Fourth Edition

Our Changing Planet

An Introduction to Earth System Science and Global Environmental Change

Fred T. Mackenzie

SCHOOL OF OCEAN AND EARTH SCIENCE AND TECHNOLOGY
UNIVERSITY OF HAWAII

Prentice Hall

Boston   Columbus   Indianapolis   New York   San Francisco   Upper Saddle River
Amsterdam   Cape Town   Dubai   London   Madrid   Milan   Munich   Paris   Montréal   Toronto
Delhi   Mexico City   São Paulo   Sydney   Hong Kong   Seoul   Singapore   Taipei   Tokyo

Acquisitions Editor: *Andrew Dunaway*
Editor in Chief, Geosciences and Chemistry: *Nicole Folchetti*
Marketing Manager: *Maureen McLaughlin*
Project Editor: *Crissy Dudonis*
Editorial Assistant: *Kristen Sanchez*
Marketing Assistant: *Nicola Houston*
Managing Editor, Geosciences and Chemistry: *Gina M. Cheselka*
Project Manager, Science: *Wendy Perez*

Art Project Manager: *Connie Long*
Art Studio: *Laserwords*
Design Director: *Jayne Conte*
Cover Designer: *Bruce Kenselaar*
Senior Operations Supervisor: *Nick Sklitsis*
Operations Supervisor: *Maura Zaldivar*
Photo Research Manager/Researcher: *Elaine Soares*
Composition/Full Service: *Laserwords Maine/Amy Saucier*

© 2011, 2003, 1998 by Pearson Education, Inc.
Pearson Education, Inc.
Upper Saddle River, New Jersey 07458

Library of Congress Cataloging-in-Publication Data

Our changing planet : an introduction to earth system science and global environmental change / Fred T. Mackenzie. — 4th ed.
   p. cm.
ISBN-13: 978-0-321-66772-4
ISBN-10: 0-321-66772-7
1. Environmental sciences. 2. Human ecology. 3. Nature—Effect of human beings on. 4. Global environmental change.
GE105.M33 2011
577—dc22   2009052220

Printed in the United States of America
10 9 8 7 6 5 4

Prentice Hall
is an imprint of

www.pearsonhighered.com

ISBN-10: 0-321-66772-7
ISBN-13: 978-0-321-66772-4

Dedication

In loving memory of our mothers, Bessie V. Mackenzie and Ella Marie Knapton Lent.
To Judy Mackenzie, who appreciates "all the general changes which it (the Earth) hath already
undergone, or is to undergo till the consummation of all things" (Thomas Burnet, 1691),
to my family Scott M., Michele, Deborah, Rebecca, Ashlyn, Jazmin, Chloe, and Scott B.,
and to my students and friends with hopes for global peace and a sustainable environment.

In fondest memory of my friend, student, and research colleague, John W. Morse.

CONTENTS

PREFACE

For about 45 years, I have been engaged in teaching undergraduate and graduate students and doing research in a field that has now come to be called Earth system science. Two of my early previous books published with my colleague and friend Robert M. Garrels were initial attempts to treat the Earth system—the solid earth, atmosphere, oceans, and living organisms—in an integrative fashion, emphasizing the interactions between the components. In addition, a third more recent book written with my colleague and friend Abraham Lerman of Northwestern University emphasized similar integrative interactions in the Earth system relative to the history and behavior of the carbon cycle and climate over time. Also, I was instrumental in developing at the University of Hawaii a Bachelor of Science degree program in Global Environmental Science. This program has grown dramatically in 12 years, demonstrating to my colleagues and me that there exists a strong interest among undergraduate students in an interdisciplinary, integrative, rigorous approach to the study of the Earth system, including the human factors involved in global environmental change.

NEW TO THIS EDITION

Because of rapid growth in the discipline of Earth system science, I have found it necessary in the 15 years since publication of the first edition of this book to update it once more.

- The text is completely revised and updated through at least the year 2008 and wherever possible through mid-2009. The text content is expanded by about 15 to 20%.
- There is a new arrangement of chapters, with a new Chapter 1 *Introduction: The Changing Planetary System*, which includes introductory material on the concept of cycling, the technosphere, the IPAT equation, the scientific method, and human dimensions of environmental change.
- There is a significant number of new figures and tables, and many revised figures and tables. URL references to satellite photos are embedded in the text and a considerable number of additional sources of information and newer reference materials have been added.
- There are new study questions, but answers are no longer provided in the back of book as in previous editions. This is to encourage students actually to work through the problems and not simply memorize the answers.
- There are new diagrams illustrating reservoirs and fluxes in the global biogeochemical cycles of C, N, P, and the trace gases of CO_2, CH_4, N_2O, SO_2, nonmethane hydrocarbons, etc. that include the most recent information on sources, sinks, and fluxes.
- Besides updating all chapters, some important changes to individual chapters include: 1. In Chapter 5, *The Fluid Earth: Hydrosphere and Air-Sea Interactions,* there is new material on interannual, decadal, and millennial scale natural climate change, including the El Niño/La Niña phenomenon. (2) Chapter 6, *Our Living Planet: Earth's Ecosphere,* has been expanded with additional material on ecological principles and the chemistry of biological systems and cells; (3) Chapters 7 and 8, *Biogeochemical Cycles of Carbon, Nutrients, and Oxygen* and *Historical Framework of Global Environmental Change,* include new information on the nature of the global biogeochemical cycles and relation to climate change and the evolution of Earth, life, and the surface environment of the planet through geological time; (4) Chapter 9, *Human Forcings on the Ecosphere: World Population, Development, and Resource Consumption,* has been significantly expanded and includes discussion of Peak Oil and Peak Coal and their ramifications and the role of human activities as a geologic force in the Earth's outer system; (5) Chapter 10, *The Changing Earth Surface: Terrestrial Vegetation,* includes the most up-to-date Global Forest Assessment (2005; see End Note, Chapter 10) on the status of the forests of the world; (6) Chapter 12, *The Changing Atmosphere: Acid Deposition and Photochemical Smog,* has additional newer material on the worldwide status of these two air quality phenomena; (7) Chapters 14 and 15, *The Changing Atmosphere: Global Warming and Stratospheric Ozone Depletion* and *Human Dimensions of Global Environmental Change in the Twenty-First Century,* include information from

the 2007 IPCC report on climate change and a critical discussion thereof, updated information on stratospheric ozone depletion, and a broad discussion of the human dimensions of global environmental change, including discussion of ethical, social, and economic issues of global warming.

Earth system science and global environmental change are subject areas of considerable interest today. The subject matter involves both the physicochemical and biological nature of change and the effects and consequences of natural and human-induced change for ecosystems, humans, and human infrastructures. The unifying theme of *Our Changing Planet* is a holistic treatment of both natural and human-induced change from the beginning of the Earth on into its future. Earth's ecosphere or exogenic system—its land, water, ice, air, sediments, and biota—has always been in a dynamic state of change. Change is probably more characteristic of the planet than constancy. However, on the time scale of human generations, change can be rapid and threatening; thus, the topic has forced itself before the world, mainly in the guise of global warming of the planet. Scientists, teachers, policymakers, economists, sociologists, lawyers, and the general public are now addressing global change topics on a regular basis.

Since publication of the first edition of this book in 1995, a great deal has happened in the field of study of Earth systems and global environmental change. Of special importance has been the modern recognition, which began in the 1970s but had its roots much further back in time, that the Earth is an intricately coupled system where the interactions between the land and its soils, oceans, atmosphere, terrestrial and marine biota, sediments, and ice are critical to an understanding of environmental change and variability on a regional-to-global scale. As a consequence of this recognition, a new interdisciplinary field has emerged, that of Earth system science or global environmental science. The emergence of this field of interdisciplinary and holistic geosciences as a legitimate scientific discipline has been coupled to an increase in scientific papers and books, an increase in the number of courses taught at both the undergraduate and graduate level in universities and colleges, and an increase in the number of textbooks and other teaching materials, including those on the World Wide Web, concerned with the subject matter. One of the ultimate challenges of Earth system science is "to develop the capability to predict (climatic) changes (and variability) that will occur in the next decade to century, both naturally and in response to human activity" (NASA, 1996). An understanding of how Earth behaves as a complex, intricately interwoven system has implications for the future pathway of economic development and the environment.

Another occurrence that has had impact on the emerging discipline was the publication in 1990, 1996, 2001, and 2007 of the volumes of the First, Second, Third, and Fourth Assessments of the Intergovernmental Panel on Climate Change (IPCC), which include the science of climatic change, as well as the human dimensions issues of change. Although some of the conclusions of this effort over the years are still being debated by scientists and nonscientists, the volumes on the science of climate change are an excellent example of the need for interdisciplinary studies to address questions of natural and human-induced environmental change. In addition, the 2005 publication of *Ecosystems and Human Well-being: Synthesis* report of the Millennium Ecosystem Assessment team documented the role of ecosystems in human well-being, the need to conserve and sustain these systems, and the necessity of recognizing linkages between human behavior and environmental change. Furthermore, the 2009 *Geovision Report* of the U.S. National Science Foundation emphasized the need in the twenty-first century for the geosciences to evolve into a truly interdisciplinary science and that they take a more holistic approach to a quantitative understanding and forecasting ability of the complex and changing behavior of our planet. All these publications have increased awareness of the complexity of natural and human-induced environmental change and the need for a more integrated and multidisciplinary approach to the Earth system and the changes it undergoes.

A third factor that has played a role in development of the discipline has been the recognition by some scientists for the need to develop research programs that are interdisciplinary. Although most ocean, Earth, and atmospheric scientists are still highly specialized, there has been greater demand by their students for some modicum of interdisciplinary training. Furthermore, these scientists are striving to communicate more effectively with the undergraduate community in colleges and universities. Earth system science has struck a chord of interest and appreciation in the students. This has been a driving force for the creation of new courses in the field. Indeed, the Geological Society of America's 1996 Annual Meeting in Denver, Colorado, had as its theme Earth Systems, and several sessions were devoted to development and teaching of curricula in the new field of Earth system science.

Although the original edition of this book was written to appeal to upper-division high school students as well as lower-division undergraduate students, the intent of this revision is for use as a text in interdisciplinary Earth, ocean, atmospheric, and ecological sciences at the undergraduate level in colleges and universities. However, both high school students and the educated layperson could benefit from reading and studying the text. The science content is in accord with National Science Education Standards that define what students need to learn to achieve scientific literacy. The text should also be useful to middle through high school science teachers in educational efforts for the professional development of these teachers.

I recognize that there will always be a need for natural resources and for development. The mining, processing, and use of resources; the construction and maintenance of transportation systems and human structures; and the activities associated with growing and distributing food are some of the human enterprises that are prone to generating pollutants and inducing environmental change. Whenever these activities occur, there is an increase in the amount of energy not available to do work (entropy) on the planet. This is an inexorable outcome of the fundamental laws of science. As a global human civilization, we must decrease the rate of production of this unavailable energy that leads to degradation of the environment and learn to manage the global ecosystem in a sustainable way. This book will provide background for students and teachers interested in protecting and managing our global commons.

The initial completion of this text and its revisions required synthesis of a large and dispersed literature; I gratefully acknowledge all of those authors from whom I have liberally acquired information. I especially express thanks to three institutions and their directors for providing space and facilities to accomplish the task of writing the first edition: Dr. Tony Knap (Bermuda Biological Station for Research), Professor Roland Wollast (Université Libre de Bruxelles), and Dr. George M. Woodwell (Woods Hole Research Center). During the writing of the second edition, I was a fellow at the Wissenschaftskolleg (Institute for Advanced Study) zu Berlin and I would like to thank its Rector, Professor Dr. Wolf Lepenies, for providing space and atmosphere for unbridled intellectual thought. My special thanks to Dr. Lei Chou (Université Libre de Bruxelles), Dr. James N. Galloway (University of Virginia), Dr. Mary Hassinger (Viterbo College), and Dr. Douglas Whelpdale (Environment Canada) for their words of wisdom; Dr. Rolf Arvidson (University of Hawaii, now at Rice University) and Dr. May Ver (University of Hawaii, now at the University of British Columbia) for their computer assistance; Ms. Michele Loujens (Université Libre de Bruxelles) for laboratory and logistical help; and Ms. Carole Frantz (Beauvoir School) and Ms. Margaret Best (Bermuda Biological Station for Research) for their critical reviews of the first complete draft of the initial edition of this work. I also thank Ms. Sue Dewing (L. P. Goodrich High School), Mr. James Rye (The Pennsylvania State University), and Ms. Dorrie Tonnis (Logan Senior High School) for their thoughtful reviews of early portions of the text and their suggestions of study questions. The reviews of the final draft of the first edition by Dr. E. Calvin Alexander, Jr. (University of Minnesota), Dr. James N. Galloway (University of Virginia), Dr. Garry McKenzie (The Ohio State University), Dr. V. Rama Murthy (University of Minnesota), Mr. James Rye (The Pennsylvania State University), Dr. Edward D. Stroup (University of Hawaii), Dr. Douglas Whelpdale (Environment Canada), and Dr. George M. Woodwell (Woods Hole Research Center) did much to improve the flow and content of the text and identify errors. Dr. Abraham Lerman (Northwestern University) and Dr. John W. Morse (Texas A&M University), through our continuous collaboration over the years, have been a source of inspiration for a number of ideas and concepts in the several editions of this book.

The revisions of the first edition of this book were written without the participation of my wife, Judy Mackenzie. The pressures of her own teaching duties prevented her from doing so. However, without her continuous encouragement, patience, and critical thinking concerning the organization of the text, there would have been no second, third, or fourth edition. During the past 15 years, I have taught the contents of this text to students in many classes in global environmental change at the University of Hawaii at Manoa and Hilo, Northwestern University, Université Libre de Bruxelles, and elsewhere. I am especially grateful to these students for insightful comments on the text. I would like to express my aloha to Dr. Jane Schoonmaker (University of Hawaii), Dr. Daniel Hoover (University of Hawaii, now at the United States Geological Survey), and Joan Hoover, who took the time to read and critically evaluate the text. Jane did so for all four editions of this book. My deepest appreciation to my student assistant, Margaret Shum (University of Hawaii), who read and edited the text and helped obtain bibliographic material. I thank Kevin Bradley of Sunflower Publishing Services, the Production Editor of the third edition. I am indebted to my past editors, Robert A. McConnin and

Patrick Lynch, who encouraged me to proceed with past revisions and supported my efforts throughout. I would like to thank Drusilla Peters, Acquisitions Editor of Geology at Pearson Prentice Hall, and Dan Kaveney for encouraging me to do the fourth edition. I thank Sean Hale, Assistant Editor, Geoscience and Environment Pearson Prentice Hall, for helping me over the hump on a few difficult occasions. Finally, and most important, I would like to express my gratitude to Crissy Dudonis, Project Manager, Geosciences and Environment Pearson Prentice Hall, for her patience, skillful advice, and friendship while helping to bring this book to fruition and to Amy Saucier, Associate Project Manager for Laserwords Maine, for her excellent copyediting skills and kindness in dealing with this author.

Some of the material in this book has been drawn from research supported by the National Science Foundation, National Oceanographic and Atmospheric Administration, and the Andrew W. Mellon Foundation.

Fred T. Mackenzie
Honolulu, Hawaii

1

Introduction: The Changing Planetary System

The Earth from here is a grand oasis in the big vastness of space.

ASTRONAUT JAMES A. LOVELL, JR. (*Apollo 8* Spacecraft)

During much of the twentieth century, the natural sciences were concerned with examining individual physical, chemical, and biological processes, or groups of processes, in the atmosphere (air), hydrosphere (water), lithosphere (land, sediments, and rocks), cryosphere (ice), and biosphere (life). For example, the processes of oceanic circulation, atmospheric circulation, plate tectonics, life cycles of plants and animals, and individual ecosystem processes have been subjects of extensive analyses and study. However, two factors have led to a reawakening of the sciences and emergence of a planetary approach to the investigation of planet Earth: (1) satellite images of the blue planet showing Earth as one entity without political boundaries and (2) concerns with environmental issues and their connection to population, resource consumption, and energy growth on regional and global or worldwide scales (e.g., Chernobyl, Bhopal, Love Canal, *Exxon Valdez* and Prince William Sound, Gulf War oil fires, deforestation, desertification, the depletion of the ozone layer, acid rain, an enhanced greenhouse effect and consequent global warming, ocean acidification, and Peak Oil and its implications). This approach uses the findings of the physical (including chemical) and biological sciences and integrates them into a broader, more global view of our planet and how it functions as an integrated unit. In this Earth system science, Earth is viewed as a complex, evolving planet that is characterized by continuously interacting physical and biological change over a wide range of space and time scales. These changes, both natural and human-induced, that occur on Earth are often referred to as global environmental change.

As an example of this integrated global concept, consider the apparently simple, naturally occurring situation of soil formation on Bermuda, an island in the western part of the North Atlantic Ocean. The soils of Bermuda have formed on limestone rock, in part from leaching of the underlying rock by rain. The soils are deep red, indicating the presence of iron oxides. However, the limestone rock is poor in iron oxide; therefore, much of this material is likely to have been derived from a source other than the limestone. The lack of localized rain and plant life and, in more recent times, poor agricultural practices in the Sahara desert and Sahel of North Africa have resulted in large dust storms (see http://earthobservatory.nasa.gov/IOTD/view.php?id=519). This dust has been blown by easterly winds across the North Atlantic for millennia, and through fallout from the atmosphere has contributed materials to the soils of Bermuda. Thus, much of the iron oxide, clay minerals, and some other materials found in the red soils of Bermuda have a source in atmospheric dust originating in North Africa.

Similarly, dust storms in the Gobi and Takla Makan deserts in China result in an increase in dust particles in the atmosphere of the North Pacific including the Hawaiian Islands, within a few days after the huge dust plumes begin on the mainland of China and move eastward across the Pacific blown by the westerly winds (see http://visibleearth.nasa.gov/view_rec.php?id=4191). These dust storms help to supply nutrients to some parts of the North Pacific oligotrophic ocean, a region of the ocean that is nutrient depleted. Such storms can also be responsible for a significant portion of the breathable particulates found in air a continent away from their source and for transport of viruses, bacteria, fungi, and toxic metals for long distances. Thus, an understanding of the formation of soils on Bermuda and of dust fallout in the North Pacific and organic productivity in the ocean necessitates scientific knowledge, covering a wide spatial range over an extended time span, of the rain, vegetation, wind patterns, and human impacts in areas that are far removed from the islands of Bermuda or Hawaii.

Another example is the change in world weather patterns, such as the dramatic changes in rainfall patterns in North and South America and Africa produced by the redistribution of heat that starts in a region as far distant as the western tropical Pacific Ocean. This phenomenon is known as the El Niño-Southern Oscillation (ENSO) (see http://www.esa.int/esaCP/GGGN8H2UGEC_FeatureWeek_0.html). To understand this phenomenon, it is necessary to have knowledge of the interactions between the atmosphere and oceans and their influence on the climate on a global scale. It is also necessary to have a large buoy system with monitoring instrumentation for temperature, salinity, etc. spread across the tropical Pacific and satellites to gather data necessary to make predictive models of the El Niño-Southern Oscillation interannual climatic phenomenon.

The interconnected processes involved in global change and their time and space scales are numerous and diverse. The natural spheres of planet Earth, such as the hydrosphere and atmosphere, have been evolving ever since the planet became cool enough for life to start, resulting in an Earth system that allowed life to flourish. The continuing evolution of the organic world and its interactions with the inorganic world led to a global Earth surface system of quasi-balanced and cyclical processes and feedbacks, an homeostatic system, that operated naturally prior to the late pre-industrial era.

SPACE AND TIME PERSPECTIVE OF THE WORLD'S PEOPLE

In general for many reasons, including social and cultural, most individuals do not have a perspective on life that includes looking far into the future or that involves geographically large areas. However, in the industrialized and industrializing world in particular, this limited view fortunately is changing somewhat. Figure 1.1 shows qualitatively the spatial and temporal perspective of most people of the world.

The period of time that it takes for a process to occur is referred to as a **time scale** and its physical dimension is termed a **space scale.** Some time scales are very short, such as the human activities that occur on the scale of a day. Longer time scales may encompass weeks, months, years, centuries, or millions of years. The short time scale of human generations is generally what is considered most important to people, because it includes events of everyday life. Space is defined by its physical extent. It may be an immediate surrounding area or any area extending outward. For our purposes, we will use the terminology of local, regional, national, continental, and global space scales. In general, people are most concerned by events that take place today, tomorrow, or less than a generation into the future. Their own land, community, state, or nation is commonly the space of most interest. An activity becomes less important the farther away it is, even though the impact in the distant future or on geographically remote locations may be severe. For example, people are more concerned with their immediate family than with families in other parts of the world. What happens this week to the environment and ourselves is of more concern than what will happen in the more distant future to the environment or to the generations of humans not yet born.

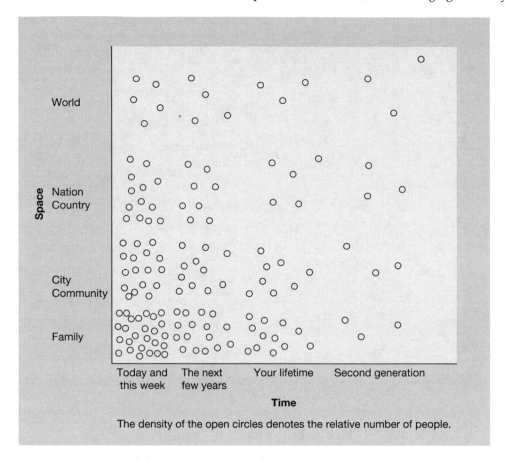

FIGURE 1.1   Space and time scale perspective of most of the world's people. The density of the open circles denotes the relative number of people.

SPACE AND TIME SCALES OF EARTH PROCESSES

Earth system processes vary over a wide range of space and time scales. Figure 1.2 shows Earth system processes grouped according to their spatial and temporal scales. The space scale is logarithmic, and the time scale is relative. Turbulent motions in the atmosphere can be a local phenomenon, arise in a matter of minutes, and cover a small area. The complete and complex mixing of the world oceans is on the order of 1000 years and occurs over distances measured in thousands of kilometers. Plate tectonic motions are slow, taking millions to hundreds of millions of years to complete, and covering distances on the order of 10,000 kilometers.

Both natural and human-induced environmental change and variability exhibit a large range of space and time scales. **Change** is often viewed as being a unidirectional property of a system, whereas **variability** implies shifts about some mean point. Often these terms are used interchangeably. For example, global climatic change involves unidirectional changes in climatic features over the vastness of the globe. Such changes may occur on decadal to centurial to millennial to longer time scales. The phenomenon of urban smog is more localized, and the events are of much shorter duration. The natural cycle of the El Niño-Southern Oscillation, as its name implies, is a phenomenon that exhibits variability and affects weather patterns in the tropical Pacific but with far-reaching effects on other regions of Earth.

Systems may exhibit both change and variability. They also may have equilibrium or set points and be stable, or they may exist in a state, known as metastable, in which a slight disturbance may cause the system to change dramatically. These "tipping points" are of great interest today in terms of the phenomenon of global warming of the planet and acidification of its oceans. The Earth system is a "complex system" characterized by processes that are overall irreversible, by phenomena that often switch from one state to another, by evolution, and by feedbacks.

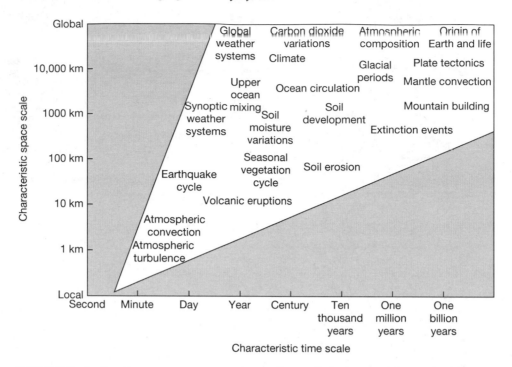

FIGURE 1.2   Earth system processes grouped according to their space and time scales. The space scale is logarithmic and the time scale is relative.   (*Source:* Graedel and Crutzen, 1993.)

A **feedback** is a process or mechanism that amplifies (positive feedback) or diminishes (negative feedback) the magnitude of a perturbation to a system. Because of these features, the behavior of "complex systems" is difficult to analyze and predict.

THE CONCEPT OF CYCLING

A **cycle** is a sequence of events that continuously reoccurs such that matter is transported from one portion of the cycle to another and returned. In a perfectly conservative cycle, no matter or energy would be lost. In nature, some matter and energy are generally lost from the cyclic system, but this is usually an insignificant amount. There are myriad cycles in the universe. The sun, Venus, Mars, and the moon have chemical cycles that are constantly active without life around. There are also biological cycles, the cycling of chemical substances that occurs within a cell or living organism. Some biological materials cycle between living and nonliving organic matter and inorganic matter. These materials are involved in biogeochemical cycles such as the carbon cycle (see Chapter 7). The physical and life processes of the Earth involve cycles on all scales of time and space. There are smaller cycles within larger global cycles. Some cycles require only a few minutes; others take

millions of years. These are the cycles that maintain life on planet Earth. An excellent example of a cyclic system on a global scale is that of water (Figure 1.3; see Chapter 5). Most (86%) of the water that evaporates from Earth's surface into the atmosphere (total evapotranspiration is 496,000 cubic kilometers per year) comes from the ocean; 14% comes from the land. However, more than 14% is returned as precipitation to the land. This surplus of water to the land surface (8%) derived from the ocean in evaporation is eventually returned via rivers to the oceans, where it is once more available for evaporation. The cycle is complete.

Natural Earth cycles generally remain *relatively* well balanced over short and long time and space scales. The life cycles of plants and animals and the water cycle are examples of well-balanced cycles. However, natural major changes do occur in the cycling of materials. For example, it is known that the periodic global ice ages (the glacial stages) of the Pleistocene with their cooler climates, low sea levels, changes in biological habitats and plant and animal species, and rates of biogeochemical cycling are part of a larger cycle that takes tens of thousands of years to complete. These ice ages are periodically interspersed with shorter and warmer periods of time (the interglacial stages) on Earth. Natural global

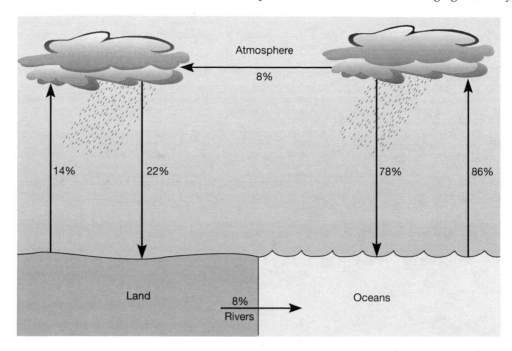

FIGURE 1.3 The global water cycle shown in percentage relative to the total water evaporated and transpired from the Earth surface annually taken as 100 percent.

environmental change, at all scales of time and space, has always been an integral part of the way in which the various systems of Earth function as part of the global ecosphere.

THE TECHNOSPHERE

Economic, political, and social growth has been an integral part of the survival of the human species. As a result, humankind has developed another sphere, the "technosphere" (agricultural, manufacturing, and industrial sectors; Figure 1.4). This sphere affects the natural surface system of Earth, the ecosphere. People are interacting with and changing the natural biogeochemical cycles of their environment by modifying the natural fluxes of materials and energy because of agricultural, industrial, transportation, and urbanization activities. Because of these activities, the rates of transfer of elements between reservoirs have changed. In some cases, these changes are modifying the composition of the natural reservoirs, such as the carbon dioxide content of the atmosphere. Also, some substances never before a part of Earth's surface system, such as chlorofluorocarbons (CFCs), are being added to the environment. As succinctly stated by Barry Commoner in 1991, the technosphere is in conflict with the ecosphere (Figure 1.4). The surface system of Earth is no longer in a natural quasi-steady

state because it is being modified rapidly by the activities of people. In fact, humankind is in the process of changing the global environment at a more rapid rate than that usually experienced by Earth throughout much of its 4.6-billion-year history. In our ignorance and simplicity, we are interfering with natural recycling processes without fully understanding what effects our perturbations will have on the natural system and, ultimately, on ourselves and other species.

As an example, consider global climatic change, a topic that is often discussed in the news media, even more so in the twenty-first century because of the problem of Peak Oil and rising oil prices. Along with the rapid economic growth of the last century, there has developed a heavy reliance on the use of fossil fuels of coal, oil, and gas as an energy source. The burning of fossil fuels releases to the atmosphere greenhouse gases, such as carbon dioxide (CO_2) and methane (CH_4), which can lead to warming of the planet. The developed, wealthy nations of the world, with about 25% of the world's population, are mainly responsible for these releases. Although the populations of the developed world are relatively stable, they experience continued economic growth. Thus, these nations are heavy consumers of resources, and their need for energy, even in 2009 despite the high price of oil per barrel, seems insatiable. The United States alone, with only 4.5% of the world's

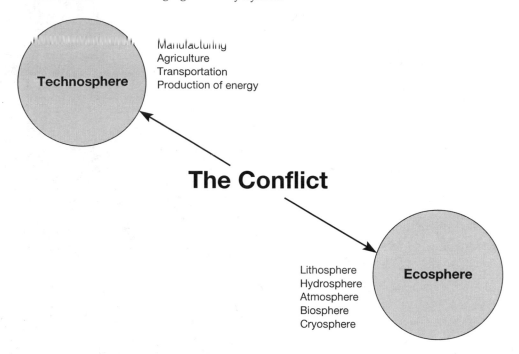

FIGURE 1.4 Schematic diagram illustrating the conflict between the ecosphere and the technosphere. (*Source:* Commoner, 1991.)

population, accounts for about 22% of the world's consumption of energy and 18% of emissions of carbon dioxide to the atmosphere from fossil fuel burning. China, with its rapid rate of economic growth during the past decade and its 1.3 billion people, now leads the United States in the total amount of CO_2 emitted to the atmosphere. In the developing, poorer countries of the world, the per capita carbon dioxide emissions from fossil fuel are relatively small, although increasing as these countries strive to become members of the developed world. Some of these nations today rely heavily on a different resource, their forests and grasslands, to provide for economic growth. The forest ecosystems are burned to provide land for urban and agricultural development or cattle grazing, or lumbered for monetary profit. Trees are cut from forests and to a large extent are not replanted. All these activities contribute to the buildup of carbon dioxide in the atmosphere (see http://earthobservatory.nasa.gov/Newsroom/view.php?old=2005091320344).

One result of the world's reliance on fossil fuels and the destruction of its forests has been the rapid accumulation of carbon dioxide in the atmosphere, particularly during the last half of the twentieth century and on into the twenty-first century. This accumulation has most likely led to an enhanced greenhouse effect and rising temperatures globally.

As a result of continued accumulation of greenhouse gases in the atmosphere, patterns of rainfall, temperature, and other environmental conditions may change rapidly, perhaps too quickly for ecosystems, agricultural systems, and human society to adjust without significant disruptions. Sea-level rise, a consequence of global warming and thermal expansion of the oceans and melting glaciers, may be accelerated and lead to flooding of coastal areas and of island nations. The acidification of our oceans is already occurring because of the release of carbon dioxide to the atmosphere by human activities and partial absorption of this gas in the ocean. Ozone depletion and acid rain are other human-induced problems that have important effects on ecosystems, people, and human infrastructures. The physical and biological changes that occur on Earth as a result of human intervention could lead to substantial changes in the social, political, and economic situation of the world.

I do not mean to imply that human-induced change is always negative. The technosphere does not necessarily need to come into conflict with the ecosphere. Careful planning in the local, regional, and global areas can be beneficial to political, economic, and social processes and at the same time lead to protection of the natural environment. It has been proposed that people "think globally, act locally." There are many examples of careful planning techniques that cross time and space scales. Terrace farming prevents regional erosion; replanting of trees and forest management provide a continual source of wood for local and world communities; conservation of forests allows for regional maintenance of water and nutrient cycles and biodiversity and the storage of carbon and thus helps to stabilize the content of carbon dioxide in the atmosphere; conservation and recycling of resources provide materials for the future; and developing new energy technologies and increasing energy efficiency save resources and promote a cleaner environment and new industries. However, any human tinkering with the natural system will probably have unforeseen side effects, some more deleterious than others.

At this stage, it is worthwhile to emphasize the distinction between growth and development. **Growth** generally means an increase in size; when something grows, it gets bigger. **Development** implies the ability to reach a more advanced or effective state and a fuller potential. When something develops, it gets better, but not necessarily bigger. Although there are limits to growth, there theoretically should be no limit to development of a sustainable nature. **Sustainability** implies development that supports human progress not simply in certain places for limited periods of time but for the entire planet far into its future. In this context, development should meet the needs of current generations without compromising those of the future. This concept applies to physical, social, economic, and political processes.

THE NATURAL SYSTEM AND ITS HUMAN DIMENSIONS

The Borromean Knot of three interlocking, but separate, rings symbolizes concerns for global change, biological diversity, and sustainable ecological systems. The integrity of the whole requires the integrity of each part. The breaking of any link can damage the whole system.

(AFTER ECOLOGICAL SOCIETY OF AMERICA, 1991.)

The ecosphere (Figure 1.5) is the thin layer of the surface of Earth that interacts with the atmosphere, hydrosphere, and lithosphere and in which life is possible. The ecosphere extends from the bottom of the oceans and from depths in rocks of a couple of thousand meters, where bacteria are found, to the mountain peaks and to the upper limits of the lower part of the atmosphere. This region of life is on the order of approximately 25 kilometers thick. If the world were the size of an apple, the ecosphere would be only about the thickness of its skin. All life on Earth inhabits this fertile skin, and as far as we know, it is the only place in the solar system that supports life, despite the discovery in recent years of many planetary bodies in the universe. This thin film of the ecosphere has evolved over 4 billion years to its characteristics of today. The ecosphere sustains life because of energy received from the sun and unique cycling processes, interactions, and feedbacks among the physical and biological systems of Earth (Figure 1.5). The physical systems (the abiotic—the hydrosphere, atmosphere, and lithosphere) include all chemical constituents and physical components of the environment of wind, ice, water, heat, air, soils, and rainfall. The biological systems include both the living organisms (the biota) and dead organic matter on Earth. The term **biosphere** has traditionally been used exclusively for the living and dead organic components of the planet. This will be the definition used in this book. The larger system that includes the biosphere and its interactions with the physical systems is referred to as the **ecosphere.**

Earth can function without humans around. The sun will burn for several billion more years. The climate will fluctuate in the future as in the past. The internal heat of Earth will drive convection currents within Earth and recycle the land. Volcanoes will erupt and continents will drift. The oceans and atmosphere will circulate and their materials cycle. The ecosystems will evolve with the planet, and the biota will adapt and influence environmental change. The planet has undergone global environmental change since its creation, and change will be a part of its future. Earth is a dynamic planet of endless change.

Prior to the existence of humans on the planet, environmental changes in the natural systems of Earth, its atmosphere, lithosphere, hydrosphere, cryosphere, and biosphere occurred over billions of years. Global environmental change is not a new phenomenon. What are new are the rates of change that the systems of Earth are currently experiencing because of human activities. Since the arrival and expansion of the human species to nearly 6.8 billion people in 2009, more rapid decadal to centurial changes are occurring to these systems compared to the rates of environmental change for most of Earth's history, with the possible exception of very large impactors [bolide (e.g., comets and meteorites)] on Earth. Many of these rapid changes are related to our

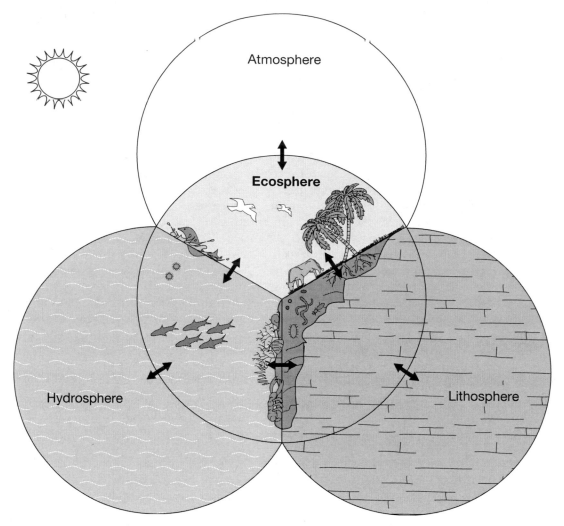

FIGURE 1.5   The ecosphere, our life-support system, showing its relationship to the other spheres of Earth. The cryosphere (not shown) is the frozen part of the hydrosphere.   (*Source:* Christensen, 1991.)

ever-increasing need for natural resources and their retrieval, production, distribution, consumption, and disposal; to short-term economic considerations; and to the growth and distribution of the human population.

Natural resources are the many different forms of available energy, minerals, and organic matter on Earth that are commercially exploitable by humans (e.g., iron, coal, oil, gas, wood, phosphate deposits). Noncommercial items (e.g., clean air, the aesthetic value of a landscape, or noncommercial organisms that make up an ecosystem) provide no monetary gain and usually are not referred to as natural resources.

Natural resources may be renewable or nonrenewable resources. Some commercial resources, such as our forests and agricultural crops, are regularly replaced on a short-term basis and are considered renewable, although replacement of these resources does not always happen. Other commercial resources, like minerals, coal, oil, and gas, are nonrenewable. They are resources that may be depleted with excessive use because of loss through non-recycling methods of disposal. They also can be essentially fixed supplies of these substances on Earth on the scale of human generations because they take millions of years to replenish by natural processes.

The people of the world have found ways to utilize the natural resources of the planet for their benefit. We have changed courses of rivers, produced bountiful agricultural lands, developed huge infrastructures to support our species, studied the stars, and traveled to the moon. However, in the twentieth and early twenty-first centuries, the world population has increased by about four times, energy use has risen 10 times, the use of water has increased nine times, and the global world product has risen 40 times, which has taken large amounts of energy, mainly the fossil fuels of coal, oil and gas, to power this growth. Because of growth in these parameters and other causes, we have also interfered in the natural dynamics of the Earth. This interference can be appreciated simply by consideration of the magnitudes of change wrought by some human activities:

1. Humans have used, co-opted, or foregone—that is, affected in one way or another—as much as 40% of terrestrial net primary productivity and several percent of marine net primary productivity.

2. Humanity now uses the equivalent of about 25% of evapotranspiration over land and 55% of the accessible water runoff from the continents.

3. Total annual discharge of sediment toward the oceans by rivers is about 20 billion tons (Gt), amounting to two times the discharge in late pre-industrial time.

4. More than 45,000 large dams are located on the world's rivers. Their reservoirs, if filled to the top, would account for nearly 300 times the amount of sediment discharged by rivers to the ocean annually.

5. About 560 Gt of carbon were emitted to the atmosphere between the years 1700 and 2008 because of fossil fuel burning and land-use practices (e.g., deforestation). In the last 50 years, fossil fuel emissions of carbon dioxide have increased more than 400%, amounting to 8.7 Gt of carbon annually in 2008, accompanied by more than 70 million tons (Mt) of sulfur and 30 Mt of nitrogen. Also, carbon dioxide emissions to the atmosphere from deforestation in this 50-year time period have increased about 200%, amounting to 1–2 (best estimate: 1.2) Gt C emitted annually in 2008.

6. Total annual mobilization of nitrogen by human activities is about 140 Mt, equivalent to two times the amount of nitrogen falling in rain each year and five times the total river discharge of dissolved nitrogen to the oceans annually. In the past 50 years, the industrial fixation of nitrogen to make synthetic fertilizers has risen about 2400%, and is a source of 85 Mt of nitrogen to the atmosphere owing to denitrification after the fertilizers are applied to croplands.

7. Total annual mobilization of phosphorus by human activities is about 25 Mt, or five times the total river discharge of phosphorus to the oceans. In the past 50 years, the global mining of phosphate ores has increased by about 600%.

8. Total annual mobilization of sulfur by human activities is about 150 Mt, equal to 70% of the sulfur rained out of the atmosphere each year. Global annual sulfur emissions decreased in the latter part of the twentieth century, partially because of regulatory measures controlling emissions in North America and Europe, but are now increasing again.

9. Almost one-third of the world's arable land has been lost due to erosion, contamination, and other processes, and cropland continues to be lost at a rate of about 10 million hectares per year.

Consequences of these human practices and human-induced changes in fluxes of materials include soil degradation and erosion, deforestation, desertification, loss of habitat and species diversity, extinction of plants and animals, eutrophication of water bodies, air pollution, acid rain, stratospheric ozone depletion, tropospheric ozone increase, the enhanced greenhouse effect and global warming, ocean acidification, toxic waste, radioactive waste, municipal waste, poverty for many, and great affluence for some. Human activities are now a force in the exogenic (outer, near surface) system of Earth, much like that of plate tectonic processes and changes in the orbital parameters of the planet but at a much more rapid rate.

The total impact on the environment of humans as a geologic force can be written in the form of a functional equation in a number of different ways. One relationship that is seen frequently is the **IPAT** formulation by John Holdren, former Director of the Woods Hole Research Center and in 2009 the U.S. President Barack Obama's chief science advisor, and Paul Ehrlich of Stanford University published in 1971 (see also Commoner, 1972):

$$\mathbf{I}_{(impacts)} = \text{function } (f) \text{ of } [\mathbf{P}_{(population)},$$

$$\mathbf{A}_{(level\ of\ affluence)}, \text{ and } \mathbf{T}_{(technology\ availability)}].$$

The consumption of natural resources can be viewed as a separate factor involved in impacts or as a variable inherent in the factors of the IPAT equation. The IPAT relationship shows that in general as population and level of affluence increase, environmental impacts increase. However, the development of environmentally friendly technologies can alleviate impacts on the environment. The removal of sulfur from coal before it is burned and the development of electric and hybrid cars are good examples. In addition, it has been shown for individual counties that as the level of affluence increases and the gross domestic product rises, a country is likely to have the financial resources necessary to implement regulations that control harmful emissions to the environment and to implement practices that are more environmentally friendly. However, for much of the Industrial Revolution as the level of affluence has increased in a country, the greater has been the country's use of natural resources.

How will the natural Earth surface system and its dynamic, quasi-balanced, cyclic pattern of processes react to the activities within the new technosphere? Although we are learning rapidly, scientists currently do not fully understand, or even agree on, what effects the activities of people will have on the physical environment of our planet. The biological effects of global environmental change are even less well understood, as are the economic, social, and political consequences of change. The ability of humankind to induce major changes in the natural system of Earth carries with it a tremendous responsibility. Earth is the only planet we know of that is capable of sustaining our form of life. People need to continue to seek an understanding of how the Earth system works to make appropriate decisions. It is necessary to focus on how to manage our planet in a way that leads to its nourishment, sustainability, and survival of all species.

The holistic approach of Earth system science is necessary to evaluate the complex functioning of the cyclic system of Earth, and the *scientific method* is an essential part of this approach. Other methodologies are prone to opinion and bias. The scientific method is a process by which scientists over time attempt to construct and present an accurate picture of phenomenon(a) that is reliable, consistent, and nonarbitrary. The method attempts to minimize the influence of bias or prejudice by the person(s) engaged in the explanation of the phenomenon(a). Although procedures vary from one field to another, the method is based on four major steps: (1) observation and description of a phenomenon or a group of phenomena by gathering observable, empirical, and measurable evidence; (2) formulation of hypotheses as explanations of the phenomenon(a); (3) use of the hypothesis to predict the existence of other phenomenon(a), or to predict quantitatively the results of new observations; and (4) performance of appropriate experimental tests of the predictions by the investigator and other groups of investigators. If the experiments and observations confirm the hypothesis, it may come to be regarded as a theory or law, for example, Isaac Newton's Law of Universal Gravitation. If the experiments and observations do not bear out the hypothesis, it must be rejected or modified. A simplified overview of the steps in the scientific method is shown in Figure 1.6. The scientific method is the methodology that should be employed in the study of Earth system science and global environmental change. Poorly supported or emotional opinion or biased conjecture should be rejected as approaches.

In this book, I discuss the Earth system prior to the intervention of humankind in its workings as well as the human impacts on the global environment. The individual and interactive cycles of Earth form a web of life. To understand this web and human interference in it, it is necessary to consider some background information dealing with the physical and derivative sciences in the context of natural and human-induced environmental change. The book chapters are arranged to do that. Following this chapter, which is a brief introduction to the subject of natural and human-induced global environmental change, Chapters 2 and 3 discuss one of the physical systems of Earth—the lithosphere, its geologic time

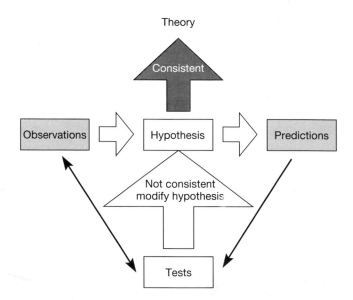

FIGURE 1.6 Simplified diagram of the steps taken in the scientific method for explanation of phenomenon(a) from the stage of asking a question usually resulting from initial observations of the phenomenon(a), followed by construction of an hypothesis, experimental tests, analysis of data, determination of the validity of the hypothesis, and final communication of results.   (From Professor Jose Wudka, http://phyun5.ucr.edu/~wudka/Physics7/Notes_www/node6.html. Courtesy of J. Wudka, UC Riverside.)

scale and building blocks of minerals and rocks, and the nature of its movement as documented in the revolutionary concept of plate tectonics initially developed in the late 1960s–early 1970s. In Chapter 4 a major portion of the fluid Earth—the atmosphere, its structure, motion, and composition—are considered. Chapter 5 examines the other major portion of the fluid Earth, the water envelope of the planet—the hydrosphere—emphasizing its structure, motion, and composition, as well as air-sea interactions in the form of the El Niño-Southern Oscillation (ENSO) phenomenon. Chapter 6 addresses our life-support system—the ecosphere—emphasizing the variety of terrestrial and aquatic ecosystems on Earth, the ecological processes operating in them, and how ecosystems function.

In Chapter 7 the biogeochemical cycles of the elements are discussed, especially those of carbon, the nutrient elements nitrogen and phosphorus, and oxygen, and their role in regulating the global environmental system through feedback processes is addressed. With the previous information as background, a historical account of the evolution of the Earth and its ecosphere from the beginning of the planet to the features of today is presented in Chapter 8. In order to gain some insight into the reasons for humans becoming a geologic force in the Earth system, the topics of world population growth, development, resource consumption, and Peak Oil are considered in Chapter 9, forming the basis for later discussion of how the ecosphere is changing owing to human activities. Based on the ecological considerations developed in Chapters 6 and 7, in Chapter 10 terrestrial vegetation and environmental problems related to land use practices, deforestation, and fertilizer and pesticide application to croplands are discussed. Following this discussion, which deals mainly

with the biosphere, in Chapter 11 the land and water environments and their interactions in a changing Earth system, including the coupled land-water ecosystem and problems of aquatic pollution, are considered. Chapter 12 investigates the mainly regional problems of a compositionally changing atmosphere addressing the phenomena of acid deposition and photochemical smog. Continuing with the theme of Chapter 8, Pleistocene and Holocene climatic and other environmental changes are discussed in Chapter 13. This serves as preparation for consideration of the global environmental problems of an enhanced greenhouse effect and global climatic change, stratospheric ozone depletion, and ocean acidification, addressed in Chapter 14. Chapter 15 deals with the human dimensions issues of global environmental change and addresses the question of sustainability.

The chapters dealing with the chemistry, biology, and physics of environmental processes are not intended to be intensive studies of the physical and biological sciences; rather, they provide important information on some aspects of these sciences and emphasize processes germane to problems of global environmental change. In addition, the material in the chapters need not be used in chapter order but, depending on the individual instructor and the students, chapters can be interleaved with those concerned mainly with human-induced global environmental change.

How did Earth evolve? How does Earth function? What will it become in the future? How can we create a sustainable planet? These are questions and challenges for humankind. The focus of this book is on the coevolution of the physical environment and life on planet Earth and, in particular, on the effect of the new, distinctive technosphere on the natural surface system of the planet and its global environment.

Additional Sources of Information

Commoner, Barry, 1972, The Environmental Cost of Economic Growth. *Population, Resources and the Environment,* Government Printing Office, Washington, D.C., pp. 339–363.
Ehrlich, P. R. and J. P. Holdren, 1971, *Impact of Population Growth.* Science, v. 171, 1212–1217.

Waggoner, P. E. and J. H. Ausubel, 2002, *A Framework for Sustainability Science: A Renovated IPAT Identity.* Proceedings of the National Academy of Science, v. 99(12), 7860–7865.

CHAPTER

2

Earth's Lithosphere: Geologic Time and Building Blocks

No vestige of a beginning—no prospect of an end.

JAMES HUTTON, 1795

Earth has four major surface reservoirs—its lithosphere, its atmosphere, its hydrosphere, and its all-encompassing ecosphere that interconnects with the other spheres. In this chapter, we consider some aspects of the lithosphere with emphasis on the concept of geologic time and the building blocks of the solid Earth—minerals and rocks. The following chapter continues with the theme of the solid Earth, focusing on the mid-twentieth century revolutionary paradigm in geology, plate tectonics, the relationship of the interior of the planet to the lithosphere, and some important processes within the crust of Earth that are germane to concepts of global change. These discussions are relevant to our consideration of the 4.55 billion year history of the Earth prior to major human interference in Earth system processes resulting from human population growth and industrialization discussed in Chapter 9.

The shape of Earth is a slightly oblate sphere, bulging a bit at the equator because of the centrifugal force of its rotation. The equatorial radius is 6378 kilometers (km), 21 kilometers longer than from pole to center. A cross section at the equator is not quite a circle, but the deviation from a circle is very small.

The mass of Earth is huge, 5.98×10^{27} grams (g). The planet has a mean density of 5.52 grams per cubic centimeter (g/cm^3), that is about twice that of average crustal rock. Measurements of gravitational force at various places on Earth's surface show that the density increase is very nearly radially symmetrical. However, if the force of gravity on a given mass is measured at Earth's surface, it may differ significantly from place to place. If the gravity readings are corrected for several factors that influence them, it is found that the differences between readings, or anomalies, tend to disappear. The overall picture is one of nearly complete isostatic adjustment between large crustal blocks of Earth. The continents "float high" because they are thicker and less dense than the materials that make up the ocean basins. However, there are regions that exhibit marked residual gravity anomalies even after corrections are made. An example is the south side of the Indonesian Archipelago where there is a marked negative gravity anomaly. This feature is due to the subduction of the oceanic crust of basalt and sediments in this region because of plate tectonic forces (see Chapter 3).

Earth is a huge magnet. Its magnetic field is like one that would result from having a bar magnet buried in the core of the planet. Molten iron in the outer core of Earth churns around the solid inner core. It is the motion of this good electrical conductor acting as a dynamo that generates Earth's magnetic field. The intensity of the field and the positions of the north and south magnetic poles change significantly on a geologically short time scale. There is good evidence from the reversal of the magnetic orientation of magnetic minerals in rocks that Earth's magnetic field has actually reversed many times in the past few millions of years.

Earth is a very energetic planet. There is continuous emission of heat from the planetary interior due to the heat released from the radioactive decay of the elements uranium, thorium, and potassium found in minerals at depth within the Earth. The average temperature gradient in drill holes in the shallow crust is about 30° centigrade per kilometer (°C/km). Exceptionally high gradients on the order of degrees per meter may be found in volcanic or hot spring areas and at hydrothermal vents on the seafloor. The temperature gradient diminishes with depth in the Earth. At the bottom of the continental crust, temperatures are estimated to be 400–600° C. This estimate gives an average temperature gradient for the continents of about 12° C/km, a value considerably less than that in the shallow crust.

The lithosphere is the outer region of solid Earth extending to a depth of about 100 kilometers. This region and its outer skin, the oceanic and continental crusts, interact strongly with the interior of the planet. The outer skin comes in contact with the atmosphere, hydrosphere, and biosphere of the planet and is in a state of continuous dynamic change.

After Earth formed, the outer crust was subjected to many forces that altered the exterior of the planet. Internal heat creates molten material (magma) within the inner Earth. The hot magma is transported toward the surface of the planet where it is intruded into the crust or flows out as lava on its surface. Over time the surface rocks are subjected to weathering and erosion and give rise to soil and detritus for rivers and the ocean. Some rocks and their alteration products eventually recycle back into the interior of the planet and may return to Earth's surface via uplift or

volcanism. These and other forces are all part of the rock cycle and continually reshape Earth's surface. Figure 2.1 is a cartoon illustrating some of the processes affecting the lithosphere that are discussed in this chapter and Chapter 3.

GEOLOGIC FRAMEWORK OF TIME

Some Historical Events

Time is determined by a sequence of events. Human beings do not find it difficult to place themselves within the temporal framework of their own existence (see Chapter 1). However, to understand ancient events is somewhat more challenging, because of the long intervals of time involved and the fact that these "happenings" are not directly observable. The geologist interprets ancient events by observing the effects that they have had on natural materials preserved in sequences of rocks. To understand Earth's history, it is necessary to have some sequential framework in which to place our observations; that is, the relative or absolute ages of the rock units must be known.

Earth's age has been determined as approximately 4.55 billion years. Earth scientists generally accept this figure, and a questionnaire circulated to them would probably reveal that only a few expect the estimate to change drastically in the future. However, if we were to plot estimates of Earth's age versus the date on which the estimate was made, we would find that every generation of students, until the late twentieth century, has been given a greater age for Earth than was conveyed to the preceding generation. Why, then, is there confidence in today's estimate of the age of the Earth?

For most of historic time, Earth's age has been determined by religious beliefs or philosophical considerations. In many cultures the age has been considered to be infinite. According to fundamentalist Christian doctrine, Earth is very young. One of the most celebrated of the specific estimates is that of Archbishop Ussher of Ireland, who, in 1658, after careful study of the Old Testament, pronounced that Earth was created at 9 A.M. on October 23, 4004 B.C. However, even during the Renaissance there were individuals (e.g., Leonardo da Vinci) who studied rocks and fossils and became convinced that Earth must be very old—at least relative to a few thousand years. However even today, there are individuals who believe the Ussher estimate and even insist that dinosaurs roamed the Earth at that time.

FIGURE 2.1 Cartoon illustrating some processes affecting the lithosphere and Earth's surface environment—A. weathering and erosion, B. sedimentation, C. burial, D. uplift, E. metamorphism and mountain building, F. volcanism, and G. seafloor spreading.   (*Source:* a display in the Royal Museum of Central Africa, Tervuren, Belgium.)

As late as the 1820s, speculation concerning the age of Earth as deduced from the rock record was tied closely to religious and philosophical beliefs. For example, one of the prevailing geologic doctrines at this time and in the late eighteenth century was that all rocks, including those we now call igneous, were deposited from an ancient ocean that once covered the entire Earth and upon recession of the sea were left in their present state. This idea is called the "Neptunist" concept and was championed by Abraham Werner of Frieburg, Saxony. Because of the biblical record of the Flood, the concept had substantial religious appeal, which probably accounted for its widespread acceptance.

The Neptunist philosophy met a slow death because of its compatibility with theological concepts; however, it soon became apparent to geologists of the early nineteenth century that the doctrine could not stand up to the pressure of field observations of the rock record. For example, basalt rocks were shown to have crystallized directly from a molten melt. In addition, in the late eighteenth century, James Hutton of Scotland, often called the Father of Geology, developed the concept of Uniformitarianism. He saw that streams and waves deposit sand, silt, and mud. He recognized that the sandstones and shales in the hills are simply hardened and uplifted deposits of ancient streams or waves. Everything he saw in the record of the rocks seemed to him to be explainable on the basis of processes like those of the present. Furthermore, his studies indicated an Earth history consisting of an endless series of cycles of deposition, burial, uplift, erosion, and deposition. He saw no evidence for a beginning to the history of Earth and no suggestion of an end. His Uniformitarianism concept that "the present is the key to the past," as envisioned today, refers to the time invariance of natural physical laws and is simply an extension of general scientific methodology to interpretation of the rock record. Nevertheless, Hutton's ideas, which were popularized by John Playfair and extended by Charles Lyell, necessitated an Earth age of much more than a few thousand, or hundred thousand, years.

In the second half of the nineteenth century, Lord Kelvin, a renowned Scottish physicist, attempted to calculate an age for Earth on the basis of its thermal history. Radioactivity was unknown at the time, so he calculated the time required for Earth to cool from an original extremely high temperature to its present state. He considered the time required as a maximum for the age of Earth. Kelvin obtained a "best estimate" of the order of 100 million years. Even this estimate was at variance with observations of the rock record and, perhaps even more significantly, did not provide the sufficiently long geologic time required by Charles Darwin's concepts of organic evolution. The origins of species through natural selection as envisioned by Darwin necessitated a much older age of Earth.

The discovery of radioactivity by Becquerel in 1896 produced almost immediate repercussions on the age problem. For one thing, Kelvin's estimate necessarily was found to be much too short, for the heat produced by the decay of radioactive elements had not been considered in his calculations. Not only did the discovery of radioactivity upset previous age estimates, it also supplied a method for determining the ages of minerals based on the degree of disintegration of radioactive elements contained in them (see Box 2.1, Radioactive Elements).

Prior to the discovery of radioactivity, only relative ages of rock units could be determined. For example, an igneous intrusive rock cutting across pre-existing strata is certainly younger than the rocks it injected. Also, younger sedimentary rocks lie above older rocks unless there has been a major disturbance of the rock sequence. Now, ancient events as

recorded in the rock record can be assigned ages in years and the age of Earth of 4.55 billion years can be determined with some confidence. With a time scale and the assumption that natural laws are invariant, the geologist is armed with the tools necessary to interpret Earth's history.

The Geologic Time Scale

Geologists have constructed a calendar of planetary history, the geologic time scale. This calendar helps us to understand the history of Earth and to gain a picture of its cycling behavior and evolution through time. The construction of the calendar was based on studies of the distribution of sedimentary rocks and the fossils they contain and determinations of the age of these materials. The relative ages of rocks can be obtained by inspection of their superposition in a thick sequence of rocks. For undisturbed sedimentary rock layers, or strata, the oldest are on the bottom and the youngest are on the top. The oldest sedimentary layers contain fossilized organisms that lived long ago, whereas the younger layers bear fossils of organisms that have inhabited the planet more recently (Figure 2.2). The absolute age of sedimentary strata

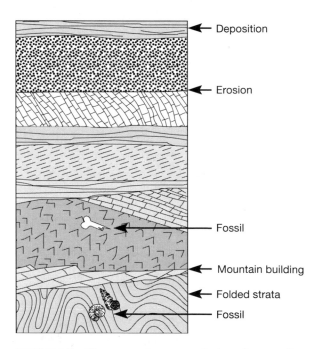

FIGURE 2.2 Diagrammatic representation of a complicated sequence of rock strata with oldest layers on the bottom and youngest layers at the top. The different strata and associated fossils and their orientation record a complex series of events of mountain building, folding, erosion, and deposition.

BOX 2.1

Radioactive Elements

The discovery by Becquerel in 1896 that the element uranium disintegrates spontaneously into other elements caused a revolution in the physical sciences. Here, we look at the process of radioactive decay to appreciate more fully the nature of decay of radioactive elements and the determination of the absolute age of Earth.

We need not be concerned here with all of the recent advances in particle physics and the fact that the Standard Model of modern particle physics has 40 species of elementary particles, but we can treat atoms as if they consisted only of a nucleus, containing protons of unit mass and unit positive charge and of neutrons of unit mass and no charge, orbited by electrons of unit negative charge and negligible mass. The chemical properties of the elements are controlled by the configuration of the electron cloud around the relatively tiny nucleus. The number of protons equals the number of electrons and is the basis for assignment of atomic number. The atomic weight equals the sum of protons and neutrons. According to this picture, the atoms should all have whole-number atomic weights, and this conclusion is true to a close approximation.

Because chemical properties depend almost entirely on the number of electrons, and hence the number of protons, atoms of a given element might and do exist with identical numbers of protons but with different numbers of neutrons. Such atoms of the same element with different masses are called isotopes.

It was many years after the discovery of the spontaneous disintegration of uranium that its decay scheme was worked out; by chance, it turned out to be one of the most complicated of the radioactive elements. In a magnificent series of early experiments, carried on in large part at the Cavendish Laboratories in England under the direction of Lord Rutherford, it was discovered that three kinds of emanations from radioactive materials were taking place: alpha particles, consisting of two protons and two neutrons (the nucleus of a helium atom); beta particles, identified as electrons; and gamma rays, electromagnetic radiation of high frequency and great energy. Later, it was found that uranium itself consists of isotopes, one with 92 protons and 146 neutrons and hence an atomic weight of 238 ($_{92}U^{238}$), and another with 92 protons and 143 neutrons and an atomic weight of 235 ($_{92}U^{235}$). The ratio of $_{92}U^{238}$ to $_{92}U^{235}$ in nature is about 138:1. A third isotope, $_{92}U^{234}$, is less than 0.006 percent of all uranium. The following table shows the decay scheme of $_{92}U^{238}$.

Decay of $_{92}U^{238}$

Element	Symbol	Emission	Original name	Half-life
Uranium	$_{92}U^{238}$	Alpha	Uranium I	4.47×10^9 years
Thorium	$_{90}Th^{234}$	Beta	Uranium X_1	24 days
Protactinium	$_{91}Pa^{234}$	Beta	Uranium X_2	1.2 min
Uranium	$_{92}U^{234}$	Alpha	Uranium II	248,000 years
Thorium	$_{90}Th^{230}$	Alpha	Ionium	80,000 years
Radium	$_{88}Ra^{226}$	Alpha	Radium	1,622 years
Radon	$_{86}Rn^{222}$	Alpha	Radon	3.8 days
Polonium	$_{84}Po^{218}$	Alpha	Radium A	3.0 min
Lead	$_{82}Pb^{214}$	Beta	Radium B	26.8 min
Bismuth	$_{83}Bi^{214}$	Beta	Radium C	19.7 min
Polonium	$_{84}Po^{214}$	Alpha	Radium C'	1.6×10^{-4} sec
Lead	$_{82}Pb^{210}$	Beta	Radium D	21 years
Bismuth	$_{83}Bi^{210}$	Beta	Radium E	5.0 days
Polonium	$_{84}Po^{210}$	Alpha	Polonium	138.4 days
Lead	$_{82}Pb^{206}$	—	Lead	Stable

A brief discussion of the first few steps in the decay scheme of uranium should suffice to indicate the consecutive processes involved when an individual atom of $_{92}U^{238}$ decays. First, an alpha particle is emitted from the nucleus, reducing the mass by 4 units and the nuclear charge by 2 units, producing the nucleus of thorium 234 ($_{90}Th^{234}$). Two electrons also are lost from the outer part of the electron cloud to balance the loss of protons from the nucleus. However, the energy change is so small in this loss, compared to the violent expulsion of the alpha particle from the nucleus, that the process can be ignored in an energy–mass balance. Next, a beta particle (electron) is expelled from the nucleus, and the reaction can be considered to be

$$Neutron = proton + e\text{-}\uparrow.$$

This process increases the nuclear charge by 1 and produces $_{91}Pa^{234}$, with the same mass as $_{90}Th^{234}$ but with a higher atomic number. Then another beta particle is expelled, and $_{92}U^{234}$ results. The final product of the long series of disintegrations and new elements is $_{82}Pb^{206}$.

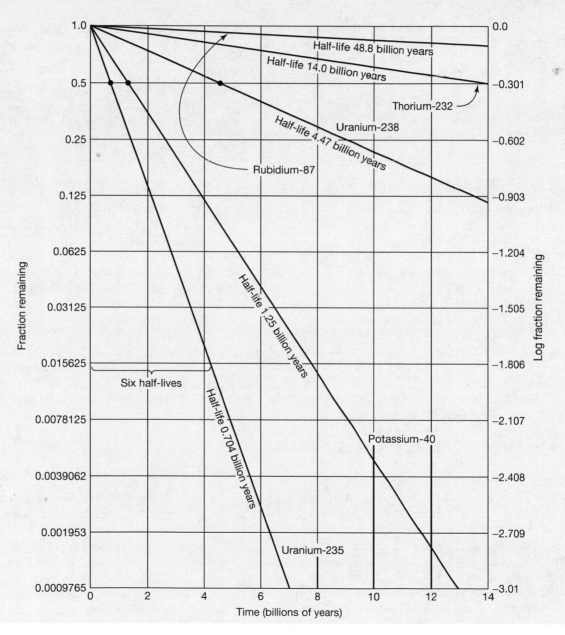

continued

If we were to crystallize today a block of pure $_{92}U^{238}$, consisting of a very large number of atoms, and then analyze it several million years hence, we would find all the elements of the reaction series in proportions determined by the relative rates of the decay reactions. At the end of a few millions of years more, the proportions of the various daughter elements would be found to have stayed the same, except for the accumulation of $_{82}Pb^{206}$, the stable end product. For any large number of uranium atoms, the fraction that disintegrates per unit time is constant; that is, the rate of loss of uranium is proportional to the amount present. It can be shown that

$$\log \frac{U}{U_0} = -kt$$

where U_0 and U are the number of uranium atoms present at time zero and time t, respectively, and k is the decay constant. Thus, a plot of the logarithm of the fraction remaining versus time for a particular radioactive element yields a linear plot, as shown in the figure on page 17.

The time required for decay of half of the original uranium atoms is known as the half-life and is a convenient way to describe the process. The half-life of $_{92}U^{238}$ is 4.47 billion years, by coincidence about the same as the current estimates of Earth's age. This means that half the $_{92}U^{238}$ atoms present at the beginning of the Earth have decayed; half the remainder will decay in the next 4.47 billion years, leaving one-quarter of the original amount. Each radioactive element has a characteristic half-life; the figure shows several elements of interest.

By using radioactive age dating techniques on minerals, the durations of the various periods and eras in the geologic time scale have been reasonably firmly established (Figure 2.3), and the age of the oldest crustal material of Earth has been estimated to be approximately 4.28 billion years, although the exact age is controversial. The origin of the solar system is about 5 billion years.

can be obtained by using dating methods that involve the radioactive decay of elements (see Box 2.1, Radioactive Elements) found in the minerals of these rocks and other rock types associated with them.

To construct the geologic time scale (Figure 2.3), the relative and absolute ages of the rocks and the fossils contained within them are arranged in order from oldest to youngest. The time scale is subdivided into a series of time periods of varying length. Most divisions of the geologic time scale mark major environmental changes, including climatic, and associated important biological changes with new species development. Biological change, as evidenced in sedimentary strata containing fossilized organisms, generally occurs slowly. However, rapid extinctions of many groups of organisms sometimes occur. These extinctions are recorded in the rocks by the absence of fossil remains of some organisms in younger rocks. Extinctions mark the boundaries of the major subdivisions of the geologic time scale. The extinctions and resultant biological changes are generally caused by natural environmental change, such as changes in temperature and precipitation patterns, atmospheric composition, and sea level. These environmental changes are due to a number of causes. They include the amount of solar radiation reaching the planet and its distribution, changes in the intensity of plate tectonics, volcanic eruptions, and the impacts of extraterrestrial bodies. More recently, human activities have been the cause of rapid and profound extinctions of

plant and animal life. These recent extinctions are on the scale of some of the larger geologic extinction events. Indeed some scientists recognize a very recent period of time extending from about 300 years in the past and into the future termed the **Anthropocene** because human activities have become a geologic force of planet Earth and led to rapid modern extinctions of plants and animals.

The geologic time scale is a standard for scientists all over the world (Figure 2.3). The calendar spans 4.55 billion years of Earth history and is modified when new information concerning the planet is obtained. The major events of Earth's history within the context of the geologic time scale are discussed in Chapter 8.

EARTH'S BUILDING BLOCKS: MINERALS

The rock cycle involves the destruction of rocks, the redistribution of the products of this destruction, and eventually their reconstitution into new rocks. Rocks are aggregates of minerals—naturally occurring, inorganic, chemical compounds (see Box 2.2, Matter: Atoms, Elements, and Compounds; some important minerals and their characteristics are given in Appendix A.) The behavior of rocks depends on the chemical, physical, and geometric properties of the individual minerals, as well as the way in which the mineral aggregates are held together.

Eon	Era	Period	Epoch	Millions of years ago
Phanerozoic	Cenozoic	Quaternary	Holocene	Today
				0.01 (10,000 years ago)
			Pleistocene	
				1.6
		Tertiary	Pliocene	
				5.3
			Miocene	
				23.7
			Oligocene	
				36.6
			Eocene	
				57.8
			Paleocene	
				65.0
	Mesozoic	Cretaceous		
				144
		Jurassic		
				208
		Triassic		
				245
	Paleozoic	Permian		
				286
		Carboniferous		
				360
		Devonian		
				408
		Silurian		
				438
		Ordovician		
				505
		Cambrian		
				545
Precambrian:				
•Proterozoic				
				2500
•Archean				
				~3800
Hadean				
				4600

FIGURE 2.3 The geologic time scale. (*Source:* Skinner and Porter, 1995.)

The elements that comprise the minerals of rocks on Earth were formed during the evolution of the universe before the planet was created. Only a dozen elements make up 99% of the chemical composition of Earth as a whole—iron, oxygen, silicon, manganese, nickel, sulfur, calcium, aluminum, chromium, sodium, potassium, and titanium. The same dozen in different relative abundance make up a comparable percentage of Earth's crust. They can be organized into five major types of compounds: elements, oxides, silicates, aluminosilicates, and sulfides. When we focus our attention on the surface environment, we need to add hydrogen and carbon to our list of elements and carbonates and sulfates to the types of mineral compounds. There are minerals made of the elements calcium, carbon, and oxygen ($CaCO_3$, calcite); aluminum and oxygen (Al_2O_3, corundum, gem variety is sapphire); sodium and chlorine ($NaCl$, common table salt); aluminum, oxygen, silicon, and hydrogen [$Al_2Si_2O_5(OH)_4$, kaolinite, a common clay mineral]; and many other combinations.

In general, minerals are natural examples of what the chemists call **solid phases** that is, they are solid substances that are homogeneous throughout and separated from other substances by a well-defined boundary. Minerals usually occur in grains ranging from submicroscopic size up to giants several meters across. However, in most rocks the maximum size is of the order of a few centimeters or less.

BOX 2.2

Matter: Atoms, Elements, and Compounds

Periodic Table of the Elements

Every object in the universe is composed of matter. Because matter can be converted to energy, it is essentially a form of energy. Matter occurs in solid, liquid, and gaseous states. For example, water, the chemical compound H_2O, exists as water vapor, liquid water, and ice.

Atoms

Modern particle physics research is currently focused on particles smaller than the atom with less structure. These particles include the atomic constituents of electrons, protons, and neutrons, particles produced by radiative and scattering processes, such as photons, neutrinos, and muons, and a range of exotic particles. All the particles and their interactions can be described by a quantum field theory known as the Standard Model, in which there are 40 species of elementary particles. However, matter in the universe can still be described in terms of the tiny atomic particles. An **atom** is the smallest unit of an element that can exist either alone or in combination and consists of a dense central nucleus containing a mix of positively charged protons and electrically neutral neutrons (except for hydrogen, with atomic weight = 1 with zero neutrons) surrounded by a cloud of negatively changed electrons. The atom is the smallest particle [on the order of angstroms (Å); $1 \text{ Å} = 10^{-10}$ meters] that can enter into a chemical reaction. Most material in the universe consists of a combination of atoms. Most atoms never change; they only combine with other atoms to make different substances. Radioactive atoms, however, do change and eventually decay into stable, nonradioactive atoms. An example is the radioactive decay of uranium atoms to atoms of lead.

Chemical Periodic Table

IA	IIA											IIIA	IVA	VA	VIA	VIIA	O
																1 **H** 1.00797	2 **He** 4.0026
3 **Li** 8.939	4 **Be** 9.0122											5 **B** 10.811	6 **C** 12.01115	7 **N** 14.0067	8 **O** 15.9994	9 **F** 18.9984	10 **Ne** 20.183
11 **Na** 22.9898	12 **Mg** 24.312	IIB	IVB	VB	VIB	VIIB	VIII			IB	IIB	13 **Al** 26.9815	14 **Si** 28.086	15 **P** 30.9738	16 **S** 32.064	17 **Cl** 35.453	18 **Ar** 39.948
19 **K** 39.102	20 **Ca** 40.08	21 **Sc** 44.956	22 **Ti** 47.90	23 **V** 50.942	24 **Cr** 51.996	25 **Mn** 54.9380	26 **Fe** 55.847	27 **Co** 58.9332	28 **Ni** 58.71	29 **Cu** 63.54	30 **Zn** 65.37	31 **Ga** 69.72	32 **Ge** 72.59	33 **As** 74.9216	34 **Se** 78.96	35 **Br** 79.909	36 **Kr** 83.80
37 **Rb** 85.47	38 **Sr** 87.62	39 **Y** 88.905	40 **Zr** 91.22	41 **Nb** 92.906	42 **Mo** 95.94	43 **Tc** (99)	44 **Ru** 101.07	45 **Rh** 102.905	46 **Pd** 106.4	47 **Ag** 107.870	48 **Cd** 112.40	49 **In** 114.82	50 **Sn** 118.69	51 **Sb** 121.75	52 **Te** 127.60	53 **I** 126.9044	54 **Xe** 131.30
55 **Cs** 132.905	56 **Ba** 137.34	57 **La** 138.91	72 **Hf** 178.49	73 **Ta** 180.948	74 **W** 183.85	75 **Re** 188.2	76 **Os** 190.2	77 **Ir** 192.2	78 **Pt** 195.09	79 **Au** 196.987	80 **Hg** 200.59	81 **Tl** 204.37	82 **Pb** 207.19	83 **Bi** 208.980	84 **Po** (210)	85 **At** (210)	86 **Rn** (222)
87 **Fr** (223)	88 **Ra** (226)	89 **Ac** (227)	104 (257)														

Lanthanum Series

58 **Ce** 140.12	59 **Pr** 140.907	60 **Nd** 144.24	61 **Pm** (147)	62 **Sm** 150.35	63 **Eu** 151.96	64 **Gd** 157.25	65 **Tb** 158.924	66 **Dy** 162.50	67 **Ho** 164.930	68 **Er** 167.28	69 **Tm** 168.934	70 **Yb** 173.04	71 **Lu** 174.97

Actinium Series

90 **Th** 232.038	91 **Pa** (231)	92 **U** 238.03	93 **Np** (237)	94 **Pu** (242)	95 **Am** (243)	96 **Cm** (247)	97 **Bk** (247)	98 **Cf** (249)	199 **Es** (254)	100 **Fm** (253)	101 **Md** (256)	102 **No** (253)	103 **Lw** (257)

Elements

Elements consist of atoms of the same kind and when pure cannot be decomposed by a chemical change. There are 118 known chemical elements; 94 occur naturally on Earth and 24 are artificially created.

Chemists arranged the elements in a table based on the properties of the elements. This periodic table of the elements was developed in the 1800s. Some elements are readily known, such as oxygen, hydrogen, gold, copper, and iron. The elements most used commercially by people in order of use are carbon (C), in the form of coal, oil, and gas; sodium (Na), in table salt and other products; iron (Fe), used in the steel industry; and nitrogen (N), sulfur (S), potassium (P), and calcium (Ca), all used in fertilizers or as soil conditioners for our food supply. Other elements are less well known or used, such as argon, antimony, bismuth, and krypton. Each element is represented by an abbreviation, a symbol, to form the periodic table. The elements are organized in the table into columns of similar properties. In the chemical periodic table shown above, the atomic number is given above the element symbol and the average atomic weight is given below.

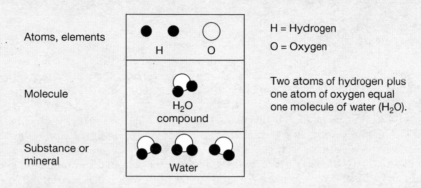

Formation of Substances

Compounds

When two or more atoms of the same or different elements are bonded in a definite proportion, a **compound** is formed. Examples of compounds are water (H_2O), carbon dioxide (CO_2), salt (NaCl), and sugar ($C_6H_{12}O_6$). The compound water (H_2O) is composed of two atoms of hydrogen (H) and one atom of oxygen (O). The compound carbon dioxide (CO_2) is composed of one atom of carbon (C) and two atoms of oxygen (O). The smallest particle of a compound that can exist and exhibit the properties of that compound is a **molecule**. The universe is composed of millions of these compounds created from the elements given in the periodic table.

Chemical Reactions

A compound is a pure substance that can be decomposed by a chemical change. The atoms in the compound may rearrange themselves. They may separate from a compound to form different compounds. These changes and interactions among compounds are called chemical reactions. A chemical equation expresses a chemical reaction process involving compounds or other chemical species. The reactants generally are shown on the left-hand side of the equation and the products on the right-hand side. Consider the decay of plant material (represented by the chemical compound CH_2O) in Earth's atmosphere containing oxygen gas (the chemical compound O_2). The simplest chemical equation representing this chemical process is

$$CH_2O + O_2 \rightarrow CO_2 + H_2O.$$

The single arrow implies that this process is irreversible—the organic matter will be completely oxidized to carbon dioxide (CO_2) and water (H_2O) in the presence of atmospheric oxygen. Some processes are highly reversible, and these are usually represented by a double arrow ($\leftrightarrow$) or an equal sign ($=$). For example, the equilibrium between $CaCO_3$ and its dissolved chemical species of Ca^{2+} and CO_3^{2-} is represented as

$$CaCO_3 \leftrightarrow Ca^{2+} + CO_3^{2-}.$$

In chemical processes, matter cannot be created or destroyed. Thus, the total number of atoms of any particular element on the left-hand side of a chemical equation is set equal to the total number of atoms of that element on the

continued

right-hand side of the equation. This is the process of balancing a chemical equation. Balancing the equation is an expression of the fact that molecules usually react in such a way as to bear simple, integral, numerical relationships to one another. These relationships are termed the **stoichiometry of the reaction.** If the stoichiometry of a chemical reaction is known, it is possible to calculate the masses of reactants and products with the use of known atomic and molecular weights. In chemical terms, the amount of a substance is expressed in moles. One **mole** of a substance is the amount that contains as many elementary entities as there are atoms in 12 grams of carbon. This number is termed **Avogadro's constant** and its value is equal to 6.022×10^{23}. In the preceding chemical equation for the equilibrium dissolution of $CaCO_3$ in water, one unit (mole) of $CaCO_3$ will dissolve in water to make one mole of Ca^{2+} and one mole of CO_3^{2-} or, in terms of mass, 100 grams of $CaCO_3$ (the gram molecular weight of $CaCO_3$) will react to give 40 grams of Ca^{2+} (the gram atomic weight of calcium) and 60 grams of CO_3^{2-} (the gram molecular weight of carbonate; gram atomic weight of C = 12 and that of O = 16). If only 10 grams of $CaCO_3$ were to dissolve, then a proportionate amount of Ca^{2+} and CO_3^{2-} would be present at equilibrium, or 4 grams and 6 grams, respectively (see also Box 2.4, Ions and Oxidation-Reduction).

Cyclic Nature of Atoms

Atoms comprise most of the matter of the universe we live in, both living and nonliving. They continuously cycle or rearrange themselves into different compounds and substances and release or absorb energy through chemical reactions. They make up the gases in the air we breathe, the land we walk on, the water we drink, and the tissues of plants and animals, including our own bodies. Because of this recycling of atoms, we might contain some atoms of extinct dinosaurs or even rocks that previously existed on Earth.

The systems of the atmosphere (air), lithosphere (soils, sediments, and rock), hydrosphere (water), and biosphere (life) are composed of organic (biotic, living, and dead) and inorganic (abiotic, nonliving) matter. Organic compounds include compounds containing carbon, particularly those specific to life forms. Most organic compounds are composed of carbon (C) and hydrogen (H) atoms (hydrocarbons). These are the compounds that make up the bodies of plants, animals, microbes, and materials derived from these organisms such as coal, oil, wood, or flour. Inorganic compounds form much of the land, air, and water and are anything that has never been alive nor has the structure or organization of living bodies. Inorganic compounds are found throughout the universe. Atoms in the form of elements and compounds are involved in the biogeochemical cycles (see Chapter 7) on the planet and are continuously being transported about the Earth surface as gases, liquids, and solids and moving from one Earth reservoir to another.

The mineral grains are usually crystalline, which is to say that they comprise repetitive groups of atoms in a fixed geometrical array internally. In some instances, where the minerals have had an opportunity to grow freely and reflect their internal structure in the development of planar faces, the grains are referred to as crystals. Also, a grain may be homogeneous throughout by any method of measurement we can devise but not consist of regularly repeated atomic groups; such minerals are called **amorphous.**

Both composition and structure are required to define a mineral; for example, the single compound SiO_2 occurs as the minerals opal, alpha quartz, beta quartz, cristobalite, tridymite, coesite, and stishovite, depending on the way in which the silicon and oxygen atoms are arranged within the grains. The structures of cristobalite, quartz, and tridymite, three of the polymorphs of silica, are shown in Figure 2.4. The individual tetrahedra are made up of a small central silicon atom surrounded by four oxygen atoms. The word **polymorphs** (often dimorphs, if there are only two known structures) is used in general to characterize minerals of the same chemical composition but of different crystalline structures. Studies of polymorphs have been of great help in investigating

Earth. Structural changes without chemical change result from changes in the temperature and pressure of the environment. Thus, knowledge of structure may give a clue to the pressure and temperature conditions of formation. Figure 2.5 and Table 2.1 give the names and chemical compositions of some common minerals. Some details of major mineral groups are given in the Appendix, Minerals.

EARTH'S BUILDING BLOCKS: ROCKS

Rocks are classified and named based in part on the minerals they contain and on other attributes, such as chemical composition and texture. There are three major classes of rocks—igneous, sedimentary, and metamorphic—each containing a spectrum of rock types.

Igneous Rocks

Some rocks are formed from molten magma generated inside Earth and intruded into the crust or extruded onto the surface of Earth through volcanoes on the seafloor or on the continents. These are the igneous rocks. The adjective *igneous* is derived from the Latin *ignis*, meaning fire. A large number of igneous rock

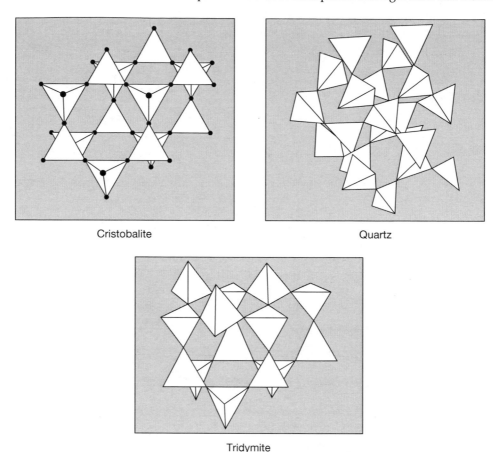

Cristobalite

Quartz

Tridymite

FIGURE 2.4 Polymorphs of silica. The tetrahedra are formed by four oxygen atoms occupying the apices of each tetrahedron. The central small silicon atom within each tetrahedron is not shown.

types exist. One classification of igneous rocks is given in Figure 2.5. This classification scheme is based on the chemistry of the rock in terms of how much silica (SiO_2) it contains, the minerals found in the rock, and whether the rock was formed by extrusive or intrusive processes.

For example, the rock granite is rich in silica and contains abundant alkali feldspar (K, Na, and Al silicate), moderate amounts of plagioclase feldspar (Na, Ca, and Al silicate) and quartz (SiO_2), and generally little amphibole (Ca, Mg, and Fe silicate containing OH^-). Granites are generally light in color, commonly pinkish, and are a major igneous rock type found on the continents. They form from molten rock intruded at depth in the crust. The subsequent cooling of the magma produces large bodies of granite rock. The Sierra Nevada Mountains of the states of California and Nevada in the United States are composed of granite-like rocks. In contrast, the rock basalt is low in silica and contains the minerals pyroxene (Mg, Ca, and Fe silicate), plagioclase feldspar, and commonly olivine (Mg and Fe silicate). Basalts are generally dark in color and are the major igneous rock type constituting the seafloor, midocean ridges, and volcanic mountains of the ocean basins. They are also found on the continents as horizontally extensive thick sheets formed from the cooling of lava as it is extruded onto the land surface. The basalt lavas found on the Columbia Plateau between the Cascades and the Rockies in the United States are an excellent example of continental basalts. Those lavas actively extruding and congealing today on the "Big Island" of Hawaii in the state of Hawaii of the United States in the Pacific Ocean and forming the massive highest mountain measured in elevation from the sea floor in the world, Mauna Loa, in Hawaii are examples of oceanic basalts.

Sedimentary Rocks

Sedimentary rocks originate from the weathering and erosion of preexisting rocks and the transportation of the debris by water, wind, and ice to areas of deposition.

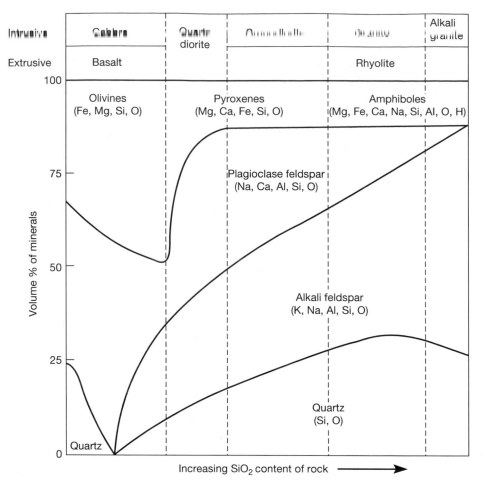

FIGURE 2.5 Classification of some common intrusive and extrusive igneous rocks according to their mineralogy and silica content. The most important elements of each mineral group are shown in parentheses. (*Source:* Clark, 1966.)

The deposition of these materials on lake and sea bottoms as gravel, sand, silt, and mud forms layers of sediment that eventually can become sedimentary rocks as these materials are buried and lithified. Some sedimentary rocks originate as chemical and biological precipitates from natural waters like seawater. The rock limestone ($CaCO_3$) is an important example. Table 2.1 gives some important characteristics of sedimentary rocks and their metamorphic equivalents.

Water, wind, and ice are responsible for the erosion of materials that eventually are deposited as sediments and converted into sedimentary rocks. The eroded materials, when deposited, are arranged such that the older materials are on the bottom and the younger sediments are on the top. Erosion, transportation, and subsequent deposition of weathered materials can lead to accumulations of sedimentary layers many kilometers thick. In the Gulf of Mexico, the thickness of the sedimentary strata reaches nearly

13,000 meters. This material has been derived from the erosion of the North American continent and transport of dissolved and particulate materials via the Mississippi River and its ancient precursors to the Gulf. Sedimentary rocks cover 80% of the present land surface of Earth.

New and freshly deposited sediments are consolidated (particles squeezed together) and cemented (spaces between particles filled by naturally occurring cements) during burial, and the sediments solidify into sedimentary rocks. For example, vast coal formations have been formed from the burial of plant materials deposited in ancient swamps and their subsequent consolidation into bituminous (soft) and anthracite (hard) coals. Sediments and sedimentary rocks contain abundant information on the evolution of life on Earth in the form of fossilized organisms. Fossilized clams, oysters, corals, and sea urchins and fossilized bones of dinosaurs, horses, and ancient

TABLE 2.1 Some Sedimentary Rock Characteristics and Their Metamorphic Equivalents

Rock	Some Sediment Equivalents	Some Metamorphic Equivalents	Origin	Grain Size	Major Minerals	Wt % of Sedimentary Rock Mass
Conglomerate	Gravel: Pebbles Cobbles Boulders	Conglomerite	Clastic particles	>2 mm	Rock fragments	<1
Sandstone	Sand	Quartzite	Clastic particles	0.062–2 mm	Quartz (SiO_2) Feldspar ($KAlSi_3O_8$) Rock fragments	10–15
Shale	Silt Mud Clay	Slate Schist	Clastic particles	<0.062 mm	Clay minerals [e.g., $Al_2Si_2O_5(OH)_4$] Quartz Feldspar	70–80
Limestone	Reef Lime mud	Marble	Biochemical precipitate, inorganic precipitate, clastic particles (shell fragments)	Variable, up to cms	Calcite ($CaCO_3$) Aragonite ($CaCO_3$) Dolomite [$CaMg(CO_3)_2$]	5–20
Evaporite	Salt	None	Precipitate from evaporating water	Variable, up to cms	Halite (NaCl) Gypsum ($CaSO_4.2H_2O$) Anhydrite ($CaSO_4$)	2–3

(*Source:* Garrels and Mackenzie, 1971.)

humans are found encased in sedimentary materials. In addition, sedimentary rocks contain chemical, mineralogical, and isotopic information that can be used to interpret the environment of the past, such as the chemistry of seawater, the temperature of the ocean, and the CO_2 content of the atmosphere.

The properties of sedimentary rocks on land can be studied by collecting samples taken from rocks exposed at the surface or by drilling deep holes into the rocks and collecting samples at depth. Exploration for oil and gas has resulted in the drilling of millions of wells all over the world and the acquisition of many data on the physical, chemical, and biological properties of sedimentary rocks.

The existing mass of sedimentary rocks preserved during the past 600 million years has been estimated to be roughly twice that of the Precambrian mass, and in general is not as complexly contorted or metamorphosed. The history of the time interval they represent is more completely preserved in these strata. From the Paleozoic on, a relatively continuous fossil record of the evolutionary history of plants and animals is found in the sedimentary strata (see Chapter 8). Abundant fossils of shelled organisms first appear in Cambrian rocks. No definite shell-bearing fossils have been found in rocks older than Cambrian. This remarkable burst of evolution gave rise to the body plans of all modern multicellular organisms. One of the great mysteries of biology and geology is the cause of this evolutionary explosion—the "Big Bang" of animal evolution, and why no fundamentally new designs for life have emerged since this event.

In general, sedimentary rocks of Cenozoic age still tend to have a "new" look about them. Porosity is high and their mineralogy is much like that of modern sediments. These rocks may be poorly cemented, and some fossil shells still retain vestiges of their original colors and composition. However, Precambrian rocks of sedimentary origin are generally contorted and metamorphosed. Fossilized remains of organisms are rare, and porosity is very low.

Bitter experience has shown that use of the degree of chemical and physical change from original sediments as a criterion of age can be an extremely dangerous practice. It is true only as the broadest kind of generalization. In some areas, sedimentary rocks 2 billion years old, by the accident of their geography and past history, are less altered than others that have survived only a few million years.

Hutton's "No vestige of a beginning—no prospect of an end" still holds in one sense for the results of studies of Earth's surface. Presumably there was a beginning of the cycles of erosion, deposition, burial, uplift, and erosion again when the first sediments were formed by weathering of primordial igneous crust, but no one has yet discovered remnants of that igneous crust. The oldest rocks we know are originally of sedimentary origin.

One of the most remarkable aspects of the sedimentary rock record is the obvious similarity of erosional and depositional processes throughout the past 3 billion years. If we do our best to eliminate post-depositional changes, it emerges that the compositions, textures, and structures of the most ancient sedimentary rocks can be matched in remarkable detail by comparable features found in modern sediments. It is perhaps this aspect of the rock record that has kept the concept of Uniformitarianism as a basic scientific principle for 200 years.

Deep-Sea Sediments and Earth History

Most sedimentary rocks started their life in the oceans as layers of sediment. Rivers transport rock particles, dissolved substances, and nutrients out to the oceans. Animals and plants in the sea use the nutrients and dissolved substances, like nitrate (NO_3^-), calcium (Ca^{2+}), and dissolved silica (SiO_2), brought in with the water to form their organic matter and inorganic skeletons. Some of the skeletal material and organic matter is deposited along with original or altered rock particles to form layers upon layers of sediments.

To study these sediments, which contain information about Earth's history, sedimentary cores are taken of the sediments at the ocean bottom, sometimes at depths of the seafloor exceeding 5000 meters. A long tube-shaped pipe is drilled into the sea bottom from a ship and then extracted. The materials encased inside the pipe are brought back to the surface, where their chemical, mineralogical, isotopic, and biotic properties can be studied. In most cases, the ages of the sediments collected in the core are determined by paleontological means or by absolute age dating methods using radioactive isotopes.

The cored sediments contain a wealth of scientific data that can provide information on the history of Earth, such as when certain kinds of plants lived and what the temperature of the ocean was long ago. For example, the sediment cores may contain fossils of the simple organisms foraminifera ("forams," protozoans) that secrete calcium carbonate shells in the form of calcite (hexagonal $CaCO_3$) when alive. These tiny creatures live in surface waters of many areas of the world ocean. When alive, they extract dissolved calcium and carbon as carbonate ion (CO_3^{2-}) from seawater to make their shells. When they die, they sink toward the bottom of the ocean, and if not dissolved in transit to the sea floor, their remains can form thick layers of sediment. The temperature of the ocean in which the forams lived can be determined by studying the isotopic composition of the oxygen contained in the $CaCO_3$ of the foram shells (see Box 2.3, Stable Isotopes). The oxygen isotopic record recorded in the tests of forams for the past 700,000 years of Earth history is shown in Figure 2.6. The periodic variations in

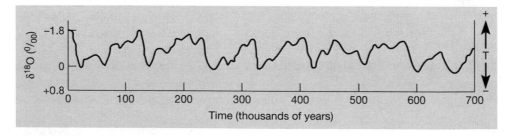

FIGURE 2.6   Generalized oxygen isotope record ($\delta^{18}O$) as recorded in the calcium carbonate of foram tests during the past 700,000 years of the Quaternary. The negative values imply warmer sea-surface temperatures, whereas the positive values indicate cooler temperatures. T is temperature.   (*Source:* Emiliani and Shackleton, 1974.)

BOX 2.3

Stable Isotopes

Because stable isotopic measurements have become a powerful tool used to interpret the environment and Earth history, it is convenient to discuss briefly some of the uses and potentials of measurements of some stable isotope ratios. Because isotopes have almost identical chemical properties but different masses, they have a tendency to become separated only by mass-dependent processes, such as evaporation or ionic diffusion. Also, many organisms tend to use one isotope selectively in their metabolic processes.

Oxygen Isotopes

Oxygen has three isotopes, $^{16}O_8$, $^{17}O_8$, and $^{18}O_8$, with relative abundances of 99.8, 0.04, and 0.2%, respectively. The $^{18}O/^{16}O$ ratios in minerals and in waters have been used extensively for their classification and for investigation of environmental processes.

Variations in the oxygen isotope ratio are currently expressed in parts per mil (parts per mil equal parts per thousand) and are given as deviations from an arbitrary standard. In the case of oxygen, the ratio of oxygen isotopes in Standard Mean Ocean Water (SMOW) or that (known as the PDB standard) of the oxygen of CO_2 produced from calcium carbonate (calcite, $CaCO_3$, originally from belemnites found in the Cretaceous PeeDee Formation of South Carolina) has been utilized as the standard. Currently, all measurements for oxygen isotope paleotemperature work are referenced to a powdered $CaCO_3$ standard supplied by the U.S. National Bureau of Standards. Deviations from the standard, in parts per mil, δ, are calculated from the relation

$$\delta^{18}O(\text{‰}) = \left[\frac{(^{18}O/^{16}O)_{\text{sample}} - (^{18}O/^{16}O)_{\text{standard}}}{(^{18}O/^{16}O)_{\text{standard}}} \right] \times 1000.$$

Consequently, positive values of $\delta^{18}O$ indicate enrichment of the sample in ^{18}O compared to the standard, whereas negative values indicate depletion of this isotope. The following figure shows the isotopic variation of oxygen in nature.

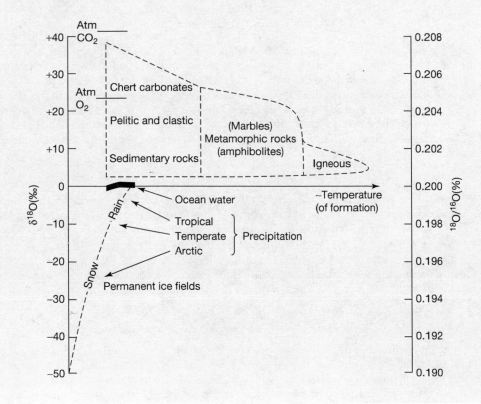

continued

Variations in the oxygen isotopic composition of materials primarily reflect the temperature and composition of the environment. Because minerals precipitated from various water bodies tend to be influenced by the isotopic composition of the water, oxygen isotopes have been used to help decide the origin of waters and whether various minerals are of marine or freshwater origin. This has been a particularly powerful tool in determination of the origin of cements in limestones. Oxygen isotopes have also been used as a thermometer. If isotopic equilibrium is established between two oxygen-containing minerals, the equilibrium constant and hence the $^{18}O/^{16}O$ ratios in the minerals are a function of temperature. Knowing the equilibrium constant as a function of temperature from experimental work and determining the $^{18}O/^{16}O$ ratio of a mineral by mass spectrometry allow one to calculate the temperature.

During evaporation, H_2O vapor derived from the tropical ocean is enriched in the light isotope ^{16}O, and thus tropical ocean water is enriched in the heavy isotope ^{18}O. Much of the water vapor is transported to higher latitudes (and higher altitudes), where it condenses and falls back to Earth as precipitation. Condensation and precipitation of water vapor favor removal of the heavy isotope ^{18}O in precipitation that leads to further depletion of the water vapor in ^{18}O as it moves to higher latitudes. The transport of water vapor toward the poles consists of many cycles of evaporation and precipitation. Because of the repeated fractionations, water vapor in the atmosphere becomes progressively depleted in ^{18}O from equatorial regions toward higher latitudes. The long-term storage of water as ice in the great continental ice sheets at high latitude will leave ocean water enriched in ^{18}O (depleted in ^{16}O).

At least as far back as Cambrian time, the oceans have been fairly constant in their $^{18}O/^{16}O$ ratio or have had a small evolutionary decrease in this ratio. However, small variations have occurred during geologic time, and these are important in interpreting the temperature of the ocean and the history of ice on the planet. For example, oxygen isotopes have been used to interpret the temperature history of the oceans during the past 700,000 years. Small, floating protozoans called foraminifera, which secrete a $CaCO_3$ skeleton, live in the surface layers of the sea. When the organisms die, their shells fall to the bottom and are incorporated into the sediments of the deep sea. Long cores of these sediments have been obtained. The forams have been separated from the sediments and the $^{18}O/^{16}O$ ratios of the calcareous tests determined by mass spectrometry as a function of depth in the cores. Shifts in the oxygen isotope ratio occur with depth in the cores and, after correction for the amount of water stored in the continental ice sheets, have been interpreted as being the result of shifts in the temperature of the surface ocean (Figure 2.6). The history of sea level during part of the Pleistocene has also been determined using the record of $\delta^{18}O$ derived from uplifted coral reefs in New Guinea and Barbados.

Hydrogen Isotopes

Hydrogen has three naturally occurring isotopes that include ^1H, $^2H_1$, and $^3H_1$. $^1H_1$ is the most abundant stable isotope and constitutes 99.98% of hydrogen. Its nucleus consists of one proton. $^2H_1$ is the other stable isotope of hydrogen and its nucleus consists of one proton and one neutron. It is known as deuterium (D). $^3H_1$ contains one proton and two neutrons in its nucleus and is known as tritium. It is radioactive and decays to $^3He_2$ by beta-minus decay (a neutron in the $^3H_1$ atom's nucleus turns into a proton, an electron, and an antineutrino) with a half-life of 12.32 years.

The deuterium-to-hydrogen ratio (D/H) of natural waters and other fluids is especially helpful in investigation of the origin and history of the fluid and in understanding the reactions between fluids and rock minerals. The D/H ratio is defined as

$$\delta D = [(D/H)_{sample} - (D/H)_{smow}/(D/H)_{smow}] \times 1000.$$

The standard in this case is Standard Mean Ocean Water (SMOW). The standard was originally defined with reference to a large volume of distilled water distributed by the National Bureau of Standards in the United States (NBS-1) such that D/H (SMOW) = 1.050 D/H (NBS-1). Hydrogen isotopes along with oxygen have been used to determine the origin and source of meteoric waters. In the figure below, δD has been plotted against $\delta^{18}O$ for a large number of meteoric waters (water that occurs in or is derived from the atmosphere) collected at different latitudes. Harmon Craig at the Scripps Institute of Oceanography first demonstrated that the isotopic compositions for water and snow exhibit a pronounced latitudinal effect, reflecting the temperatures of condensation. The δD and $\delta^{18}O$ values are linearly related according to

$$\delta D = 8 \delta + 10.$$

Some authors employ an intercept of 5. This relationship is the result of the isotopic fractionation of both the isotopes of oxygen and hydrogen during evaporation of water from the ocean and the subsequent condensation of the water vapor in clouds, which leaves fresh water generally depleted in ^{18}O and D relative to seawater. This fractionation process is a form of Rayleigh distillation. The trend line is often referred to as the global meteoric water line for natural terrestrial waters. In the figure, waters from closed basins and certain lakes with excess evaporation plot off the meteoric line. δD,

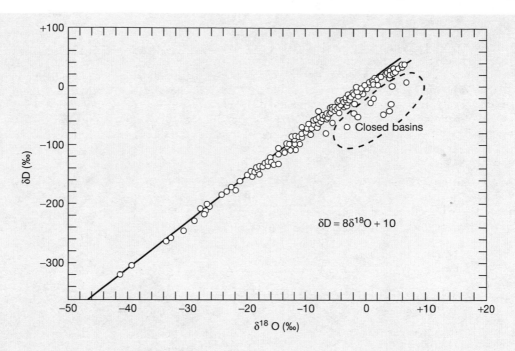

as with $\delta^{18}O$ mentioned above, can also be used to determine the temperature at which snow and ice formed in long ice cores, like the Vostok and the Dome C European Project for Ice Coring in Antarctic cores discussed in Chapter 13.

Carbon Isotopes

Much work has been done with the $^{13}C_6$ isotope in relation to the abundant $^{12}C_6$. ^{12}C accounts for more than 99% of the carbon present on Earth, with ^{13}C accounting for most of the rest. Variations of the ratio of the two isotopes as much as 10 to 12% occur. The standard for carbon isotope measurements was originally the carbon in the $CaCO_3$ from belemnites collected from the Cretaceous PeeDee Formation. The $^{12}C/^{13}C$ ratio of this carbon is 88.99. The U.S. National Bureau of Standards now supplies the $CaCO_3$ standard for use as a common reference point. Differences in the carbon isotopic composition of various substances are expressed as

$$\delta^{13}C(\permil) = \left[\frac{(^{13}C/^{12}C)_{sample} - (^{13}C/^{12}C)_{standard}}{(^{13}C/^{12}C)_{standard}} \right] \times 1000.$$

The following figure shows the carbon isotope ratios of some naturally occurring carbon-bearing materials.

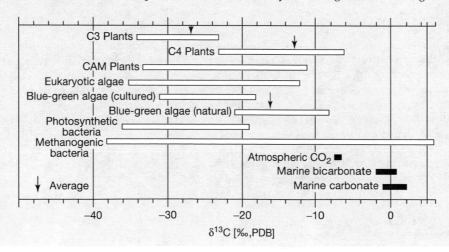

continued

Photosynthesis, during which CO_2 from the atmosphere is utilized, is attended by marked selection of the "light" carbon isotope ^{12}C, that is, plant tissue is depleted in ^{13}C. Atmospheric CO_2 has a $\delta^{13}C$ value of -7, whereas plants are much "lighter," having $\delta^{13}C$ ratios varying from an average of about -12 to -27. Among the problems that have been studied by the aid of $^{13}C/^{12}C$ ratios is the origin of petroleum. In the laboratory, progressive distillation of oily materials yields fractions with increasing $^{13}C/^{12}C$ ratios. In the field, attempts have been made to follow the migration routes of hydrocarbons through rocks to form oil accumulations by mapping out the spatial variation of the isotope ratio. Also, $^{13}C/^{12}C$ ratios of petroleum are compatible with an organic origin of this important energy source. In fact, carbon isotope ratios indicate that petroleum is probably derived from the lipid (fat and waxes) fraction of pelagic marine organisms.

Sulfur Isotopes

The two chief isotopes of sulfur are $^{32}S_{16}$ and $^{34}S_{16}$. $^{32}S_{16}$ is by far the more abundant; the average ratio in earth materials is 22.6:1. $^{33}S_{16}$ and $^{36}S_{16}$ also exist but make up only 1% of all sulfur. The mass difference of some 6% between ^{32}S and ^{34}S is sufficient for marked fractionation in many Earth-surface processes. In the case of sulfur, the ratio in meteorites (Canyon Diablo troilite meteorite; $^{32}S/^{34}S = 22.22$) is used as the standard, and the deviations from the standard are expressed as

$$\delta^{34}S(\permil) = \left[\frac{(^{34}S/^{32}S)_{sample} - (^{34}S/^{32}S)_{standard}}{(^{34}S/^{32}S)_{standard}} \right] \times 1000.$$

The following figure is a summary diagram of the $^{32}S/^{34}S$ relations of some sulfur-containing materials.

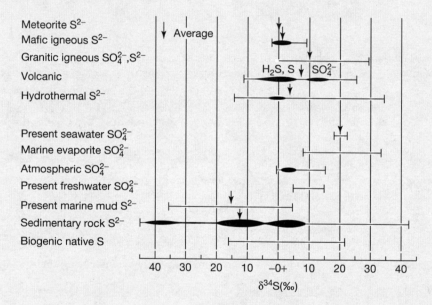

Sulfur in sulfates is "heavier" than that in sulfides. This difference is caused by organisms; sulfur-reducing bacteria selectively utilize ^{32}S when they convert sulfate into sulfide. These bacteria live within sediments that are depleted of oxygen (anoxic) and oxidize organic material as they reduce sulfate. The hydrogen sulfide generated usually reacts with minerals containing iron to produce iron sulfides, most of which eventually become pyrite, FeS_2, or is released to the atmosphere as hydrogen sulfide gas (H_2S) where it is oxidized to sulfate. Sulfate can be reduced inorganically under sedimentary conditions only at a vanishingly small rate; fractionation is therefore evidence of biochemical processes.

The sulfate of the present-day oceans averages about $\delta^{34}S = +20\permil$, and sedimentary sulfides about $-20\permil$; therefore, at present about half of the total sulfur of the sedimentary rock–ocean system is present as sulfate, and most of the other half resides in sedimentary rocks as iron sulfides. The isotopic composition of sedimentary sulfate deposits varies as a function of their geologic age. Because the precipitation of gypsum ($CaSO_4 \cdot 2H_2O$) or anhydrite ($CaSO_4$) from seawater causes little fractionation, this variation implies that the distribution of sulfide and sulfate sulfur between rocks and oceans has changed markedly from time to time.

Sulfur isotope ratios have provided information on many other aspects of geologic history; for example, they have been used to determine that certain economic deposits of copper sulfide gained their sulfur from sedimentary sources and that certain others have been derived from the mantle or the deep crust. Also, sulfur isotope ratios have been used to determine the sources of atmospheric sulfur in some urban areas.

isotopic values reflect changes in the temperature of surface waters of the ocean for part of Pleistocene time. The more negative values of $\delta^{18}O$ (Figure 2.6) represent times of warmer ocean temperatures (less continental ice) during interglacial stages. The more positive values record times of cooler ocean temperatures (more continental ice) during glacial stages. The ultimate cause of the oscillations in the oxygen isotope values is discussed in Chapter 13.

The oldest sediments obtained by drilling the ocean floor have an age of about 200 million years (Jurassic Period). Thus, these sediments and those of younger age recovered in this drill core contain environmental information representing the time that dinosaurs inhabited the planet. Drilling continues in the deep sea under auspices of the Integrated Ocean Drilling Program (IODP) with the lead funding agencies being the U.S. National Science Foundation and Japan's Ministry of Education, Culture, Sports, Science and Technology (MEXT). Studies of sedimentary rocks and deep-sea sediments are providing us with many clues to the history of our planet and can help us to forecast its future.

Metamorphic Rocks

Metamorphic rocks are formed when igneous and sedimentary rocks are subjected to high temperatures and pressures. These conditions occur in the deeper portions of subduction zones or in regions of continental collision where rock strata are squeezed and deformed at depth. Under such conditions, the mineralogy and structure of the original rocks may be altered substantially. On the continent, there are regions that have undergone deep erosion. In these regions, it is possible to follow sedimentary rocks that outcrop at Earth's surface and see them grade into rocks of different mineralogy and structure. One such transition is that from shale, a fine-grained sedimentary rock, through slate to schist. The latter two rock types are metamorphic. Also, with increasing temperature and pressure, limestone is converted to marble and sandstone to quartzite. However, even within the highly altered metamorphic rocks, it is still possible in many cases to recognize features of sedimentary origin.

The term **metamorphism** has been used for well over a century to describe the changes that occur in the conversion of an original rock type to a metamorphic rock. Because the various metamorphic minerals form at different temperatures and pressures, the mineralogy of metamorphic rocks is a good indicator of the environmental conditions of the metamorphism. Hence, metamorphic rocks can provide information concerning the depth of burial, temperature, and subsurface water compositions experienced by sedimentary rocks in their conversion to metamorphic rocks. Table 2.1 gives the names of some metamorphic rock types that are produced by the metamorphism of sedimentary rocks.

SOILS

Soils are made up of matter existing in solid (organic and inorganic), liquid (water), and gaseous states. Gases, derived from the chemical reactions and biological activity in soils, accumulate in air spaces within the soil. Much of the carbon dioxide in soils is produced by the biological decay of organic matter in soils. Other gases, such as nitrogen and oxygen, come directly from the atmosphere. The material discussed below on the nature and character of soils and controls on soil formation is germane to consideration of the impacts of humans on soil ecosystems found in Chapter 11.

Soils have a wide range of physical, chemical, and biotic complexity. They are the solid products of weathering of minerals and rocks and decomposition of plants and animals that form a mantle (the regolith) of alteration products over the underlying rocks. The weathering products are formed mainly by the attack of low pH, carbon dioxide (CO_2)–charged soil, and ground waters on the rocks, producing dissolved and solid alteration products. The carbon dioxide is derived from atmospheric CO_2 dissolved in rainwater percolating through the soils and rocks into the subsurface and from root respiration and the decay of plant and animal remains in the soil by oxygen transported into the soils from the atmosphere. An example reaction is

$$2KAlSi_3O_{8solid} + 2CO_2 + 11H_2O =$$
$$Al_2Si_2O_5(OH)_{4solid} + 2K^+ + 2HCO_3^- + 4H_4SiO_4^0.$$

Soils are generally arranged in layers of different composition or horizons with the most altered and oxidized (see Box 2.4, Ions and Oxidation-Reduction) zone near the top and the least altered part at the base. In the upper A horizon, organic matter, composed of living and dead plant and animal material, mixes with the subsoil to form humus or topsoil—a homeland for many living species (Figure 2.7). The topsoil is generally the most fertile area of soils. In the B horizon, the fresh rock is broken down by water and acids that descend through the rock. Subsoil is formed. Little organic matter exists in the subsoil. In the C horizon, the underlying, little altered rock forms the solid inorganic layer of the soil.

It may take hundreds to thousands of years for topsoils to form that are capable of supporting substantial plant life. Depending on the climate and type of bedrock, 200 to 1500 years are required to form just 2.5 centimeters of topsoil from bedrock. This slow development of soil is evident when looking at the

volcanic flows of the Hawaiian Islands located in the North Pacific Ocean. The ages of the islands increase progressively toward the northwest (see Chapter 3). The southeastern-most island of Hawaii is youngest and is an area of present-day volcanism. The new volcanic rock that is forming on Hawaii will not support plants for many years. Initially, the new plants that colonize the fresh lava surfaces are found in small nooks and crannies, where water may accumulate. The rocks gradually weather over the years, and a thin soil layer is eventually formed. On the older island of Kauai northwest of Hawaii, thick red soils

BOX 2.4

Ions and Oxidation-Reduction

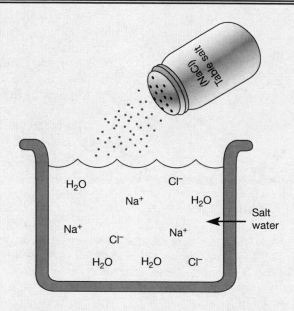

Salt Solution

The concepts of ions and oxidation-reduction are closely linked to the valence state of an element. When many compounds, like common table salt (NaCl), are put in water, they react with water and dissociate; that is, they break down into charged entities. Ions are formed as shown below. There are positively charged ions (cations), like those of calcium as Ca^{2+} and sodium as Na^+, and negatively charged ions (anions), like those of chlorine as Cl^- and sulfur as SO_4^{2-}. The type and amount of charge depend on the valence state of an element, which is related to the number and distribution of electrons in the electron cloud surrounding the nucleus (containing protons and neutrons) of an atom. *Valence* is the number of electrons that an atom can gain or lose and is equivalent to the number of hydrogen atoms or twice the number of oxygen atoms with which an atom of interest will combine. For example, the valence of nitrogen in the chemical compounds NH_3 (ammonia), N_2O (nitrous oxide), NO (nitric oxide), and NO_2 (nitrogen dioxide) is 3, 1, 2, and 4, respectively. If hydrogen is assigned a valence of +1 and oxygen that of −2, then the valence of nitrogen in the above compounds is −3, +1, +2, and +4, respectively. As another example, carbon in the chemical compound carbon dioxide gas (CO_2) has a valence of +4, whereas according to the above convention, its valence is zero in the compound glucose ($C_6H_{12}O_6$).

The Daniell Cell

The valence of an element is one reason for its position in the periodic table of the elements (Box 2.2). Metal-like elements are on the left of the table, nonmetals are on the right, and the transition elements occupy the middle of the table.

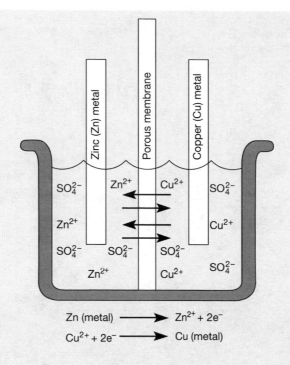

$$Zn \text{ (metal)} \longrightarrow Zn^{2+} + 2e^-$$
$$Cu^{2+} + 2e^- \longrightarrow Cu \text{ (metal)}$$

Metals tend to lose electrons, and nonmetals tend to gain electrons. In the Daniell cell shown above, Zn (metal) loses two electrons ($2e^-$) and becomes the cation Zn^{2+}, whereas the cation of copper (Cu^{2+}) gains the two electrons and becomes copper metal. The overall chemical reaction is

$$CuSO_4 \text{ (in solution)} + Zn \text{ (metal)} \rightarrow ZnSO_4 \text{ (in solution)} + Cu \text{ (metal)}.$$

This reaction is an example of an oxidation-reduction process. The substance oxidized by losing electrons is zinc metal. The cation reduced by gaining electrons is Cu^{2+}. In other words, Zn metal acts as a reductant, reducing Cu^{2+} to metallic copper. Copper ion (Cu^{2+}) acts as an oxidant, oxidizing zinc metal to the cation Zn^{2+}.

Photosynthesis is an important other example of an oxidation-reduction reaction. In the photosynthetic reaction

$$CO_2 + 2H_2O \rightarrow CH_2O + H_2O + O_2$$

carbon, which has a valence of +4 in carbon dioxide, is reduced to organic matter (CH_2O) in which the valence of carbon is zero. The oxygen of the water, which has a valence of -2, acts as the reductant providing the four electrons to reduce the carbon and in itself being oxidized to oxygen gas, in which the valence of oxygen is zero.

have developed on the old and deeply weathered volcanic flows.

Soil Classification

There are many types of soils. Government soil maps and surveys classify soils on the basis of series and by topsoil texture. The series designation reflects the ordering and type of soil layers in the subsoil below the topsoil. The series name is taken from the locality where the series was first studied and is used wherever soil of this nature is found. For example, the term **Miami loam** is applied to soils far removed from Miami County, Ohio, in the United States where the term was first applied to a soil.

The texture of a soil is a measure of its degree of fineness or coarseness and the way in which the soil breaks apart. Gravel, sand, silt, and clay are textural size terms applied to soils and other materials (Table 2.1). Particles greater than 2 millimeters in diameter are termed gravel, whereas particles less than 0.002 millimeter are termed clay. Silt (0.002–0.062 millimeter) and sand (0.062–2 millimeters) are intermediate-size terms. The textural types of soils from coarse to fine are termed sandy, sandy loam, loam, silt loam, clay loam, and clay. Loam is a loose soil of clay and sand containing organic matter. These textural terms are added to the soil series designation; thus, we may have a silt or clay Miami loam. The principal types of soils of the United States and other countries have been mapped, and soil maps

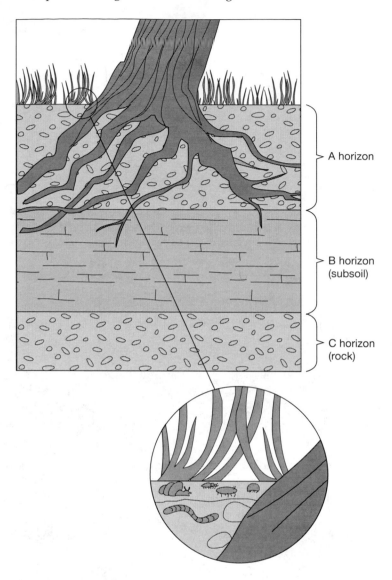

FIGURE 2.7   Weathering profile showing gradation upward from fresh rock to a well-developed, humus-rich, earthy soil zone. Inset shows soil organisms that reduce plant detritus to humus, mix soil, and help in development of soil structure.

are available for most countries. Figure 2.8 is a map of the distribution of the principal soil groups of the world. The various groups are discussed in the following section in the context of the relationship between climate and soil type.

Climate and Soils

Although a number of factors affect soil type, Russian scientists before World War I demonstrated that given sufficient time for soil development, the principal factor determining soil type is the climate of a region. This conclusion to some extent is supported

by a comparison of the world map of major soil groups (Figure 2.8) with the world map of the principal climatic zones of the planet (Figure 2.9). Notice how the patterns of major climatic zones mimic to a first approximation the patterns of different soil groups. For example, the humid subtropics and tropics (Figure 2.9) are generally characterized by red soils (latosols) and red-yellow soils (Figure 2.8). This correspondence in patterns reflects the heavy rainfall and warm temperatures of this climatic zone, leading to intense weathering and leaching of soluble materials from the underlying bedrock. Deep reddish soils with abundant insoluble iron and

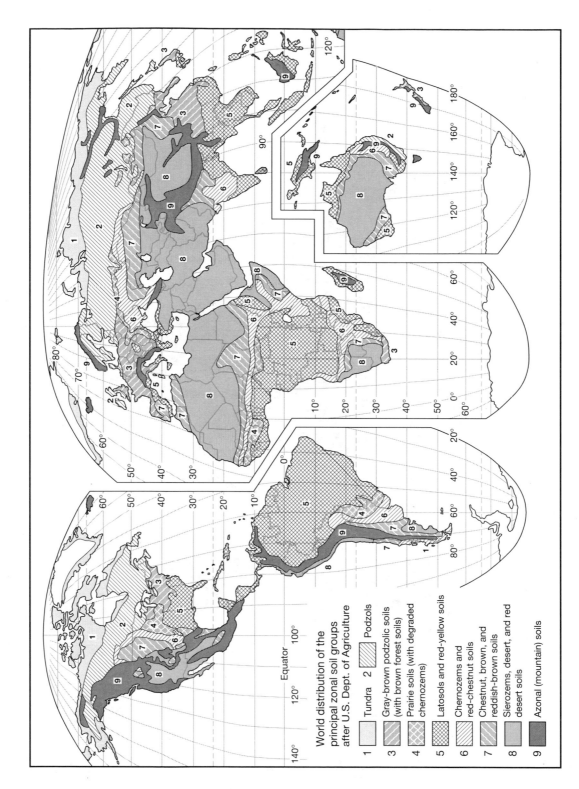

World distribution of the
principal zonal soil groups
after U.S. Dept. of Agriculture

1 — Tundra   2 — Podzols

3 — Gray-brown podzolic soils
    (with brown forest soils)

4 — Prairie soils (with degraded
    chernozems)

5 — Latosols and red-yellow soils

6 — Chernozems and
    red-chestnut soils

7 — Chestnut, brown, and
    reddish-brown soils

8 — Sierozems, desert, and red
    desert soils

9 — Azonal (mountain) soils

FIGURE 2.8 World map of principal zonal soil groups. (*Source:* Strahler, 1963.)

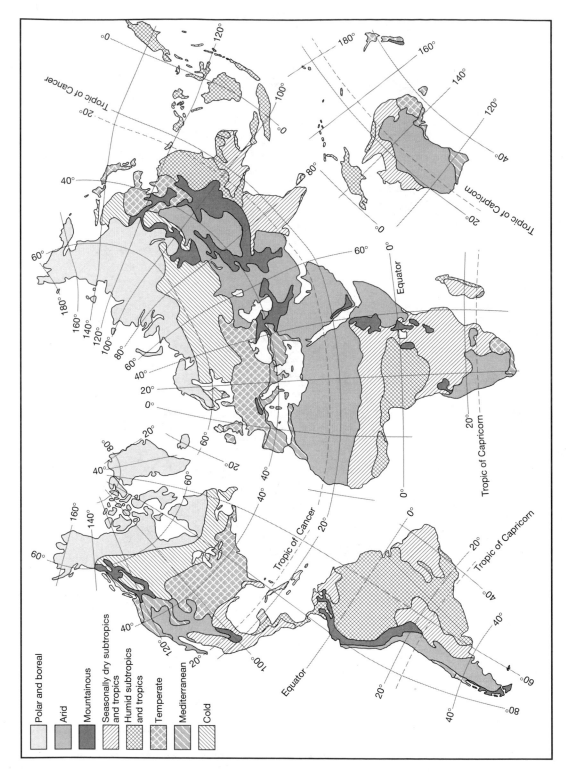

FIGURE 2.9 Major climatic zones of the world. (*Source:* FAO, 1991.)

Polar and boreal

Arid

Mountainous

Seasonally dry subtropics
and tropics

Humid subtropics
and tropics

Temperate

Mediterranean

Cold

aluminum oxyhydroxides develop as a result of the weathering process and can lead to the formation of aluminum-rich ore deposits termed bauxites. Furthermore, it will be shown in Chapters 6 and 10 that the patterns of the global distribution of types of vegetation are strongly related to climate and hence soil patterns.

Polar-climate or tundra soils of high latitudes are frozen (permafrost) except for a thin upper horizon that thaws in the summer. Because of the frozen subsoil, leaching of materials is restricted to the thin upper layer of the topsoil, and the subsoil has the same dirty gray color as its substratum. Brown humus derived from decomposition of mosses and lichens characterizes the A horizon (Figure 2.7). These tundra soils are reacting to the present rapid warming of the high northern latitudes in that the frozen layer at the top of the soil melts earlier in the spring and the melt layer is thickened. The thicker melted layer allows more carbon dioxide and methane produced by the decay of organic material in the soil to be released to the atmosphere. Recent increases in the methane content of the atmosphere may be a result of permafrost melting owing to global warming of the planet.

In warmer wet and dry climates, the soil undergoes deep thawing, and the soil changes with time and climate. In wet climate soils, soluble nitrogen compounds and calcium are leached from the topsoil to the subsoil. Loss of the calcium salts results in an acidic topsoil. Such a soil is called a pedalfer. In contrast, in a dry climate after each rain, the calcium leached from minerals is transported upward by capillary action, and stringers and layers of calcium carbonate are deposited in the topsoil. The soil is termed a pedocal or alkali soil. In the United States in the more humid climate of the eastern region of the country, there is the tendency for acid pedalfer soil formation, whereas in the west, basic or alkaline pedocal soils are more common. The acidic soil composition of the eastern part of the United States is thought to compound the problem of acid deposition on forests, whereas in the western United States, the alkalinity of soils is a problem for croplands (see Chapters 11 and 12).

In cool, moist temperate climates where conifer forests are present, a podzol soil forms. This soil is acidic because of leaching of calcium from the upper soil horizons and the acidic nature of decaying conifer needles and wood. The soil is also poor in nitrogen, because the long winters depress the metabolism of nitrogen-fixing bacteria that transfer nitrogen from the atmosphere to the soil. Podzol soils generally have an upper topsoil comprised of two layers. Raw humus material overlies ashy or salty-looking inorganic and organic debris. The underlying subsoil is often stained reddish brown because of iron oxide compounds and decaying organic matter. This type of soil is extremely poor for farming.

In contrast, a good farming soil, gray-brown podzol, is formed in milder temperate climates where deciduous trees intermingle with or supplant conifers. The deciduous trees provide basic materials to the soils upon decay. The longer summers of this climatic region promote more nitrogen fixation and nitrogen-rich soils. The developing podzol soil exhibits a humus-rich upper layer overlying a gray to brown layer and a deeper subsoil stained brown by decaying humus and iron oxides.

In warm temperate climates, loss of calcium and important oxidation of iron-rich soil minerals and organic matter give rise to red and yellow soils. In areas of heavy tropical rainfall, like the Amazon and Congo river basins, the soils are well leached, lack nutrients, and are deep red in color because of the oxidation of iron-bearing soil minerals to iron oxides. In such systems, the tropical forest trees get much of their nutrients from decaying vegetation on top of the soil and from the dead understory below the forest trees. This land is generally very infertile for any vegetation except the tropical forest itself (see Chapter 10).

The types of dry climate soils depend on the amount of rainfall they receive. In deserts the soil is patchy and is light-gray to whitish-yellow in color, and is termed gray desert soil. With scant rainfall, gray soil, or sierozem, may develop and is characterized by brownish streaks of humus derived from decay of sparse desert grasses and other plants. As rainfall increases in dry climates, so does the amount of plant life and decaying vegetation. This leads to the development of brown and dark-brown or chestnut soils. Chernozem, or black soil, develops in the subhumid climatic regime with slightly less than 50 centimeters of rain per year. Such soil characterizes the wheat belt of southern Russia and the Ukraine. In the United States under the wetter climatic conditions characterizing the prairie of the Midwest, dark-brown prairie soil with a deep brown humus layer is found. This soil, because of the higher rainfall of the region, is even better for farming than chernozem.

Concluding Remarks

Plate tectonic forces (see Chapter 3) give rise to the formation of the Earth's crust, affect continental positions, and produce changes in global topography. The surface processes of weathering and erosion that destroy crustal rocks can be thought of as steps in a rock cycle. These processes lead to alteration of the crust of Earth and begin the movement of materials via river and atmospheric transport to the oceans. In the oceans these materials are deposited and then buried. Later in their history they are uplifted to be exposed once more to the processes of weathering and erosion. The exogenic (near surface) cycle is complete. This near surface cycle is part of the greater rock cycle involving the formation of crystalline (igneous plus metamorphic) rock and its subsequent destruction and the exchange of materials with the mantle.

Weathering and subsequent erosion of the weathered rocks shape the exterior of the continental crust. Weathering is a complex physical, chemical, and biological process that leads to the disintegration of rocks and their preparation for erosion. The products of weathering can be dissolved substances, like sodium as Na^+ ion and chlorine as Cl^- ion, or solid materials, like original rock particles or minerals formed by alteration of the original rock. Erosion is the set of processes whereby these solid particles and dissolved substances are transported to another area by the agents of water, wind, and ice. Intense weathering and erosion have resulted in the creation of immense valleys, some the size of the Grand Canyon and its miniature version, the Waimea Canyon on the island of Kauai, Hawaii, and the Samari Gorge in Crete, the longest gorge in Europe.

The nutrient elements of nitrogen and phosphorus are also released from rocks and organic detritus in soils by weathering. The nutrients are leached by rainwater from soils as dissolved substances into groundwaters and rivers, or are transported as suspended inorganic and organic solids in rivers to different environments. An excellent example of the dispersion of nutrients as dissolved and suspended materials is the flooding of the Mississippi River in the United States in 1993 and in more recent years. This flooding led to the transport of materials into the Mississippi Basin and the Gulf of Mexico. The plume of fresh water generated by the 1993 flood was observed as far away as the western margin of the North Atlantic Ocean. The flooding spread nutrients, mud, and silt throughout the floodplain of the Mississippi River.

Thus, the building blocks of solid Earth, its minerals and rocks, are continuously being altered and transformed in contact with air and water to new materials at the surface of the planet. The processes generally proceed at relatively slow geologic rates but during times of major floods, large quantities of both dissolved and solid materials may be removed quickly from their sites of origin and transported long distances before deposition. The chemical compositions of the major river waters of the world are determined to a significant extent by the reactions of weathering.

The altered zone of soils provides a refuge for organisms and a substrate for the growth of plants. In turn, plants help to break up the rocks to form soils. Biochemical reactions in the soils lead to transfer of gases into and out of them. These gases then influence atmospheric composition. The slow, inexorable wearing down of the land surface of Earth would lead to its demise and reduction in height to sea level in less than 20 million years owing to weathering and erosion, if it were not for the return of materials to the continents via the plate tectonic forces discussed in the next chapter. In addition, human activities are leading to the movement of huge amounts of soil and subsoil inorganic and organic materials about Earth's surface and soils are being lost from the continental surface to be deposited in lakes or the coastal ocean or to be trapped behind dams built by humans. River flow and chemistry are also being modified by human activities (Chapter 11). Thus, the land surface of Earth is being rapidly altered because of human behavior and the natural rates of weathering, transport, erosion, and deposition are also changing due to human activities.

Study Questions

1. The Earth has a mass of 5.98×10^{27} g and an average density of 5.52 g/cm^3. What is the volume of the Earth in cubic kilometers?
2. Why do the continents "float high" relative to the ocean basins?
3. If the near-surface temperature gradient of 30° C/km continued to the base of the continental crust at 40 km, what would be the temperature at this depth? Is this the estimated temperature at this depth? If not, what is the temperature and why the difference between estimates?

4. Why was James Hutton's concept of the vastness of geologic time not generally accepted by both scientists and nonscientists for much of the eighteenth, nineteenth, and the early part of the twentieth centuries?

5. What are the major generic subdivisions of the geologic time scale?

6. How does one determine the relative ages of a sequence of sedimentary rocks? Absolute ages?

7. What are the two major attributes that define a mineral?

8. What are polymorphs?

9. Define the term *rock*. How many major classes of rocks are there and what are they?

10. In what kind of environmental setting would you expect to find marine evaporite deposits forming?

11. The reaction $Ca^{2+} + SO_4^{2-} + 2H_2O \rightarrow CaSO_4 \cdot 2H_2O$ represents the precipitation of the evaporite mineral gypsum. If 2.5 moles of gypsum precipitated from an evaporating brine, how many grams of sulfate would this represent?

12. Latosols are relatively rich in iron, aluminum, and silicon and depleted in sodium, potassium, calcium, and magnesium. Why?

13. The podsols of Figure 2.8 are principally a product of what kind of climate?

14. What are the two major types of weathering products?

15. What effect does photosynthesis have on the fractionation of carbon isotopes?

16. What effect does the formation of great continental ice sheets at high latitude have on the $\delta^{18}O$ of surface ocean water?

17. How and why do both the $\delta^{18}O$ and δD of precipitation change from the tropical climatic zone to the polar?

18. Uranium-235 has a half-life of 0.705 billion years. What is the fraction of $_{92}U^{235}$ uranium remaining after four half-lives?

Additional Sources of Information

Brady, N. C. and Weil, R. R., 1996, *The Nature and Properties of Soils.* Prentice Hall, Upper Saddle River, N.J., 740 pp.

Eicher, D. L., 1976, *Geologic Time*, 2nd edition. Prentice Hall, Englewood Cliffs, N. J., 150 pp.

Ernst, W. G., 1969, *Earth Materials.* Prentice Hall, Englewood Cliffs, N. J., 150 pp.

Ernst, W. G., 2000, *Earth Systems: Processes and Issues.* Cambridge University Press, New York, N.Y., 566 pp.

Press, F. and Siever, R., 1985, *Earth.* W. H. Freeman, San Francisco, 656 pp. 5th edition: http://bcs .whfreeman.com/understandingearth5e/default.asp?

Tennissen, A. C., 1974, *Nature of Earth Materials.* Prentice Hall, Englewood Cliffs, N. J., 439 pp.

3 Earth's Lithosphere: Plate Tectonics

There's nothing constant in the universe,
All ebb and flow, and every shape that's born
bears in its womb the seeds of change.

OVID, *METAMORPHOSES*, XV (A.D. 8)

The theory of plate tectonics, first advanced in the early 1960s, revolutionized the Earth sciences. Plate tectonic theory proposes that the lithosphere (the upper 100 km of the Earth) is an assembly of units or plates resembling the shell of a cracked egg. At least seven major plates and a number of smaller ones move relative to one another around the surface of Earth (Figure 3.1).

In the late nineteenth and early twentieth centuries, most geologists assumed that vertical movements of the crust explained geologic features, such as mountain ranges, continents, and ocean basins. The early ideas of the theory of geosynclines in which deep linear subsiding basins filled with sediment formed by compression of the Earth's crust were also invoked to explain the features of Earth's surface. It was the hypotheses of Alfred Wegener formulated first in 1912 and later fully developed in his book *The Origin of Continents and Oceans* (1915) that were influential in changing this perspective and led to the theory of plate tectonics. Wegener pointed out that the present-day continents probably once formed a single landmass with, for example, present-day South America connected to Africa. This large continental mass, which was to become known as Pangaea, then broke into two major parts, the supercontinents of Laurasia and Gondwana, and these supercontinents drifted and floated like icebergs of primarily low-density granitic material on a sea of more deeply buried denser basalt. A major impediment to the acceptance of the Wegner hypothesis of continental drift was that the calculated forces required to move the large continental masses were too great. The hypothesis remained a focus of much scientific debate for several decades. It was not until 1920, when the English geologist Arthur Holmes suggested that the junctions of the moving crustal blocks might actually be within the oceans, and Holmes' subsequent 1928 suggestion of convection currents within the mantle as a driving force for moving the blocks, that the hypothesis of continental drift began to be accepted almost everywhere.

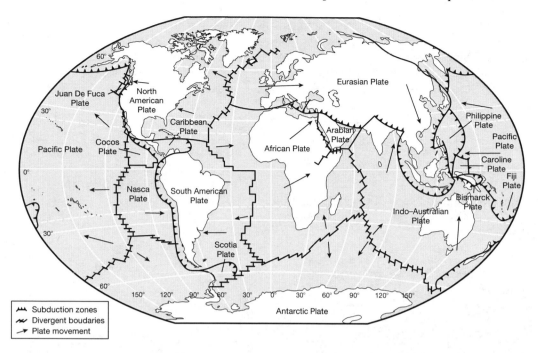

FIGURE 3.1 Major lithospheric plates of the Earth and source regions of generation and spreading of the crust (midocean ridges) and regions of destruction and sinking of the crust (subduction zones). Primarily horizontal motions along transform faults are shown as short lines approximately perpendicular to the ridge axes. The relative directions of movement of the plates are shown by the arrows.

In the 1950s oceanographic exploration of the sea floor led to a much better understanding of the origin of ocean basins and their basaltic and sediment floor. The discovery of zebra stripe–like magnetic patterns parallel to the midocean ridges of the ocean floor was among the new oceanographic findings (Figure 3.6b). Field reversals had already been demonstrated for magnetic rocks on the continents; the first examination of the timing of the reversals was done by Motonori Matuyama as early as the 1920s. Two British geologists, Frederick Vine and Drummond Matthews, and Lawrence Morley of the Canadian Geological Survey, hypothesized in 1963 that this sea floor magnetic striping was produced by repeated reversals of the Earth's magnetic field and not, as earlier thought by some scientists, by changes in the intensity of the magnetic field. This interpretation supported Robert Dietz's and Harry Hess' ideas that seafloor spreading was occurring at midocean ridges. A logical next step was to see if the continental magnetic reversals might be correlated in geologic time with the oceanic magnetic striping. They did correlate, and a team of U.S. Geological Survey scientists, geophysicists Allan Cox and Richard Doell, and isotope geochemist Brent Dalrymple, reconstructed the history of magnetic reversals for the past four million years

using the potassium argon radiometric dating technique (Box 2.1, Figure 3.6b).

Thus, although initially thought to be a result of an expanding global crust, the movement of the plates and hence the continents was later shown to be due to spreading at midocean ridges, a consequence of new basaltic rock upwelling from the Earth's interior, subduction of the lithosphere at the edges of ocean basins, and to conservatively behaving transform faults. Harry Hess at Princeton University and others derived the precise mechanisms responsible for the upwelling of molten rock from the Earth's interior. These realizations avoided the need for an expanding crust. Thus, Wegener's continental drift theory moved from being radical to mainstream science and plate tectonics quickly became broadly accepted as a theory with extraordinary explanatory and predictive power.

The theory of plate tectonics has revolutionized the Earth sciences and led to explanations of a diverse range of geological phenomena and their implications. The fact that the Earth has plate tectonics involving differential slow horizontal and vertical motions is unique among the planets of our solar system. This mobility of an apparently rigid planetary body explains why the continents stand high and are old parts of the crust and the ocean basins are low-riding and young. It explains why we find fossil seashells in the rocks of the great mountain ranges, volcanoes in the seas, like Mauna Loa and Kilauea in Hawaii, and the Ring of Fire around the Pacific Ocean. It also offers a mechanism for the drifting of the continents and a variety of other geological phenomena. Plate tectonic processes can be understood by considering first the subsurface of Earth.

EARTH'S INTERIOR

The interior of Earth has been subdivided into zones. There are two major systems of zonation. Historically,

the first approach to subdivision of Earth's interior was based on composition and included the zones of core, mantle, and crust (Figure 3.2b,c). The inner part of Earth, the core, is principally metallic iron with a light-alloying element. The thick middle portion, the mantle, is made of magnesium and iron silicate minerals. The thin outer portion, the crust, is rich in silicates of aluminum, potassium, and sodium. The zonation of the planet and the types of minerals and rocks found in its interior are determined in part by the patterns that earthquake waves make when they travel through the interior of Earth (Box 3.1: Structure and Composition of Earth). For example, the contact

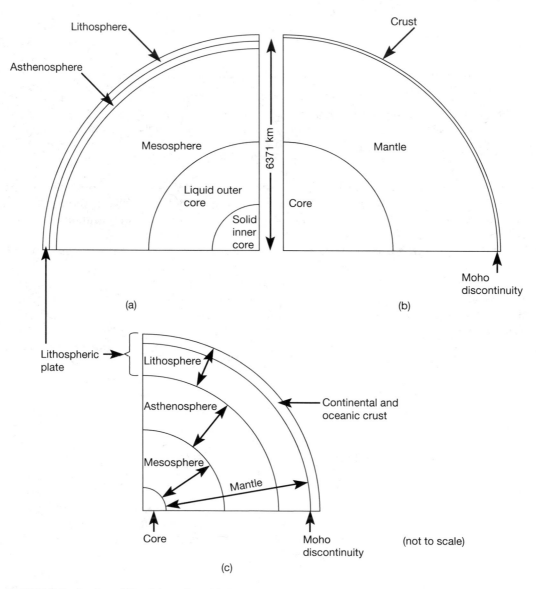

FIGURE 3.2 Section of Earth based on (a) dynamical and (b) compositional properties; (c) shows an expanded view.

between the crust and the mantle is defined by a sharp increase in the velocity of earthquake waves and is called the Moho (Mohorovičić) discontinuity. Data obtained from experiments done on minerals and rocks at high temperatures and pressures and from the composition of meteorites, and additional theoretical and observational considerations, also provide clues as to the nature of Earth's interior.

Since the advent of plate tectonic theory, the interior of Earth has been subdivided into the regions of lithosphere, asthenosphere, and mesosphere based on dynamical properties of the planet (Figure 3.2a,c). The outer, roughly 100 km lithosphere is thought of as a rigid surface layer overlying the warmer, softer, and more ductile asthenosphere. The mesosphere is the hot, high-pressure, interior portion of Earth that is little involved in plate tectonic activity near the surface but contributes to it owing to rising plumes of molten rock from perhaps as deep as the core–mantle boundary. The new dynamical subdivision of the interior of Earth is also based on earthquake wave patterns and is confirmed by a variety of other information. This information includes knowledge of the movement of large areas of the surface of Earth, the magnetic properties of rocks of the seafloor, the distribution of submarine mountain ranges, island-arc volcanoes, and earthquake sources, and a variety of other data. The dynamical subdivision is an integral part of the new and unifying plate tectonic theory of planet Earth.

The energy for plate movement comes from the mantle of the planet that produces an immense amount of heat energy from the radioactive decay of elements found there and from the core. This interior furnace gives rise to convection cells that function as an escalator or conveyor belt, carrying materials upward from below and perhaps helping to force the lateral and sinking motions of the lithospheric plates. This motion is similar to the motions of the atmosphere full of moisture or a pot of boiling water, when hot vapor or water rises and relatively cooler water condenses or sinks, only to be reheated and recycled upward (Figure 3.3). In the interior of Earth, the convection process is much slower, and motions are measured in centimeters per year. Rising, hot mantle plumes of molten magma also play an important role in motion at the surface of the planet.

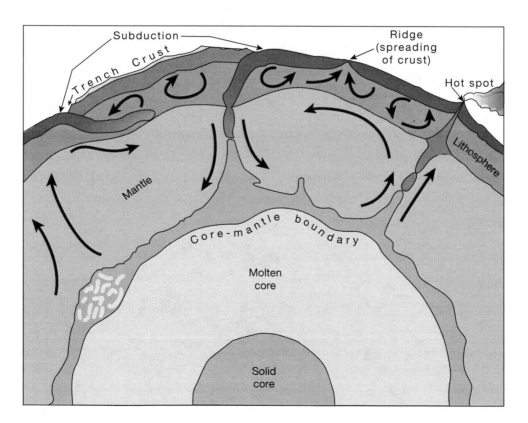

FIGURE 3.3 One scheme of convection cells postulated for the interior of Earth leading to upwelling currents below midocean ridges and downwelling currents at subduction zones (e.g., trenches) (not to scale).

BOX 3.1

Structure and Composition of Earth

Most of the interpretations of the internal structure of Earth come from studies of earthquake-wave velocities. When a liquid or a gas is subjected to a shock, energy is transmitted through it in waves. Because liquids and gases have no shear strength, only compressional waves in which particles are displaced parallel to wave propagation are formed. In solids, shear waves with particle displacements at an angle to the direction of wave propagation also are set up.

The important point is that earthquakes set up compressional and shear waves that travel through Earth. Wave velocities are dependent on rock properties such as compressibility, density, and resistance to shear. Rocks can be tested in the laboratory to see the effects of increasing temperature and pressure on their physical properties and the resultant effects on wave velocities. Based in part on the results of these experiments, various models can be constructed for the nature and properties of the material within Earth. Inevitably, because wave velocities are not unique functions of a single property such as density, density profiles within Earth are not uniquely determined by velocity distributions.

The following figure shows velocities of P (compressional) and S (shear) waves as a function of depth within Earth. Interpretations of density, pressure, and temperature relations within the planet are also shown. The important features, as velocities change with depth, are as follows. Velocities beneath the continents, for a depth of about 40 kilometers, increase irregularly downward, approximately as would be expected from the effects of increasing temperature

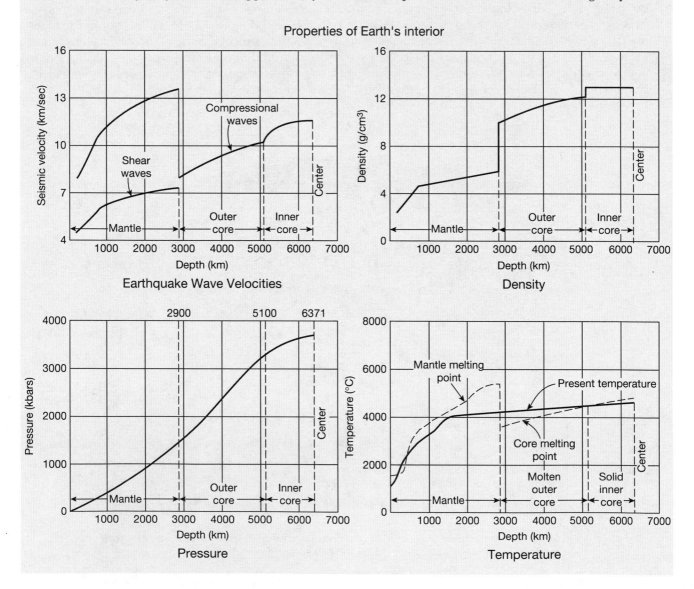

Properties of Earth's interior

and pressure on various mixtures of rock materials we can observe at the surface. Beneath the bottoms of the oceans, a similar increase is observed, but only for a depth of about 10 kilometers. Then at both the 40- and 10-kilometer depths, there is an abrupt increase in velocity over a short vertical distance, the Mohorovičić discontinuity. This velocity discontinuity is a result of a change in both the composition of the materials and the minerals present.

The P or compressional waves are transmitted through liquids and gases, as well as through solids. The S or shear waves pass only through solids. The absence of shear-wave transmission below about 2900 kilometers is the main evidence for a liquid core. The presence of a core is dramatically illustrated by the absence of S waves and the abrupt drop in P-wave velocity. The P-wave velocity increase at about 5100 kilometers indicates an inner core that is solid.

The Core of Earth

The conclusion that Earth's core is dominantly metallic iron is long-standing. This conclusion finds support in the fact that the average density of Earth is 5.52 g/cm^3, whereas the average density of crustal material is only about 2.7 g/cm^3. Thus, the interior of Earth must contain denser materials than its outer regions. The choice of metallic iron for core composition over other dense materials is based on indirect evidence that mainly has been accumulated during the last half century.

For example, meteorites have been a major source of data helpful in deducing the origin and history of Earth and the solar system. Meteorites are relatively large pieces of solid matter that formerly traveled through our solar system. Upon entering the atmosphere of Earth, the solid matter was not completely vaporized and fell to Earth's surface as a meteorite. Many, but not all, meteorites that reach Earth come from the asteroid belt between Mars and Jupiter. Some come from Mars itself. One controversial find on Earth may contain evidence for prokaryotic life having existed on Mars.

Meteorites are thought to have formed from the fragmentation of larger planetary bodies up to several hundreds of kilometers in diameter. Meteorite compositions have been subdivided into three main categories: the irons (4% of meteorite finds), which consist primarily of iron metal with important concentrations of nickel; the stones (95% of meteorite finds), consisting of silicates with little metal; and the stony irons (1% of meteorite finds), containing abundant metal and silicates. The stony meteorites have been further subdivided into the chondrites and achondrites. The chondrites are so named because of their characteristic texture consisting of an aggregate of rounded grains. Some investigators consider these grains as actual original particles that agglomerated to form the parent body of the meteorite. The carbonaceous chondrites, containing about 5% carbon as organic carbon (Table 3.1), are the most primitive group of meteorites. Their composition suggests that they were formed by accretion of the dust phase of the solar nebula (see Chapter 8) into a small parent body, which was subjected to a very mild degree of metamorphism.

The implication is that the chondrites were derived from a parent body too small to have differentiated into a core and mantle. Their compositions are thought to represent "cosmic dust." The achondrites and irons seem to have been derived from a planetary body big enough to have melted and differentiated into the two types. Note in Table 3.1 that the chondritic composition could reasonably give rise to the irons and the silicate portion of chondrites. Thus, meteorites composed predominantly of iron provide evidence that parts of other bodies in the solar system, presumably similar in origin to Earth, were composed of metallic iron. The evidence from meteorite compositions and origin lends support to the conclusion that Earth's core is primarily composed of metallic iron.

Experimental studies of various materials, in which pressure-temperature conditions presumably comparable to those in the center of Earth (about 4500° C and 3700 kilobars of pressure) have been achieved, have strongly reinforced the concept of a metallic iron core. Of all the materials that have been suggested as possible components of the core, only iron has properties of density and P-wave velocity that correspond closely to those previously deduced for the core. In addition, the melting point of iron, under the estimated pressure conditions of the core, is just about at the deduced temperature of the core. This makes the assumption of the presence of both solid and liquid in the core reasonable.

To fit the earthquake wave velocities observed for the core and its density, it appears that an alloy of iron with some quantities of silicon, oxygen, and sulfur best satisfies our estimate of core composition at this time.

The Mantle of Earth

Just as experiments have shown that almost without doubt the core is dominantly iron, experiments also

TABLE 3.1 Comparison of the Chemical Composition of Earth with that of the Sun and Meteorites (Weight Percent of Element)[a]

Element	Sun	Earth				Meteorites			
		Whole Earth	Continental Crust	Oceanic Crust	Mantle	Average Iron	Average Silicate[b]	Average Chondrite[b]	Carbonaceous Chondrite
Fe	0.00032	35.0	5.63	8.56	5.3	90.78	9.88	27.24	18.75
O	0.078	30.0	46.40	43.8			43.7	33.24	41.93
Si	0.0027	15.0	28.15	24.0	22.0		22.5	17.10	10.58
Mg	0.0021	13.0	2.33	4.5	22.7		18.8	14.29	9.56
Ni	0.00007	2.4	0.0075	0.015		8.59		1.64	1.02
S	0.0017	1.9	0.026	0.025				1.93	6.09
Ca	0.00012	1.1	4.15	6.72	2.0		1.67	1.27	1.11
Al	0.00014	1.1	8.23	8.76	2.2		1.60	1.22	0.87
Na	0.00017	0.57	2.36	1.94	0.4		0.84	0.64	0.56
Cr	0.000014	0.26	0.01	0.02	0.1		0.38	0.29	0.24
Co	0.000004	0.13	0.0025	0.0048		0.63		0.09	0.04
P	0.000019	0.10	0.105	0.14			0.14	0.11	0.18
K	0.000004	0.07	2.09	0.83	0.2		0.11	0.08	0.06
Ti	0.000004	0.05	0.57	0.90	0.1		0.08	0.06	0.04
Mn	0.000007	0.22	0.095	0.15			0.33	0.25	0.16
H	86.0		0.14	0.2					0.21
He	13.0								3.97
C	0.045								5.53

[a] *Source:* Cameron, 1966; Ringwood, 1966; Mason, 1966; Taylor, 1964; and Anderson, 1989.
[b] Stony meteorites.

have shown that the mantle, which makes up two-thirds of the mass of Earth, has properties consistent with a silicate composition and cannot be composed of metals. This interpretation also is in agreement with meteorite compositions. The stony meteorites (Table 3.1) have properties appropriate for mantle material. One estimate of the approximate composition of the mantle is given in Table 3.1.

Major debate still exists concerning the compositions and structures of the individual compounds that are presumed to be in the mantle and the relationships of mantle materials to rocks of the crust. On the basis of knowledge to date, it is thought that the elements are perhaps combined as iron-magnesium silicates in the upper mantle and as silicates and oxides in the lower mantle.

Using the diamond-anvil high-pressure cell that can duplicate the pressures and temperatures of the deep Earth, some experimenters have concluded that a single high-pressure mineral probably dominates the lowermost mantle. The mineral is a dense form of magnesium iron silicate $[(Mg, Fe)SiO_3]$ called perovskite. Magnesium oxide (MgO) and iron oxide $(FeO$, the mineral wustite) may also be present in a combined mineral composition—magnesiowustite.

The chemical differences within the upper mantle are thought to be small, although important for interpretations of its behavior. Changes of earthquake wave velocities with depth appear to be consistent with those expected from differences in physical conditions. Uranium, thorium, tantalum, barium, strontium, zirconium, hafnium, beryllium, and the rare Earth elements are thought to be relatively more concentrated in the upper mantle.

THE CRUST AND LITHOSPHERE OF EARTH

Convection processes in the inner Earth deliver hot molten lava toward Earth's surface and may aid in moving the lithospheric plates and the continental and oceanic crusts attached to them. The continental crust extends to a depth of 20 to 70 kilometers and is less dense than the oceanic crust. The denser oceanic crust lies under the oceans and is only about 10 kilometers thick. These two types of crust have different compositions, and hence densities, because of different processes that led to their formation (Figure 3.4).

The composition of the crust has received a great deal of attention for many years. A fair proportion of the crust can be sampled directly by investigation of outcrops of rock at its surface and by drilling into both the continental and oceanic crusts. Only oxygen, silicon, aluminum, iron, calcium, sodium, potassium, and magnesium have abundances in the crust greater than 1% by weight. The crust is enriched in the elements silicon, oxygen, calcium, aluminum, sodium, and potassium relative to the Earth as whole

(Table 3.1). The crust is so heterogeneous that it is difficult to classify the various parts without a long and detailed discussion. However, two chief kinds of crust can be distinguished: continental and oceanic (Table 3.1). The composition of continental crust is higher in its concentration of oxygen, silicon, sodium, and potassium and depleted in iron, magnesium, calcium, and aluminum relative to oceanic crust. Also, continental crust has a lower average density than does oceanic crust.

Oceanic Crust

RIFT ZONES—SPREADING OF THE OCEANIC CRUST
The composition (Table 3.1) and density (3 g/cm^3) of the oceanic crust are a result of the processes that led to its origin. Upward-moving convection currents from the asthenosphere and deeper rise toward the ocean floor and diverge, leading to spreading of the lithospheric plates. Long cracks or rifts are created, and molten rock from the inner Earth intrudes the oceanic crust and oozes and spreads out onto its surface (Figure 3.5). These accumulations of dense basalt rock (basalt rock is rich in iron and magnesium silicate minerals) form submarine mountain ranges (midocean ridges or rises) that encircle the globe today for a total distance of 32,000 kilometers (Figure 3.1). The ridge system is broken by a series of fracture zones (transform faults) oriented roughly perpendicular to the trend of the ridge system. Transform faults permit parts of the ridge system to slip horizontally past one another. The faults

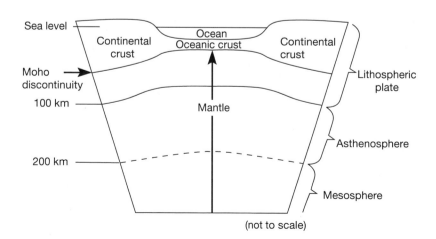

FIGURE 3.4 Schematic diagram showing thin, young oceanic crust and thick, old continental crust. The former is formed at midocean ridges by volcanism and by sedimentation in the oceans and mostly destroyed at subduction zones; the latter is formed by accretion of oceanic materials and thickening throughout geologic time.

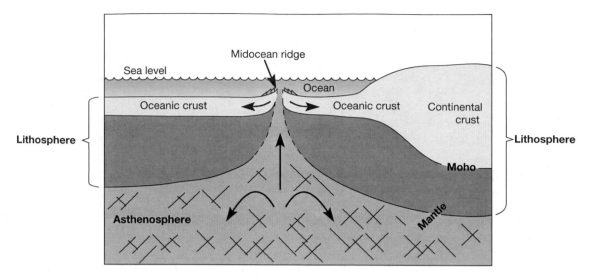

FIGURE 3.5 Diagram illustrating the process of upwelling of molten rock below the midocean ridge system and spreading of the crust.

form a type of boundary to plate motion, along with the divergent boundary of spreading of the seafloor at ridge systems. Fracture zones generated by transform motion on the continents form the great fault zones of the San Andreas of California and the Anatolian of Turkey. Both these zones are areas of repetitive and often destructive earthquake activity.

The midocean ridges with their associated rift zones are regions of relatively high heat flow (see Box 3.2: Heat Flow) compared to that seen for much of the ocean basins and the continents (Figure 3.6a). The high heat flow reflects the rising of hot basaltic magma beneath the ridge crests. Heat flow values up to 350 milliwatts per square meter (mW/m^2) have been observed compared to some areas on the continents and deep-sea floor of less than 60 mW/m^2. For comparison, the amount of solar radiation reaching the top of Earth's atmosphere (the solar constant) is about 1370 W/m^2, almost 4000 times the heat escaping from the hottest areas of the midocean ridges!

Perhaps one of the strongest arguments for the spreading of the lithospheric plates (the seafloor-spreading hypothesis) is the observation that magnetic profiles across some oceanic ridges show a series of anomalies that trend parallel to the ridge and are symmetrical about the ridge axis (Figure 3.6b). The explanation for this observation is that as mantle material wells up beneath the ridges and cools below its Curie temperature (the temperature above which a material cannot be magnetized), it is magnetized in the direction of the magnetic field of

Earth (the geomagnetic field). The magnetic anomalies are deviations from the regional average magnetic strength of the present-day magnetic field and represent reversals in Earth's magnetic field. Belts of low magnetic strength are correlated with times when the geomagnetic field of Earth was reversed. Those of high magnetic strength represent times of a normal field like today. The magnetic profile across a ridge system forms a series of peaks and troughs representing, respectively, regions of normal magnetic polarity and reversed magnetic polarity. The polarity is set, "locked in," in the zone of magma injection and cooling underneath the ridge (Figure 3.6b).

Based on paleomagnetic data and radiometric age dating of rocks from the continents, it has been shown that Earth's magnetic field has reversed itself many times in the past. From these data a time scale of reversals of Earth's magnetic field has been developed and applied to the reversal pattern observed for seafloor basalts, which had their origin along midocean ridges. An excellent correlation exists between the age of the magnetic reversal intervals for the basalts and their geographic distance from the spreading-ridge center. One can calculate the rate of seafloor spreading simply by dividing the distance of the basalt from the ridge crest by its age. This rate comes out to be about one to a few centimeters per year, consistent with movement rates of large blocks of Earth's crust obtained from other relationships.

The crests of the ridges rise about 3.5 kilometers above the deep-ocean floors. The Mid-Atlantic Ridge has grown above sea level in the North Atlantic Ocean

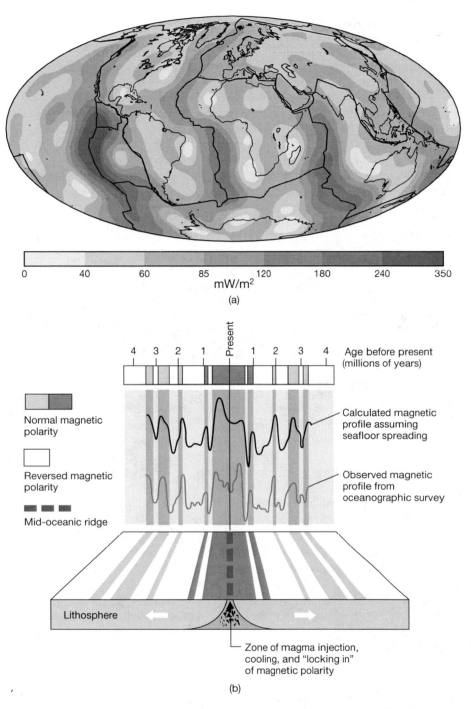

FIGURE 3.6 (a) Model calculation of the distribution of global heat flow. Notice the high heat flow values found along midocean ridges where molten magma upwells from the interior of Earth. (After Pollack, et al., 1993.) (b) Magnetic anomalies on the seafloor along a section perpendicular to a ridge axis. Rocks having normal, or present-day, magnetic polarity are shown in dark gray; rocks of reversed polarity are shown in light gray. The symmetrical pattern of the magnetic stripes and their correlation with the time scale of magnetic reversals derived for lava flows on land provide a means of dating the age of the stripes and are evidence for seafloor spreading. The calculated and observed magnetic profiles showing normal and reversed magnetic polarity are also shown. Ages of magnetic reversals in millions of years. (*Source:* Press and Siever, 1974.)

BOX 3.2

Heat Flow

The conclusion that there is a difference in the temperature gradient between the continents and ocean basins came as a result of studies of heat flow from the interior of the Earth. From measured temperature gradients and knowledge of the thermal conductivities of rocks, it is possible to calculate the heat released per unit area of Earth's surface. For a linear gradient in temperature, the heat loss can be calculated from the equation

$$q/t = \frac{KA(T_2 - T_1)}{d}$$

where q is the heat in calories, t is the time in seconds, K is the thermal conductivity of the rock, A is the cross-sectional area of interest in centimeters squared, T_2 and T_1 are the temperature in degrees centigrade at two depths in a drill hole in the crust, and d is the distance between them in centimeters. As a specific example, for a 30° C/km gradient in a particular granite with a thermal conductivity of 7.9×10^{-3}, the heat flow per second per square centimeter is

$$q = \frac{7.9 \times 10^{-3} \times 1^2 \times 30}{10^5} = 2.4 \times 10^{-6} \text{ cal/cm}^2/\text{sec}$$

or 2.4 microcalories (μcal)/cm^2/sec. Heat flow measurements are also expressed in units of milliwatts per square meter. Several thousand measurements of heat flow have been made over the world. Figure 3.6a shows the global pattern of heat flow on Earth and illustrates the high heat flow over midocean ridges, reflecting steep temperature gradients beneath these features.

and forms the island of Iceland. A large rift system is exposed on Iceland. Roughly one-half of Iceland is on the Eurasian plate, and the other half is on the American plate. Consequently, Iceland is an environment of intense volcanic activity. Hot-water springs are found at the surface and are used for bathing, and the geothermal heat is used to heat buildings.

Investigations of ridge systems using deep-diving submersibles have led to a number of exciting discoveries. Parts of the ridge system have been shown to be sources of dissolved constituents, such as calcium and silica, for the ocean. At the same time, they act to remove some substances, such as magnesium, from seawater. The processes controlling the input and removal of elements are related to the circulation of seawater through ridges. Cold seawater enters ridges and off-ridge areas through cracks and crannies and circulates downward. In its journey, the heat escaping from molten rock at depth heats the seawater. Temperatures may reach several hundreds of degrees Celsius. The seawater reacts with basaltic rock at these high temperatures. The chemical composition of the circulating seawater is modified during these reactions. Because it has been heated, the density of the seawater decreases. The water becomes buoyant enough to rise through openings in the basalt rock back toward the seafloor. It exits the seafloor in hot-water springs or in large chimney-like structures termed "black smokers." The hot-water springs are different in chemical composition from

that of the original seawater that entered the ridge. These hydrothermal (hot water) vents can be sites of prolific growth of life that is adapted to the high pressures of the deep sea and to the warmth of the environment surrounding the vents. The organisms found near vents include blood-red, tubular worms that extend a meter or so above the seafloor. Crabs, clams, and abundant bacteria also colonize these vent areas (see Figure 3.7). Most midocean ridge hot spring deposits discovered so far are formed from iron- and sulfur-containing minerals. However, at 30 degrees north on the Mid-Atlantic Ridge, at a site dubbed the *Lost City*, huge chimney structures towering 60 meters above the sea floor and composed of carbonate and silica minerals were discovered on a submersible *Alvin* dive during a cruise of the oceanographic vessel *Atlantis* on December 5, 2000. The low to moderate-temperature hydrothermal fluids associated with these structures have chemistries different from those of the high temperature vent smokers. The aqueous environment is extremely alkaline compared to the acidic fluids of the high-temperature vents. Dense mats of microbial communities are found in the trapped pools of warm water emanating from the sides of the structures, apparently living off the methane and hydrogen found in the vent fluids. Transparent or transluscent shrimp and crabs are also found living in the *Lost City* vent field.

A portion of the midocean ridge system, the East Pacific Rise, has been used systematically for the

FIGURE 3.7   The Genesis hydrothermal vent at 13° N along the East Pacific Rise, showing lush populations of tube worms (*Riftia pachyptila* and *Tevniajerichonana*) and brachyuran crabs (*Bythograea thermydron*). Note the characteristic bright-red "plume" (dark areas in this picture) of *R. pachyptila.*   Photo by R. A. Lutz.

study of seafloor-spreading processes. In 1989, photographs of the distribution of hydrothermal vents and animal communities on the East Pacific Rise were obtained using a visual imaging system employed by the deep-diving submersible *Alvin.*

During March–April 1991, further investigations were conducted in the area previously photographed. Recent volcanic activity was evident. The type and distribution of the biota around the vents and the location of the vents had changed over the two-year period. Upward-moving molten rock had solidified and created new crust at the boundary line of the two spreading plates—American and Pacific. This series of photographs documented the dynamic and now active nature of volcanism and hydrothermal activity along ridge systems.

SUBDUCTION ZONES—SUBMERGING PLATES The ocean floor spreads as new material is added to the lithospheric plates along the midocean ridge system. The lithospheric plates glide over the underlying, softer asthenosphere as the laterally moving convection currents carry them along. The rate of spreading ranges from less than 1 to greater than 10 centimeters per year. This rate has been confirmed by satellite observations of continental motion.

The plates eventually return to the interior of the planet or Earth would expand. There must be a sink, as well as a source, for the plates. The areas where the plates return to the mantle are along boundary lines termed subduction zones (Figure 3.8) where two large plates meet. At these zones of convergence, one plate bends and slides beneath the other. The oceanic crust is a relatively thin layer of dense, young basaltic rock and sediments, the former recently regurgitated from the interior of the planet along ridge axes. Because of its density, the oceanic crust is subducted as part of the downward moving lithospheric plate toward the interior of Earth, to depths greater than 650 kilometers. The driver for seafloor spreading is not directly the magmatism at

the ridge/rise system. This is considered to be a passive process, which is a result of the plates being pulled apart under the weight of their own cool, dense, subducting slabs.

Deep trenches in the seafloor may be formed in these regions of plate collision and subduction. The Marianas Trench located east of the Philippine Islands is one example. The depth of the trench seafloor is 11,033 meters and is the greatest ocean depth measured. Most of the world's most active volcanoes are found near these subduction zones. The margin of the Pacific Ocean exhibits frequent volcanic activity and is called the Ring of Fire (Figure 3.9). The volcanoes are often explosive and much volcanic debris and gas, like carbon dioxide and sulfur oxides, are injected into the atmosphere. The eruption of the volcano Pinatubo in the Philippines in 1991 and that of El Chichón in Mexico in 1982 are excellent examples of this type of volcanism.

Subduction zones are also defined by intense, repetitive, and often disastrous earthquake activity. The movement of lithospheric plates caused the devastating Alaskan earthquake of 1964 near Anchorage. The island nation of Japan, which lies next to a trench, has repeatedly felt the power of these forces of the planet. In addition, the San Andreas Fault in California is a focus of earthquake activity, the result of the lateral movement of the Pacific plate as it makes its journey toward the subduction zone off the Aleutian Chain in Alaska.

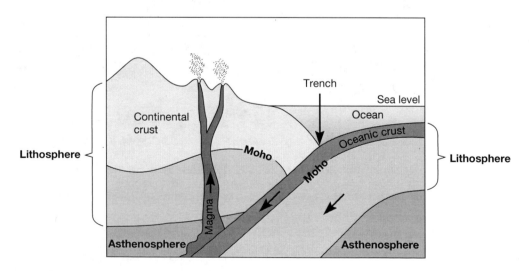

FIGURE 3.8 Diagram illustrating the process of subduction of the lithosphere at subduction zones and volcanism associated with the subduction process. Oceanic crust has been pulled down by the descending plate and partially melts, leading to the generation of rising magma that feeds the island volcanoes above. Arrows indicate direction of flow.

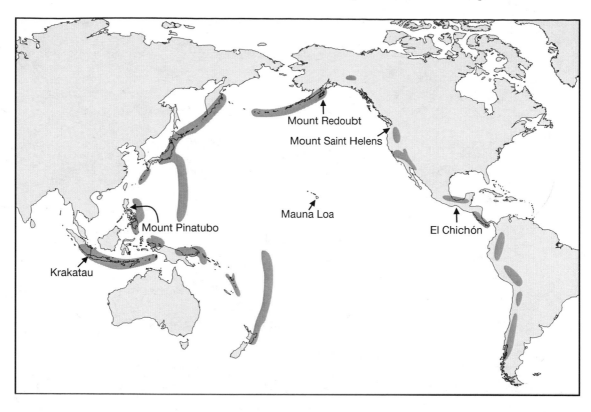

FIGURE 3.9 An illustration of the Ring of Fire, the volcanic region surrounding the Pacific Basin. Darkened areas are regions of major earthquakes and volcanoes.

At times, tsunami waves are generated in the waters overlying these zones of subduction. The waves are caused by the shifting of the plates and subsequent movement along faults (cracks) of the seafloor that initiate disturbances of the waters above. This process generates waves internal to the ocean of very long wavelength and small amplitude. These waves are unnoticeable at sea, but upon reaching shallow water near coasts, they can build up to heights exceeding 20 meters and flood coastal areas. Crustal movement in the vicinity of the Aleutian Trench and along the western boundary of South America has caused tsunamis to propagate across the Pacific Ocean and cause coastal damage and loss of life in areas within the Pacific Basin. The 2004 Indian Ocean undersea earthquake that occurred with its epicenter (point on the Earth's surface that is directly above the hypocenter or focus of the earthquake—the point underground where the earthquake originates) in the subduction zone off the west coast of Sumatra, Indonesia had a magnitude of 9.1 to 9.3 on the logarithmic MMS (moment magnitude scale) and is the second largest earthquake ever recorded by seismograph. The earthquake and the resulting tsunami led to the death of more than 225,000 people in 11 countries and widespread damage of infrastructure, shortages of food and water, and economic damage. Nations all over the world provided more than seven billion U.S. dollars in aid.

HOT SPOTS The Hawaiian Islands and the Emperor Seamounts are another example of plate tectonics in action. They are a chain of volcanic mountains in the Pacific Ocean that were formed in the middle of a moving plate. The Emperor Seamounts are the oldest and most northerly of the chain and consist of submerged limestone islands and volcanic mountains (seamounts). The Hawaiian Archipelago, consisting of islands, atolls, and submerged volcanic structures, trends northwest across the Pacific and intersects the trend of the Emperor Seamounts at their southern limit (Figure 3.10). The southern islands of the Hawaiian Archipelago are the youngest and comprise part of the state of Hawaii. The state's northern islands of Niihau and Kauai are 4.9 million to 5.1 million years old. Farther south, the islands of Oahu and Molokai have ages of 2.6 million to 3.7 million years and 1.8 million to 1.9 million

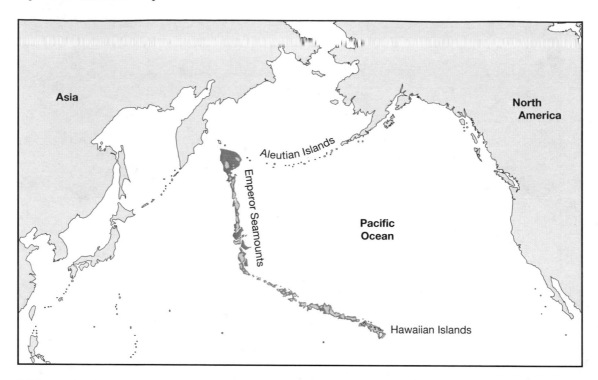

FIGURE 3.10 Hawaiian-Emperor chain of seamounts and islands. The southernmost part of the chain constitutes the well-known Hawaiian islands.

years, respectively. The more southern islands of Maui, Lanai, and Kahoolawe have ages of 0.8 million to 1.9 million years. The island farthest south in the chain is the big island of Hawaii, which has been experiencing nearly continuous spectacular volcanic activity in the form of lava flows since 1983. The seamount Loihi is currently forming under the waters south of the island of Hawaii and may become an island in a few million years. It is a seismically active area and a source of warm hydrothermal waters for the ocean.

The dates of formation and the differing intensities of erosion of the islands and seamounts along the Hawaiian-Emperor chain indicate that this submarine and subaerial mountain system was formed as the Pacific plate moved northward over a fixed (immobile) hot spot in the Pacific Ocean. The hot spot is a source of magma located deep below the Pacific lithospheric plate. As the plate moves northward, volcanic islands are formed by the rise of magma and its extrusion onto the oceanic crust. The lava piles up into a submarine volcanic mountain, and eventually, into a subaerial volcanic mountain. For example, the summit of the volcanic island of Hawaii rises 9000 meters above the seafloor. It is the tallest mountain on the face of Earth when measured from the depth

of the ocean floor to the summit of the mountain. Once a lava mountain is carried away from the hot spot by the northwesterly movement of the Pacific plate, the mountain is eroded and eventually submerged. Someday, some of these existing seamounts may be carried into the trench at the subduction zone near the Aleutian Islands of Alaska (Figure 3.11). The change in direction of trend of the Emperor Seamounts relative to that of the Hawaiian Archipelago represents a change in the direction of motion of the Pacific plate about 40 million years ago. Figure 3.12 shows the global distribution of hot spots and the inferred movement of the lithosphere relative to the fixed hot spots.

Continental Crust

CONTINENTAL DRIFT   Earth's continental crust has a different thickness and composition (Table 3.1) from that of the oceanic crust. The continental crust is older, thicker, less dense ($3.0\,g/cm^3$ or less), and composed of a mixture of sedimentary, metamorphic, and igneous rocks. Because these rocks are generally less dense than the basaltic rocks constituting much of the oceanic crust, the continental crust floats higher within the lithosphere. The continental crusts,

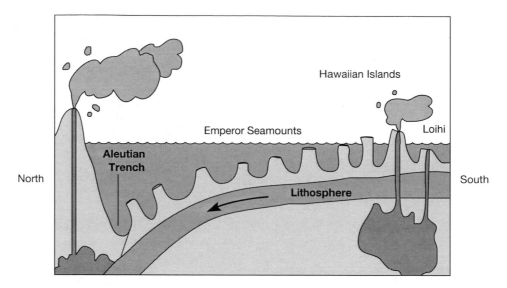

FIGURE 3.11 Schematic diagram illustrating how the Hawaiian-Emperor chain of islands, atolls, and seamounts has formed from slow movement of the Pacific plate initially in a north, then a northwest, direction across a fixed (immobile) hot spot. The hot spot is the source of basaltic lava that forms the volcanic structures of the chain, and the Aleutian Trench is the region of destruction of the chain by subduction.   (*Source*: Skinner and Porter, 1995.)

like the oceanic crusts, are attached to lithospheric plates and ride with them. This movement of the continents is known as continental drift (Figure 3.13). When a lithospheric plate with its affixed continent reaches a subduction zone, the continent is not subducted, because it is lighter and floats like a cork in water. Moreover, when two continental blocks attached to separate lithospheric plates collide, neither can be subducted because of their lightness. Thus, their collision results in the bending, folding, and rupturing of sedimentary and other rock types caught between them. The analogy can be made to the closing of the jaws of a vise on a piece of hardened, but still plastic, clay. Indeed flat, horizontal plates of clay have been used in laboratory experiments to mimic the formation of natural structures and the stresses involved in creating them. The sheet of clay is squeezed from one end and held rigid at the other, leading to compression of the clay and its distortion into folds and faults. In this manner, but in a more complicated series of movements, the Himalayan Mountains were formed when the Indian plate crashed into the Asian plate during the late Cenozoic, about 50 million years ago, carrying upward its cargo of sediments and fossil seashells.

Plate tectonic motions result in the movement of continental crust around the world (see Chapter 8 for details of Phanerozoic continental drift). These motions began sometime in the Precambrian, continued throughout the Phanerozoic, and will continue into the future (Figure 3.14). The formation and subsequent breakup of the supercontinent of Pangaea are an excellent example of continental drift and the concept of plate tectonics. Pangaea was a supercontinent that existed from about 300 to about 200 million years ago. The proto-ocean surrounding Pangaea has been termed Panthalassa. The supercontinent included most of the continental crust of the Earth at the time. The present continents have been derived from the fragmentation and displacement of continental blocks that were originally part of Pangaea. The supercontinent was intact at the close of the Paleozoic Era. It began its breakup in the late Triassic with the formation of the supercontinents of Laurasia and Gwondana. Lithospheric plates like those of today (Figure 3.1) were generated and set in motion. New continental blocks formed and drifted apart. As a result of these plate tectonic motions, we find oil in the Arctic and fossilized shells of marine organisms in sedimentary rocks outcropping on Mount Everest. Also, there exist similarities in the fossilized flora and fauna and rock types between continents that were once joined in Pangaea and are now separated. However, the separation of Australia and New Zealand from Pangaea, and their subsequent isolation from other continents, led to

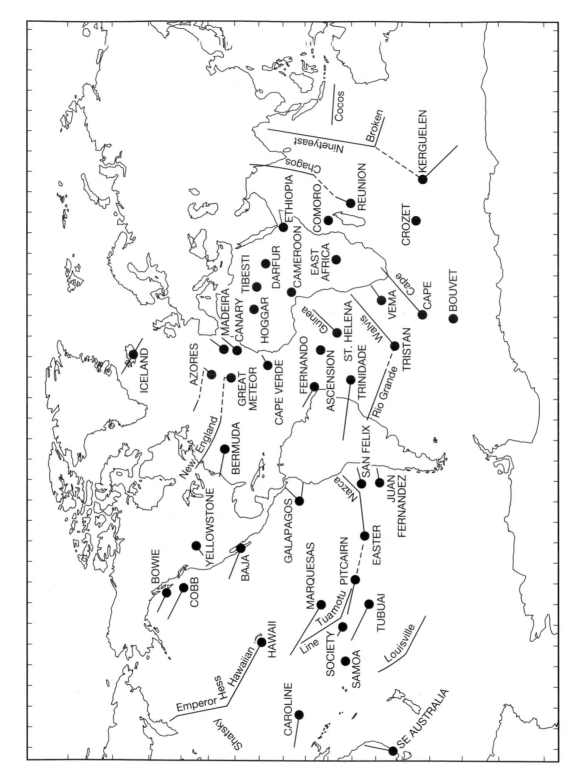

FIGURE 3.12 Global distribution of hot spots and tracks showing motion of the plates relative to the hot spots. (*Source:* Crough, 1993.)

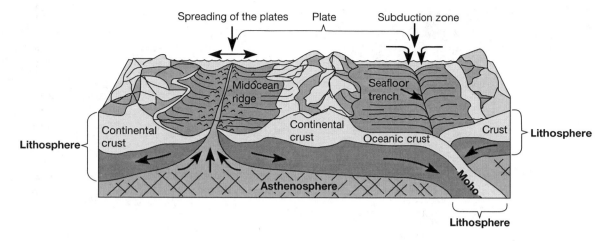

FIGURE 3.13 Cross section of outer layers of Earth showing two continental crusts located on separate lithospheric plates. Movement of the lithosphere away from its source at midocean ridges leads to continental drift.

the development of flora and fauna on these continents along different lines. This separation gave rise to the evolution of species not found on the other continents (e.g., the koala bear). Because of vast oceans separating them from other continents, the southern lands were isolated, prohibiting the migration of animals and plants to or from the other continents.

CONDITIONS WITHIN THE UPPER CRUST We are all familiar with the irregular cracked surface of Earth, underlain by the debris of billions of years. Here is the zone of intersection of the atmosphere with the rocks, where rain falls and accumulates into rivers, or

sinks into the ground and percolates through pores and fractures. Rocks are broken up and form soils, which are eventually swept into the sea. Other rocks rise to take their place, with the same eventual result.

The conditions of the shallow crust are hard to describe, because they are so varied and the rock units so discontinuous. Let us look at some of the processes that transfer materials within the upper few thousands of meters of solid Earth. These processes are particularly important to the movement of fluids in Earth, including groundwaters and their dissolved chemical compounds and pollutants, oil, and gas, and to concepts discussed in this book.

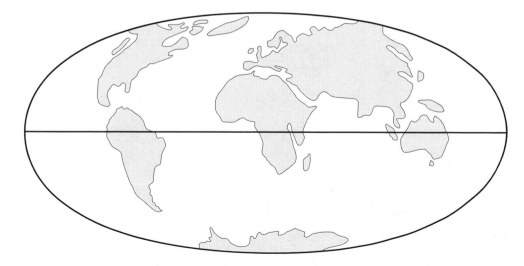

FIGURE 3.14 A possible configuration of continents in the future resulting from plate motion and concurrent continental drift. Note how Australia has moved northward to an equatorial position.   (*Source:* Christensen, 1991.)

Pressure-temperature Gradients in the Shallow Crust
In the upper few kilometers of the crust, rocks have the strength to maintain openings. The rocks and the waters within their pores can be considered two separate interpenetrating systems. Pressures in the water-filled pores usually are higher by a factor proportional to the ratio of rock density to water density.

A 100-meter column of water exerts a pressure of about 10 bars (approximately 10 atmospheres, or 147 pounds per square inch). The hydrostatic (water) pressure of the rock column depends on the density of the rocks at a particular place. Sedimentary rocks have densities of the order of 2.5 g/cm^3; metamorphic and high-silica igneous rocks, such as granite and granodiorite, have densities of about 2.7 g/cm^3; and for low-silica igneous rocks, such as basalts, densities are a little over 3 g/cm^3. Therefore, the lithostatic (rock) pressure increases by about 250 to 300 bars/km. Figure 3.15 shows hydrostatic and lithostatic pressure-temperature-depth relations for various temperature gradients in the crust. The boiling curve for water is also plotted and shows that for shallow depths in the crust, water remains liquid

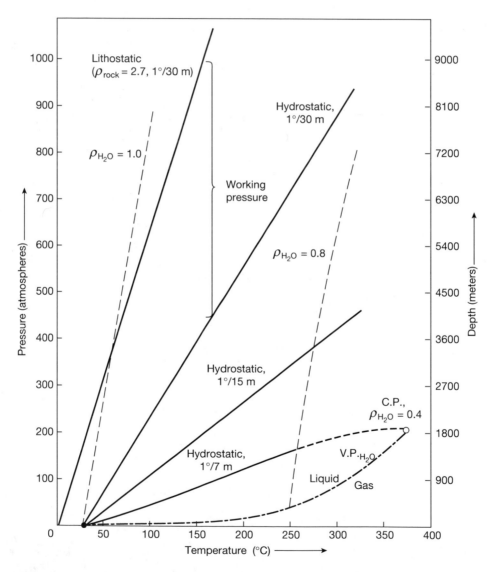

FIGURE 3.15 Relationships among hydrostatic and lithostatic pressures, temperature, and depth for various gradients of temperature in the crust. Two lines for the density of water (ρ) as a function of pressure and temperature and the vapor pressure curve (V.P.) for water are also shown. C.P. is the critical point of water.

under almost all temperature gradients likely to be encountered. The production of steam or the presence of geysers requires somewhat unusual conditions since in general the hydrostatic pressure curves for most temperature gradients in the crust do not intersect the vapor pressure curve for H_2O (Figure 3.15). Furthermore, down to depths of several kilometers, the density of liquid water lies between 0.8 and 1.0 g/cm^3. The critical point of water (the temperature and pressure at which the density of liquid H_2O equals the density of H_2O vapor) occurs at a H_2O density of 0.4, which for a one degree Celsius per meter temperature gradient is at a temperature of 375°C and a depth of about 1800 meters (Figure 3.15).

The pressure difference for any given temperature between the typical lithostatic and hydrostatic pressure curves is labeled "working pressure" in Figure 3.15. This difference in pressure is the maximum that can be achieved underground for water to flow. If the pore-water pressure exceeded lithostatic pressure, the overlying rock column would be lifted and fractured. Because working pressure increases with depth, the maximum potential for water flow through rocks increases linearly with depth.

Although the usual situation encountered in the shallow Earth is the expected hydrostatic pressure owing to the weight of the water column, water pressures approaching lithostatic are observed in a few places. This leads to a situation in which the weight of the overlying rock column is almost entirely supported. Under these conditions, lateral sliding of large masses of rock under the influence of very small forces may occur at depth. The presence of nearly horizontal, thick, plastic, impermeable, shale layers is probably necessary for high water pressures to occur. In other types of rocks, the water would escape through fractures. Indeed, overpressured sandstones encapsulated in shales are found at depth in sedimentary basins. In some cases these sandstones may contain fluids other than water, like oil and gas.

Porosity   Below a depth of 5 or 6 kilometers, the pressure exerted by the overlying rock column is so great that open spaces in rocks are eliminated. The porosity is only a small fraction of 1% of the rock volume. **Porosity** is the proportion of total rock volume occupied by pores. Transportation of water, oil, or gases in important quantities can take place only with the development of forces sufficient to break the rocks, permitting movement along fractures, or by

ionic or molecular diffusion (see p. 62) in microscopically thin films along the boundaries of mineral grains.

In the upper few kilometers of the crust, the confining pressures are less, and most rocks have the strength to maintain openings. The maximum size of these openings diminishes from huge caves near the surface of Earth to grain boundary films at depth.

Igneous rocks and some highly altered metamorphic rocks (crystalline rocks) almost never have significant intergranular porosity. In many fine-grained extrusive igneous rocks, the grains are typically so tightly intergrown that the rock density is almost exactly that of the individual minerals. Intrusive igneous rocks are almost as tightly put together. A granite may have 0.1% pore space. Most of the pore space occurs where the intergrowth of the original grains left small irregular cavities. Consequently, transport of water, which is the chief agent of material transfer in the shallow crust, takes place in crystalline rocks through major cracks and fractures. The degree to which crystalline rocks are fractured near the surface is extremely variable. In some places closely spaced fractures subdivide the rock into fragments a few centimeters in diameter. Elsewhere, the fracture spacing may leave great blocks many meters across. When the rock is heavily fractured, water can pass through it easily and rapidly. However, even in such rocks, the percentage of open space is small so that they do not contain much water per unit of rock volume.

Sedimentary rocks, which make up 10% to 20% of the entire crust and underlie about 80% of the continental surfaces, range from materials with no significant intergranular porosity to those with 30% to 40% pore space. The clastic sedimentary rocks, those deposited mechanically as grains, contain from 30% to 90% by volume of water at the time of deposition. As these sediments are compacted by the weight of subsequent deposits, the water is squeezed out until the grains make sufficient contact with each other to form a structural framework that can support the rock load. In the absence of any later cementing material, sandstones and shales average about 30% porosity. Average actual porosities are about 15% for sandstones and shales, but differences between sandstones and shales, or from one part of a rock to another, are large.

The typical history of a sedimentary rock comprises its original deposition and compaction in seawater; burial to depths ranging up to several kilometers; uplift above the sea, with little or great distortion and fracturing; and eventual exposure at

Earth's surface. During this history the original sea-water present in the pores of the rocks may be in part retained, apparently with selective loss of water and salts with increasing compaction of the sediments. The water also may be completely displaced by freshwaters percolating downward from the surface or by waters moving from adjacent rocks or from deep within the basin of deposition. The original porosity may be increased by solution of grains by the moving waters, or it may be eliminated by precipitation and cementation from such waters. Also, the fluids of oil and gas may move many hundreds of kilometers from their source in fine-grained, organic-rich rocks before they are trapped in a porous and permeable sediment from which they can be exploited by drilling into the sedimentary formation.

At any rate, rain falling on Earth's surface tends eventually to reach the sea. Part of it is carried in streams; part of it sinks into the ground and circulates downward wherever it can, its movement dominated by those rocks through which it can move easily and continuously. In addition, the percolating rainwater leaches dissolved constituents from the surface and subsurface minerals, rocks, and organic matter, including potential pollutants like nitrate (NO_3^-) and phosphate (PO_4^{3-}) derived from the application of fertilizers to croplands; toxic trace metals, like arsenic (As); and synthetic organic compounds, like the pesticide dichloro-diphenyl-trichloroethane (DDT) and the chlorinated hydrocarbon insecticide dieldrin. These natural and pollutant dissolved constituents can move through the subsurface via groundwater movement and into streams and rivers and to other aquatic systems, as described below.

Figure 3.16 shows an idealized situation involving an upland region, a stream, and a valley in which the subsurface material is uniformly porous and in which the force causing water movement is the height of the zone of saturation under the upland above that in the valley. Figure 3.16 can be used to indicate the sequence of events due to the pulsating drive of percolating rain followed by dry weather. The pore spaces of the rocks tend to fill up to the ground surface during the rain, when addition of water is faster than its flow through the rock. This increase in potential energy beneath the upland tends to disappear between rains, causing continuous flow. The surface of saturation (water table) is clearly a function of the

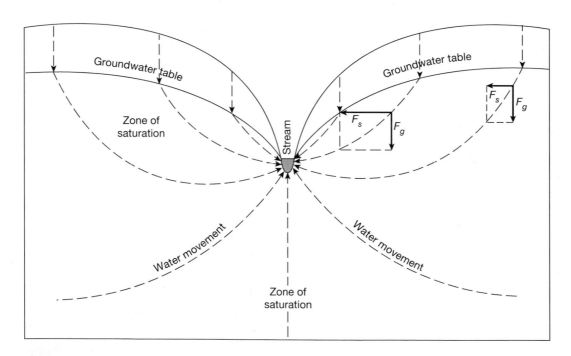

FIGURE 3.16 Diagram illustrating the movement of water from the land surface into a stream. The path traveled by the water particles through homogeneous rock materials is shown by dashed lines. The path traveled is the resultant of the force of gravity (F_g) and the "pulling" force (F_s) of the stream. In the unsaturated zone where water does not pervade the rock material, only the force of gravity operates. (*Source:* Garrels, 1951.)

amount and frequency of rainfall and of the water-carrying characteristics of the rocks. In arid areas evaporation might exceed rainfall, and no zone of saturation might exist. In areas of heavy continuous rain, the saturation surface remains at the topographic surface. In general, the zone of saturation in the United States, exclusive of the desert areas, is encountered at an average depth of 6 to 12 meters and appears as a subdued image of the topography.

The most effective water conduits (aquifers) are the sandstones. The pore spaces in these rocks are relatively abundant and large, which make it easy for water to pass through them. Also, sandstones commonly occur as continuous thin sheets over many thousands or tens of thousands of square kilometers. The source of water for many wells drilled beneath the Great Plains of the United States is a subsurface sandstone (the Cretaceous Dakota Sandstone) that is recharged through outcrops of the formation in the Rocky Mountains far to the west.

Permeability A measure of the ease of transmission of water through a rock is called its **permeability.** Movement of water is driven by a pressure gradient, commonly gravitational, and is retarded by friction with the grains of the rocks. The permeability of a rock can be tested by determining the rate at which the water will flow through a given cross-sectional area of a rock sample of a given length,

$$\frac{q}{t} = \frac{K(P_2 - P_1)A}{d}$$

where q is the quantity of water delivered in cubic centimeters, t is time in seconds, $P_2 - P_1$ is the drop in pressure across the sample in atmospheres, A is its cross-sectional area in square centimeters, d is its length in centimeters, and K is a factor termed the coefficient of permeability. A permeability of 1 darcy is possessed by a rock that transmits 1 cubic centimeter of water in 1 second through an area of 1 square centimeter down a length of 1 centimeter under a pressure gradient of 1 atmosphere. Few rocks have permeabilities as high as 1 darcy. The millidarcy (darcy/1000) is in common use. The difference in permeability of several orders of magnitude between sandstones and shales is one major reason sandstones are the dominant carriers of water underground.

The situation in the Great Plains of the United States is fairly typical of what occurs in areas underlain by slightly inclined sedimentary rocks (Figure 3.17). Rain infiltrates the sandstone in its catchment area. The underground waters (groundwater) move downward through the sandstone beneath an impervious layer. Where a drill hole pierces the rock capping

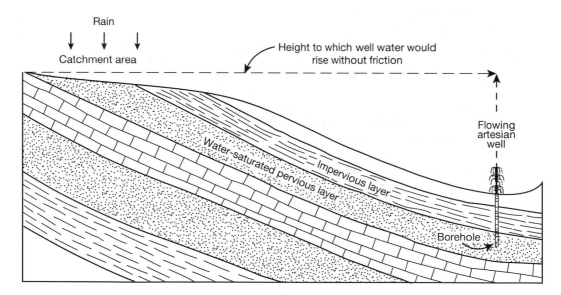

FIGURE 3.17 Water movement through slightly dipping sedimentary rocks of varying lithology. Rainwater percolates into the pervious layer, where it intersects the land surface. The water then flows through the pervious layer below the impervious layer. If the water-saturated pervious layer is drilled into, the pressure is released and the water flows upward in the form of an artesian well. (*Source:* Garrels, 1951.)

the sandstone, the pressure in the sandstone is released and the water flows upward (artesian flow). If the permeability of the sandstone were 10 millidarcies, the head of water 500 meters, and the distance from catchment area to discharge 200 kilometers, the flow per year per square centimeter would be (100 meters of water = 10 atmospheres):

$$\frac{q}{3.15 \times 10^8} = \frac{(0.01)(50)(1)}{2 \times 10^7}$$
$$q = 7.9 \text{ cm}^3 \text{ of water.}$$

In other words, the water would be expected to move only 8 centimeters during a year. Such a flow rate is not atypical. Groundwater flow rates of several centimeters to a few tens of meters per year are usual, and rates as high as 1000 meters per year or more are attained in unconsolidated gravelly sediments. Even a slow rate of movement of 8 cm/year becomes significant on a geologic time scale. A liter of water would be transmitted through each square centimeter of rock in 125 years, amounting to 8000 liters in a million years. Groundwaters in rocks are highly variable in their chemical composition, but about 0.1 grams per liter of dissolved solids in sandstone waters is not uncommon. Therefore, about 800 grams of dissolved material could be transported through each square centimeter of sandstone in a million years. In addition, groundwaters can contain elevated concentrations of some dissolved constituents, like nitrate derived from leaching of nitrogenous fertilizer applied to croplands.

In the history of sedimentary rocks, the tendency is to displace the original seawater with rainwater in a highly irregular way, controlled by the permeability and porosity of the rock, its geometry, its relation to topography and to adjacent rocks, the climatic conditions of the region, and a host of other variables. However, for any given mass of rocks, it is apparent that sandstones will be the chief water carriers, along with some of the limestones. Shales will transmit little if any water, even over geologic periods. The older a rock, the more chance it has had to be flushed out, with accompanying effects of solution and alteration of mineral grains, as well as cementation and pore filling.

About 0.40×10^{20} grams, or 40 trillion metric tons, of freshwater flow down the rivers into the oceans each year. All this water has at least a short period of contact with surface materials, and perhaps 25% to 35% percolates through these materials before draining into streams. Most of the percolating water moves close to the surface. The amount of this water diminishes rapidly with depth in the crust. The total amount of water in the pores of rocks is estimated to be as much as 20% to 25% of that in the oceans. Some of the molecules of this pore water have not seen the surface of Earth since the day of their entrapment. Others have been residents of many different rocks at many different times.

In addition to the gravity drive for groundwater flow, for which energy is derived ultimately from the sun, water circulation is driven by Earth's internal heat. The geysers and hot springs of Iceland and of Yellowstone National Park in the United States are extreme examples of water circulation driven by heat. There are many other areas, without obvious surface expression, where a steepened thermal gradient drives water upward through the rocks, including the hydrothermal vent fields found along the midocean ridges of the deep sea.

Diffusion   Bulk flow of water with its contained solutes is the chief medium of material transfer in the upper crust. Differential movement of water and solutes by ionic and molecular diffusion is of secondary importance in moving large quantities of materials long distances but is one of the dominant factors in controlling reactions between waters and rocks. For example, if a groundwater is moving through a fractured rock and the water is not in equilibrium with the rock, the rock tends to be altered at the contacts between the mineral grains of the rock and water. An altered layer is created around the fresh mineral grains. The alteration process involves transport of dissolved ions and molecules from the water through the altered material at the surfaces of the mineral grains. These dissolved constituents then react with the fresh mineral grains inside the alteration layer. Soluble reaction products are released and move outward through the alteration layer and into the water. This process of diffusion is important in the weathering and alteration of minerals and rocks and can be expressed by the following equation that is analogous to those for heat flow and water flow:

$$q/t = \frac{D(C_2 - C_1)A}{d}$$

where, if q is expressed in equivalents of ions, t is in days, D is the diffusion coefficient of the chemical species in question in square centimeters per day, $C_2 - C_1$ is the concentration gradient in equivalents per cubic centimeter, A is the effective cross-sectional area in square centimeters, and d is the distance in centimeters over which the diffusion is occurring. D is specific to each species and ranges from less than $1 cm^2/day$ for ions like Ca^{2+}, Mg^{2+}, and Fe^{2+} to as much as $8 cm^2/day$ for the hydrogen ion. C_2 represents a fixed concentration at the boundary between the fresh mineral or rock and altered material, and C_1 represents a constant concentration at the outer boundary of the alteration zone. The effective cross-sectional area is that fraction of the actual cross-sectional area that behaves as if it were a straight solution-filled pore. In most rocks, it is about two-thirds of the total porosity.

For example, using the preceding diffusion equation, it can be calculated that it takes only about ten years to convert half of a 1-centimeter potassium feldspar grain in a rock undergoing weathering by freshwater percolating through soils at 25° C to an altered layer of the mineral kaolinite. The process represents the transport of about 0.6 grams of potassium from the feldspar grain into the soil water. In some places, where water has moved through fractures in rocks, diffusion-controlled alteration may extend for a meter or to more than a few meters outward from the fracture. The alteration may represent tens of thousands or hundreds of thousands of years of water flow in the fracture during which time the diffusion of ions into and out of the rock lining the fracture was taking place.

Concluding Remarks

The rock cycle was introduced in Chapter 2. Rudimentary allusions to the cyclic concepts of geology can be found in writers of antiquity. However, James Hutton (1726–1797) was the first post-antiquity geologist to recognize the importance of crustal denudation and sediment transport as ongoing geological processes on a global scale and the general concept of a rock cycle. Since then considerable attention has been focused on the global sedimentary cycle, including the cycling of salts in the ocean, as a result of William Thompson's (later Lord Kelvin) early estimate of the age of the Earth as no older than 100 million years, made between 1864 and 1899, and John Joly's estimate of the age of the ocean in 1899 as 99.4 million years based on the rate of accumulation of sodium brought in by rivers. In the 1980s C. Bryan Gregor of Wright State University summarized and discussed in detail the geological arguments proposed in the second half of the 1800s and the early 1900s for the recycling of oceanic sediments after their deposition and for the existing sinks of dissolved salts in ocean water brought in by rivers, such as their removal by absorption on clays, entrapment in sediment pore waters, and the formation of evaporite deposits. The removal and recycling of the oceanic salts is contrary to the older Joly and others' idea of the ocean continuously filling up with dissolved salts. In 1971 Robert M. Garrels and Fred T. Mackenzie presented the concepts of the sedimentary cycling of materials that had lain dormant for some years in a book entitled *Evolution of Sedimentary Rocks*, and in 1972 these two authors developed a quantitative model of the complete sedimentary rock cycle. Most materials derived from human activities, whether an element or compound that originally was transported in the cycle or a synthetic compound, are added to and flow within the natural sedimentary cycle. These anthropogenic emissions perturb the cycle, change the composition of reservoirs in the cycle—like the land, biosphere, atmosphere, and aquatic systems—and lead to modern environmental problems.

A generalized diagram of the rock cycle and its circulation illustrating the major units (reservoirs) and rates of transfer of materials (fluxes) is shown in Figure 3.18. The lithosphere is an integral part of the ecosphere. It is being renewed constantly and rearranged through plate tectonics. The land surface is acted upon continually by the biota and by the air and water systems. The processes of weathering and erosion of sedimentary and crystalline rocks modify the land surface and are the beginning of the rock cycle. The products of weathering and erosion on land reach the ocean through the agents of wind, water, and ice. The rivers are the main agent of delivery of materials to the ocean. In the ocean, the products are deposited and accumulate as sediments of gravel, sand, silt, mud, and organic and skeletal debris on the seafloor. Materials derived from hydrothermal exhalations at midocean ridges and the products of submarine erosion of midocean ridge basalts are added to the materials derived from land. The sediments may be buried to great depths because of subduction or by other means. During burial the sedimentary materials encounter increasing pressures and temperatures and different subsurface fluid compositions. They are compacted and lithified to rock. Some are converted to metamorphic rocks,

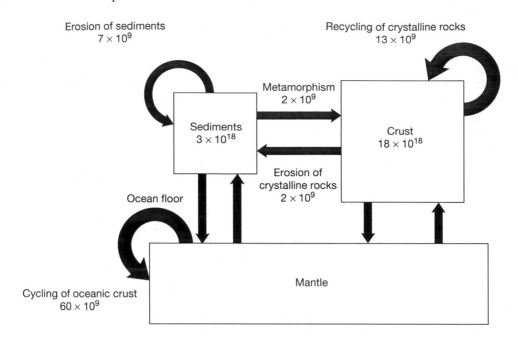

FIGURE 3.18 A generalized diagram for the steady-state rock cycle. Sediments and continental crystalline crust masses are in units of metric tons. Erosion of sediments, metamorphism, erosion of crystalline rocks, recycling of crystalline rocks, and cycling of oceanic crust are fluxes in units of tons per year. Total sedimentation rate is 9×10^9 tons annually. (*Source:* Gregor, 1988.)

and others are so altered that they resemble igneous rocks. Later in their history, plate tectonic forces uplift these rocks. The materials are once again subjected to the processes of weathering and erosion. The rock cycle is complete.

Within the great rock cycle (Figure 3.18), the sedimentary cycle of weathering, erosion, transportation, deposition, burial, lithification, and uplift has repeated itself many times during Earth's history. Gregor, Garrels and Mackenzie, and others have made quantitative estimates of sediment recycling rates, based on the distribution of sediments and sedimentary rocks through geologic time. The distribution of sediment mass per unit time as a function of age of the sediments roughly follows an exponential decay curve (see half-life, Box 2.1; Radioactive Elements) diminishing with increasing age of the sediments. The total sedimentary mass has a mass half-age of 500 to 600 million years; that is, half of its present mass—about 3×10^{18} tons—will disappear in about 600 million years owing to the processes of

weathering, erosion, and transport of materials to future basins of deposition and the formation of "new" sediments and sedimentary rocks. The total amount of sediments deposited during Earth's history is about 18×10^{18} tons, and the present erosion rate of sediments and crystalline rocks is about 9 billion tons per year. In 1988 Jan Veizer showed that the recycling rates of the sedimentary lithosphere and the various rock types within it are mainly a function of the recycling rates of the tectonic realms, such as active margin basins, oceanic crust, and continental basement, in which the sediments were accumulated. Thus, as James Hutton recognized in 1795, there is still no "vestige of a beginning—no prospect of an end." However, in the eighteenth century, human activities had not as yet become a major player in the ecosphere. Since then and particularly in the twentieth and twenty-first centuries, human activities have become a "geologic force" in the sedimentary cycle, leading to major modifications of the cycle as a whole and of its individual components.

Study Questions

1. What is the lithosphere, and what are the three major boundaries of lithospheric plates?
2. What are the two major systems of subdivision of Earth's interior, and what zones do they contain?
3. Why is it difficult to subduct the continental crust of Earth?
4. How does Earth's crust spread?
5. Why must plates be destroyed at subduction zones?
6. What are the major processes affecting the lithosphere and the surface of Earth?
7. How do the processes of weathering and erosion relate to the formation of soils and sedimentary rocks?
8. What are some evidences for an Earth core of metallic iron?
9. What are the major chemical differences among the core, mantle, and crust?
10. What is the main evidence for the conclusion that part of Earth's core is molten?
11. The Gilbert reversal in magnetic anomaly patterns of the seafloor has an upper age limit of about 3.3 million years (Figure 3.6). The basaltic lava representing this time is found 100 km from the present ridge axis. What has been the average spreading rate of this portion of the oceanic crust?
12. What is the significance of the term *working pressure* in Figure 3.15?
13. What is the permeability of a rock layer through which $80 \, cm^3$ of water flow in one year through a cross-sectional area of $1 \, cm^2$ a distance of 200 km? The catchment area is located 1000 meters above the point of discharge of the water in a spring.
14. (a) Write a chemical reaction for the weathering of calcium carbonate ($CaCO_3$) by water (H_2O) and carbon dioxide gas (CO_2) in soil to dissolved calcium (Ca^{2+}) and bicarbonate (HCO_3^-).
    (b) Where does the carbon dioxide come from to chemically weather $CaCO_3$?
15. What is the rock cycle?

Additional Sources of Information

Garrels, R. M. and Mackenzie, F. T., 1971, *Evolution of Sedimentary Rocks*. Norton, New York, 397 pp.

Gregor, C. B., Garrels, R. M., Mackenzie, F. T., and Maynard, J. B., 1988, *Chemical Cycles in the Evolution of the Earth*. John Wiley, New York, 276 pp.

Kearey, P. and Vine, F. J., 1990, *Global Tectonics*. Blackwell Scientific Publications, London, 302 pp.

LeGrand, H. E., 1988, *Drifting Continents and Shifting Theories*. Cambridge University Press, Cambridge, 313 pp.

Skinner, B. J. and Porter, S. C., 1995, *The Dynamic Earth: An Introduction to Physical Geology*, 3rd Edition. John Wiley, New York, 567 pp.

Wilson, J. T., 1976, *Continents Adrift and Continents Aground: Readings from Scientific American*. W. H. Freeman, San Francisco, 230 pp.

Wyllie, P. J., 1976, *The Way the Earth Works*. John Wiley, New York, 296 pp.

The Fluid Earth: Atmosphere

No one doubts the reality of the natural greenhouse effect, which keeps us over 20° C warmer than we would otherwise be. The science of it is well understood; it is similar science which applies to the enhanced greenhouse effect.

SIR JOHN HOUGHTON, 1997

The lithosphere, as discussed in the previous chapters, interacts extensively with the atmosphere and hydrosphere. These interactions have helped to produce an environment in which life could develop and organic evolution could take place. In turn, the development of the ecosphere has produced a surface environmental system (exogenic system) that has been relatively stable and resilient, but changeable, throughout the 545 million years of Phanerozoic time. The stability and resilience of the system are maintained by biological and physical processes, fluxes (flows, rates of exchange), and feedbacks (processes that aid or abet change) that involve the major environmental subdivisions of the near surface of Earth.

Continuing the theme of Chapters 2 and 3, in this and the following chapter, the air and water envelopes of the planet are discussed as part of this interactive exogenic system. Natural climatic interactions between the atmosphere and ocean are also considered to provide some background for discussion of Pleistocene and Holocene climate change in Chapter 13 and global warming and stratospheric ozone depletion in Chapter 14.

First let us consider some facts concerning the atmosphere and hydrosphere as background for this and the following chapter. The surface area of Earth is 510×10^{12} square meters (m^2). The oceans, the largest reservoir of water in the hydrosphere, constitute 70% of this area, or $360 \times 10^{12}\,m^2$, and the land surface, 30%, or $150 \times 10^{12}\,m^2$. The Northern Hemisphere has about 50% land surface, much of which exists in the zonal wind belt of the prevailing westerlies. The Southern Hemisphere in contrast is 75% sea surface, with a marked absence of land at and near 60°S latitude. The average height of the land is 0.8 kilometers (km), and the average depth of the oceans is 3.8 kilometers. The vertical relief between the highest point on land (Mt. Everest, 8935 meters) and the deepest depth of the oceans (Marianas Trench, 11,000 meters) is 19,935 meters, or about 64,624 feet. The mass of the ocean is about $14,000 \times 10^{20}$ grams (g), or 1.4 billion billion metric tons, about 270 times that of the atmosphere of 52×10^{20} grams. The salt alone in the ocean has a mass about 10 times that of the entire atmosphere, or 500×10^{20} grams. The average density of seawater is 1.026 g/cm^3, whereas pure water has a density of 1.0 g/cm^3, and air near the base of the troposphere, 0.00122 g/cm^3. The average molecular weight of air is 29 g/mole; thus, the total number of moles of atmospheric gas is 52×10^{20} g/29 g/mole = 1.8×10^{20} moles (1 mole or gram molecular weight of a gas contains 6.02×10^{23} molecules and occupies a volume of 22.4 liters at 0°C and 1 atmosphere pressure).

The pressure of the air at sea level is 1 atmosphere, which is approximately equal to 1 bar, which in turn equals about 1000 g/cm^2 or 14.7 lb/in^2 or 29.92 inches of mercury. The pressure in the ocean increases approximately 1 atmosphere for every 10 meters of depth. At the average depth of the ocean of 3800 meters, the pressure is 380 atmospheres. The pressure of the atmosphere halves every 5 kilometers above Earth's surface. Thus, for jet aircraft flying at 10 kilometers above the surface, three-quarters of the mass of the atmosphere is below the plane. The mass of air in a given volume decreases with increasing height at a rate less than that for water vapor; thus, the upper atmosphere has a very low water content. The upper atmosphere actually acts as a trap for water vapor. Because of the very cold temperatures of the upper atmosphere, water vapor condenses into rain droplets or freezes and falls as precipitation, thus preventing it from escaping into space. Now we will consider some of the characteristics of the thin gaseous envelope of Earth, the atmosphere.

THE ATMOSPHERE

The global atmosphere extends 500 kilometers above the surface of Earth from the lower zones of the troposphere and stratosphere to the upper zones of the mesosphere, thermosphere, and exosphere (Figure 4.1). The atmosphere has been evolving since Earth was formed. It is the most rapidly changing and dynamic of the three physical systems, but it is a well-balanced system. It is the heat engine (along with the ocean) that distributes heat throughout the globe and drives the climate system of the planet. Climate and weather have much in common, but they are not identical. **Weather** is an everyday experience and represents the sum total of atmospheric variables in a particular region for a short period of time. **Climate** is a lasting feature of the atmosphere and represents a composite

of the day-to-day weather conditions and of atmospheric variables in a region over a long period of time. The atmospheric variables, or elements, are principally solar energy (sunshine), humidity and precipitation (moisture), and winds. Atmospheric pressure may also be thought of as an element of weather and climate. Variations in atmospheric pressure determine the characteristics of the other variables. Atmospheric pressure differences determine the speed and direction (velocity) of the winds. The winds, in turn, move atmospheric air masses of different temperature and moisture conditions from one region to another.

The major source of energy for driving atmospheric and surface processes on Earth comes from the sun (see Box 4.1: Energy). The surface temperature of the sun is about 5480°C, and the light energy or radiation from this stellar body easily reaches our atmosphere. Some of the incoming solar radiation is reflected back to space in the outer reaches of the planetary atmosphere. Most of the energy reaching the outer boundary of our atmosphere is visible, shortwave, and ultraviolet radiation (Figure 4.2). In the atmosphere each atom and molecule has a different efficiency and wavelength region for the absorption of radiation. The most important absorbers of incoming solar radiation in Earth's atmosphere are molecular oxygen (O_2) and ozone (O_3). High in the atmosphere, molecular oxygen absorbs photons with wavelengths less than about 240 nanometers (nm, 10^{-9} meters) and deeper in the atmosphere, molecular ozone absorbs wavelengths mainly in the range of 200 to 300 nanometers. The incoming intense solar radiation actually breaks the bonds holding the oxygen atoms together of both O_2 and O_3 molecules; for diatomic oxygen, the reaction is

$$O_2 + h\nu \rightarrow O + O$$

where $h\nu$ represents incoming solar radiation. The free oxygen atoms each reunite with other diatomic oxygen molecules to form ozone. For triatomic oxygen, the reaction is

$$O_3 + h\nu \rightarrow O_2 + O.$$

The oxygen atom reacts with O_3 to make two O_2 molecules. Thus, ozone is produced and destroyed in the atmosphere through a natural cyclic process known as the Chapman cycle, named after Sidney Chapman of Oxford University who in 1930 proposed the destruction reaction for atmospheric ozone. Most of the ozone in the atmosphere occurs at an altitude of 19 to 48 kilometers in the stratosphere. This altitude region of increased ozone concentrations is known as the stratospheric ozone layer (see Chapter 14).

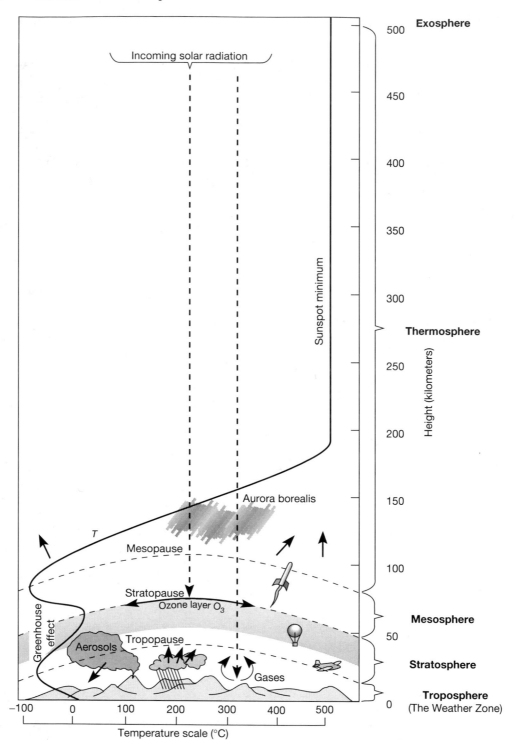

FIGURE 4.1 The layered structure of Earth's atmosphere. The structure is defined on the basis of the way temperature varies with height. The warm stratosphere acts as a lid for the lower atmosphere; thus, weather systems develop and circulate only in the troposphere. The temperature of the thermosphere varies with the intensity of solar ultraviolet radiation, and this, in turn, varies with the sunspot cycle and solar activity in general. Sunspots are dark areas on the face of the sun visible with telescopes or the partially shielded naked eye. Sunspots occur in a definite cycle, with periods of more or less sunspots. The periodicity of the sunspot cycle is approximately 11.2 years. The temperature of the thermosphere in this figure is for a period of time of minimum sunspot activity.

BOX 4.1

Energy

The Electromagnetic Spectrum

Energy on Earth has led to the evolution of life. It is vital for the existence of our planet and for ourselves. The major source of energy for the surface of the planet is radiant energy that reaches Earth from the sun. The sun runs on atomic energy; the source is the fusion of hydrogen in the stellar interior.

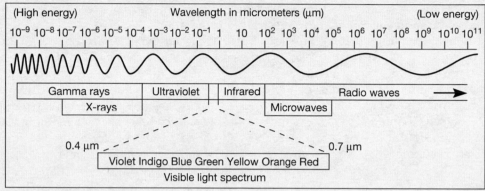

(*Source:* Johnson, 1990.)

Energy from the sun travels through space in the form of electromagnetic radiation. One measure of the character of radiation is its wavelength. The wavelengths of energy are analogous to water waves. The wavelength of a water wave is defined as the distance from one high (crest) or low (trough) point on the wave to the next high or low point on the following wave. There are water waves of long and short wavelengths. Electromagnetic radiation can also be thought of as a series of energy waves passing a fixed point and, thus, their wavelengths can be determined.

Electromagnetic radiation has a wide range of wavelengths forming a continuum from short to long. This continuum is the electromagnetic spectrum. The amount of energy in radiation depends on its wavelength. More energy is contained in electromagnetic radiation that is short in wavelength than that in long wave radiation. Hot bodies tend to emit energetic, short-wavelength radiation, whereas cool bodies emit longer wavelength, less energetic radiation. For example, most of the sun emits radiation in the visible part of the spectrum, whereas Earth, with a much lower surface temperature, emits less-intense radiation with a longer wavelength spectrum than that of the sun.

The surface temperature of the sun is 5480°C. Objects of this temperature give off radiation with wavelengths mainly in the range of 0.2 to 3.0 micrometers (μm). The sun emits approximately 95% of its energy as ultraviolet, visible, and shortwave infrared radiation. Additionally, the sun gives off gamma rays, X-rays, microwaves, and radio waves.

Gamma rays are extremely powerful, shortwave radiation. They can penetrate most objects and are deadly to most life. X-rays are almost equally powerful. Both of these types of radiation are absorbed by gas molecules in the outer reaches of the atmosphere of Earth. Ultraviolet radiation is shortwave, high-energy radiation that is dangerous to living organisms and can

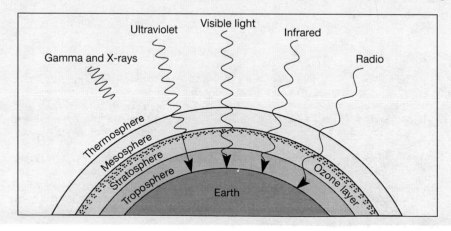

be the cause of skin cancer and sunburns. This radiation is mostly absorbed in the stratosphere with just a small percentage actually reaching Earth's surface. Visible light waves reach Earth's surface, and our eyes are sensitive to this radiation. Visible light consists of a range of electromagnetic radiation. The color violet is on the short-wavelength end of the color spectrum, and the color red is on the long-wavelength end. Infrared radiation also reaches the surface of Earth but is not visible. We feel it as heat. There are photographic films that are sensitive to this radiation and can image sources and intensity of infrared radiation. Microwaves are high-frequency radio waves. This radiation is used in microwave ovens. Radio waves span a large range of the electromagnetic spectrum. All of these different radiation wavelengths make up the electromagnetic spectrum.

The absorption of solar radiation by stratospheric ozone is the source of the heat for the stratosphere and much of the mesosphere (Figure 4.1). The ozone also provides a screen for ultraviolet radiation of wavelengths shorter than about 300 nanometers. This radiation is responsible for biological mutations, sunburn, and other ecosystem and physiological effects (see Chapter 14). As solar radiation continues to penetrate the atmosphere, closer to the Earth's surface, air molecules, clouds, and various types of particles scatter, reflect, and absorb the radiation.

Of the incoming solar radiation, approximately 30% is reflected back to space by clouds and Earth's surface (the planetary albedo is 30% or 0.30), 25% is absorbed by the atmosphere and reradiated back to space, and 45% is absorbed by the surface of land and water. The energy reaching the surface is used to evaporate water; photosynthesize organic matter; drive winds, currents, and waves; generate rising thermals;

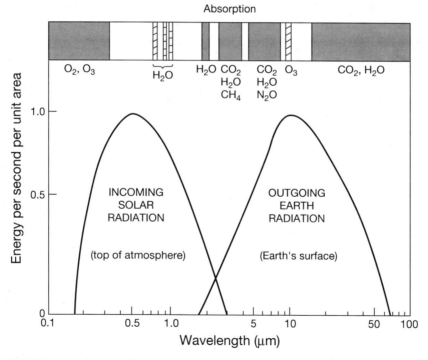

FIGURE 4.2   Diagram illustrating the relationship between the intensity of radiation in arbitrary units and the wavelength of radiation in μm (micrometers, 10^{-6} meters) of incoming solar radiation at the top of Earth's atmosphere and outgoing Earth radiation. Much of the incoming short-wavelength ultraviolet solar radiation is absorbed by the gases of oxygen (O_2) and ozone (O_3) in the upper atmosphere (black wavelength regions at top of diagram). The outgoing long-wavelength Earth radiation is partially (hatched regions) or totally (black regions) absorbed by the greenhouse gases of water vapor (H_2O), carbon dioxide (CO_2), methane (CH_4), nitrous oxide (N_2O), and tropospheric ozone (O_3). Today, the synthetic chlorofluorocarbons (CFCs) produced by human society are also present in the atmosphere and absorb outgoing Earth radiation. The solar curve is scaled to equal Earth's curve at the peak.   (*Source:* Streete, 1991.)

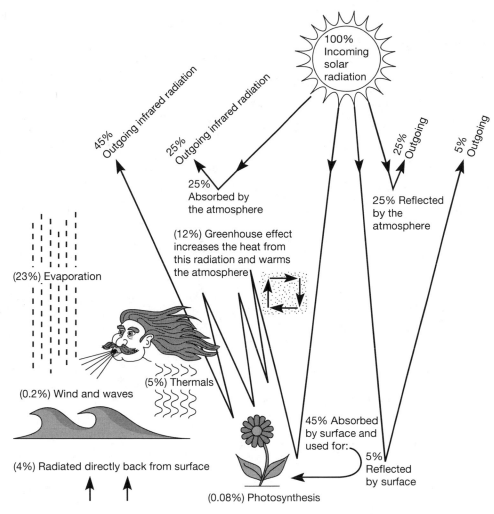

FIGURE 4.3 Earth's radiation budget. Incoming solar radiation is shortwave, ultraviolet, and visible radiation; outgoing Earth radiation is long-wave infrared radiation.

and is absorbed by greenhouse gases before being radiated back toward space (Figure 4.3).

The temperature of the surface of Earth and lower atmosphere is higher than would be expected for a planet of our distance from the sun. This is because of the greenhouse gases. When short-wavelength radiation from the sun that is not intercepted by the outer atmosphere or the ozone layer penetrates to the surface of the planet, it is reradiated back as energy of a longer wavelength (infrared radiation, Earth radiation, Figures 4.2 and 4.3). Carbon dioxide and other greenhouse gases absorb and trap this longer wavelength radiation on its rebound from the planet, leading to a natural warming of Earth's surface and the lower atmosphere (Figure 4.2).

The amount of carbon dioxide residing in the atmosphere affects the amount of heat retained in the atmosphere, and this heat retention in turn affects the climate of Earth. The greater the carbon dioxide level, the warmer the climate. Nitrous oxide (N_2O), water vapor (H_2O, the most important greenhouse gas), methane (CH_4), and other gases have effects similar to those of carbon dioxide in controlling the amount of heat retained in the atmosphere (Figure 4.2). This overall process has been called the natural greenhouse effect, although the trapping of heat within the glass enclosure of a greenhouse actually is due to a different process. Without greenhouse gases, the planetary surface temperature would be $-18°C$. This temperature is $33°C$ cooler than the Earth's present average temperature of $15°C$ (see Box 4.2: Solar and Earth Radiation and the Greenhouse Effect).

The atmosphere is a mixture of gases constituting air and is currently composed of 78.1% nitrogen

BOX 4.2

Solar and Earth Radiation and the Greenhouse Effect

The principal source of energy at Earth's surface to drive biological, geophysical, and geochemical processes is from the sun in the form of ultraviolet, visible, infrared, and other forms of radiation. Only small amounts of energy come from the interior of the planet. The basic mechanism involving the absorption or emission of radiation is the transition of an atom or molecule between a higher and lower energy state. When a molecule gains energy, the electrons jump to a higher energy state, and as the molecule loses energy, energy is emitted as radiation.

Radiation has been described both in terms of particles (photons) and waves. A basic relationship between the two is

$$E = hc/\lambda = h\upsilon$$

where E is the energy of the photons, λ is the wavelength of the radiation, h is Planck's constant with a value of 6.626×10^{-34} joule-second [1 joule (J) = 0.239 calories; 1 watt (W) = 1 joule per second], υ is the frequency of the radiation, and c is the velocity of light (2.998×10^8 m/sec). It can be seen from the preceding equation that photons with short wavelengths have higher energies than photons with long wavelengths. The quantity of radiation emitted by a radiating body can be described by Planck's law:

$$\Psi = a\lambda^{-5}\exp(-b/\lambda T)$$

where Ψ is the energy flux (amount of energy, or number of photons, in an electromagnetic wave passing perpendicularly through a unit surface per unit time) at a given wavelength of radiation λ, a and b are constants, and T is the temperature in Kelvin (K). From the preceding relationship, Wien's displacement law can be obtained as

$$\lambda = 2898/T$$

where λ is the wavelength of the radiation in micrometers, and T is the temperature in K. This law holds true for blackbodies, objects that emit or absorb electromagnetic radiation with 100% efficiency at all wavelengths. The radiation emitted by a blackbody is called **blackbody radiation**. For the sun with an effective surface temperature of 5780 K, the maximum in its radiated energy is at 0.5 micrometers (µm) or 500 nanometers (nm) in the middle of the visible electromagnetic spectrum (Box 4.1, Figure 4.2). For Earth with an effective surface temperature of 288 K, its peak emission in radiation is at 10 micrometers, well into the infrared region of electromagnetic radiation (Box 4.1; see Figure 4.2). The term **effective radiating temperature** is the temperature that a true blackbody would need to have to radiate the same amount of energy as an object radiates.

It can also be shown from Planck's law that the energy flux emitted by a blackbody is related to the fourth power of the body's absolute temperature in K; that is,

$$F = \alpha T^4$$

where F is flux, T is temperature in K, and α is a constant with a numerical value of 5.67×10^{-8} watts per square meter per K^4 (W/m^2/K^4). This relationship is known as the Stefan–Boltzmann law. Our sun with a surface temperature of 5780 K has an energy flux per unit area of 6.3×10^7 W/m^2. If we treat Earth as a blackbody with an effective radiating temperature of T_E, the total energy emitted by Earth is

$$\alpha T_E^4 \times 4\pi R_E^2 \text{ (the surface area of Earth)}$$

where T_E is the effective radiating temperature of Earth, $\pi = 3.1416$, and R_E is the radius of Earth. The total energy absorbed by Earth is

$$\pi R_E^2 \times S \times (1 - A)$$

where S is the solar flux of energy, and A is the albedo (percentage of incoming solar radiation reflected back to space) of Earth. The planetary energy balance to a first approximation implies that the energy emitted by Earth is equal to the energy absorbed by Earth or

$$\alpha T_E^4 = S/4\,(1 - A).$$

By solving for T_E in the preceding equation, we obtain

$$T_E = \sqrt[4]{S/4\alpha \times (1 - A)}.$$

This is the effective radiating temperature of Earth. Using the known values of S of 1370 W/m^2, A of 30% or 0.30, and an α of 5.67×10^{-8} W/m^2/K^4, the Earth's effective radiating temperature is calculated to be a very cold −18° C (255 K). The actual surface temperature of modern Earth is 15°C (288 K). The difference between the two temperatures, 33° C, is due to the greenhouse effect of the Earth's atmosphere. In other words, without greenhouse gases in the atmosphere, Earth would be approximately 33°C colder than it is presently. The naturally occurring greenhouse gases of H_2O vapor, CO_2, CH_4, N_2O, and tropospheric O_3 absorb part of the infrared Earth radiation radiated upward from Earth's surface and reemit it in all directions. A consequence of this greenhouse effect is warming of the surface of the planet. It would be anticipated that increasing greenhouse gas concentrations in the atmosphere would eventually lead to warming of the Earth's surface and decreasing concentrations would lead to cooling.

(N as N_2), 20.9% oxygen (O as O_2), 0.93% Argon (Ar), 0.03% carbon dioxide (CO_2), and a number of trace gases (Table 4.1). The water vapor content of the atmosphere varies from several percent in hot, humid environments to tens of parts per million in cold, dry conditions. The trace gases, although small in concentration, are of importance. For example, O_3 in the stratospheric ozone layer shields life from detrimental ultraviolet (UV) radiation, and CH_4 (methane) and N_2O (nitrous oxide, "laughing gas") contribute to the natural greenhouse effect that moderates climate.

One can gain an impression of the sources and removal processes (sinks) of various substances in the atmosphere by looking at the vertical profiles of the concentrations of substances in the atmosphere. Figure 4.4 shows examples of the distribution of some atmospheric components with height above Earth's

surface. Substances that remain in the atmosphere for a reasonably long time tend to be well mixed throughout the atmosphere. An example is the gas carbon dioxide. Although its concentration in dry air of nearly 386 ppmv (parts per million parts of air by volume) in 2008 does vary about 10 ppmv seasonally and is higher by several ppmv in the Northern than in the Southern Hemisphere, the gas is rather uniformly distributed with height above Earth's surface (Figure 4.4). The major source of carbon dioxide in the atmosphere on a short time scale is the respiration of plants, the decay of rotting vegetation, and the exchange of the gas between the ocean and atmosphere.

On the contrary, the gradient of the atmospheric gas ozone reflects its formation in the upper atmosphere (stratosphere; see the following) and its mixing down into the lower atmosphere (troposphere; see the following) where it is converted back to oxygen gas (Figure 4.4). Carbon monoxide gas shows a profile opposite to that of ozone. It is generated naturally in the lower atmosphere by reaction of methane with hydroxyl radical (OH^*) and at Earth's surface through the decay of rotting vegetation, and is converted to carbon dioxide in the upper atmosphere by reaction with hydroxyl radical (Figure 4.4). Water vapor has a profile similar to that of carbon monoxide in that its concentration in air decreases with increasing elevation above Earth's surface (Figure 4.4). The source of water vapor for the atmosphere is the evaporation of water from aquatic systems and the transpiration of plants. It is removed from the atmosphere at high elevations because it is frozen out of cold, high air.

The concentration profile of dust and aerosols in the atmosphere (Figure 4.4) is very informative. Both dust and the precursors of atmospheric aerosols have as their source the Earth's surface. Thus, their concentrations are high near the surface and decrease with increasing elevation above the surface. The decrease in concentration upward reflects the fact that the particles settle out of the lower atmosphere relatively rapidly because of gravity and are swept from the atmosphere in rainfall. The maximum in concentration in the stratosphere stems from sulfur gases generated at Earth's surface (see Chapter 7), which

TABLE 4.1 Gaseous Composition of Dry Air

Constituent	Chemical Symbol	Mole Percent
Nitrogen	N_2	78.084
Oxygen	O_2	20.947
Argon	Ar	0.934
Carbon dioxide	CO_2	0.0370
Neon	Ne	0.001818
Helium	He	0.000524
Methane	CH_4	0.00017
Krypton	Kr	0.000114
Hydrogen	H_2	0.000053
Nitrous oxide	N_2O	0.000031
Xenon	Xe	0.0000087
Ozone*	O_3	trace to 0.0008
Carbon monoxide	CO	trace to 0.000025
Sulfur dioxide	SO_2	trace to 0.00001
Nitrogen dioxide	NO_2	trace to 0.000002
Ammonia	NH_3	trace to 0.0000003

*Low concentrations in troposphere; ozone maximum is found at 30 to 40 km above Earth's surface in the equatorial region.

(*Source:* Warneck, 1988; Anderson, 1989; and Wayne, 1991.)

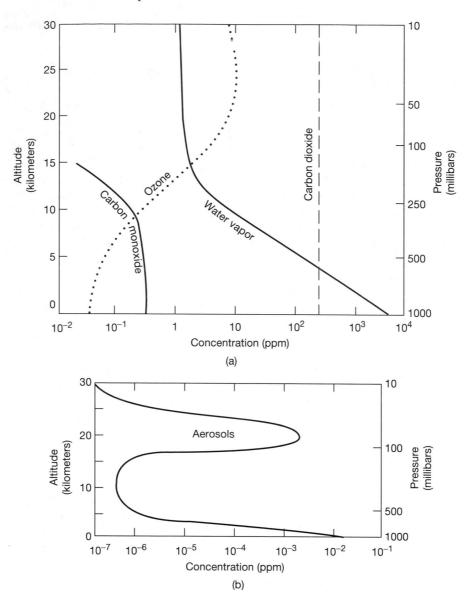

(a)

(b)

FIGURE 4.4 Generalized vertical concentration profiles of some atmospheric components. (a) Water vapor and gases of carbon dioxide, carbon monoxide, and ozone in units of micrograms per gram of air (ppm). (b) Fine-grained terrigenous dust and aerosol in units of micrograms per gram of air (ppm). Notice the concentration maxima at the surface and in the stratosphere. In the stratosphere, the maximum reflects high concentrations of sulfate aerosol in the Junge layer (the sulfate veil); in the troposphere, sea-salt and sulfate aerosols and fine-grained dust particles derived from the land surface are the cause of the maximum. (*Source:* Newell, 1971.)

are converted to small aerosol particles of sulfate in the upper atmosphere. Such particles may reside in the stratosphere for several years before falling back into the troposphere because of the stability of the stratosphere and its slow rate of mixing with the lower atmosphere. During volcanic episodes that eject sulfur gases high into the stratosphere, the concentration of sulfate particles in the stratosphere increases. This stratospheric zone of relatively high sulfate aerosol concentrations that encircles the globe is known as the Junge layer or the sulfate veil (see Chapter 7). The veil of particles acts to cool the planet by scattering and reflecting incoming solar radiation (see Chapter 14).

STRATOSPHERE

The stratosphere extends from about 12 to 48 kilometers above Earth (Figure 4.1). Harmful incoming ultraviolet radiation from the sun is absorbed by ozone (O_3) in the top two-thirds of the stratosphere between approximately 24 and 48 kilometers above Earth's surface. This region is the ozone layer that protects life on the planet from UV radiation. When UV light from the sun reaches oxygen molecules in the stratosphere, the molecular oxygen (O_2) is broken down into free oxygen (O) atoms. The necessary energy is supplied by the absorption of solar UV energy. The oxygen atoms unite with other oxygen gas molecules to produce ozone ($O_2 + O = O_3$).

The overproduction of ozone in the natural system is kept in check by the destruction of ozone by solar radiation and by gases, such as nitrous oxide (N_2O), emitted from Earth's surface that diffuse throughout the stratosphere. If this destruction were not so, ozone would continue to build up in the stratosphere and soon overwhelm atmospheric composition. Before human intervention in the natural system, the production and destruction of ozone were in "nature's" balance (see Chapter 14).

Absorption of ultraviolet light by ozone produces a warming of the stratosphere, such that its temperature is above that of the troposphere immediately below it. The warmer stratosphere creates a stable situation with less-dense air on top of more-dense air and acts like a lid, trapping the cooler weather system of the planet in the troposphere below. Materials that do reach the stratosphere, like gases and aerosol particles from volcanic eruptions (e.g., the eruption of El Chichón in Mexico or Mt. Pinatubo in the Philippine Islands), nuclear explosions, or high-flying airplane exhausts, may stay trapped there for years and affect the temperature of Earth.

Jet airplane trails (contrails) can be observed in the relatively still air of the region between the stratosphere and troposphere—the tropopause. These trails give some feeling for how turbulent the air is on a particular day. The more turbulent this region, the more convoluted the contrail, and the more rapidly the ice and water vapor of the trail are dispersed. Planes often take advantage of rapidly moving jet streams that meander between the stratosphere and the upper troposphere.

Figure 4.5 is a cross section of the atmosphere from pole to pole and shows wind velocities as a function of latitude, altitude, and season. Notice the strong seasonal pattern of wind velocities, with stronger velocities in both the troposphere and stratosphere during the winter seasons in the Northern and Southern Hemispheres. The stratosphere is particularly windy and cold ($-80°C$) during the austral winter above the South Polar Region. An intense polar vortex of winds and thin, wispy polar stratospheric clouds (PSCs) develop during the winter season. The clouds consist of a collection of droplets of a mixture of water and nitric acid. The surfaces of the particles act as sites for chemical reactions, including those involved in the destruction of stratospheric ozone (see Chapter 14). A less intense polar vortex and PSCs also develop above the North Polar Region during the Northern Hemisphere winter season.

TROPOSPHERE

The lowermost atmospheric region, the troposphere, extends from about 12 kilometers above Earth down to its surface (Figure 4.1; see also Figure 5.10). The tropospheric boundary layer is the lower one kilometer or so of the troposphere where mixing and frictional effects between the land and atmosphere are most dramatic. The troposphere is the well-mixed region of the atmosphere. It is a turbulent region because it is mainly heated by the sun's energy at its base, the planet's surface. In the lower troposphere, the air temperature is generally warmer near the surface and cooler above. This situation is very unstable. The additional warming near the surface can make the lower layers of the troposphere less dense than the overlying air and cause convective motions with warm air rising and cold air sinking.

In the troposphere, the weather system of clouds, surface winds, and water vapor circulates around the planet, capped by the stratosphere. Weather is a product of the several weather elements (variables) previously mentioned: temperature, moisture in the air, movement of the air, and pressure of the air. The amount of solar energy reaching Earth is not uniformly distributed. The equatorial belt of the planet is more perpendicular to the sun's rays and, when the sun is directly overhead, receives more radiation per unit area than is received at the polar regions. This occurs twice a year during the times of the equinox when the sun crosses the equator and day and night are everywhere of equal length. As the Earth moves about the sun during the year, this latitudinal band of intense net solar energy flux toward Earth migrates from hemisphere to hemisphere (Figure 4.6). This seasonal fluctuation in the distribution of solar energy affects the amount of heating of the surface and lower atmosphere in the Northern and Southern Hemispheres. The seasonal fluctuations in the amount of

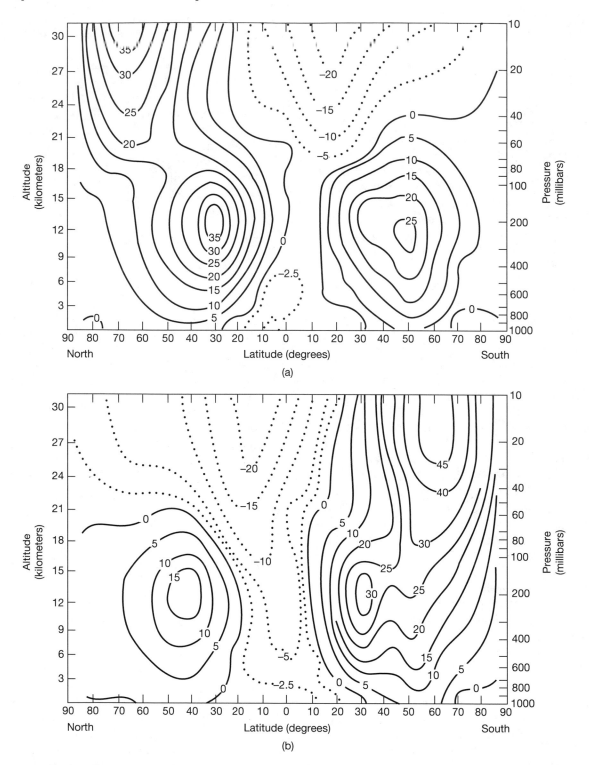

FIGURE 4.5 The velocity of the average prevailing wind in meters per second as a function of latitude, altitude, and the season of the year. (a) Northern hemispheric winter/Southern hemispheric summer (average for December–February). (b) Northern hemispheric summer/Southern hemispheric winter (average for June–August). Notice the distinct shifts in wind patterns caused by the seasonal distribution of solar energy, and that the seasonal changes in the lower atmosphere are larger in the Northern Hemisphere than in the Southern. Solid lines denote westerly winds; dotted lines, easterly winds. (*Source:* Newell, 1971.)

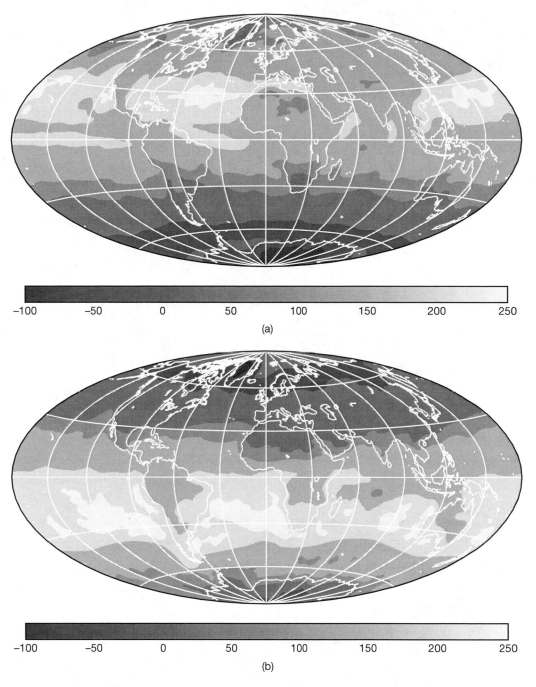

(a)

(b)

FIGURE 4.6 Distribution of the net total flux of heat to Earth's surface in watts per square meter in (a) July 1983 and (b) January 1984. [*Source:* The International Satellite Cloud Climatology Project (ISCCP) and the Earth Radiation Budget Experiment (ERBE), after Darnell et al., 1992.]

heating lead to a seasonal shift in the distribution of the global isotherms (lines of equal temperature) from hemisphere to hemisphere (Figure 4.7). In turn, the global distribution patterns of water vapor, cloudiness, and wind intensities change with the seasons. These changes, coupled with interactions between the atmosphere and the land and sea surface, lead to our changing weather patterns and, in the longer run, climatic patterns.

Because air has weight, it presses down upon Earth. At sea level the pressure on average is 1013.2 millibars and decreases logarithmically, roughly 50%

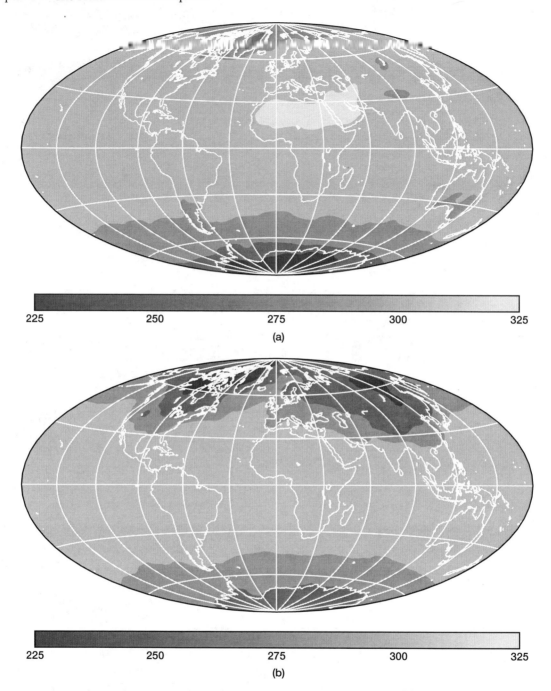

FIGURE 4.7   Distribution of temperature at the surface of Earth in Kelvin in (a) July 1983 and (b) January 1984.   (Ibid., after Darnell et al., 1992.)

for every 5 kilometers of altitude, according to log pressure = 0.06 × altitude in kilometers. As air warms, it becomes less dense as the molecules of air expand and, in turn, rises, and colder heavier air sinks, creating convection currents. Because warm air weighs less than cool air, it forms areas of low pressure, whereas cool air is associated with regions of high pressure. Winds are the movement of air from a high to a low pressure area. As warm air rises, it carries water vapor upward into the atmosphere. Warm air can hold more water vapor than cooler air. Humidity is a measure of the concentration of water vapor in the atmosphere. Commonly, we refer to the humidity as high or low. The relative humidity of air is the percent of water vapor present in the air relative to the maximum it can hold at a given temperature. When water vapor reaches the dew point or the point at which air is cool enough for water to condense out of the air at the critical temperature, clouds are formed, and eventually it may rain or snow.

Land masses and their elevations and oceans distribute heat differently and interact and interfere with air circulation patterns. Land masses tend to heat quickly during the day and cool rapidly at night. Air that travels over land tends to heat and cool with the diurnal cycle. Water heats slowly during the day, and cools only slightly at night. Air that moves over water tends to change little in temperature from day to night. As a result, winds may blow from the water to the land during a hot day and from the land to the sea at night.

The high flux of solar radiation at the equator results in warm equatorial air rising and moving toward the poles and cooler air from the poles moving toward the equator. The warm, moist, rising air contains a great deal of water vapor that condenses and forms large cumulus clouds extending up into the high troposphere. These clouds are sources of precipitation. On the way toward the poles, the air cools and begins to sink. The cooler air descends and moves toward the equator, creating the trade winds of both hemispheres. As the air journeys to the equator, it begins to warm, rise, and starts its cycle again. Part of the global atmospheric circulation pattern, known as the Hadley cell (Figure 4.8; see also Figure 5.10), is completed.

There is a phenomenon produced by rotating bodies like Earth termed the Coriolis effect (discussed

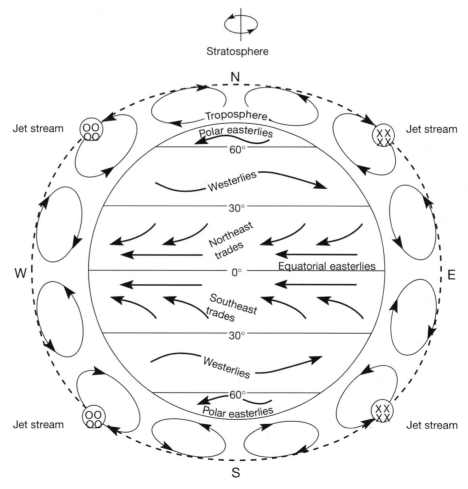

FIGURE 4.8 A view of the planet and atmosphere showing locations of major zonal wind directions and jet streams. The jet streams meander from mid to high latitudes in both hemispheres and flow predominantly from west to east. The Earth revolves west to east, making one full anticlockwise revolution about its N–S axis in 24 hours. (*Source:* Schneider and Londer, 1989.)

in Chapter 5 in more detail) that modifies what would have been simple wind circulation patterns. This effect (actually an acceleration) is due to the rotation of the planet and causes air (and water) to veer to the right of the direction of the force that is causing the motion in the Northern Hemisphere and to the left in the Southern Hemisphere. Instead of traveling due north or south toward the equator, the trade winds approach the equator from the northeast (northeast trade winds) and the southeast (southeast trade winds). Other global wind systems are the prevailing westerlies and polar easterlies (Figure 4.8). Most of the heavily industrialized nations of the world lie in the wind belt of the Northern Hemisphere westerlies. These westerlies are a region of intense cyclonic (counterclockwise) and anticyclonic (clockwise) motion, due in part to the Coriolis effect. The winds tend to transport atmospheric pollutants from west to east in the industrialized nations of the north (Figure 4.8). For example, Bermuda, an island located in the North Atlantic Ocean, receives materials from the North American seaboard blown out to sea by westerly winds.

Air Masses

Winds are essentially air in motion. Great masses of air move, generating winds. An **air mass** is an extensive portion of air in a horizontal direction with similar temperature, pressure, moisture content, and stability in the atmosphere. Drafts are vertical motions of air. Air masses are formed over the land and the ocean in both hemispheres and are named according to their source region. There are about 20 regions in which these air masses form. Five major air masses are recognized: maritime tropical, continental tropical, maritime polar, continental Arctic or Antarctic, and continental polar. For example, masses of very cold and dry air are formed in the interior portions of Eurasia and North America in winter above about 55°N latitude. These continental Arctic air masses may penetrate far south out of the interior regions into lower latitudes. In North America, outbreaks of extremely cold air originally with temperatures of −55 to −35°C may penetrate as far south as Florida and the coast of the Gulf of Mexico. In contrast, continental tropical air masses are formed in the subtropical deserts of the Sahara and Australia, especially in summer. These are very warm masses of air with temperatures of 30 to 42°C. Moist, 2 to 14°C, maritime polar air formed over the Pacific Ocean in the summer and traveling in the prevailing westerlies may deliver heavy rainfall to the states of Oregon

and Washington. The mountain ranges of western North America inhibit the deep penetration of this air mass into the continental interior. In contrast, in Europe, because of the lack of such a topographic barrier, maritime polar air travels very far into the heart of the continent.

In terms of the weather machine of the troposphere, air masses are important in three ways. First, the elements of air masses are determined by the characteristics of their source regions—an air mass forming in the Caribbean (maritime tropical air mass) will be warm and wet; one forming in Arctic lands will be cold and dry.

Second, as air masses move, they encounter different conditions of the land and ocean over which they are traveling. The contrast in conditions leads to instabilities that may generate rain, snow, or wind. For example, when warm air undersaturated with water vapor encounters a mountain range, it rises. The rising air expands (volume increases) and cools (temperature decreases). Little heat is added to or removed from the rising parcel of air because the parcel of air rises faster than it can exchange heat with its surroundings. The process is termed *adiabatic* because there are no exchanges of heat between the air parcel and its external environment. The rising air cools because of adiabatic expansion. The cooling leads to the condensation of water vapor contained in the air mass, and rain or snow may ensue. Sinking air actually warms because of adiabatic compression.

When air masses of two different temperatures (and hence density) meet, they do not easily mix. A transition zone of temperature and other environmental conditions develops between them, and a front is created. A warm front may form when warm air overrides the gentle slope of a cold air mass moving in the same direction as the warm air. A cold front develops when a cold air mass collides with the rising air of a warm air mass. When very cold air jams itself beneath warm and humid air, very dangerous weather may develop, including conditions that lead to severe thunderstorms and tornadoes. Fronts are zones of steep gradients in temperature and consequently regions of strong winds aloft above them. They are associated with regions of low atmospheric pressure and cyclones. Jet streams are commonly found above frontal zones.

Within the broad zonal wind belt of the prevailing westerlies with wind speeds of 50 to 100 km/hr, there are relatively narrow corridors of much stronger westerly winds. These features are known as jet streams (Figure 4.8) and encircle the globe. Aircraft often fly at the altitude of the jet streams, and if flying east, may be "pushed" along by the winds. The jet streams are caused by the fact that the temperature of the atmosphere does not increase uniformly from the poles to the tropics. There are regions associated with frontal zones in which there are large horizontal temperature gradients. These gradients are a result of the fact that atmospheric surfaces of equal pressure (isobaric surfaces) above the cold polar regions to the north are lower in the atmosphere than those to the south in the warm tropical air. The isobaric surfaces slope downward to the north. The greatest downward slope is found above frontal zones. A meandering polar jet is found near the tropopause at an elevation of about 10 kilometers (varies from 8 to 15 km) at the northern boundary of the prevailing westerlies. This jet is approximately coincident with the polar front that separates the warm westerlies from the cold polar easterlies. Large undulations occur in the path of the jet stream. A second jet, the subtropical, is located farther toward the equator at an elevation of about 12 kilometers above a frontal zone. Wind speeds within the jet streams may reach 150 to 300 km/hr, with extremes of 400 km/hr. Cyclones and

anticyclones (see the following) commonly form below these jet streams and move with them.

Third, the air found in air masses moving principally horizontally within the atmospheric boundary layer is commonly spinning, creating "highs" and "lows" (Figure 4.9). Cool air weighs more than an equal volume of warm air. Because of its weight, it sinks and more air piles up on top of it. A zone of convergence at higher elevation is generated and one of divergence in the tropospheric boundary layer. A dome or mountain of cool air builds up in the atmosphere—a region of greater weight or air pressure. Because of the spin of Earth and hence the Coriolis effect, in the Northern Hemisphere, the air cell slowly revolves in a clockwise, or anticyclonic direction, with the winds in the boundary layer blowing somewhat outward from the center of the high-pressure cell. In contrast, in a "low" there is a convergence of air at the base of the tropospheric boundary layer and a divergence aloft (Figure 4.9). The whirlpool of air revolves in an anticlockwise, or cyclonic, direction. The winds in the boundary layer

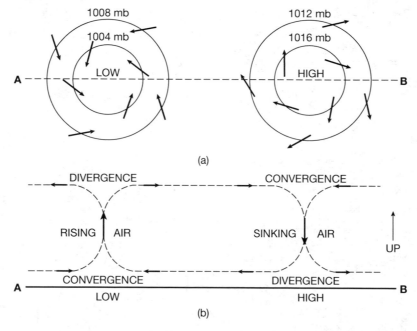

FIGURE 4.9 The motion of air in a low (low-pressure cell) and a high (high-pressure cell) in the lower atmosphere in the Northern Hemisphere. (a) Horizontal section through a low and high showing the counterclockwise and slightly inward motion of winds in a low and the clockwise and slightly outward wind motion in a high. (b) Vertical cross section along the line AB in (a). In a low there is convergence of air at lower elevations and a rising motion and divergence at higher elevations. In a high there is divergence of air near the surface and convergence and a sinking motion at higher elevations. In the Southern Hemisphere, rotational motions in highs and lows are reversed from those in the Northern Hemisphere, but the patterns of convergence and divergence are the same.   (*Source:* Trewartha and Horn, 1980.)

blow somewhat inward toward the center of the low-pressure cell. A low-pressure cell forms when two high-pressure cells encounter one another, and waves of air are created, generating a low-pressure area between them. Because "lows" mark the meeting place of two air masses of different temperatures and moisture contents, they usually give rise to periods of bad weather.

Large-scale upward motions of air, called atmospheric disturbances, may give rise to tropical cyclones, also called hurricanes in the North Atlantic and typhoons in the western Pacific. These disturbances may form along the Intertropical Convergence Zone. The disturbances, for example, may move off the African continent and Central America into the Atlantic and Pacific Oceans, respectively, at about 10 to 15°N latitude. The pattern of convection in these stormy disturbances may become well organized with a definite low-pressure warm core and cyclonic circulation. If wind speeds within the cyclonic circulation exceed 120 km/hr (72 mi/hr) the storm is called a hurricane. Figure 4.10 shows the well-developed hurricane *Iniki* as it passes over the island of Kauai in the southeast Hawaiian Islands on September 11, 1992. Notice the well-developed eye of the storm and the strong cyclonic motion of the low-pressure area.

Hurricanes form where the sea surface temperature (SST) exceeds 27°C, primarily during the period of midsummer to early fall in the Northern Hemisphere. However, in the western part of the North Pacific Ocean, the water can be warm enough to spawn a typhoon in winter. The warm tropical ocean water supplies abundant moisture to the hurricane system that can lead to thunderstorms and heavy

FIGURE 4.10 The eye and aerial extent of the hurricane *Iniki* as it passes over the island of Kauai in the southeastern Hawaiian Islands at 3:00 P.M. HST on September 11, 1992. (*Source:* NOAA, 1992, http://www1.ncdc.noaa.gov/pub/data/images/hurricane_iniki_1992_avhrr.gif.)

rains. The release of large amounts of latent heat (**latent heat** is the quantity of heat absorbed or released by a substance undergoing a change of state, such as water vapor changing to liquid water) from the condensation of atmospheric water vapor into rain warms the atmosphere. Because warm air weighs less than cold air, the air pressure drops, and an initial low-pressure disturbance may intensify. As the low develops, the convergence of air in the tropospheric boundary layer leads to the rise of more warm air, increased thunderstorm activity, and increased release of latent heat. A full-fledged hurricane may develop. This is an example of feedback—in this case, a positive feedback in which the initial disturbance is intensified by processes related to the disturbance.

When natural global climatic change occurs, the distribution of the elements of weather and climate change. The mean global temperature of Earth has changed in the past because of natural causes. As a consequence, certain regions of the land and oceans became warmer while others became cooler. Precipitation and evaporation patterns changed, leading to drying out of certain regions of the continents while other regions became wetter. The positions of the jet streams changed. The frequency and intensity of hurricanes and the variability and intensity of El Niño–Southern Oscillation events changed (see Chapter 5). Such changes certainly have happened during the history of the planet. Examples are the Medieval Warm Period of 800 A.D. to 1300 A.D. and the Little Ice Age of 1350 A.D. to 1850 A.D (see Chapter 13). A concern today is that the activities of humans may lead to modifications of climate and weather patterns at rates that exceed those of most of the geologic past (see Chapter 14).

Clouds

The evaporation of water involves the escape of those water molecules that have gained enough speed through heating of the water surface to break away from that surface. The amount of heat required to evaporate a given mass of water depends on its temperature. For water at the mean global temperature of the planet of 14°C, it would take about 590 calories to evaporate 1 gram of the liquid. This value is known as the "latent heat of evaporation or vaporization." The heat is taken from the air. This is the reason that the skin feels cool when perspiration is being evaporated from its surface. When the evaporated water condenses into a liquid again, the heat absorbed in evaporation is evolved. The surroundings are warmed. Thus, when warm moist air cools, and the

air becomes saturated with water vapor and rain occurs, the surrounding air is warmed. Figure 4.11 illustrates the global distribution of atmospheric water vapor and the seasonal change in its distribution pattern.

Warm air holds more water vapor than the same amount of cold air. When warm air cools, the water vapor in the air condenses, and clouds may form. The warm air may be cooled by convection, by being thrust upward over an advancing cold air mass, and by being pushed upward over mountains on Earth's surface. The latter *orographic effect* can result in rainfall near the summit and leeward side of mountain ridges. All clouds are liquid or frozen water. Rain or frozen precipitation in the form of ice or snow occurs when a cloud has sufficient moisture to approach a saturated condition. In the presence of tiny particles of dust and aerosol in the clouds, the water vapor will condense on the particles and rain, ice, or snow may form. The water droplets or ice crystals are initially very small. It would take millions of them to form a single raindrop or snow crystal. However, the small droplets join together, or coalesce, into larger ones, as more and more of them condense from the water vapor in the cloud. Coalescence produces droplets heavy enough to overcome the upward turbulence in the cloud, and the drops fall.

If the drops have formed in clouds above the freezing point, they will fall as rain. In colder air, water may fall as ice particles formed in the clouds from water vapor that has been quickly frozen at temperatures below freezing (supercooled). These icy particles melt in their descent when they reach the warmer temperatures close to the surface. Snow results from supercooled water vapor that has crystallized around particles in a cloud and fallen through freezing air all the way to the surface. Sleet is produced from warm rain that falls through freezing air on its way to the surface. Hailstones generally form in towering thunderstorm clouds through which frozen raindrops have fallen part way and then have been resuspended by the high winds in the cloud. The continuous circulation of the frozen droplets allows them to collect more and more of the supercooled moisture and grow in the cloud. Eventually they become heavy enough to overcome the upward turbulence in the cloud and they fall to Earth.

Clouds are important to the heat balance of the planet. They have both heating and cooling effects. Clouds tend to brighten the planet as seen from space. Thus, incoming solar radiation is reflected by clouds back to space, leading to a cooling of the planet.

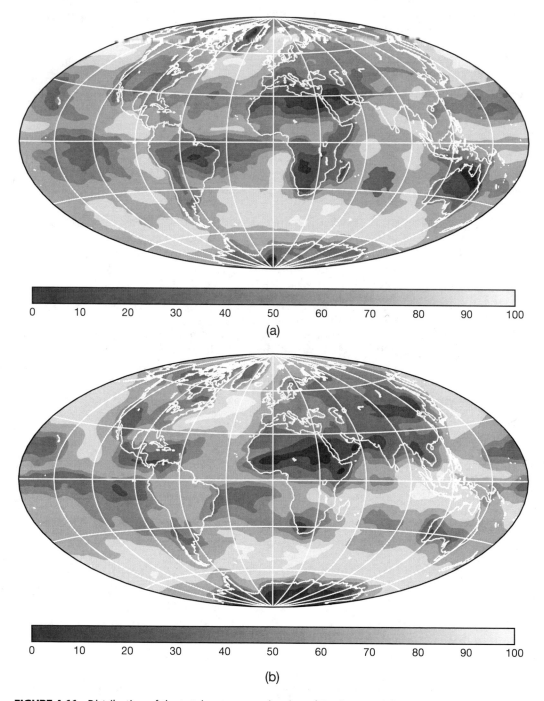

FIGURE 4.11 Distribution of the total water vapor burden of Earth's atmosphere in millimeters of water vapor available for precipitation in (a) July 1983 and (b) January 1984.   (Ibid., after Darnell et al., 1992.)

The potential cooling due to reflection lowers the planet's surface temperature by 20°C. The reflectivity, or albedo, of Earth is 30%; that is, 30% of the incoming solar radiation is reflected back to space (see Figure 4.3). On a cloudless Earth, the albedo would drop to 10%, and the planet would be much warmer. Clouds also warm Earth by trapping outgoing, long-wave, infrared Earth radiation. High, thin clouds are the most effective warmers. The degree of warming can be appreciated by consideration of an Earth cloaked in a high, thin, uniform blanket of cloud. On such a planet, the global mean surface temperature would be about 42°C. On balance, it appears that clouds naturally cool the planet. The

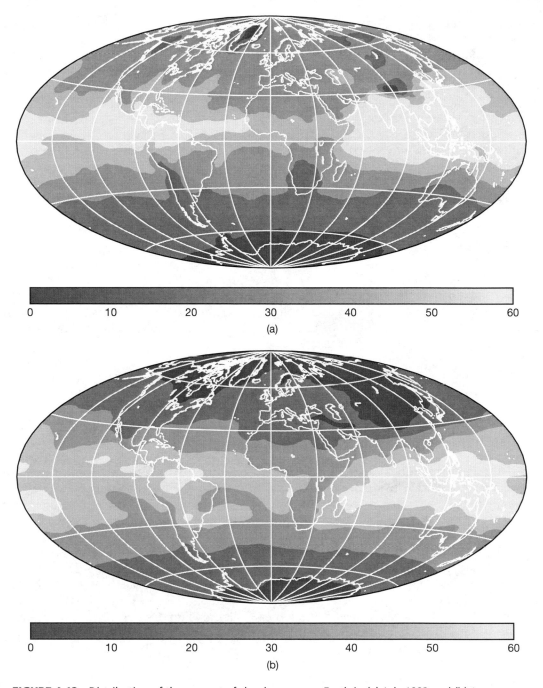

FIGURE 4.12   Distribution of the percent of cloud cover over Earth in (a) July 1983 and (b) January 1984.   (Ibid., after Darnell et al., 1992.)

net natural cooling effect is equivalent to about 13 W/m^2. The cooling effect is about 50% greater than the warming effect. Figure 4.12 shows the cloudiness of Earth and changes in global and regional cloudiness with season. With global climatic change, the patterns of cloudiness will change, as well as those of heat flux, temperature, and water vapor (Figures 4.6, 4.7, and 4.11). Clouds and their role

in global climatic change are further discussed in Chapter 14.

PREDICTION OF WEATHER AND CLIMATE

The zonal distribution of solar radiation over the surface of Earth, the distribution of land masses and their elevations, and ocean and atmospheric circulation and interaction, among other causes, control the global patterns of winds, the balance between precipitation and evaporation, and humidity, cloudiness, air pressure, and temperature. These latter factors determine both global and regional weather and climate. It is a very complicated dynamic system, one that weather forecasters have long tried to comprehend and predict. Computer modeling of climate systems has helped with forecasting the weather and climate, but precise forecasting on any time or space scale is difficult because of the many variables involved.

It is appropriate to consider here some of the systems set up to study weather, climate, and thus global environmental change (see Box 4.3: Remote Sensing). To gain some basis for the prediction of climate, many scientists and organizations are trying to determine how the climate system functions. One international effort in place is known as the World Weather Watch (WWW) under the auspices of the World Meteorological Organization (WMO) of the United Nations. The WWW consists of the Global Observing System (GOS) (data-gathering facilities and instruments), the Global Data-Processing System (GDPS) (computer centers used to analyze information), and the Global Telecommunications System (GTS) (exchange of data and rapid dispersal of information). The GOS currently receives information from land- and space-based observational systems.

Land-based information is received from many countries around the world. Information is obtained from approximately 9500 surface-based stations that send up weather balloons, several weather ships, 7000 merchant ships, 350 automated weather buoys, commercial aircraft, and ground weather-radar stations.

The first weather satellite was launched in 1960 and provided space-based observations. There are two types of satellite systems: polar orbiters and geostationary satellites. Polar orbiters orbit 850 kilometers from the surface of Earth observing every point on Earth every 2 days as the planet rotates below

them. Geostationary satellites orbit Earth above the equator at an elevation of about 36,000 kilometers. They have the same orbital period of 24 hours as that of the revolving Earth; thus, they remain stationary above one point on Earth. These satellites are able to keep continuous track of an area or a large weather system, for example, a hurricane in the Atlantic Ocean. Future Earth-observing satellites are designed to be launched in the coming years.

The GDPS has three nerve centers located in Moscow, Russia; Melbourne, Australia; and Washington, D.C. They analyze the immense amount of weather information that arrives and, using sophisticated weather models, produce about 500 weather forecasts daily. These forecasts are rapidly distributed by the vast GTS computer systems to all parts of the world.

The WMO's World Climate Programme has climate-observing programs in existence and planned for the near future. The overall plan is to monitor and model stratospheric ozone levels, global cloud fields, world ocean and atmosphere circulation patterns, global energy and water cycles, and the land surface and climatic interactions within specific biomes. In addition, the international Joint Global Ocean Flux Study (JGOFS), the International Global Atmospheric Chemistry Program (IGACP), and the International Geosphere Biosphere Program (IGBP) are programs set up to investigate Earth systems in light of global environmental change.

Satellite observations related to global change, for example, can be found in the book *Atlas of Satellite Observations Related to Global Change* edited by R. J. Gurney, J. L. Foster, and C. l. Parkinson (1993), and in *Climate Change 2007: The Physical Science Basis* edited by S. D. Solomon et al. (2007). The National Aeronautics and Space Administration Goddard Space Flight Center (http://www.nasa.gov/centers/goddard/home/index.html) is an especially good source of satellite data.

Concluding Remarks

There has been a substantial research effort put into understanding the composition and dynamics of the atmosphere since the eighteenth century. In recent decades, partly as a response to potential environmental problems associated with interannual, decadal, and global climate change, the effort has increased dramatically. We know the chemical composition of the atmosphere well, with more than 99.9% of the air molecules (exclusive of highly variable amounts of water

vapor) of Earth's atmosphere being made up of N_2, O_2, and chemically inert noble gases, primarily argon (Ar). The remaining less than 0.1% of the gases in the atmosphere are the trace gases, such as CO_2, CH_4, N_2O, and O_3. These gases are very important to atmospheric phenomena and to Earth processes. Atmospheric carbon dioxide is used in photosynthesis and is the most important greenhouse gas in the modern atmosphere next to water vapor; CH_4 is a

BOX 4.3

Remote Sensing

Components of a Remote Sensing System

Remote sensing is the art of obtaining information through measurements made from a distance. Our five senses have been the usual way of obtaining information about the environment. However, these senses provide only limited information about what is actually occurring around us. We cannot see harmful X-rays or hear radio waves without the use of an X-ray detector or a radio, respectively. Remote-sensing technology detects and analyzes electromagnetic radiant energy. This technology is used extensively in satellites.

Global Circulation Models (GCMs) require global initial estimates of the three-dimensional structure of the atmosphere in order to make accurate multiday weather predictions and to make future climate predictions. Over most of the globe, these estimates are obtained using satellite observations. The instrumentation (sounders) on board the satellites measure infrared and microwave radiations (see Box 4.1: Energy) leaving the Earth's atmosphere, which depend on the surface temperature; surface emissivity (ratio of the radiation emitted by Earth's surface to that emitted by a blackbody at the same temperature; see Box 4.2: Solar and Earth Radiation and the Greenhouse Effect); and surface reflectivity; vertical temperature, humidity, and ozone profile distributions; and a number of cloud parameters.

Remote-sensing satellites focus on the target of interest, and instruments on board record electromagnetic information. Radiance in the atmospheric windows (the wavelength spectrum of absorption of energy) of 11 μm to 3.7 μm are most sensitive to surface parameters; radiance in the 15 μm to 4.3 μm CO_2 absorption bands and the microwave 56 gigahertz (GHz, millions of hertz; 1 hertz is a cycle of frequency of one cycle per second) O_2 band are sensitive to atmospheric temperature; the 6.7 μm water vapor absorption band depends on the temperature and water vapor content of the atmosphere; and the 9.6 μm O_3 band depends in part on the distribution of O_3 in the atmosphere.

The satellite data from various instrumental arrays are beamed back to Earth and picked up by a receiver. Computers are used to analyze the data and produce an image or picture. Many of the beautiful pictures of planet Earth from

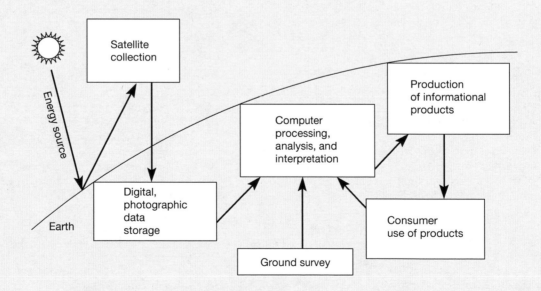

space have been obtained this way. Many disciplines employ the use of remote-sensing devices in their studies of climatic change, land-use activities, mineral resource locations, ocean currents and plankton distributions, plate tectonic movement, distribution and changes in agricultural land, and changing population densities. This new technology has enabled people to view the planet Earth as one system, one entity. (After a display in the Denver Museum of Natural History.)

greenhouse gas and enters into photochemical reactions in the high atmosphere that lead to the formation of water vapor; N_2O is a greenhouse gas, inert in the troposphere but reactive in the stratosphere, and is partially responsible for the destruction of stratospheric O_3; and tropospheric O_3 is a greenhouse gas, and in the stratosphere, the gas shields the planetary surface from intense ultraviolet (UV) radiation. The general chemistry of reactions involving these gases in the atmosphere is reasonably well known and the physics involving the greenhouse effect of these gases well established.

The physics governing the dynamics of the global atmosphere and heat transport therein are also reasonably well known. The year-round energy transport from low to high latitudes by air (and ocean) currents is a consequence of the fact that more energy from the sun is received at tropical latitudes than is given off in Earth radiation. It is this imbalance in the distribution of solar radiation at Earth's surface that drives the general circulation of the atmosphere. The circulation of the atmosphere is responsible for the transport of gases and aerosols.

The vertical distribution of temperature to a large extent governs the exchange of air in the vertical direction. Because the atmosphere is heated at its base by solar radiation, there tends to be a mixture of stable and unstable atmospheric conditions in the troposphere, but the tendency will be for warm surface air to rise to the top of the turbulent planetary boundary layer. Unlike the troposphere, the stratosphere is a region of continuous stability, in part because of the heating of the stratosphere by absorption of solar energy by stratospheric O_3. Temperatures generally increase with height in the stratosphere, resulting in a temperature inversion above the troposphere. Materials injected into the stratosphere from below, as from volcanic eruptions, or at altitude from aircraft, or the products of chemical reactions in the stratosphere will mix slowly and can remain in the stratosphere for months to years.

The composition of the atmosphere has changed over geologic time (see Chapters 7, 8, and 13) and on longer time scales of centuries to millennia will naturally change in the future. However, atmospheric chemical composition is changing today at a rapid rate as trace gases and particles from human activities of coal, oil, and gas combustion, land use change, deforestation and biomass burning, agricultural practices, and industrial processes are vented to the atmosphere. It is likely that the change in composition has already modified climate and atmospheric processes, and if emissions from human activities remain unchecked, the compositional changes of the future will lead to substantial modifications of the ecosphere.

Study Questions

1. Why is it said that the troposphere is unstable, and the stratosphere stable?
2. Why do temperatures in the troposphere decrease upward?
3. (a) Why do temperatures in the stratosphere increase upward?
   (b) What does this temperature gradient have to do with the stability of the stratosphere?
4. Total precipitation over Earth's surface is 496,000 cubic kilometers of water per year. The amount of water vapor in the atmosphere is 13,000 cubic kilometers. If there were no replenishment of water vapor in the atmosphere, how long would it take to remove all the water from the atmosphere?
5. Dust is susceptible to being rained out of the atmosphere. From your answer to question 4, would you expect dust to be evenly mixed throughout the atmosphere?
6. (a) What are the four major gases in the atmosphere?
   (b) List three trace gases that have Earth's surface as their source for the atmosphere.
7. Why is the long-wave, infrared radiation that is reradiated from Earth's surface so important to climate?

8. The average molecular weight of the gases in the atmosphere is 29 g/mole and the mass of the atmosphere is 52×10^{20} grams.
   (a) What is the total number of moles of gases in the atmosphere?
   (b) Carbon dioxide gas makes up 0.038 mole % of the atmosphere. How many moles of CO_2 are there in the atmosphere? How many grams?
9. Water vapor in the atmosphere averages about 0.2 weight %.
   (a) What is the total mass of water vapor in the atmosphere?
   (b) How many moles of H_2O vapor are there in the atmosphere?
10. Earth's winds are zonal in that in general they tend to blow along latitudinal lines around the planet. The major wind belts are the polar easterlies, westerlies, trade winds, and equatorial easterlies in both the Northern and Southern Hemispheres. The winds drive the surface currents of the ocean. Cool air that has descended at midlatitudes moves toward the equator as the northeast and southeast trade winds in the Northern and Southern

Hemispheres, respectively. These winds exert a force on the sea surface, and currents are generated that flow toward the equator. Both the winds of the atmosphere and the surface currents of the ocean converge near the equator. The warm equatorial air rises and moves north and south toward the poles. The converging ocean currents generate westward-flowing equatorial currents.

(a) Based on this, would you expect a pollutant gas like carbon monoxide with major anthropogenic sources in the Northern Hemisphere and a short atmospheric residence time to be evenly distributed in the troposphere? Explain.

(b) Is there an equatorial barrier to the dispersal of floating tar balls produced in the petroleum tanker shipping lanes of the North Atlantic by accidental and deliberate spillage of oil?

(c) Does such a barrier exist for the deep waters of the ocean?

11. The very finest particles of airborne dust transported off the Sahara Desert by winds travel in the troposphere for long distances westward across the Atlantic Ocean. These particles are deposited on the ocean surface and settle out at a rate of 500 cm/yr. How long would it take such particles to reach the bottom of the Atlantic Ocean at 4 km?

12. The primary energy sources for Earth are (1) solar radiation, $0.5 \text{ cal/cm}^2/\text{min}$ (about 343 W/m^2); (2) heat flow from the interior of Earth, $0.9 \times 10^{-4} \text{ cal/cm}^2/\text{min}$; and tidal energy, $0.9 \times 10^{-5} \text{ cal/cm}^2/\text{min}$. About 45% of the solar radiation is absorbed by Earth's surface and reradiated to space as long-wave, infrared radiation.

(a) What is the percentage contribution of heat from the interior of Earth and from tidal energy to the energy sources of the planet?

(b) In units of W/m^2, how much solar radiation reaches Earth's surface and is absorbed there? What does this energy do?

(c) Of the 390 W/m^2 of long-wave radiation emitted by Earth's surface, 324 W/m^2 are absorbed by water vapor, CO_2, and other greenhouse gases in the atmosphere and reradiated back to Earth. What happens to this energy, and what effect does it have on Earth?

Additional Sources of Information

Graedal, T. E. and Crutzen, P. J., 1993, *Atmospheric Change: An Earth System Perspective.* W. H. Freeman and Company, New York, 446 pp.

Gurney, R. J., Foster, Foster, J. L., and Parkinson, C. L., 1993, *Atlas of Satellite Observations Related to Global Change.* Cambridge University Press, U.K. and New York, 470 pp.

IPCC, 2007, *Climate Change 2007: The Physical Science Basis. Contribution of Working Group 1 to the Fourth Assessment Report of the International Panel on Climate Change.* Solomon, S., D. Qin, M. Manning, Z. Chen, M. Marquis, K. B. Averyt, M. Tignor, and H. L. Miller (eds.). Cambridge University Press, Cambridge, U.K. and New York, 996 pp.

Streete, J., 1991, *The Sun–Earth System.* University Corporation for Atmospheric Research, Global Change Instructional Program, Boulder, CO, 25 pp.

Trewartha, G. T. and Horn, L. H., 1980, *An Introduction to Climate.* McGraw-Hill, New York, 416 pp.

5 The Fluid Earth: Hydrosphere and Air–Sea Interactions

And Noah he often said to his wife when he sat down to dine, "I don't care where the water goes if it doesn't get into the wine."

G. K. CHESTERTON

The planet has had a hydrosphere for more than 4 billion years. Without it, life on Earth as we know it would not exist. Throughout this time, the size and shape of the ocean basins and ocean circulation patterns have changed because of seafloor spreading and continental drift, which rearrange the configuration and distribution of the land and ocean over the globe. Continuous changes have occurred in the composition of the atmosphere, including its water vapor and carbon dioxide content, and there have been changes in the type and distribution of clouds and land vegetation. The sun's luminosity has increased over geologic time, and the amount of solar radiation received by the Earth has varied on several different time scales. These changes have affected the temperature and climate of Earth and indirectly affected the volume of the cryosphere (ice). Sea level has risen and fallen during Earth's history as plate accretion rates waxed and waned. On a shorter time scale, the great Pleistocene continental ice sheets repeatedly moved forward and retreated across the planetary surface, and sea level consequently fell and rose as the cryosphere repeatedly grew and shrunk (see Chapter 13).

Precipitation and evaporation distribution patterns have changed over the surface of the Earth, as have glaciated areas and the extent of sea ice, and floods and droughts have been events that continuously reoccur on the planet. In the late Holocene, the centurial scale Medieval Warm Period of 800 A.D. to 1350 A.D. was a time of relatively mild climate in the North Atlantic and Europe but the following Little Ice Age of 1350–1850 was an interval of time of cooler weather that followed the Medieval warming (see Chapter 13). During both of these periods, the atmosphere and hydrosphere and their interactions were significantly affected, with some regions of the world becoming drier and other regions wetter. The ships of the Norsemen were able to sail the Northern Atlantic to colonize Greenland and go as far as North America because of decreased sea ice in the Arctic during the Medieval Warm Period. However in the American West, droughts were a common occurrence due to the warming and people went hungry and died. In contrast, during the Little Ice Age, times of markedly cooler climates in Western Europe led to starvation and death (see Chapter 13).

In the modern world, the enhanced greenhouse effect and global warming (see Chapter 14) have led to rising sea levels, melting of mountain glaciers and ice caps, melting of Arctic sea ice and permafrost, breakup of the western ice margins of Antarctica, and increased worldwide precipitation. Thus, it is necessary to understand the major features of the hydrosphere and the important natural processes operating within it to provide a framework for consideration of human-induced global and regional environmental change.

THE WATER CYCLE

Water circulating through the ecosphere is part of a continuous hydrologic cycle that makes life on Earth possible. It is a dynamic system, with water stored in many places at any one time. The water cycle (Figure 5.1 and Chapter 1, p. 5) involves the transfer of water in its various forms of liquid, vapor, and solid (ice and snow) through the land, air, and water environment. Matter and energy are involved in the transfer. Heat from the sun warms the ocean and land surfaces and causes water to evaporate. The water vapor enters the atmosphere and circulates with the air. Warm air rises in the atmosphere and cooler air descends. The water vapor rises with the warm air. The farther from the warm planetary surface the air travels, the cooler it becomes. Cooling causes water vapor to condense on small particles (cloud condensation nuclei) in the atmosphere and to precipitate as rain, snow, or ice and fall back to Earth's surface. When the precipitation reaches the land surface, it is evaporated directly back into the atmosphere, runs off or is absorbed into the ground, or is frozen in snow or ice. Also, and perhaps most important, plants require water and absorb it, retaining some of the water in their tissue. The rest is returned to the atmosphere through transpiration.

Over the land, precipitation is balanced by evaporation plus runoff, but for the ocean the situation is different. Much of the water evaporated from the ocean returns there directly; however, a small amount (about 8% of that evaporated) is carried by atmospheric winds over the continents where it precipitates. Once on the ground, the water finds its way to streams, lakes, or rivers in runoff or by percolation into and through groundwater. In due time, the water will return to the ocean mainly in stream and river flows. This return flow balances the net loss from the ocean surface by evaporation.

Snow and ice may remain on the land for a long time before the water in these forms evaporates to the atmosphere or returns via rivers or as direct input via glacial melt water and icebergs to the oceans. The snow and ice that feed glaciers may remain locked up in the cryosphere for thousands of years, but finally the ice will melt, and the water will travel to another part of the hydrologic cycle.

One can gain an idea of how much heat is involved in the water cycle by calculating the amount of heat necessary to evaporate water from Earth's surface annually. Each year 496,000 km^3 of water are evaporated from Earth (Figure 5.1). The total heat, assuming a latent heat of water of 590 calories/g of water, associated with this evaporation is

$$496,000 \text{ km}^3 \times 10^{15} \text{ g/km}^3 \times 590 \text{ calories/g}$$
$$= 2.9 \times 10^{23} \text{ calories of heat.}$$

This amount of heat is equivalent to 23% of the incoming solar radiation of 342 W/m^2 that pass through a clear atmosphere (Figures 4.3 and 14.3). Of the energy used for evaporation, 85% is used in the evaporation of ocean water, amounting to two meters of water evaporated from the tropical ocean each year. On average, 78 W/m^2 of the incoming solar radiation is lost to the atmosphere during evapotranspiration. When the evaporated water condenses into precipitation, all of the latent heat is released to the atmosphere. This process warms the atmosphere. The heat involved in the water cycle and its distribution globally are major variables in the development of weather patterns and climate.

WATER RESERVOIRS

Water on our planet is found in the reservoirs of oceans, glacial ice (cryosphere), groundwater, lakes, soils, atmosphere, and rivers (Table 5.1). Approximately 99% of the total water on Earth is located in the oceans and glacial ice, and only about 1% is found in all the groundwater, lakes, soils, rivers, and atmosphere on the planet. Water is continually being shifted (recycled) from one of these reservoirs to another in the water cycle. The total amount of water in the different reservoirs remains nearly constant with time on a short time scale, but it can change for various reasons. These changes can have profound effects on the ecosphere. For example, it is known that the temperature of Earth can fluctuate on time scales varying from annually to interannual, centurial to millennial, and

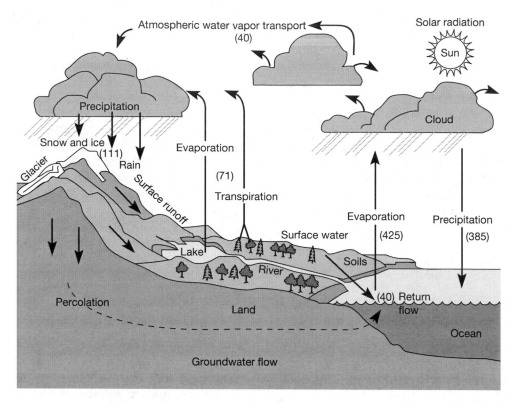

FIGURE 5.1 The hydrologic cycle showing the major processes of water movement. Numbers in parentheses show the water budget in units of thousands of cubic kilometers per year. (*Source:* Maurits la Rivière, 1990.)

to millions or more years (see Chapter 13). Therefore, both alpine and continental glaciers have decreased and increased in size as a result of regional and worldwide climatic change. A consequence of these fluctuations in the cryosphere is that the amount of

water in each reservoir of the hydrologic cycle has changed over time. On the time scale of the Pleistocene glacial-interglacial stages, growth of continental glaciers took water from the ocean and sea level fell; the water reservoir of the cryosphere (ice) grew at

TABLE 5.1 Distribution of Water in the Ecosphere

Reservoir	Volume (10^6 km^3)	Percent of Total
Ocean	1370	97.25
Cryosphere (ice caps and glaciers)	29	2.05
Groundwater	9.5	0.68
Lakes	0.125	0.01
Soils	0.065	0.005
Atmosphere	0.013	0.001
Rivers	0.0017	0.0001
Biosphere	0.0006	0.00004
Total	1408.7	100

(*Source:* Berner and Berner, 1996.)

the expense of the seawater reservoir of the oceans. The converse was true when the giant glaciers receded (see Chapters 2 and 13).

Oceans

This section discusses some aspects of the structure, circulation, and composition of the ocean. Later, similar aspects of other water reservoirs will be discussed. About 1.4 billion km^3, or 97% of the water on Earth, resides in the oceans. The oceans blanket 71% of the surface of Earth, giving the planet its beautiful blue countenance when viewed from outer space. About 60% of the area of the Northern Hemisphere is covered by water, whereas in the Southern Hemisphere 80% is covered. The average depth of the oceans is 3.8 kilometers (see Figure 6.1 and Table 6.6). The oceans reach a maximum depth of 11 kilometers in the Marianas Trench associated with the subduction zone near the island of Guam in the western Pacific Ocean. The oceans receive their heat from the sun, mainly in the equatorial region, and this heat is distributed around the planet by ocean and atmospheric currents (see Figure 4.6).

The oceans can be thought of as a two-layer system, with a thin, generally warm and less dense surface layer on top of a thick, cold and more dense deep layer. The boundary between the two layers is the thermocline, a region of rapid decrease in temperature with increasing depth. The upper layer is turbulent and usually well mixed by winds blowing across the sea surface, whereas the deep ocean is a region of relative calm and slowly moving currents. This stable layering of the ocean makes the transfer of substances, like gases and other dissolved materials, between the two layers a slow process.

WAVES AND TIDES   As early as Aristotle (384–322 B.C.), a relationship between wind and waves was recognized. However, even today, because of the complexity of the dynamics of wave formation and growth, our understanding of waves is not complete. The most common force for creation of water waves

is the wind. Wind blowing across a smooth, quiet water surface creates friction between the air and the water and disturbs the water surface. Energy is transferred from the air to the water. The disturbance is propagated through the water without any substantial overall motion of the water itself. Most of the transferred energy results in the formation of waves but a small proportion of the energy is used to generate wind-driven currents. The transfer of energy from the air to the water surface results in stretching of the water surface, and small wrinkles or ripples form. These small waves are known as capillary waves and have wavelengths less than 1.7 centimeters. As the water surface becomes rougher, it makes it easier for the wind to grip the surface and add more energy to the surface, thus increasing the size of the waves. Surface tension, an elastic property of the water surface, causes the water to return toward the undisturbed state. However, as the waves become larger, the force returning waves toward a level state changes from surface tension to gravity. Waves maintained by the force of gravity and of wavelength greater than 1.7 centimeters are known as gravity waves. Figure 5.2 shows the anatomy of a surface wave in profile, with the descriptive terms used to characterize a wave. **Wave height** (H) is the overall change in height between a wave crest and trough. The wave height is twice the wave amplitude (A). **Wavelength** (L) is the distance between two successive wave crests (peaks) or troughs. The **period** of a wave is the time interval that it takes for two successive peaks or troughs to pass a fixed point and is usually measured in seconds. The **frequency** is the number of peaks or troughs that pass a fixed point per second.

There are several types of waves. Waves that travel through material are known as body waves. The P and S waves discussed in Chapter 3 are examples. Surface waves occur at the interface between two mediums, for example, the atmosphere and ocean. Wind-driven waves, long-period tsunami, seiche, and storm surge waves and ordinary tide waves of fixed period are surface gravity waves.

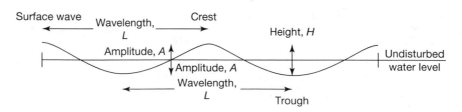

FIGURE 5.2   Anatomy of a surface wave in profile.

Internal waves in the ocean occur at the interface between two layers of ocean water of different densities. A standing wave, for example, the wave generated by plucking a guitar, is the sum of two waves of equal dimensions traveling in opposite directions and generated by energy moving across or through the material.

A moving waveform in the open ocean sets particles in motion. The particles describe an orbital path of rising and moving forward as the crest of the wave advances and falling and moving backward as the trough of the wave passes. The diameter of the orbit of the water particles at the surface is determined by the height of the wave. With increasing depth below the surface, the orbital diameter decreases, as there is less energy of motion available. At a depth of one-half the wavelength of the wave, motion ceases. The speed at which individual surface waves move at sea, that is their celerity, can be calculated from

$$C = L/T$$

where C is speed in meters per second, L is wavelength in meters, and T is period in seconds. Thus, a wave with a wavelength of 150 meters and a wave period of 10 seconds would travel at a speed of 15 m/sec. In the open ocean where the water is commonly deeper than one-half the wavelength of a wave, orbital particle motion does not reach the seafloor, and the wave is considered a deep-water wave. The wavelength of a deep-water wave is

$$L = (g/2\pi) \times T^2 = 1.56\,T^2$$

where g is the acceleration due to gravity, 9.81 m/sec^2, and $\pi = 3.1416$. Its speed is

$$C = 1.56\,T.$$

The average energy of a wave is directly related to wave height. The height and period of waves formed by storms at sea are related to the wind speed, the duration of time that the wind blows, and fetch, the distance over water that the wind blows in a single direction. In the open ocean, the speed of the wind is the most important factor. For an average wind speed of about 20 m/sec, the maximum wave height can be on the order of 4.4 meters.

As deep-water waves move into shallow water, the orbits of the water particles become flattened circles or ellipses. The waves begin to "feel the bottom," and the forward speed of the waves is reduced due to interactions with the bottom. The wavelength decreases and the height of the waves increases. At a depth of one twentieth of the wavelength, C and T depend on depth only, and the height of the wave increases, sometimes dramatically. For such a shallow-water wave, the speed and wavelength are

$$C = 3.13D, \text{ and}$$
$$L = 3.13D \times T$$

where D is depth in meters.

As shallow-water waves approach the shoreline, they may be refracted or bent and diffracted, changing the angle at which the waves approach the shoreline or pass a barrier. The ancient Polynesians who sailed to Hawaii in canoes learned to recognize these patterns of wave motion in the open sea and refraction patterns around islands and used them as navigational tools to make their long-distance voyages. In the shallow area of the coast, the surf zone, wave speeds slow greatly, the waves steepen, break, and disappear in the turbulence and generation of sea spray as the energy of the breaking wave is dissipated. Harbors and constructions near the shoreline must be designed to account for the motions of shallow-water waves to avoid damage to moored vessels or erosion of the shoreline. Surfers watch closely for the waveforms, their breaking, and the timing of incoming wave crests before and after they enter the water.

Another wave type of interest is the tsunami, or seismic sea wave. **Tsunami** is the Japanese term for "harbor wave." Tsunamis are generated when an underwater area of the Earth's crust is displaced, perhaps due to an earthquake. This displacement can cause a sudden rise or fall in the sea above the area of displacement, generating extremely long waves. Tsunamis have wavelengths of 100 to 200 kilometers and periods of 10 to 20 minutes. When a tsunami leaves its point of origin, its height is small, perhaps 1 to 2 meters distributed over its long wavelength. Vessels in the open ocean probably would not see the wave. When the wave reaches the coast, its height builds rapidly, and the wave acts like a shallow-water wave. In 1946, an earthquake occurred in the Aleutian Trench and generated a tsunami that struck Hawaii, killing more than 150 people. The tsunami took only about five hours to travel from the Aleutians to Hawaii. Tsunamis can also be generated close to coastal areas. In July 1998, an earthquake close to the shore of Papua, New Guinea, generated a vertical displacement of the seafloor of about 2 meters. This resulted in 7- to 15-meter tsunami waves striking the coast and the destruction of four fishing villages. Two thousand people were killed or reported missing. In another example of the devastation that can be caused by tsunamis, the December 26, 2004 Indian Ocean undersea earthquake of magnitude 9.1 to 9.3 occurred near the island of Simeulue off the western coast of northern Sumatra at a depth of 30 km (19 mi) and was the second largest earthquake ever recorded on a seismograph. The sudden vertical rise of the seabed of several meters generated a series of tsunami waves that stuck the coastlines of land throughout much of the Indian Ocean. There is evidence that one wave on coming ashore reached 24 m (78 ft) at Aceh in Sumatra. The death toll resulting from the earthquake and the ensuing tsunami was on the order of 250,000 people.

To conclude this section, a few words concerning the tides are appropriate. The force of gravity acting between any two bodies is proportional to the product of the masses of the bodies and inversely proportional to the square of the distance between the bodies:

$$F = G \times (m_1 \times m_2)/r^2$$

where F is the magnitude of the gravitational force between two masses in newtons (N, a newton is the amount of force required to accelerate a one-kilogram mass at a rate of one meter per second); G is the gravitational constant, equal to 6.673×10^{-11} N m^2/kg^2; m_1 is the mass of one body in kg; m_2 is the mass of the second in kg; and r is the distance between the center of the two masses in meters.

Periodic tides on Earth are caused by the gravitational effects of both the moon and sun on Earth, but the moon is most important in terms of tides because it is closer to the Earth. The gravitational effects are modulated by the rotation of the Earth and the shape of the basin affected by the tide. The extent of the tide is largely determined by the difference in the gravitational attraction between the moon and Earth on either side of the Earth (Figure 5.3). Water is pulled by gravitational attraction toward the moon on the side of the Earth facing the moon, creating a tidal bulge. On the opposite side of the Earth, the gravitational force is at a minimum, and this minimum, coupled with the spin of the Earth, creates a net excess of centrifugal force (a fictitious outward force on a particle rotating about an axis and having an equal and opposite direction to the centripetal force, defined as $F = m \times \omega^2 \times r$, where F is force, m is mass, ω is the angular velocity, and r is the distance between the mass and the center of rotation), leading to the development of a tidal bulge away from Earth.

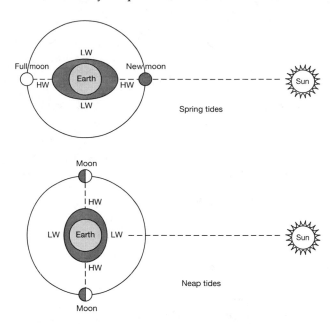

FIGURE 5.3 The alignment of the sun and moon during the development of spring and neap tides showing high-water (HL) tidal bulges and low-water (LW) tidal depressions.

These two bulge regions are areas of high tide. Tidal depressions, low tides, exist on parts of the Earth between the bulges. Ideally, there are two low and two high tides each tidal day because the moon passes over any point on the Earth's surface every 24 hours and 50 minutes, or once each tidal day. Because the moon's position relative to the equator shifts from 28.5°N to 28.5°S, and thus gravitational attraction is not uniform over Earth, the relative heights of high and low water differ geographically.

Tides may be semidiurnal, with two approximately equal high tides and two approximately equal low tides each day; diurnal with a single high tide and low tide each day; and mixed, where the semidiurnal and diurnal components of the tide combine to give two high tides and two low tides each day of unequal tidal ranges. Maximal tidal ranges or spring tides occur when the sun, moon, and Earth are in line, whereas when the sun, moon, and Earth form a right angle, neap tides occur with minimal tidal range (Figure 5.3).

Tides are especially important in coastal areas where tidal heights and tidal currents are greater than in the open ocean, usually because of physical constriction of the tide in bays or funnel-shaped basins, such as estuaries. Tidal action is especially important in certain estuaries, inducing mixing and vertical exchange of estuarine waters. In the Bay of Fundy, an inlet of the Atlantic Ocean between New Brunswick and Nova Scotia, Canada, the tidal range can exceed 10 meters! Unusually high tides commensurate with hurricane-force winds in an onshore direction can generate exceptionally strong storm surges that penetrate far inland.

SURFACE CURRENTS   The atmospheric winds are the driving force for surface ocean currents (Figure 5.4). The winds blow across the water and literally drag it along. The direction of wind currents is influenced by the rotational spin of Earth, the **Coriolis effect** (see p. 79 and the following). There is little exchange of air masses between hemispheres. The Coriolis effect and coastal boundaries cause the surface currents of the world oceans in both hemispheres to flow in a generally circular pattern, like great gyres in the sea. Winds blowing from easterly directions in the trade wind belts of the two hemispheres force water toward the equator. Here, the water flows converge and then move westward as the great North and South equatorial currents of the world oceans. On the western boundaries of the oceans, the equatorial water flows separate and move north and south into the Northern and Southern hemispheres, respectively. The continents on the western boundaries of the major oceans form barriers to the direction of motion of the surface currents and help create great "rivers in the sea," like the Gulf Stream off the east coast of the United States and the Kuroshio Current east of Japan. The Gulf Stream alone transports 4500 times more water than the Mississippi River does annually. These surface ocean currents move poleward and carry warm and salty tropical and subtropical waters to the colder polar regions of both hemispheres. The north and south transport of heat by the surface currents of the oceans and by the atmosphere (see Chapter 4) is called the **meridional heat transport.** The oceans carry about one-half of this heat to the high latitudes of the Northern and Southern hemispheres, and the atmosphere is responsible for the rest. In the prevailing westerly wind belts, eastward current flow is helped along by these winds. To complete the circle, waters return equatorward from the polar regions along the eastern margins of the ocean basins. The return of the cooler waters is not in strong flows like the Gulf Stream but is more diffuse.

When a wind blows across the surface of the ocean, energy is transferred from the wind to the sea surface. This energy is used in the generation of surface waves and in driving currents. The force of friction acting on the sea surface across which the wind is blowing is known as the **wind stress.** In general, the greater the wind speed, the greater the wind stress, and the stronger the current.

In the 1890s, a Norwegian explorer and scientist, Fridtjof Nansen, permitted his ship to be frozen in ice in the Arctic. The ship drifted with the ice for more than a year. During this time, Nansen observed that ice flows do not move parallel to the wind but at an angle of 20° to 40° to the right of the direction of the wind. Later in the decade V. W. Ekman developed a theory to explain these observations. The substance of his theory is shown in Figure 5.5.

The motion of a wind blowing on the ocean surface is transmitted downward because of internal friction within the upper ocean. The friction in the moving fluid stems from the transfer of momentum between different parts of the fluid, where momentum is the product of mass times velocity. In the ocean, the friction results from turbulence in the water. Flow is not laminar where momentum transfer occurs owing to the transfer of water molecules between adjacent layers. In the ocean, parcels of water, rather than individual molecules, are exchanged between one part of the moving fluid and another. Thus, the internal friction that develops in a turbulent fluid is greater than that in a fluid flowing in a laminar fashion. The internal friction that develops is known as **eddy viscosity.**

Ekman considered the balance between frictional forces and a force previously mentioned, the Coriolis force, for an infinite number of water layers making up the upper ocean. The Coriolis force is the force experienced by a moving particle because of its being on a rotating Earth. The force is proportional to the sine of latitude (ϕ) and is defined as

$$\text{Coriolis force} = m \times 2\Omega \sin \phi \times u$$

where Ω is the angular velocity of the Earth about its axis, m is the mass of a parcel of water, and u is its speed. Notice in Figure 5.5 that the speed of the wind-driven current induced by a wind blowing across the surface ocean decreases with depth. The direction of the current at the surface in the Northern Hemisphere is 45° to the right of the wind direction. The angle of deviation from the wind increases with increasing depth. Thus, the current vectors representing both speed and direction form

FIGURE 5.4 Average direction of motion of the surface currents of the world's oceans and names of major currents. The currents exhibit the distinctive pattern of curving to the right in the Northern Hemisphere and the left in the Southern Hemisphere. The currents are determined by the force of the prevailing winds, by Earth's rotation, and by the location of land masses that block current motions.   (*Source:* Munk, 1955.)

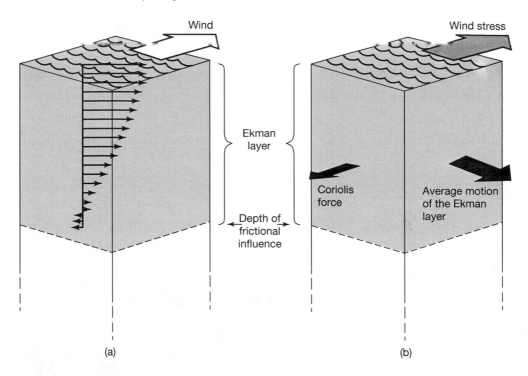

FIGURE 5.5 Ekman circulation pattern in the oceans formed from the balance between the force of the wind and the Coriolis force. The Ekman layer extends to a depth of about 100 meters, below which friction from the surface wind stress is minimal. (a) The Ekman spiral current pattern. The arrows show the direction and relative speed of the wind-driven current with depth. (b) Balance of the wind force and the Coriolis force leading to net transport of water 90° to the right of the wind in the Northern Hemisphere. The surface current theoretically moves at 45° to the right of the wind in the Northern Hemisphere—actual observed motions in the ocean are at an angle somewhat less than this.

a spiral pattern downward. The theoretical pattern is known as the **Ekman spiral.** For the layer of water as a whole affected by the wind (the **Ekman layer**), the force due to the wind is balanced by the Coriolis force, and the average motion of the Ekman layer is at 90° to the direction of the surface wind. This motion in response to prevailing wind fields is very important in terms of the general circulation of the ocean.

For example, along the coast of Peru, the southeast trade winds blow roughly parallel to the coastline. This motion sets up surface ocean currents that move at an angle of about 45° to the left (we are in the Southern Hemisphere) of the direction of the wind, that is, offshore. The net Ekman transport of water is offshore. To replace this water, water from deeper in the oceans rises (upwells) along the coast of Peru. This upwelling is important in delivery of nutrients from depth to the surface waters and hence phytoplankton production and the anchovy fishery that is found along the Peruvian coastline (see further

sections in this chapter). As we will see later, El Niño–Southern Oscillation events can significantly modify the depth and intensity of the upwelling offshore Peru.

Besides the wind-driven currents described by Ekman theory in which there is a balance between wind stress and the Coriolis force, there is another important type of current motion in the ocean. This current motion arises because, as in the atmosphere, there are horizontal pressure gradients in the ocean. Just as winds blow from regions of high to low pressure, water in the ocean tends to flow in a direction that smooths out the horizontal differences in pressure. These horizontal pressure differences may be caused by a current encountering the boundary of a continent. The "piling up" of water against the boundary may create a sloping sea surface and hence a horizontal pressure gradient. If the pressure gradient were the only horizontal force acting, current motion would occur in the direction of the pressure gradient. Boundaries are much more important in the ocean to current flow than in the atmosphere because of the presence of continents that impede flow.

When the Coriolis force acting on a moving water body is balanced by a horizontal pressure-gradient force, the current generated as a result of these interacting forces is said to be in geostrophic equilibrium, and the current is described as a geostrophic current. Geostrophic equations for the velocity of the current are a function of the pressure gradient, the Coriolis parameter f ($f = 2\ \Omega \text{sine } \phi$), and density. For example, let us imagine a region in the North Atlantic Ocean where water has "piled up" so that there is a slope in the sea surface. Imagine the slope is down toward the west. The slope is not great, only perhaps a few meters in 1000 kilometers. The slope may be created by water piling up against a continental coastal boundary, by lateral variations in density caused by differing temperature and salinity conditions in the ocean, or by variations in the pressure of the overlying atmosphere. Whatever the case, in the situation described here, there is a horizontal pressure force acting from east to west. Water moving westward under the influence of this pressure force will be deflected to the north by the Coriolis force. If an equilibrium situation is eventually reached between the pressure force and the Coriolis force, the water will flow northward. The horizontal pressure-gradient force acting toward the west will be balanced by the Coriolis force acting toward the east. Thus, the resulting geostrophic current will flow at a right angle to the horizontal pressure gradient instead of down the pressure gradient.

The real ocean and atmosphere are very complicated systems. Spatial and temporal variations in the speed and direction of flows in both systems are complex. Actual wind or ocean current directions as seen on one particular day may be quite different from the average conditions shown in Figures 4.8 and 5.4. However, many of the characteristics of the wind and current fields in both fluids may be appreciated by consideration of the Coriolis force and pressure forces and the interactions between the atmosphere and the sea surface.

DEEP CURRENTS The deep currents of the ocean are a part of the temperature- and salinity-driven (thermohaline) circulation of the oceans and are related to the movement of great parcels of water termed **water masses.** The motion of deep waters is very complex and not well understood. Similar to air masses in the atmosphere, there are volumes of water in the ocean that have unique properties. These water masses are defined on the basis of the properties of salinity and temperature, and hence density, which is related to these properties. Water masses obtain their characteristics at or near the sea surface but may eventually sink and move to great depths in the ocean. These water masses might travel long distances and during transit retain their salinity and temperature properties without appreciable mixing. There are five commonly recognized water masses in the ocean produced by the stratification of the masses due to density differences. These include the surface and upper water mass layers, which are under the influence of the wind, and the deeper masses of intermediate, deep, and bottom waters. Figure 5.7b shows the distribution of some water masses in the Atlantic Ocean.

The deep currents of the oceans are not directly affected by winds but are driven by changes in temperature and salinity. If the temperature of the water does not change, seawater becomes denser as its salinity increases and will sink below less dense water. Also, warm water may be less dense than cooler water and will rise, while cooler, heavier water sinks. The deep currents of the oceans are driven by these processes, and in contrast to surface currents, they are able to flow between hemispheres. The flow reminds one of a conveyor belt but is much more complex (Figure 5.6); nevertheless, the global flow has been termed the **global conveyor belt,** which results in interhemispheric exchange of warm water, most vigorously in the Atlantic. The pattern of flow in the Atlantic Ocean is occasionally called the **meridional overturning circulation (MOU),** which is a more accurate and well-defined term and not necessarily equivalent to the conveyor belt circulation. Cold winds from northern Canada cool the waters of the North Atlantic and initiate the movement. Dense, cold, salty waters sink east of Greenland in the Greenland and Norwegian Seas and flow southward along each side of the Mid-Atlantic Ridge at great depth as North Atlantic Deep Water (NADW) into the South Atlantic. This flow constitutes part of the lower limb of the conveyor belt circulation. In the South Atlantic, the NADW meets and mixes with very cold and denser, northward-flowing Antarctic Bottom Water (ABW) formed mainly in the Weddell Sea off Antarctica (Figure 5.7b). The deep-water flow then turns east around the tip of Africa and, continuing its deep passage, heads toward the Pacific Ocean. It flows into the Pacific along the western margin of that ocean basin, where it is deflected eastward by the presence of the

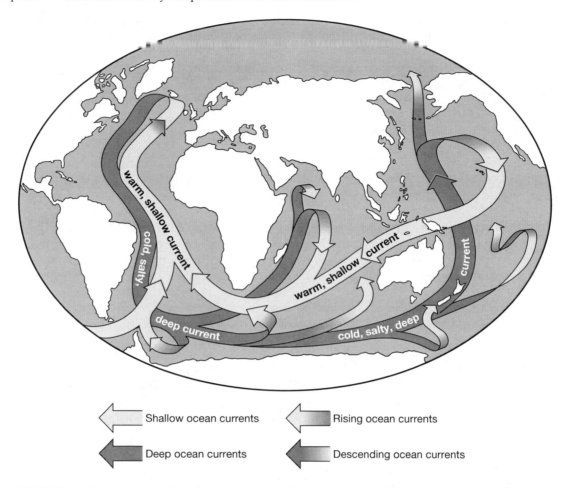

Shallow ocean currents

Rising ocean currents

Deep ocean currents

Descending ocean currents

FIGURE 5.6 The conveyor belt circulation pattern of the world's oceans. Cold, salty water in the North Atlantic sinks to the deep ocean and moves southward to resurface and be warmed in the Indian and Pacific Oceans. Surface currents then return the water to the Atlantic. A complete passage takes about 1000 years. Currently, this conveyor belt circulation pattern is driven to some extent by an imbalance between the loss of water from the Atlantic by evaporation and its gain by precipitation and continental runoff. (*Source:* Broecker, 1995.)

Asian continental margin and the rotation of Earth. No water mass similar to the North Atlantic Deep Water or Antarctic Bottom Water forms in the high latitudes of the Pacific Ocean, because the Bering Strait prevents the free exchange of cold Arctic water with the North Pacific Ocean. However, in the western tropical Pacific, there is also northerly water flow in the upper kilometer of the water column as in the Atlantic. This water flow is mainly balanced by export of warm North Pacific water through the Indonesian Seas and, less important, through the Drake Passage between Antarctica and South America. This is part of the upper limb of the conveyor belt circulation.

The upper limb of the conveyor belt circulation carries water northward into the North Atlantic in the upper kilometer of the water column. The Gulf Stream is an important part of this northerly flow. In the high latitudes of the North Atlantic, the cold winds of Canada cool the northward flow of salty Gulf Stream waters, the waters become sufficiently dense to sink as NADW, and the trip begins anew. This cycling of deep ocean waters takes on average about 1000 years and is driven mainly by North Atlantic Deep Water formation at the rate of 15 to 20 Sv (Sverdrups; 1 Sverdrup = 1 million cubic meters of water flow per second). The heat associated with the upper limb of this circulation in the North

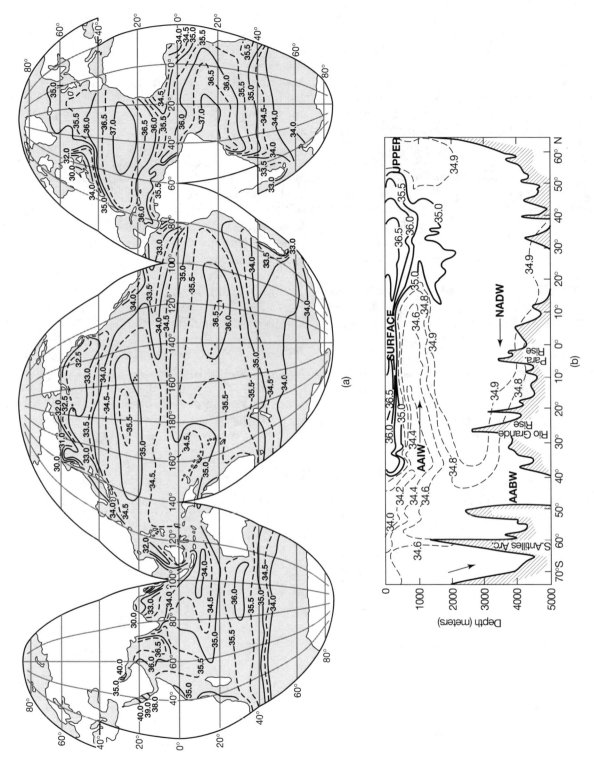

(a)

(b)

FIGURE 5.7 Salinity of the world's oceans. (a) Surface salinity of the oceans in the summer in the Northern Hemisphere. (*Source:* Sverdrup, et al., 1945.) (b) Cross section of the Atlantic Ocean showing the vertical distribution of salinity and major water masses. Relative motions of deeper water masses shown by arrows. Surface, surface water mass; Upper, upper water mass; NADW, North Atlantic Deep Water; AAIW, Antarctic Intermediate Water; AABW, Antarctic Bottom Water. (*Source:* Von Arx, 1962, and The Open University, 1989c.)

Atlantic helps to maintain a more equable climate for Greenland, and in the Gulf Stream's surface return flow along the eastern margin of the North Atlantic, for Great Britain and Europe. Shifts in the intensity of this conveyor belt circulation are one cause of natural climatic change and may have influenced the climatic patterns of the Younger Dryas, the Medieval Warm Period, and the Little Ice Age (see Chapter 13).

COMPOSITION Seawater is an electrolyte solution composed of water (H_2O) and dissolved salts. Common table salt (NaCl) is the most abundant dissolved substance in seawater, followed in abundance by salts of every naturally occurring element (Table 5.2). The salt content of seawater is expressed in terms of salinity (S), where **salinity** is the total weight in grams of solid matter in 1 kilogram of seawater after evaporation to dryness. The average salinity of seawater is about 35 parts per thousand (ppt), or 3.5% by weight. Because seawater salinity is now generally measured using conductimetric salinometers, it is common not to use units in expressing the salinity of seawater. Thus, a seawater of 35.5 ppt is simply noted as 35.5. Figure 5.7 illustrates the surface salinity of the world's oceans during the northern summer and the salinity at depth

in the western part of the Atlantic Ocean. Notice that the salinity of seawater does not vary greatly with geographical position or depth in the ocean. The chlorinity of seawater is a measure of the most important dissolved element in seawater—chlorine as chloride ion. Salinity divided by the factor 1.805 approximately equals the chlorinity of a seawater sample.

The major elements of calcium, magnesium, sodium, potassium, sulfur, and chlorine make up more than 99% of the salts in the ocean. The ratios of the concentrations of the dissolved chemical species of these elements in seawater are remarkably constant with geographical position and depth in the ocean, except near major river mouths or in semi-enclosed seas. In other words, as the salinity of seawater varies, the concentrations of the major dissolved ions vary accordingly, but their proportions remain constant, and their concentrations depend only on the salinity of seawater.

This observation forms the basis of the concept of the constancy of composition of seawater. The concept was first formulated by Marcet in 1819 on the basis of the chemical analysis of 14 samples of seawater and expanded on by Forchhammer in 1865. In 1884 it was further developed by Dittmar, who analyzed 77 seawater samples from the *Challenger*

TABLE 5.2 Major Dissolved Components of Seawater

Constituent	Chemical Symbol	Concentration (g/kg seawater of salinity 35)	Percentage of Salts
Chloride	Cl^-	19.353	55.29
Sodium	Na^+	10.760	30.74
Sulfate	SO_4^{2-}	2.712	7.75
Magnesium	Mg^{2+}	1.294	3.70
Calcium	Ca^{2+}	0.413	1.18
Potassium	K^+	0.387	1.11
Bicarbonate	HCO_3^-	0.142	0.41
Bromine	Br^-	0.067	0.19
Carbonate	CO_3^{2-}	0.016	0.05
Strontium	Sr^{2+}	0.008	0.02
Silica	SiO_2	0.006	0.02
Boron	B	0.004	0.01
Fluorine	F	0.001	0.003

(*Source:* Wilson, 1975.)

expedition of 1872–1876—the first major oceanographic expedition covering the oceans of the world. The reports of the *Challenger* expedition were published in 50 volumes and mark the birth of the scientific study of the oceans.

The fact that the ratios of the major dissolved ions are nearly constant and depend only on salinity implies that the distributions of these major constituents of seawater are controlled mainly by physical processes of mixing in the ocean—advection and diffusion. Advection is the bulk flow of water, as in a river, whereas diffusion in turbulent fluids, as in the atmosphere and oceans, implies the movement of small parcels of fluid. Biological or inorganic chemical processes are not important in controlling the concentration of the major dissolved ions in present-day seawater. The major elements of seawater are commonly referred to as **conservative elements.**

This does not hold true for the concentrations of many of the minor and trace elements found in seawater. For example, the concentrations and distributions of the chemical species of carbon (C), nitrogen (N), phosphorus (P), silicon (Si), and oxygen (O) in the ocean are not simply related to salinity, and hence mixing in the ocean. Besides sulfur

and hydrogen, carbon, nitrogen, phosphorus, and oxygen are the most important elements found in organic matter. They are actively involved, for example, in the photosynthesis of aquatic plants and are products of the respiration and decay of organic matter (see Chapter 6 for an expanded discussion of biological processes). Silicon is used to build the skeletons of certain marine organisms, like diatoms and radiolarians. Thus, the concentration and distribution of these elements in the ocean depend on both biological and physical processes. These elements are usually referred to as **nonconservative elements.**

Figure 5.8 illustrates the generalized vertical distributions of dissolved inorganic carbon (DIC), nitrate (NO_3^-), phosphate (PO_4^{3-}), dissolved Si (most of the Si present in the ocean is as monomeric silicic acid, H_4SiO_4), and oxygen (O_2) in the central portion of an ocean basin, like the North Pacific. It serves to show how biological processes are involved in determining the concentration and vertical distribution of these biologically important compounds in seawater. Dissolved inorganic carbon (DIC) is the sum of the concentrations of the chemical compounds bicarbonate (HCO_3^-), carbonate (CO_3^{2-}), and carbonic acid (H_2CO_3).

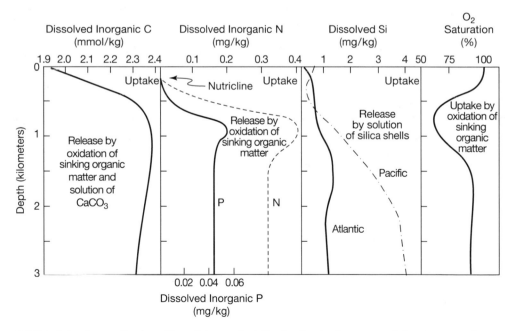

FIGURE 5.8 Generalized profiles of the vertical distribution in the ocean of the concentrations of dissolved inorganic C, N, P, Si, and O_2 saturation. In deep water the concentrations of C, N, and P are nearly uniform with depth. Notice the peak in inorganic N and P concentrations and the minimum in oxygen saturation at depths of about 800 and 500 meters, respectively, reflecting the oxidation of sinking organic detritus as it falls through the upper ocean.

Concentrations of nitrate and phosphate in the upper, lighted zone of the ocean, the euphotic zone, are nearly imperceptible, and those of DIC and dissolved silica are vanishingly small. This is due principally to two biochemical processes going on in the near-surface waters of the oceans: (1) photosynthesis of organic matter in the form of plant plankton and (2) extraction of carbon and silicon from the water to make the inorganic shells of marine plankton (see Chapter 6). The chemical equations representing these processes may be written as follows:

$$106CO_2 + 16NO_3^- + PO_4^{3-} + 19H^+ + 122H_2O$$
$$\rightarrow (CH_2O)_{106}(NH_3)_{16}H_3PO_{4(\text{marine plankton})}$$
$$+ 138O_2,$$
$$Ca^{2+} + CO_3^{2-} \rightarrow CaCO_{3(\text{shell})}, \text{ and}$$
$$H_4SiO_4 \rightarrow SiO_{2(\text{shell})} + 2H_2O.$$

Notice that the formation of the organic matter and shell material of the plankton removes carbon, nitrogen, phosphorus, and silicon from the water column and through the production of phytoplankton releases oxygen. Because vertical mixing processes in the upper ocean generally are not sufficiently rapid or continuous to replace the dissolved constituents of DIC, NO_3^-, PO_4^{3-}, and Si at a rate commensurate with their rate of uptake, concentrations of these chemical compounds are lowered substantially in the euphotic zone. The nutricline is the depth range in the ocean where there is a sharp gradient in nutrient concentrations from nutrient-poor surface waters to nutrient-rich waters at depth (Figure 5.8). The top of the nutricline usually coincides with the depth of the mixed layer, the wind-stirred upper region of the ocean, or the euphotic zone, whichever is deeper. Because the major source of oxygen for the surface waters of the ocean is from the atmosphere, the concentration of this gas approximates that in equilibrium with the concentration of oxygen in the atmosphere above the sea surface. However, surface waters may contain a little more oxygen than predicted owing to the release of oxygen to the water during photosynthesis. When such a situation exists, the water is considered to be oversaturated with respect to oxygen. Waters in equilibrium with atmospheric oxygen are said to be saturated; those containing less oxygen than that predicted at equilibrium with the atmosphere, perhaps because respiration exceeds photosynthesis, are said to be undersaturated with respect to oxygen.

After death of the phytoplankton and the animal plankton that feed on them, the organic matter and shells begin to sink through the water column toward the bottom of the ocean. As they do so, the organic matter is used as food by bacteria in the ocean. The organic matter is decomposed, and the products of decomposition are released back to the water. The photosynthesis reaction described previously is reversed to some extent. Nitrate, phosphate, and dissolved inorganic carbon are returned to the water column, and oxygen is consumed in the process. The shells of silica dissolve, and dissolved Si is released to the water. Much of this decomposition and solution takes place in the upper water column, and the nutrients of nitrate, phosphate, and silica and the dissolved inorganic carbon are used again in phytoplankton production. Thus, in the upper water column, much of the nutrients and carbon simply recycle (see Chapter 7).

Some portion of the dead organic matter and shell material escapes the upper ocean and settles into the depths. Here the organic matter continues to be decomposed by biochemical reactions involving bacteria. Also at depth, the shells dissolve and calcium, silica, and dissolved inorganic carbon go back into solution in the deeper ocean. Thus, the deeper ocean is enriched relative to the upper ocean in the dissolved chemical species of DIC, NO_3^-, PO_4^{3-}, and Si (Figure 5.8). The peaks in the inorganic nitrogen, phosphorus, and carbon concentrations at depths of 800 to 1000 meters are in part a result of the decomposition of the sinking detritus. The reduction in the oxygen concentration at about this depth range is in part due to the utilization of oxygen by the bacteria involved in the process of decay of organic matter. The depth region of the world ocean at which the dissolved oxygen content of the ocean is low is known as the **oxygen minimum zone,** which occurs at an average depth of 800 meters but varies from ocean basin to basin.

Deep-water concentrations of inorganic C, N, P, and Si are variable within a particular ocean basin, and each ocean differs significantly from the others in the average concentrations of these elements at depth. This is because the deep waters of the ocean basins have important source areas in the high latitudes of the North Atlantic Ocean (see Figure 5.7b). As these waters travel at depth through the ocean basins, they age and pick up the products of decay and solution. The waters increase in their concentrations of dissolved chemical compounds of carbon, nitrogen, phosphorus, and silicon. Deep waters of the North Pacific are very old, and the concentrations of carbon and nutrients in them are high relative to the deep waters of the Atlantic or Indian Oceans. When intermediate depth or deep waters upwell, they carry with them these high concentrations of carbon and nutrients. The carbon and nutrients are utilized again in photosynthesis and shell production in the euphotic zone. Some of the debris of the organic and inorganic components of the plankton escape the ocean and are deposited on the sea floor to accumulate in marine sediments. In the late Holocene, in terms of organic carbon, this sediment accumulation was equivalent to $120-144 \times 10^{12}$ g C/yr, <0.01% of the net primary production of organic carbon (see Chapter 6) in the euphotic ocean of the world.

Atmospheric gases are also major components of seawater. All atmospheric gases dissolve in seawater. Some, like oxygen (O_2), are moderately soluble. Most of the oxygen in the ocean is derived from the atmosphere by absorption of the gas and a minor amount is produced in the surface water column by photosynthesis (see equation, p. 104). Thus for a body of water to remain oxygenated, there must be strong mixing of the gas from the atmosphere into the water body. This does not always occur. During the summer due to temperature stratification of some water bodies, particularly estuaries, lagoons, and bays, oxygen cannot mix downward through the water column and the deeper parts of the water body may go anoxic as the oxygen initially present is consumed by the microbial respiration (oxygenation) of the dead organic matter suspended in the water body (see Chapter 11).

Atmospheric carbon dioxide (CO_2) is an environmentally relevant example of a gas that exchanges with freshwaters and seawater and is produced by both natural and human activities. The 2008 concentration of carbon dioxide in the atmosphere was approximately 386 ppmv (386 parts of carbon dioxide per million parts of air by volume). In the ocean, 1 kilogram of seawater contains about 160 milligrams of dissolved inorganic carbon (equivalent to about 160 parts of dissolved inorganic carbon per million parts of seawater). Some of this carbon, about 0.50 mg/kg of seawater, is in the form of aqueous CO_2, like carbon dioxide gas dissolved in bottled carbonated soda. Atmospheric carbon dioxide gas that is absorbed by the surface waters of the ocean enters into a series of chemical reactions in which the CO_2 gas dissolves into the seawater to form uncharged aqueous CO_{2aq} [a small percentage of the dissolved CO_2 is present as carbonic acid (H_2CO_3)]. The CO_{2aq} reacts with water

(H_2O) to make a HCO_3^- ion and a proton (H^+), and this is followed by dissociation ("breakdown") of HCO_3^- ion into another proton and carbonate ion (CO_3^{2-}). The net overall chemical reaction is:

$$CO_{2aq} + H_2O + CO_3^{2-} = 2HCO_3^-.$$

In this net reaction, as the CO_2 gas enters the ocean, the DIC content increases, the total alkalinity (Alk_T) remains the same, and the pH (the negative logarithm of the concentration of H^+) decreases (the acidity of seawater increases). The total alkalinity of seawater is a measure of its ability to neutralize acids. It is the sum of all bases that can be titrated by hydrochloric acid. The total alkalinity of seawater is mainly a function of the concentrations of the HCO_3^-, CO_3^{2-}, $B(OH)_4^-$ (borate), and OH^- (hydroxyl) ions in seawater. Some chemical details of the marine carbon dioxide-carbonic acid-carbonate system are given in Box 5.1.

The carbon dioxide in the ocean and that in the atmosphere can undergo rapid exchange, leading to a situation in which the concentration levels in the lower atmosphere and the waters near the sea surface are very nearly in equilibrium. The concept of chemical equilibrium is only an approximation of the true state of the natural system because of several factors (see Box 5.1), including variation of ocean surface temperatures with latitude, variation in the solubility of CO_2, which increases at lower temperatures and salinities, and hence latitude, and biological productivity. Tropical warm and salty equatorial ocean surface waters tend to contain a little more dissolved CO_2 than the atmosphere because of the upwelling of CO_2-rich waters. This generally leads to a sea-to-air flux of CO_2; therefore, large parts of tropical equatorial waters are sources of CO_2 to the atmosphere. On the contrary, high latitude, cold and less saline surface waters of large parts of both hemispheres are regions of air-to-sea transfer of CO_2, hence sinks of the gas. In the modern world, because of emissions of CO_2 to the atmosphere due to fossil fuel burning and deforestation practices, the ocean surface waters absorb some of this CO_2 and act as sink of anthropogenic CO_2 emissions. Globally there is a net transfer of CO_2 into the oceans of 2.3 billion tons of carbon per year (see Chapter 14).

It would take more than several thousand years to remove all the carbon dioxide gas that has accumulated in the atmosphere during the last century because of fossil fuel burning and land use activities, if the oceans were the ultimate main repository of the

BOX 5.1

The CO_2-Carbonic Acid-Carbonate System in Seawater

The carbon dioxide-carbonic acid-carbonate system is one of the most studied of seawater chemical systems and is very important in terms of modern environmental problems involving the carbon cycle on Earth (see Chapters 7 and 14). Dissolution of carbon dioxide in water is the first step that enables photosynthetic production of organic matter in aquatic systems, precipitation of carbonate minerals, and chemical weathering of the crust of the Earth. Carbon dioxide dissolves in water and reacts with water (H_2O), producing negatively charged bicarbonate (HCO_3^-) and carbonate (CO_3^{2-}) ions. The electrical charges of these ions are balanced in pure water by the hydrogen ion (H^+) or in a natural solution like seawater by other metal cations and the hydrogen ion. Absorption of the gaseous species $CO_{2(g)}$ in seawater is represented by the following reaction

$$CO_{2(g)} = CO_{2(aq)}. \tag{1}$$

The aqueous uncharged CO_2 chemical species also includes the carbonic acid (H_2CO_3) species, which constitutes only a small fraction—about 1/400—of CO_2 in solution.

To further describe relations in the CO_2-carbonic-acid-carbonate system, it is necessary to introduce the concept of an equilibrium constant touched upon in Chapter 2, Box 2.2. When the forward rate of a chemical reaction equals the backward rate, the reaction is said to be at equilibrium; this equilibrium can be represented by an equilibrium constant. In seawater the usual type of equilibrium constant employed to describe chemical relationships is the apparent or stoichiometric constant, which is a function of temperature, pressure, and salinity. An apparent constant is defined as the product of the concentrations of the products, each product concentration raised to the power of its stoichiometric constant in the reaction, divided by a similar expression for the reactants in the reaction. For reaction (1) above, the apparent equilibrium constant for the reaction in seawater may be written as

$$K'_0 = [CO_{2aq}]/P_{CO2}$$

where K'_0 is the apparent constant of the reaction, the brackets denote concentration, and P_{CO2} is the partial pressure of CO_2 gas in the atmosphere. The value of K'_0 at 25° C, an average seawater salinity of 35, and a total pressure of 1 bar ($=\sim1$ atmosphere) is 2.84×10^{-2}.

After absorption of the CO_2 gas, the aqueous CO_2 reacts with water

$$CO_{2(aq)} + H_2O = H^+ + HCO_3^-. \tag{2}$$

Under the environmental conditions specified above,

$$K'_1 = [H^+][HCO_3^-]/[CO_{2aq}] = 1.39 \times 10^{-6}.$$

The HCO_3^- ion in reaction (2) then dissociates ("breaks down")

$$HCO_3^- = H^+ + CO_3^{2-}, \tag{3}$$

and

$$K'_2 = [H^+][CO_3^{2-}]/[HCO_3^-] = 1.19 \times 10^{-9}.$$

As mentioned previously, the values of the K's vary with the salinity, temperature, and total pressure of the seawater environment.

Because of the formation of the aqueous ions HCO_3^- and CO_3^{2-} from CO_2 in solution, the total concentration of dissolved inorganic carbon (DIC) in seawater is the sum of the concentrations of all the inorganic carbon species, $CO_{2(aq)}$, HCO_3^-, and CO_3^{2-}, and is defined as

$$DIC = [CO_{2(aq)}] + [HCO_3^-] + [CO_3^{2-}].$$

The concentration unit for *DIC* is usually in moles of carbon per 1 kilogram of solution; in the case of seawater, the total moles of the *DIC* chemical species per kilogram of seawater.

For seawater the difference between the charges of the conservative cations and anions is equal to the algebraic sum of the charges of the H^+-dependent ions. This difference is called the **total alkalinity** of the solution, denoted Alk_T:

$$2[Ca^{2+}] + 2[Mg^{2+}] + [Na^+] + [K^+] - [Cl^-] - 2[SO_4^{2-}] = [HCO_3^-] + 2[CO_3^{2-}] + [B(OH)_4^-] + [OH^-] - [H^+]$$

and

$$Alk_T = [HCO_3^-] + 2[CO_3^{2-}] + [B(OH)_4^-] + [OH^-] - [H^+].$$

The Alk_T is usually given in units of mol-equivalent/kg seawater. The determination of Alk_T and DIC in ocean waters is a powerful tool to use to trace water masses (Alk_T) in the ocean and to obtain estimates of primary productivity and the dissolution and precipitation of carbonate minerals (DIC and Alk_T). Also if you know the values of these two variables, you can calculate the pH and pCO_2 of seawater.

It is also possible to write reactions and equilibrium constants for the solution (dissolution) and precipitation of minerals in aqueous solutions. The carbonate minerals are especially important minerals found in the ocean as the skeletons and tests of marine organisms and as cements in marine sediments. Calcite (hexagonal $CaCO_3$), aragonite (orthorhombic $CaCO_3$), and dolomite [hexagonal $CaMg(CO_3)_2$] are important marine minerals (see Appendix: Minerals). The latter mineral is not found in great abundance in the modern oceans but it is an important carbonate mineral found in ancient rocks deposited in the marine environment. The reaction representing both the dissolution of calcite and aragonite in seawater may be written

$$CaCO_3 = Ca^{2+} + CO_3^{2-},$$

and at 25° C, S = 35, and 1 bar total pressure, the apparent equilibrium constants are

$$K'_{cal} = [Ca^{2+}][CO_3^{2-}] = 4.27 \times 10^{-7}, \text{ and}$$
$$K'_{arag} = [Ca^{2+}][CO_3^{2-}] = 6.48 \times 10^{-7}.$$

Despite the two minerals having the same composition, they differ in their atomic structure, and thus the K' values for the two minerals differ. These K's basically represent the solubilities of the two minerals. The larger K' value for aragonite over calcite implies that aragonite is a more soluble mineral than calcite and is not stable relative to calcite under the environmental conditions specified. However, for these minerals to dissolve, the seawater in which the minerals are bathed must be undersaturated with respect to the two minerals. This implies that the product of the concentrations of Ca^{2+} and CO_3^{2-} (termed the ion concentration product, ICP) in the seawater must be smaller than the K' values for the minerals. The converse is true for the minerals to precipitate, where $Ca^{2+} + CO_3^{2-} = CaCO_3$.

Both calcite and aragonite contain other elements in minor or trace amounts and this affects their solubilities. Strontium (Sr) is especially important in aragonite and magnesium (Mg) in calcite. Indeed Mg can be so plentiful in biologically and inorganically produced calcite that a special term has been applied to the calcites containing more than about 1 weight % Mg, the **magnesian calcites,** which generally have greater solubilities than pure calcite, and for some magnesian calcite compositions, even aragonite.

The general relationships discussed here for the CO_2-carbonic acid-carbonate system are true for all natural waters. The major difference is that for other water compositions, the K's have different values. These concepts involving this chemical system in seawater will be especially important to the consideration of the problems of how rocks weather chemically and remove CO_2 from the atmosphere in the process (see Chapter 7) and of rising atmospheric concentrations of CO_2 and the acidification of the world's oceans (see Chapter 14).

gas. This slow rate of removal of atmospheric CO_2 is principally due to the stable layering of the ocean, with less dense warm water on top of cold and more dense water. Gases entering the surface layer of the ocean do not rapidly penetrate the thermocline to mix downward into deeper waters, except in limited areas of the ocean.

COASTAL MARGINS   Because the margins of the continents and islands are regions of important interactions among the land, ocean, and atmosphere, it is important to consider some of the biological, chemical, and physical characteristics of estuaries, bays, seas, and gulfs adjacent to the shoreline and of the margins of islands and atolls. Furthermore,

pollutants and natural substances are carried into the coastal oceans via the atmosphere, rivers, and groundwaters. How these coastal environments function is important to considerations of the dispersal and fate of these substances and also to the problem of the air–sea exchange of atmospheric CO_2 and the acidification of the oceans (see Chapters 11 and 14).

Several general types of water circulation can develop in restricted coastal margin basins. Two major types are estuarine and lagoonal circulation. In the former type, less dense freshwater flowing into the estuary from rivers flows out over denser seawater. This produces a circulation pattern like that seen in Figure 5.9a. The degree of stratification of the

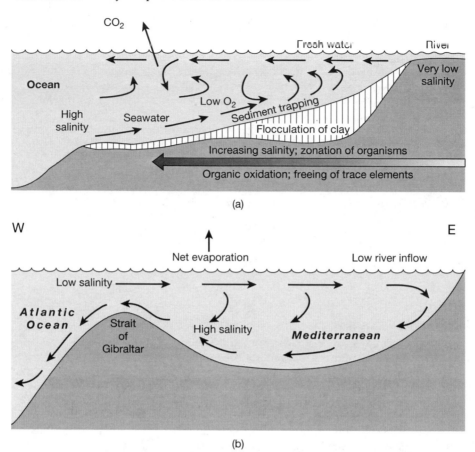

FIGURE 5.9   Generalized circulation patterns in an estuary and in the Mediterranean Sea. (a) Estuarine circulation in which there is a strong flow of freshwater from a river into the estuary and out to the open ocean along the surface of the estuary. The degree of mixing between freshwater and seawater varies from estuary to estuary, leading to estuaries with well-defined surface and seawater layers to those with nearly uniform vertical salinity gradients. (b) Idealized flow pattern in the Mediterranean Sea in which seawater enters at the Strait of Gibraltar between North Africa and Europe, flows eastward, increases in density because of evaporation, sinks, and returns at depth out through the Strait into the Atlantic Ocean. See text for discussion.   (*Source:* Garrels, et al., 1975.)

less dense fresher water on top of seawater varies from estuary to estuary. In lagoonal circulation, which characterizes circulation in some atolls at latitudes where evaporation exceeds precipitation and in the large basin of the Mediterranean Sea, less dense water flows in along the surface and more dense water flows out as an underflow along the bottom of the basin (Figure 5.9b).

In estuarine circulation, freshwater brought in by rivers mixes at the interface, with seawater coming into the estuary at deeper depths. The mixing is driven in part by winds and tidal currents. Because of the strong gradient in salinity along the longitudinal axis of an estuary, a diversity of benthic and planktonic

animal and plant species adapted to the various salinities of the estuarine waters is found along the salinity gradient. Estuaries are highly productive ecosystems (see Chapter 6). As the river water mixes with seawater and salinity increases, a salinity is reached where the suspended particles brought in by the rivers flocculate and settle rapidly to the bottom of the estuary. The process of flocculation involves the clotting together of very fine-grained (colloidal) particles to form larger aggregates that settle at much faster rates than the smaller materials. The flocculation of the very fine-grained river suspended particles entering an estuary is caused by the salts in seawater. As the salinity of estuarine waters increases, the dissolved salts of seawater alter the electrical charges of the colloidal particles so that the particles attract rather than repel one another as they do in freshwater. Flocculation plays an important role in estuaries and hence in the oceans. The flocculation of particles leads to the accumulation of inorganic and organic matter on the estuarine seafloor, thus preventing suspended clays from being dispersed widely throughout the oceans. Sediment is trapped within the estuary. The suspended clays may absorb elements from estuarine waters or release them from the particulate form to the water. More than 80% of the river suspended load of 150–180 × 10^{14} g/yr is deposited in near-shore coastal and shelf environments and does not make it out to the open ocean.

Organic materials entering estuaries may be oxidized and decomposed, and gases like carbon dioxide (CO_2), methane (CH_4), and hydrogen sulfide (H_2S) released to the waters and then to the atmosphere. Nutrient elements like nitrogen in the form of nitrate (NO_3^-) and phosphorus in the form of phosphate (PO_4^{3-}) and trace elements such as molybdenum (Mo), zinc (Zn), and copper (Cu) may reach estuaries via rivers or through the atmosphere. In the estuary, some elements will be used in the production of plant matter and some portion of the organic matter produced deposited on the seafloor. Some elements will be scavenged by inorganic suspended particles to accumulate in estuarine sediments or released to the pore waters of the sediments and escape to the overlying water or enter into biochemical reactions and remain trapped in the sediments in the pore water or in solid sediment phases.

An example of lagoonal circulation is shown in Figure 5.9b. In the Mediterranean Sea, located in a climatic region where evaporation exceeds precipitation, surface waters of salinity 35 enter from the North Atlantic Ocean through the Strait of Gibraltar. As the flow moves east, evaporation increases the salinity of the water to 38. In the winter, intense cooling and higher than normal evaporation of the waters occur because of the cold, dry mistral wind from southern Europe that blows across the water surface in the Mediterranean. The density of the surface waters is increased so much so that there is vertical mixing, or convection, of the waters to a depth of 2000 meters. The water mass formed has a salinity of 38.4 and a temperature of 12.8°C. This water flows west as an undercurrent at depth in the Mediterranean Sea and out through the Strait of Gibraltar.

It is known as Mediterranean Water. The Mediterranean Water enters the North Atlantic Ocean and spreads out across the basin at a depth of about 1000 meters. Mediterranean Water can be recognized at about this depth throughout much of the Atlantic Ocean because of its signature of high temperature and high salinity.

The Mediterranean contains about 3.8 × 10^6 km^3 of water. Water flows into the Mediterranean Basin through the Strait of Gibraltar at a rate of 1.75 × 10^6 m^3/sec. Thus, all the water in the Mediterranean Sea could be replaced by incoming North Atlantic Ocean water in

$$(3.8 \times 10^6 \, \text{km}^3 \times 10^9 \, \text{m}^3/\text{km}^3)$$
$$\div (1.75 \times 10^6 \, \text{m}^3/\text{sec})$$
$$\times (3.15 \times 10^7 \, \text{sec}/\text{yr}) = 70 \, \text{years}.$$

This value is the residence time (see Chapter 7) of water in the Mediterranean Basin. Because of the magnitude of the value of residence time, some dissolved chemical pollutants transported into the Mediterranean via rivers, for example, may remain in Mediterranean waters for several hundreds of years and be transported into the deep waters of the basin before exiting through the Strait of Gibraltar.

The estuaries and other coastal environments of the world's oceans are being heavily impacted by human activities. Some of these impacts are discussed in Chapter 11.

Cryosphere

Another major reservoir of water is the cryosphere. However, the mass of water in this reservoir is small compared to that in the oceans, with only about 2% of the water on the planet residing in snow, ice, and frozen ground.

Glacial ice covers 10% of the land area of the world today. There are many types of glaciers, such as crevice, valley, mountaintop ice cap, ice field, ice sheet, and ice cap. Glaciers are formed when moist air turns to snow in the winter, and the snow falls on the ground. Because it is so cold at high latitudes or at high elevations of mountain ranges, snow can accumulate in these environments and lead to the formation of ice. The ice can build up to great thicknesses and eventually a mountain or continental glacier can develop. These thick ice accumulations will flow downhill because of the pull of gravity. The rate of flow generally equals a few centimeters to tens of meters per year. Glaciers are responsible for the erosion of large U-shaped valleys and the transport and deposition of fine grained to boulder sized rock debris.

Ninety-five percent of all glacial ice is found in the polar regions in the form of ice sheets. They are the largest glaciers on Earth and spread out over the continents. Ice shelves that float on the oceans may be attached to the continental glaciers. In the past, ice sheets covered large portions of North America and Europe. The only ice sheets left today are in Greenland and Antarctica. The Greenland ice sheet is up to 3000 meters thick. The Antarctic ice sheet consists of two thick ice masses the size of Canada and most of the United States combined. These expansive ice sheets comprise 84% of the glacial ice mass covering the world. Ice in Antarctica reaches a thickness of 3600 meters or more; the melting of this ice would raise the level of the oceans about 60 meters. If all of the rest of the ice on the planet were melted, the sea level would rise only an additional 6 meters.

In contrast to the Antarctic ice sheet, the Arctic Ocean in the Northern Hemisphere is covered by only a thin layer of sea ice. Sea ice differs from glacial ice in that it is frozen seawater. It does not move downhill. When it melts or when it forms, sea level is not affected. In some areas of the world, sea ice development is seasonal, whereas in other areas it is a permanent feature. The degree of freedom with which ships can move in and out of the harbor of Vladivostok in Russia, and some other high-latitude ports, depends on the extent of development of seasonal sea ice. Polar bears in the Arctic wait each year on the shores of Hudson Bay in Canada for the sea to freeze, so that they may venture out on the sea ice to hunt for seals. The melting of the Arctic sea ice and permafrost (permanently frozen ground) as a result of global warming is discussed in Chapter 14.

The temperature of Earth's surface and the size and amount of water stored in ice on the continent varied during the Pleistocene Epoch. These changes in water storage led to changes in sea level. Continental glaciers grew and advanced as temperatures fell during glacial times. More water was locked in ice and sea level fell. During times of glacial retreat, as temperatures warmed, the cryosphere shrunk and as water was added to the oceans, the sea level rose (see Chapter 13).

Because glaciers are immense and heavy, continental masses with huge glaciers resting on them will be depressed into the underlying mantle, while continents free of ice sheets will rise higher. Today, in areas of the world like Scandinavia, the continent is rebounding because of the melting of the ice sheet that once covered the region during the last ice age, which climaxed 18,000 years ago. Because of this uplift, ancient beaches that were deposited at sea level are now found at higher elevations.

Minor Water Reservoirs

Only 0.68% of water in the hydrosphere is found encased in the pores of rocks of Earth. On the average, groundwater occurs within 750 meters of the land surface. It is very scarce below this depth, although the deepest subsurface water found was at a depth of 11 kilometers on the Kola Peninsula in the former Soviet Union. Groundwater travels very slowly in the subsurface. It enters the subsurface through soils as rain that has fallen on Earth's surface and then seeps into the ground. This water, in its travels in the subsurface, may encounter rock strata through which it can pass easily. The water is capable of leaching elements from the soil and rock that it passes through. It can remain in rock strata for hundreds of years and travel hundreds of kilometers. Eventually, the water originally derived as evaporation from the ocean and that fell as precipitation on land to enter groundwater can return to the ocean in rivers and flowing springs (see Chapter 3).

Some groundwater can return to the land surface in the form of springs. Also, the strata containing the water can be drilled into, and the water can be recovered by pumping or by free flow from an artesian well. Water entering rock strata at the surface in Colorado can be obtained below ground more than 1000 kilometers away in Kansas by drilling (see Chapter 3). In Hawaii, an extensive groundwater lens provides water for the island of Oahu. The water that is pumped out for consumption is approximately 20 years old. Groundwater is a major source of water for human consumption even though it comprises a small percentage of the hydrosphere. It is a natural resource that may be rapidly depleted and easily contaminated by the waste products of human society (see Chapter 11).

The small remainder of water on Earth is found in lakes (0.01%), soils (0.005%), atmosphere (0.001%), rivers (0.0001%), and the biosphere (0.00004%). These water reservoirs, although small, are an integral part of the hydrologic cycle, and their waters are generally relatively fresh, although variable in chemical composition. Rain contains dissolved salts and gases only at the parts per million level, while the average river water has about 0.01 weight % of dissolved salts, mainly as dissolved calcium (Ca), carbon (C), and silicon (Si) chemical species. Soil waters have salt concentrations like those of rivers, but the waters of lakes are very variable in composition, ranging from freshwater like the Great Lakes to salty brines like the Great Salt Lake. Rivers are the major source of water and dissolved and solid materials for the ocean. The atmosphere acts as a conveyor of water from the ocean back to the land. The soils collect water from the atmosphere as rainfall and are temporary repositories for water that will make its way into deeper groundwaters. Large lakes, like the Great Lakes, are temporary storage basins for waters on their way to the oceans and modulate regional climate. As with groundwaters, soil, lake, and river waters are easily modified by human activities and contaminated by effluents from industrial, agricultural, and household sources (see Chapter 11).

THE AIR–SEA INTERACTIONS

Intertropical Convergence Zone

Air–sea interactions are important to the transfer of energy and mass between the oceans and atmosphere. Both atmospheric and ocean circulations are driven by low-latitude heating and high-latitude cooling. Heat is absorbed by the planetary surface at low latitudes and is transferred poleward in both the Northern and Southern hemispheres. The equatorial region is a barrier to the exchange of materials between the atmosphere of the Northern and Southern hemispheres. It is also a barrier to water, salt, and heat exchange at the surface of the ocean. The large solar input to the tropics leads mainly to heating of the ocean surface. In turn, the air above the surface ocean is heated, expands, and becomes less dense. This reduction in density causes the air to rise owing to convection. A region of low pressure develops because the mass of the overlying atmosphere in the tropics is reduced. The "void" left by the rising, warm, moist air is replaced by air that moves toward the equator. A region of intense convergence in the lower atmosphere is created near the equator. This region is called the **Intertropical Convergence Zone** (ITCZ) (Figure 5.10). It is the zone along which the wind systems of the Northern and Southern hemispheres meet.

Heating of the tropical ocean surface causes evaporation of water. The water vapor rises and cools at higher elevations in the atmosphere, leading to the formation of clouds. Thus, the region of the ITCZ is characterized by cloudiness and heavy precipitation. The zone of the ITCZ is particularly intensely developed in the western Pacific. Here a warm water pool of surface seawater is found with mean temperatures of about 31°C. The warm pool extends nearly halfway around the world into the Indian Ocean and is known as the Indian Ocean/West Pacific Warm Pool (Figure 5.11). On average, the region of the warm pool around Sumatra, Java, Borneo, and New Guinea and into the central Pacific Ocean is also a region of intense atmospheric convection and the wettest part of the tropics. The surface ocean warm pool and the associated zone of heavy rainfall are important to the dynamics of El Niño–Southern Oscillation events (ENSO, discussed later).

The ITCZ shifts position seasonally (Figure 5.12). During the northern summer, the ITCZ shifts northward as the Asian continent is warmed more than the adjacent ocean (Figure 5.12a). The warm continental air rises, and air is drawn from the ocean toward the land. The zone of heavy rainfall expands northwestward from Indonesia into India and Southeast Asia. Southerly winds blowing from the oceans toward India and Southeast Asia are dominant at this time (Figure 5.12a). This is the time of the Southwest Monsoon (May to September). During the northern winter, the reverse is true (Figure 5.12b). The air above the Asian continent becomes very cold. An intense high-pressure region develops in the atmosphere above continental Asia. The flow of air is from the continent toward the ocean. This is the time of the Northeast Monsoon (November to March).

During these shifts in the ITCZ and monsoon direction, the currents of the ocean also change position. During the northern winter, the North Equatorial Current (Figure 5.4) flowing generally from east to west is the dominant current in the northern part of the Indian Ocean. However, during the austral winter, the Southwest Monsoon Current generally flowing from west to east replaces the North Equatorial Current in the Indian Ocean. This change reflects primarily the switch in the dominant direction of monsoon winds.

As mentioned previously, the global wind field and density differences in seawater from place to

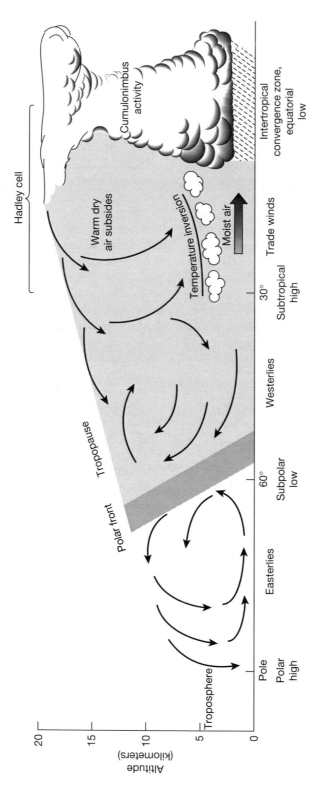

FIGURE 5.10 Cross section through the lower 20 kilometers of Earth's atmosphere from the polar region to the equator showing general vertical air circulation patterns. The Intertropical Convergence Zone (ITCZ) is the region near the equator where the wind systems of the Northern and Southern Hemispheres converge. This is a low-pressure region, one of tropical cloud formation and intense thunderstorm activity. Notice the return flow of the Hadley cell circulation in the upper part of the troposphere and the slope of the tropopause toward the pole. Compare with Figure 4.8, showing the zonal wind pattern of the planet.   (*Source:* The Open University, 1989c.)

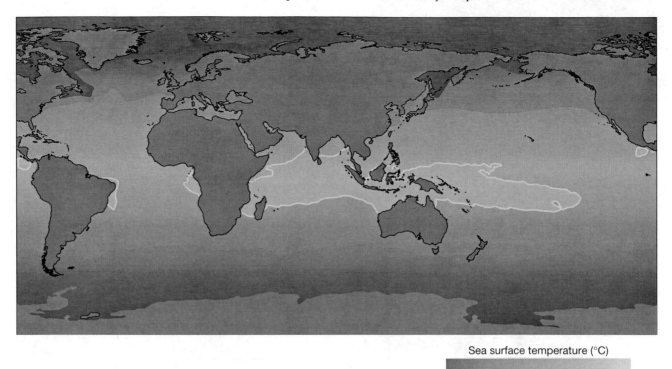

Sea surface temperature (°C)

−3   35

FIGURE 5.11   The Indian Ocean/West Pacific Warm Pool, showing its nearly halfway extension around the globe in the equatorial ocean. The waters of the pool reach an average temperature of 31°C but can reach nearly 35°C. The warm pool has significant effects on the climate of the surrounding land. Its slow fluctuations in size and heat content are probably linked to ENSO events.   (http://earthobservatory.nasa.gov/Features/WarmPool/.)

place drive oceanic circulation. In turn, both of these variables are strongly related to conditions in the atmosphere. For example, the trade winds are initially cool winds formed by the sinking (subsidence) of air that originally diverged poleward at the ITCZ (Figure 5.10). The warm, moist air rising at the ITCZ cools and loses moisture as it moves north and south in the atmosphere. The air subsides at about 30°N and S latitudes (Figure 5.10). The descending air results in an increase in the mass of the overlying atmosphere and thus an increase in atmospheric pressure at the surface. As the air moves toward the equator in the return flow of the trade winds, it sinks and warms. The warming during descent prevents the formation of clouds because the air can hold more water vapor at higher temperatures. Thus, the midlatitudes are characterized by a lack of cloud cover and low rainfall. However, the trade winds do pick up moisture from the sea as they move toward the equator. This increases the salinity of surface seawater at midlatitudes to the point where the seawater is dense enough to sink. At the ITCZ, the trade winds rise, causing the cloudiness and precipitation in this region and lowering the surface salinity of seawater.

The cloud cover also decreases the amount of solar heating at the equator. The pattern of vertical air movement from tropics to midlatitudes and return is known as the Hadley cell circulation (Figure 5.10).

Similarly, air sinking through the troposphere at 30° is veered to the right (east) due to the Coriolis force and forms the zonal winds of the prevailing westerlies. In turn, these winds drive ocean currents to the east (Figure 5.4). Figure 5.13 shows a cross section of the atmosphere from pole to pole, illustrating the relationships between the latitudinal variations in precipitation and evaporation, net transport of water in and out of the ocean, surface seawater salinity, and major wind belts.

El Niño–Southern Oscillation (ENSO)

A change in established currents, both air and water, can have an extreme impact on regional and global climate. The El Niño–Southern Oscillation (ENSO) phenomenon in the Pacific Ocean and the resultant intrusion of warm and low salt content water along the South American west coast are an excellent example. In 1892, Captain Camilo Carrillo, a Peruvian sea

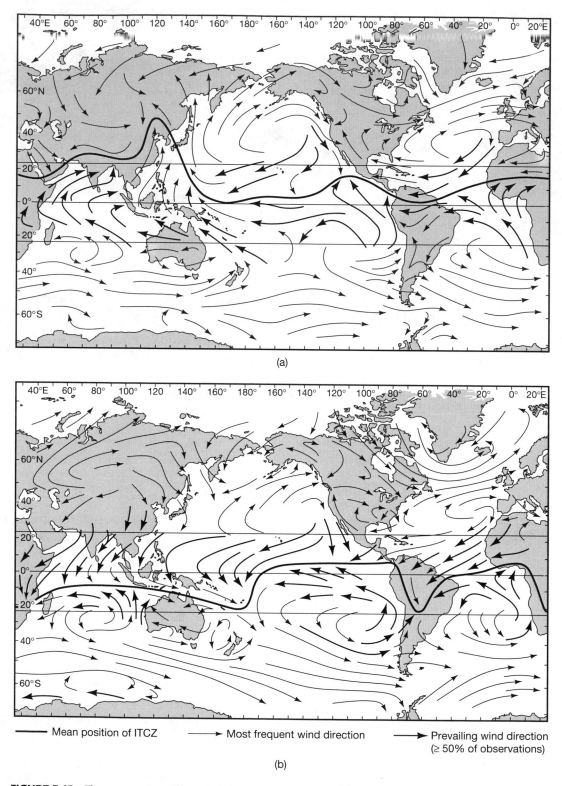

(a)

(b)

——— Mean position of ITCZ   ———→ Most frequent wind direction   ———→ Prevailing wind direction
                                                                              (≥ 50% of observations)

FIGURE 5.12 The average prevailing wind directions at Earth's surface during the Northern hemispheric summer (a, July) and winter (b, January). The solid black line shows the position of the Intertropical Convergence Zone (ITCZ) and its change with season. Note in particular the seasonal development of the strong low during the summer and high during the winter over Siberia, the seasonal movement of the ITCZ in the vicinity of India, and the seasonal switch from summer southwesterly to winter northerly monsoon winds in the northern Indian Ocean and over the subcontinent of India and southeast Asia.   (*Source:* The Open University, 1989c.)

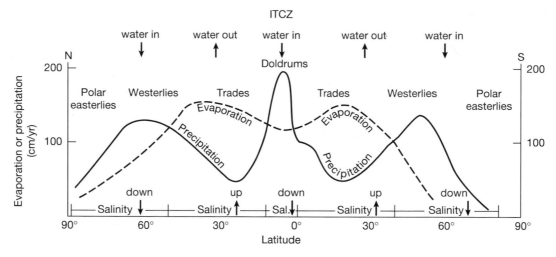

FIGURE 5.13 Latitudinal variation of global precipitation and evaporation and generalized relationship to zonal wind belts and the salinity of the ocean. (*Source:* Garrels, et al., 1975.)

captain, published a paper in the Bulletin of the Lima Geographical Society describing a tropical counter-current offshore Peru known by the local fishermen as **El Niño,** meaning "the Christ Child" in English, so named because the phenomenon usually begins around Christmastime. In the 1920s, British meteorologist Sir Gilbert Walker discovered the atmospheric phenomenon of the Southern Oscillation (SO). The meteorologist, Jacob Bjerknes, then connected the Southern Oscillation to El Niño (EN) events, hence the term El Niño–Southern Oscillation (ENSO).

Normally, cool currents flow north along the coastline of Peru and Chile. These currents are the Peru Oceanic Current, which extends to a depth of 700 meters and may be as wide as 600 kilometers, and the Peru Coastal Current, running close to the coastline with a depth of 200 meters and a width usually not exceeding 200 kilometers. Beneath these two currents lie the southward-flowing Peru Undercurrent and to the north the Peru Countercurrent, which is an extension of the eastward-flowing Equatorial Countercurrent (Figure 5.14a). The southeast trade winds usually blow approximately parallel to the Peruvian coastline. Thus, because Peru is in the Southern Hemisphere and the Coriolis force deflects surface currents to the left of the wind direction, surface water moves offshore along the entire Peruvian coastline at a direction somewhat less than 45° to the prevailing wind direction, that is, to the west. The net Eckman transport of water is offshore and to the west.

The movement of the surface water offshore from the Peruvian coast leads to its replacement by vertical water motions (upwelling) from a depth of 40 to 80 meters at a speed of 1 to 3 meters per day

(Figure 5.14b). This upwelling water is rich in nutrients and results in stimulation of biological productivity in the surface waters (see Chapter 6). The upwelled water comes from below the nutricline. It is this rich fertilization of the surface waters by nutrient-rich upwelling water that accounts for the spectacular yields of the Peruvian coastal anchovy fishery. Figure 5.15 shows the Peruvian anchovy biomass and fish catch from 1950 to 1998, the latter year coinciding with the large El Niño event of 1997–1998. Notice the large variability in fishery yields and in particular the low yields coinciding with the early 1970s, early 1980s, and the 1997–1998 El Niño events. One cause of these large fluctuations in fish catch appears to be the El Niño phase of the El Niño–Southern Oscillation phenomenon.

About every 3 to 7 years, an El Niño occurs that lasts 1 to 2 years, and sometimes longer. El Niño is a time when the warm waters of the western Pacific extend far to the east and bathe the coast of central South America in warm, nutrient-poor water. The water may be as much as 10°C warmer than the original coastal water. The water is transported to the Peruvian coast by the Peru Countercurrent. This warm, low-salinity, nutrient-poor water overrides the Peru Coastal Current because its density is less than that of the coastal current. A nutrient-poor water layer comes to reside along the coast of Peru. The surface-water layer is on the order of 30 meters in thickness. Because of its thickness, upwelling water no longer comes from below the nutricline but from within the nutrient-poor water itself. Hence, if upwelling continues during an El Niño event, the upwelled waters are low in nutrients, and biological productivity in

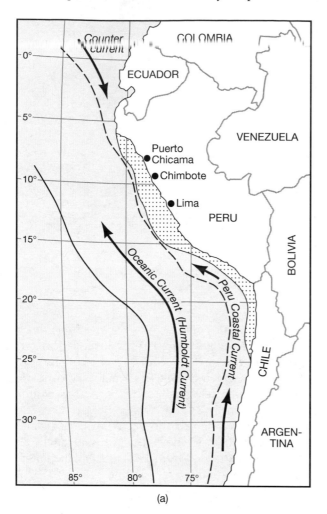

(a)

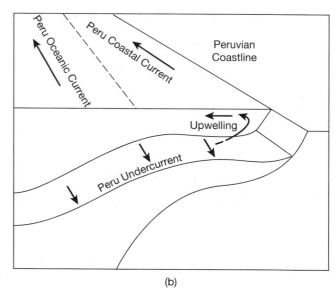

(b)

FIGURE 5.14 Shallow movements of water offshore western South America. (a) Shallow current system; the dashed line is the approximate boundary between the Peru Coastal Current and the Peru Oceanic Current (the Humboldt Current). During an El Niño event, the countercurrent to the north extends farther south, transporting warm water to the coast of Peru. The stippled area is the region of the Coastal Current system in which the anchovy fishery is found. (*Source:* Idyll, 1973.) (b) The relationship among the coastal upwelling along the coast of Peru, the surface currents, and the Peru Undercurrent. Southeasterly winds prevail along the coast of southern Ecuador and Peru and blow in both normal and El Niño years. The wind blowing on the surface leads to current flow in a northwesterly direction and upwelling of nutrient-rich waters during normal years. (*Source:* Laws, 1992.)

the surface waters is substantially reduced. At the same time, the trade winds usually slacken, and this may cause upwelling to cease completely.

The warming of the coastal waters adversely affects the anchovy populations that thrive in the waters offshore of Peru. With the cessation or disruption of coastal upwelling of nutrient-rich waters, nutrients do not reach the shallow euphotic zone along the coast of Peru. Plant plankton (phytoplankton) productivity slows dramatically. The phytoplankton and the animal plankton (zooplankton) that feed on them virtually disappear. The small anchovy fish prefer the cool and plankton-rich waters of non–El Niño years. Thus, the anchovies nearly disappear with the

warming of the waters during El Niño events. The fish may actually die or simply disperse to deeper waters or far offshore. The local fishing industry suffers.

Since 1970 the catches of adult anchovies have declined during El Niño years. In part these declines reflect the changing environment of El Niño years but also previous overfishing and fish recruitment failures during the years preceding El Niño events. From 1970 through the El Niño years of 1972 to 1973, the anchovy catch dropped from 13 million metric tons to about 2.5 million metric tons (Figure 5.15). This global loss of protein was partially responsible for driving up the price of another protein source, soy, from

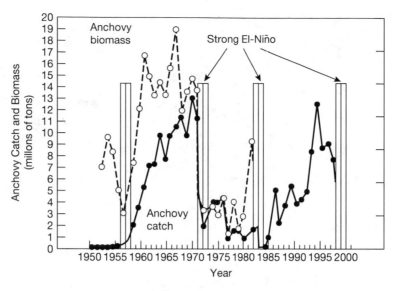

FIGURE 5.15 The annual catch of anchovy (dark black line) and the anchovy biomass (light line) offshore Peru in millions of tons between 1950 and 1998. Vertical bars denote El Niño events. Arrows show times of strong El Niño events. (From *Climate Change and Long-term Fluctuations of Commerical Catches*. © FAO 2001. Reprinted by permission of the Food and Agricultural Organization of the United Nations.)

US$100 to almost US$300 per ton during the same period of time.

Without anchovies, the staple food supply of the local guano bird populations is lost, and the birds die off or leave the area for other food supplies. In mild El Niño years, there may be enough food for the adult birds to survive. However, during such events, the adult birds usually vacate their nests, leaving the young birds to starve. During severe El Niños, significant numbers of adult birds also die. In 1957, during an El Niño event, the guano bird population dropped from 27 million to about 6 million, recovering to 17 million just prior to the El Niño of 1965.

Rainfall in the Sechura Desert of Peru increases dramatically during an El Niño event. During the El Niño of 1982–1983, rainfall in the desert increased from near zero to 40 millimeters per hour in some rain events. Such torrential rains in Peru during the extended El Niño of the early 1990s and that of 1997–1998 led to floods, landslides, destruction of crops and homes, and death in Peru.

When the more usual situation of the cool currents offshore Peru return (normal condition or La Niña events), upwelling returns, biological productivity increases, the anchovies return, the birds feed, the fish and bird stocks increase, and the fishermen fish. Atmospheric/oceanic changes accompanying El Niño events can affect circulation patterns and weather around the world and are responsible for warm, dry winters in the northern United States and wet winters in the southern United States, torrential rains and flooding in coastal South America and parts of Asia, droughts in Africa, Australia, and the Hawaiian Islands, unusual tropical cyclone activity, and failure of the monsoon in Asia (Figure 5.16). The cold phase of the 1988 **La Niña** (the "sister" of El Niño) event was probably the cause of the drought in the United States during that year.

Why do we have El Niño and the cold extreme of ENSO, La Niña, events or phases? This is a difficult question to answer, although in recent years much has been found out about the air–sea interactions that lead to these interannual variations. We know the ENSO phases are a result of both atmospheric and oceanic changes. As mentioned, the western equatorial South Pacific is a region of unusually high sea-surface temperature (as high as 31°C) known as the West Pacific Warm Pool. Associated with the warm pool is a low-pressure region in the atmosphere of intense convective atmospheric motions, storminess, and rainfall. During the warm phase of an El Niño–Southern Oscillation (an El Niño), the West Pacific Warm Pool and the atmospheric convection center migrate to the east along the equator, spreading into the eastern Pacific. In addition, waves in the ocean are generated. These waves, known as **internal waves,** have their maximum amplitude below the surface of the ocean and are affected by the Coriolis force. The internal waves of interest here are the Kelvin and Rossby waves. Kelvin waves cross the Pacific Ocean

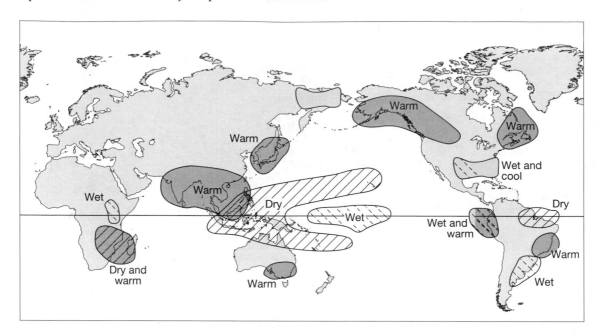

FIGURE 5.16 Rainfall and temperature patterns associated with an El Niño–Southern Oscillation (ENSO) event. The diagram shows conditions during a winter season in the Northern Hemisphere. (*Source:* Ropelewski, 1992.)

in 2 to 3 months and lead to a rise in sea level along Peru and Ecuador, as the Peru Countercurrent carries the warm water from the equator southward. The thermocline in the equatorial ocean deepens and upwelling along the equator slows. As sea level rises and warm water accumulates in the eastern Pacific, Rossby waves are generated that move west across the Pacific. The time it takes for these waves to cross the Pacific is strongly dependent on the latitude at which the wave is traveling. Rossby waves near the

equator cross in about 9 months; those at a latitude of 12° take nearly 4 years. On reaching the western Pacific, the Rossby waves travel toward the equator as coastal Kelvin waves, which on reaching the equator turn east and begin another crossing of the Pacific.

During the developing phase of an El Niño, a pressure anomaly in the atmosphere known as the **Southern Oscillation** decreases (Figures 5.17 and 5.18). As a result, the trade winds weaken and westerlies tend to blow along the equator. The pressure

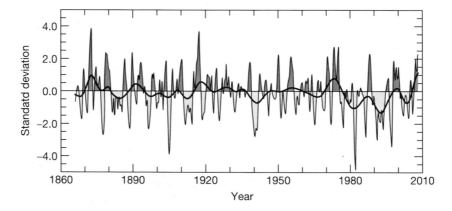

FIGURE 5.17 The Southern Oscillation Index (SOI) from the mid-nineteenth century to 2008 computed using monthly mean sea level pressure anomalies at Tahiti and Darwin, Australia, derived using monthly values of the index. The smoothed black curve represents the index when fluctuations of less than 8 months are excluded. The indices represent deviations (anomalies) from the mean state of the system. (From Climate and Global Dynamics Division, National Center for Atmospheric Research, http://www.cgd.ucar.edu/cas/catalog/climind/soi.html. Copyright © 2010 University Corporation for Atmospheric Research. Used by permisssion of UCAR.)

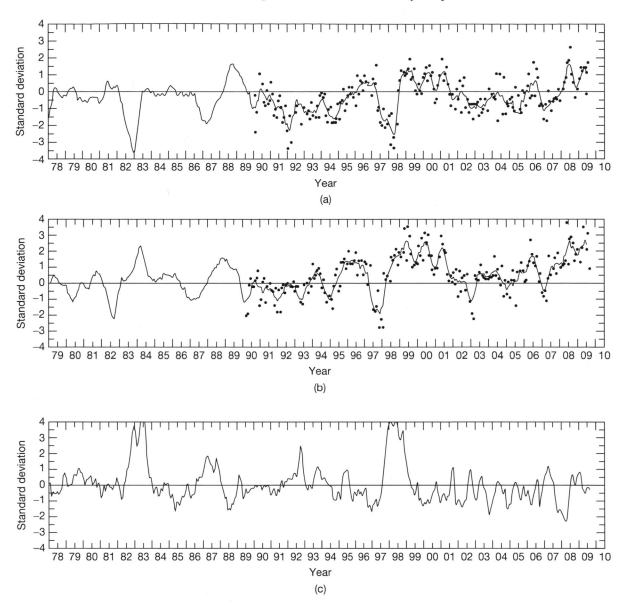

FIGURE 5.18 Some trends in variables associated with the El Niño–Southern Oscillation ENSO) phenomenon for the period of the late 1970s to 2009. (a) The Southern Oscillation Index (SOI), a measure of the sea-level pressure differential between the equatorial western Pacific and the eastern Pacific (now commonly reported in terms of the difference in atmospheric pressure between Darwin, Australia, and Tahiti; earlier in the history of the derivation of the SOI, Easter Island pressures were used to represent the eastern Pacific). The SOI is plotted as departures (standard deviation) from the 1951–1980 base-period means. (b) Near-surface zonal wind averaged over the area 5°N−5°S, 135°E−180° and plotted as departures (standard deviation) from the 1979–1995 base-period means. (c) Equatorial sea-surface temperature (SST) anomaly in degrees Celsius in the eastern Pacific near the coast of Peru plotted as departures from the 1950–1979 period. The 1982–1983 and the 1997–1998 El Niño events show up very clearly in changes in these variables: a sharp drop in the SOI (implying weakening of the Walker cell circulation) and the zonal wind anomaly (implying a decrease in the strength of the equatorial trade winds), and a rise in SST near Peru of more than 4°C. For (a) and (b), black dots represent individual monthly means. As this text was being completed in mid-2009, a transition from La Niña conditions to ENSO neutral conditions followed by El Niño conditions was underway.   (*Source:* various U.S. Department of Commerce Climate Diagnostics Bulletins, http://www.cpc.noaa.gov/products/CDB/CDB_Archive_pdf/pdf_CDB_archive.shtml.)

anomaly is an index of the intensity of the Walker cell circulation in the atmosphere. The **Walker cell circulation** involves large-scale convective motion in the Pacific atmosphere. As the trade winds blow across the Pacific, they pick up heat and water vapor from the tropical ocean. In the western Pacific near Indonesia, the air rises and cools. Water vapor condenses from the air, and rain ensues. This is the region of intense convection in the western Pacific referred to previously. The rising air aloft moves toward the east, radiating heat to the surrounding atmosphere, cooling, and eventually sinking in the eastern Pacific to begin its journey to the west again. This pattern of circulation is known as a Walker cell. The air that sinks in the eastern Pacific is more dense than that rising in the western Pacific because it has lost heat and water vapor. Thus, there is a small atmospheric surface-pressure difference between the eastern and western Pacific. The eastern Pacific has the higher pressure. This sea level pressure differential is referred to as the **Southern Oscillation Index** (SOI). The SOI in the past was taken as the pressure difference at sea level between Darwin, Australia, and Easter Island, but it is now calculated from the pressure difference between Darwin and Tahiti at a latitude where the pressure differential is maximum. Figure 5.17 shows the Southern Oscillation Index from the mid-nineteenth century to the year 2008. The negative values denote conditions of weaker Walker cell circulation and trade winds and the positive values the reverse. The relationships among the SOI, zonal equatorial winds

between 5° N and 5° S latitude, and the sea-surface temperatures off the coast of Peru are shown in Figure 5.18. Drops in the Southern Oscillation Index and the wind anomaly index are associated with high sea-surface temperatures in the eastern equatorial Pacific and are indicative of El Niño events. The drop in the wind anomaly index reflects weakening of the Walker cell circulation, less intense trade winds, and westerly winds blowing along the equator in the central Pacific. It is possible that El Niño events are triggered by bursts of westerly winds that last for a period of only a week and cover a few hundred kilometers of area in the western Pacific.

The second set of Kelvin waves that crosses the Pacific leads to the lowering of sea level, and sea-surface temperatures decline in the eastern Pacific. The conditions off Peru begin to return to normal. However, the lowering of sea-surface temperatures increases the pressure at sea level, leading to an increase in the SOI and intensification of the Walker cell circulation and the trade winds. This shift in winds sends Rossby waves westward back across the Pacific. Once again, these waves on reaching the western Pacific travel toward the equator as coastal Kelvin waves, which turn east and move across the Pacific. This final set of Kelvin waves raises sea level in the eastern Pacific, and the El Niño cycle is complete. Figure 5.19 shows the motion of the Kelvin and Rossby waves as they cross the Pacific Ocean.

There has been considerably less research done on the cold extreme (La Niña) of the ENSO cycle and

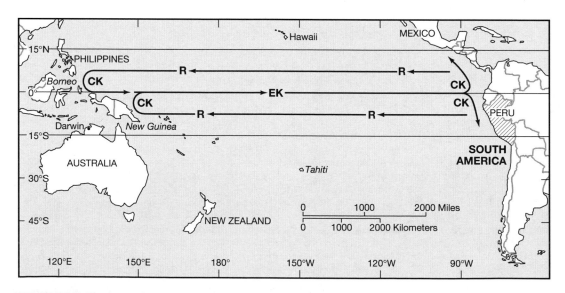

FIGURE 5.19 The internal wave system in the ocean that leads to the generation and relaxation of an El Niño event. EK, equatorial Kelvin waves; CK, coastal Kelvin waves; R, Rossby waves. The complete El Niño cycle requires about 3 to 5 years, reflecting the time it takes for two sets of both Kelvin and Rossby waves to cross the Pacific. One set raises sea levels in the direction they are traveling and the other lowers sea levels.   (*Source:* Laws, 1992.)

its climatic implications compared to El Niño. There are several possible reasons for this, including the fact that there have been only a few major cold events that preceded the 1998–2000 La Niña event. There are three possible states for sea-surface temperatures of the equatorial Pacific. These are a range of extreme warm conditions (El Niño), a range of extreme cold conditions (La Niña), and a range of temperatures around an average that can be referred to as normal. These three phases of differing sea-surface temperature, thermocline depth, and location of the western Pacific zone of intense atmospheric convection are shown in Figure 5.20. As the 1997–1998 El Niño warm event decayed rapidly in May 1998, the ocean's surface and subsurface temperatures fell, passed through average, and then continued to cool rapidly (Figure 5.18). This indicated the onset of La Niña cold-phase conditions in the Pacific central and eastern equatorial waters. Since the 1997–1998 strong El Niño event, the eastern equatorial Pacific has remained near normal, with a tendency toward cooler waters. During 2007 and 2008, a distinct La Niña event occurred followed by a slight warming trend. As of the time of this writing in 2009, there is a tendency toward El Niño conditions across the equatorial Pacific Ocean.

The impacts of cold events on meteorological conditions throughout the world depend on the intensity of the La Niña event, but are less well known than the impacts of El Niño events. Different researchers have different opinions about the intensity of past La Niña events. For example, some classify the 1988–1989 event (Figure 5.18) as a strong one, whereas others classify its intensity as only moderate. Figure 5.21 represents a composite of meteorological conditions of several La Niña events to provide some idea of the rainfall and temperature impacts of La Niña events. A comparison with Figure 5.16 provides some clue as to how the impacts differ geographically between El Niño and La Niña events. For example, notice that generally during an El Niño event, the southwestern Pacific is warm and dry, and during a La Niña, it is cool and wet.

It appears that the ENSO has been a feature of the equatorial Pacific for some time. During the early Pliocene about four million years ago, there is evidence for a greatly reduced meridional temperature gradient between the equator and the subtopics, implying a poleward expansion of the ocean tropical warm pool. The contrast in temperature between the equator and 32°N was only about −2°C at this time and evolved to −8°C today. The reduced temperature gradient of the Pliocene led to a slowdown in the atmospheric Hadley cell circulation and distinct El Niño–like

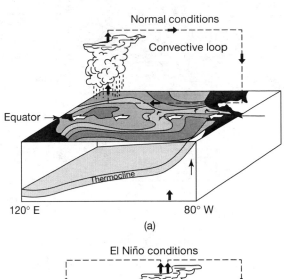

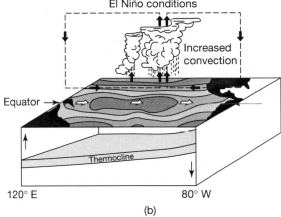

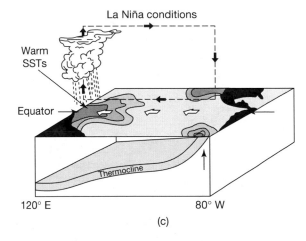

FIGURE 5.20 Schematic diagrams showing sea-surface temperatures (SSTs), atmospheric convection, and the position of the thermocline during (a) normal, (b) El Niño, and (c) La Niña conditions in the equatorial Pacific. The dark, thick arrows represent atmospheric circulation in the convective loop of the Walker cell. The white, thick arrows show the direction of movement of the warm water of the West Pacific Warm Pool (the darker the shading, the warmer the water). The thin arrows show the relative movement of the oceanic thermocline. (*Source:* Glantz, 2001.)

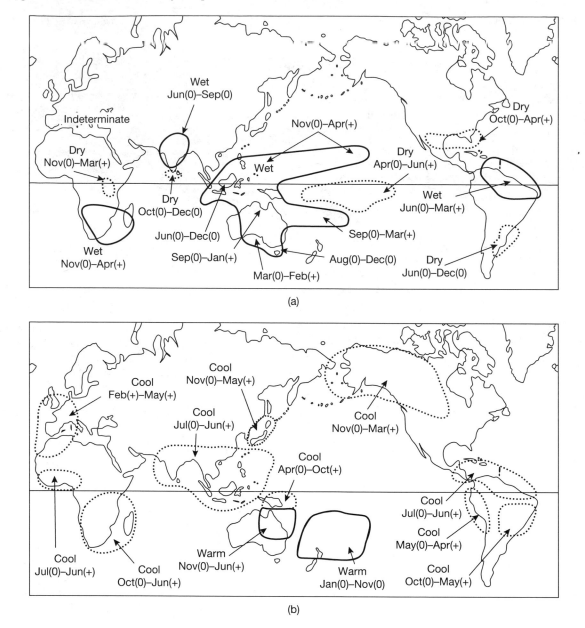

FIGURE 5.21 The (a) potential rainfall and (b) temperature impacts from La Niña events. The (0) represents the conditions for the year of the onset of La Niña, and the (+) represents conditions for the year following the onset of La Niña. (*Source:* Ropelewski and Halpert, 1987.)

conditions in the equatorial Pacific. In addition, there are geologic records of El Niño events attesting to the fact that the modern expressions of these air–sea interactions have been around for about 5000 years. The record suggests that prior to this time there was a warm background climate without El Niño episodes (the Holocene Climatic Optimum; see Chapter 13). Since then the cooler background climate of the past 5000 years has been conducive to the spawning of El Niño events. Historical records for Peru document four major events of flooding and rebuilding of settlements between 2150 B.C. and 1770 A.D., and 87 El Niño events between 1690 and 1987. A question that is of concern to scientists studying the current environmental issue of global climatic change is: What might happen to the El Niño phenomenon if there is a global warming induced by the accumulation of greenhouse gases in the atmosphere from industrial, transportation, and other sources related to human activities? There is some evidence that global warming might lead to conditions in the Pacific region like that of an extended El Niño (see Chapter 14).

The Atlantic and Pacific Decadal Oscillations

The ENSO is an interannual climatic phenomenon involving air–sea interactions; there are also natural oscillations in the coupled ocean–atmosphere system affecting climate that take place on other time scales. Two of importance are the North Atlantic Oscillation (NAO) and the Pacific Decadal Oscillation (PDO). Much like the ENSO Southern Oscillation index, indices are available for these two phenomena (Figure 5.22a, c).

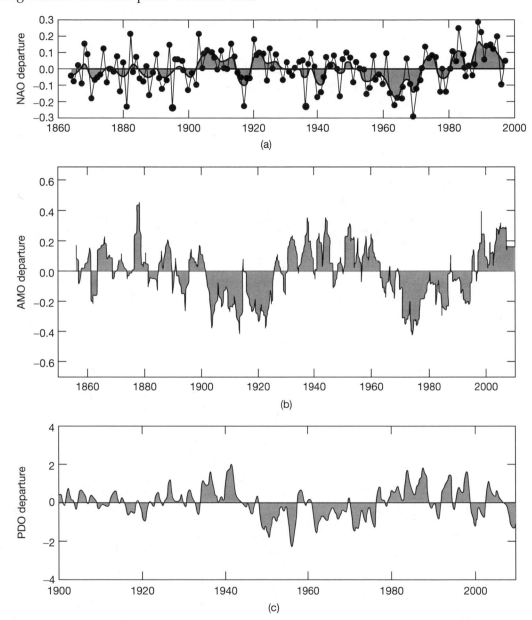

FIGURE 5.22 Time series of expressions of three major air–sea interactions that affect climate. (a) The North Atlantic Oscillation (NAO) index plotted yearly (black dots) and as a three-month running mean (patterned area) from 1863 to 1995. (shttp://www.ldeo.columbia.edu/res/pi/NAO/.) Later data are available at http://www.cpc.noaa.gov/data/teledoc/nao.timeseries.gif. (b) The Atlantic Multidecadal Oscillation (AMO) index from 1856 to 2008 plotted as monthly values. (http://www.cdc.noaa.gov/data/timeseries/AMO/.) (c) The Pacific Decadal Oscillation (PDO) index from 1900 to 2008 plotted as monthly values. (http://jisao.washington.edu/pdo/.) For both the AMO and PDO, shaded areas above the zero line indicate SSTs above the mean, and shaded areas below the zero line, SSTs cooler than the mean state of the system. All indices are plotted as standardized values of the departures from the mean state of the system. (From Lemont-Doherty Earth Observatory, www.ldeo.columbia.edu/res/pi/NAO/ by Dr. Martin Visbeck. Used by permission of Dr. Martin Visbeck.)

The **North Atlantic Oscillation** (NAO) was discovered in the 1920s by Sir Gilbert Walker and is an air–sea interaction and climatic phenomenon involving mainly the atmosphere in the North Atlantic Ocean. It is expressed as an irregular seasaw in the difference of atmospheric pressure at sea level between the Icelandic low-pressure region and the Azores high-pressure region. The NAO is a dominant mode of winter climatic variability in the North Atlantic region ranging from central North America to Europe and into much of Asia, due to the oscillatory motions of the Icelandic low and the Azores high controlling the strength and direction of westerly winds and storm tracks across the North Atlantic. The NAO index varies from year to year and exhibits a tendency to remain in one phase for intervals of time that may last for several years or even decades. A positive NAO index (Figure 22a) implies a stronger than usual subtropical high-pressure center and a deeper than normal Icelandic low. The increased pressure difference results in more and stronger winter storms crossing the Atlantic Ocean that have a northerly track. This results in warm and wet winters in Europe, cold and dry winters in northern Canada and Greenland, and mild and wet winter conditions in the eastern United States.

A negative NAO index (Figure 5.22a) phase implies a weak subtropical high and a weak Icelandic low. The reduced pressure gradient results in fewer and weaker winter storms crossing the Atlantic on a generally west to east pathway. This brings moist air into the Mediterranean and cold air to northern Europe, and the east coast of the United States experiences more cold spells and hence snowy weather conditions. Bitterly cold air flows south and west from the North Pole and Siberia.

The NAO index generally has been on the positive side since the 1970s. This has had marine ecological effects including enhancing the growth of populations of Labrador Sea snow crabs that thrive under the colder conditions in the northwestern Atlantic Ocean associated with a positive NAO. In addition, the peak in the NAO in the 1990s may have been part of the reason for the collapse of the Newfoundland cod fishery.

The behavior of the NAO depends on several complex factors, including sea surface temperatures in the Atlantic, the flow of the warm waters of the Gulf Stream, and the intensity of the downwelling of the lower limb of the conveyor belt circulation in the North Atlantic. There are certainly links between the NAO phase shifts and ENSO but they are not well defined. From a historical perspective, tree ring drought reconstruction data from Morocco and speleothem (cave deposit)–based precipitation proxy data for Scotland have shown that the Medieval Warm Period was characterized by persistent positive NAO conditions. In addition, based on climate model results and proxy data, a shift to weaker NAO conditions has been demonstrated for the Medieval Warm Period transition into the Little Ice Age. The Medieval Warm Period–Little Ice Age climatic transition probably was coupled to prevailing La Niña–like conditions amplified by an intensified Atlantic meridional overturning circulation during the Medieval Warm Period (see Chapter 13).

The NAO should not be confused with the **Atlantic Multidecadal Oscillation** (AMO, Figure 5.22b) that is reflected primarily in changes in sea surface temperatures on a decadal time scale in the North Atlantic Ocean and is correlated to changes in air temperatures, rainfall, and drought over much of the Northern Hemisphere. In addition, changes in the AMO seem to be tied to the frequency of severe hurricanes in the Atlantic. Cool and warm phases of the AMO may last 20 to 40 years; these changes have been occurring naturally for at least 1000 years. The variability in the AMO may be related to small changes in the intensity of the North Atlantic part of the conveyor belt circulation.

In contrast to the NAO and AMO, the **Pacific Decadal Oscillation** (PDO) is a decadal temperature fluctuation pattern in the Pacific Ocean (Figure 5.22c) and has been described as a long-lived El Niño–like pattern of Pacific climatic variability. The PDO was named by Steven R. Hare of the International Pacific Halibut Commission in Seattle, Washington, and the pattern of variability was also described at about the same time by Yuan Zhang at the University of Washington. The PDO is detected as warm or cool surface waters north of 20°N and waxes and wanes every 20 to 30 years. During a cool or negative phase, lower than normal sea surface heights and temperatures characterize the eastern equatorial Pacific; the opposite pattern is true of a positive warm phase. The changes in cold and warm water influence the path of the jet stream, which flows farther to the north during cool phases resulting in reduction in rainfall, e.g., in the western North America. The **Interdecadal Pacific Oscillation** (IPO) is a variant of the PDO, with a cycle of 15 to 20 years, but it affects both the north and south Pacific regions. The PDO signal has been indentified in tree-ring chronologies (see Chapter 13) from the Baja, California area going back nearly 350 years. Northeast Pacific marine ecosystems are affected by phase changes in the PDO. Positive warm phases result in enhanced coastal ocean biological productivity in Alaska and decreased productivity off the west coast of the contiguous United States. During negative cold PDO times, the opposite pattern in marine ecosystem productivity is seen. The causes for the PDO and its phases lasting years to decades are not well understood. Currently we are in a cool phase, implying less rainfall and more drought over western North America for two to three decades.

From a societal standpoint, recognition of the natural phenomena of ENSO, NAO, AMO, PDO, and IPO and their causes are important because they show that climatic conditions can vary from what we humans call "normal" on a time scale of a generation or less. Thus, these climatic variations have important societal and ecological consequences discussed here, and later in this book. The natural climatic phenomena also play a role in the climatic history of the planet during the past 150 years, the time scale of the recent warming of the planet attributed mainly to the release of greenhouse gases to the atmosphere by human activities (the problem of global warming; see Chapter 14).

Concluding Remarks

Processes in the atmosphere and hydrosphere are interconnected. The ocean and atmosphere redistribute the heat received at the planetary surface from the sun. Warm air and water are transported toward the poles from the equatorial region by atmospheric and oceanic circulations, respectively. In the process, heat and water are exchanged between the two large fluid reservoirs of Earth. Cool air and water are returned toward the equator by low-level atmospheric winds and surface ocean currents, respectively. The winds drive the surface currents of the ocean, and in turn, heat exchange between the ocean and atmosphere helps propel the wind systems of Earth.

The atmosphere is an easily perturbed system. Many gases in the atmosphere are found there in trace concentrations. These trace gases have sources at Earth's surface. On one hand, natural variations in the gas fluxes associated with these sources can lead to changes in atmospheric composition on a relatively short time scale. On the other hand, fluxes of these trace gases to the atmosphere from human activities can also rapidly modify the chemical composition of the atmosphere. Such changes in atmospheric composition can result in climatic change.

The water cycle of Earth is complicated. Water may be salty or fresh; it may be warm or cold; it may be dense or less dense; it may be clear or opaque; and it may be teeming with life or sterile. Each of these factors affects the movement of water and its influence on Earth. These effects include a range of time and space scales of interaction, involving water in small, local, and transient showers to water locked up in continental-scale glaciers for moderately long periods of time. Stream flow can add an amount of water to a lake equivalent to the volume of the lake in less than 100 years. However, it would take 40,000 years to replace all the water in the oceans by river input. Both these values are residence times, where **residence time** is defined as the mass of a substance in a reservoir divided by the flux of the substance into the reservoir (see Chapter 7).

During one residence time of a substance in a reservoir, the material may be stirred or mixed repeatedly. Small lakes may be stirred many times in 100 years by currents generated by winds blowing across their surfaces. However, in 40,000 years, the ocean would be stirred only about 25 times. Surface-water systems are very susceptible to contamination by chemicals derived from human activities; the extent of contamination depends to some degree on how fast the waters of a lake, river, or bay are renewed. Atmospheric chemical changes and climatic change can also affect the properties of surface waters by modifying their temperature, circulation, and chemistry.

There are several different natural air–sea interactions and climatic phenomena that exhibit variability on interannual to decadal time scales. These natural phenomena can result in changes in weather and climatic patterns throughout much of the world. The changes, particularly those of an extreme nature, include changes in regional rainfall, flood, and drought patterns, storm tracks, global wind patterns, cooling of some regions and warming of others, and perhaps increased hurricane intensity and frequency. The pattern of *global surface warming* of the past 150 years of Earth history is affected by these natural climatic phenomena and by volcanic explosions and changes in solar radiation. These natural climatic fluctuations can modify any upward trend in the global warming pattern; global warming in turn may affect the natural climatic phenomena.

Study Questions

1. Discuss briefly the pattern of the trade winds, their origin, and their relationship to ocean currents.
2. What are the seven most abundant elements in seawater?
3. Average seawater is primarily a sodium chloride solution. Seawater contains about 35 grams of salt per kilogram of seawater, of which about 19 grams are chloride ion (Cl^-). What is the weight percent of chloride in seawater salt?
4. Several major shipping lanes cross the North Atlantic Ocean. If ships accidentally or deliberately discharge oil or plastics in the ocean, how would you expect these materials to circulate?
5. What is the conveyor belt circulation pattern of the world's oceans?
6. The concentration of dissolved potassium (K) in the ocean is 390 mg/kg. The atomic weight of potassium is 39.
   (a) What is the concentration of K^+ in seawater in moles kg^{-1}?
   (b) What is its concentration in parts per million by volume of seawater?
7. The average depth of the ocean is 3.8 kilometers and the average upwelling rate of deep water into the surface open ocean and into coastal environments is 4 m/yr.
   (a) About how long would it take to upwell the volume of the ocean?
   (b) The average nitrogen content of deep water is about 40×10^{-6} moles/L. What is the annual rate of addition of nitrogen to the surface water due to upwelling?
   (c) At a molar ratio of C:N of 106:16, what is the productivity in the global surface ocean in g C/m²/yr sustained by the upwelling of nitrogen?
8. One source of the deep water of the world's oceans is in the Norwegian Sea. Here, water is sufficiently dense to sink to the bottom. This water mainly forms from the cooling by evaporation of water carried northward by the Gulf Stream. The Gulf Stream carries heat to the high latitudes of the North Atlantic Ocean, which helps to moderate the climate of Europe. There is considerable interest in the rate of formation of North Atlantic Deep Water because of the role it plays in climate. Possible changes in the rate of its formation have been cited as the cause of rapid climate change in the past. Any global warming could modify the rate of deep-water formation.
   (a) How would you expect the residence time of the deep water to change from the Atlantic Ocean to the Pacific Ocean?
   (b) How does the deep water return to the surface?
   (c) If the continental glaciers were to begin melting because of a global warming, what would you expect might happen to the rate of deep-water formation and the climate of Europe?
9. During an El Niño year, how do sea-surface temperatures (SSTs), the depth of the thermocline, and rainfall differ from normal conditions over the eastern Pacific near Peru?
10. During an extreme La Niña year, how do the SOI, SSTs in the eastern Pacific, and the strength of the trade winds change from normal conditions?
11. A tsunami wave has a wavelength of 150 kilometers; what is its speed?
12. If all the ice in the Earth's cryosphere were melted, how high would sea level rise?
13. What are three ecological responses involving the biota to an El Niño event in the eastern Pacific near Peru?
14. Contrast the coupled ocean–atmosphere conditions of the natural oscillations of the NAO, AMO, and PDO.
15. Calculate the pH and DIC of seawater of salinity (S) 35 at 25°C and 1 atmosphere in equilibrium with the PCO_2 (partial pressure of CO_2 in the atmosphere) of the late pre-industrial atmosphere of roughly $10^{-3.5}$ atmosphere. You will need the following equations:

$$[H^+] = 10^{-14}/[OH^-] \qquad (1)$$

$$[H_2CO_3] = 10^{2.84 \times 10^{-2}}[PCO_2] \qquad (2)$$
(Henry's Law equation, H_2CO_3 is carbonic acid)

$$[H^+][HCO_3^-] = 10^{1.39 \times 10^{-6}}[H_2CO_3] \qquad (3)$$
(First dissociation constant for H_2CO_3, HCO_3^- is bicarbonate ion and CO_3^{2-} is carbonate ion)

$$[H^+][CO_3^{2-}] = 10^{1.19 \times 10^{-9}}[HCO_3^-] \qquad (4)$$
(Second dissociation constant for H_2CO_3)

$$[H^+] = [HCO_3^-] + 2[CO_3^{2-}] + [OH^-] \qquad (5)$$
(The charge balance equation)

You now have five equations with five unknowns. **Hint:** the master equation is the charge balance or charge conservation equation. Set this up in terms of solving for $[H^+]$. You will obtain a cubic equation. If you cannot solve the cubic equation, assume that the carbonate ion and hydroxyl ion in solution are negligible in concentration and approximate the solution.

Additional Sources of Information

Dudley, W. C. and Lee, M., 1998, *Tsunami!* 2nd ed. University of Hawaii Press, Honolulu, HI, 362 pp.

Glantz, M. H., 2001, *Currents of Change: The Impacts of El Niño and La Niña on Climate and Society.* Cambridge University Press, Cambridge, UK, 252 pp.

Laws, E. D., 1992, *El Niño and the Peruvian Anchovy Fishery.* University Corporation for Atmospheric Research, Global Change Instructional Program, Boulder, CO, 61 pp.

The Open University, 1989a, *The Ocean Basins: Their Structure and Evolution.* Pergamon Press, Oxford, UK, 171 pp.

The Open University, 1989b, *Ocean Chemistry and Deep-Sea Sediments.* Pergamon Press, Oxford, UK, 134 pp.

The Open University, 1989c, *Ocean Circulation.* Pergamon Press, Oxford, UK, 238 pp.

The Open University, 1989d, *Seawater: Its Composition, Properties and Behavior.* Pergamon Press, Oxford, UK, 165 pp.

The Open University, 1989e, *Waves, Tides and Shallow-Water Processes.* Pergamon Press, Oxford, UK, 187 pp.

Philander, S. G., 2004, *Our Affair with El Niño: How We Transformed an Enchanting Peruvian Current into a Global Climate Hazard.* Princeton University Press, Princeton, NJ, 275 pp.

Weiher, R. F. (ed.), 1999, *Improving El Niño Forecasting: The Potential Economic Benefits.* U.S. Department of Commerce, National Oceanic and Atmospheric Administration, Office of Planning and Strategic Planning, Washington, D.C., 57 pp.

CHAPTER

6 Our Living Planet: Earth's Ecosphere

Nothing can survive on Earth unless it is a cooperative part of a larger global life.

BARRY COMMONER, 1991

The ecosphere is the thin film around the planet where the biota interacts with the atmosphere, lithosphere, and hydrosphere in a complex system involving biological, geological, and chemical cycles (see Chapter 7). This biogeochemical system is powered by energy from the sun, with a small amount of energy derived from processes within the Earth. Matter circulates and energy flows throughout the ecosphere on local, regional, and ultimately global scales. This complex system is unique in that it supports the only known life in the universe, our biota.

The ecosphere exhibits large differences in weather and climatic regimes, length of daylight hours, topography, depth of the oceans, and elevation of the land surface. These physical factors influence the types of life that can live within a particular environment. The majority of life is restricted to a zone between 200 meters below the surface of the oceans to 6000 meters above sea level (Figure 6.1). In this narrow band, the biota sequesters carbon in organic (see Box 6.1: Organic Compounds) and inorganic compounds and, along with the lithosphere, atmosphere, and hydrosphere, helps to maintain the balance of carbon and other biologically reactive elements, like N, P, S, O, Ca, Si, and a number of trace metals. This dynamic system involving processes of nutrient transport, utilization and recycling, carbon uptake, respiration and decomposition, and elimination of wastes within a relatively narrow and stable range of environmental variables, e.g., solar radiation, temperature, and water availability, enables the maintenance and continuous evolution of life on Earth.

The number of biotic species in an area is a measure of the diversity of life on the planet, its biodiversity, and can be viewed as one measure of the "health" of the Earth system. The health of the ecosphere should be a primary concern of this and future generations, as the ecosphere is undergoing an assault on its integrity due to human activities that is unique in the history of the Earth.

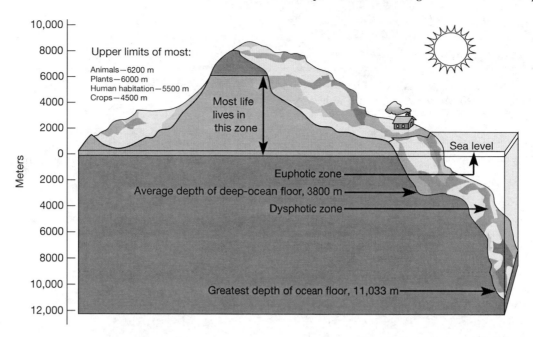

FIGURE 6.1 Vertical dimension of the ecosphere. Most life is restricted to the zone between 200 meters below sea level and 6000 meters above sea level. The thin euphotic zone (~100 meters) of the ocean is lighted and supports plant life, whereas the thick dysphotic zone is without the abundance of life found in the euphotic zone, although it is an important habitat for bacteria and nekton (fish).

CHEMISTRY OF BIOLOGICAL SYSTEMS AND CELLS

Biochemistry is that branch of chemistry dealing with the chemical properties, composition, and biologically related processes of complex substances in living systems. The subject matter forms an important part of the basis of gene science and genetics. Life processes are studied using biochemical tools. Biochemical processes are both influenced by chemical species in the environment and determine the nature of these species, their synthesis, and degradation. Molecules comprising organic matter (biomolecules) are often polymers with molecular masses on the order of a million or even larger. The biomolecules can be divided into the categories of carbohydrates, proteins, lipids, and nucleic acids. Proteins and nucleic acids consist of macromolecules; lipids are generally very small molecules, whereas carbohydrates range from small sugar molecules to high molecular mass macromolecules such as cellulose. Proteins are nitrogen-containing organic compounds that are the basic building blocks of living systems. Proteins are composed of amino acids linked together in long chains. Amino acids contain the carboxylic group, $-CO_2H$ and the amino group, $-NH_2$. Two amino acids illustrating their C-H-N-O configuration are shown in Figure 6.2a.

There are approximately 100 amino acids available in nature, of which the human body contains 20. Two major types of amino acids are recognized: those that are essential and must be obtained from food intake, and nonessential amino acids that are manufactured within the organism. Cells use amino acids to build enzymes and structural proteins. Carbohydrates have the relatively simple chemical formula CH_2O. Glucose, as an example, is shown in Figure 6.2b.

Lipids are substances that can be extracted from plant or animal tissue by organic solvents such as chloroform. Lipids are defined mainly by their physical characteristics. Common lipids are fats and oils composed of triglycerides formed from the alcohol glycerol and a long-chain fatty acid. **Deoxyribonucleic acid (DNA)** and **ribonucleic acid (RNA)** are nucleic acids that store and transport genetic information controlling reproduction and protein synthesis. The molecules of DNA are huge, with a molecular weight greater than 1 billion. DNA has the structure of the famous double helix. The DNA structure was figured out by James D. Watson and Francis

BOX 6.1

Organic Compounds

Carbon-containing compounds of which life is composed are organic chemicals. Other than carbon, virtually all organic compounds contain hydrogen and have at least one carbon–hydrogen (C–H) bond. The three-dimensional shape of an organic molecule is particularly important because this molecular geometry determines in part the behavior of the organic compound. There is an enormous number of organic compounds because of the ability of carbon atoms to bond covalently to each other through single (two shared electrons), double (four shared electrons), and triple (six shared electrons) bonds found in an endless variety of straight and branched chain and ring structures. Most organic compounds can be divided into six classes—hydrocarbons, oxygen-containing compounds, nitrogen-containing compounds, sulfur-containing compounds, phosphorus-containing compounds, and organohalides—or combinations of these compounds. The following figure shows examples of the structures in two dimensions of the major types of organic compounds that belong to the hydrocarbon class of organic compounds containing only carbon and hydrogen atoms.

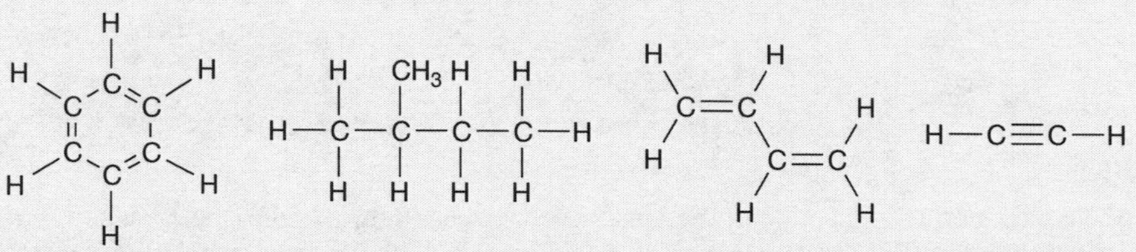

Benzene (aryl compound)   2-Methylbutane (alkane)   1,3-Butadiene (alkene)   Acetylene (alkyne)

It is not possible to discuss in detail the great variety and many classes of organic compounds. Suffice it to look at one class and compounds in that class as an example of the chemistry of an environmentally important organic compound that is implicated in the destruction of stratospheric ozone (Chapters 4 and 14). The organohalides are organic compounds that consist of halogen-substituted hydrocarbon molecules, each of which contains at least one atom of fluorine (F), chlorine (Cl), bromine (Br), or iodine (I). One important manufactured product in this class is the chlorinated hydrocarbons that are widely distributed and regarded as environmental pollutants or hazardous wastes. Chlorofluorocarbons (CFCs) are volatile 1- and 2-carbon organohalide compounds that contain Cl and F bonded to carbon. These compounds are highly stable and nontoxic. The most widely manufactured of these compounds are CCL_3F (CFC-11, Freon-11), CCL_2F_2 (CFC-12, Freon-12), $C_2Cl_3F_3$ (CFC-113), $C_2Cl_2F_4$ (CFC-114), and C_2ClF_5 (CFC-115). As an example, the structure of Freon-12, dichlorodifluoromethane, follows.

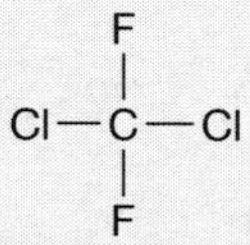

Dichlorodifluoromethane
(Freon-12, boiling point = –29°C)

In past decades, the CFC compounds were widely used in the fabrication of flexible and rigid foam materials and as fluids for refrigeration and air-conditioning units. It is estimated that more than 85% of the freons (the fluorocarbons) produced were released to the atmosphere. Halons are organohalide compounds related to CFCs that contain bromine and are used in fire extinguisher systems. They include $CBrClF_2$ (Halon-1211), $CBrF_3$ (Halon-1301), and $C_2Br_2F_4$ (Halon-2402). Production of CFCs and halocarbons in the United States was curtailed starting in 1989 after the United States Environmental Protection Agency developed regulations in accordance with the 1986 Montreal Protocol on Substances That Deplete the Ozone Layer (Chapter 14).

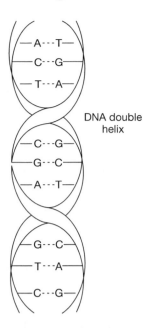

DNA double helix

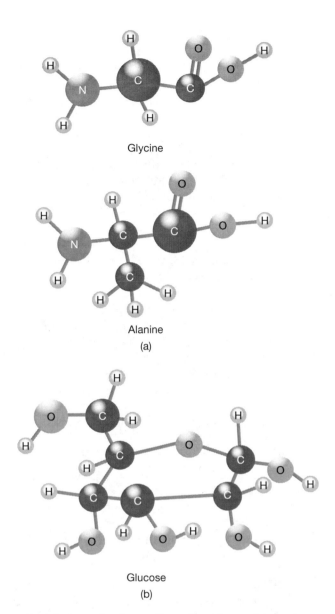

Glycine

Alanine

(a)

Glucose

(b)

FIGURE 6.2 Molecular configuration of two of the amino acids: glycine and alanine (a) and that of the simple sugar glucose (b).

FIGURE 6.3 Structure of the deoxyribonucleic acid (DNA) double helix. DNA consists of two long polymers of simple nucleotide units with backbones made of alternating sugar and phosphate groups joined by ester bonds. The bases of adenine (A), guanine (G), cytosine (C), and thymine (T) are attached to the sugar/phosphate molecules to form the complete nucleotide of base, sugar, and phosphate.

Crick, for which they received the Nobel Prize in 1962. Basically, the structure of DNA is that of two spiral ribbons counterwound around each other. A representation of the double-helix structure of DNA is shown in Figure 6.3. The nitrogen-containing base pairs held together by hydrogen bonding (dashed lines) of adenine (A), cytosine (C), guanine (G), and thymine (T) are shown between the phosphate/sugar polymer backbones of the two strands of DNA.

The double-helix structure of DNA enables storage and replication of genetic information. The genetic information contained in and transmitted by nucleic acids depends on the sequence of bases of which the nucleic acids are composed. The DNA double helix may unravel, producing a strand of RNA. The RNA then travels from the nucleus of a cell out into the cell and regulates the synthesis of new protein. This is the way in which DNA regulates the function of a cell and acts to control life processes. Whenever a new cell is formed, the DNA of its nucleus must be accurately reproduced from the parent cell. The DNA in a single cell can be responsible for controlling the synthesis of more than 3000 proteins. Directions for protein synthesis are contained in a segment of DNA called a gene. A **gene** is the basic unit of heredity in a living organism and is a segment of nucleic acid that when taken as a whole specifies a trait in an organism. In the cell, a gene is a portion of the DNA containing sequences that code what a gene does, and noncoding sequences that determine when the gene is active. DNA molecules may be modified, leading to mutations. Among other factors, chemical substances and radiation from X-rays and radioactivity may alter DNA and lead to mutations.

At the microscopic level, life is composed of cells. In the human body, there are 10 trillion cells divided into about 200 different types, including cells of the muscle, the liver, and the skin. The largest human cells are about the diameter of a hair but most are much smaller, on the order of 10 micrometers (microns). A bacterium is a very simple organism when viewed at the cellular level being a single, self-contained living unit, the cell (Figure 6.4). Thus, a bacterium makes a good template for understanding the parts of a cell.

The bacterium species *Escherichia coli* (*E. coli*) consists of an outer wrapper, cell membrane, within which is a fluid called the cytoplasm containing nearly 70% water. The remaining 30% consists of about 1000 enzymes and smaller molecules like amino acids, glucose molecules, and ATP (adenosine triphosphate, a multifunctional nucleotide acting as a coenzyme transporting chemical energy within cells for metabolism). Within the *E. coli* cell membrane, near its center, is a ball of DNA. Stretched out, the ball of DNA would be 1000 times longer than the three-micrometer length of the bacterium. *E. coli* propels itself by use of the long strands of flagella attached outside the cell to the cell membrane. Cells of higher trophic level organisms, like humans, are much more complex and contain a greater variety of additional membranes and structures than those of *E. coli*.

Enzymes are very important to life processes. An enzyme is formed from amino acids strung together in a specific and unique order and hence is a protein. Enzymes allow the cell to carry out chemical reactions at rapid rates; a pure inorganic chemical reaction of like nature would progress less quickly, if at all. The enzyme catalyzes the reaction, and the amino acids that constitute the enzyme are folded into a unique shape which enables the enzyme to carry out a specific chemical reaction acting as a catalyst in that reaction. For example, maltese is an enzyme and is shaped in such a way that it can break apart the two glucose sugar molecules found in the sugar maltose. This activity is specific to maltese and it is the only thing the maltese can do as an enzyme.

CLASSIFICATIONS OF THE BIOTA

The rich abundance of different life-forms found on this planet is commonly referred to as **biological diversity** or **biodiversity.** Hundreds of thousands of different species have been described to date, and the number is continuously increasing. This diversity has necessitated arranging these organisms in some sort of order. Classification, or **taxonomy,** is the science of arranging organisms according to some sort of scheme of differences and likenesses among the various groups. Life-forms can be classified on the basis of their genealogical (ancestry) and evolutionary relationships, common morphologic and genetic characteristics or qualities, and their function in the ecosphere. The following section discusses a classification based on genealogy and evolutionary relationships. Later, the way in which organisms function is presented in the context of their role in ecosystems.

Genealogy and Evolutionary Relationships

One classification of organisms is given in Figure 6.5. In this figure, life is divided into three principal types. One group of organisms is the viruses. The other two groups, the prokaryotes (bacteria, cyanobacteria) and the eukaryotes (all other life-forms), are further subdivided into kingdoms.

VIRUSES   Viruses are acellular organisms consisting of only a small amount of protein and genetic material in the form of complex organic substances of high molecular weight RNA (ribonucleic acid) and DNA (deoxyribonucleic acid). Viruses act as parasites and live on materials of other life. They are the smallest infectious agents capable of replicating themselves in the living cells of their host. The majority of viruses are found within the size range of about 0.02 to 0.25 micrometer (one micrometer = one millionth of a meter). Viruses are the cause of a number of diseases and infections including types of influenza like swine flu, mumps, smallpox, poliomyelitis, and many more. All life other than viruses is cellular.

PROKARYOTES: KINGDOM MONERA   The prokaryotes, bacteria and cyanobacteria, are the only members of the kingdom Monera. They have a cell

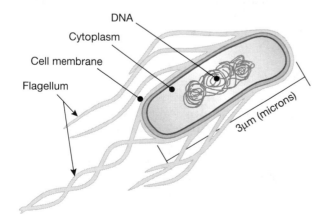

DNA
Cytoplasm
Cell membrane
Flagellum

3μm (microns)

FIGURE 6.4   The cell of the *Escherichia coli* bacterium showing its cellular components.

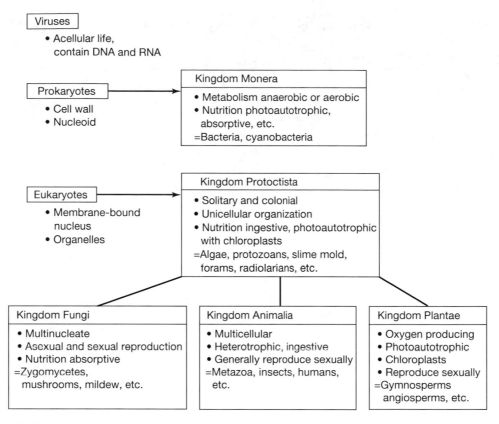

FIGURE 6.5 The biological classification of organisms; compare with Figure 6.6. (*Source:* Stolz, et al., 1989.)

wall but the nucleus of the cell is not bound by a membrane, as it is in eukaryotic organisms. The prokaryotes are exceedingly important organisms in that they take part in a variety of biogeochemical processes that cycle elements about the surface of Earth (see Chapter 7). These processes involve a number of biochemical pathways including (1) **autotrophy** in which inorganic carbon in the environment is converted into organic materials, (2) **heterotrophy** in which organic substrates are used to make organic matter, and (3) **mixotrophy** in which both organic and inorganic compounds are used by an organism to manufacture organic tissues. For carbon, the biogeochemical processes include (1) the fixation of carbon dioxide from the atmosphere during photosynthesis and its release during respiration, (2) the production and oxidation of methane by bacteria, and (3) the fermentation of sugars to carbon dioxide by bacteria (Table 6.1).

Prokaryotic bacteria are also involved in the processes of reduction of sulfate to sulfide and in the oxidation of sulfide to elemental sulfur and sulfate

(see Box 2.4: Ions and Oxidation-Reduction). For nitrogen, the processes of nitrogen fixation, nitrification, and denitrification (Table 6.1) are all mediated by bacteria. These processes occur in both terrestrial and aquatic environments. Human interference in the cycling of chemical substances in these environments is affecting the ability of prokaryotes to carry out their functions in the natural system.

In the 1980s and 1990s, there was what could be called a revolution within the larger molecular and genetic revolution in biology. This is the increasingly rapid capacity to sequence proteins and nucleic acids. The sequencing approach provides a new and powerful means for determination of evolutionary relationships among organisms. It has led to the development of the universal phylogenetic tree (Figure 6.6) that apparently encompasses all of modern life-forms. Although still somewhat controversial, the tree demonstrates the recognition that the Monera are considerably more complex phylogenetically than shown in Figure 6.5. On the basis of the development of RNA patterns, this kingdom has been divided into

TABLE 6.1 Some Reactions Involving Organic Matter (Simply Taken as CH_2O), Including Photosynthesis and Remineralization in Which Oxygen, Nitrate, Iron Oxide, Sulfate, and Water Act as Oxidizing Agents

Process	Reaction
1. Photosynthesis organic matter	$CO_2 + H_2O = CH_2O + O_2$
2. Remineralization organic matter	$CH_2O + O_2 = CO_2 + H_2O$
3. a. Denitrification	$5CH_2O + 4NO_3^- + 4H^+ = 5CO_2 + 2N_2 + 7H_2O$
b.	$2CH_2O + 2NO_3^- + 2H^+ = 2CO_2 + N_2O + 3H_2O$
c.	$2CH_2O + 2NO_3^- = N_2O + 2HCO_3^- + H_2O$
4. a. Fe^{3+} iron reduction	$CH_2O + 4Fe(OH)_3 + 8H^+ = CO_2 + 4Fe^{2+} + 11H_2O$
b.	$CH_2O + 4FeOOH + 8H^+ = CO_2 + 4Fe^{2+} + 7H_2O$
c.	$CH_2O + 7CO_2 + 4Fe(OH)_3 = 4Fe^{2+} + 8HCO_3^- + 3H_2O$
d.	$2Fe(OH)_3 + H_2 = 2Fe(OH)_2 + 2H_2O$
5. a. Sulfate reduction	$3CH_2O + 4H^+ + 2SO_4^{2-} = 3CO_2 + 5H_2O + 2S$
b.	$2CH_2O + 2H^+ + SO_4^{2-} = 2CO_2 + 2H_2O + H_2S$
c.	$2CH_2O + SO_4^{2-} = 2HCO_3^- + H_2S$
6. a. Methanogenesis	$2CH_2O = CO_2 + CH_4$
b.	$3CH_2O + H_2O = 2CO_2 + CH_4 + 2H_2$
c.	$CH_3COOH = CO_2 + CH_4$

(*Source:* Mackenzie and Lerman, 2006.)

two major groups, the Archaea and the Bacteria and, in turn, many phyla are recognized within these groups. The Archaea appear to have given rise to both the Bacteria and the Eucarya but the common ancestor of all three groups has not been described. Bacteria that live in extremely salty and hot environments (the extreme halophiles and extreme thermophiles, respectively) and methanogenic bacteria (those that convert organic matter to methane, Table 6.1) are examples of Archaea. In 1982, a species of Archaea, *Methanococcus jannaschii,* was found at a hydrothermal vent 2.5 kilometers deep in the Pacific Ocean off Mexico. In 1996, it was determined that at least half of its 1738 genes are unknown in modern plants, animals, or bacteria.

The Bacteria include the photosynthesizing cyanobacteria (blue-green algae), the photosynthesizing green and purple sulfur bacteria that oxidize reduced sulfur to sulfate (SO_4^{2-}) (e.g., *Chlorobium*), and those that reduce sulfate to hydrogen sulfide (H_2S Table 6.1) (e.g., *Desulfovibrio*). On the early Earth

devoid of oxygen, the Archaea and the Bacteria probably flourished in the harsh environmental conditions of the time (see Chapter 8). In the modern world, both the sulfur-oxidizing and the sulfate-reducing bacteria live in habitats that are anaerobic (without oxygen), as do the methanogenic bacteria.

EUKARYOTES Eukaryotes are organisms in which most of the genetic DNA is found within a membrane-bound nucleus of the cell. Outside the nucleus, organ-like structures called organelles are found. Mitochondria that produce energy for the cell and chloroplasts that are sites of photosynthesis and starch formation are examples of organelles. The eukaryotes are divided into four kingdoms. They are the Protoctista, Fungi, Plantae, and Animalia kingdoms. Some characteristics of organisms in these kingdoms are given in Figure 6.5.

Kingdom Protoctista The Protoctista have a diverse number of ways in which they obtain materials

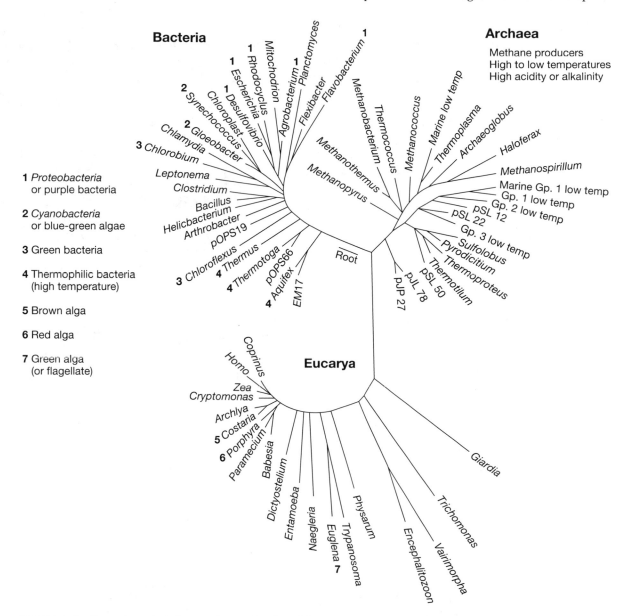

Bacteria

Archaea
Methane producers
High to low temperatures
High acidity or alkalinity

1 *Proteobacteria*
  or purple bacteria

2 *Cyanobacteria*
  or blue-green algae

3 Green bacteria

4 Thermophilic bacteria
  (high temperature)

5 Brown alga

6 Red alga

7 Green alga
  (or flagellate)

Eucarya

FIGURE 6.6 The universal phylogenetic tree developed from sequencing a specific type of RNA (ribonucleic acid), ribosomal ribonucleic acid (rRNA). rRNAs at present are the most useful and most used of the molecular chronometers, a molecule whose sequence changes randomly in time. (*Source:* Pace, 1997.)

for energy and growth. They absorb substances from the environment, as well as ingesting them or producing them from inorganic substances and light energy in the process of **photoautotrophy.** These life-forms include algae, protozoa, primitive fungi with flagella, and slime molds. Foraminifera and radiolarians are important protozoans that are part of the plankton of the sea. The former build shells of calcium carbonate and the latter construct intricate skeletons of spicules made up of silica. These protists are important to the cycling of carbon and silica in the ocean.

Kingdom Fungi The Fungi are heterotrophic, living on decaying organic matter and absorbing substances from their environment. They include mushrooms, bread molds, and mildews. Some fungi that create microscopic channels in hard, dense materials (bore into them), in cooperation with boring algae, aid in weathering and erosion of hard substrates like rock surfaces on land and corals growing in the sea. They also are important in the decomposition of plant and animal litter found on forest floors and the cycling of nutrient substances produced by the decay process.

Kingdom Plantae The Plantae are principally photoautotrophic. This kingdom includes the mosses and the vascular angiosperm and gymnosperm plants. These organisms are especially important in the cycling of water and the bioessential elements of carbon, nitrogen, phosphorus, and sulfur about Earth's surface. They are involved in the processes of transpiration and oxygen-producing photoautotrophy. They also help to prevent the erosion of soils by stabilizing them with their root systems. Furthermore, plants play a role in climate through their effects on the water cycle and by affecting the amount of radiant energy reflected from the surface of Earth, the planet's albedo.

Kingdom Animalia The kingdom Animalia includes exclusively multicellular eukaryotes. These life-forms are heterotrophic and ingest organic compounds as a source of energy and for growth. Animals are an important component of food chains, generally always occupying the upper trophic levels (discussed later). They range in size from small crustacea to whales and elephants. They are important in the process of biomineralization in which elements like calcium and phosphorus are used to make bone and calcium and carbon are used to make the skeletons of corals and mollusks. *Homo sapiens* (modern human) is a major life-form of the Animalia and is responsible for many modifications of life and element cycles involving organisms of the rest of the ecosphere.

Although nearly every day we learn more about human evolution, it appears that human lineage began at least 8 million years ago. At that time, there was the divergence of the chimpanzees and the australopithecines from a common ancestor. The australopithecines are a complex group of archaic hominids from which the genus *Homo* diverged. *Homo habilis*, a hominid form that could walk upright, emerged more than 3 million years ago, probably in the Rift Valley of eastern Africa. The ultimate emergence of *Homo* began with a tectonic event about 8 million years ago that led to the development of the Rift Valley and to mountains that constitute the present-day western rim of the valley. Prior to this time, with no Rift Valley and mountains present, equatorial Africa was nearly climatically and biologically homogeneous from west to east, from the Atlantic to the Indian Ocean. With the development of the Rift mountains, differences emerged between western and eastern equatorial African climate and hence ecosystems. The west remained humid and wet and covered by forests and woodlands, while the east became dry and a region of open savanna grasslands with sparse trees. It is possible that the population of the common ancestor of the chimpanzees and the hominids also found itself divided. The western population eventually led to the development of the chimpanzees, and their relatives adapted to life in a humid, wet, arboreal environment, while the eastern population adapted to the aridity and the openness of the savanna environment. In the latter milieu, *Homo* eventually emerged, including *Homo neanderthalensis* and *Homo erectus*. The emergence and growth in the population of modern humans, *Homo sapiens*, and its ability to manipulate and manage the natural environment have been unprecedented features of organic evolution. How this species can avoid actions that are destructive to the biosphere and at the same time preserve its own interests are at the heart of the environmental issues of modern society.

Classification within a Kingdom

Table 6.2 gives the terminology used in describing the hierarchy of the taxonomic units of classification within the kingdoms. The fundamental category of taxonomy is the species. A **species** is a single kind of plant or animal in which any variations among individuals are not regarded as affecting the essential sameness that distinguishes them from other organisms. This means that there is free gene flow between individuals of a species under natural conditions. In other words, all the healthy individuals of a species can breed with other individuals of the opposite sex

TABLE 6.2 The Hierarchy of the Taxonomic Units of Classification

Larger units ↑ — Kingdom, Phylum, Class, Order, Family, Genus, Species — Larger number of shared characteristics ↓

of that species, and they do not breed with members of other species.

The next higher level of biological classification is that of genus, which is the major subdivision of a family or subfamily of organisms and usually consists of more than one species. The genus designation is the first part of the scientific name of a species. For example, *Acropora palmata* is the scientific name for the elkhorn coral, one species within the genus *Acropora*. The next higher level is the family, followed by order, class, phylum, and finally kingdom. In the hierarchy of taxonomic units, the subdivisions from species to kingdom represent increasingly larger units with a decreasing number of shared characteristics.

ECOSYSTEMS AND THEIR DYNAMICS

To gain a firmer understanding of the workings of the ecosphere, let us look at organisms and the relationships they have with each other and their physical environment. Where an organism lives is its habitat, and what it does or how it interacts with its habitat is its niche. As a practical example, a lecture hall may be the habitat for students attending a physics colloquium. What role each student plays in the lecture hall is his or her niche. This role includes how one interacts with other students, where one sits, and how one performs or responds during the colloquium. There are specific niches and habitats for some organisms. However, some other organisms are generalists (e.g., cockroaches) and may occupy a variety of habitats or niches.

Organisms of the same species living in a specific area are populations of that particular species. Populations never live in isolation but interact with other populations. A group of plant and animal populations living together in the same region is a community. Specific types of plants and animals live together within a particular community. For example, deer and mountain lions can coexist. In contrast, cacti and reindeer never share the same community.

Communities of organisms that interact with one another, as well as with their physical and chemical environment, in such a way as to sustain a system are collectively termed an **ecosystem**. These dynamically balanced systems are found from the tallest peaks to the floors of the deepest oceans. Ecosystems come in many sizes, exhibit variations in life-forms, and have distinct chemical and physical properties, such as those found in desert, forest, grassland, tide pool, stream, or pond. Earth has a multitude of naturally balanced ecosystems. As one

progresses toward the equator, the complexity of ecosystems increases from a few species and communities in the polar regions to multiple, diverse communities at the equator. Ecosystems tend to merge gradually into one another geographically. The diffuse boundary between ecosystems is termed an **ecotone**. An ecotone may create an unique habitat of its own and contain a mixture of plant and animal species that do not match those of the ecosystems on either side of the boundary. For example, a deciduous forest ecosytem contains forest tree species that lose their leaves every winter and a grassland ecosystem contains a variety of grass species. The boundary between the two, the grassland–forest ecotone, could contain a mixture of species from the two ecosystems, plus perhaps some species not found in either ecosystem. Another example is the transitional zone between a terrestrial ecosystem and a lake ecosystem in which the transitional zone is a marshland ecotone with its own unique species. The many ecosystems and ecotones coupled together sustain the larger, complex, and intricately interlinked global ecosystem—the ecosphere.

Components of Ecosystems

Every ecosystem, in order to function properly, must have an energy and nutrient source and maintain a relationship between its biotic and abiotic components that allows the processes of energy flow and nutrient cycling to go on efficiently. These dynamically balanced components lead to sustainability of the ecosystem.

ENERGY FLOW Energy is defined as the ability or capacity to do work. There are many forms of energy such as kinetic (energy of motion) and potential (stored energy). The movement of energy is described by basic scientific observations called the Laws of Thermodynamics. The first law states that energy can neither be created nor destroyed, but is only transferred from one form to another. The second law states that when energy changes from one form to another, part of it is always converted into unusable waste heat. As a result, a constant supply of energy is needed to sustain a system. The waste heat lost in each transformation must be replaced.

Energy flows through an ecosystem in a series of transformations. The light energy of the sun enables plants to make organic tissue from carbon dioxide, water, and inorganic nutrients through the process of photosynthesis. The light energy is changed to chemical energy in plant cells and used for growth or stored in the plant tissue. When the plant dies and decomposes, or is eaten by a consumer, energy stored in the plant is transferred.

The ultimate source of energy for animals is the plant. Animals require energy to convert nutrients from their food into body tissue because they cannot make body tissue directly from the energy of the sun. When plants are eaten, a small portion of the energy stored in plants is transferred to animals for growth, maintenance, and performance of activities. When animals are food for other animals, another transfer of energy occurs. With each transfer, energy is lost to waste heat and ultimately radiated back into space as infrared radiation.

When animals use energy stored in their bodies, inorganic compounds are released through body excretions. When an animal dies, decomposition processes convert its organic compounds into inorganic compounds. These inorganic compounds are one source of nutrients that are used by plants.

NUTRIENT CYCLING Elements and compounds are cycled through the organic and inorganic systems. Their cycling is essential for life to exist. The complexities of the biogeochemical cycles and their involvement in the sustainability of life are topics of great scientific interest and practicality. Knowledge of these topics is necessary to evaluate the effects of global change on the ecosphere.

Of the 94 naturally occurring elements on Earth, 17 are essential for the growth of most plants (Table 6.3). Life could probably not exist without each of these elements. Six of the 17 elements—nitrogen (N), carbon (C), hydrogen (H), oxygen (O), phosphorus (P), and sulfur (S)—represent approximately 95% of all matter in plants, animals, and microorganisms. In addition, the essential elements must be present in a form that the organism can use. These compounds are nutrients. For example, the carbon (C) in carbon dioxide (CO_2) is the only major form of carbon that plants can use. In general, they cannot use carbon directly from other compounds. Water (H_2O) provides the only form of hydrogen (H) that a plant can use. The deficiency or lack of any one of the 17 nutrients can be a limiting factor for organic growth in an ecosystem. The one nutrient, the **limiting nutrient**, that is not provided, or is deficient, may limit the development of an organism in some way.

Nutrient cycling is a critical function of life systems. The nutrients are cycled throughout the ecosphere in a series of interwoven biogeochemical cycles. These highly important cycles are discussed more thoroughly in Chapter 7.

RELATIONSHIP BETWEEN ABIOTIC AND BIOTIC COMPONENTS An ecosystem has abiotic and biotic components that interact with each other (Figure 6.7). The abiotic components of an ecosystem are all the chemical and physical parts of the environment, such as nutrients and their availability, soils, temperature, water, and sunshine. Every organism must adapt to the abiotic factors of an area or it cannot live in that area. Change in a single abiotic factor may lead to collapse of the system. This factor may be a limiting factor for an ecosystem. For example, the availability of water determines the types and extent of plant life in a desert, while the supply of nitrogen or phosphorus as nutrient compounds determines the extent of plant growth in the oceans and in many terrestrial environments. Disruptions in the supply of these life-essential materials can lead to ecosystem degradation and ultimately to collapse.

The members of the biota may be categorized by their function in an individual ecosystem and include producers, primary and secondary consumers, and decomposers (Figures 6.7 and 6.8). These organisms form part of a cycle, a food chain. A food chain results in the transfer of energy and nutrients from one organism to another within a particular ecosystem. A more complex intertwining of individual food chains in an ecosystem is a food web. The feeding level occupied by an organism in a food chain is called its **trophic level** (Figure 6.9). The first trophic level contains the producers, the plants. Primary consumers, or plant eaters (herbivores), occupy the second trophic level, and the secondary consumers, the meat eaters (carnivores), occupy the higher trophic levels. Humans occupy the higher trophic levels, existing as carnivores, as herbivores (vegetarians), and as omnivores (eating both meat and plants).

Primary Producers First, there are the primary producers (autotrophs; e.g., the green plants; Figure 6.8). The autotrophs evolved before animals because these organisms can get nourishment from inorganic compounds to form their tissues. Animals generally cannot extract sufficient nutrients directly from soil and water in order to survive. Land and aquatic plants have this ability in the process of photosynthesis (Figure 6.10; see Box 6.2: The Photosynthetic Process for details of the photosynthetic process).

Land plants are able to make their own food (carbohydrates—organic carbon compounds) from

(text continues on Page 144)

TABLE 6.3 The Essential Elements Necessary for Plant Growth

% of Plant Matter	Essential Element	Available from Nutrients	Present in	Role in Plants
95%	Carbon, C	Carbon dioxide, CO_2	Air or dissolved in water	Essential in the structure of all organic molecules
	Hydrogen, H	Water, H_2O	Water	Essential in the structure of all organic molecules
	Oxygen, O	Carbon dioxide, CO_2	Air	Essential in the structure of most organic molecules
	Nitrogen, N	Nitrate, NO_3^- Ammonium, NH_4^+	Some soil minerals or dissolved in water	Essential in the structure of all proteins, nucleic acids, and some other organic molecules
		Nitrogen gas, N_2, but only by nitrogen fixation	Air	
	Sulfur, S	Sulfate, SO_4^{2-}		Essential in the structure of most proteins
	Phosphorus, P	Phosphate, PO_4^{3-}		Essential in the structure of nucleic acids and in energy transfers
	Potassium, K	Potassium ions, K^+		Essential in maintaining water balance; necessary in certain enzymes
5%	Calcium, Ca	Calcium ions, Ca^{2+}		Essential in maintenance of membrane function; essential in cell walls of most plants
	Magnesium, Mg	Magnesium ions, Mg^{2+}		Essential in chlorophyll molecules
	Iron, Fe	Iron ions, Fe^{2+}, Fe^{3+}	Some soil minerals or dissolved in water	Essential in photosynthesis and in energy releasing reactions
	Manganese, Mn	Manganese ions, Mn^{2+}		
	Boron, B	Boron ions, B^{3+}		
	Zinc, Zn	Zinc ions, Zn^{2+}		
	Copper, Cu	Copper ions, Cu^{2+}	(Mn, B, Zn, Cu, Mo, Co)	All essential for the function of certain enzymes
	Molybdenum, Mo	Molybdenum ions, Mo^{2+}		
	Cobalt, Co	Cobalt ions, Co^{2+}		
	Chlorine, Cl	Chloride ions, Cl^-		Essential in photosynthesis

(*Source:* Nebel, 1981.)

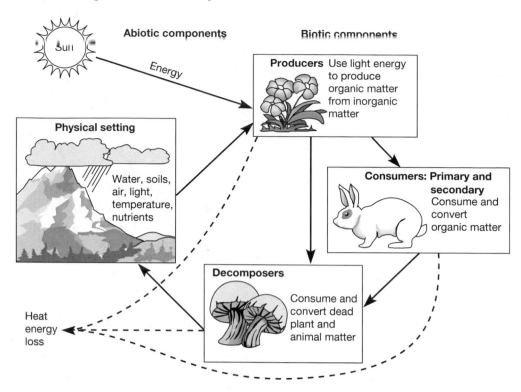

FIGURE 6.7 The biotic and abiotic components constituting an ecosystem structure and their interactions. The relationship among producers, consumers, and decomposers is critical to functioning of the system. Light energy is degraded as it flows through the system.   (*Source:* Nebel, 1981.)

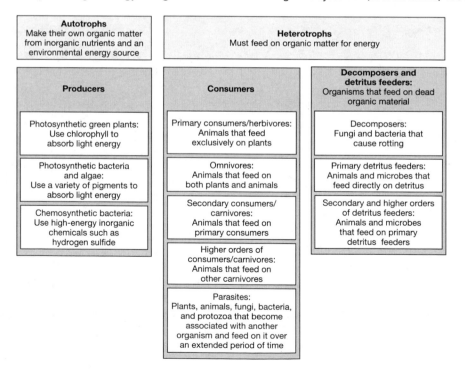

FIGURE 6.8 Autotrophs and heterotrophs as producers, consumers, or decomposers of organic matter. The characteristics of each trophic group is also shown.   (From *Environmental Science: The Way the World Works, 6E* by Bernard J. Nebel and Richard T. Wright, © 1998. Electronically reproduced by permission of Pearson Education, Inc., Upper Saddle River, New Jersey.)

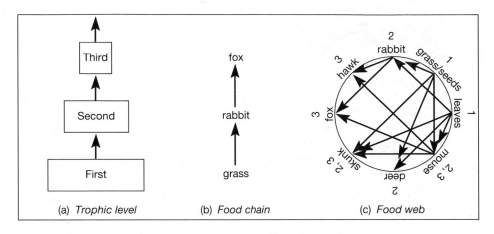

FIGURE 6.9 The transfer of nutrients and energy is represented by (a) trophic levels, (b) a food chain, and (c) a food web. In (c), numbers refer to the trophic levels occupied by each member of the food web.

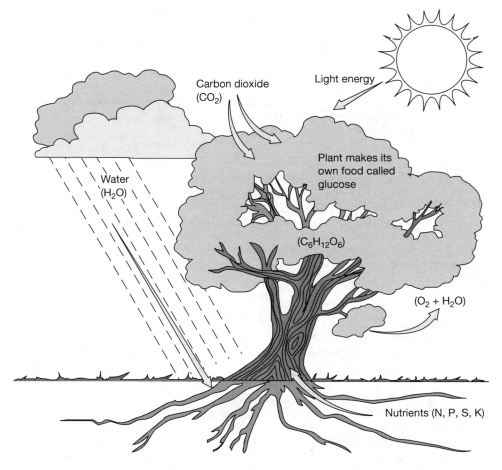

FIGURE 6.10 The process of photosynthesis, in which carbon dioxide from the air and water and nutrients from the soil react in the presence of light energy from the sun to make organic matter in land plants ($6CO_2 + 12H_2O \rightarrow C_6H_{12}O_6 + 6O_2 + 6H_2O$).

BOX 6.2

The Photosynthetic Process

There are few biological mechanisms on Earth that can reduce inorganic carbon to organic carbon. On a global basis, photosynthesis is the most important and is one of the principal achievements of biological evolution. The reverse reaction of respiration and decomposition is complementary to photosynthesis in that it returns carbon and nutrients to the environment to be used again in production of organic matter. The production of organic matter by freshwater and marine planktonic organisms involves carbon, nitrogen, and phosphorus in the atomic ratio $C:N:P = 106:16:1$, named the Redfield ratio after its discoverer Alfred Redfield. The reaction involving CO_2, nutrients of N and P, and H_2O is

$$106CO_2 + 16NO_3^- + HPO_4^{2-} + 18H^+ + 122H_2O = (CH_2O)_{106}(NH_3)_{16}(H_3PO_4) + 138O_2.$$

Oxygen is a by-product of the photosynthetic reaction. The reverse reaction, respiration or oxidation of organic matter, proceeds from the right to the left, utilizing O_2 and producing CO_2, nitrate, phosphate, water, and hydrogen ions. In land plants, the average $C:N:P$ atomic ratios vary significantly. Land-plant photosynthesis produces organic matter with a relatively much higher concentration of carbon than found in the aquatic plants, in such a reaction as, for example:

$$1000CO_2 + 8NO_3^- + HPO_4^{2-} + 19H^+ + 2074H_2O = (CH_2O)_{1000}(NH_3)_8(H_3PO_4) + 2091O_2.$$

Photosynthesis in terrestrial and aquatic plants involves net transport of carbon from the surrounding medium into the plant cell. For terrestrial plants, the medium is the atmosphere with its gaseous CO_2; for aquatic plants, the medium is the water with its dissolved inorganic carbon. The site of carbon fixation in any eukaryotic plant cell is the chloroplast and its smaller morphological units of grana, thylakoids, and stroma surrounding the chloroplast and where CO_2 is stored. Inorganic carbon may cross the membranes simply by diffusion along a positive diffusion gradient (see Chapter 3) from the external medium to the site of carbon fixation, the chloroplast. In this case, because CO_2 is the only major carbon species that can freely cross the chloroplast membranes due to the fact that it is an uncharged molecule, the gas must dissolve in the membrane lipid phases, diffuse across the membranes, and dissolve back into the aqueous phase of the chloroplast. In terrestrial plants with openings in the plant epidermis, known as stomata, the rate of gaseous diffusion to the cell walls of the mesophyll is enhanced through the stomata and is higher than their total area would suggest; hence the concentration of CO_2 in the intercellular spaces of the mesophyll is usually very close to that of the ambient air for most plants.

In aquatic plants the slow rate of equilibration between CO_2 and the chemical species H_2CO_3 and HCO_3^- is rapidly catalyzed by an enzyme, carbonic anhydrase, which promotes the conversion of HCO_3^- to CO_2 for photosynthesis. Carbonic anhydrase is a group of related enzymes, all of which use Zn in their active site. It is one of the most catalytically reactive enzymes known. The activity of the enzyme is located mainly on the plasmalemma and as a soluble, extracellular enzyme. Although in some unicellular algae and macrophytes, the rates of diffusive fluxes of CO_2 seem adequate to support photosynthetic demands, it appears that in the majority of aquatic plants, there are carbon-concentrating mechanisms (CCMs) that actively transport carbon to the site of fixation. The location and mechanism of this active carbon pump have been difficult to pinpoint; however, the plasmalemma appears to be the only possible location and semicrystalline arrays of Rubisco associated with carbonic anhydrase activity appear to be involved in the carbon-concentrating mechanism.

Photosynthesizing organisms use preferentially the light carbon isotope ^{12}C in making their organic matter (see Chapter 2, Box 2.3: Stable Isotopes), which is thought of as an evolutionary adaptation that uses the lighter-isotope compounds where bond energies are lower and reaction rates faster than in their heavy-isotope equivalents. A typical value of biological fractionation is, for example, the difference between $\delta^{13}C$ of atmospheric CO_2 and that of the C_3 land plants: fractionation factor $\varepsilon^{13}C \approx (-7\text{‰}) - (-27\text{‰}) = 20\text{‰}$. Plants build their tissue from CO_2 that is converted to the basic organic form of $C_6H_{12}O_6$ (glucose) by three main photosynthetic pathways. The more widespread C_3 pathway (called C_3 because the initial stable product of this photosynthetic pathway is a 3-carbon compound 3-phosphoglycerate) characterizes 85% to 96% of known plant species. The C_3 plants are mainly trees, shrubs, and cool-climate grasses. Forests, shrub land, and prairies are generally C_3 plants that respire up to 50% of the photosynthetically fixed carbon. The C_4 pathway (where the first stable end product is a 4-carbon compound oxaloacetate) is common to 3% to 14% of plant species, and the crassulacean acid metabolism pathway (CAM) is exhibited by only 1% of plant species. About 7500 species of flowering plants or 3% of all the land plant species use C_4 photosynthesis. The figure below shows the essential features of CO_2 fixation in the C_3 and C_4 pathways.

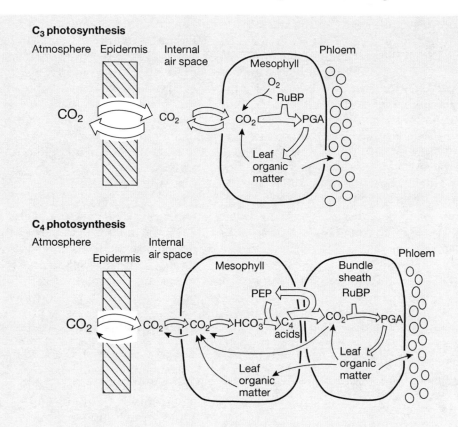

The C_3 photosynthetic mechanism characterizes all the plants, but the C_4 pathway, also known as the Krantz or Hatch-Slack pathway, is an addition to some of the plant groups. The smaller group of C_4 plants includes tropical and subtropical grasses, and marsh and wetland plants. They retain most of the photosynthetically produced carbon, respiring only about 1% of it. C_4 photosynthesis in the terrestrial environment probably became widespread about 6 to 8 million years ago with the expansion of grasslands dominated by C_4 species in response to declining atmospheric CO_2 concentrations. C_4 plants have evolved a unique cell architecture called Krantz leaf anatomy and have developed very effective mechanisms for modification of their internal CO_2 concentrations in both gas and aqueous micro-environments. The C_4 plant is capable of lowering its internal CO_2 gas concentration well below atmospheric thus creating a substantial CO_2 gradient between the atmosphere and mesophyllic cells. However at the same time, the CO_2 concentration in the aqueous phase of the bundle sheath cells of the plant is increased well above the atmospheric CO_2 gas–water equilibrium value for subsequent assimilation of carbon by the C_3 pathway. Thus CO_2 is not rate-limiting for C_4 plant photosynthesis, as it is for C_3 plants, and C_4 species would be expected to respond less than C_3 plants to increasing atmospheric CO_2 concentrations. C_4 plants include some of our most productive crops, including corn, sugarcane, and sorghum, but also weeds such as purple nutsedge, Bermuda grass, and pigweed.

The CAM pathway appears to be a metabolic adaptation to drought or otherwise dry conditions. The adaptation has resulted in these plants closing their stomata during the day and opening them during the night when evaporative transport is low. Another possibility is that CAM is an adaptation to low daytime CO_2 levels because many, if not most, CAM plants are tropical epiphytes that live in forest canopies that have low atmospheric CO_2 concentrations during the day when photosynthesis is maximal. However, the stomata of the plants are open at night when atmospheric CO_2 concentrations are relatively elevated due to respiration. CAM plants fix carbon derived from HCO_3^- into organic acids such as malate, which accumulates as malic acid during the night. During the day malic acid is decarboxylated (splitting off of a carboxyl group, $-COOH$, from the acid as carbon dioxide) to free CO_2 and a 3-carbon compound. Because the stomata are closed during the day, the CO_2 is trapped and water retained, and the CO_2 is assimilated by the C_3 pathway. CAM plants respond positively to elevated CO_2 levels but generally display lower photosynthetic rates than C_3 or C_4 plants.

(After Mackenzie and Lerman, 2006.)

water (H_2O), nutrients [nitrogen (N), phosphorus (P), sulfur (S), and trace quantities of other essential elements] removed from the soil by the roots of plants, carbon dioxide (CO_2) from the air, and light energy from the sun, which is trapped by chlorophyll. In the process, plants discharge oxygen (O as O_2), heat energy, and water vapor into the air while storing carbon in their tissue as glucose and other carbohydrates. In the aquatic realm, the photosynthetic process is similar, except in this case the "plants," for example, the microscopic phytoplankton of the ocean, take up carbon and nutrients directly from dissolved forms of these substances in their water environment. The organic compounds produced by photosynthesis, and the energy stored in them, are the basis for the food chains and food webs of the land and oceans. This same process initially provided Earth with its oxygenated environment.

Respiration is the reverse process of photosynthesis. It is a process common to producers, consumers, and decomposers. It is the burning, through the use of oxygen, of the stored carbohydrates in plants and animals. Respiration produces energy for organisms to move, grow, and maintain existing body structures. Respiration has as its by-products CO_2 and H_2O. In plants, these by-products are released into the atmosphere through the stomata of leaves. **Transpiration** is the process whereby water is released and evaporated by plants. The processes of photosynthesis and respiration are common to plants, algae, cyanobacteria, and certain bacteria. They are the organisms that form the foundation for the entire living world.

Primary and Secondary Consumers The consumers (heterotrophs, Figure 6.8) are organisms living on land or in water that feed directly or indirectly on the producers. They cannot make organic compounds within their bodies from inorganic materials. They must feed on plants or other animals, who initially fed on plants, to get their nutrients and energy. Primary consumers are organisms that feed directly on plants and include deer, cows, or horses on land and microscopic crustaceans in the sea. Secondary consumers feed on the organisms that feed on the primary consumers. Humans, sharks, and cougars are examples of secondary consumers. The secondary consumers keep a system in balance by maintaining the primary consumers at population levels for which the ecosystem has sufficient space, food, and other materials to sustain them.

Decomposers Every living organism in the world is born, lives, and then dies. Decomposers are

heterotrophic organisms (e.g., fungi and bacteria, Figure 6.8) that are essential in the process of decay of dead organisms. These organisms derive energy and nutrients from dead organisms and change organic compounds back into inorganic compounds. This decomposition completes the cycle of organic matter, because inorganic compounds are now available for plants to use again. Decomposers seem like an unpleasant group, but without them organic materials would rot exceedingly slowly (in outer space where bacteria do not live, organic materials may persist and not decay), nutrients would not be replenished, and life would cease.

The more general term **detritivore** includes the decomposers but also includes animals that live on the refuse of an ecosystem. Crabs, jackals, and vultures are examples. Bacteria that consume dead organic matter are also considered detritivores.

Ecological Pyramids

The type, variety, and abundance of the members of the biota must be in balance to sustain an ecosystem. There can be only a certain total number of each type of organism, fulfilling the roles of producer, consumer, or decomposer in an environment.

Biomass is the mass of organic matter and usually refers to the mass created by plants and other photosynthetic organisms that is passed up the food chain. It represents an enormous amount of stored, or potential, energy in the chemical bonds that bind the various chemical compounds of biomass together. Generally, water is excluded when determining biomass; thus, biomass is the total combined dry weight of any specific group of organisms. The biomass of the first trophic level is the total dry weight of all producers in an area. The biomass of the second trophic level is the total dry weight of all herbivores, and the third trophic level is the total dry weight of secondary consumers, the carnivores, and so on. A phenomenon common to most ecosystems is a decrease of biomass with an increase in the trophic level (Figure 6.11). This is due to losses of energy as biomass is transferred from a lower trophic level to a higher one.

Energy enters an ecosystem as light from the sun. Less than 0.1% of the total solar energy absorbed by the surface of Earth is used in plant photosynthesis (see Figure 4.3, Chapter 4.) All of the energy absorbed by plants is not available to higher trophic levels because much of the plant matter produced is not consumed by herbivores, but is consumed and digested by decomposers and detritus feeders

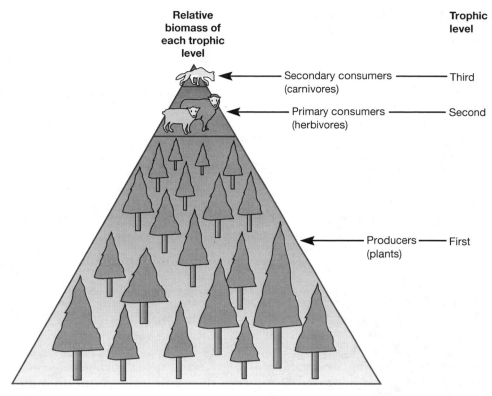

Relative biomass of each trophic level

Trophic level

Secondary consumers (carnivores) —— Third

Primary consumers (herbivores) —— Second

Producers (plants) —— First

FIGURE 6.11 An ecological pyramid. Biomass in successively higher trophic levels decreases due to losses of energy originally introduced at the base of the pyramid in the form of sunlight. Bacteria may constitute more biomass than that of all other life on the planet. They are involved in myriad processes, including nitrogen fixation, denitrification, nitrification, O_2 metabolism of organic matter, sulfide formation, sulfate reduction, methanogenesis, methanotrophy, and iron oxidation and reduction. (see Chapter 7).

instead. Furthermore, each time energy from food is transferred from one organism (and trophic level) to another, a large portion of the available energy is lost through bodily excretions and waste heat as work is performed.

An ecological pyramid (Figure 6.11) is established because of the loss of energy available for use by each subsequent trophic level. As a result, only about 10% to 20% of energy and biomass is actually transferred upward at each level, resulting in an approximately 90% decrease in potential energy available for each succeeding trophic level. Thus, the organisms at the bottom of the food chain, the plants, contain more biomass and generally are more abundant. Plants are more abundant than herbivores, and there are more herbivores than carnivores.

An ecosystem remains quasi-balanced when the structure of the ecological pyramid is satisfied and the type and relative abundance of members of the biota are in certain proportions. In fact, the "**balance of nature**" refers to this stability of plants and

animals in relationship to their environment. Humans frequently interfere with and alter the natural balance of ecosystems. An example of the way in which the balance of an ecosystem can go awry and nearly collapse is that of plant–deer–cougar relationships in the southwestern United States. In the early 1900s, Arizona placed a bounty on large predators (wolves, cougars, and coyotes) in an effort to protect and increase the deer population. This led to a massive decline in these secondary consumers. As a result, the deer population (primary consumers) swelled as desired due to no natural force to cull the herd. As the growing deer population searched for food, vegetation (producers) in the area was stripped. Eventually, many deer starved to death after much of their food source was depleted. The deer population ultimately decreased in number to less than that prior to the extermination of the predators. However, since then, predators have been reintroduced into their natural habitat and they, the deer, and the vegetation have made a comeback.

TERRESTRIAL AND AQUATIC ECOSYSTEMS

There are several different ways to subdivide the ecosphere into units. An **ecosystem** (as described previously) is the smallest unit of the ecosphere that has all the characteristics necessary to sustain life. The ecosystem is a functional unit in which living and nonliving components interact dynamically and exchange materials. A **biome** is a group of ecosystems within a geographical region exposed to the same climatic conditions and having dominant organisms with a similar life cycle, climatic adaptations, and physical structure. Sometimes ecosystems and biomes are referred to by the same terminology. The term biome simply represents a large geographical subdivision, whereas the definition of an ecosystem specifically includes organism–environment interactions within an area, whether they be interactions within a large biome or those within a smaller region.

The ecosphere can be divided into both terrestrial and aquatic biomes. A terrestrial biome generally receives its name from the main type of vegetation in the region. A perusal of the literature shows that a large number of names, descriptions, locations, and areas have been used to characterize the major biomes. An example of some of the major subdivisions of the terrestrial realm is that of polar ice, tundra, taiga (the transition region between tundra and temperate boreal), temperate boreal forests, temperate coniferous and temperate deciduous forests, grassland, desert, chaparral, savanna/woodland, mountain, tropical rain forest, and tropical deciduous forests (Figure 6.12). The freshwater and marine aquatic realm also can be divided into distinct biomes such as river, marsh, lake, estuarine realm (wetlands and estuaries), continental shelf, coral reef, and deep ocean areas.

The boundaries between different biomes are not sharp, but tend to grade into one another. Thus, transition areas may contain species from adjacent areas. In addition, the major biomes contain within them a variety of physical conditions and plant and animal species. These variations of physical and organic components make it difficult to subdivide the ecosphere into distinct biomes. Different researchers have different opinions on subdivisions. As a result, there is confusion concerning the specific boundaries

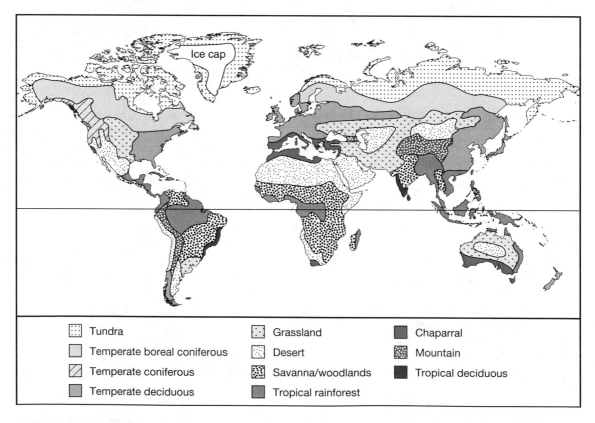

FIGURE 6.12 Distribution of the major biomes of the terrestrial realm. The correlation of the terrestrial biomes with soil type (Figure 2.8) and climate (Figure 2.9) is strong.

of biomes and the use of consistent terminology for the subdivisions.

Terrestrial Biomes

Various terrestrial biomes are generally recognized, from a region like the tundra, which is characterized by a lack of trees, a cold and dry climate, and permanently frozen subsoil, to that of a tropical rain forest with substantial amounts of rain throughout the year, warm temperatures, and soils of low organic content (Figure 6.12, Table 6.4). In the tundra region, only a few species live year-round, such as mosses, lichens, the snowy owl, the musk ox, and the arctic hare. However, the tundra is also the feeding ground for other animals and many species of migratory birds that spend their summers in the relative warmth of the region. The taiga area, with its hardy coniferous trees, serves as a protective wintering ground for some of these animals. In the temperate boreal forests of the northern subpolar zone, timber wolves and moose live among coniferous trees conditioned to withstand freezing weather. The temperate forests farther south on the North American and Euro-Asian continents contain coniferous and deciduous trees, birds, bears, squirrels, owls, and deer. The grasslands of the Great Plains in North America are home to prairie dogs, sage grouse, and deep-rooted grasses. Many original grassland areas are now used to graze cattle and sheep and to grow crops. Former grassland areas in the central United States constitute the breadbasket of the plains, where wheat and corn now grow. The emu lives in the Australian grasslands, and rheas live in small flocks on the grassland of the Pampas in South America, where they graze with cattle. Zebras, lions, and elephants roam the grassy and tree-studded savanna regions in Africa. Hot and dry deserts contain plants such as cacti and succulents, which are able to tolerate low annual precipitation. In this region, lizards, birds, snakes, and mammals all seek refuge from the noonday sun to survive. Tropical forests are characterized by tropical moist-deciduous trees and evergreen rain forests. They contain the largest number of plant and animal species on the planet, from tiny poisonous frogs to huge bats.

Aquatic Biomes

Aquatic ecosystems (Table 6.5) are of two principal types: freshwater and marine. Lakes, rivers, marshes, and swamps are typical aquatic biomes found on continents. In the marine realm, continental shelves, coral reefs, and the open ocean are specific biomes. The estuaries of the world are unique in that they overlap both terrestrial and marine biomes (Figure 6.13).

In aquatic ecosystems, there are two major groups of organisms. Benthic animals and plants live on the bottom, while pelagic organisms float or swim in waters. Benthic organisms may live attached to rocks or dwell on sediment surfaces (epifauna and epiflora) or live within sediments (infauna). Benthic organisms living in shallow waters of the ocean include corals, green algae, coralline algae, and sea grasses. Bivalves, brittle stars, and worms that burrow into the sediment are found from the shallowest to the deepest parts of the ocean. Specialized clams, tube worms, and crabs inhabit the warm areas near hydrothermal vents at great depths on the ocean floor. Lake bottoms are often covered with benthic plants firmly rooted in the sediments and a variety of molluscan fauna.

Zooplankton, phytoplankton, and nekton are the major types of pelagic flora and fauna. **Neuston** are very small pelagic animals and plants inhabiting the upper few centimeters of a water body. They are generally less than 0.5 millimeter in size. The **phytoplankton** and **zooplankton** are plants and animals, respectively, and vary in size from 0.005 millimeter to about 3.0 centimeters. Diatoms are important representatives of the phytoplankton of both freshwater and marine ecosystems. Crustacea are important zooplankton in most aquatic systems. Foraminifera (protozoans), pteropods (gastropod mollusks), and Coccolithophoridae (algae) are abundant plankton groups in the ocean. The **nekton** are generally much larger, ranging in size from 3.0 centimeters to marine organisms as large as tunas, sharks, and whales. The phytoplankton are limited to the shallow, lighted portions of water bodies because they require the sun's energy to photosynthesize. Zooplankton, nekton, and bacteria are widely distributed with depth throughout aquatic systems.

FRESHWATER BIOMES: LAKES, RIVERS, AND STREAMS
The major freshwater biomes are lakes and rivers. On the scale of human lifetimes, lakes are permanent features of the landscape. However, in terms of geologic time, they are transient features. They act as temporary repositories for water and sediment before being filled in with organic and inorganic matter derived from within the lake or from stream runoff. The lakes of the world contain 125,000 cubic kilometers of water. These waters may be very fresh, like those of the Great Lakes of the United States and Canada and of Lake Victoria and Lake Tanganyika in Africa, or they may be very salty, like the Great Salt

TABLE 6.4 Characteristics of Terrestrial Biomes

Biome	Location	Climate	Plants	Animals
Tundra	High latitudes of Asia and Europe below the frozen Arctic and above the boreal forest to the south.	Treeless plain, extremely cold and dry. Precipitation as rain or snow less than 10 cm per year. Permafrost (permanently frozen subsoil). Organic-rich soils.	Grasses, mosses, lichens, and rushes.	Permanently dwelling musk ox, lemming, arctic hare, and snowy owl; migratory birds, caribou, and insects.
Taiga	Transition zone between the tundra and boreal.	Soil and climate transitional.	Grasses, mosses, and conifers (spruce).	Caribou and birds.
Temperate Forests				
Located mainly between the Arctic Circle, latitude 66°N, to 23.5°N but also in the Southern Heimisphere.				
Temperate (coniferous boreal)	Located below the tundra area on the continents of North America, Asia, and Europe.	Cold winters. Subsoil thaws in the summer. Acidic soils.	Coniferous trees (jack pine, black and white spruce).	Moose, wolves, bears, insects, and migratory birds.
Temperate (coniferous) Montane, Subalpine, Coastal Forests	Northwestern North America.	Cool and moist. Acidic, nitrogen-poor soils.	Coniferous trees (cedar, bristlecone pine, fir, and redwood).	Elk, bears, mountain lions, and wolves.
Temperate (deciduous)	Eastern United States, western Europe, and northeast China.	Winters are cold to mild. Precipitation is high. Soil is rich in organic and inorganic nutrients.	Broadleaf trees (maple, hickory, oak, elm, beech, and birches).	Foxes, rabbits, deer, squirrels, raccoons, rodents, and many birds.
Grassland	North American prairies, Russian steppes, Argentinian pampas, and large portions of Australia.	Hot summers, cold winters. Sparse precipitation. Soil is rich in organic and inorganic nutrients.	Grasses.	Bison, antelope, coyotes, and wolves in North America.
Chaparral (Mediterranean)	Borders the Mediterranean Sea, and in southwestern United States and Australia.	Hot and dry summers. Cool wet winters.	Shrub brush and olive trees.	Lizards and coyotes.
Desert	Southwest United States, Sahara, and central China.	Low rainfall. Extremely arid with short, intense rains at times. Little nitrogen or organic nutrients in soils.	Cacti, succulents, mesquite, acacia, and yucca.	Snakes, lizards, kangaroo rats, and birds.
Savanna/Woodlands	Africa, South Asia, and South America.	Long dry seasons.	Open grasslands with sparse trees, grasses, acacia, thorn scrub, and baobab trees.	Giraffes, zebras, and lions in Africa.

Biome	Location	Climate	Plants	Animals
Tropical Forests				
Located in an area around the equator in Africa, South America, and Southeast Asia.				
Tropical (rain forest)	South America, Africa, and Southeast Asia.	Wet all year. Warm temperature, nutrient-poor soil.	Epiphytes, tall trees, and many different species of plants.	An abundance of birds, insects, and mammals.
Tropical (deciduous)	Southeast Asia and South America.	Monsoon weather, wet with periods of dry weather.	Trees that lose their leaves during the dry period.	Colorful birds, snakes, and monkeys.

Lake of the United States and the Dead Sea of the Middle East. The sediments of lake bottoms are composed of the debris of pelagic and benthic organisms and organic and inorganic materials from land runoff and atmospheric deposition.

Solar heating of lake waters frequently produces permanently or temporarily stratified lakes having strong gradients in temperature and density with depth of the lake. This stratification may result in

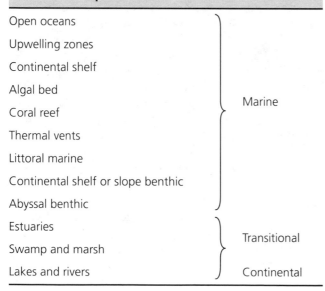

TABLE 6.5 Some Terms Applied to Biomes of the Aquatic Realm

Open oceans	
Upwelling zones	
Continental shelf	
Algal bed	
Coral reef	Marine
Thermal vents	
Littoral marine	
Continental shelf or slope benthic	
Abyssal benthic	
Estuaries	
Swamp and marsh	Transitional
Lakes and rivers	Continental

lighted, warm, oxygenated, and less dense water (the epilimnion) overlying cooler, denser, deeper waters (the hypolimnion) that are devoid of oxygen. In temperate regions, there is typically a seasonal overturn of lake waters. In the summer, stratification sets in owing to a high degree of solar insolation, and in the fall, cooling causes the lake to be circulated from top to bottom. This situation is important to biological and biogeochemical processes occurring in the lake. The products of organic production in the epilimnion accumulate in the hypolimnion because of the settling and decay of dead organic matter. The waters become rich in nutrients and oxygen deficient (anoxic). On overturn due to cooling of the surface waters and turbulent mixing, the nutrient-rich and anoxic water from the hypolimnion is mixed throughout the lake. This mixing restores nutrients to surface waters, resulting in a bloom of phytoplankton in the fall. During the winter, temperate lakes that freeze go through another period of stratification. During winter stratification, the warmest water is at the bottom of the lake rather than at the top. This layering occurs because water has its maximum density at 4°C. As the lake cools to this temperature, the 4°C water sinks to the lake bottom and is protected against further cooling. During the spring, partial overturn of the lake water occurs again. Phytoplankton blooms are again observed at this time.

Rivers and streams drain lakes and watersheds upstream of lakes. These conduits of water connect with other streams to form eventually the great rivers

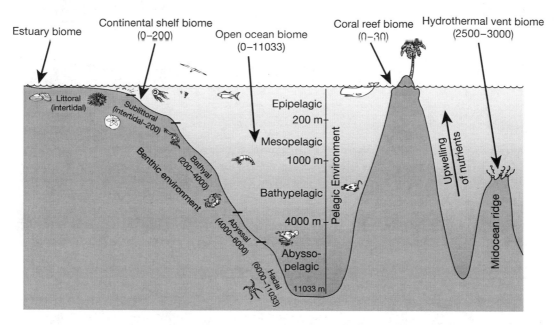

FIGURE 6.13 Classification and distribution of benthic and pelagic ecological zones in the world's oceans. Several of the larger marine biomes discussed in the text are also shown. (*Source:* Hedgpeth, 1957.)

of the world that empty into ocean basins. Both the Amazon and the Orinoco River of South America and the Zaire River of Africa drain great floodplains in the tropics and alone account for more than 30% of the freshwater discharge to the oceans. At any one time, rivers contain 1700 cubic kilometers of water. This water is flowing to the ocean at a rate of 40,000 cubic kilometers per year. Thus, river waters are very ephemeral. In one year, the water of the world's rivers is replaced nearly 24 times by precipitation.

Rivers transport dissolved constituents, suspended organic and inorganic materials, and coarse sediment in the bed of the river. The dissolved substances are mainly bicarbonate (HCO_3^-) calcium (Ca^{2+}) and silica (SiO_2). The solids are particulate organic matter and rock minerals and their alteration products eroded from the land surface. On a global basis, rivers transport annually toward the ocean about 4.3 billion tons of dissolved materials and 18 billion tons of solid matter. A significant percentage of the suspended solid matter is currently being caught behind dams and does not make its way to the coastal ocean. On entering the ocean, the solid and dissolved materials settle to the bottom because of gravity or are biologically or inorganically precipitated out of ocean water, respectively.

As with the larger aquatic biomes of the ocean, lakes and rivers contain planktonic and benthic plants and animals. The plankton typically include algae (diatoms), protozoa, rotifers, copepods, cladocera, and phantom midge larva. Unlike in the ocean, none of the larger invertebrates of freshwater ecosystems have temporary planktonic stages. The plankton in these systems are often kept suspended mainly by the turbulence of the water. Plankton in lakes show a great deal of variability spatially and with season of the year. Phytoplankton populations change rapidly, with the dominant species governed by the ever-changing conditions of light, temperature, nutrient availability, and other factors. The majority of zooplankton in freshwater systems are herbivores and feed on algae.

Rooted aquatic plants are major photosynthetic organisms in freshwater systems. They are often the most visible organisms found in rivers (riparian vegetation of river margins) and comprise much of the organic production of these ecosystems. In lakes, rooted plant production varies with lake age. The older a lake and the more it has been filled by sediment, the more important the benthic plants are to total plant production in the lake. These plants have long life cycles compared to those of the phytoplankton. Consequently, they are capable of tying up nutrients for long periods of time and contributing more organic matter to the lake bottom than the phytoplankton. This situation is quite different from that of most oceanic waters, where plankton detritus is the major source of organic matter for bottom sediments.

TABLE 6.6 Characteristics of Major Benthic Ecological Zones of the World's Oceans

Characteristics	Littoral and Sublittoral	Bathyal	Abyssal	Hadal
Area of sea bed (%)[a]	8%	16%	76%	1%
Depth (meters)	0–200 m	200–4000 m	4000–6000 m	6000–11,033 m
Pressure (atmospheres)	1–21 atm	21–401 atm	401–601 atm	601–1104 atm
Temperature (°C)	25 − 5°C	15 − 5°C	<5°C	<3.5°C
Light	bright–dim	dim–dark	total darkness	total darkness

[a] Does not total to 100% due to rounding.

Very commonly benthic plants in lakes show a zonation from the shoreline to deeper waters. This zonation is not only true for the plants but for the sediment types and the invertebrates that live on or within the sediments. Emergent plants, insects, and molluscs inhabit the shallow, oxygenated waters of lakes and their sandy bottoms. Floating leaf plants characterize waters farther offshore, followed by submersed plants in deeper waters, including mosses and branched algae. In the deep, cold, light- and oxygen-deficient realms of lakes, the bottoms are often muddy and plant and animal assemblages are lacking. Midge larvae and clams are representative of the larger animals found in these areas.

Fish are the dominant nekton of fresh waters. Trout, bass, pike, pickerel, burbot, and many other species are found in freshwater ecosystems. In lakes fish tend to spawn in shallow waters but live both offshore and in the nearshore zone among the benthic plants. Rivers form the spawning and nursery areas of migratory fish species such as salmon that migrate great distances upriver from their ocean habitat to spawn and die.

When a freshwater ecosystem undergoes change induced naturally or by human activities, organisms other than plankton and nekton also may be affected. The habitats of vertebrates like frogs, salamanders, turtles, snakes, and various aquatic birds, such as herons and kingfishers, which depend on the resources of these systems, may be disturbed. In Chapter 12, examples are presented of this situation due to anthropogenic acid deposition.

MARINE BIOMES Benthic and pelagic organisms in the ocean live in a number of different environments. These environments can be grouped into a series of ecological zones. As an example, the benthic environment of the ocean is subdivided into the littoral and sublittoral, bathyal, abyssal, and hadal zones (Table 6.6). The environments range from the shallow waters of the intertidal or littoral zone, to the deeper sublittoral waters of the continental shelves, to the hadal zone of the deep sea. A similar classification has been developed for the pelagic environment. The relationships between the benthic and pelagic ecological zones of the world's oceans are shown in Figure 6.13. A number of marine biomes can be found within these zones or coextensive with them.

In this section, we consider some examples of marine biomes, emphasizing their general characteristics and flora and fauna. These biomes include the estuarine realm, continental shelves, coral reefs, and the open ocean (Figure 6.13).

Estuarine Realm The estuarine realm includes coastal wetlands and estuaries. These are highly productive areas teeming with aquatic life. **Wetlands** are periodically or continually flooded lands. Inland wetlands are found adjacent to bodies of freshwater. Coastal wetlands are located along coastlines and include mangrove swamps, marshes, and parts of bays and lagoons. Wetlands regulate water flow and act like a sponge, holding back floodwaters. They filter out sediments from incoming water and reduce sedimentation in streams. They also effectively trap excess nitrogen and phosphorus and the potential pollutants delivered to them by rivers.

Estuaries are the main avenues of input of riverine materials to the coastal zone and their area is less than 1% of total ocean area, or one million square kilometers. They are important habitats for aquatic life, and along with wetlands, are the breeding grounds for many organisms that migrate to the open ocean.

An **estuary** is a region where river water meets and mixes with seawater (see also Chapter 5). The region of mixing is a zone of salinities intermediate between that of river water and that of seawater. This gradient in salinity is responsible in part for the distribution of sediment, biological activity, and species diversity in an estuary. As the river water mixes with the saltwater, suspended particles clump together (flocculate) and settle rapidly to the floor of the estuary, and thus are not carried out to the open ocean. Commonly, these particles accumulate on the floor of the estuary in a relatively restricted zone. Metals and organic matter transported by the river to the estuary also tend to settle out in this zone of flocculation. The sediments may build up on the floor of the estuary such that the water becomes shallow. In estuaries used for shipping, extensive dredging may be necessary to maintain channels for ship navigation. This dredging may remobilize materials originally deposited with the sediment and release them into the estuarine water. These substances include metals and organic compounds, some of which may be toxic to organisms.

Continental Shelves The continental shelves are an important region of major fisheries and of exploration and exploitation of petroleum resources. The continental shelves occupy an area of 27 million square kilometers, or about 7.5% of ocean area. They extend outward from the shorelines of the continents to a depth of roughly 200 meters. On the eastern margins of modern continents shelves tend to be wide, whereas on western margins they are generally narrow, dropping off quickly to the oceanic slope below. About 90% of the suspended sediment delivered by rivers to the ocean generally is deposited very close to the coastline in bays, lagoons, estuaries, and the nearshore area of continental shelves. However, some fine-grained materials often bypass shelves because of strong currents on the shelves. These materials are transported to the deep sea.

The continental shelf is essentially coincident with the sublittoral ecological zone of the ocean (Table 6.6). This zone is usually well nourished by nutrients transported to the shelf by rivers and in certain regions by nutrients upwelled from deeper waters of the continental slope. The latter process globally is the most important source of nutrients for **new production** (primary production not supported by regenerated nutrients) on shelves. Also, the death of pelagic organisms in the surface waters of the shelf provides a rain of organic detritus for the regeneration of nutrients at the seafloor below. Consequently, the continental shelf is an area rich in a diversified collection of benthic flora and fauna. The sublittoral environment includes a diversity of plant and animal communities. In some areas, sea stars, brittle stars, and feather stars are important organisms. In other areas, sea urchins dominate the submarine landscape. Coral reefs, discussed next, are also an important ecosystem found in the sublittoral, continental shelf environment.

Coral Reefs Coral reefs are the ocean's analogy to tropical forests on land. They occupy an area of 284,300 square kilometers, or about 1% of the area of the shallow continental shelf, equivalent to about half the area of France. Nearly 45% of coral reef area is found in Indonesia (17.9%), Australia (17.2%), and the Philippines (8.8%). Despite their small area, two-thirds of all marine fish species are associated with coral-reef environments, and these systems exhibit a wide diversity of invertebrate life. Reefs are largely limited to tropical waters and are not found in waters that are too deep, too muddy, too diluted by freshwater, or too hot or cold. Major coral reefs do not occur in waters that are less than 8°C or exceed 30°C for extended periods of time. However, reefs are found at high latitudes, for example, off the southern coast of Alaska and off the western coast of Norway. Many of the species living in reef environments are near their upper thermal limit for survival, at least during some months of the year. This is one reason that during the warming of surface marine waters during an El Niño event, when the cyanobacterial symbiotic zooanthellae leave the coral polyps, coral morbidity and mortality can be high.

Coral reefs are dominated by corals and coralline algae. Corals belong to the large and varied phylum Coelenterata, which are simple, multicellular invertebrate animals. The coralline algae are red algae (phylum Rhodophyta) that secrete a hard calcareous skeleton. Corals and coralline algae form the living and dead framework of coral reefs, which are usually cemented by inorganic and biologically produced cements. The coral reef framework is infilled by the skeletal and organic matter debris of corals, coralline algae, mollusks, echinoderms, green algae, foraminifera, soft fleshy algae, and other organisms. Coral-reef area comprises 617,000 square kilometers, or 0.2%, of total ocean area, and 15% of the shallow seafloor area of 0 to 30 meters in depth. Reefs account for at least one-third of all the calcium carbonate deposited in the oceans of the world.

There are three major types of coral reefs. Fringing reefs are found bordering coastlines or are separated from them by a very narrow band of water. Barrier reefs, like the Great Barrier Reef of northern Australia, parallel the coast but are located some distance from it. A large lagoon generally separates a barrier reef from the neighboring land. An atoll is a ring of coral reefs surrounding a central lagoon. Hundreds of atolls, like Eniwetok and Bikini, dot the Pacific Ocean.

Coral reefs are an important ecosystem on a global basis. They protect coastlines from erosion and supply calcareous sediment to nearshore areas. They are recreational areas for millions of people. The potential yield of fish from coral-reef ecosystems is about 10% of the total commercial oceanic fish landings. As will be seen later, coral reefs are threatened ecosystems worldwide because of human activities.

Open Ocean   The open ocean includes nearly 90% of ocean area, or 325 million square kilometers. The open ocean starts at the edge of the continental shelf. Its waters cover with increasing depth and decreasing bottom slope the outer submerged margins of the continents, the continental slope and rise. In the large central portions of the ocean basins, the waters overlie the deep abyssal floor at depths commonly exceeding 4.5 kilometers. The sea bottom of the open ocean is covered mainly by fine-grained, clay- and silt-sized sediments derived from the land, the atmosphere, the settling of the skeletal and organic debris of pelagic organisms, hydrothermal emanations, and weathered and eroded basalts. Sediment types include materials derived from the skeletons of pelagic organisms. Radiolarian (protozoan) and diatom skeletons characterize the clay- and silt-size particles of siliceous deposits (siliceous oozes). Foraminifera, pteropod, and Coccolithophoridae skeletons comprise the deposits of calcareous oozes. Fresh basaltic rock is found along the midocean ridges. In the deepest portion of ocean basins, little skeletal debris is found. The major sediment type is red to brown mud.

The open ocean includes the bathyal, abyssal, and hadal ecological zones of the world's oceans (Table 6.6). In these deeper benthic zones, light is not sufficient for plant growth. As a result, the amount of organic material in these zones is much less than that found in shallower environments. Animals living on the seafloor of the open ocean rely primarily on sinking organic detritus from above, including dead phytoplankton, zooplankton, and fish, and occasional large falls of food such as dead whales. Sponges, soft corals, sea lilies, sea stars, sea cucumbers, crabs, shrimp, and fish are characteristic animal life found in the bathyal zone. Even in the abyssal zone, which

comprises 80% of the seafloor, life is diverse. The diversity of life may be comparable to that found in shallow tropical seas. Sea cucumbers, brittle stars, sea stars, and bizarre fish characterize the sea bottom of the abyssal zone. In the hadal zone of the long, narrow oceanic trenches, large predators are absent and the concentration of life may be only 1/1000 of that found on the seafloor of the shelf. Sea cucumbers, sea anemones, huge protozoans, and exotic worms characterize this deep environment.

Biomass and Productivity

Biomass and **productivity** are measures, respectively, of the amount of organic matter in a certain area and the rate at which that organic matter is formed. The amount or weight of terrestrial and aquatic plant (phytomass) and animal (zoomass) matter found in a particular area of Earth's surface can be measured and is referred to collectively or separately as biomass, standing crop, or standing stock.

Phytomass comprises about 99% of all biomass on Earth, with only about 1% accounted for by all animal life. Because of the small percentage accounted for by animal life, the term **biomass** generally refers to plant matter or phytomass. The units of biomass are variable. For plankton and nekton, typical units are micrograms per liter ($\mu g/L$), milligrams per cubic meter (mg/m^3), grams per square meter (g/m^2), kilograms per hectare (kg/ha), and so forth, where the weight is given as wet or dry weight or as carbon. For benthic marine communities, like algal beds or coral reefs, and terrestrial biomes, biomass is usually given in kilograms of dry matter (DM) per square meter (kg/m^2).

DISTRIBUTION OF PHYTOMASS   The distribution of phytomass in the global biosphere is shown in Figure 6.14. The total world phytomass is about 1250 billion tons of dry matter. Most of this phytomass is continental; about 0.3% is marine. There are about 563 billion tons of carbon tied up in the total world phytomass. About 560 billion tons are found on the land surface and about 3 billion tons in the oceans. The atmosphere contains 800 billion tons of carbon as carbon dioxide gas. Thus, only a slight change in the carbon stored on land in terrestrial ecosystems could affect significantly the content of carbon dioxide in the atmosphere. This is one reason there is so much concern about the burning of the world forests, which releases carbon dioxide into the atmosphere (see Chapters 10 and 14).

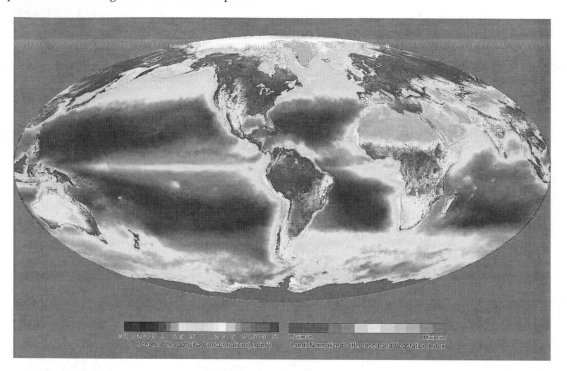

FIGURE 6.14 A view of the global biosphere showing its distribution on land and in the ocean. The ocean chlorophyll a pigment concentration in mg/m^3 is a measure of the biomass of phytoplankton in the surface waters of the ocean. The normalized difference land vegetation index is a measure of the density of vegetation on land. Notice the large areas of tropical and very productive temperate forest biomass on land, the fertile coastal regions of the ocean, and the large areas of the central ocean basins with relatively low productivity. The image is generated from data obtained from the SeaWIFS instrument aboard the Seastar satellite.   (e.g., http://earthobservatory.nasa.gov/Features/WorldOfChange/biosphere.php.)

The average specific phytomass (phytomass per unit area) of aquatic ecosystems varies by a factor of 15,000, if swamps and marshes are included (Figure 6.15). For example, the average specific phytomass of the small area of upwelling zones of the ocean, like that off the coast of Peru, averages 0.02 kg/m^2, resulting in a standing stock of 8 million tons of dry matter globally. In contrast, the large area of the open ocean has an average specific phytomass of only 0.003 kg/m^2 but represents a total of 1000 million tons globally. In this case, the larger surface area of the open ocean more than makes up for its lower specific area phytomass. Although small in area, estuaries are important marine ecosystems. They account for about 35% of all marine phytomass. Coral reefs and algal beds account for another 30% of marine phytomass; the remainder is principally found in the open ocean. The average specific phytomass of lakes and streams is 0.02 kg/m^2, like that of oceanic upwelling areas. However, lakes and streams have

an area more than five times larger than that of upwelling zones, producing a total phytomass of 50 million tons of dry matter. Tropical swamps and marshes also represent significant phytomass, equivalent to 15 kg/m^2 or 23 million tons of dry matter (Figure 6.15).

There is also a large range in the phytomass of the different terrestrial biomes. Tropical humid rain forests have a specific phytomass of 42 kg/m^2, representing 420 billion tons of dry matter, the largest of any ecosystem. In contrast, hot and dry deserts have a specific phytomass of 0.06 kg/m^2, amounting to 480 million tons of dry matter (Figure 6.16).

The amount of plant matter stored on the terrestrial surface increases as one moves from the polar biomes toward the equator. Of particular importance is the large amount (530 billion tons of dry matter) of phytomass in the tropical rain forests and tropical deciduous forests. These biomes alone contain 43% of the total terrestrial phytomass of Earth. They are

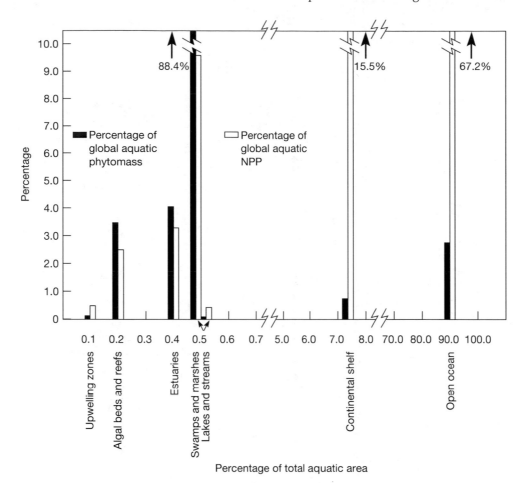

FIGURE 6.15 Relationships among the percentage of aquatic area occupied by an aquatic biome, the percentage of the global living plant matter (phytomass) that occupies an aquatic biome, and the percentage of global net primary production (NPP) of an aquatic biome. Note changes in scale. (*Source:* Whittaker and Likens, 1975.)

home to an abundance and great variety of plant and animal species and represent a renewable resource of vast dimension. Public concerns over deforestation and species extinction occurring in the tropical forests are an expression of the value of this vast resource and are more thoroughly discussed in Chapter 10.

NET PRIMARY PRODUCTIVITY   Gross primary productivity (GPP) is the total amount of organic matter produced by photosynthesis in a defined area in an interval of time. Some of the organic matter produced is used in respiration within the cells of plants. The remainder is available to plant growth and is termed **net primary productivity (NPP).** It is the net amount of organic matter produced per unit area per unit time. Net primary productivity is on the order of one-half of gross primary productivity. The units used in the measurement of NPP are the

same as those used for phytomass when expressed per hour, day, or year. The net production of an ecosystem is the NPP times the total area of the ecosystem. The NPP and consequently the net production of terrestrial and aquatic ecosystems are very variable.

The average NPP of marine systems varies from 2500 grams of dry matter per square meter per year ($g/m^2/yr$) for algal beds and coral reefs to $125\ g/m^2/yr$ for the open ocean (Figure 6.15). However, much of the productivity of the open ocean is found in restricted areas where nutrient- and carbon-rich waters upwell from depth. These include the equatorial and high-latitude regions of the major oceans. The great central interiors of the open ocean are relatively unproductive. Some authors term them the "deserts of the sea." However, because of the large area of the ocean interior, the total annual net primary production of this area is still large, amounting to about 42 billion

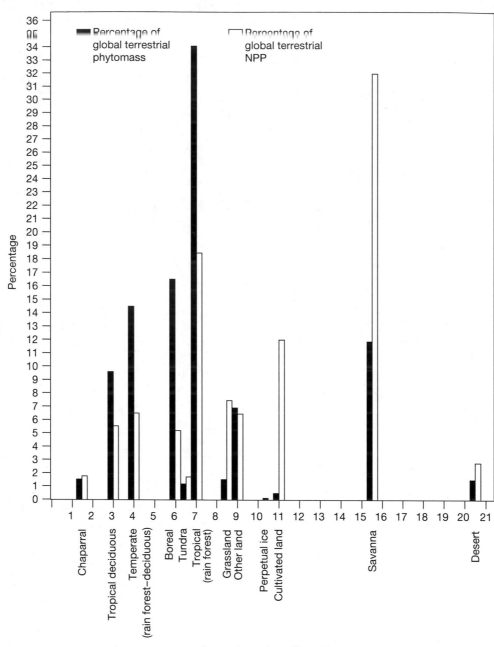

FIGURE 6.16 Relationships among the percentage of land area occupied by a terrestrial biome, the percentage of the global living plant matter (phytomass) that occupies a terrestrial biome, and the percentage of global net primary production (NPP) of a terrestrial biome. (*Source:* Atjay, et al., 1979.)

tons of carbon as organic matter, or about 76% of total oceanic net production.

For terrestrial biomes, the range in NPP is even greater. The NPP of tropical swamps and marshes is 4000 g/m^2/yr and that of hot and dry desert is 10 g/m^2/yr. The total yearly production of dry organic matter by world forests is about 49 billion tons, while that of cold and dry deserts is only 50 million tons (Figure 6.16).

Global annual NPP and total production of plant material (terrestrial and marine) are about 0.37 kg/m^2 and 189 billion tons of dry matter, respectively. Average terrestrial NPP is nearly six times greater than that of marine, and the total production of plant matter on land exceeds that in the ocean by perhaps as much as two times. Thus, the terrestrial biosphere takes up more carbon each year than the marine ecosystem in total organic production. However, carbon used in biological production in the ocean turns over more rapidly than that on land. This is because of the essential differences between marine and terrestrial ecosystems that are discussed in the following section.

Fundamental Differences

Some fundamental differences exist between terrestrial and marine ecosystems. The major difference lies in the fact that terrestrial ecosystems have a large storehouse of structural materials in wood and bone. In the terrestrial realm, organisms must first and foremost adapt to the strains of gravity. In contrast, marine organisms grow up bathed in the fluid of the sea. Pelagic organisms that float or swim in the sea are close to being neutrally buoyant in their medium of seawater. Because of this difference, plants and animals of the terrestrial realm must expend a great deal of energy making wood and bone to survive. In contrast, the need for large bones among most marine organisms is greatly reduced, and cellulose, the material of wood, and calcium phosphate, the material of bone, are very minor components of marine plants and animals. Furthermore, when in motion, terrestrial animals must expend a great deal of energy to overcome gravity, whereas the business of swimming or floating is more effortless and energy efficient.

The predominant organic material of terrestrial organisms is carbohydrate, whereas that of marine organisms is protein. Carbohydrate is a vast store of energy, and this energy is used slowly in the development of terrestrial plants. The mean residence time of carbon in terrestrial vegetation is 10 years. In contrast, marine plants use much of their energy to grow, and there is little energy stored for a rainy day. Marine plants (phytoplankton) grow rapidly to maturity, are quickly grazed by herbivores, and generations rapidly succeed one another, resulting in rapid turnover of nutrients and carbon. The mean residence time of carbon in marine phytoplankton is only 26 days.

Concluding Comment

Terrestrial and marine ecosystems have evolved over geologic time. Certainly, no rain forest ecosystem

existed 3 billion years ago, and dinosaurs are no longer a part of the fauna of modern ecosystems. However, prior to major human interference in Earth processes, ecosystems were fairly stable on the time scale of many human generations, as are some ecosystems today. Biological and physical variables—for example, temperature, precipitation, and abundance of plants or animals—did change from day to day, but on the whole they remained more or less the same. There was a state of stability or dynamic equilibrium, in which a change in one variable resulted in a change in another, bringing the system back into balance. In other words, ecosystems are complex self-regulating systems, involving a form of regulation known as **negative feedback.** This does not mean that all parts of an ecosystem operate in perfect harmony or that conditions are absolutely constant. It does mean, however, that the abundance of species remains much the same from year to year, that the same species are present each year, and that the population sizes of species remain about the same from year to year.

There is no doubt that because of human activities, the characteristics of many ecosystems are changing globally. Habitat and species have been lost from the rain forests of Brazil. Coral reefs of Southeast Asia are losing species because of unsustainable rates of fishing and pollution from suspended sediments, pathogens, pesticides, and nutrients mobilized by humans in the watersheds of rivers that enter coastal areas with living corals. Estuaries in many parts of the world are undergoing changes for the same reasons. If one wishes to consider the lower atmosphere as an integral abiotic part of some ecosystems, then demonstrated changes in atmospheric composition during the past 300 years due to human activities are leading to ecosystem change. In later chapters in this book, we discuss many of the environmental issues that are due to an impingement of human interests and activities on other living organisms.

EXTINCTION

Life first emerged about 4 billion years ago and got off to a slow start. In the latter part of the Permian Period, after more than 3.5 billion years of development, there were only 350,000 species on the planet, most of which were marine. From its first appearance until today, life has continued to evolve and to produce new and more varieties of its theme.

While life has also continued to evolve over time, producing increasing diversity, the extinction of species also has been a natural part of life and life cycles (Figure 6.17). Over 90% of the more than one-half billion species that have occupied Earth at one time or another have become extinct.

Several mass extinctions of life-forms have occurred during the Phanerozoic Eon, some more dramatic than others. Six major global extinction events stand out during the Phanerozoic as times when an unusual abundance of life-forms disappeared from the planet forever: the Cambrian-Ordovician boundary (510 million years ago), the Ordovician-Silurian boundary (extinction of 12–20% of all families approximately 440 million years ago), the Devonian-Carboniferous boundary (14–21% of all families of marine organisms lost 365 million years ago), the Permian-Triassic boundary (54–57% decline in all marine families 245 million years ago), the Triassic-Jurassic boundary (12–23% of marine and nonmarine animal families lost 208 million years ago), and the Cretaceous-Tertiary boundary (the dinosaur extinction; loss of 11% of all families of marine organisms 65 million years ago). The Permian-Triassic and the Cretaceous-Tertiary boundaries are characterized by the most devastating extinction events. A more recent minor extinction event occurred at the end of the Pleistocene

(10,000 years ago) when many large mammals, such as the saber-toothed tiger and the mammoth, passed into obscurity.

Regardless of the occasional high extinction rates during the Phanerozoic, the variety and number of life-forms have steadily increased to an all-time high since the time the dinosaurs became extinct. The planet has been richer in life-forms during the current interglacial period, the last 10,000 years, than at any period in its history. No one is sure of the number of species on Earth today, but so far about 1.59 million species have been identified, including estimates of about 297,000 plants, 1,200,000 invertebrates, 60,000 vertebrates, and nearly 29,000 lichens, mushrooms, and brown algae in 2007 (compare with E. O. Wilson's earlier estimates in 1992, Figure 6.18). The biological diversity of bacteria is poorly known but there are certainly about 70,000 species or more (E. O. Wilson's estimate in 1992 was only 5000). Bacterial biomass may exceed the mass of all the rest of life combined. The total number of life-forms present on the planet today is estimated to be anywhere from 5 million to 30 million. Nature, if left on its own, would probably continue on its course of expansion in the development of new varieties of life forms.

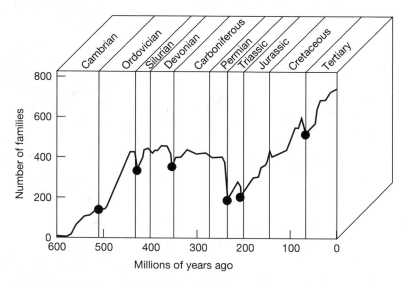

FIGURE 6.17 The pattern of biological diversity through Phanerozoic time, as expressed by the number of families present in the fossil record. Biodiversity has increased slowly over time, interrupted occasionally by massive extinction events shown as black dots on the figure.   (*Source:* Wilson, 1990.)

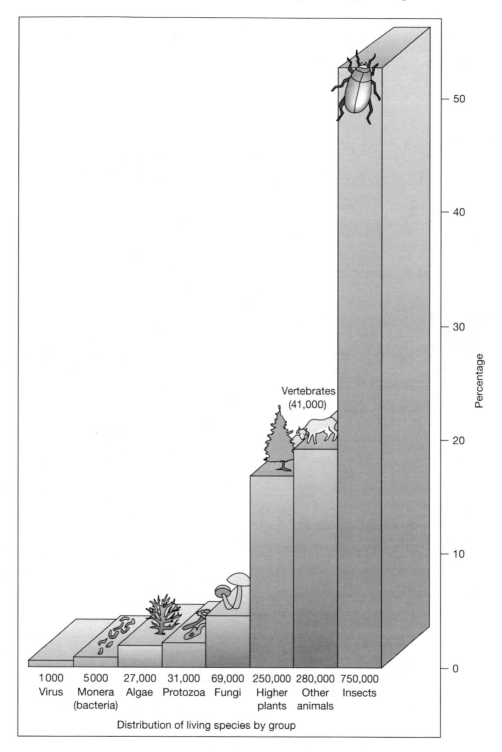

FIGURE 6.18 Percentage distribution of living species by group. Approximately 1.4 million living species have been identified to date. The total number of life-forms is estimated to be between 5 million and 30 million. The number of species of Monera is poorly known and may be considerably larger than shown. (*Source:* Wilson, 1992.)

Concluding Remarks

To deal with problems of global environmental change, it is necessary to know the sizes, distribution, and composition of the various ecosystems of the world. These data are difficult to obtain on a global scale and are a subject of controversy among scientists. For example, it is important to know how much carbon is stored in living and dead plants in the ocean and on land (the biomass of carbon) and at what rate this carbon is incorporated into living plant matter (the productivity). These are not well-known values, even today.

Why is this important? For one reason, current estimates of the size of the various terrestrial ecosystems of the world vary greatly. We will see in Chapters 10 and 11 that the human activities of deforestation and other land use changes are modifying the terrestrial ecosystems of the world, but by how much? To answer this question we require accurate estimates of the size of these systems and the carbon stored in them. Do we have them? The answer is, probably not. This lack of knowledge makes it difficult

for scientists to evaluate how damaged the global terrestrial ecosystems are, and to some extent the magnitude of loss of biological diversity, and the role the ecosystems play in global climatic change.

The human species, a newly dominant species, is one of the late arrivals on the planet. Until the agricultural revolution 10,000 years ago, the human population was small and had little effect on the plants and animals of the world but had already begun to modify the landscape. With the cultivation of crops and domestication of livestock, our ancestors multiplied. We are the only species to have attained the ability to influence comprehensively the natural ecosystems of the planet. Unfortunately, human pressures on natural ecosystems have caused an ongoing wave of extinction of other species. According to some scientists, species are being extinguished at a rate that exceeds that at any time in the history of Earth because of the current dominant role of the human species in the environment. Figure 6.19 shows

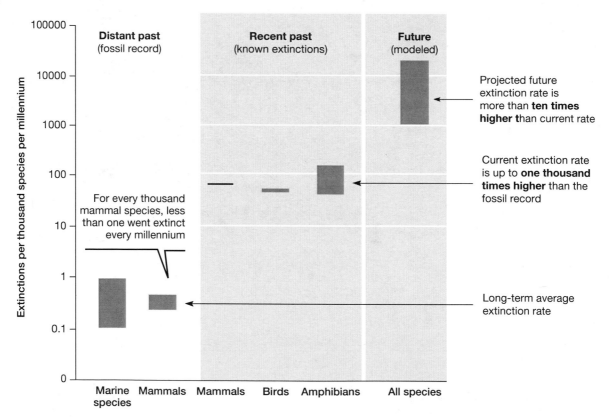

FIGURE 6.19 Comparison of rates of mass extinctions during the distant geological past and the recent past with future model extinction rates. The predicted future rates are 10,000 times higher than the long-term average extinction rate. (From Millennial Ecosystem Assessment, 2005 – *Ecosystems & Human Well-Being: Synthesis*. Island Press, Washington, DC. Copyright © 2005 World Resources Institute. Reprinted by permission of World Resources Institute.)

the 2005 Millennium Ecosystem Assessment estimates of the number of extinctions per thousand species per millennium for the fossil record, for the recent past, and the modeled rate for the future, which is ten times higher than the current rate and 10,000 times higher than the rate for the fossil record.

This rapid loss of biological diversity threatens the future organic evolutionary processes of the planet. In the six large extinction events of the past, when millions of species were lost in certain groups, other species continued to survive and multiply. For example, the dinosaurs became extinct at the end of the Cretaceous, but most mammals and plants survived. Unfortunately, the global extinction of today is one that is affecting many different groups of plants and animals and is due largely to the land use changes associated with human activities and to some extent with climatic change.

Study Questions

1. What is the depth and altitude range of most life on Earth?
2. In what three ways may life be classified on Earth?
3. In what major way do viruses differ from all other life-forms?
4. What is the difference between *autotrophy* and *heterotrophy*?
5. What are three biogeochemical processes performed by prokaryotes?
6. Organisms in the kingdom Plantae are *photoautotrophic*.
   (a) What does this term mean?
   (b) Write a chemical equation demonstrating this process.
7. The simplest chemical expression for the production of organic matter in plants is

$$\downarrow CO_2 + H_2O \rightarrow CH_2O + O_2\uparrow.$$

The chemical compound CH_2O is organic matter; CO_2 is carbon dioxide gas; O_2 is diatomic oxygen gas; and H_2O is water. The atomic weights of the elements carbon (C), hydrogen (H), and oxygen (O) are, respectively, 12, 1, and 16.
   (a) What are the gram molecular weights of the compounds of CH_2O, CO_2, O_2 and H_2O?
   (b) What is the weight of 1 mole of each of these compounds?
   (c) If 10 moles of plant matter have been produced, how many moles of CO_2 did it take to produce the plant matter? How many grams of CO_2?
   (d) If the molar C:N ratio of the plant material is that of marine phytoplankton of 106:16, how many moles of nitrogen were needed to produce the plant matter?
   (e) If the source of the nitrogen were nitrate (NO_3^-) in the euphotic zone of the ocean, how many grams of nitrate were consumed in the process?
8. What is an ecosystem? How does it differ from a biome?
9. For an ecosystem to function in a sustainable manner, what properties must it possess?
10. What are the three major functional groups of organisms, and how do they perform in an ecosystem?
11. What are the two major generalized biomes of the ecosphere?
12. How does the tropical rain forest biome differ from that of the tundra?

13. What are the major types of pelagic flora and fauna, and how do they differ?
14. Lakes and streams have an average global biomass of 0.02 kg dry matter/m^2. Their total area is 2×10^6 square kilometers.
   (a) What is their total biomass?
   (b) Why is their biomass so much larger than that of oceanic upwelling zones with the same specific area biomass?
15. Tropical rain forests have a total area of $17 \times 10^{12} \, m^2$, and estuaries have an area of $1.4 \times 10^{12} \, m^2$. Their mean net primary productivity is 2000 g dry plant matter m^2/yr and 1800 g dry plant matter m^2/yr, respectively. Their mean plant biomass in kg C m^2 is 20 and 0.45, respectively. Forty-five percent of dry plant matter is carbon.
   (a) What is the total net primary production of tropical rain forests and estuaries in metric tons of plant dry matter and carbon per year?
   (b) What is the total plant mass of these ecosystems in metric tons of plant dry matter and carbon?
   (c) The tropical rain forests of the world lost 9% of their area due to cutting in the 1980s. How much plant dry matter does this cutting represent? How much carbon?
   (d) If all of the carbon in the cut trees of question 15c were emitted to the atmosphere by burning and slow oxidation, how many grams of CO_2 would this represent? What fraction of the atmospheric CO_2 reservoir of 800 billion tons of carbon does this represent?
   (e) Estuaries receive 1.5×10^{12} g of pollutant dissolved phosphorus (P) annually. This amount of P could support how much additional plant productivity as dry matter and as carbon per square meter per year?
   (f) If all the additional plant matter in 15e were buried in the sediments of the estuaries, would this flux qualify as a biological feedback to the accumulation of CO_2 in the atmosphere? Explain. What percentage of the atmospheric flux of 8.7×10^{15} g C from fossil fuel burning in 2008 does the additional total plant production in estuaries represent?

16. The net primary production of Earth's surface is 0.37 kg dry matter/m^2/yr.
    (a) With a total area of 510×10^6 km^2 what is the total production of organic matter?
    (b) With a conversion factor from dry matter to carbon of 0.45, what is the global NPP in terms of carbon?
17. Total production on land exceeds that in the ocean by perhaps as much as two times. Based on your answer to question 16, what is the annual total production on land and in the ocean?
18. What are the differences between the photosynthetic pathways of C_3 and C_4 plants? Why would C_3 plants be expected to respond more to increasing atmospheric CO_2 levels than C_4 plants?
19. At least five compounds can act as oxidizing agents in remineralization of organic matter. What are they?
20. Extinction events have occurred in the geologic past. Why the concern with the present rate of extinction?

Additional Sources of Information

Coppens, Y., 1994, East Side Story: The Origin of Humankind. *Science*, v. 270 (May), pp. 88–95.

Curtis, H., 1983, *Biology*, 4th ed. Worth Publishers, New York, 1159 pp.

Levington, J. S., 2001, *Marine Biology: Function, Diversity, Ecology*, 2nd ed. Oxford University Press, New York, 515 pp.

Margulis, L. and Olendzenski, L. (Eds.), 1992, *Environmental Evolution: Effects of the Origin and Evolution of Life on Planet Earth*. The MIT Press, Cambridge, MA, 405 pp.

Millennial Ecosystem Assessment, 2005, *Ecosystems & Human Well-Being*. Island Press, Washington, D. C., 160 pp.

Rambler, M. B., Margulis, L. and Fester, R. (Eds.), 1989, *Global Ecology: Towards a Science of the Biosphere*. Academic Press, Boston, 204 pp.

Raup, D. M., 1991, *Extinction: Bad Genes or Bad Luck?* W. W. Norton & Co., New York, 210 pp.

Spalding, M. D., Ravilious, C., and Green, E. P, 2001, *World Atlas of Coral Reefs*, University of California Press, Berkeley, CA, 424 pp.

Southwick, C. H., 1996, *Global Ecology in Human Perspective*. Oxford University Press, New York, 392 pp.

Wilson, E. O. (Ed.), 1988, *Biodiversity*. National Academy Press, Washington, DC, 521 pp.

7 Biogeochemical Cycles of Carbon, Nutrients, and Oxygen

The integrity of the whole requires the integrity of each part.

ECOLOGICAL SOCIETY OF AMERICA, THE SUSTAINABLE BIOSPHERE INITIATIVE, 1991

The ecosphere is a highly interactive system with matter and energy flowing between and within individual ecosystems in interconnected biogeochemical cycles. For various reasons, including those involving nutrient cycling and limitation and climate control, it is essential that the integrity of all biogeochemical cycles on Earth is preserved. Most elements and compounds on Earth are involved in these cycles. For example, Figure 7.1 shows the many gaseous compounds that are produced biologically in terrestrial ecosystems and exchanged with the atmosphere. These compounds enter into biogeochemical cycles. In their transport through the atmosphere, they may react to form other compounds before returning to Earth's surface.

Biogeochemical cycles of elements and compounds are generally portrayed in the form of box models. In **biogeochemical-cycling models**, the major compartments or reservoirs (boxes) that contain the substance of interest are defined, and the mass of the substance in each box determined. Examples of reservoirs on a global scale are the atmosphere, ocean, and biosphere. After the reservoirs in the cycle are defined, the processes or mechanisms that lead to transport of the substance between reservoirs, that is, transport paths, are determined. An example is the process of production of organic matter by terrestrial plants that results in removal of carbon dioxide from the atmosphere and uptake in land vegetation, a reservoir in the biogeochemical cycle of carbon. Finally, the rates (fluxes) at which the substance of interest moves between boxes must be known to develop a quantitative model of a biogeochemical cycle. An example is the rate at which carbon as carbon dioxide is removed from the atmosphere and incorporated in land vegetation. This flux is 63 billion tons of carbon per year. A following section describes in more detail the construction of global biogeochemical cycling models.

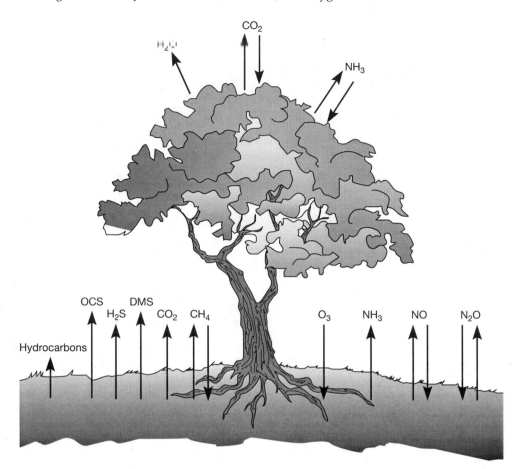

FIGURE 7.1 Exchange of gases between the terrestrial realm and the atmosphere. These gases in their transit through the atmosphere may react to form other compounds before returning to Earth's surface. (*Source:* Mooney et al., 1987.)

Individually, or in a complex web of interaction, biogeochemical cycles serve many purposes. For example, processes occurring in the nitrogen and phosphorus cycles shuffle these elements between organic and inorganic materials and maintain production of organic matter in aquatic and terrestrial ecosystems. The carbon and sulfur cycles are an important part of Earth's regulatory system for climate: Carbon dioxide is a greenhouse gas and warms the planet, and sulfur, as sulfate (SO_4) aerosol, in the troposphere and stratosphere helps to cool the planet. To provide an introduction to how cycles operate, a few biogeochemical cycles, or parts thereof, are discussed in this chapter. Emphasis is on the biologically important element carbon (C) and the elements nitrogen (N), phosphorus (P), oxygen (O), and sulfur (S) that are tied intimately to it. Additional cycles are treated in more detail in Chapter 12 on acid deposition and photochemical smog and Chapter 14 on climate.

CONSTRUCTION OF GLOBAL BIOGEOCHEMICAL CYCLING MODELS

The factors involved in the construction of biogeochemical cycling models are considered in more detail in this section. These factors include the following:

1. Definition of the boundaries of the system and subsystems (reservoirs or boxes);
2. Prediction and evaluation of transport paths;
3. Problems of evaluating fluxes; and
4. Mathematics of modeling of global systems.

Reservoirs

In developing a biogeochemical-cycling model, it is necessary to separate the system of interest from its natural surroundings. In general, the boundaries of a natural system are defined by the scale of the phenomena of interest and by previous knowledge of possible interactions between the system and its surroundings.

Chapter 7 • Biogeochemical Cycles of Carbon, Nutrients, and Oxygen **165**

Global biogeochemical cycling models involve consideration of phenomena on a worldwide scale. The Earth is divided into a number of physically well-defined spheres referred to as "boxes" or "reservoirs." The term **box model** is commonly applied to models of this type.

The number of reservoirs considered in modeling the global movement of a substance depends on previous knowledge of the way in which the substance of interest is distributed about Earth's surface. A **transport path** is a directional property of the system, providing the route by which a substance moves or is moved by some agent from one reservoir to another. The **flux** of a substance is a measure of the mass of the substance transferred along a transport path between reservoirs. Flux is usually defined in terms of mass per unit time (e.g., grams or moles of a substance per year). Fundamentally, the way a substance is transferred from one box to another depends on its physical and chemical properties. A substance with a high solubility in water, for example, sodium chloride, would be expected to move about Earth's surface via agents, such as rivers, involving water movement. However, a substance with a high solubility in fatty tissue, for instance, DDT (dichlorodiphenyl-trichloromethane), would be expected to be involved in biological processes. For example, DDT accumulates in the fatty tissue of some fish and is transported by fish from place to place.

If it can be demonstrated that there is a means of transporting a substance from one Earth sphere to another, then the spheres are considered boxes or reservoirs in a global model. If there are several sources within a reservoir for transport of materials from that reservoir to another, but there are no known transport paths for these sources, then the individual sources within a reservoir are considered as compartments in a global model. This distinction between reservoir and compartment is shown subsequently.

In Figure 7.2, three reservoirs of a hypothetical substance A are shown with transport paths between reservoirs. The ocean reservoir is subdivided into the compartments of water, biota, and sediment. The land reservoir includes the compartments of plants, animals, and soil. The masses of the substance A in these compartments are known. However, although transport paths among the compartments are suspected, no values for the fluxes are known for these paths. The transport of substance A involves (1) gaseous transport from the land and sea surface to the atmosphere

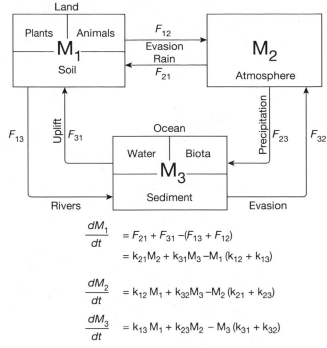

Schematic Global Biogeochemical Cycle of Substance A

$$\frac{dM_1}{dt} = F_{21} + F_{31} - (F_{13} + F_{12})$$
$$= k_{21}M_2 + k_{31}M_3 - M_1(k_{12} + k_{13})$$

$$\frac{dM_2}{dt} = k_{12}M_1 + k_{32}M_3 - M_2(k_{21} + k_{23})$$

$$\frac{dM_3}{dt} = k_{13}M_1 + k_{23}M_2 - M_3(k_{31} + k_{32})$$

FIGURE 7.2 Schematic diagram of the global biogeochemical cycle of a hypothetical substance A. See text for discussion.

(evasion, F_{12} and F_{32}) and return in rainfall (F_{21}) and precipitation (F_{23}); (2) river transport of dissolved and solid materials (F_{13}); and (3) return of the substance to the land via uplift of sedimentary rocks (F_{31}). Fluxes of substance A are known for these transport paths. The quotient dM/dt is simply the calculus notation for the change in mass of a reservoir per unit of time, the rate of change of mass. In a steady-state system, it is equal to zero. In a non–steady-state system, it differs from zero.

Transport Paths and Fluxes

Transport paths represent the major agents responsible for movement of materials about Earth's surface. In global models of the movement of a substance, these agents must describe the transport paths among reservoirs. Their nature might have various origins. They may be physical, chemical, or biological processes. The quantitative evaluation of their contribution is probably the most critical point of model building. It necessitates both a good knowledge of all the interactions possible between the reservoirs and the dynamics of the physical, chemical, and biological processes involved. The complexity of the task increases rapidly if subsystems are considered. In this case, it is necessary to understand the physical processes of movement (mixing by advection and diffusion; see Chapters 4 and 5), as well as biological processes, within the reservoirs. Although it might be easy to estimate the transport of a dissolved substance from the land to the oceans via rivers, it is more difficult to estimate the substance's movement within the ocean by advective and diffusive flows interacting with the biological processes of organic production and decomposition.

Once transport paths are defined in a model, they must be quantified. This step involves estimates of rates of transfer (fluxes) involved with the transport paths. Herein lies a major problem. The root cause is that in many cases basic data and an understanding of the processes operating in the system of interest are lacking. Many material cycles have innumerable processes and fluxes. Some are better known than others, and some are quantitatively unknown. Our knowledge of many biogeochemical cycles, particularly their dynamics and potential for feedback during perturbations, for example, natural or human-induced climatic change, is at a similar stage of development.

Modeling

Figure 7.2 is a model of a cycle consisting of three reservoirs. Each reservoir receives fluxes from the other two and delivers fluxes to them. If this cycle is thought of as representing a biogeochemical cycle of an element, then the chemical or biological form of the element may differ from reservoir to reservoir. Any transformations or chemical reactions, such as oxidation or reduction and solution or precipitation, are not represented in the diagram of the cycle. The flows between the reservoirs involve the masses transported. The rates of transport (fluxes) are measured in units of mass per unit of time. The nature of the biogeochemical processes that are responsible for the transformation of the components from one form to another, and the physical and biological processes that are responsible for the flows, determine in detail the rates of transport.

Knowledge of the fundamental processes and of the driving forces behind the flows may not be sufficient to enable development of a quantitative relationship between them and the material fluxes. Thus, the fluxes of materials (denoted F_{ij} in the figure for the flux from box i to box j) are commonly measured and related to the conditions in the system according to some chosen model. The two simplest flux models

are of zeroth- and first-order fluxes. The zeroth-order flux is a constant

$$F_{ij} = \text{constant}.$$

A first-order flux is one that is proportional to the reservoir mass

$$F_{ij} = k_{ij}M_i.$$

M_i is the mass of a substance in a reservoir i, and k_{ij} is a rate parameter for the flux going from reservoir i to j. In general, k_{ij} may vary with reservoir size and time, and may be a function of environmental conditions within a system. However, in the preceding equation, it is treated as a constant.

In a **steady-state** (unchanging) system, reservoir concentrations or masses of a substance do not change with time. This requires that the input and output fluxes for every reservoir are equal. If one of the fluxes changes, the steady state of the system becomes perturbed. The system is no longer at steady state, and the system is described as transient or **non–steady-state**. Such a condition can result in changes in all the reservoir masses.

The **residence time** λ of a substance in a reservoir is an important concept. It provides some clue as to the reactivity of a substance within a reservoir. It is defined as the ratio of the mass of the reservoir (M_i) to the sum of either input or output fluxes of the substance at steady state. The residence time is equal to 1 divided by the rate constant k_{ij}.

A perturbation of a biogeochemical cycle caused by a change in an input flux will result in a change in reservoir mass. For a fixed change in input, up or down, the reservoir mass will come to within 5% of a new steady-state value after three residence times have elapsed. Thus, most of the change (95%) caused by a perturbation in input would be completed in a time period equal to three times the residence time. Consequently, reservoirs of short residence times respond rapidly to external perturbations. In reservoirs of long residence time, perturbations require more time to work their way through.

Feedback is an important concept in biogeochemical cycles and environmental change. **Feedback** is a self-perpetuating mechanism or process of change and response to that change. For example, you may experience threats from another person as the result of something that you did to that individual. You react to that threat and modify your actions. In turn, the threatening person modifies his or her behavior and thus the subsequent feedback that you receive. A loop of behavior has been set up between you and the threatening individual. Natural systems behave in a similar manner, reacting to perturbations in a positive or negative fashion. In a positive feedback loop, the effects of a perturbation are amplified; in a negative feedback loop, the effects of the disturbance are diminished. Feedback is discussed later in the book with respect to climate change and other issues.

CARBON

Carbon comprises approximately 50% of all living tissues and in the form of carbon dioxide is necessary for plants to grow. Another function carbon dioxide serves is helping to sustain an equable climate on Earth. The concentration of carbon dioxide in the atmosphere has varied during the geologic past, but has remained within limits that permit life to exist on Earth. Carbon dioxide is cycled throughout the spheres of Earth on different time scales. For convenience, we can refer to these scales as short-, medium-, and long-term (Figure 7.3). Figure 7.4 is a simplified model of the biogeochemical cycle of carbon. A discussion of the

behavior of the carbon cycle over a range of time scales follows. Further details of the cycle are discussed in Chapter 14.

Short-Term Cycling: Photosynthesis/Respiration

Photosynthesis is a process that is part of the short-term carbon cycle (Figure 7.3a). We can look at the short-term cycling of carbon as carbon dioxide by beginning with the producers of organic carbon, the plants. Carbon in the form of atmospheric carbon dioxide is removed from the air by plants. This removal occurs on land, for example, in forests, grasslands, and aquatic systems, and in the surface waters of the oceans. The primary producers, the photosynthetic phytoplankton in the oceans and plants on the terrestrial surface, transform inorganic carbon as carbon dioxide into organic carbon within their tissue. A simplified chemical reaction representing the photosynthetic reaction is

$$6CO_2 + 6H_2O \rightarrow C_6H_{12}O_6 + 6O_2$$
$$\text{(carbon dioxide)} + \text{(water)} \rightarrow \text{(organic matter)} + \text{(oxygen)}$$

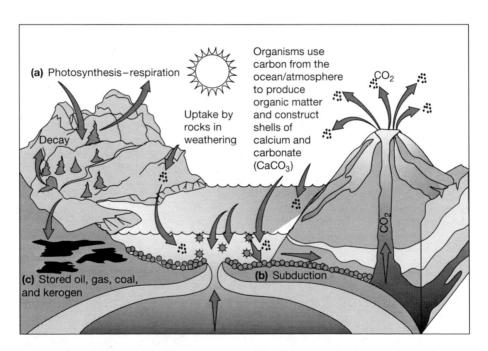

FIGURE 7.3 Diagram of the biogeochemical cycle of carbon prior to human interference showing (a) the short-term photosynthesis–respiration part of the cycle; (b) the long-term cycle involving accumulation of organic carbon and $CaCO_3$ in marine sediments, their subduction, alteration, and return of carbon dioxide via volcanism; and (c) the medium-term situation involving storage of carbon in organic materials in sedimentary rocks. This diagram does not show the important anthropogenic perturbations of the carbon cycle—those of fossil fuel burning, cement manufacturing, and deforestation.

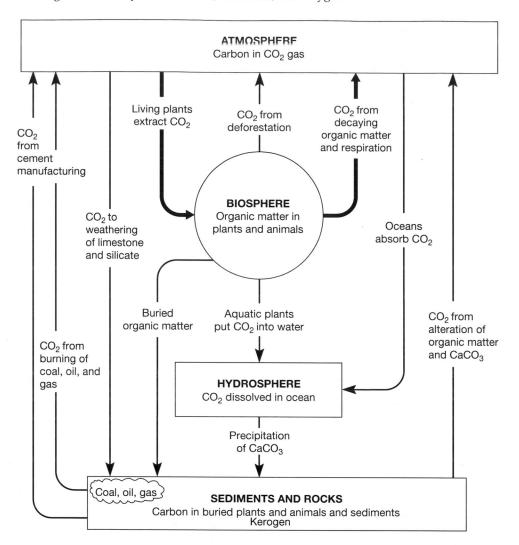

FIGURE 7.4 A diagram of geometric shapes, termed **boxes**, showing the major reservoirs and flows in the biogeochemical cycle of carbon. The arrows represent the processes and their direction that transfer carbon from one box to another. One can conceive of the cycle as representing a series of interlocking circuits in the reservoirs of biosphere, hydrosphere, atmosphere, and crust. The cycle would be in balance if it were not for human interference by burning of fossil fuels, cement manufacturing, and deforestation. (*Source:* Skinner and Porter, 1987.)

Energy and nutrient substances, like phosphate (PO_4^{3-}) and nitrate (NO_3^-), are necessary for this reaction to occur. In the photosynthetic process, some of the energy from sunlight is used in the growth of plants, and some energy remains stored in the tissues of plants as carbohydrates.

Plants remove about 200 billion tons of carbon as carbon dioxide from the atmosphere each year in gross primary production, which is about 26% of the world's total atmospheric carbon. Most, but not all, of the carbon dioxide taken from the atmosphere during photosynthesis in the terrestrial and oceanic realms is returned to the atmosphere during cell respiration and decay of plants (Table 7.1). The reaction is the reverse of that of photosynthesis; oxygen is consumed, and energy, carbon dioxide, and nutrients are released. The yearly removal rate of atmospheric carbon dioxide in photosynthesis is slightly larger on land than in the ocean.

After photosynthesis, carbon may be transferred next to a consumer organism when plants are eaten for food. The carbon stored in the tissue of the plant enters an animal's body and is used as energy or stored for growth. Land animals, such as cows and

TABLE 7.1 Fluxes of Carbon, Nitrogen, Phosphorus, Sulfur, and Oxygen in their Global Biogeochemical Cycles

Flux	Element (Million Metric Tons of Element Per Year)				
	C	N	P	S	O
1. River dissolved[a]	400	40	3	115	
2. Net primary production					
Land	63,000	580	320	265	168,000
Ocean	45,000	7925	1097	1925	120,000
3. Respiration and decay					
Land	61,400	560	310	260	163,700
Ocean	45,200	7960	1100	1930	120,500
4. Nitrogen fixation					
Land		270			
Natural		130			
Anthropogenic		140			
Ocean		40			
5. Denitrification					
Land		115			
Ocean		70			
6. Combustion fossil fuel	8700	30		80–100	
7. Land use activities[b]	1400	15–46	20 (mining)	1–4	
8. Burial and uplift[c]	400	15	3	40	
9. Metamorphism and volcanism	120			10	
10. Weathering	220				380

[a]Inorganic flux to ocean.
[b]Deforestation and biomass burning.
[c]Steady-state flux.

deer, are primary consumers of this stored energy. Aquatic plants are eaten by zooplankton and larger consumer animals. Whales, for example, feed directly on the animal plankton krill. When an animal breathes, some of the carbon taken up is released from the body as carbon dioxide gas in the process of energy expenditure (respiration). Aside from carbon stored in above-ground living and dead vegetation, carbon is also present in the root systems of terrestrial plants. When plants die, some of this carbon may be released as carbon dioxide or methane gas to the soil atmosphere, or may accumulate in the soil as dead organic material. This dead organic matter may be ingested by consumers, such as insects and worms living in the soil.

Some of the organic carbon generated in land environments is weathered and eroded, and the organic debris is transported by streams to the ocean. In the ocean, some of this debris, along with the organic detritus of marine plants, is sedimented on the ocean floor and accumulates in the sediments. However, some of the terrestrially derived organic matter is respired in the ocean to carbon dioxide; this carbon dioxide leaves the ocean and is transported over the continents where it is utilized again in the production of land plants.

Long-Term Cycling: The $CaCO_3$–SiO_2 Connection

Long-term cycling of carbon dioxide involves a series of processes dating back to the beginning of plate tectonics and includes not only land and ocean reservoirs but that of limestone rocks (Figure 7.3b). Limestone rocks are mainly composed of calcium carbonate ($CaCO_3$) and are the fossilized skeletal remains of marine organisms or, less commonly, chemical precipitates of calcium carbonate. Limestones are great storage containers for carbon; most of the carbon near Earth's surface is found in these rocks or in fossil organic matter in sedimentary rocks (Table 7.2). Weathering and erosion of Earth's surface results in the leaching of dissolved calcium, carbon, and silica (SiO_2) from limestones and rocks containing calcium silicate ($CaSiO_3$). The chemical weathering reactions are

$$CaCO_3 + CO_2 + H_2O = Ca^{2+} + 2HCO_3^-$$

and

$$CaSiO_3 + 2CO_2 + H_2O = Ca^{2+} + 2HCO_3^- + SiO_2.$$

The dissolved substances produced by these weathering reactions are transported to the ocean by rivers. In the ocean, the dissolved compounds are used to form the inorganic skeletons of some benthic organisms and plankton, which are composed of calcium carbonate and silica. The plankton include protozoans, such as foraminifera and radiolarians, and algae, such as coccolithophores and diatoms. During formation of the calcium carbonate skeletons, the carbon dioxide derived from the weathering of limestone is returned to the atmosphere in a chemical reaction that is the opposite of that for the weathering of limestone.

When marine animals and plants die, their remains settle toward the seafloor, taking the carbon stored in their bodies with them. En route, much of their organic matter is decomposed by bacterial decomposers, just as on land, and some shells may dissolve, thus turning animal and plant organic and skeletal matter back into dissolved carbon dioxide, nutrients, calcium, and silica in the ocean (see Chapter 5). This carbon dioxide is stored in the deeper waters of the oceans for hundreds to a thousand or so years before being returned to the atmosphere by the upwelling of deep-ocean waters.

Some of the animal and plant plankton sink to the bottom, where the carbon in the organic matter and

TABLE 7.2 Reservoir Masses in the Global Biogeochemical Cycles of Carbon, Nitrogen, Phosphorus, Sulfur, and Oxygen

Reservoir	Element (Billion Metric Tons of Element)				
	C	**N**	**P**	**S**	**O**
1. Atmosphere	800 and growing	3,950,000	0.00003	0.003	1,216,000
2. Ocean	38,400[a]	570[b]	80[c]	1,248,000[d]	4100[e]
3. Land biota	600	10	3	2.5	800
4. Marine biota	3	0.5	0.07	0.1	4.2
5. Soil organic matter	1600	190	5	95	850
6. Sedimentary rocks	78,000,000	999,600	4,030,000	12,160,000	1,250,000,000

[a]Dissolved inorganic carbon.
[b]NO_3^-.
[c]PO_4^{3-}.
[d]SO_4^{2-}.
[e]Dissolved O_2.
One billion metric tons = 10^{15} grams.

shells escapes degradation and becomes part of the sediment. As the seafloor spreads through plate tectonic processes, the sediments containing the remains of marine plants and animals are carried along to subduction zones, where they are transported down into the mantle. At the severe pressures and high temperatures in the subduction zones, organic matter is decomposed and calcium carbonate reacts with the silica found in the subducted rocks and forms rocks containing calcium silicate. This metamorphic reaction is the reverse of that of weathering of calcium silicate. Similar "reverse weathering" reactions (Ebelman-Urey reactions) occur at the base of huge piles of sediments deposited in sedimentary basins like the Gulf Coast of the United States and during collision of lithospheric plates and subsequent deformation and mountain building (see Chapter 3). During this process of metamorphism, carbon dioxide is released and makes its way into the atmosphere in volcanic eruptions and via hot spring discharges. Once in the atmosphere, carbon

dioxide can then combine with rainwater. The rainwater falls on the land surface and seeps down into the soils, where it picks up more carbon dioxide from decaying vegetation in the soils. This water, enriched in carbon dioxide, is very aggressive. It weathers and dissolves the compounds of calcium and silica found in rocks of the continents. The cycle begins again. This series of processes has been active for at least 600 million years, since the advent of the first organisms that made shells, and was significant even earlier in Earth's history when calcium carbonate was deposited in the ocean by inorganic processes (see Chapter 8).

One outcome of changes in the rates of processes in the long-term biogeochemical cycling of carbon is that atmospheric carbon dioxide has varied in a cyclic fashion during the last 600 million years of Earth's history. Robert A. Berner of Yale University, his colleagues, and others have developed models of the carbon cycle to calculate these variations. Figure 7.5 shows a comparison of two model calculations.

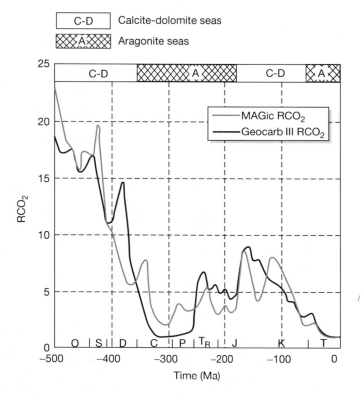

FIGURE 7.5 Two model calculations of atmospheric carbon dioxide during Phanerozoic time. RCO_2 is the concentration of CO_2 in the atmosphere relative to average Holocene levels. Abbreviations indicate geologic periods shown in Figure 2.3. Aragonite and calcite-dolomite seas are times of dominant precipitation of these minerals from the ocean and are explained further in Chapter 8.   (*Source:* Berner and Kothavala, 2001, Geocarb III and Arvidson et al., 2006 (revised), MAGic.)

The high atmospheric carbon dioxide levels of the Cretaceous and mid-Paleozoic are a result mainly of intense plate tectonic activity, a time of increased metamorphism of limestone and release of carbon dioxide to the atmosphere from volcanoes. These high carbon dioxide periods are termed Hot Houses (Greenhouses) and are described in more detail in Chapter 8. The lower atmospheric carbon dioxide levels of the early Phanerozoic, the Carboniferous through early Triassic, and much of the Cenozoic are an outcome of less intense plate tectonic activity and enhanced removal of carbon dioxide from the atmosphere by weathering. These three intervals in time are three of the Ice Houses described in Chapter 8. Other factors, like the evolution of land plants and their effect on weathering and changes in the amount of solar radiation received by the planet with time, also play a role in regulating atmospheric carbon dioxide levels over the long term. The large drawdown of atmospheric CO_2 in the later part of the Paleozoic can be attributed to the evolution of vascular plants in the Devonian Period leading to an acceleration of CO_2 uptake by weathering and burial of this new type of organic carbon appearing on Earth. The drawdown in the late Cretaceous into the Cenozoic is also due to a biological factor, the geographical migration and diversification of the angiosperms (flowering plants) with their extensive root systems, leading to an enhancement in weathering rate.

The modeling work (Figure 7.5) has been confirmed to some extent by proxy data for atmospheric CO_2 concentrations involving, for example, the density of stomata in fossilized plants, which is related to atmospheric CO_2 concentrations. The important point is that atmospheric carbon dioxide concentrations have varied by a factor of about 20 during the last 600 million years of Earth's history. This variation certainly has had climatic implications because carbon dioxide is an important greenhouse gas. In fact, for much of its Phanerozoic history the planet had a different atmospheric composition (and ocean composition; see Chapter 8) and a more equable climate than that of today.

Medium-Term Cycling: The Organic Matter–Oxygen Connection

Medium-term cycling of carbon dioxide involves organic matter in sediments, coal, oil, and gas, and atmospheric oxygen (Figure 7.3c). It commences, as with short-term cycling, with the removal of carbon dioxide from the atmosphere by its incorporation into plants and the accumulation of the dead plant

and animal carbon in sedimentary organic matter. Sedimentary organic matter is the dead and fossilized remains of plants and animals. When dispersed throughout a sedimentary rock, this material is termed **kerogen**. Shales are very fine-grained sedimentary rocks that are often rich in kerogen. Coal, oil, and gas deposits are also the remains of the soft tissues of plants and animals, but represent large and segregated accumulations of these materials in an altered condition.

Coal is derived mainly from terrestrial plant material, which is often deposited in swampy environments. The plant material is altered during burial of the swamp sediments. If buried deep enough, it may be changed substantially owing to the increased temperatures and pressures at depth. Different types of coal are formed because of varying conditions of temperature and pressure. Anthracite is a hard, dense coal formed by alteration of plant material at a relatively high temperature and pressure, whereas bituminous coal is formed under less intense conditions. Peat used as a fuel for fires is little-altered plant material containing high amounts of carbon that has not been buried deeply. Peat is an important component of tundra areas in the Northern Hemisphere.

Oil and gas represent highly altered organic matter, principally microscopic plant matter, the altered remains of marine phytoplankton that have been sedimented on the seafloor. During burial, these organic materials are broken down at elevated temperature and pressure, forming oil and gas. The oil and gas may travel many miles in the subsurface before coming to rest in large accumulations in the voids of rocks. Often oil and gas are formed in shales, but leave these rocks and migrate to more coarse-grained rocks like sandstones and limestones. It is in these latter rocks that the great Cretaceous and Cenozoic oil and gas reservoirs of the world are found, like those of the Persian Gulf.

These deposits of coal, oil, and gas represent organic carbon that has escaped respiration and decay, and thus carbon dioxide that has been removed from the atmosphere. The same is true of the kerogen dispersed as fine-grained materials in the sedimentary rocks. Because of their burial, the oxygen normally used to decay these materials remains in the atmosphere. The carbon in the coal, oil, and gas deposits and the kerogen are recycled back into the atmosphere to return carbon dioxide to that reservoir. At the same time, the oxygen that accumulated in the atmosphere is removed. All of this is accomplished when these fossil fuel deposits and kerogen are uplifted by plate tectonic forces after millions of years of burial and exposed to the atmosphere. When this occurs, the oxygen that previously accumulated in the atmosphere reacts with the coal, oil, gas, and kerogen. The reaction involves the oxidative decay of these organic materials. This results in the removal of oxygen from the atmosphere and the return of carbon dioxide to the atmosphere. The ongoing dynamic cycle is complete.

Fossil fuel is a nonrenewable energy source, because coal, oil, and gas deposits take millions of years and specific environmental conditions to form, and the mining of these deposits brings these materials back to the surface much more rapidly than natural processes. The stored energy from the long-dead organisms is released in the form of heat when the coal, oil, and gas are burned. This fossil fuel energy keeps us warm, powers our cars, and moves the machinery of industry. It is also a main cause of environmental pollution because a by-product of fossil fuel burning is the release of gases and particulate materials into the environment. Climatic change is an important global environmental consequence of the release of carbon and other gases to the atmosphere by combustion activities (see Chapter 14).

The Methane Cycle: The Wetland-CH_4 Connection

During the past decade as research in paleoclimate and modern climate has progressed, we have come to appreciate more the long-term changes in atmospheric composition. This has led to an understanding of the Phanerozoic behavior of atmospheric methane. Figure 7.6 shows the methane biogeochemical cycle prior to human influence (the modern cycle influenced by human activities is illustrated and discussed in Chapter 14). Notice in particular that natural wetlands are a very important source of methane to the atmosphere, with the oceans and termite metabolic processes being less important sources. The sinks of natural CH_4 emissions are mainly oxidation in the atmosphere by hydroxyl (OH^*) radical and uptake in soils. As mentioned, the organic matter deposited in natural wetlands and its alteration are the source material for coal deposits. The size and geographical distribution of natural, swampy wetlands and hence coal deposits have varied through geologic time, with the Permo-Triassic and much of the Jurassic, Cretaceous, and parts of the Tertiary being times of extensive tropical swamplands that resulted in major coal accumulations.

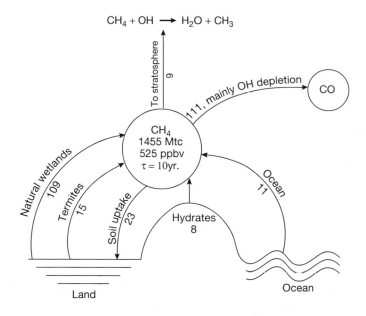

FIGURE 7.6 Prehuman biogeochemical cycle of methane. Fluxes are in units of millions of tons of carbon as CH_4 per year. The residence time (λ) of CH_4 in the atmosphere is 10 years. Notice the important flux of methane to the atmosphere associated with microbial degradation of organic matter in natural wetland sediments. (*Source:* Beerling et al., 2009.)

These extensive swamplands led to increased emissions of CH_4 to the atmosphere and increased atmospheric methane concentrations (Figure 7.7). So as we see with Phanerozoic atmospheric CO_2 concentrations, there is also a cyclic pattern of atmospheric CH_4 concentrations, although fundamentally for a different reason. Because methane is a greenhouse gas, it also contributes to the natural global warming or cooling of the planet. Also, because methane and carbon dioxide show cyclic patterns in atmospheric concentrations through the Phanerozoic that in part overlap, it is likely that atmospheric nitrous oxide (N_2O) also exhibited cyclic behavior. All three greenhouse gases played a role in climatic fluctuations through the Phanerozoic but CO_2 was most important.

Summary

Carbon is found in all four spheres or reservoirs. It is essential to every life-form, occurring in all organic matter in the ecosphere. It is found in the atmosphere, as the gases carbon dioxide, methane, and several other compounds, and it occurs in the hydrosphere as carbon dioxide dissolved in lakes, rivers, and oceans. In the crustal part of the lithosphere, it is found as calcium carbonate that was

originally deposited on the seafloor, as kerogen dispersed in rocks, and as deposits of coal, oil, and gas. It is because carbon is stored in the large sedimentary reservoirs of limestone and fossil organic carbon, and not in the atmosphere, that life on Earth is possible. If all this carbon were stored in the atmosphere, the result would be a heating of the atmosphere because of the increased absorption of the sun's energy; the greenhouse effect would be very strong, and Earth would probably have a temperature similar to that of Venus!

OXYGEN

The biogeochemical cycle of oxygen is diagrammed in Figure 7.8. Presently oxygen comprises 20% of the gases of the atmosphere. The cycling of oxygen is strongly coupled to that of carbon. Oxygen is produced by plants during photosynthesis, when carbon dioxide is consumed, and it is removed by respiration and decay, when carbon dioxide is produced.

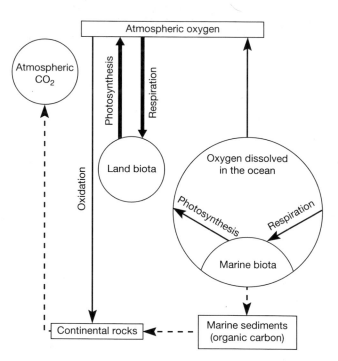

FIGURE 7.8 The biogeochemical cycle of oxygen. This cycle is strongly coupled to that of carbon in Figure 7.4. The geometric shapes, termed **boxes**, represent the major reservoirs of carbon and the arrows the flows, or fluxes, of oxygen from one box to another. The heavier the arrow, the larger the flux. The broken line shows the flow of carbon in sedimentation on the ocean floor, burial in sediments, and uplift by plate tectonic processes. When uplifted, organic carbon is oxidized by oxygen in the atmosphere and carbon dioxide is released back to the atmosphere. (*Source:* Andreae, 1987.)

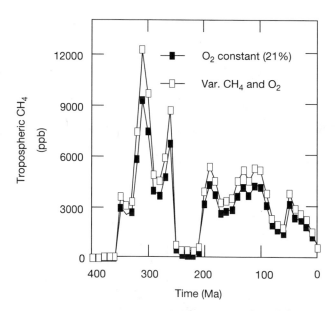

FIGURE 7.7 Model calculation of the variation of tropospheric methane concentrations during much of Phanerozoic time. Notice the cyclic pattern as with atmospheric CO_2 in Figure 7.5. The model calculations were done holding atmospheric O_2 concentrations constant and varying them closely following Figure 7.9. (*Source:* Beerling et al., 2009.)

This is a short-term, nearly balanced cycle on land because the amount of oxygen produced yearly by land plants is about equivalent to the amount used in the processes of respiration and decay in the terrestrial realm (Table 7.1). It takes about a decade for oxygen and carbon dioxide to recycle through living plants. However, there is a little leakage of non-respired organic matter from the land to the ocean. This organic matter, the leaves and trunks of trees and smaller size organic debris, is carried to the oceans by rivers as particulate and dissolved organic carbon (POC and DOC), where some of it is deposited in the sediments accumulating in the ocean and some is respired.

In the oceans, the production of oxygen by phytoplankton exceeds slightly consumption of this gas in respiration and decay. As a result, oxygen is released to the atmosphere. The organic carbon not decayed by this oxygen, as well as some of that brought to the oceans by rivers, is deposited on the seafloor and accumulates in the sediments of the ocean. If this accumulation process continued too long, all the carbon dioxide in the atmosphere would disappear in less than 10,000 years, and the oxygen content of the atmosphere would double in less than several million years. Changes of this magnitude did not occur, at least on these time scales, during the Phanerozoic. If they had occurred, they would have

been detrimental to life on Earth. Loss of atmospheric carbon dioxide would inhibit the photosynthetic process, and a large increase in oxygen would lead to the massive burning of plant life in forest and grassland fires. Fortunately, as mentioned previously, oxygen is removed by weathering of fossil organic carbon and other materials (particularly iron sulfide minerals; e.g., pyrite, FeS_2) found in rocks exposed on land, and during this process carbon dioxide is returned to the atmosphere. The overproduction of oxygen in the oceans and uptake by rocks on land balance the cycle.

One outcome of changes in the rates of processes involved in the medium-term cycling of carbon is that increased burial of organic carbon in sediments implies the possibility of accumulation of oxygen in the atmosphere. Times of high organic carbon burial in the past should have given rise to high atmospheric oxygen levels. This seems to be the case. Figure 7.9 is a model calculation of atmospheric oxygen concentration variations during the Phanerozoic. Just before the Carboniferous, vascular plants evolved and spread over the continents. Their organic remains were a new source of organic matter resistant to degradation by atmospheric oxygen. During the Carboniferous and Permian, large quantities of vascular plant organic matter were buried in the vast coastal lowlands and swamps of the

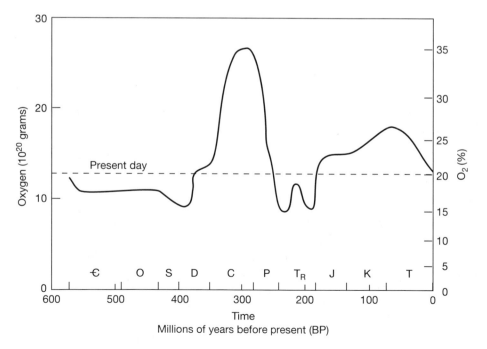

FIGURE 7.9 Model calculation of atmospheric oxygen during Phanerozoic time. Abbreviations indicate geologic periods shown in Figure 2.3. (*Source:* Berner and Canfield, 1989.)

time. This material became the coal deposits mined from rocks of Carboniferous and Permian age today. This large accumulation of organic matter gave rise to the high atmospheric oxygen levels of the late Paleozoic. Coal deposits are also important in Cretaceous and early Cenozoic-age rocks, another time of high atmospheric oxygen concentrations, and as discussed, methane concentrations.

Once again we see that atmospheric composition has not remained constant throughout Earth's history. However, until humans evolved, the changes in atmospheric composition were driven by changes in the rates of natural processes involving the biogeochemical cycles of materials. Now, the activities of human society are rapidly changing the composition of the atmosphere because the natural biogeochemical cycles cannot cope with the rates at which substances are entering these cycles from anthropogenic activities.

NITROGEN

Nitrogen is important to life because it forms part of the molecules that make up living things, such as amino acids, the building blocks of proteins, and DNA. The nitrogen in proteins acts to bond together various amino acids to form the protein structure. The atmospheric abundance of nitrogen is very large compared to that found in the oceans or in rocks (Table 7.2). Approximately 79% of the atmosphere is composed of diatomic nitrogen gas (N_2). Plants and animals cannot use nitrogen directly from the air. It must first be changed into the compounds ammonia (NH_3) or nitrate (NO_3^-) by certain bacteria that live in soils or water before it can be used by plants and incorporated into their tissue. Nitrogen fixation is the process of converting atmospheric nitrogen to usable nitrogen. One reaction representing the nitrogen-fixation process in which the bacterium *Azobacter* plays a role is

$$2N_2 + 6H_2O = 4NH_3 + 3O_2.$$

The ammonia (NH_3) reacts with water to make ammonium ion (NH_4^+), which is then incorporated into plant tissue. Ammonium may also take part in another bacterially mediated process, that of nitrification. In nitrification, ammonium is converted to nitrite (NO_2^-) and then to nitrate (NO_3^-) by bacteria. The nitrate can then be used in plant production.

The roots of land plants take up nitrogen from the soil, and the nitrogen is used as a nutrient element for growth. In the ocean, phytoplankton (producers)

obtain their nitrogen for growth from the water. They have complicated methods to use the nitrogen dissolved in ocean water. Animals (consumers) receive the nitrogen they need when plants are eaten. When plants or animals die, bacteria and fungi (decomposers) decompose the remains. In the process of denitrification, microbes use nitrogen as a food source and produce nitrogen gas (N_2) and nitrous oxide (N_2O) as by-products. An example of a denitrification process is that of the production of diatomic nitrogen gas by the bacterium *Pseudomonas*:

$$4NO_3^- + 2H_2O = 2N_2 + 5O_2 + 4OH^-.$$

Decomposition releases nitrogen back into soils or waters which, if converted to ammonia by nitrogen-fixing bacteria, can be a fertilizer for plant growth. Some of the nitrogen is transferred back to the air as nitrogen and nitrous oxide gases, and the cycle is complete.

A newly discovered important reaction pathway involving nitrogen in the environment is that of ammonium oxidation that involves anammox bacteria discovered in wastewater sludge in the 1990s and only identified in 2003 in the marine environment. The process, along with the conventional denitrification process, contributes to nitrogen loss from oxygen-deficient environments. The anammox reaction is performed by a specialized group of bacteria belonging to the *Planctomycetes* phylum. The reaction is:

$$NH_4^+ + NO_2^- = N_2 + H_2O.$$

The anammox reaction seems to occur most importantly in oxygen-deficient zones of the ocean, like the oxygen minimum zone offshore of northern Chile, the Benguela upwelling zone offshore of western Africa, and the suboxic zone of the Black Sea. However, annamox facilitating bacteria have also been found in permafrost soils, freshwater lakes, and subtropical wetlands.

Much of the nitrogen used in productivity in both terrestrial and aquatic systems is recycled. Little new nitrogen is added via streams and the atmosphere. Of the total nitrogen used annually in terrestrial production, only about 25% of it is derived directly from the atmosphere in nitrogen fixation; the rest is recycled. In the ocean, recycling is even more efficient, with only about 1% of the nitrogen consumed yearly coming from river runoff or the atmosphere. The rest enters the productive shallow waters of the ocean through the upwelling of deep nitrogen-enriched water. These deep waters have been enriched in nitrogen by the sinking of dead organic matter out of the euphotic zone of the ocean and its decomposition in deep waters.

The global biogeochemical cycle of nitrogen is very complex, involving more than seven major gaseous and aerosol species of nitrogen in the atmosphere, three major species in aquatic systems, and nitrogen in organic matter and absorbed on minerals in rocks. Thus, there are many reservoirs, processes, and fluxes in the nitrogen cycle when viewed in detail. To give a feeling for the complexity of the system, Figure 7.10 shows a part of the biogeochemical cycle of nitrogen in the land environment. It illustrates the major chemical species of nitrogen in soil and their conversion from one species to another, principally by bacterial processes.

The natural biogeochemical cycle of nitrogen is being heavily impacted by anthropogenic activities to the extent that the global flux of nitrogen through its natural cycle has at least doubled because of

human activities. Figure 7.11 illustrates schematically that both natural and human sources emit nitrogenous gases to the atmosphere and dissolved and particulate nitrogen compounds to rivers that carry the nitrogen to aquatic systems, including the ocean. In their cycling through the Earth system, the human sources of increasing nitrogen fluxes have environmental impacts on people and terrestrial and oceanic ecosystems. Nitrogen is stored in the Earth in the form of dispersed organic matter, coal, oil, and gas. When the fossil fuels are burned, nitrogen gases may be released to the environment. In the exhaust streams of cars, nitrogen oxide gases are generated from atmospheric nitrogen and constitute an important component of urban smog and contribute to acid rain. Agricultural crops use nitrogen as a nutrient in growth. When these crops are harvested and taken to stores for our consumption or used to feed domestic animals, some nitrogen is removed with the crop. The nitrogen taken up during

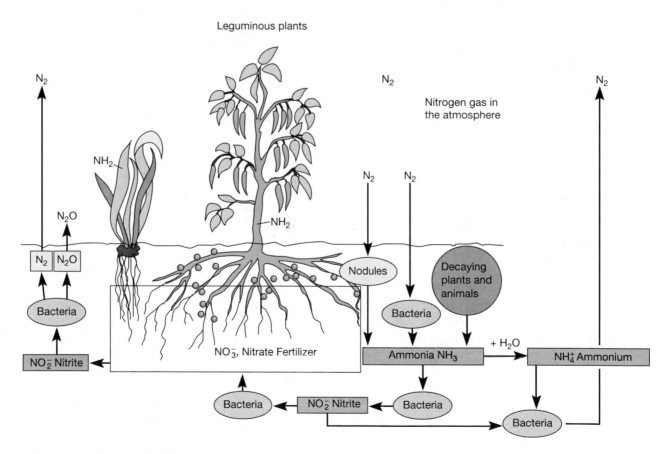

FIGURE 7.10 The simplified biogeochemical cycle of nitrogen in the land environment. This cycle is very complex, involving many reactions in which bacteria take part. These reactions are described in the text.

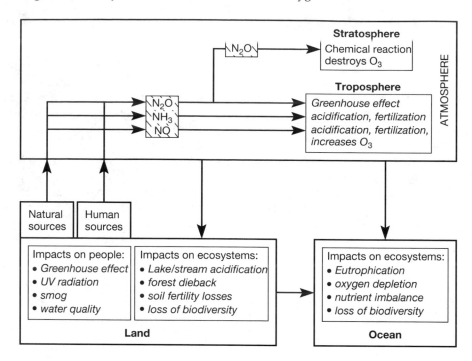

FIGURE 7.11 Diagram illustrating the gaseous nitrogen compounds emitted by natural and human sources to the atmosphere and the effects these compounds have on the atmosphere. Also shown are the impacts of deposition of these compounds or nitrogen-bearing reaction products on people, terrestrial ecosystems, and oceanic ecosystems. Rivers deliver both natural and anthropogenic nitrogen compounds to the ocean. (J. N. Galloway, personal communication.)

crop growth is not directly returned to the soil during decomposition of the crop plants. This nitrogen is not available for future plant growth. Artificial fertilizers of urea and ammonium nitrate made in the Haber-Bosch industrial process or, in some cases, nitrogen in sludge from waste treatment plants or in animal and human manure are spread on the land surface to replace the missing nitrogen. Frequently, excessive nitrogen is used on the fields, and this excess may get washed into streams and rivers, causing pollution. Also, the decomposition of nitrogen fertilizers applied to the land surface produces nitrous oxide, a greenhouse gas. Production of nitrous oxide also occurs during the decomposition of sewage released into the environment and in areas of eutrophication (see Chapter 11). Later chapters discuss further the impacts of human sources of nitrogen on the environment. The environmental issues related to excess nitrogen added by human activities to the natural biogeochemical cycle of nitrogen are perhaps as significant as those of global warming (see Chapter 14) and are not totally independent of this latter problem.

SULFUR

The global biogeochemical cycle of sulfur involves three major atmospheric components, those of reduced gaseous forms of sulfur, like dimethylsulfide (DMS) and carbonyl sulfide (OCS), sulfur dioxide (SO_2) gas, and sulfate (SO_4) aerosol. In aquatic systems, dissolved sulfate (SO_4^{2-}) is the major chemical compound found, and in sediments, sulfur occurs in the minerals pyrite (FeS_2) and gypsum ($CaSO_4 \cdot 2H_2O$) or its dehydrated form anhydrite ($CaSO_4$), and in organic matter.

A portion of the global biogeochemical cycle of sulfur is shown in Figure 7.12, that of the ocean and its role in sulfate aerosol and cloud formation. Dimethylsulfide is produced by algae in the surface waters of the ocean. The DMS concentration in the ocean is very low, but the dissolved gas is found nearly everywhere near the sea surface, where it may escape into the atmosphere. It is a trace gas comprising much less than 1% of the gases in the atmosphere. Once in the atmosphere, it takes only a few days before DMS is oxidized to sulfate and, along with other chemical species in the atmosphere,

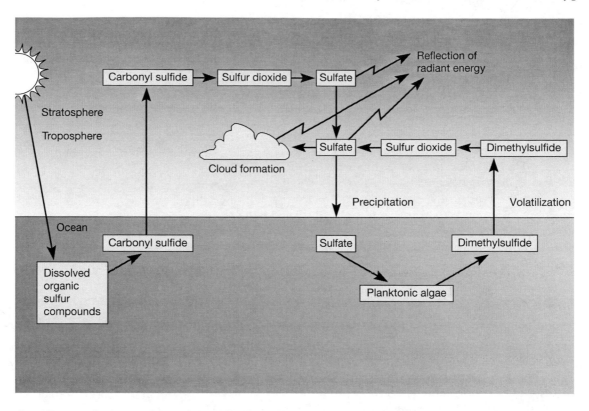

FIGURE 7.12   The biogeochemical cycle of sulfur as dimethylsulfide $[(CH_3)_2 S]$ and carbonyl sulfide (OCS) in the ocean–atmosphere system. This illustrates the biological and inorganic processes involved in the production of these gases in the ocean and their conversion to sulfate aerosols in the atmosphere. This aerosol may lead to the formation of clouds in the troposphere. The reflection of radiant energy back to space from sulfate aerosols and clouds may help cool the planet.   (*Source:* Andreae, 1987.)

condenses into small aerosol particles. These atmospheric sulfate aerosols act as nuclei on which rain droplets form, thereby facilitating formation of clouds and rain. The aerosol particles formed in the troposphere produced by oxidation of dimethylsulfide by themselves and acting as seeds for cloud formation affect the radiant energy balance of the oceanic atmosphere. Cloud formation over the oceans may lead to reflection of incoming solar radiation and have a cooling effect on the troposphere and surface of Earth. Furthermore, these aerosol particles interact with sunlight and water and produce some acidity in rain.

As seen in Figure 7.12, another reduced sulfur compound, carbonyl sulfide, is produced in the ocean from dissolved organic sulfur compounds and escapes the sea surface to the atmosphere. This gas is chemically inert in the troposphere, but in the stratosphere, it too is oxidized to sulfate. This process has contributed to a blanket of sulfate aerosols, the sulfate veil or Junge layer, in the stratosphere surrounding Earth. This veil reflects incoming solar radiation

and helps to cool the planet, thus helping to maintain an equable climate for Earth.

Two more features of the biogeochemical cycle of sulfur are important to climatic change (see Chapter 14). Volcanic explosions, like that of Mt. Pinatubo in the Philippines in 1991, can lead to the injection of sulfate aerosol into the stratosphere, resulting in a short-term cooling of Earth's climate. Also, sulfur dioxide gas released from fossil fuel burning can form atmospheric sulfate aerosol, particularly in the Northern Hemisphere—where most heavily industrialized nations are located—and potentially act as a cooling agent for the planet. Thus, aside from the role it plays in the acidity of rain, sulfur is an important player in regulating climate.

Eventually sulfur that is released into the atmosphere from the surface of Earth is returned to the ocean and land surface in precipitation. The cycle is complete.

The global sulfur cycle has been dramatically perturbed by the fossil-fuel-burning activities of human society. The flux of sulfur dioxide to the atmosphere from the burning of fossil fuels in certain regions of the world greatly exceeds natural fluxes of the gas. On a global scale, the principal flux involved in the exchange of sulfur between Earth's surface and the atmosphere is that of fossil fuel burning (Table 7.1). This flux has been so large compared to natural fluxes of sulfur that the net transport of sulfur between the land and the ocean has been reversed. Prior to human interference in the biogeochemical cycle of sulfur, the net transport of sulfur via the atmosphere was from the ocean to the land. It is now from the land to the ocean because of fossil fuel burning emissions of sulfur from the land. One environmental problem related to this large fossil fuel flux of sulfur is that of acid deposition, discussed in Chapter 12.

PHOSPHORUS

Phosphorus is important to life because it is an integral part of the RNA and DNA of living organisms. It is found both in the organic tissue and skeletons of organisms as phosphate (PO_4^{3-}). Unlike carbon and nitrogen, phosphorus is not present in significant quantities in the atmosphere as a gas. Only small amounts are transferred about Earth's surface

in gaseous form as phosphine gas (PH_3). The greenish glow seen at times in the air above swamps is a result of the oxidation of this reduced phosphorus gas escaping from the muddy, organic-rich sediments of the swamp. The glow on some occasions has mistakenly been attributed to an Unidentified Flying Object (UFO).

Nutrient phosphorus is leached by rain from rocks containing phosphate minerals, absorbed by plants from soil waters, and then passed on to animals when they feed. It may be recycled from excretions of animals and from dead plants and animals through organic decay processes and returned to the soil. After leaching and organic matter decays some phosphate is also transported by rivers to the oceans, where benthic organisms and marine plankton utilize it. Some phosphate settles to the ocean bottom in dead organic matter and inorganic particles and becomes part of the sediments. These sediments are eventually uplifted and phosphorus is returned to the land surface to begin its trip again. Figure 7.13 shows that part of the global biogeochemical cycle of phosphorus involving the land, ocean, and sediments.

Similar to nitrogen, most of the phosphorus consumed in organic production in terrestrial and aquatic ecosystems is recycled. Phosphorus is considered to be a limiting nutrient in most continental

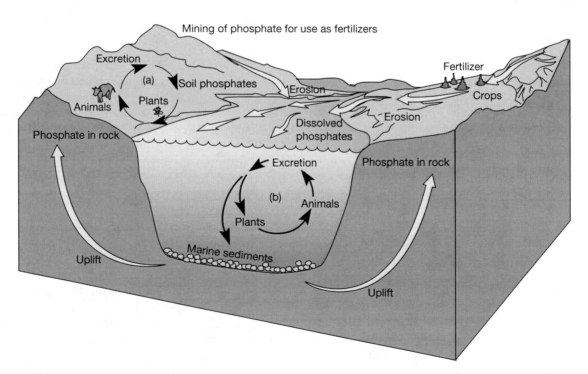

FIGURE 7.13   The global biogeochemical cycle of phosphorus involving the land, ocean, and sediments. The major processes moving phosphorus about Earth's surface are shown, as well as the internal cycling processes occurring within the (a) land and (b) ocean environments.

ecosystems. The reason for its low availability is that phosphorus ultimately must be derived from the slow process of weathering of surface rocks and is often bound to solid phases in sediments. There is no large atmospheric reservoir of phosphorus as there is for nitrogen. Whether phosphorus is also a limiting nutrient in the ocean is controversial. Phosphorus, nitrogen, iron, and other elements have been proposed as limiting nutrients in the ocean. For the global ocean over a long time, it is likely that phosphorus is the limiting nutrient, if for no other reason than that nitrogen can be obtained from the atmosphere by nitrogen fixation.

Phosphorus as inorganic phosphate compounds is an important element used in fertilizing agricultural crops. When the crops are harvested, the nutrient must be returned to the soils for the growth of future crops. Phosphorus is mined from the ground, where it occurs in commercial deposits of calcium phosphate rock and is added to cropland as fertilizer. Generally an excess of phosphorus is added to agricultural land to stimulate production of crops. It, like fertilizers of nitrogen, eventually may be washed into aquatic systems. Because phosphorus can be a limiting nutrient in rivers, lakes, and coastal marine environments, phosphorus additions to these systems can act as a potent fertilizer. Rapid growth of plants in both freshwater and marine environments may occur and lead to problems of eutrophication (see Chapter 11).

Up to this point, only portions of the biogeochemical cycles of certain elements have been discussed. To give the reader an appreciation of the complexity of the element cycles, the detailed, natural, global biogeochemical cycle of phosphorus is shown in Figure 7.14. The rates of transfer (fluxes) associated with processes involving the natural flow of phosphorus in the Earth system are also shown.

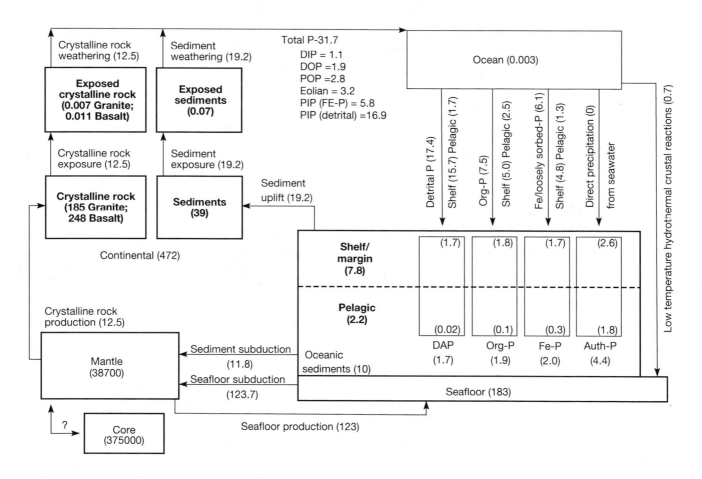

FIGURE 7.14 The geologic long-term biogeochemical cycle of phosphorus. All fluxes are in units of 10^{10} mol P/yr and masses are in units of 10^{18} moles P. To obtain masses and fluxes in grams, multiply the values on the diagram by the atomic weight of P = 31. (*Source:* Guidry and Mackenzie, unpublished.)

The weathering of both crystalline and sedimentary rocks leads to the delivery of various chemical forms of phosphorus to the ocean. The chemical forms include dissolved inorganic phosphorus (DIP), dissolved organic phosphorus (DOP), particulate organic phosphorus (POP), and detrital and iron oxyhydroxide particulate inorganic phosphorus (PIP). These materials are deposited in continental margin and deep sea sediments as detrital apatite [$Ca_5(PO_4)_3$ (F, Cl, OH)], phosphorus in organic matter, and phosphorus contained in iron oxyhydroxide compounds. In the sediments, diagenetic reactions occur that lead to the authigenic formation (new mineral formed *in situ* in the sediment) of a phosphorus mineral termed carbonate fluoroapatite. Hydrothermal reactions at the sea floor between basalts and seawater also add phosphorus to deep-sea sediments. The sediments containing the phosphorus-bearing minerals are then uplifted by tectonic forces to the Earth's surface to be chemically and mechanically weathered, releasing phosphorus once more to the environment. Deep-sea sediments and seafloor basalts may be subducted into the mantle, where they can be metamorphosed or partially melted and eventually returned as crystalline rock to the Earth's surface to be weathered. It is particularly the shallow ecosphere part of this great cycle and the internal processes within aquatic and sediment systems that are being impacted by the human activities of mining phosphate rock and the manufacture and application of phosphate fertilizers to the landscape.

C-N-P-S-O CYCLES AND HUMAN INTERFERENCE

A summary diagram of the movement of the key biological elements of carbon, nitrogen, phosphorus, sulfur, and oxygen in their biogeochemical cycles is shown in Figure 7.15. Oxygen is intimately associated in its biogeochemical cycle with the other elements through the processes of photosynthesis and respiration and decay, and its ubiquitous presence in the atmosphere and aquatic systems. Carbon, nitrogen, phosphorus, and sulfur are the elements that are continually cycling between inorganic compounds and life and, along with oxygen, hydrogen, and bioessential trace elements, are responsible for sustaining life on Earth. The movements of C, N, P, S, and O about Earth's surface involve transfers owing to bacterial processes in soils and waters, as well as those from power plants that release substantial quantities of C, N, and S into the regional and global environment. The principal environmental issues confronting humankind result from human interference in the biogeochemical cycles of the biological elements.

Some data on the sizes (masses) of reservoirs (boxes) and fluxes involved in the global biogeochemical cycles of carbon, nitrogen, phosphorus, sulfur, and oxygen are given in Tables 7.1 and 7.2. The inspection of a few quantitative data for biogeochemical cycles should give the reader an appreciation for the magnitudes of values involved in the global circulation of the elements. As a summary of earlier material in this chapter, some highlights of these element reservoir masses and fluxes in the context of human interference in the global biogeochemical cycles of the elements are discussed in this section.

Let us start with carbon. Notice that the mass of carbon tied up in sedimentary rocks is considerably larger than that in any other reservoir. This statement is also true for phosphorus, sulfur, and oxygen (Table 7.2). Furthermore, the atmospheric mass of carbon as carbon dioxide gas is smaller than that of any other surface ocean or land reservoir with which it may exchange carbon, except for the mass of carbon tied up in marine plants. For example, there is about 50 times more inorganic carbon dissolved in the ocean and three times more carbon stored on land in the form of living plants and soil humus than there is in the atmosphere. This large difference in masses implies that any change in the carbon content of the surface ocean, land biota and humus, and sedimentary rocks could significantly affect the concentration of CO_2 in the atmosphere.

The land and ocean reservoirs exchange large amounts of carbon with the atmosphere because of biochemical and physical processes (Table 7.1). Thus, changes in the carbon content of these reservoirs can affect the CO_2 concentration of the atmosphere on a time scale as short as that of a year. In contrast, for the large sedimentary carbon reservoir, changes in its carbon content affect atmospheric CO_2 concentrations on the much longer time scale of millions of years. Weathering of minerals and rocks on land consumes about 220×10^{12} grams of carbon annually. Thus, in only several thousands of years, the atmosphere would become depleted of CO_2, if there were no way in which the CO_2 used in weathering was returned to the atmosphere. An important part of this return is through emissions of volcanic and metamorphic CO_2 (Table 7.1), as we saw earlier in this chapter.

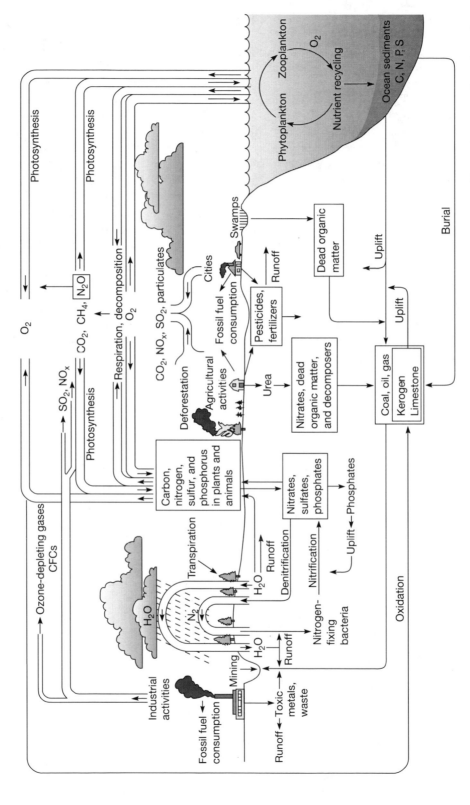

FIGURE 7.15 A summary diagram of the biogeochemical cycles of the key elements carbon, nitrogen, phosphorus, sulfur, and oxygen. The cycling of these elements is critical to the maintenance of life.

In terms of human interference in the carbon cycle, note that present emissions of CO_2 to the atmosphere from the burning of coal, oil, and gas (fossil fuels) and from deforestation and other land use activities amount to about 12% of the total flux of CO_2 to the atmosphere from natural respiration and decay processes on land. This is a significant perturbation of the global biogeochemical cycle of carbon, which is discussed in detail in Chapter 14.

Nitrogen and phosphorus play key roles as nutrients fueling organic productivity on land and in the ocean. The global biogeochemical cycle of nitrogen differs from those of the other elements in that the major reservoir of nitrogen is the atmosphere (Table 7.2). The element is present in the atmosphere primarily as diatomic (N_2) nitrogen gas. Major exchanges of the element between the atmosphere and the oceans and land occur via the processes of nitrogen fixation and denitrification (Table 7.1). This element cycle has been significantly perturbed by human activities. The rate of nitrogen fixation by industrial, combustion, and rice cultivation practices of about 140 million tons of nitrogen per year is now slightly more than the natural rate of biological fixation on land. As a consequence of this nitrogen fixation by humans, rivers transport to the ocean perhaps as much nitrogen from human sources as they do from natural sources. This is a major anthropogenic perturbation of a global element cycle.

The phosphorus cycle is considerably different from those of C, N, and S. Although there is transport of phosphorus through the atmosphere in aerosol and dust, there is no major gas of phosphorus in the atmosphere. The phosphorus in the atmosphere is in particulate form and represents a very small reservoir of the element (Table 7.2). Therefore, the major reservoirs of phosphorus are the land, ocean, and sedimentary rocks and the organic materials within them. The mining of phosphorus ores adds about 20 million tons of phosphorus to the land surface annually, mainly in the form of fertilizer. The transport of dissolved phosphorus in rivers to the ocean derived from sewage and by leaching of phosphorus from fertilizers added to croplands is about 1.5 million tons of phosphorus per year, equivalent to about the flux derived from natural sources. We shall consider in later chapters the effects of the human interferences in the phosphorus and nitrogen cycles.

Sulfur is present in the atmosphere in both chemically reduced and oxidized forms. The ocean is an important reservoir of this element, where it is present primarily as dissolved sulfate ion (SO_4^{2-}) (Table 7.2). Combustion of fossil fuels is an important source of sulfur dioxide to the atmosphere. This flux of about 60 million tons per year exceeds any natural flux of the element to the atmosphere. The result is that about 35% of the total global deposition of oxidized sulfur on Earth's surface from the atmosphere in rainfall and in particulate matter comes from sources related to human activities. This is a very significant perturbation of the global sulfur cycle, which is discussed in Chapter 12.

The global biogeochemical cycle of oxygen is strongly coupled to the cycles of C, N, P, and S. Oxygen oxidizes reduced chemical forms of C, N, and S and is an active agent of weathering, affecting the availability of phosphorus on land and, through runoff, the ocean. The reservoir of atmospheric diatomic oxygen (O_2) is large, amounting to 12×10^{20} grams of oxygen. However, as with carbon, phosphorus, and sulfur, most of the mass of oxygen is tied up in minerals in sedimentary and other rock types (Table 7.2). Oxygen is mainly exchanged between the atmosphere and Earth's surface via the processes of photosynthesis, which adds oxygen to the atmosphere, and respiration and decay, which abstract oxygen from the atmosphere (Table 7.1).

The oceans also exchange large quantities of oxygen, as well as carbon, with the atmosphere because of physical processes. Some years ago concern was voiced that the burning of fossil fuels would significantly reduce the oxygen content of the atmosphere. This opinion stems from the fact that for each new CO_2 molecule that appears in the atmosphere from the burning of fossil fuels, one O_2 molecule disappears. However, a simple calculation shows that the statement concerning oxygen depletion is not true. There are about 5×10^{12} metric tons of carbon tied up in the fossil fuel reserves of the world. If these reserves were burned completely, their burning would require about 11×10^{12} metric tons of oxygen, less than 1% of that available in the atmosphere. It is interesting to note that the small annual depletion of oxygen in the atmosphere of about 0.001% because of fossil fuel burning is observable with modern analytical techniques.

Concluding Remarks

Chemical, biological, and biogeochemical cycles are found in all "spheres" of planet Earth. These cycles interact with one another and are essential for life on the planet. The activities of people have significantly impacted these cycles and Earth's surface system. These human-induced changes in biogeochemical cycles can be the root causes of global environmental changes on the time scale of human societies. The extent to which these cycles can accommodate change brought about by fluxes from human activities and their influence on and response to global environmental change are subjects of considerable scientific interest and practical importance today. Later in this book we shall explore in greater detail these human impacts on biogeochemical cycles and global environmental change induced by human activities.

Study Questions

1. The total carbon in the atmosphere currently is about 800 billion tons as carbon dioxide gas. Plants remove 100 billion tons in net primary production each year. How long would it take the atmosphere to run out of carbon dioxide if there were no means of replenishment of the gas?

2. On the time scale calculated in question 1, how is the carbon dioxide of the atmosphere replenished?

3. Where is most of the carbon in Earth's surface system stored?

4. What are the two major processes controlling atmospheric carbon dioxide levels on a long time scale? Write a balanced set of chemical equations demonstrating these processes.

5. During the Phanerozoic, how has atmospheric carbon dioxide content varied on a long time scale?

6. What is the connection between organic matter and atmospheric oxygen?

7. Why is fossil fuel a nonrenewable resource?

8. During the Phanerozoic, how has atmospheric oxygen varied on a long time scale?

9. What is the process of nitrogen fixation?

10. By what process is nitrogen returned to the atmosphere? Write a balanced chemical reaction demonstrating this process.

11. What are the three major forms of atmospheric sulfur, and where are they produced?

12. (a) What is a major difference in the reactivity of DMS and carbonyl sulfide in the atmosphere?
    (b) How might these gases affect climate?

13. How can volcanic eruptions like that of Mt. Pinatubo affect global climate?

14. What is the major difference between the biogeochemical cycle of phosphorus and those of carbon, nitrogen, and sulfur?

15. (a) Why is there concern for the leaching of agricultural fertilizers of nitrogen and phosphorus into aquatic systems?
    (b) Why are these elements able to change the biological characteristics of an aquatic ecosystem?

16. Construct a simple box model of the global biogeochemical cycle of water. Use the information in Chapter 5, Figure 5.1, and Table 5.1, and the reservoirs of atmosphere, land, and ocean.
    (a) What is the residence time of water in the ocean with respect to the flux of water via the rivers to the ocean?
    (b) If the concentration of dissolved calcium is 400 ppm in the ocean and 15 ppm in average river water, what is the residence time of calcium in the ocean?

17. A nearly rectangular-shaped lake is 5 kilometers long, 2 kilometers in width, and 100 meters deep and contains 0.001 mg/L of dissolved mercury. A river discharges 2×10^{12} L/yr of water with a concentration of mercury of 0.0005 mg/L into the lake.
    (a) What is the water volume in the lake?
    (b) What is the total mass of mercury in the lake?
    (c) What is the residence time of mercury in the lake?
    (d) An accidental discharge of mercury into the lake occurs from an industrial plant located near the mouth of the river. How long will it take for most (95%) of the mercury contamination of the lake to disappear?

18. Write balanced equations for the three major processes in the nitrogen cycle of nitrogen fixation, nitrification, and denitrification.

19. (a) The Earth has a surface area of 510×10^{12} m^2. The weight of air above each cm^2 of the Earth's surface is 1,031 g/cm^2. What is the mass of the atmosphere?
    (b) The average molecular weight of air is 29 g/mole. How many moles of gas are there in the atmosphere?
    (c) SO_2, a major atmospheric pollutant in photochemical smog, has an average concentration of 0.2 ppbv (parts per billion by volume) in remote air. How many moles of SO_2 are there in the background atmosphere?
    (d) What is the residence time of SO_2 in the remote atmosphere assuming a natural flux of 1×10^8 metric tons of S per year into the atmosphere? What does the value of this residence time tell you qualitatively about how far from a fossil fuel combustion

source (e.g., coal-fired electrical generating plant) SO_2 released into the atmosphere may travel?

20. Very commonly the relationship between flux (F) and mass (M) in a biogeochemical box model is considered to be first-order. (See box on residence time in your book.) Construct (diagram) a three-box ocean model for the oceanic phosphorus biogeochemical cycle with the reservoirs of Surface Ocean Water (M_1, $2{,}710 \times 10^{12}$ g P), Oceanic Phytoplankton (M_2, 138×10^{12} g P), and Deep Ocean (M_3, $87{,}100 \times 10^{12}$ g P). NPP (F_{12}) removes $1{,}040 \times 10^{12}$ g P/yr from surface ocean water and 998×10^{12} g P/yr are returned in oxidation and decay processes (F_{21}). 42×10^{12} g P/yr in organic matter sink into the deep ocean (F_{23}) and upwelling (F_{31}, 58×10^{12} g P/yr) and downwelling (F_{13}, 16×10^{12} g P/yr) processes lead to exchange of P between the deep ocean and the surface ocean.

(a) What is the residence time of P in the oceanic phytoplankton calculated with respect to NPP? In one year, how many times does P turnover in the oceanic phytoplankton reservoir?

(b) Write the 5 equations showing the fluxes between reservoirs assuming a first-order relationship between mass and flux.

(c) Calculate the constants (the rate constants) for each transfer process, assuming the P cycle is at steady state.

(d) The major controls of organic production in the present oceanic system are generally the availability of the limiting nutrients of nitrogen and phosphorus. Let us hypothesize that some other control takes over the photosynthetic rate globally, say iron as a limiting nutrient, changes in light availability due to particulates in the atmosphere, warming of the sea surface, etc. Furthermore, let us assume that the effect is instantaneous, and that the photosynthetic rate is cut in half overnight; that is, k_{12} is changed from 0.384/yr to 0.192/yr.

Qualitatively, what should happen? The rate of production of the biomass will be approximately cut in half, but there is no reason to assume that the rate of decay, the rate of sinking of organisms, the rate of upwelling, or the rate of downwelling will be affected. The biomass should shrink; the flux of decay should increase; and the flux of sinking organisms should diminish. Because the total system is a closed one, that is $M_1 + M_2 + M_3$ equals a constant, there should be a redistribution of phosphorus among the reservoirs. As the biomass declines in phosphorus content, the surface and deep waters should be enriched in P. A new steady state would be achieved, probably with lower total P in the biomass and higher total P in the waters. This is qualitatively what might happen.

Now solve this problem quantitatively using the new value of 0.192/yr for k_{12}. Remember the system is closed and only the k for the flux F_{12} has been changed. You need to set up a new three-box model with five fluxes. As a check on your calculations, you should end up showing a lowering of the biomass by about 2/3 of its original size and an increase of dissolved P in the ocean surface waters of about 50% owing to the instantaneous lowering of the photosynthetic rate by 50%.

Additional Sources of Information

Berner, E. K. and Berner, R. A., 1996, *Global Environment: Water, Air, and Geochemical Cycles.* Prentice Hall, Upper Saddle River, NJ, 376 pp.

Berner, E. K. and Lasaga, A. C., 1989, Modeling the geochemical carbon cycle. *Scientific American,* March, pp. 74–81.

Butcher, S. S., Charlson, R. J., Orians, G. H. and Wolfe, G. V. (Eds.), 1992, *Global Biogeochemical Cycles.* Academic Press, London, 379 pp.

Few, A. A., 1991, *System Behavior and System Modeling.* Global Change Instruction Program, University Corporation for Atmospheric Research, Boulder, CO, 57 pp.

Mackenzie, F. T., 1997, *Global Biogeochemical Cycles and the Physical Climate System.* Global Change Instruction Program, University Corporation for Atmospheric Research, Boulder, CO, 69 pp.

Mackenzie, F. T. and Lerman, A., 2006, *Carbon in the Geobiosphere: Earth's Outer Shell.* Springer, Dordrecht, The Netherlands, 402 pp.

Schlesinger, W. H., 1997, *Biogeochemistry: An Analysis of Global Change,* 2nd ed. Academic Press, New York, 588 pp.

Smil, V., 1997, *Cycles of Life: Civilization and the Biosphere.* Scientific American Library, W. H. Freeman, New York, 221 pp.

8 Historical Framework of Global Environmental Change

If the Lord Almighty had consulted me before embarking on the Creation,
I would have recommended something simpler.

ALFONSO X OF CASTILE

Before we can discuss how the ecosphere is influenced by human activities, it is necessary to recognize how the system has changed naturally throughout geologic time. Geologists, oceanographers, biologists, chemists, astronomers, climatologists, theologians, and others have applied their professions to find answers to the puzzle of how the universe and Earth evolved, how the planet works today, and what its future will be. In effect, they study Earth system science and global environmental change. It is the accumulation of the wisdom and research involving scientists in many disciplines that is helping to piece the puzzle together. The puzzle is far from complete, but each new piece of information added from knowledge of Earth's past provides a new perspective on the phenomenal world in which we live.

In this chapter the broader aspects of the evolution of the universe and the solar system are presented along with an encapsulated view of the evolution of Earth's surface system, its exogenic system and ecosphere. The discussion involves some of the biggest mysteries of our planet for which there is still controversy. Further details of the history of the planet in terms of its biogeochemical cycling and Pleistocene and Holocene environmental and climatic change are given in Chapters 7 and 13.

"THE BIG BANG"—THE EVOLVING UNIVERSE

This period of time spans the evolving universe from creation at about 13.7 billion years ago to 4.6 billion years ago.

The vastness and complexity of the universe (everything that physically exists) make it difficult to comprehend. The **universe** is the entirety of space and time, all forms of matter, energy and momentum, and the physical laws and constants that govern them. It is filled with an array of celestial energy, matter, and space (Box 2.2, Matter: Atoms, Elements, and Compounds). The universe was probably created from a very small, dense conglomeration of matter and energy. Modern physical theory suggests there may be multiple universes and that our modern universe may be the most recent in a series of preceding universes that formed from small, dense conglomerations of matter and energy, expanded, then collapsed, and re-expanded. It is anticipated that when the European Union of Nuclear Research CERN (Centre Européen de Recherche Nucleaire) Large Hadron Collider, built in the French village of Cessy near the Swiss city of Geneva, is fully operational that experiments will be performed shedding light on universe origin and evolution (a beam of protons was first fired within the collider in September 2008). Whatever the case, data collected from the light of distant stars reveal that matter currently is moving away from an initial center of mass, with the more distant bodies moving at faster velocities. This observation and other data, including recently acquired microwave data from the Cosmic Background Explorer (COBE) satellite, are consistent with the idea of an explosion, occurring 10 to 15 billion years ago, of the dense center of matter into a titanic fireball. This explosion is called the **Big Bang**.

One way of estimating the age of the universe is from a cosmological model and the **Hubble constant** that gives the rate of recession of distant astronomical objects per unit distance away. There is some degree of uncertainty in this age estimate that depends on the value taken for the constant. The most recent age estimate of the universe of 13.73 billion years (±120 million years) is based on precision measurements of the oldest light in the universe obtained from cosmic microwave fluctuations using the NASA satellite known as the Wilkinson Microwave Anisotropy Probe and a cosmological model.

One confirmation alluded to above for the age of the universe comes from the relationship between the distance from Earth to a distant galaxy or group of galaxies and the velocity at which the galaxy is receding. Such observations depend on several different techniques, involving trigonometric, spectroscopic, and stellar luminosity considerations, and are fraught with difficulties. The red shift first reported by Edwin Hubble is part of this story. Hubble observed a shift toward red in the spectra of light reaching Earth from stars in distant galaxies. His

explanation of this red shift was that the distant galaxies were moving away from ours at incredible speeds. When the observations for the various galaxies are plotted on a graph, they exhibit a linear relationship (Figure 8.1). This is to be expected if the universe formed as a result of the Big Bang and was flying outward due to the explosion. If we divide the distance in centimeters of any galaxy from Earth shown in Figure 8.1 by its retreat velocity in centimeters per second, we obtain the time in seconds that it has taken for the galaxy to recede from us to its present position. This value is approximately 4.6×10^{17} seconds, or roughly 15 billion years. However, the uncertainty in this method of calculation of the age of the universe is several billion years.

The value of the Hubble constant is constantly being refined from observations made with the Hubble telescope and instrumentation on satellites. The range in values of the constant accommodates several different hypotheses concerning the origin and fate of the universe—will the universe continue to expand or collapse back on itself? As of this writing, the former seems most likely, but as mentioned, there may have been a universe that preceded the modern one.

When the universe was only 10^{-35} seconds old, it was still incredibly small, about the size of a pea. It had a temperature of 10 billion billion billion (10^{28})

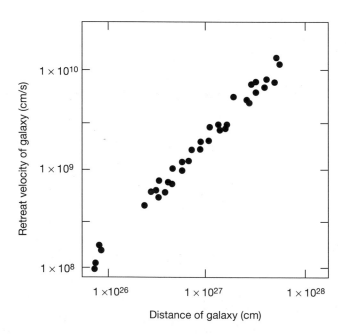

FIGURE 8.1 Relationship between the distance to a galaxy or group of galaxies and the velocity at which the galaxies are moving away from ours. Notice that the data are plotted on a logarithmic scale. (*Source:* Broecker, 1985.)

degrees! After the fiery Big Bang, the universe was slow to cool. After a million years of evolution, its temperature was still about 3000°C. In the enormous, outward-expanding, gaseous cloud created when the universe began, smaller clouds of hydrogen and helium gas formed. These clouds of gas condensed under gravitational pull and contracted into swirling, tabular masses of material that eventually would become galaxies (Figure 8.2). The spinning cosmic clouds fragmented further, and smaller, individual, swirling masses (protostars) formed within them. During their evolution, protostars contracted under gravitational pull and became denser. Their temperatures rose and nuclear fusion reactions began within them. At this time, the inward collapse of the protostars stopped. Energy-producing stars were born. Billions of stars were created in this way. Even today stars are still being formed in the galaxies. Within reach of our telescopes, there are estimated to be more than 100 billion billion (10^{20}) stars.

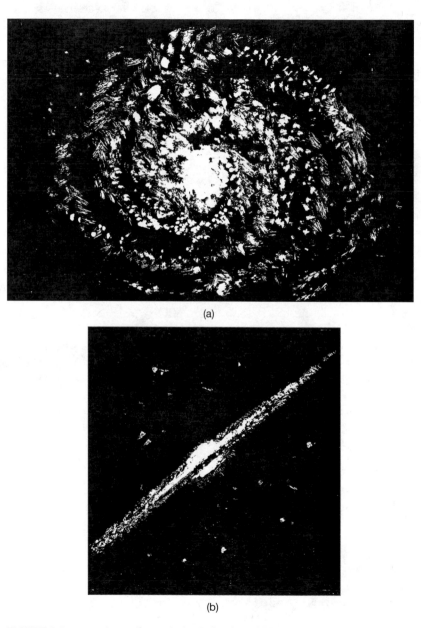

(a)

(b)

FIGURE 8.2   Two views of a typical spiral galaxy. (a) Perpendicular to plane. (b) Side view.

Within the first 20 minutes of the life of the universe, the Big Bang created by nucleosynthesis most of the atoms and masses of the elements hydrogen (H) and helium (He) in the ratio of about 3H:1He by mass. The other heavier elements (in 2008 there were 117 known elements of which 94, including H and He, naturally occur on Earth) were created through the birth and death of several generations of stars. First-generation stars formed from contraction of a large mass of gas under its own gravity. Inside these stars, hydrogen was fused in thermonuclear reactions to make helium, which in turn led to the creation of the heavier elements neon, oxygen, silicon, and iron. Later in the life of these massive stars, their temperatures rose to billions of degrees. This was due to the continual compression of their cores from the force of gravity. The compression resulted in the death of these stars in supernova explosions. The elements in the stars and new heavier elements like uranium and lead created by their explosions were spewed out into the neighboring gas cloud and seeded interstellar space. Generations of supernova explosions were necessary to produce the additional 92 elements heavier than hydrogen and helium. These explosions generated the gas, dust, and water in interstellar space needed to form our solar system and other planetary bodies.

Supernova explosions continue today. Massive stars, red giants, explode at the rate of about one per galaxy per century. The Crab Nebula is the debris from one such supernova. Chinese astronomers observed this nebula forming in A.D. 1054. Today, 99% of the mass of the universe remains as hydrogen and helium. Only 1% is heavier elemental matter produced from supernova explosions.

Our own galaxy of the Milky Way started to form approximately 13.2 billion years ago, nearly the age of the universe; the oldest stars in the galaxy are of this age. Interstellar matter makes up about 10% of the total mass of the galaxy, equivalent to 10 billion stars. There are about 100 billion stars in the present galaxy. In the outer rim of the Milky Way Galaxy, our solar system started to form about 4.6 billion years ago from a solar nebula consisting of a swirling cloud of gas and dust that condensed and collapsed under its own weight. The result was a thin rotating disk with most of the mass of the material at the center of the disk. Through frequent collisions of material, the matter heated up and hotter regions characterized the interior of the disk and cooler regions the outer regions of the disk. Within about 50 million years, the pressure and temperature at the center of the disk increased sufficiently so that the fusion of hydrogen could take place and an inner star, which today is our sun, formed from the solar nebula. This protostar initially was cooler than today's sun. It had a composition similar to that of the original cloud and contained 99% of its weight in hydrogen and helium. The remaining 92 elements comprised the rest of its mass. A small amount of matter of the solar nebula formed outer, rocky and icy smaller systems swirling and revolving around the inner star. Within about 100 million years, the accretion of smaller bodies led to the solar system planets of today, surrounded by a rim of nonaccreted material (Figure 8.3). The planets attained compositions different from that of the star (for example, see Table 3.1 showing the difference between the composition of the sun and the Earth).

The planets closest to the sun developed as rocky, terrestrial planets from accretion of solid matter termed "**planetesimals.**" The material of these planets consists primarily of four heavy elements—iron, magnesium, silicon, and oxygen—originally released during supernova explosions. This matter forms much of the mass of the inner terrestrial planets: Mercury, Venus, Earth, and Mars. Mercury, closest to the sun, is extremely hot; Venus, second planet from the Sun, is hot; and Mars, the fourth, is cold. Earth, the third planet from the sun, is located at a distance that enabled life and a hydrosphere and an unique oxygen-containing atmosphere to evolve together. This coevolution produced the only known life-support system in the solar system. Recently acquired controversial evidence for fossilized bacteria and organic compounds indicative of life, and found in a meteorite on Earth that originated on the surface of Mars, may point to life existing, or having existed, on Mars.

The planets farthest away from the sun formed mainly as huge gaseous and icy bodies. The outer planets of Jupiter, Saturn, Uranus, and Neptune are composed of frozen gases like hydrogen (H_2), helium (He), methane (CH_4), and ammonia (NH_3). Because of its distance from Earth, relatively little is known about rocky and icy Pluto. Recent photographs obtained from the Hubble telescope are revealing more about the constitution of Pluto. It should be pointed out that Pluto is no longer recognized as a true planet in the solar system but is considered the largest body in a region known as the Kuiper belt. The Kuiper belt is a region in the solar system extending from the orbit of Neptune at 30 astronomical units (AU; 1 AU is equivalent to the distance from the sun to Earth, approximately 93,000,000 miles

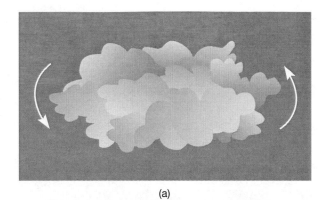

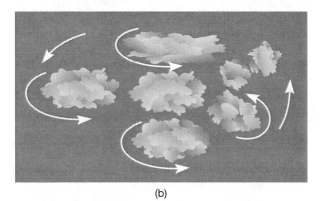

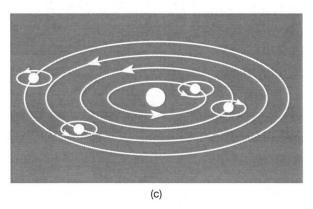

FIGURE 8.3 Stages in the evolution of the solar system containing Earth. (a) Initial swirling cloud of gas and dust. (b) Early stage in separation and formation of the sun and protoplanets. (c) Parts of the present solar system with the sun and the planets Mercury, Venus, Earth, and Mars. (*Source:* Mintz, 1972.)

or 149,597,871 kilometers) to approximately 55 AU from the sun. Like the asteroid belt of small bodies of rock and metal between Mars and Jupiter, the Kuiper belt consists of small bodies also but is composed mainly of frozen methane, water, and ammonia ice. Both the material of the Kuiper belt and the asteroid belt are remnants of early solar system formation. The

present characteristics of the planets are given in Table 8.1.

The planets, with the exception of Pluto, orbit the sun in nearly circular paths in a plane that includes the equator of the sun. With the exception of Venus and Uranus, they also spin and revolve in the same direction as the sun (Figure 8.4). The effect somewhat resembles that of a flat plate, like that of the galaxies.

The present solar system consists of a star we call the sun, its eight discovered planets, their 166-plus named moons (not including Pluto's three moons), myriad comets, and innumerable asteroids. Ninety-nine percent of the total mass of the solar system still resides in the sun, resulting in less than 1% left over for all other orbiting matter. The solar system is the only well-documented planetary system in the universe. The total number of stars in the universe is estimated at $10^{22}-10^{24}$ and there are more than 400 known planets. In 2008 the first ever direct-imaging pictures were released by the National Aeronautics and Space Administration (NASA) and the Lawrence Livermore National Laboratory of four newly discovered planets orbiting stars outside our solar system. Considering the size of the number of stars, there may be as many or more planets than there are stars!

We have briefly considered the origin of the universe and solar system in light of modern evidence for their formation. Now we will take a tour through geologic history to understand how planet Earth evolved prior to human arrival.

EVOLUTION OF PLANET EARTH

The surface environment of Earth has dramatically changed since its formation from a hot, oxygenless, and lifeless planet to an oxygenated world with a diversity of life forms. We are unable for most of geologic time to sample directly atmospheric or hydrospheric composition, and we must rely on the fossil record to provide clues as to organic evolution and past biodiversity. Thus, much information about the evolution of the ecosphere comes indirectly in the form of biological, chemical, mineralogical, and isotopic data primarily from sedimentary rocks, so called proxy data. The data are not uniformly distributed through geologic time and become more scant with increasing age of the rock record because of the selective preservation of younger sedimentary rocks and probably the lack of large quantities of sediments deposited early in Earth history. The sedimentary

TABLE 8.1 Characteristics of the Principal Planets of the Solar System

		Terrestrial Inner Planets					Great Outer Planets			
	Sun	Mercury	Venus	Earth	Mars	Jupiter	Saturn	Uranus	Neptune	Pluto
Distance from sun (millions of km)		58	108	150	228	778	1430	2870	4500	5900
Mass (units of 10^{21} tons)	1,984,000	0.3586	4.900	5.976	0.657	1900	568.8	86.89	102.5	0.012
Density (g/cm^3)	1.41	5.44	5.25	5.552	3.94	1.33	0.69	1.27	1.64	2
Volume (units of 10^{12} km^3)	1,410,000	0.066	0.933	1.083	0.167	1429	824	68.4	62.5	0.006
Equatorial radius (km)	696,000	2439	6052	6378	3397	71,492	60,268	25,559	24,764	1140
Surface temperature (°C)	5480	350	460	15	−50	−103	−153	−214	−214	Approximately absolute zero
Number of satellites	Many	0	0	1	2	16	18	15	8	1
Gases in atmosphere	Many	None	CO_2, N_2	Many, primarily N_2 and O_2	CO_2, H_2O, N_2	H_2, He, NH_3, CH_4	H_2, He, NH_3, CH_4	H_2, He, CH_4	H_2, He, CH_4	CH_4, other compounds

(*Source:* Wayne, 1991; Boyce and Maxwell, 1992.)

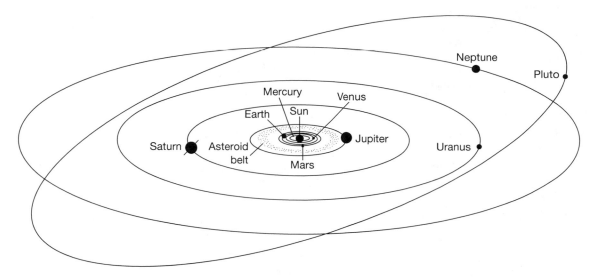

FIGURE 8.4   The present solar system containing Earth showing positions and orbits of the planets. Terrestrial planets are Mercury, Venus, Earth, and Mars. The outer planets of Jupiter, Saturn, Uranus, and Neptune are gaseous. Pluto is composed of methane ice and other compounds.

rock mass preserved per unit increment of geologic time decreases roughly exponentially with increasing age of the mass. The 3.3 billion years of the Precambrian Eon are represented by only about one-third of the total sedimentary mass. More than two-thirds of the mass is found in the younger Phanerozoic Eon of 545 million years' duration. Thus, any interpretation of the geologic history of the exosphere is fraught with difficulties.

We next discuss the evolution of Earth with emphasis on the ecosphere. In Chapter 13, the more recent history of the Earth is described in more detail.

Hadean Eon: Earth's Violent Beginning

Hadean *is Greek for "beneath Earth." The eon spans 4.6 billion to about 3.8 billion years* B.P. *(before the present).*

No rocks that would give a complete and accurate picture of the earliest evolution of Earth remain on the planet from this eon. However, the oldest confirmed date for a rock comes from the Acasta gneiss in Canada at 4.03 billion years; zircon ($ZrSiO_4$) mineral grains dating about 4.2 billion years old have been found in the Jack Hills Formation of Western Australia. The debatable finding of a rock age of 4.28 billion years for the Nuvvuagittuq greenstone belt of northern Quebec, Canada (dark metamorphosed basaltic rock that once was deep sea lava containing various green minerals), would imply an even earlier formation of Earth's primitive continents. The early surface of the planet was

transformed completely because of bombardment of planetesimals, tectonic movements, and erosion by wind and water. This has made it impossible to determine by observation of the rocks of Earth exactly what our planet was like when it formed. However, evidence of the nature of formation of the early Earth is partially derived from rocks found on the moon. Also, the composition and history of other planets that formed at the same time as Earth and the chemistry of their atmospheres, experimental information on minerals and rocks, telescopes and other instruments on space platforms examining processes in the universe, and the composition of meteorites provide clues as to the nature of the early Earth. An understanding of the earliest history of our planet has been obtained from these and other sources of information. Figure 8.5 shows a summary of major environmental trends and events and biological evolution during the Hadean and Precambrian eons.

During formation of the solar system, Earth accreted from the agglomeration of kilometer-sized and bigger planetesimals. "Time zero" for the formation of the solar system is generally agreed to be 4.567 billion years and by 4.55 billion years as much as 65 percent of Earth's mass had accreted. The proto-Earth formed in a very hot nebular region of the solar system where materials were depleted in the lighter elements of hydrogen (H), carbon (C), nitrogen (N), and oxygen (O). During the early growth history of the proto-Earth, it was stuck by a large Mars-sized protoplanet. The collision removed whatever primordial

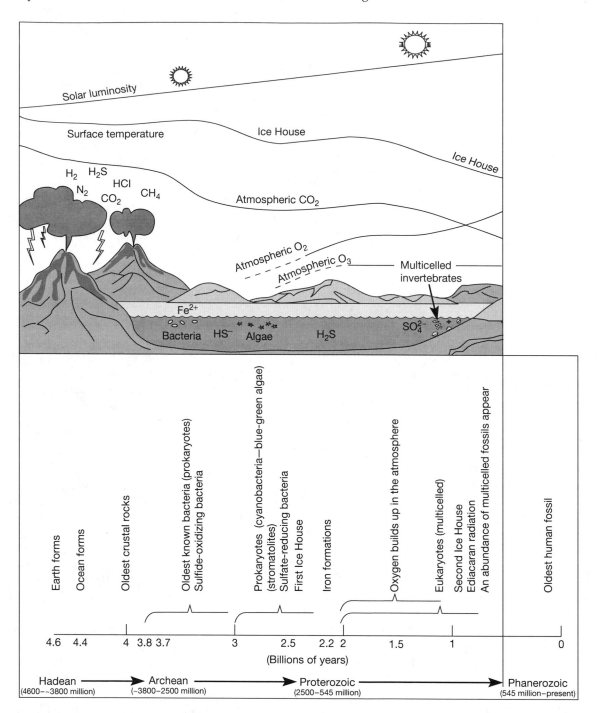

FIGURE 8.5   Some important biological and physical trends and events in the history of Earth's surface environment during the Hadean and Precambrian.

ocean and atmosphere that existed at the time and further depleted the planet in light volatile elements, leaving the reformed Earth depleted in light elements. The collision most likely also led to the formation of the moon by agglomeration of some of the

debris released into space by collision of the proto-Earth with the Mars-sized planetary body.

It is now believed that the early Earth was in a molten condition nearly to its surface because of the energy released in the impact and agglomeration of large planetesimals. The radioactive decay of the elements uranium (U), thorium (Th), and potassium (K) contained in the planetesimals and the production of heat by this process also added energy to the planet. As the planet accreted, a series of hierarchically larger planetesimal bodies collided with the primordial surface of Earth. These large impacts created so much energy that nearly the entire planet was molten at the end of its formation. Thus, it is likely that an early molten magma ocean existed at Earth's surface.

The early accreting planetesimals were probably materials formed at relatively high temperatures in the solar nebula during its cooling. These materials were rich in iron, an element that would eventually make up most of the interior core of Earth (Figure 8.6). The later materials of accretion from the nebula were probably enriched in volatile elements like hydrogen (H), helium (He), and sodium (Na). Icy planetesimals, asteroids, comets, and interplanetary dust were probably the source of replenishment of the lighter volatile elements and of organic compounds as the primordial solar system cleared itself of materials. Sodium and other elements, which eventually would make up the crust of Earth (Figure 8.6), form solids at lower temperatures than iron. Therefore, they condensed out of the solar nebula later in its cooling history.

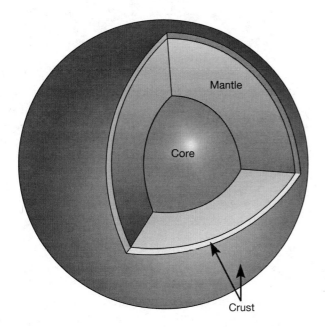

FIGURE 8.6 The zonation of Earth into a dense, iron-rich core, a thick iron-magnesium-silicon-oxygen–rich mantle, and a thin sodium-potassium-oxygen-hydrogen–rich crust.

The molten state of the early Earth led to core formation nearly simultaneously with the accretion of the planet. The planet very early in its history became zoned into core, mantle, and crust (Figure 8.6). Dense materials sank inward and formed the iron core of primitive Earth, leaving less dense materials to accumulate in a thick zone surrounding the core. This region is the mantle. At the surface, an outer, thin, discontinuous skin of islands of an original igneous rock crust developed as the planet cooled. The islands eventually formed proto-continental size masses and they and the rest of Earth continued to undergo heavy bombardment by meteors from 4.5 billion to 3.8 billion years ago. Also, the surface of the planet was racked by cometary impacts and volcanic eruptions. It is likely that if early life formed at this stage in Earth's evolution, it was killed off several times before it survived.

During accretion of primitive Earth, its heat energy escaped into space, along with some volatile materials. Earth is still depleted in some elements, like hydrogen (H), argon (Ar), nitrogen (N), neon (Ne), and helium (He), relative to solar matter. This depletion indicates that these elements and other volatile constituents probably were lost from the planet early in its evolution. As mentioned, any primitive atmosphere was probably blown away by strong solar winds and the giant Mars-sized impact and other impacts. Ninety-nine percent of Earth's original noble gas (helium, neon, argon, krypton, xenon, and radioactive radon) inventory was lost to space during these catastrophic impacts.

Above the crust, after the loss of some of the most volatile of the constituents of the planet, a secondary atmosphere developed from gases of nitrogen (N) and sulfur (S), carbon dioxide (CO_2), and water vapor (H_2O) emitted (degassed) from the mantle via volcanoes and hot springs. These gases were called "**excess volatiles**" in a classic paper in 1951 by W. W. Rubey. In small quantities, these gases are still reaching Earth's surface because of volcanism. A substantial portion of these gases was probably added to Earth from space in relatively cool planetesimals and other bodies during the later stages in the planet's accretionary history. Water, an essential compound for supporting life, is formed from two of the three most abundant elements in the universe (hydrogen and oxygen) and is ubiquitous in interstellar space, in our solar system, and on Earth. Most of it probably accumulated in Earth during formation of the last half of its mass from wet planetary embryos. On impact the water was degassed to

Earth's surface where it would dissolve in the early magma ocean, and eventually, once the planet had cooled sufficiently, much of it would constitute the primitive hydrosphere.

These late accreting planetesimals may also have been the source of original organic compounds for the planet. Another source may have been interplanetary dust particles floating around when Earth was forming. Table 8.2 lists some compounds detected in outer space. If compounds of this nature were transported to Earth during its early history, they could serve as building blocks for life. Lightning was probably a prominent phenomenon of the sky at this time and may have been a source of energy for conversion of simple organic compounds into the building blocks of life, RNA (ribonucleic acid) and amino acids. Indeed in 1953 Stanley L. Miller and Harold C. Urey, working at the University of Chicago, used some of the compounds listed in Table 8.2 (methane, ammonia, hydrogen, and water) and put them in a closed vessel simulating a primitive atmosphere and ocean. Miller then generated electrical sparks in the primitive atmosphere simulating lightning storms, which could have been common on the early Earth. The result of this experiment at the end of one week was that 10%–15% of the carbon originally present in gaseous form was now in the form of organic compounds. Two percent of the carbon was in the form of amino acids, the building blocks of proteins, and many amino acids were made in the process. Although this experiment has its detractors, it demonstrated that organic compounds, such as amino acids, could be

synthesized from inorganic compounds believed to exist on the early Earth. Another experiment performed by Juan Oro in 1961 using other gases (Table 8.2, hydrogen cyanide and ammonia) found in outer space in an aqueous solution produced the amino acid adenine. Adenine is one of the four bases in RNA and DNA (diribonucleic acid) and is a component of adenosine triphosphate (ATP). ATP is a major energy-producing compound in cells. Furthermore, in 2008 a scientific team led by Stanley Miller's former graduate student Jeffrey Bada, now a Professor at Scripps Institute of Oceanography at the University of California in San Diego, analyzed dried material stored in vials from Miller's original experiments. Bada and biochemist Adam Johnson of Indiana University in Bloomington, Indiana, discovered that some of the vials' contents contained dry matter created in the presence of a stream of water vapor. The reconstituted dried material contained 22 amino acids, 10 of which had not been identified in the original Urey and Miller experiments. In addition, more than 90 amino acids have been identified from the September 28, 1969, meteorite fall at Murchison, Australia, suggesting that organic compounds and amino acids could have been added to Earth by planetary infall from extraterrestrial space.

Thus, the early Earth had all the ingredients necessary for the beginnings of life, which probably originated in a series of steps, each building on one another, and certainly did not originate fully formed from some primordial soup. Simple organic molecules were highly likely the first step in formation: most likely nucleotides (molecules that comprise the structural units of RNA and DNA) followed by RNA and DNA, the genetic material for all life, formed from long chains of nucleotides. The next step would be the evolution of replicating molecules because all living organisms reproduce, copy their genetic material, and pass it on to future generations. Self-regulation would open the door for natural selection to take place. This ability probably arose in an RNA molecule that could copy itself, an RNA replicator. It has been assumed that RNA acted as a precursor of both DNA and proteins and that the early planet Earth was a RNA World. This hypothesis has run up against a number of stumbling blocks. However, keeping in mind the limitations of the RNA World hypothesis, the next step in the origination of life would probably be the encasement of replicating molecules in an envelope, a cell membrane, which would keep the products of the genetic material close by and maintain the internal environment of

TABLE 8.2 Some Compounds Detected in Outer Space by Spectral Analysis

Compounds	
Hydroxyl, OH	Formic acid, HCOOH
Ammonia, NH_3	Carbon sulfide, CS
Water, H_2O	Formamide, NH_2CHO
Formaldehyde, H_2CO	Carbonyl sulfide, OCS
Carbon monoxide, CO	Acetonitrile, CH_3CN
Cyanogen, CN	Methylacetylene, CH_3C_2H
Hydrogen, H_2	Acetaldehyde, CH_3CHO
Hydrogen cyanide, HCN	Thioformaldehyde, H_2CS
Cyanoacetylene, HC_3N	Hydrogen sulfide, H_2S
Methanol, CH_3OH	Sulfur monoxide, SO

the protocell different from its exterior environment. Finally, the cell would begin to evolve metabolic processes resembling those of modern organisms. The ideas concerning the origin of life still have a way to go and are changing at a rapid rate.

The Hadean Earth was a hostile planet for life. It was an environment that contained no free diatomic oxygen (O_2), a gas that enables most life to exist on Earth today; it was hot; and the planet had a dense atmosphere of vastly different composition from that later in Earth's history. Atmospheric pressure at this time was much greater than that of today. The atmosphere contained large quantities of water vapor, carbon dioxide, sulfur and nitrogen gases, and probably hydrogen chloride (HCl) gas.

Figure 8.7 shows one interpretation of the pressure and temperature history of the early Earth atmosphere and its gases. As Earth cooled, water vapor in the atmosphere reached the critical point of water and condensed to make rain. At 283° C, more

than 76% of the water vapor had condensed, leaving an atmosphere of 64 bar total pressure with mainly CO_2, N_2, HCl, and H_2S as gases. At about 56 bar, most of the H_2S and HCl gas was removed into the water phase and eventually at lower temperatures, N_2 and CO_2. Even today, because of its low solubility in water and chemical inertness, most nitrogen gas resides in the atmosphere and not in the water of the ocean and lakes or in sediments.

The rainwater and the gases contained in it chemically weathered the primordial igneous crust of the planet. If the rounded detrital zircon ($ZrSiO_4$) grains 4.2 to 4.4 billion years old found in gravelly deposits of the Jack Hills Formation of Western Australia stand up to scientific scrutiny as having been deposited in a cold, wet environment, then a primitive cool hydrosphere could have existed 150 to 350 million years after Earth's formation. Thus, life on Earth could have a longer history than previously thought.

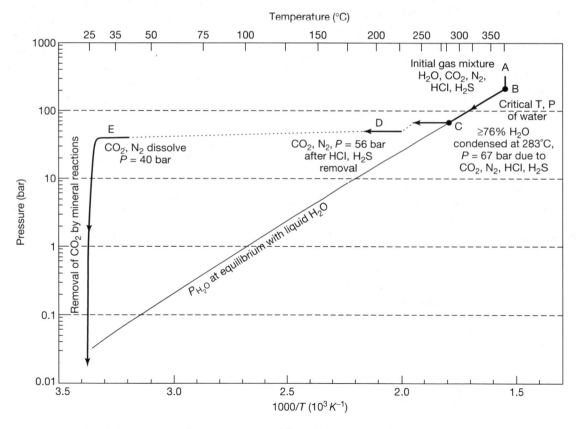

FIGURE 8.7 One interpretation of the prebiotic evolution of the temperature and atmospheric pressure of the Earth's early surface environment and atmosphere. As the Earth's total atmospheric pressure fell and its gaseous composition changed from a prebiotic carbon dioxide–rich atmosphere to one in which life could evolve, the planet cooled with time.   (*Source:* Mackenzie and Lerman, 2006.)

Carbon, nitrogen, sulfur, and chlorine gases when dissolved in water make carbonic (H_2CO_3), nitric (HNO_3), sulfuric (H_2SO_4), and hydrochloric (HCl) acids, respectively. These acids falling in rain reacted with the primordial igneous rock minerals of the crust. The minerals, because of their compositions, acted as bases to neutralize the acids. Thus, a gigantic acid–base titration (the **Goldschmidt reaction**), like a sour stomach with some bicarbonate of soda added, was an important feature of the Hadean evolution of Earth. The solid alteration products of this titration became the sediments of early Earth. The waters and dissolved products derived from the reactions became the early hydrosphere (oceans, rivers, and lakes). The formation of sediments and the cycling of gases between the Hadean crust and the early atmosphere started to change the composition of the atmosphere toward that of today. Earth was on the threshold of life.

In the Hadean environment, where did life start? The answer to this question is still a matter of debate. An old argument is that of Charles Darwin, who suggested that life began in a "warm little pond" containing a rich broth of organic compounds. Another suggestion is that life began in the deep sea in the vicinity of hydrothermal vents, where it was protected from the continuous and devastating bombardment of the planetary surface by meteors and other objects. In this environment, the microorganisms at the base of the food chain obtain their energy by metabolizing sulfur compounds and methane (CH_4) supplied to them by the hot waters surrounding the vents. These organisms are archaic and probably a close living link to the first organisms on Earth. Other environments suggested for the origin of life are surface hot springs, like those of Yellowstone National Park, United States, and other volcanically active areas, and organic-rich bubble foams that formed on the surface of the primordial ocean. A fortuitous bolt of lightning may have provided the energy that led to the formation of amino acids, the building blocks of proteins necessary for life, in an environment containing precursor chemical compounds (Table 8.2). It is possible that life emerged in several different environments. Indeed, although most scientists believe that atmospheric CO_2 and perhaps CH_4 as greenhouse gases kept the early Earth warm, some envision an early global winter in which life had its start in an ice-capped ocean. Whatever the case, it is likely that the first microorganisms belonged to the Archaea and Bacteria phylogenetic groups (Chapter 6).

Precambrian Eon: Radical Atmospheric Changes and the Beginnings of Life

The **Precambrian Eon** *is subdivided into the* **Archean Eon,** *Greek for "ancient rocks," extending from about 3.8 billion to 2.5 billion B.P., and the* **Proterozoic** *Eon, Greek for "early life," dating from 2.5 billion to 545 million B.P.*

The early Archean atmosphere still contained no oxygen. However, it is likely that much of the mass of the excess volatiles had been neutralized by reaction with minerals of the primordial crust. The sun's luminosity was perhaps 25% to 30% less than that of today (Figures 8.5 and 8.8). This **faint young sun** has led to a paradox. There is no direct evidence from the scant rock record of the Archean that the planetary surface was frozen. However, if Earth had no atmosphere or an atmosphere of composition like that of today, the amount of radiant energy received by Earth from the sun would not be enough to keep it from freezing. The way out of the dilemma is to have an atmosphere present during the early Archean that was different in composition from that of today. There are several gases that could have been in greater atmospheric abundance during the Archean—for example, water vapor (H_2O), carbon dioxide (CO_2), methane (CH_4), ammonia (NH_3), and nitrous oxide (N_2O). These gases are all greenhouse gases. They affect the solar radiation balance of the planet, and consequently climate. They produce a warming effect on climate termed the **greenhouse effect,** which is a result of the absorption by these gases of energy radiated from Earth's surface (see Chapters 4 and 14). For a variety of reasons, it has been concluded, although still debated, that the most likely gases present in greater abundance in the

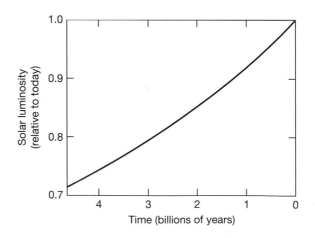

FIGURE 8.8 Change in solar luminosity relative to today during geologic time. (*Source:* Gough, 1981.)

Archean atmosphere were carbon dioxide, water vapor (the most important greenhouse gas), and perhaps methane. It is possible that if the CH_4:CO_2 ratio were high enough, an organic hydrocarbon haze may have existed in the Archean atmosphere. The presence of these greenhouse gases warmed the atmosphere and planetary surface and prevented the early Archean Earth from being frozen. The atmosphere at this time may have had 100 times or more the carbon dioxide content of today's atmosphere. If so, it is likely that the surface of the planet was even warmer than later in Earth's history. However, the surface had certainly cooled to a temperature favorable for life. Some evidence suggests that early Archean surface temperatures could have been as high as 60°C. At such a temperature, halophilic (salt adaptive) and thermophilic (heat adaptive) archaeabacteria (see Chapter 6) could evolve and thrive.

Furthermore with the high atmospheric CO_2 and vanishingly small O_2 levels of the Hadean–earliest Archean, ocean chemistry would have been much different from that in the latter part of the Precambrian. For example, the ocean would have been moderately acidic, with a pH (see Chapter 12) of 5.6 rather than its present surface water pH of about 8.2. The total alkalinity (a measure of a solution to neutralize acids) of the ocean and its dissolved inorganic carbon [DIC, dissolved $(CO_2 + HCO_3^- + CO_3^{2-})$] content (see Chapter 5) would have been significantly higher and the dissolved calcium content lower (Figure 8.9). Dissolved reduced iron (Fe^{2+}) and dissolved reduced sulfur chemical species (H_2S, HS^-), instead of dissolved sulfate (SO_4^{2-}) as is abundant in the present ocean, would be abundant in the seas of the time. What a different atmosphere and ocean!

REMOVAL OF ATMOSPHERIC CARBON DIOXIDE The removal of carbon dioxide from the atmosphere is critical to the temperature history of the planet. If the apparently high levels of carbon dioxide (as well as water vapor) in the atmosphere of the Hadean–early Archean were not removed by some process, Earth could have undergone a runaway greenhouse. It may have become as hot as Venus. The cooling of the planetary atmosphere is attributed to processes that remove carbon dioxide from the atmosphere. These processes are the deposition of limestone ($CaCO_3$) and organic matter (CH_2O) and storage of this carbon in sedimentary rocks. Consider what happened when calcium silicate ($CaSiO_3$) compounds ($CaSiO_3$ is used as an approximation of the Ca-, Si- and Al-bearing silicate mineral compositions of the igneous crust) of

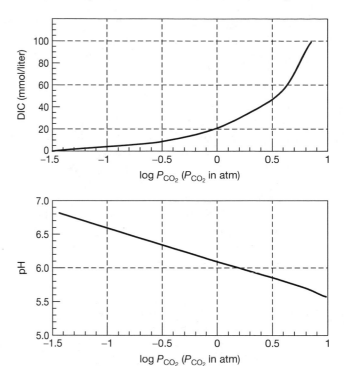

FIGURE 8.9 An interpretation of the trends in the dissolved inorganic carbon composition (DIC, $HCO_3^- + CO_3^{2-} + H_2CO_3$) and pH (measure of hydrogen ion content) of the ocean as a function of CO_2 at a temperature of 85°C. As atmospheric CO_2 decreased in the Hadean atmosphere, and hence temperature, the originally acidic pH of the ocean rose and the total inorganic carbon content decreased. In addition, the calcium content of seawater also may have risen. (*Source:* Morse and Mackenzie, 1998.)

the primordial igneous crust were attacked by the very acidic, carbon dioxide–rich rainwater of the Hadean–early Archean. The chemical weathering of these minerals removed carbon dioxide from the atmosphere. Dissolved calcium (Ca^{2+}), silica (SiO_2), and carbon as bicarbonate (HCO_3^-) were produced. These substances were transported by the ancient rivers to the ocean, where they were deposited as limestone and solid silica. After life evolved, carbon also accumulated as dead organic matter in the sediments of the primitive ocean. This process also removed carbon dioxide, which was originally taken up in the production of organic compounds, from the atmosphere. As these processes continued, the amount of carbon dioxide in the primordial atmosphere fell, and the amounts of limestone and organic matter on Earth increased. The planet cooled and more and more water vapor condensed from the atmosphere into the waters of the time. However, this set of processes cannot simply continue without any set of processes to return carbon dioxide to the atmosphere.

The present-day mass of carbon stored in sedimentary rocks is 7.78×10^{22} grams or $77,800 \times 10^{12}$ metric tons. All this carbon had its origin in the atmosphere. However, it can be easily shown that for any reasonable rate of chemical weathering, it would take only several hundreds of millions of years to remove a quantity of carbon from the atmosphere equivalent to that found in sedimentary rocks. Therefore, the very high atmospheric carbon dioxide contents of the early Earth did not persist for a long time. The total mass of carbon in limestone and organic matter in sedimentary rocks grew relatively rapidly. This mass attained a value close to its present value early in Earth's history, perhaps as early as the late Archean, although the mass grew somewhat through the Proterozoic, as evolutionary events took place. If so, then there must be some set of processes that returns carbon dioxide to the atmosphere. Without such processes, growth of these carbon masses would lower atmospheric carbon dioxide concentrations to vanishingly small levels. These processes began to be understood as early as the nineteenth century and especially in the scientific revolution that occurred in geology in the early 1960s, that of plate tectonic theory. Plate tectonics involves motions of large portions of the outer Earth and was explained in detail in Chapter 3.

What happens to limestone and organic matter deposited on the seafloor? Because of plate tectonic motions, these materials may be deeply buried along the margins of continents. They also may be moved away from their sites of deposition in the deep sea and transported down (subducted) into the interior of the planet. In either case, because of the high temperatures and pressures encountered by the limestone, silica, and organic matter during burial or subduction, these materials become chemically reactive. The limestone ($CaCO_3$) and silica (SiO_2) form calcium silicate ($CaSiO_3$), the original mineral compound weathered. Carbon dioxide is released in the process. The reaction is a type of metamorphism (see Chapter 2). Also, the decomposition of the organic matter releases carbon dioxide. The carbon dioxide eventually makes its way back to Earth's surface through volcanism, hot spring emanations, and other processes, where it is used again in weathering and in the production of organic matter.

Thus, by the early Archean, a cyclic system involving atmospheric carbon dioxide was established.

On a long time scale, it is this system, along with other factors, that regulates atmospheric carbon dioxide concentrations. Imbalances in the system produced by changes in rates of weathering and release of carbon dioxide by metamorphism and volcanism cause changes in atmospheric carbon dioxide concentrations over geologic time (see Chapter 7). The changes in rates are intimately connected to variations in plate tectonic activity through geologic time. The phenomenon of plate tectonics and the position of Earth in the solar system relative to the sun make it a habitable planet. Because of differences in plate tectonic history and position in the solar system, Mars is cold and frozen, Venus is hot, and neither planet appears suitable for life, but Earth is just right. The question why this is so is sometimes called the "**Goldilocks paradox**" of comparative planetology. The zone where environmental conditions might potentially support life has been called the "**Habitable Zone.**" The definition of the Habitable Zone first encompassed the orbits of Venus to Mars close enough to the sun for solar energy to drive the evolution of life and life processes but not close enough to boil off water or break down the organic molecules that life depends on. However, the Habitable Zone may be larger than originally proposed, if for no other reason than life might exist in very extreme environments. The recently acquired controversial evidence for microorganism life on Mars and the possibility of liquid water and volcanic activity on Europa, one of the moons of Jupiter, have led to modification of our ideas concerning the Habitable Zone, and if the requisite ingredients of life of water, energy, and organic molecules exist beyond our solar system, there will be further modifications of the concept.

EARLY LIFE—PROKARYOTES It is unknown when life first appeared on Earth. Possible evidence for the emergence of life is found in approximately 3.8 billion year old banded iron formation from the Isua supercrustal belt in western Greenland and a similar formation from nearby Akilila Island (Figure 8.5). Carbonaceous inclusions in calcium phosphate minerals (apatite) found in these sedimentary sequences have carbon isotopic compositions that are indicative of biological activity (depleted in ^{13}C; see Chapter 2). In addition, some of the tiny 0.3 mm across zircon crystals found in the Jack Hills Formation of Western Australia contain microdiamonds with a light carbon signature. If their interpretation of origin holds up to scientific scrutiny, this would push back the appearance of life to beyond 4.25 billion years.

It is likely that the first living cells on Earth were primitive microbes (prokaryotes, "before nuclei"; see Chapter 6 for discussion of taxonomic groups and their life processes), most likely mainly archaeabacteria [Archaea, single-celled microorganisms found in hot springs, salt lakes, soils, oceans, and marshlands today (see Chapter 6)]. *Disputable* biomarkers of hopanes, a class of chemical compounds produced by cyanobacteria and some other microorganisms, as well as streranes, which are produced only by eukaryotes (organisms whose cells contain nuclei), have been reported from 2.7 billion year old black shales from Australia. The oldest known fossil cyanobacteria are about 2.15 billion years in age and the oldest eukaryotic fossils date at about 1.7 billion years in age.

Perhaps the first good scientific evidence for life is found in rocks of the Dresser Formation and Stelley Pool Chert in the Pilbara region of Western Australia. Here rocks dating at 3.4 to 3.5 billion years contain laminated or layered structures known as **stromatolites**. Modern stromatolites are found in the present-day ocean in shallow-water areas like Shark Bay, Australia, and the Bahamas. The modern structures are formed by symbiotic microbial communities, especially cyanobacteria, that create mats and bulbous structures on the sea floor and whose secretions of mucus trap sedimentary grains and cement them into layers. The upper part of the mat requires sunlight for the cyanobacteria to photosynthesize aerobically (in the presence of sunlight, convert carbon dioxide and water into food, with a waste product being oxygen) and thus the mats continue to grow upward, if there is space to accommodate them. The Pilbara region rocks have structures and shapes that resemble modern stromatolites and these features and isotopic evidence suggest they too are of biological origin. Because there appears to be no evidence of complex life in stromatolites until about 2.7 billion years ago, the organisms that constituted the Pilbara microbial communities are unknown and might not have included oxygen-producing organisms. It should be pointed out that some scientists contest the biological origin of the 3.5 billion year old stromatolites and believe the structures were formed chemically under unusual environmental conditions on Earth at the time.

Some of the earliest primitive organisms were likely **chemolithoautotrophs**, nonphotosynthetic autotrophic (food manufactured from inorganic raw materials) microorganisms. These bacteria do not use light, and in their life processes, a variety of inorganic metabolites, like sulfur compounds and methane, are absorbed from the environment and combined with oxygen in their cells, releasing energy and a variety of inorganic by-products. Water and carbon dioxide are the inorganic raw materials used in subsequent food manufacture. The process of food manufacture is known as **chemosynthesis**. Among other primitive bacteria, the sulfur bacteria that can oxidize reduced sulfur (sulfide, S^{2-}) to oxidized sulfur (sulfate, SO_4^{2-}) and the methane bacteria that can oxidize methane (CH_4) to carbon dioxide (CO_2) were probably early groups to evolve (see Box 2.4: Ions and Oxidation-Reduction). These bacteria are found in hot sulfur springs and are associated with hydrothermal vents today. Other early organisms were the **photolithoautotrophs** in which the external energy source is light. Photosynthetic microalgae, cyanobacteria, and the photosynthetic purple sulfur bacteria and green sulfur bacteria are examples of this group, in which one or more varieties of chlorophyll are present in the organism to trap solar energy. The hydrogen source of the photoautotrophic microalgae and cyanobacteria is water (H_2O), which is split into hydrogen and oxygen by the light energy. Oxygen is released as a by-product of the reaction and the hydrogen liberated as a result of decomposition of H_2O reduces the carbon dioxide to carbohydrate. This compound is used as food for the organism. The hydrogen source of the photoautotrophic bacteria is not water and in their metabolism, oxygen is not liberated. These bacteria today are adapted to live in sulfur springs where reduced sulfur (H_2S) is available as a hydrogen source. The bacteria able to oxidize reduced sulfur were probably responsible for the initial concentrations of dissolved sulfate in the hydrosphere of the Archean environment. Oxygen is toxic at high concentrations to these primitive bacteria. The primitive microorganisms and their environment had to evolve to a stage where oxygen was produced and emitted to the atmosphere in order for more advanced forms of life to evolve.

As mentioned, evidences of *probable* single-celled, blue-green algae, the cyanobacteria (prokaryotes), are found in 3.5 billion year old sedimentary rocks from Australia. The evolution of these organisms changed the process of photosynthesis. The blue-green algae were able to emit oxygen into their surrounding environment as they used carbon dioxide from the atmosphere. Nevertheless, oxygen in the early Archean was still not abundant enough to lead to an oxygenated hydrosphere–atmosphere. However, the new life form of cyanobacteria altered the environment. With the release of oxygen into the atmosphere from blue-green algae photosynthesis, a protective ozone (O_3) layer began to develop in the atmosphere, blocking intense, shortwave, ultraviolet radiation from the sun (Figure 8.5). The development of the ozone layer allowed for the expansion of blue-green algae communities. These organisms emerged from shaded areas where they were shielded from the intensity of the sun's ultraviolet light. With this expansion, oxygen was more rapidly emitted to the atmosphere. The primitive bacteria, intolerant to oxygen, retired to parts of the environment that had no oxygen, such as the fetid muds of the oceans. As mentioned, the cyanobacteria still live in parts of the world today, growing in small mounds along the ocean shoreline. Their ancient fossil remains (stromatolites) often are found as structures resembling towering columns and bulbous laminations in rocks.

ICE HOUSES, HOT HOUSES, AND SULFATE-REDUCING BACTERIA The surface of Earth was also changing over time. Continental crust had been forming initially in the form of large island masses and about 2.6 billion years ago, there was a major tectonic event that led to significantly more amounts of crust formation. It is believed that the volume of the crust at the end of the Archean was perhaps as much as 80% to 90% of that of today. The first of four well-documented ice ages is thought to mark the boundary between the Archean and Proterozoic some 2.5 billion years ago (Figure 8.5). One possible cause of this ice age may be related to the intensity of plate tectonic activity. Before the ice age, plate tectonic activity may have been decreasing. Such a decrease in activity may have led to less carbon dioxide being emitted to the atmosphere from metamorphism and volcanism. Because of these decreased emissions, atmospheric carbon dioxide may have fallen. The planet may have cooled because of a less effective greenhouse. This initial cooling might have permitted other factors, such as changes in the amount of heat received from the sun and its distribution (see Chapter 13), to enhance the cooling and lead to an ice age. Thus, it is likely that this 2.5 billion year old ice age and those that followed were initiated because the planet was in an extended period of unusual coolness. These prolonged cool periods are termed "**Ice Houses.**" They are probably the result of the waning of plate tectonic activity. Intervening warmer and longer periods in the history of Earth have been termed "**Hot Houses**" ("Greenhouses").

The long transitions from an Ice House to a Hot House and vice versa involve not only the waxing and waning of plate tectonic activity and resultant changes in atmospheric carbon dioxide concentration, but the water cycle on Earth. For example, the development of a Hot House implies the potential of more water vapor in the atmosphere. It also means increased rainfall over land. The higher temperatures and increased rainfall cause an increase in the rate of chemical weathering of calcium silicates. This change, in turn, leads to increased deposition of the weathering products in the ocean as limestone and silica. The increased rate of weathering would lead to an increased rate of removal of carbon dioxide from the atmosphere. This would tend to cool the planet because of a decreased greenhouse effect. An Ice House could develop. With cooling, the rate of weathering would decrease and slow the removal of carbon dioxide from the atmosphere. The increased volumes of limestone and silica deposited on the seafloor during the previous Hot House would be buried or subducted to depths where these materials would be converted back to calcium silicate and carbon dioxide. The carbon dioxide would be returned to the atmosphere in volcanism. As a result of decreased weathering rates and enhanced emissions of volcanic carbon dioxide, atmospheric carbon dioxide would rise again. The planet would warm and enter another Hot House. This series of events is an excellent example of **negative feedback** in which an initial perturbation of a system gives rise to a change in the system that relieves the perturbation.

Another important event took place at least as early as 2.5 billion years ago. Bacteria evolved that could reverse the process of bacterial oxidation of sulfide (Figure 8.5). These sulfate-reducing bacteria are capable of taking dissolved sulfate in water and reducing it to sulfide. In the process, they oxidize organic matter back to carbon dioxide and water, and the mineral pyrite (FeS_2) is formed. The evolution of these bacteria necessitated sulfur in the environment as sulfate. The early concentrations of sulfate in the hydrosphere probably came from reactions involving the sulfide-oxidizing bacteria that had evolved earlier. The availability of sulfate and perhaps even the higher atmospheric CO_2 contents of the time, and hence higher temperatures, were important factors in the environment that led to the evolution of the sulfate-reducing bacteria. The emergence of the sulfate-reducing bacteria was a major event in the evolution of Earth's surface environment. Their emergence meant the establishment of carbon and sulfur cycles with processes much like those of today. The sulfate reducers, like the sulfide oxidizers, are also found today in the fetid muds of the seafloor and in waters that are devoid of oxygen.

IRON FORMATIONS During the period of 2.2 billion to 1.6 billion years ago, large volumes of sedimentary rock rich in iron (iron formations) were deposited, probably during an extended Hot House (Figure 8.5). These formations are unique, never to be found again in such quantities in younger rocks. They are the source of 90% of the world's mineable iron ore. Their origin is controversial. However, it is likely that the deposition of these sediments required low oxygen concentrations in the global atmosphere–hydrosphere of this time. Such a global environmental condition would mean that most of Earth's waters were anoxic (without oxygen), but the surface waters of some oceans and lakes may have contained small quantities of dissolved oxygen. The anoxic waters would have high concentrations of iron in soluble, reduced form (Fe^{2+}) derived from weathering of Earth's crust and hydrothermal vent emanations at the sea floor. The transportation of this iron in anoxic river waters to the ocean and in the sea itself and its subsequent deposition in iron minerals in more oxygenated environments of the ocean may have led to the formation of the vast iron deposits of the late Archean–early Proterozoic. Today most iron is transported about Earth's surface in solid form as particles of iron oxide (Fe_2O_3) because our present atmosphere–hydrosphere is well oxygenated and oxidized iron (Fe^{3+}) is insoluble.

The iron deposits of the late Archean–early Proterozoic actually consumed oxygen, generated by photosynthesis, in chemical reactions that led to their deposition. Consequently, the accumulation of oxygen in the atmosphere slowed. This period of iron formation is transitional between an atmosphere with virtually no free oxygen and one in which oxygen was more abundant. However, even with free oxygen in the atmosphere, it is likely that much of the volume of the ocean was not fully oxygenated until the late Proterozoic.

EMERGENCE OF THE EUKARYOTES Approximately 2 billion years ago new, complex, aerobic (oxygen tolerant), photosynthetic cells (eukaryotes, "membrane-bound nuclei") evolved (Figure 8.5). This evolutionary event is considered a milestone in the history of the organic world, although its timing is still questionable. Early forms of eukaryotes were single-celled algae and

were followed by the development of multicelled algae. The early eukaryotes were more efficient than blue-green algae at generating oxygen, and their appearance led to a more rapid and massive buildup of oxygen in the atmosphere, especially after 1.6 billion years ago. Eventually, these organisms developed into two broad groups, one evolved into plants and the other into animals.

It had taken Earth more than 3 billion years to reach this stage in its development. Nearly another billion years would pass as oxygen continued to increase in abundance in the atmosphere, and atmospheric carbon dioxide levels continued to fall. With the accumulation of atmospheric oxygen, Earth's environment approached conditions like today at the end of the Precambrian, with a less oxygen deficient and sulfide and ferrous iron–rich ocean and a more oxygenated and less carbon dioxide–rich atmosphere, and life-forms began to expand rapidly. Invertebrate metazoans evolved in the late Proterozoic.

The earliest likely animal fossils, collectively called the **Ediacaran fauna**, were first found in late Proterozoic rocks from the Ediacara Hills in southern Australia more than 50 years ago. These organisms seem to have been floating, gelatinous forms that lived along the subtidal zone of the sandy beaches of the late Proterozoic sea. Some of the organisms are flat, others have a quilted texture, and still others are intricately textured. They were probably large protoctists (see Chapter 6) and animals. Similar soft-bodied forms of similar age have been found in rocks in England, Greenland, Siberia, Namibia, and 20 other locations. The Ediacarans may have been ancestors to the animals preserved in early Cambrian sediments (e.g., the Burgess Shale of Canada), or they may have been a "false start" in organic evolution. Whatever the case, it is likely that the Ediacaran radiation of organisms in the late Proterozoic was coupled to a complex series of events involving tectonic, climatic, and biogeochemical changes. One likely important change was an increase during this metozoan radiation in the oxygen content of the atmosphere—probably to levels of 10% or more of that of today.

Earth's environment was substantially cooler 850 million to 600 million years ago as the Precambrian drew to a close. Planet Earth entered a second major Ice House. The sedimentary record of this period of time records intervals of extensive glaciation. Indeed, some scientists contend that near the close of the Proterozoic, a nearly completely ice-covered world existed, a "**Snowball Earth**."

The Precambrian Eon closed about 545 million years ago. About seven-eighths of the entire history of Earth is recorded in the scant fossil and rock record of this eon. Only one-third of the existing sedimentary rock mass is from this eon; the rest is from the much shorter Phanerozoic Eon.

Phanerozoic Eon: Life Multiplies and Wanders

Phanerozoic is Greek for "visible life"; the eon spans 545 million years ago to the present and includes the Paleozoic, Mesozoic, and Cenozoic eras.

PALEOZOIC ERA—THE RAPID BLOOMING OF LIFE

Paleozoic is Greek for "old life"; the era spans 545 million to 245 million years B.P. This era consists of the Cambrian, Ordovician, Silurian, Devonian, Carboniferous, and Permian periods.

The volume of the continental crust of Earth 545 million years ago was probably similar to that of today. The continental masses were scattered in positions around the equator and adrift because of plate tectonic motions (Figure 8.10). They had reached their positions in the Cambrian following breakup of a great supercontinent, Rodinia, that existed between 1,100 and 750 million years ago, into several continental masses. Oxygen had become plentiful, and atmospheric carbon dioxide had reached levels like those to be found throughout the rest of Earth's history. The start of the era was a time of low plate tectonic activity, leading to small midocean ridge volume and hence relatively low global sea level (Figure 8.11), low atmospheric carbon dioxide levels, and cool surface temperatures. However, during much of the Paleozoic Era the planet was in a Hot House, although 450 million years ago, there probably was a continental-scale glaciation.

Plate tectonic activity gradually increased during the Paleozoic, and volcanism and metamorphism associated with this activity led to high atmospheric carbon dioxide levels. The planet was warm because of the stronger greenhouse effect produced by the elevated levels of atmospheric carbon dioxide. The great midocean ridge system gradually increased in size, resulting in the displacement of water onto the continents. The oceans flooded the continents, and sea levels were high (Figure 8.11). The continents wandered during this era from early positions around the equator to a position by late Paleozoic time in which they were collected into the great supercontinent of Pangaea (Figure 8.10).

At the beginning of the Cambrian, there was an organic evolutionary explosion with the appearance of shelled forms of life representing most of the existing phyla on Earth today. For the first time, the rocks contain fossilized, well-preserved, hard body parts. A proliferation of eukaryotic life forms had occurred. Marine invertebrates, such as trilobites, sponges, snails, clams, and corals, dominated the sea. By 400 million years ago, during the Silurian Period, plant life occupied the land for the first time and dispersed rapidly over the landscape. Beginning in the Devonian, invertebrates emerged on land. Insects, spiders, and scorpions thrived during this period. Cockroaches are one of the oldest surviving insects from this time. They have not changed substantially in characteristics for 320 million years. Fish species, arising in mid-Ordovician time, became abundant, along with sharks, during the Devonian (Age of Fishes). The Coelacanth "lungfish" that evolved in mid-Devonian time persists today in deep-ocean waters off the island of Madagascar. Vertebrates appeared on land in the late Devonian, 360 million years ago. Amphibians lived in the waters of the time and ventured onto the land and evolved into reptiles. Mammal-like reptiles evolved. Land plants, from tree-size conifers to low-lying ferns, proliferated. Swamps existed and produced many of the future coal deposits of Earth. Much of the coal on Earth, a product of terrestrial plant matter and its decay, formed during the Carboniferous Period, at the end of this warm climate of the Hot House. The deposition of such vast deposits of organic matter likely was one cause of high atmospheric oxygen concentrations during the Carboniferous. Atmospheric methane levels also were relatively high during the latter part of the Paleozoic due to increased emissions of CH_4 associated with the formation of the large coal basins. (see Chapter 7).

The rapid development of life in the early and middle Paleozoic was also accompanied by periods of extinction followed by rapid increases in new life forms. Fossil remains of some life forms disappeared from the rock record and new forms appeared. The Cambrian–Ordovician and the Ordovician–Silurian boundaries and the Late Devonian mark periods of mass extinctions of some life forms and the arrival and persistence of others. The exact causes of these periods of mass extinction are poorly known. A number of different causations leading to the destruction of habitats were probably involved, including climatic change, volcanism, and meteorite impacts. As the environment changed through time, life forms found suitable habitats and survived, adapted, and proliferated or became extinct.

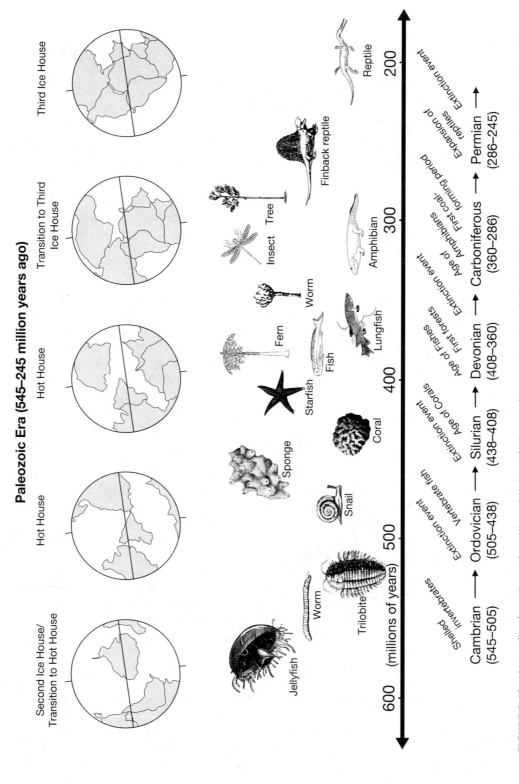

FIGURE 8.10 Generalized continental positions, biotic evolution, and temperature trends during the Paleozoic Era. (Continental positions after Calder, 1983.)

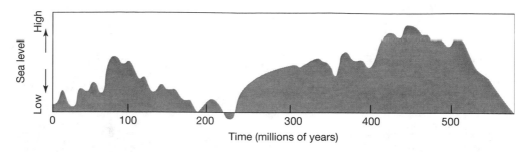

FIGURE 8.11 Generalized pattern of relative sea level change during the Phanerozoic. (*Source:* Hallam et al., 1983; Haq et al., 1987).

During the latter part of the Paleozoic, plate tectonic activity slowed. The spread of lowland and eventually highland plants led to storage of atmospheric CO_2 in vegetation. The midocean ridge system decreased in volume and ocean water receded from the continents (Figure 8.11). The release of carbon dioxide to the atmosphere by volcanism and metamorphism gradually decreased and, with storage of carbon in terrestrial vegetation and enhanced plant weathering, atmospheric carbon dioxide levels fell. The reduced concentrations of carbon dioxide in the atmosphere caused surface temperatures to fall because of a decreased greenhouse effect. During the Permian, at the close of the Paleozoic Era, these changes led to the development of a third major Ice House. There is evidence of a continental-scale glaciation 300 million to 200 million years ago. Because of smaller midocean ridge volumes and because water was locked up in glacial ice, many shallow continental seas and their biotic occupants disappeared. Two-thirds of the surface of Earth became one great body of water, whereas the other one-third became the supercontinent of Pangaea (Figure 8.10). Pangaea was oriented on Earth's surface such that it was spread out in a north–south direction from pole to pole. Life forms easily migrated because of the interconnection of the great landmasses.

By the end of the Paleozoic Era, 245 million years ago, a gradual disappearance of Permian life forms occurred as plant (flora) and animal (fauna) forms were replaced by new varieties of life. The fossilized remains of many organisms of this era are not found in rocks younger in age. The Permian-Triassic boundary is recognized as the most devastating extinction of Phanerozoic life. Fifty-seven percent of marine families disappeared, never to be found in the fossil record again.

MESOZOIC ERA—EXTINCTION AND TRANSFORMATION
Mesozoic *is Greek for "middle life." This era, extending from 245 million to 65 million years* B.P., *consists of the* Triassic, Jurassic, *and* Cretaceous periods.

We are beginning to gain some insight into what caused the massive extinction of species near the termination of the Paleozoic. The cause was complex but involved a reorganization of the global carbon and other biogeochemical cycles (see Chapter 7) and perhaps a return of the ocean to the nearly anoxic state of the Precambrian accompanied by high atmospheric H_2S and CH_4 concentrations and lower ozone levels. Whatever the cause, the demise of many life forms at the close of the Paleozoic allowed room for new life forms to emerge and proliferate (Figure 8.12). By the Late Triassic, the huge continent of Pangaea was breaking up, and over the following millions of years, its parts drifted around the surface of the planet, ferrying its diverse life forms. Australia and Antarctica separated from Africa to form one continent 150 million years ago, then divided again 50 million years ago. This separation allowed for independent development of life forms on these continental masses. South America and Africa, previously joined together, broke apart about 100 million years ago, during the Cretaceous, to evolve independently until South America joined up with North America. During the Mesozoic, the subcontinent of India began its journey northward.

The Earth was cycling into a warm period again. For much of the Mesozoic Era, the environment consisted of a stable, warm climate (Hot House) (see Chapter 13, Figure 13.13). Surface temperatures were relatively high. Adequate oxygen and water existed on the planet. Atmospheric carbon dioxide and global sea level rose continuously, but erratically, during this era, reaching a maximum in the late Cretaceous and then declining into the Cenozoic Era as the intensity of plate tectonics decreased (Figures 7.5 and 8.11). The ecosphere of Earth was well suited for higher life forms of eukaryotes during the Mesozoic Era.

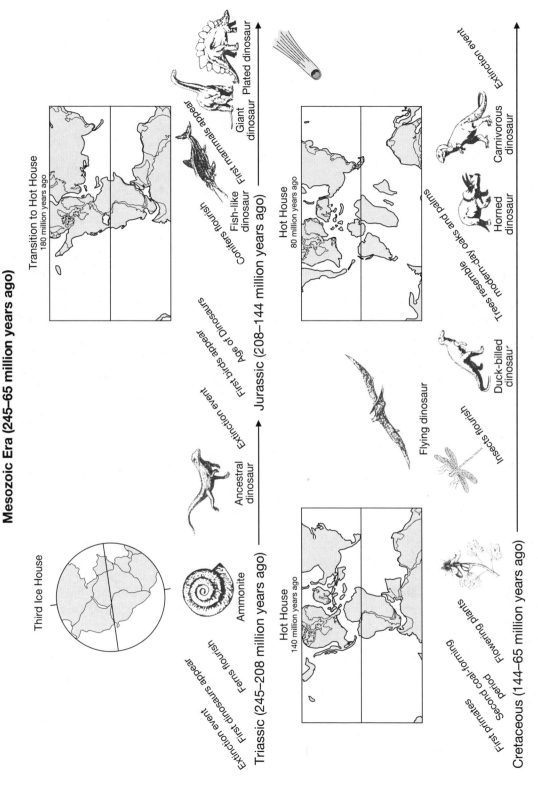

FIGURE 8.12 Generalized continental positions, biotic evolution, and temperature trends during the Mesozoic Era. (Continental positions after *Eclogae geologicae Helvetiae*, 1981.)

Mesozoic Era (245–65 million years ago)

Third Ice House

Triassic (245–208 million years ago)

Extinction event
First dinosaurs appear
Ferns flourish

Ammonite

Ancestral dinosaur

Extinction event
First birds appear
Age of Dinosaurs

Hot House
140 million years ago

Conifers flourish
Fish-like dinosaur
First mammals appear

Fish-like dinosaur

Transition to Hot House
180 million years ago

Jurassic (208–144 million years ago)

Giant dinosaur   Plated dinosaur

Hot House
80 million years ago

Flying dinosaur

Duck-billed dinosaur

Insects flourish

Trees resemble modern-day oaks and palms

Horned dinosaur   Carnivorous dinosaur

Extinction event

First primates
Second coal-forming period
Flowering plants

Cretaceous (144–65 million years ago)

207

Many new plants and animals emerged in this new, mainly Hot House environment. The descendants of mammal-like reptiles and amphibians had lived on through the early part of the Triassic. The end of the Triassic, about 200 million years ago, saw another wave of extinctions when 23% of marine and land families disappeared. This event ushered in the dinosaurs and other new life forms. By the beginning of the Jurassic (Age of Dinosaurs) and continuing into the Cretaceous, dinosaurs roamed the land and swam in the waters of Earth. Their fossils have been found on all continents of the world, including Antarctica. Marine invertebrates inhabited the sea and flying dinosaurs patrolled the sky. Conifers flourished on the land. At times, extensive swamps existed in the warm climate of the Mesozoic. During the Cretaceous, flowering plants evolved. The Mesozoic Era was the time of a second great coal-forming period, beginning in the early Jurassic, extending through the Cretaceous, and continuing on into the Cenozoic. Atmospheric methane levels were probably elevated during much of this coal-forming time. In addition, a significant amount of all known petroleum reserves is found in rocks of Cretaceous age.

The dinosaurs and their entourage lived on Earth for more than 140 million years during this Hot House. Then at the end of the Cretaceous, 65 million years ago, they suddenly disappeared. It is very likely that a catastrophic environmental disaster was brought about at this time by a meteorite collision with Earth. As a result, Earth's ecosphere was greatly modified. The highly specialized dinosaurs, who survived the initial impact, were ill suited to exist in such a drastically changed environment, could not adapt, and died off. The cause of the demise of the dinosaurs is a hotly debated scientific topic today. However, whatever the reasons, as in the past, another massive extinction of many life forms occurred at the end of the Cretaceous. New types of animals and plants, replacing those that perished, began to flourish on Earth. These new life forms were very similar to the flora and fauna of today. The Age of Mammals had arrived.

CENOZOIC ERA—A MODERN EARTH

Cenozoic *is Greek for "recent life"; it spans 65 million years ago to the present and consists of the Tertiary and Quaternary Periods. The Tertiary Period lasted from 65 million to 1.6 million years B.P. and consists of the Paleocene,* *Eocene, Oligocene, Miocene, and Pliocene epochs. The Quaternary Period has lasted from 1.6 million years ago to the present day. This period consists of the Pleistocene Epoch, comprising the time interval between 1.6 million and 10,000 years B.P., and the Holocene Epoch, which occupies the last 10,000 years of Earth history.*

The beginning of the Cenozoic was still a warm time (Hot House), as in the Mesozoic Era (Figure 8.13; see Chapter 13, Figure 13.13). Most of the flora and fauna as we know them today arose and modern ecosystems developed. Oil deposits were formed in marine sediments, and today one-half of our oil reserves are found in Cenozoic rocks. Mammals became abundant, and as a result the Cenozoic has been called the Age of Mammals.

The continents were still drifting around the world after the breakup of Pangaea. India had broken away from Antarctica, moved north, and eventually crashed into Tibet and other parts of Asia to form the Himalayan Mountains about 50 million years ago. Australia broke away from Antarctica 50 million years ago, and life on Australia evolved isolated from all other continents. The continents eventually reached their present-day positions in the Quaternary (Figure 8.13), forming the global pattern that we are familiar with today. The continents continue to drift at rates of several centimeters per year.

The Paleocene–Eocene boundary at about 56 million years ago was marked by a significant climatic and extinction disturbance termed the Paleocene–Eocene Thermal Maximum (PETM). The event was characterized by temperatures rising globally about 6°C in 20,000 years, a higher level of atmospheric carbon dioxide, probably a more anoxic and acidic ocean, a higher sea level, presumably due to the expansion of ocean water from thermal heating, and mass extinction of marine benthic foraminifera and changes in mammalian life on land. The cause(s) of the event is still being debated and investigated but it may have been due to the decomposition of methane ice deposits termed clathrates in marine sediments (see Chapter 14) owing to a previous warming trend and the release of methane gas (a greenhouse gas) to the atmosphere and its oxidation to CO_2 (a greenhouse gas). It also may be related to an increase in volcanic activity that led to significant releases of CO_2 to the atmosphere or due to hot magma intruding carbon-rich sediments leading to the release of CH_4 to the atmosphere. Recovery of the Earth system from this event took about 150,000 to 30,000 years.

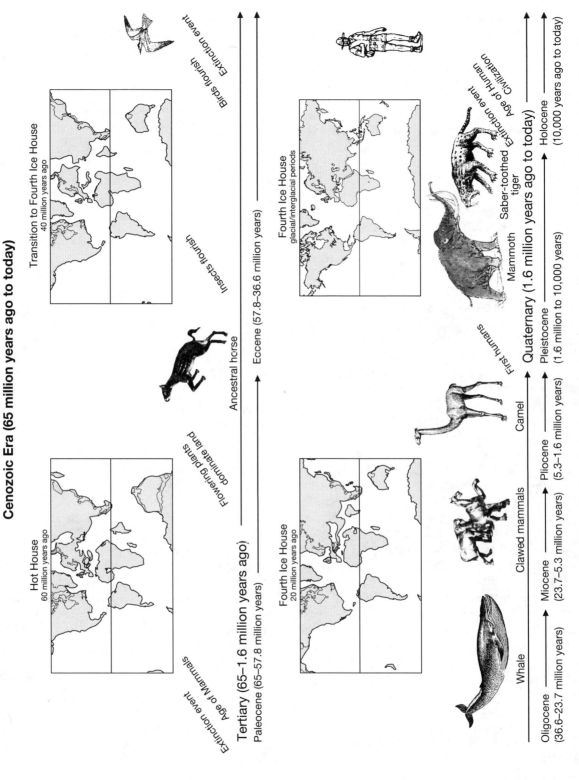

FIGURE 8.13 Generalized continental positions, biotic evolution, and temperature trends during the Cenozoic Era. (Continental positions after *Eclogae geologicae Helvetiae*, 1981.)

The mid-Tertiary, some 35 million years ago, was the beginning of a progressive, but erratic, cooling of the global environment, and the formation of the first Cenozoic continental glaciers. Earth entered a fourth, well-documented Ice House that culminated in the Pleistocene Epoch. Sea level, which had attained a maximum elevation of several hundred meters above today's level in the warmer late Mesozoic, receded from the continents to reach about its present-day level just prior to the Pleistocene Epoch (Figure 8.11).

The Pleistocene Epoch includes a series of cooler and warmer episodes termed, respectively, **glacial** and **interglacial** stages. Glacial stages are characterized by global average low temperatures (about 6°C degrees lower than that of today), low atmospheric carbon dioxide, methane, and nitrous oxide (N_2O) levels, and low sea levels. Higher temperatures, sea levels, and atmospheric carbon dioxide, methane, and nitrous oxide concentrations, conditions more similar to the environment of today, are representative of interglacial stages. In the last 1.6 million years, there have been 10 major and 40 minor periods of glacial and interglacial conditions. The most recent glacial period reached its climax about 18,000 years ago. The Holocene Epoch, a relatively short period of time of 10,000 years, during which the human population grew rapidly, appears to be part of an interglacial interlude in an extended series of glaciations and interglaciations (see Chapter 13, Figure 13.13). Broadly speaking, during this period of interglacial time, the global climate has warmed several degrees, sea level has risen about 120 meters, and atmospheric carbon dioxide, methane, and nitrous oxide concentrations have climbed. Chapter 13 discusses late Pleistocene and Holocene environmental change in more detail.

There were two other occurrences of extensive extinctions of species during the Cenozoic—the Eocene–Oligocene boundary and at the termination of the Pleistocene. It is likely that the causes of these extinctions were associated with climatic change, a result of changes in ocean circulation and concurrent temperature change. The Pleistocene extinction affected mainly large mammals, such as the mammoth and saber-toothed tiger. Homines (primate species including modern humans), evolving in the Cenozoic Era, adapted and survived these glacial–interglacial environmental changes. The human species evolved, proliferated, and multiplied in the late Pleistocene and Holocene. Human history is discussed in more detail in Chapter 9.

Climate, Glaciations, and Atmospheric CO₂ During the Phanerozoic

On the long time scale of tens to hundreds of millions of years in the Phanerozoic, atmospheric CO_2 and climate appear to go hand in hand (Figure 8.14). Major continental-scale glaciations occurred when atmospheric CO_2 (and to some extent CH_4 concentrations and probably N_2O concentrations), radiative forcing from CO_2 and other greenhouse gases, global temperatures, and worldwide sea levels were relatively low. The Earth was cold enough for large ice sheets to form and spread over large areas. At these times, it was also the case that seawater chemistry was more similar to the more recent geological past with relatively high dissolved $Mg^{2+}:Ca^{2+}$ and dissolved $SO_4^{2-}:Ca^{2+}$ ratios, and pH (more basic ocean water), and relatively low dissolved inorganic carbon (DIC) and total alkalinity. Biological and inorganic carbonate precipitates and sediment accumulations in the ocean were dominated by the $CaCO_3$ mineral aragonite (the aragonite seas). These are the Ice Houses of the Phanerozoic, which contrast sharply with atmospheric and seawater conditions of the Hot Houses. During the Hot Houses, atmospheric CO_2, temperatures, and sea level were relatively high, and seawater $Mg^{2+}:Ca^{2+}$ and $SO_4^{2-}:Ca^{2+}$ ratios, and pH were low (extended times of ocean acidification), and the DIC and total alkalinity of seawater relatively high. Biological and inorganic carbonate precipitates from seawater were dominated by calcite and dolomite, as were the carbonate sediments formed from these precipitates (the calcite/dolomite seas). Thus, we see that fluctuations in climate are tied to atmospheric greenhouse gas concentrations at this long geologic time scale. What is also important to understand is that similar connections exist at the time scales of Pleistocene glaciations and interglaciations (see Chapter 13) and the recent period of Earth history under human influence, the Anthropocene (see Chapter 14). These connections will be discussed later in this book.

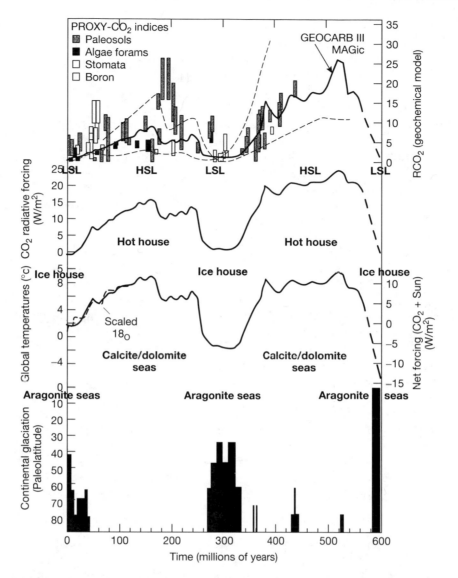

FIGURE 8.14 Climate, glaciations, and atmospheric CO_2. Notice the oscillatory pattern in all parameters and that the three major glaciations occurred at times of relatively low atmospheric CO_2, hence radiative forcing and temperature. Various types of proxy data for atmospheric CO_2 concentrations are compared with the composite model curve for Geocarb III and MAGic (see Chapter 7). RCO_2 is the concentration of CO_2 in the atmosphere relative to average Holocene levels. The dashed lines in the upper panel are upper and lower bounds of CO_2 estimates. HSL is high sea level; LSL is low sea level. Aragonite seas are times when the mineral aragonite was especially abundant as a chemical and biological precipitate from ocean water. Calcite/dolomite seas are times when these minerals were abundant as precipitates from ocean water. (See text for further explanation; modified from Crowley and Berner, 2001.)

Concluding Remarks

The universe is believed to have formed about 13.7 billion years ago, Earth 4.55 billion years ago. A comparison between the length of time since the creation of Earth and a 46-year time span, as depicted by Crispin Tickell in 1977 in *Climate Change and World Affairs,* can be used to put into perspective the events that have taken place since the planet formed. If Earth were 46 years old, or one year per hundred million years of its approximately 4.6 billion year existence, the major ice ages occurred approximately 25, 7, and 3 years ago and 7 days ago (2 million actual years ago). The dinosaurs would have existed for 1.5 years until they died out six months ago. Humankind emerged about a week ago, and in the last minute of the 46 years (150 actual years ago), the Industrial Revolution began. John McPhee in 1980 provided another analogy. An English yard is measured by using the distance from the tip of one's nose to the end of one's middle finger. If one used this as a measure of the span of Earth history, then a stroke of a file on the nail of the middle finger would remove humankind's existence from this planet. Life as we know it has existed only for a speck of time since the creation of the planet. During this brief moment, humans and their activities have become a powerful geologic force for environmental change on the planet. Many of these changes are occurring at rates that exceed those of the entire history of planet Earth.

The physical environment and life on Earth have evolved over geologic time. However, environmental conditions, particularly the chemistry of the atmosphere and hydrosphere and climatic factors, as interpreted from the sedimentary rock record, have been repeated over time. There is evidence for a cyclic pattern of development. The periodicity in Hot Houses and Ice Houses and glacial–interglacial stages is evidence of this cyclicity, as is the rock cycle (see Chapter 3).

There is a certain degree of coevolution of the physical environment and life. Certainly, the evolution of the cyanobacteria with their ability to produce oxygen led to nearly simultaneous changes in the chemistry of Earth's surface environment. Changes in chemistry, in turn, provided the necessary conditions for further organic evolution.

In the 1970s, J. E. Lovelock and L. Margulis presented the Gaia hypothesis, a statement about how the surface environment is regulated. They claimed that the environment and climate of the surface of Earth were controlled only by the biota in such a way as to be favorable for life. This hypothesis about Gaia has led to a great deal of discussion and controversy since its introduction.

There is little doubt that organisms play a strong role in processes operating at Earth's surface. Witness the many reduced gases produced and consumed by organisms that are exchanged between the surface and the atmosphere and the effect of some of these gases on climate (see Chapters 7 and 13). There is also little doubt that inorganic processes such as metamorphism and volcanism are important in determining the composition of Earth's atmosphere and the climate of the planet over geologic time. It seems likely if Earth's surface environment is regulated, the stability, resilience, and changes in the environment are a function of a complex system of organic and inorganic processes. It is this system, a homeostatic system, which is being disrupted by the human species.

Study Questions

1. How did the universe originate?
2. Where did the atmospheric oxygen of Earth come from?
3. What are the major differences in environmental conditions between Hot Houses and Ice Houses with respect to temperature, atmospheric gas composition, sea level, seawater composition, carbonate precipitates, and glaciations?
4. Explain the difference between Hot/Ice Houses and glacial/interglacial periods.
5. Briefly describe the pattern of evolution of life during the Precambrian.
6. About 200,000 million tons of carbon are removed from the atmosphere each year in the gross primary production of organic matter on the surface of Earth. The atmosphere contains 800,000 million tons of carbon as carbon dioxide (CO_2) gas. How long before the carbon dioxide in the atmosphere would be exhausted if there were no way to compensate for the removal of carbon dioxide?
7. On this time scale, how is the carbon in the atmosphere replenished?
8. On the longer geological time scale, what processes control the content of carbon dioxide in the atmosphere?
9. What is the basis for subdivision of the Phanerozoic Eon into three major eras?

10. Create an analogy comparing the length of time since the creation of Earth until today, putting into general perspective some events that have taken place since the planet formed.
11. Explain briefly the Goldilocks and Faint Young Sun paradoxes.
12. On the long-term geological time scale, how does the temperature feedback to rising atmospheric CO_2 and temperature work?
13. A pure limestone ($CaCO_3$) undergoes weathering according to the following reaction:

$$CaCO_3 + CO_2 + H_2O = CaCO_3 + 2HCO_3^-.$$

    (a) What happens to the dissolved constituents of this reaction?
    (b) Where does the CO_2 come from to weather the $CaCO_3$?
    (c) If two moles of CO_2 are used in the weathering reaction, how many moles of HCO_3^- are produced? How many grams?
14. The BIFs are an important sedimentary rock type found in the late Archean and early- to mid-Proterozoic. What are the BIFs? What do these sedimentary rock types tell you about the levels of atmospheric O_2 at the time of their formation?
15. Why on the early Earth must you have had some atmospheric O_2 before development of an ozone shield (see Chapters 4 and 14)? What effect did the development of the O_3 shield on the Earth have on the cyanobacteria and the development of stromatolites?

Additional Sources of Information

Broecker, 1986, *How to Build a Habitable Planet.* Eldigio Press, New York, 299 pp.

Cowen, R., 1995, *History of Life.* Blackwell Scientific Publications, Boston, 462 pp.

Kasting, J., Toon, O. B., and Pollack, J. B., 1988, How climate evolved on the terrestrial planets. *Scientific American,* February, pp. 90–97.

Kump, L. R., Kasting, J. F., and Crane, R. G., 1999, *The Earth System.* Prentice Hall, Upper Saddle River, NJ, 351 pp.

Levinton, J. S., 1992, The big bang of animal evolution. *Scientific American,* November, pp. 84–91.

Margulis, L. and Sagan, D., 1995, *What Is Life?* Simon & Schuster, New York, 207 pp.

Mottl, M. J., Glazer, B. T., Kaiser, R. I., and Meech, K. J., 2007, Water and astrobiology. *Chemie der Erde,* v. 67, pp. 253–282.

Schopf, J. W. (Ed.), 1983, *Earth's Earliest Biosphere: Its Origin and Evolution.* Princeton University Press, Princeton, NJ, 543 pp.

9 Human Forcings on the Ecosphere: World Population, Development, and Resource Consumption

One generation passeth away, and another generation cometh:
but the Earth abideth forever.

ECCLESIASTES 1:4

In Chapter 1, the major factors involved in human impacts (*I*) on the environment were discussed and shown to be population growth and distribution (*P*), affluence (*A*), and technology (*T*), where *I* is a function of (*P, A,* and *T*). Consumption of resources is a hidden additional factor in the IPAT equation and can be considered as a separate factor or one that is related to the three other factors. As preparation for discussion of global environmental change related to human activities in the Anthropocene in the following chapters, the increasing world population and consumption of natural resources as major factors contributing to environmental degradation are considered in this chapter.

The world in the early twenty-first century has been split in two. The northern developed countries are most of the industrialized, technological societies and have the affluence and resource consumption rates that are symbols of such societies. The developing countries have most of the people of the world; a large majority of them currently exist in poverty or just above the poverty level and have been deprived of the wealth of the North. However, several of the developing nations are changing rapidly as their populations increase and they industrialize along the lines of western societies during the Industrial Revolution. This pathway of development leads, as history has shown, to considerable resource consumption and major problems related to the environment of the planet. We will travel back in history to gain some understanding of the disparity in population and resource consumption and then consider in detail human population and industrial trends, and energy and mineral resource consumption.

BRIEF HISTORICAL REVIEW

Earth evolved for more than 4 billion years prior to the arrival of the first hominid, *Australopithecine,* some 4 million years ago. The early *Homo sapiens* species came on the scene about 400,000 years ago (see Chapter 6). The Earth was well prepared for the needs of the species. It was stocked with all the natural resources that allowed *Homo sapiens* to multiply and prosper. Food, minerals, forests, and fresh water were readily available. These early humans were hunters and gatherers. They used fire and later made tools, but their population was small, and as a result they had minimal impact on the ecosphere of Earth and evolutionary processes of ecosystems. Modern humans, Cro-Magnon, evolved 30,000 to 50,000 years ago. These humans developed efficient hunting skills and exhibited more control over their environment. The human species proliferated and slowly migrated and spread around the world. During their expansion, the ability of the human species to control the environment grew, as did its population. It is estimated that the human population stood at 250,000 to 5 million people at the end of the Pleistocene Epoch, some 10,000 years ago (Figure 9.1). At that time, human use and control of natural resources began to play an important role in the evolution of the environment of the Earth.

During the warming trend on Earth that began at the end of the Pleistocene, the major Agricultural Revolution had its beginning. The human population had learned to manipulate the natural environment to the benefit of the species by settling into small communities in which farming and eventually animal husbandry were important activities. This event marked the first of a series of human population explosions and the beginning of major, human-induced impacts on the environment and resources of Earth. Slash-and-burn techniques were employed to clear forests and grasslands to provide fields for crops. Animals were domesticated to supplement the dietary requirements of the expanding human population. Agriculture and the domestication of animals provided a better and more reliable supply of food. As a result, the first major permanent human settlements were established more than 5000 years ago (Figure 9.1). Agricultural communities developed in Pakistan, China, Africa, Egypt, and the Tigris-Euphrates Valley in Iraq.

The human population had increased, owing to the shift from hunting and gathering to the cultivation of food and the domestication of animals. This change allowed for more people to live in smaller areas and to establish larger communities. In this settled pattern of life, people could perform activities other than that of growing food. Time was available for the development of labor-intensive manufacturing activities such as craft making and the production of goods. Simple manufacturing techniques evolved. In all parts of the world, civilizations developed, advanced, or passed into obscurity during the next few thousands of years as evidenced by the history of the Egyptian, Mayan, Incan, Roman, Han, Grecian, and Khmer empires. During this time, the natural environment of areas adjacent to the larger population centers was slowly altered by the activities of humans. Forests were felled, cities grew, and the mining of mineral resources developed and expanded.

The environmental impact of the Agricultural Revolution resulted in a change in the natural ecosystems of those areas affected by human civilization.

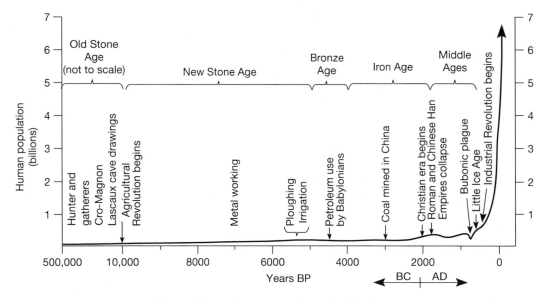

FIGURE 9.1 The growth of the world population over the past half million years. Some major historic ages and events in the history of the human population are also shown.

For example, the Mediterranean region has experienced the rise and fall of several civilizations. It now no longer contains its once vast primordial forests. It is difficult to determine what the original natural vegetation was like several thousands of years ago because of extensive modification of the Mediterranean ecosystems. In addition to political and social change and changes evoked by nature, such as climatic change, another probable cause of decline of some of these ancient civilizations was related to human-induced regional depletion of natural resources, such as forests, that once sustained these societies. It is during the Agricultural Revolution that the first significant modifications of natural carbon dioxide and other gas emissions to the atmosphere probably occurred because of human activities. As Jared Diamond has articulated in his book *Collapse* (2005), the demise of great ancient civilizations was the result of the intersecting factors of rapid population growth, unstable trade partners, pressures from enemies, and—of importance in the context of this book—climate change and environmental damage.

Until the last few centuries, most environmental changes as a result of human activities had been mainly a regional phenomenon and had occurred at a relatively slow pace. In the nineteenth century, the Industrial Revolution had its beginning in Europe and North America (Figure 9.1). The development of energy-intensive machinery resulted in mass production of goods and supplies. Energy in the form of wood, coal, and oil was used in abundance in the development of new industries. The steam and internal combustion engines provided power to replace the previous labor-intensive and simple manufacturing techniques that human society had used for thousands of years. More people and natural resources were needed to sustain the new industrial system and the needs of growing societies. Additional forests were converted to farmlands and pasture. Energy and food production increased. In the past, birth and death rates were nearly balanced. However, new effective medicines and better disease controls and sanitation methods led to a lower death rate, but the birth rate remained the same. As a result, the population increased and the demand for conversion of natural resources into energy and commodities for the expanding population increased.

In the twenty-first century, an adequate, if not high, standard of living exists for the majority of people in the industrialized world, but the developing world still strives to improve its living conditions and economies. The latter few decades of the twentieth century and on into the twenty-first century in the developed world could be characterized as the Technological Revolution, fueled by production and use of seemingly abundant energy, mainly from fossil fuels of coal, oil, and gas, modern advanced agricultural techniques, modern medicines, products of biotechnology, computerized systems, and reliable and efficient communication, informational, and transportation systems. Since World War II, the consumption of resources has mushroomed, and the rate of anthropogenic change dramatically increased. Today the stresses on our natural resource base are great. For example, oil, natural gas, and coal are rapidly being consumed. Some estimates of world oil reserves and consumption rates lead to the conclusion that there could be major oil shortages in the early to middle part of the twenty-first century. In addition, forests were and are being depleted of their stocks. The North and South American Atlantic forests were cut and cleared in the seventeenth century. More recently, according to the 2007 Food and Agricultural Organization (FAO) of the United Nations' estimates, tropical forest area was lost at a rate of roughly 0.8% per annum during 1981 to 1990. In addition, between 1980 and 1995, the extent of the world's forests decreased by about 180 million hectares (1 hectare = 10,000 m^2), an area equivalent to about the size of Indonesia or Mexico. From 1990 to 2000, the net forest loss was 13 million hectares per year slowing to 7.3 million hectares per year from 2000 to 2005, an area the size of Sierra Leone or Panama. Furthermore, water resources are being used and polluted at increasing rates. In the late twentieth century, global water use amounted to about half of the annual accessible runoff from the land. The scarcity and overuse of water resources in the Middle East are certainly one reason for the political problems of the region. In the United States, certain minerals and building materials are scarce and necessitate imports. For example, domestic sources of ores of tungsten, mercury, and chromium are in short supply. Lithium used in lithium batteries is scarce worldwide. In the future, the increasing demand for these elements will lead to increased imports, recycling, or substitution of other materials.

The Technological Society is rapidly phasing into an Informational Society. In the twenty-first century, new materials, products, and manufacturing techniques will most likely be developed. There will be a strong need to develop methods to manage the vast amount of information confronting society and to develop the necessary computerized systems that will operate the agricultural, industrial, and transportation sectors. The stresses on the Earth's natural resource base continue to increase as the developed nations of the world maintain their growth patterns of the past 150 years and the developing nations of the world, many exhibiting strong population growth, seek a higher standard of living and industrialize along the same pathway that the developing world took since the advent of the Industrial Revolution.

INDUSTRIAL AND HUMAN POPULATION TRENDS

Sustainable development implies economic progress, investment in human resources, stable population growth at replacement levels, and a technology that does not significantly degrade the environment or deplete the natural resource base. Sustainable development is a difficult and confusing concept. The meaning of each aspect of sustainability and the value attached to it differs from culture to culture and from country to country. The nations of the world vary in their political, social, and economic structure. In many instances, processes within these structures have led to growth without due attention to one or more aspects of sustainable development. It appears that none of the countries of the world have been entirely successful at achieving a reasonable level of sustainability.

Data on mineral resources, energy production and consumption, population, economic indicators, and other parameters are available from a number of institutions, including the United Nations (UN), Food and Agricultural Organization of the United Nations (FAO), World Bank (WB), Organization for Economic Cooperation and Development (OECD), Energy Information Administration (EIA), and World Resources Institute (WRI). The data are available in many cases for specific nations as well as regions. The regions considered vary from source to source but generally include subdivisions of the developed and developing world.

For example, as defined by the United Nations, the First World, the industrialized, affluent, developed countries particularly, but not exclusively, in the Northern Hemisphere, includes the United States, Canada, Australia, New Zealand, Japan, the former USSR, and countries of Western Europe. Recently, these countries have been referred to collectively as the North. The European community has created a unified European Union (EU). The East bloc nations of Europe, prior to Mikhail Gorbachev's presidency, had been referred to as the centrally planned economies of the Second World. The world situation has changed dramatically with the demise of the Soviet Union and the rise of the newly independent countries of Eastern Europe and Central Asia. Data on population, resources, and other parameters for these countries were just becoming available in 1997.

The other nations of the world are usually considered to be developing countries (also referred to as the Third World or the South), and many are located in the Southern Hemisphere. These nations are relatively poor but are striving toward development of their economies. They generally have a greater fertility rate than the developed countries. Nations of South and Southeast Asia, Latin America, and Africa fall under this heading.

Industrialization

The nations of the world have industrialized at different rates and along different paths. The industrialized countries have created an advanced society; however, it has been a consumptive society that has been built on the natural resource base of other countries and the planet as a whole. Currently, these nations are attempting to become more globally responsible, efficient in using energy and other natural resources, and adept at reducing pollution and recycling waste.

In contrast, the industrialized but newly independent countries of central Europe have a different and extremely difficult challenge. They have spent years in political systems that did little to advance human resources or sustain economic development, but led to environmental degradation of huge proportions. The still heavily polluted waterways and urban smog of much of eastern Europe are dramatic testimony to their present environmental situation.

Developing countries are in different stages of industrialization and economic development, and it is somewhat inappropriate to lump them all together as developing countries of the South. The poorer countries of the developing world are the worst off. At present, they generally lack the political will or environmental infrastructures necessary to lead them along a pathway of sustainable resource and economic development and care for the environment. For example, of the developing countries, the sub-Saharan nations of Africa are making minimal progress industrially, are still experiencing relatively large population increases, and are suffering economically. Poverty, environmental degradation, and social and political unrest are occurring in several of these nations.

Other developing countries of the world are rapidly becoming industrialized, such as Indonesia, Malaysia, Thailand, Mexico, Brazil, and Chile. In

Asia, the countries of South Korea and Taiwan and the city-states of Hong Kong and Singapore experienced an even more dramatic phase of industrialization and economic expansion in the latter decades of the twentieth century. These four strong economies are known as the four tigers of Asia. Following Japan's lead, they have concentrated on the development of high-quality educational systems and scientific infrastructures that include extensive research and technological development activities. This effort has rapidly advanced their social, economic, and political development. China, with more than 1.33 billion people in 2008, experienced strong economic growth in the late twentieth century and on into the twenty-first century. India with more than 1.14 billion people in 2008 is not far behind. For all these countries, growth has had an environmental cost. Air and water pollution and land degradation have been an outcome of the economic expansion effort in most regions.

Developing countries face a great challenge in the future if they wish to achieve rapid economic development, avoid environmental degradation, and build a sustainable environment and economy. To attain sustainability, many of these countries need to combat poverty, provide an adequate standard of living for their people, build a skilled workforce, and protect their natural resources that serve as the basis for their economic development.

Population Growth

The world population is still increasing rapidly, but growth has slowed in the twenty-first century. **Demography** is the study of population and causes for changes in population and its distribution. Estimates of world population in the early stages of human development on the planet are difficult to make. The range of estimates for total world population at the dawn of the Agricultural Revolution, about 10,000 years ago, is 250,000 to 5 million people. At the time of Christ, the world population was probably around 200 million people, and 500 million by A.D. 1650. It was not until about 1850 that the world population reached the mark of 1 billion people. Eighty years later, by 1930, the population doubled to 2 billion people. It doubled again 45 years later when it reached the 4 billion mark in 1975. In just another dozen years, in 1987, another billion people lived on the planet, bringing the total world population to 5 billion people! The human population reached the 6 billion mark late in the year 1999, and in the year 2009, there were an estimated 6.756 billion people on Earth (Figure 9.1). It is predicted that the world's population will reach 7 billion people in 2011.

The future of the world's population is difficult to project because it depends on a number of factors that in themselves are difficult to predict, not least of which are the fertility rate and the mortality rate. The growth rate of the world's population peaked in the late 1960s at 2.1% per year and has since fallen to about 1.2% per year in 2008. The annual absolute increment of people added to the world population peaked at 87 million per year in the late 1980s and is now about 72 million per year (Figure 9.2). However, even today about 27 people are added to the world's population every 10 seconds. Projections of the world population until the year 2100 under different scenarios of mortality and fertility and other factors are shown in Figure 9.3. Approximately 8 billion people may be on the Earth in the year 2020 and perhaps more than 10 billion by the year 2050. Notice that several of the scenarios suggest a peak in the world population around 9.5 billion people during the third quarter of the twenty-first century and a slow decline thereafter. The vast majority of these people will live in urban areas of the world and especially the urban centers of the developing countries.

The doubling time for a projected world population in the year 2020 of 8 billion people would be about 45 years (Table 9.1). The doubling time with a growth rate of 1.2% per year in 2008 is 58 years. About three babies are currently born every second. This doubling of human population is an example of exponential growth. If you double the value of a penny and then continue to double the value of the sum of your pennies at a constant rate, the amount you have is not worth much for a long period of time. Then, very rapidly, the total value will increase, and you will have a large amount of money over a short period of time. Another example: If you have a chessboard and put one penny on the first square, two pennies on the second, four on the third square and eight on the fourth and so on, by the time you reach the twenty-first square, you would have more than a million pennies. The twenty-second square would give you more than 2 million pennies, and so forth. A short time to doubling is a characteristic of exponential growth. Such growth starts very slowly, but soon, large numbers are created very quickly.

Exponential growth at a constant growth rate (r) can be described by the following relationship:

$$N_{(t)} = N_{(0)} e^{rt}.$$

For population growth, $N_{(t)}$ is the number of people at time t, $N_{(0)}$ is the number of people at some starting time 0, t is time, and e is the natural logarithm

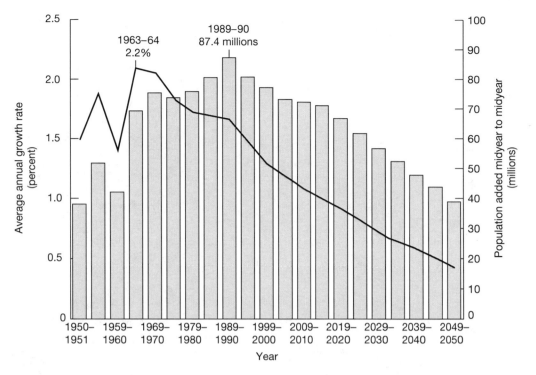

FIGURE 9.2 Annual additions to the world population and the annual growth rate of the global population from 1950 projected to 2050. In 1963–1964, the average annual percentage growth rate of the world population reached 2.2% and has been in decline ever since. In 1989–1990, the pace of the absolute annual increase to global population began its decline. Bars represent population added and the dark line, average annual growth rate.   (*Source:* United States Census Bureau, International Programs Center, International Data Base.)

with the value of 2.718. In exponential growth, a simple relationship exists between the rate of growth in percent and the time it takes to double a quantity like population. The doubling time of the population equals $N_{(t)}/N_{(0)} = 2 = e^{rt}$, and taking the natural logarithm of both sides gives $0.69 = rt$, or $0.69/r$ equals the doubling time t. Thus, the doubling time is approximately 70 divided by the growth rate in percent. For example, at the present and constant growth rate of the world population of 1.2% per year, the time to double the population would be about 58 years.

The projected increase in world population will be distributed unevenly among the nations of the world. The more affluent countries generally are experiencing slow rates of population growth, whereas the poorer nations are experiencing the greatest gains in population (Figure 9.4). In 1950, 33% of the world population resided in the industrialized, developed countries as compared to 23% at the close of the twentieth century. In 1995, 77% of the world population, or 4.4 billion people, lived in the developing world and 1.3 billion resided in the industrialized world. By the

year 2025, the percentage of world population attributable to the developed world is projected to drop to 16%. The developed countries have the smallest birth rate and, in fact, several of them are approaching zero population growth. In contrast, the developing countries generally are still experiencing a relatively rapid increase in population. The industrializing nations of the developing world are experiencing modest gains in population. The largest increases in birth rates are in the developing countries with the least amount of industrial growth. Declining birth rates tend to accompany economic affluence.

The regional aspects of population change, such as that of the developed and developing world, are diverse. For example, between 1995 and 2025, some countries of eastern Europe and the former USSR are projected to lose population, including the Russian Federation, Belarus, Bulgaria, Hungary, Romania, and Ukraine. The populations of Denmark, Greece, Italy, Spain, and Portugal are also expected to decline in the early part of the twenty-first century. On the contrary, the population of the United States is expected to grow from 278 million in the year 2000 to

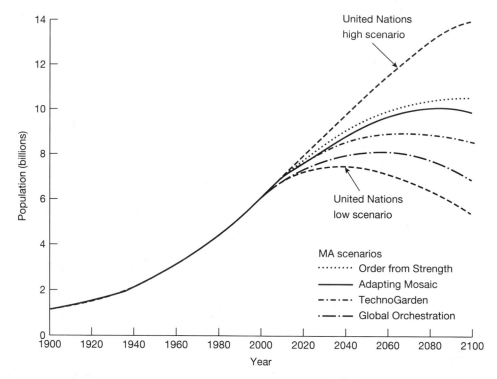

FIGURE 9.3 Alternative projected paths of global population growth. The various millennium assessment (MA) scenarios are shown along with the United Nations high and low trend estimates for global population growth. The four MA scenarios are based on different assumptions about the driving forces of change, like economic growth and how management of ecosystems is achieved, and their interactions. (From Millennial Ecosystem Assessment, 2005 – *Ecosystems & Human Well-Being: Synthesis*. Island Press, Washington, DC. Copyright © 2005 World Resources Institute. Reprinted by permission of World Resources Institute.)

325 million by the year 2025, mainly due to immigration. By 2050 the five largest countries of the world in order of population are projected to be India (1.747 billion people), China (1.437 billion people), the United States (420 million people), Indonesia (297 million people), and Pakistan (295 million people).

The population of sub-Saharan Africa is projected to grow from 640 million in 2000 to one billion in 2025. Asia could grow during the same period from 3.42 billion to 4.31 billion people. By the year 2025, the populations of Asia and Africa could be more than 6 billion people and exceed the global population on Earth today. Some 85% of the world population by the year 2025 will live in the developing nations, in the regions of Africa, Latin America, and Asia. This rapid increase in population growth will have important and perhaps serious ramifications for the world in terms of energy and material resource production, distribution, consumption, and waste disposal. Also, this disparity in population growth rates has important implications for any attempt to develop a sustainable global environmental and economic situation.

TABLE 9.1 The Doubling Time of the World Population Throughout Human History

Population Size, Billions	Year	Time Required
~1	1850	30,000 to 50,000 years (all of modern human history)
2	1930	80 years
4	1975	45 years
8 (projected)	2020	45 years

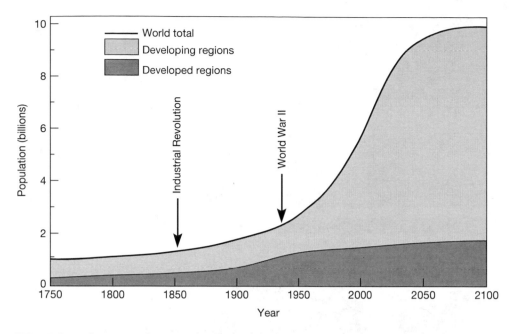

FIGURE 9.4 The past and projected growth of the human population in the developed and developing regions of the world from 1750 to 2100, assuming population stability is reached in the latter part of the twenty-first century. Notice the large expansion in the population of the developing world after 1950.

For example, on one hand, the increasing populations of the developing world will need energy sources for transportation, home heating, and cooking, and, if they choose so, industrialization. The choice of energy for their future needs is critical. If they choose fossil fuels, they, as the industrialized nations of today, will contribute progressively more to carbon dioxide emissions to the atmosphere and to changes in atmospheric gas composition. On the other hand, without energy conservation and increased efficiency and development of alternative energy sources, increased affluence in the developed world, even with stabilized population, will lead to increased per capita consumption of fossil fuels and increased gas emissions to the atmosphere.

URBANIZATION The world is quickly becoming urbanized as people migrate to the cities in droves. Although urban areas in the year 2008 occupied only 5% of the Earth's land area, they are home to nearly 50% of the world's population. Urban areas with more than 500,000 inhabitants accounted for 49.1% of the world's urban population in 2003. Approximately 75% of the people in the developed nations and 50% of the population of the developing world live in these urban centers (Figure 9.5). In 1950, less than 29% of the world population lived in cities. By 1985

this value had increased to 41%, and, it is projected that by the year 2030, about 60% will live in urban areas. The urban populations estimated in 2008 for the more developed nations and the less developed nations were 915 million and 2,443 million, respectively, and are projected to increase to 1,016 million (11% increase between 2008 and 2030) and 3,949 million (61.6% increase between 2008 and 2030), respectively, in 2030. In 1970, only the cities of London, New York, Tokyo, and Shanghai had populations greater than 10 million. In the year 2000, 24 cities reached this population level; 18 are located in the developing nations. In 2009, five cities had populations exceeding 20 million people and by 2030 12 cities will exceed this population size. Figure 9.6 shows some examples of city growth from 1970 to 2030. The concentration of the urban population undoubtedly has a profound effect on the ability of city systems to provide adequate services for their residents. For example, it is estimated that 40% of the urban population in the developing world currently lives in slum areas. In the future, providing for water, health services, pollution control, transportation, employment, and education in the megacities will be a staggering task. In other words, the "footprint" of these cities in terms of their effects on the environment will be even larger than one might expect simply from their area or population.

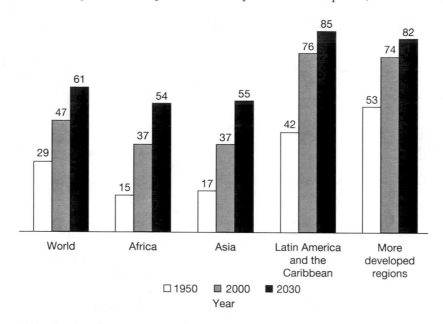

FIGURE 9.5 Trends in percent growth of the urban population of the world between 1950 and 2030. Notice, in particular, the more rapid growth of the urban population in less developed regions. (*Source:* United Nations, 2004.)

Coastal regions will be particularly endangered by this urban growth. At least 40% of the world population lives within 100 kilometers of the coastline. Nearly 40% of the cities with a population larger than 500,000 people are located on the coast. These urban and near-coastal populations depend heavily on river water and groundwater resources and near-coastal and marine environments for food, construction sites, transportation, recreation, and waste disposal. The World Resources Institute estimates that about 50% of the world's coastal ecosystems are at significant risk of degradation because of developmental and other activities within the regions or in the watersheds of rivers flowing to coastal regions (see Chapter 11).

DEMOGRAPHIC TRANSITION To determine how fast a population is growing, demographers must have information about birth and death rates. The crude birth rate is the number of live births per 1000 people. The birth rate is calculated as follows: (number of live births) ÷ (total population at midyear) × 1000. The crude death rate is calculated in a way similar to the crude birth rate. It is the number of deaths per 1000 people and is calculated as (number of deaths) ÷ (total population) × 1000. Thus, the natural rate of increase of the population, its growth rate, is the surplus of births over deaths in a population

at a given time. For a declining population, the death rate would exceed the birth rate.

Demographic transition is a model that describes population changes over time for the world or individual regions or countries. It is based on the ideas first set forth in 1929 by the American geographer Warren Thompson that the observed changes, or transitions, in birth and death rates in industrialized countries during the last century or so follow a general pattern of development as shown in Figure 9.7. Demographers have recognized four stages in the historical pattern of human population growth. The first stage is that of high birth and high death rates and no increase in population. During most of the history of the human species, high birth and death rates were balanced. Families had many children, but a limited food supply and disease curtailed population growth. The second stage is transitional and is defined as a time of high birth and low death rates resulting in a large growth in population. During the Industrial Revolution, an adequate food supply, improved medicines, and disease control led to a lower death rate, while the birth rate remained high and thus the population grew. The third stage occurred later during the industrialization period when a lower birth rate accompanied the already lower death rate and population growth slowed. Nations that had developed industrially and economically provided

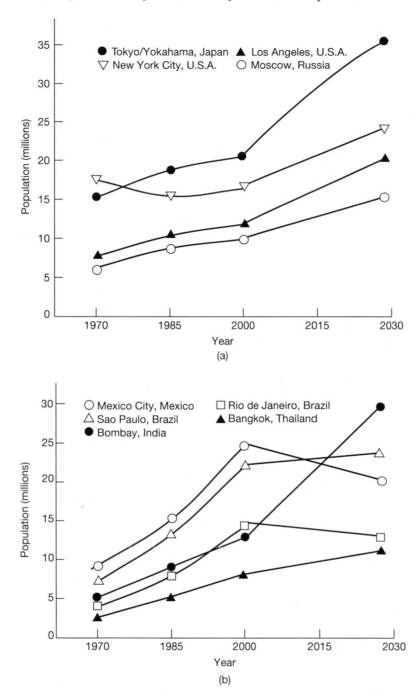

FIGURE 9.6 Examples of growth in the populations of cities located in (a) developed and (b) developing countries between 1970 and projected to 2030. Notice the different patterns of growth of the urban centers with most growing, but Mexico City and Rio de Janeiro predicted to shrink modestly. The city Bombay is now called Mumbai.   (*Source:* United Nations, 1987 and Demographia, 2009.)

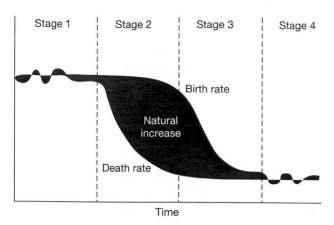

FIGURE 9.7 Generalized pattern of stages in demographic transition. Natural increase in population is produced from the excess of births over deaths.

more jobs and money and better food, medicine, and education. More people lived past childhood. The affluence level increased, as did the demand for the use of natural resources. The necessity and desire for large families decreased and thus the birth rate was lower. The fourth and last phase of the population pattern is the post-industrialized stage. Low birth and death rates result in slow population growth. When the birth and death rates are balanced, the population is stabilized. This latter stage has not been reached by the world population but has been nearly achieved by some countries in the industrialized world where population growth rates are low. Indeed, Europe as a whole actually had a rate of population growth in 2007 of −0.1%. This negative growth has become a concern. However, a number of industrialized countries are experiencing important population growth because of immigration. Figure 9.8 compares the world demographic transition with patterns of some individual countries. It can be seen that patterns of transition differ from country to country.

It is interesting and encouraging to note that infant mortality rates have fallen worldwide during the past half-century. Furthermore, life expectancy at birth has risen during this same period. For the world as a whole, infant mortality rates have fallen from about 160 deaths per thousand births in 1950 to close to about 50 deaths per thousand births in 2008, while life expectancy has increased from about 46 years to 70 years in 2008 (see, for example, https://www.cia.gov/library/publications/the-world-factbook/rankorder/2102rank.html), with a projection to 72.4 in 2025. These overall trends are similar for both developed and developing countries, although the rates of change differ from country to country (Table 9.2).

TABLE 9.2 Infant Mortality and Life Expectancy for Selected Countries, 2007*

Country	Infant Mortality (deaths per thousand births)	Life Expectancy (years)
Albania	20.0	77.6
Angola	184.4	37.6
Australia	4.6	80.6
Austria	4.5	79.2
Bangladesh	59.1	62.8
Brazil	27.6	72.2
Canada	4.6	80.3
Chile	8.4	77.0
China	22.1	72.9
Costa Rica	9.5	77.2
Cyprus	6.9	78.0
Czech Republic	3.9	76.4
Denmark	4.5	78.0
Ecuador	22.1	76.6
Egypt	30.1	71.6
Finland	3.5	78.7
France	4.2	79.9
Germany	4.1	79.0
Greece	5.3	79.0
Guatemala	29.8	69.7
Hungary	8.2	72.9
India	34.6	68.6
Iran	38.1	70.6
Ireland	5.2	77.9
Israel	6.8	79.6
Italy	5.7	79.9
Japan	3.2	81.4
Kenya	57.4	55.3
Korea, South	6.1	77.2
Mexico	19.6	75.6
Mozambique	109.9	40.9
New Zealand	5.7	79.0
Nigeria	95.5	47.4
Norway	3.6	79.7
Pakistan	68.5	63.8
Panama	16.0	75.2
Peru	30.0	70.1
Poland	7.1	75.2
Portugal	4.9	77.9
Russia	11.1	65.9
Slovakia	7.1	75.0
South Africa	59.4	42.5
Spain	4.3	79.8
Sri Lanka	19.5	74.8
Sweden	2.8	80.6
Switzerland	4.3	80.6
Syria	27.7	70.6
United Kingdom	5.0	78.7
United States	6.4	78.0
Venezuela	20.9	74.8
Zimbabwe	51.1	39.5

*Source: http://www.infoplease.com/ipa/0004393.htm/

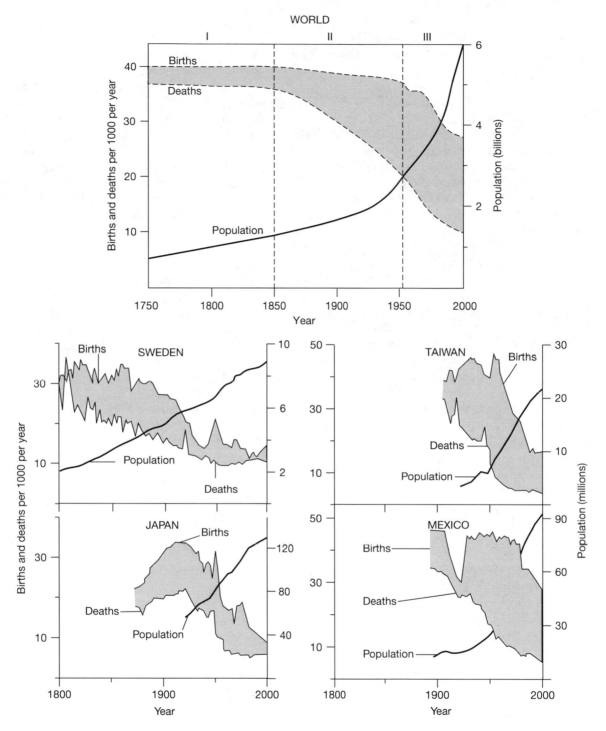

FIGURE 9.8 Comparison of world demographic transition with some examples of demographic transitions in industrialized (Sweden and Japan) and less industrialized (Taiwan and Mexico) countries. The shaded area between births and deaths is the rate at which the population grows. The solid curve shows population growth. I, II, and III are the first three stages in world population growth. Sweden's demographic transition occurred over 200 years, whereas Japan will undergo transition in less than a century. Gaps between birth and death rates for the less industrialized countries are large, and these countries have less time to undergo transition.

Many poorer countries of the world are currently in the second stage of population growth and are experiencing high birth and low death rates. Originally, a number of the countries of the Northern Hemisphere colonized many countries of the world. Much of the wealth of the colonized countries was transported back to the North. The northern countries, in turn, exported their knowledge of disease-control methods and efficient agricultural practices throughout the world. In the formally colonized countries, eradication of major world diseases, improved medicines and health care, modern agricultural techniques, and higher sanitation standards reduced the death rate, but the birth rate remained high. The majority of the people, however, remain illiterate and poor. Rapid population growth and increased poverty have been characteristic of many countries in the Southern Hemisphere in the twentieth and early twenty-first century. The sheer task of providing for this rapidly growing population hinders the ability of the poorer developing nations to educate their people and to develop commercially and industrially, both prerequisites for an adequate modern standard of living and an eventual stabilized population.

Influence of Fertility Rates   Many factors contribute to the birth rate. These factors involve complex individual and family decision-making processes that differ among cultures and socioeconomic groups. Fertility is one factor. In comparison to the birth rate, the total fertility rate is the average number of children that women will have in their lifetime. Figure 9.9 shows the world and regional differences in fertility rates and the fact that these rates have fallen during the past 50 years of the twentieth century and into the early twenty-first century. In 2008, the estimated total fertility rate for the world was 2.61, with a range of 1.7 for the more industrialized countries to 4.7 for the least developed countries. The developing countries had a fertility rate of 2.8. The fertility rate for Africa alone was 4.61, with Niger and Mali having estimated rates for 2009 of 7.75 and 7.29, respectively (see, for example, https://www.cia.gov/library/publications/the-world-factbook/rankorder/2127rank.html and http://data.un.org/Data.aspx?d=PopDiv&f=variableID%3A54).

Figure 9.10a, b shows population growth and total fertility rates for various countries of the world for the early part of the twenty-first century. Although both population growth rates and total fertility rates have fallen in much of Africa and the Middle East, notice the relatively high rates for these regions compared to most of the world. Also, note the low rates for much of the developed and industrialized world. The present rates of change of these population variables plus decreased infant mortality rates in Africa and the Middle East are reasons for the projected continued strong rate of growth of the world's population, at least through the first half of the twenty-first century. Also, countries like China and India, with populations exceeding 1 billion even with modest growth rates of their populations, add substantial numbers of people to the world population each year.

A woman's education level plays an important role in lowering fertility (Figure 9.11a). Fertility rates fall when the educational level and status of women in a community are increased. Educated women can be employed outside the family and tend to marry at a later age and thus delay childbirth. Women who do not have access to education and paid employment and whose primary employment is unpaid family labor tend to have high fertility rates.

Another factor that is believed to affect fertility rates is the influence of family-planning programs (Figure 9.11b). Such programs have the potential to enable couples to constrain their fertility rates within their ability to have and care for children through their growth years. In 2001, about 62% of married couples used modern contraceptive procedures in the world. The United States prior to 1985 was a major contributor to the United Nations Population Fund. This fund provides alternatives to abortions through education and health care for women and children of the world. Because of the domestic political debate over birth control in the United States, the United States in the late 1980s cut off funding through the United Nations of worldwide family-planning programs. The United States has recently reestablished this funding. Obviously, family-planning programs of a national or global nature are strongly influenced by social and political decisions.

A change in the role of children within the family labor force also may affect population growth. Higher fertility rates occur when children are necessary to the family labor force or to bring in income. As the cost of raising children increases, the birth rate drops. In addition, urbanization tends to play a role in decreasing birth rates because city life is generally conducive to smaller families.

A final factor of importance in declining fertility rates is the wealth of a country as expressed in its gross domestic product per capita (GDP; see discussion later in this chapter) (Figure 9.12). It appears that the "richer" the country, that is the higher a country's GDP, the lower its total fertility rate, approaching the replacement fertility rate (Figure 9.13) of about 2.1 at US$5,000 to $10,000 per capita.

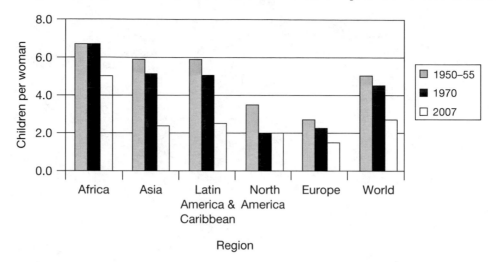

FIGURE 9.9   Fertility levels of the world and various regions of the world from 1950–1955 to 2007. The replacement fertility rate is 2.1 children per woman.   (*Source:* The United Nations.)

The fertility rate of the world has generally been declining. Women, on the average, are having fewer children. However, there is a greater population of women in the world having babies. As a result, the decrease in the fertility rate does not compensate for the large number of women having children, and the population of the world continues to increase, to 6.757 billion people in mid-2009.

Population Pyramid   Another important element of demography is the age–sex structure of the population. This structure can be represented in the form of a population pyramid. The pyramid shows how many people of each sex there are in various age groups. The horizontal axis of the pyramid is usually the percentage of the total population or the number of people, and the vertical axis is generally the age grouping or year of birth. The age and sex distribution of people in the world as a whole is compared with various regions and countries in Figure 9.14 (http://www.census.gov/ipc/prod/wp02/wp-02.pdf). There is a great deal of variation in the population pyramids. Notice the strongly pyramidal shape of the age–sex distribution of the population of sub-Saharan Africa as contrasted to the world as a whole and that of the developed world. A significant percentage of the population of women in sub-Saharan Africa is in their childbearing years or younger. Furthermore, nearly 50% of the male population is below the age of 40 and capable of fathering children. Except for China, Eastern Europe, and the Newly Independent States

(NIS) of the former Soviet Union, India and other developing regions of the world tend to exhibit somewhat pyramidal shapes in their population pyramids. In contrast, the population pyramid for people in the industrialized countries shows rather similar percentages of males and females of various age groups up until an age of 40 to 45 years. Age groups older than 45 years show diminishing percentages of both males and females; however, with increasing age in the industrialized countries, females are proportionately a greater percentage of the population.

The shape of the population pyramid of people in the developing countries shows why, despite the decreasing total fertility rates for most regions of the world, the global population in the twenty-first century will continue to grow at a rate that could lead to a world population of slightly less than 7.5 billion to slightly more than 10 billion people in the year 2050 (Figure 9.3). It is simply a reflection of the fact that there are many women of childbearing age or younger in the population, as well as potential young fathers. When the younger people mature to become parents, they will contribute to a considerable growth in population, despite the fact that they may have fewer children than their parents (lower total fertility rates). Contributing to the overall future world population growth will be (1) the reduction in infant mortality from 104 to 56 deaths of infants under age one per 1000 live births from the 1960s to the early twenty-first century, (2) the increase in life expectancy from the 1950s to the early twenty-first century of 9.7 and

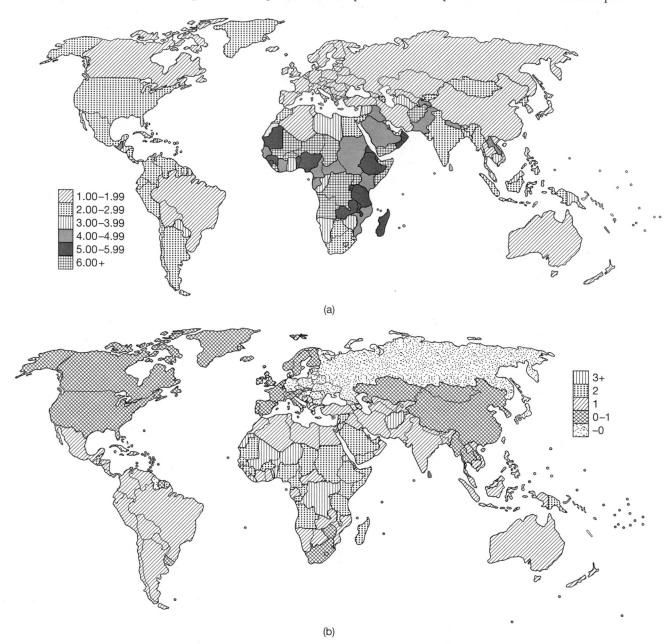

FIGURE 9.10 World maps of the early twenty-first century fertility rates in children per woman (a) and population growth rates in percent for various countries (b). (*Source:* The United Nations; modified from Fertility Rate World map.PNG-Wikimedia Commons and Population_growth_rate_world_2005-2010UN.PNG.)

22.4 years for the more developed countries and the less developed countries, respectively, and (3) the overall aging of the world population from a percentage of people older than 65 years of 7% in 2007 to 26% in the year 2050. The AIDS epidemic, although for some select countries of the world like South Africa having the potential to be devastating by early in the twenty-first century, will not have a dramatic effect on the growing world population.

Gross National and Domestic Products (GNP and GDP)

One measure of the economic activity of a nation is the dollar value of all goods and services produced by a nation's economy in a year, including goods and services produced abroad, and is known as the **Gross National Product (GNP)**, also called **Gross National Income (GNI)** by some organizations.

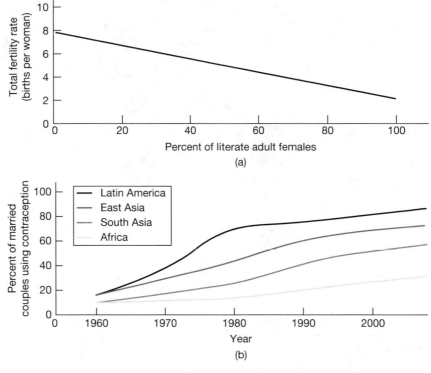

FIGURE 9.11   The generalized trend between fertility rates and female literacy for the countries of the world (a) and the percent of married couples using contraception (b) from 1960 to 2007 for various regions of the world.   (*Source:* http://www.un.org/esa/population/ and http://www.prb.org/Publications/Datasheets/2008/2008wpds.aspx.)

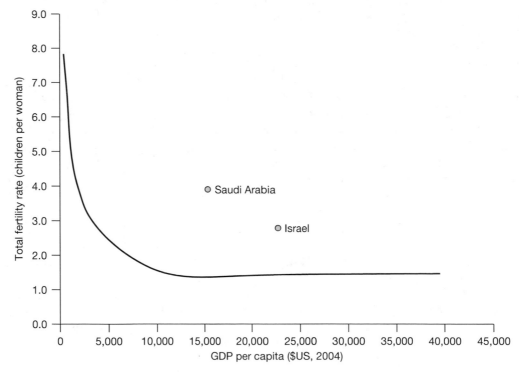

FIGURE 9.12   Total fertility rates vs. GDP per capita for countries with over 5 million people in 2004. Israel and Saudi Arabia stand out from the overall trend.   (*Source:* Central Intelligence Agency, *CIA World Fact Book,* www.cia.gov.)

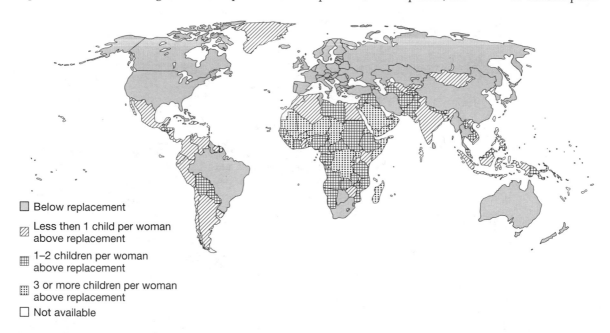

Below replacement

Less then 1 child per woman above replacement

1–2 children per woman above replacement

3 or more children per woman above replacement

Not available

FIGURE 9.13 Total fertility rates relative to replacement levels for the countries of the world in 2002, at which time the global average replacement level was 2.3 children and fertility rates were highest in sub-Saharan Africa and the Near East. (*Source:* United States Census Bureau International Programs Center, International Data Base, http://www.census2010.gov/ipc/www/publist.html.)

Until recently this index has been the main criterion used to evaluate the economic situation of a nation and broadly distinguish the different degrees of economic development of nations. Another measure of economic activity is **Gross Domestic Product (GDP)**, which is the total value of goods and services produced by a nation within the nation's borders in a year. It is calculated by adding together the market values of all the final goods and services produced during the year. This index is being used more frequently as a true indicator of a nation's wealth but not necessarily its overall quality of life. Changes in GDP from one year to the next reflect changes in the output of goods and services and changes in their prices. Real GDP may also be calculated and changes in this index are more meaningful in providing a better understanding of what is occurring in a nation's economy. The changes in real GDP show what has actually happened to the quantities of goods and services, independent of changes in prices. The average annual GDP growth rate in percentage for the world's countries is shown in Figure 9.15 for the period 1990 to 2003. This rate has slowed in the latter part of the first decade of the twenty-first century because of the worldwide financial crisis.

When the GNP or GDP is divided by the population of the country, you have the per capita GNP or GDP per year, one measure of the average potential affluence of each person in a country. Figure 9.16 shows the relationship in 2008 between GDP and GDP per capita for some selected countries. Notice in particular the great disparity in wealth between the industrialized world and the developing world. For example, in 1997 the United States had a per capita GDP of $28,651 as compared to 20 countries with a GDP per capita of less than $300. In 2008 the International Monetary Fund estimated the United States GDP as $46,859 international dollars and that of the Congo as $328. In the early twenty-first century, there still exist large differences in GDP and, more significantly, GDP per capita between developed and developing countries.

One may also calculate the gross value of all the goods and services produced by the world as a whole and per capita termed the world GDP and GDP per capita, respectively, or the **Gross World Product (GWP)** and the Gross World Product per capita. The world had a GWP of about 33 trillion United States dollars and a per capita GWP equivalent to $5,700 in 2000. In 2008 the GWP was $60 trillion and the GWP

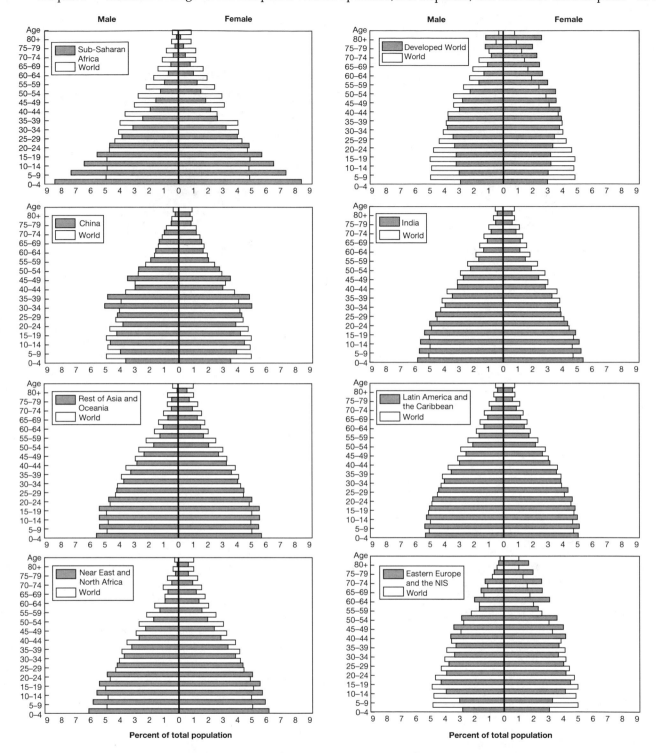

FIGURE 9.14 Population pyramids showing the age–sex structure of the population in various regions and selected countries compared with the world as a whole in 2002. Fertility has fallen in the industrialized countries, although it has not reached the stage of simple replacement fertility, the number of children that parents need to have over a generation to replace themselves. Zero population growth cannot be achieved in the industrialized world until the baby-boom generation ages and dies. Also, there are significant migrations of people into several of these countries, including the United States, which will delay zero population growth. The developing countries are poised for continuous population growth. There are many more young people than older ones. When these young people age and become parents, they will contribute significantly to population growth despite the fact they may have lower fertility rates than their parents. (*Source:* United States Census Bureau International Programs Center, International Data Base, http://www.census2010.gov/ipc/www/publist.html.)

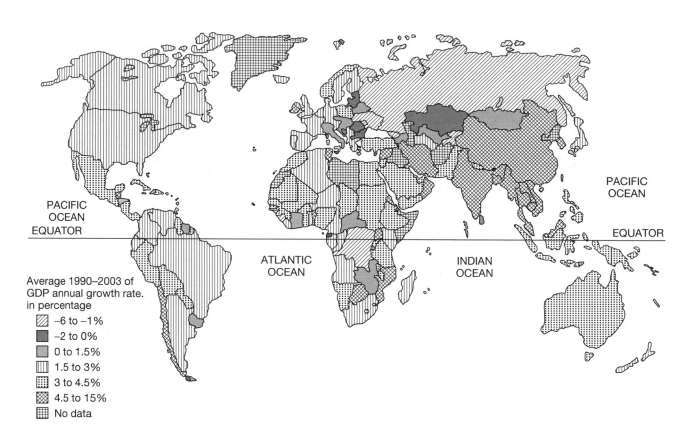

FIGURE 9.15 The percent average annual growth from 1990 to 2003 of the Gross Domestic Product for various countries of the world. (*Source:* http://www.worldbank.org/.)

per capita was $8,800 (Figure 9.17). The GWP grew by more than 1000% between 1950 and 2008, and the per capita GWP by about 340%. The difference in percentage growth reflects the fact that for the world, the population during this approximately 60-year span had grown faster than GDP. Also notice the sharply exponential trend in GWP and GWP per capita from the late 1990s to 2008. To a very significant degree, the energy for this strong growth has been derived from the combustion of the fossil fuels of coal, oil, and gas. This reliance on fossil fuels as the energy source to drive the world's economic growth and that of most nations has led to emissions of hydrocarbons and other organic compounds, nitrogen- and sulfur-bearing trace gases, carbon dioxide, methane, trace metals, and particulates to the environment. These emissions are the root cause of several environmental issues discussed later in this book.

GDP is generally related to the rate of resource consumption by a country. As GDP rises, resources are usually consumed more rapidly, with an accompanying increase in pollution and waste residues. In recent decades, this relationship has been broken somewhat in the industrialized world partially because of conservation of resources and increases in energy efficiency. However, national values of GDP generally do not include environmental "costs of doing business." For example, the costs associated with the loss of soil by erosion that would eventually make land unproductive for future farmland are not considered in determining GDP, nor are future replacement costs. Likewise, costs associated with the destruction of the tropical forests and the eventual loss of these natural resources are also not included in GDP estimates. These environmental costs of doing business have been allowed to accumulate over time; for example, consider the projected costs of the United States Environmental Protection Agency's Superfund Clean-up, about $1 billion.

When natural resources near depletion, the potential for ecological, social, and economic disaster

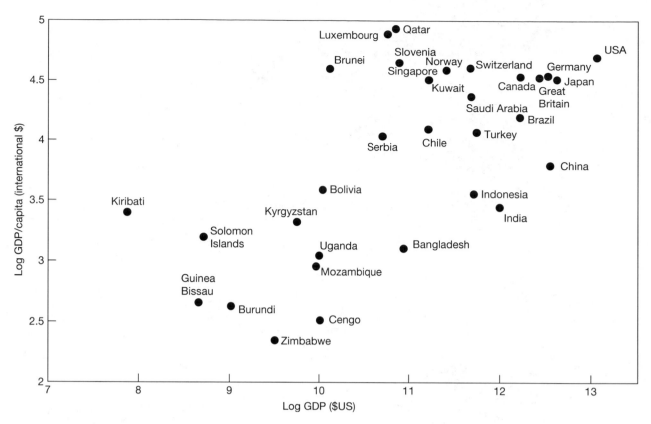

FIGURE 9.16   Gross Domestic Product (GDP) per capita plotted against GDP illustrating the differences between some countries in 2008. A log scale is used because of the large spread of the values. For example, the United States had a GDP/capita and a GDP of $47,580 and $14.2 trillion, respectively, whereas Burundi had a GDP/capita and GDP of $140 and $1.16 billion, respectively, in 2008. The data for Belgium includes Luxembourg.   (Data for 2008 from http://web.worldbank.org/.)

exists. An example of the results of the cost of doing business as usual scenario is that experienced by the inhabitants of the island of Nauru in the South Pacific. Nauru had ample mineable phosphate rock, a commonly used fertilizer. The phosphate rock was mined and exported and brought prosperity to the inhabitants of Nauru and revenues to them and foreign investors. However, large environmental and social costs resulted from this mineral exploration and exploitation program. The phosphate rock is nearly completely mined out, and mining revenues on Nauru have fallen drastically. Four-fifths of the island have been devastated by mining operations that were mainly controlled by the foreign investors. The remaining land is incapable of sustaining vegetation and is essentially unrecoverable. The reestablishment of vegetation is impractical because of the lack of soil and the costs associated with importing soil. In 1993, the International Court of Justice ruled that reparations were due Nauru because of the

environmental damage stemming from the mining operations.

Developed countries are great consumers; generally, the more affluent a nation, the more it consumes. There is a disparity between the developed and developing nations in terms of consumption patterns of many materials. The developed countries generally have very high GDPs. The developing countries have low GDPs but obviously desire the affluence that exists in the United States and elsewhere. However, debt, unequal wealth distribution within the country, population growth, and poverty interfere with the fulfillment of this desire. Some of the rapidly industrializing countries in Asia and South America are finding ways to acquire a piece of the pie and are experiencing significant economic growth along with a corresponding increase in consumption of resources. Other developing countries find it more difficult to obtain a share of the resources of the planet.

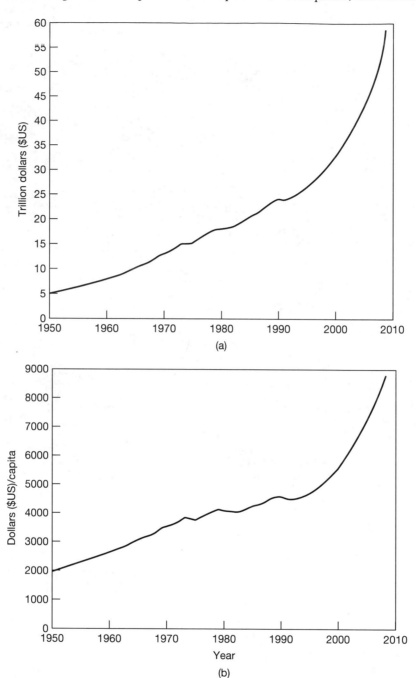

FIGURE 9.17 Gross world product (a) and gross world product per person (b) in $US from 1950 to 2008. (*Source:* Worldwatch Institute, 2001.)

In some developing countries, natural resources such as forests and minerals are exploited in an attempt to provide for the burgeoning population. The South trades its natural resources to buy energy and goods from the North. As a result, there is a net transfer of natural resources from poor countries of the South to the wealthy countries of the North. However, this trading of resources on the world market does not always cover the cost of the imported energy and goods for some Third World nations. There is a net cash flow of approximately $50 billion a year from the poorer countries of the world to the wealthy. The late chancellor Willy Brandt of Germany equated this transfer to that of giving a blood transfusion from the weak to the healthy. This imbalance forces the poorer nations into debt to foreign countries and, as a result, the developing countries continue to deplete their natural resources. This debt factor alone prevents many developing countries from improving their standard of living. In addition, inequitable distribution of wealth within nations, such as in the countries of Brazil, Zambia, Kenya, and Nigeria, adds to this problem. Other countries of the South, with little natural resources left to exploit, such as the Congo and Madagascar, have actually experienced a decline in their GDP per capita.

It is very obvious that the economic level of the industrialized nations is not shared by all. A person born in a developed country consumes more and has a greater impact on the natural resources of Earth than a child born in a developing country. In fact, 3 billion people today, or 45% of the world population, consume very little and live on the edge of the poverty level or below, and 1 billion people, or 15% of the world population, live in extreme poverty. The gap between the richer and poorer nations, the wealthy and the poor, and the haves and the have-nots is expanding.

RESOURCES: ENERGY AND MINERALS

The sun was the primary source of energy available to early Stone Age humans. It indirectly provided food, through photosynthesis, for energy for physical labor. As primitive industries developed, energy beyond muscle power was needed for the growing demands of society. Wood, wind, water power, and fossil fuel coal, oil, and gas energy were employed in increasing amounts. As early as 2500 B.C., the Babylonians used petroleum. Coal and natural gas were mined in China by 1100 B.C. Coal came into use on a large scale as an energy source three hundred years ago, and the expanded use of oil and natural gas followed. The improved use of these new energy sources allowed for the refining of metals, such as gold, silver, copper, lead, tin, and iron, by many societies. The commercial goods that are produced from energy consumption are used to provide the basic needs of people, such as food, water, shelter, clothing, and health care. In addition, transportation, communications, and the many luxuries of more affluent societies require an energy source. This control and use of energy enabled humans to use minerals and other resources of the world to advance the development of modern societies.

Energy Resources

The fossil fuels of coal, oil, and gas, along with nuclear power and hydropower, are called **commercial fuels.** These fuels are traded in the commercial marketplace. Global production and consumption figures for commercial energy are easy to obtain and are generally widely quoted in a number of sources of research data. One of the best sources is the Energy Information Administration (EIA; http://www.eia.doe.gov/). Of the commercial fuels, fossil fuels account for 80% to 90% of global commercial energy, and approximately 10% is derived from nuclear and hydropower (roughly, oil 40%, coal 30%, gas 20%, nuclear 5%, hydroelectric 5%) (Figure 9.18). Led by China, the developing nations of the world now consume more energy than the developed nations. The developed nations, with 25% of the world's population, used about 48.8% of the global energy budget in 2008 and the developing countries, with 75% of the world population, used 51.2%. Figure 9.19 shows the rapid rise in commercial energy consumption by the developing countries.

At the global level, energy needs increased about 1.4% in 2008, the smallest increase since 2001, mainly because of the high costs of petroleum and the economic crisis of the first decade of the twenty-first century. With a surge in demand of 7.2% in 2008, China alone accounted for nearly three-quarters of the world increase in that year. India and the Middle East also saw increases of 5.6% and 5.9%, respectively. The United States and the European Union exhibited decreases of 2.8% and 0.5%, respectively. The United States, with a population of about 304 million in 2008, or 4.5% of the world's population, consumed, mainly in the form of fossil fuels (Figure 9.20), approximately 22% of the total global energy produced. Figure 9.21 shows the primary energy consumption by source and sector for the United States in 2008, illustrating that 90% of energy use came from petroleum, natural gas, and coal and the transportation and industrial sectors used 49% of the energy consumed. The rapid more recent growth of energy consumption in the developing world is illustrated by the fact that China and India with 17% of the world's population, used about 19% of the world's total commercial energy consumption in 2008, an increase of 90% over 1990.

In the developing world, **noncommercial fuels** (fuels not traded in the commercial marketplace), collectively defined as biomass (firewood, charcoal, and animal and crop residues), are used extensively. More people in the world depend on

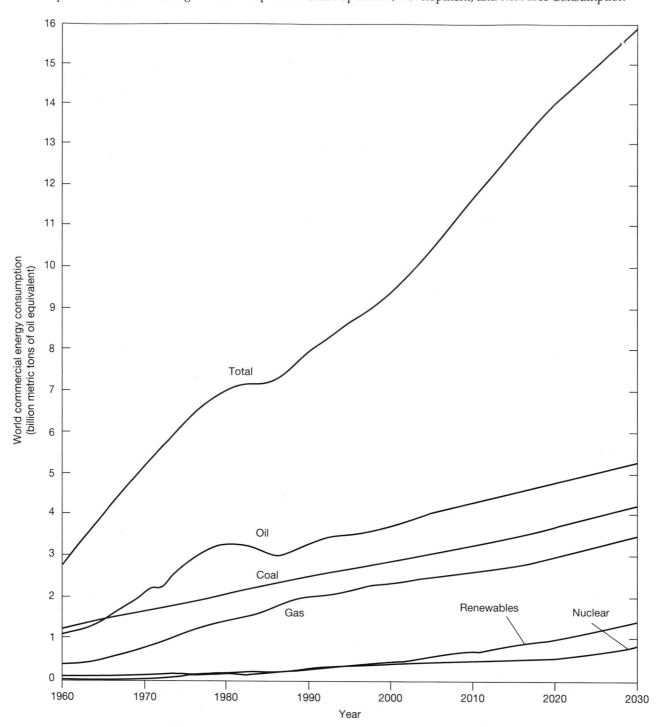

FIGURE 9.18 World consumption of commercial energy between 1960 and projected to the year 2030 by fuel type. (*Source:* International Energy Agency, 1993, 1999 and http://www.eia.doe.gov/oiaf/ieo/world.html.)

biomass than any other source of energy. It is difficult to calculate the exact amount of energy acquired from biomass burning in the developing nations, and, in fact, most energy data compilations exclude all forms of energy except those traded commercially. However, it is possible to get a rough estimate of energy derived from biomass use. Of total energy consumption including both commercial and biomass, biomass in the early twenty-first century accounted for about 14% of energy consumed in the

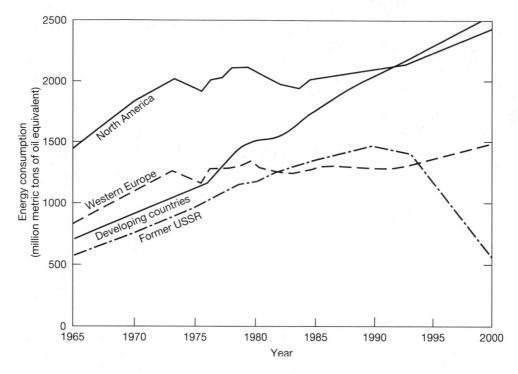

FIGURE 9.19 Commercial energy consumption, 1965–2000, by region. (*Source:* United Nations Statistical Office, 1991, 1993; World Resources Institute, 2000, http://www.wri.org/.)

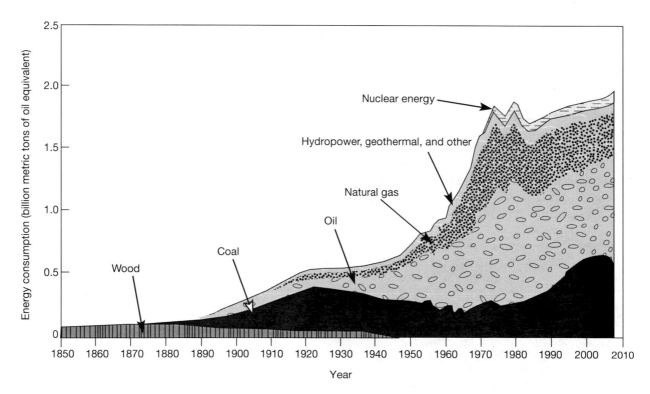

FIGURE 9.20 United States consumption of commercial energy, 1850–2007, by fuel type. (*Source:* Chiras, 1988; Energy Information Administration, 1994; http://www.eia.doe.gov/cneaf/solar.renewables/page/trends/rentrends.html.)

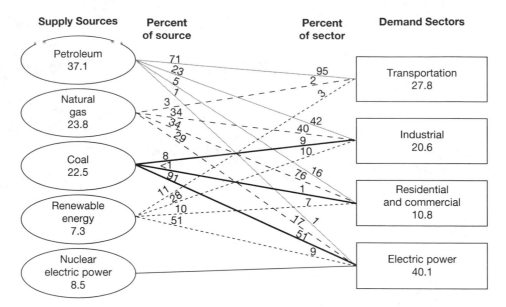

FIGURE 9.21 Primary energy consumption for the United States according to the supply source of the energy and the demand sector in which it was used in 2008 in quadrillion Btu (1 barrel of oil equivalent $= 5.8 \times 10^6$ Btu). For example, petroleum (oil and gaseous fuels) accounts for 37.1 quadrillion Btu of the energy used in the United States. Seventy-one percent of the petroleum is used in the transportation sector, accounting for 95% of the 27.8 quadrillion Btu of energy used in the transportation sector. (*Source: Energy Information Administration/Annual Energy Review 2008; http://www.eia.doe.gov/emeu/aer/pdf/aer.pdf.*)

world. This energy is the main source of energy for 2.5 billion people, or roughly 42% of the world's population.

NONRENEWABLE RESOURCES Large amounts of energy are needed to sustain the expanding population of the world and its desire for an adequate standard of living. However, there is a limit to the natural resources of coal, oil, and gas that provide us with energy. These energy sources are nonrenewable because coal, oil, and gas formed many millions of years ago. It would take an additional millions of years for Earth processes to form new reserves to replace those currently being depleted. These reserves will be depleted in a few hundred years with current consumption patterns. Indeed, at the current global rate of oil consumption of 31 billion barrels per year, if it could be maintained to depletion, the global proven reserves of 1342 billion barrels of oil would be exhausted in approximately 43 years. Furthermore, the uranium for nuclear power will eventually be depleted as it is mined and consumed.

Fossil Fuels Figure 9.22 compares the carbon content of proven and estimated total fossil fuel reserves with the other major organic carbon reservoirs.

Notice that the proven fossil fuel reserves are about equivalent to the carbon content of the atmosphere and that a major carbon reserve is that of methane hydrates (icy compounds of water and methane, see Chapter 14), which could be burned as an energy source if they could be safely and commercially extractable. Burning of all the total reserves of fossil

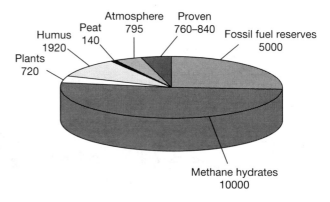

FIGURE 9.22 Carbon content in billions of metric tons (gigatons, or Gt C) of the major organic carbon reservoirs and as CO_2 in the present-day atmosphere of Earth. Fossil fuel reserves amount to 5000 Gt C, of which 760 to 840 Gt C are proven reserves. (*Source:* Mackenzie and Lerman, 2006.)

TABLE 9.3 The Elemental Composition of Oil, Natural Gas, Wood, and Coal
All values are in weight percent.

Element	Oil	Natural Gas	Wood	Peat	Lignite	Bituminous	Coal Anthracite
C	82.2–87.0	65–80	49.6	55.4	72.9	84.2	93.5
H	11.7–14.7	1–25	6.2	6.3	5.2	5.6	2.8
N	0.1–1.5	1–15	0.9	1.7	1.3	1.5	0.97
O	0.1–4.5	—	43.2	36.6	20.5	8.7	2.7
S	0.1–5.5	trace–0.2	—	—	—	0.1–6	

(*Source:* Clarke, 1924; Levorsen, 1956.)

fuels of 5000 Gt C would increase the carbon content of the atmosphere by 42 times over its late pre-industrial level, assuming that about half the carbon remains in the atmosphere.

Fossil fuels are derived from the remains of dead plants and animals that were encased in sedimentary rocks many millennia ago. Plants and animals initially trap energy from the sun for their growth and store carbon in their bodies. After they die, some of the organic matter decays and some is accumulated and stored in rocks. This organic matter formed the fossil fuels: the oil shales, coals, oil, and gas deposits of Earth. When ignited, the stored energy in the fossil fuels is released and produces heat and energy. The burning of the fuel releases carbon dioxide, sulfur dioxide, and other substances into the atmosphere. These emissions are the major cause of air pollution and regional acid rain problems. They are also the

largest source of anthropogenic greenhouse gases for the atmosphere, and their accumulation in the atmosphere leads to an enhanced greenhouse effect and global warming.

Coal is the most abundant of the fossil fuels. It is a black, combustible, sedimentary rock, composed primarily of the elements of carbon, oxygen, and hydrogen (Table 9.3). Land plants evolved during the Silurian Period. Coal deposits are formed from the remains of terrestrial vegetation that lived mainly in coastal, lowland swamps. The greatest coal formations are those of the Carboniferous and Permian age. These deposits are located mainly in the Northern Hemisphere in the United States and Europe. The second most important coal deposits formed during the Cretaceous Period and are found mainly in the former USSR and China (Figure 9.23). Coal occurs in layers or seams called strata. These

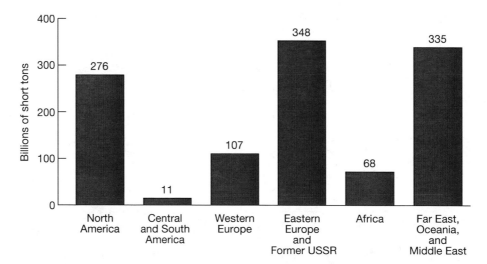

FIGURE 9.23 Geographical distribution of the world's coal reserves. Notice, in particular, the large (jointly about 55%) reserves in North America and the former USSR and Eastern Europe. The Far East, Oceania, and the Middle East have about 30% of the reserve. One short ton equals 2000 pounds equals 0.909 metric ton. (*Source:* Energy Information Administration, 1994.)

coal strata may be from 3 to 30 meters thick and cover extensive areas.

Coal is the second largest source of energy in the world, supplying 30% of commercial global energy. It is very abundant on Earth, low in price, and is likely to increase in use throughout the world. The world recoverable coal reserves were estimated at 800 to 900 gigatons (Gt) as of 2006. At the current extraction rate, the reserves could last about 132 years. Coal is a viable option to many nations of the world with a limited availability of petroleum. Coal is a major energy source in the eastern European countries of Poland, the former East Germany and Czechoslovakia, and Hungary. Coal, however, is considered a dirty fuel. Large amounts of carbon dioxide, sulfur, and soot are released into the atmosphere when coal is burned. Eastern European countries have suffered devastating damage to the environment and to the personal health of their populations owing to many years of intensive coal burning. Regardless of the environmental costs, but because of economic and social demands and needs, other coal-rich nations, such as China, have every intention to continue burning coal as a major source of energy at least into the near future. The United States is also expanding its use of coal because of the perceived need for increased coal use as an energy source due

to lack of internal reserves of oil and the fact that the United States is relying on costly imports of oil from regions of the world that are not especially politically stable or friendly. The United States, several European countries, and Japan are conducting research into methods that would produce a cleaner burning coal fuel that is also more energy efficient and of sequestering the CO_2 emissions from coal-burning electrical generating plants. However, no totally satisfactory economically feasible method has yet been developed to prevent releases of carbon dioxide to the atmosphere from fossil fuel burning.

Figure 9.24 shows the history of coal extraction in the major coal-producing countries. Since the 1890s, total production has shifted from the United Kingdom to the United States, then briefly to the former USSR, and now China is a major producer of coal and will probably remain so in the future. The extensive use of coal in China is one of the reasons that China is experiencing air pollution problems and becoming a major contributor of carbon dioxide to the atmosphere from the combustion of fossil fuels.

Petroleum, including oil and natural gas, forms from the decomposition of microscopic marine plant and animal life and is also composed mainly of the chemical compounds of carbon, hydrogen, and oxygen (Table 9.3). Cenozoic and Mesozoic rock strata contain

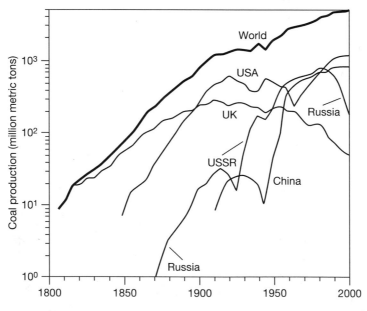

FIGURE 9.24   A history of coal extraction between 1800 and 2000 in millions of metric tons of coal. Notice the increase in the use of coal by both the United States and China during the past few decades. Total world coal production is projected to be about 7 billion metric tons in 2030. The increase will be due in large part to the demand for coal by China and the United States.   (*Source:* Smil, 1997.)

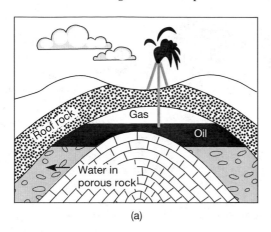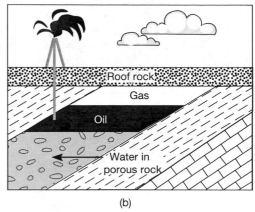

(a)   (b)

FIGURE 9.25 Two ways in which oil is trapped in sedimentary rocks. Gas overlies the oil, which floats on water. The hydrocarbons are held in place by a capping of nonporous and nonpermeable roof rock; (a) a structural trap, (b) a stratigraphic trap. (*Source:* Skinner and Porter, 1987.)

60% of all the oil discovered so far. Petroleum is usually found accumulated in pools in underground porous and permeable sedimentary rocks. The oil floats on a bed of water, and natural gas overlies the oil. These deposits are commonly overlain by a cap rock that encases the oil and gas and prevents their escape (Figure 9.25).

Petroleum provides 60% of the total world use of commercial energy and is used mainly in the transportation sector. Other major uses of petroleum are to heat homes, fuel the fires of industry, and produce chemical fertilizers, Styrofoam products, and plastics. An industrialized nation's many petroleum products are generally viewed as necessities rather than luxuries.

Petroleum reserves are widespread throughout the world, but are highly concentrated in specific areas. Ninety-five percent of known reserves are found in 20 countries, with the Organization of Petroleum Exporting Countries (OPEC) countries of Saudi Arabia, the United Arab Emigrate, Kuwait, Iraq, and Iran controlling more than 65% of the world total reserves of oil and the FSU and the Mideast the site of 72% of the gas reserves (Figures 9.26 and 9.28). Over 500 billion barrels (oil and gas calculated as oil equivalent) have already been consumed by the world economies. The total reserve plus undiscovered producible petroleum resource is estimated as 1.5 trillion to 3 trillion barrels (oil and gas). At present rates of consumption, petroleum cannot meet the long-term energy needs of the world societies because the resources will last less than 100 years.

The use of oil has been steadily increasing worldwide. Oil, by itself, provides about 40% of commercial global energy. Present global reserves of oil are about 990 billion barrels, and the average

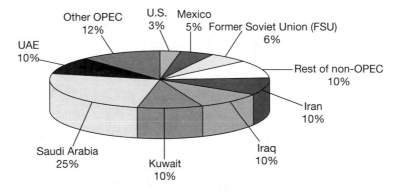

FIGURE 9.26 Percentage of proven oil reserves of major OPEC and non-OPEC nations. OPEC nations have 76% of the reserves and the non-OPEC nations 24%. (*Source:* Energy Information Administration, 2008; http://www.eia.doe.gov/.)

estimate of undiscovered producible oil is about 550 billion barrels for a total reserve plus resource of 1.5 trillion barrels. In the 1990s, the world used between 20 billion and 30 billion barrels of oil annually. At the present global rate of consumption of 22 billion barrels annually, the oil reserve would be exhausted in about 40 years and the reserve plus undiscovered resource in 60 years. The present consumption rate amounts to 4.5 barrels of oil per person per year or about 0.4 tons of carbon emitted to the atmosphere annually. Figure 9.27 illustrates the various regions where oil is produced compared to where the oil is consumed. Of the oil produced, North America and Europe consume 52% of the production and if Australia and Asia are included, 88% of the production is consumed in these four regions. All of Latin America consumes only 6%.

The main importers of oil are the United States, Japan, and Western Europe, with the exception of the United Kingdom and Norway, which export oil. More than one-half of the world demand for oil is by the developed nations, and the United States alone consumes one-half of that. In 2008 the United States paid 425 billion dollars for its oil imports. Seventy-one percent of the petroleum in the United States is used in the transportation sector (Figure 9.21), mainly for gasoline in cars, while transportation in Western Europe accounts for 40%, Japan 25%, developing countries 50%, and more than 50% for the former USSR. A large number of developing countries are finding their need and demand for oil increasing rapidly. The price of oil has created a heavy financial burden for these countries and an important but lesser burden for the developed nations, particularly in the twenty-first century. Developing nations spend a great deal of their foreign exchange currency on oil imports. This expenditure is partly the cause of their heavy foreign debt. In the late twentieth century, the developing nations spent between 37 and 61% of their export earnings on oil imports. This dependency makes them very vulnerable to the political and economic pressures of oil-exporting countries.

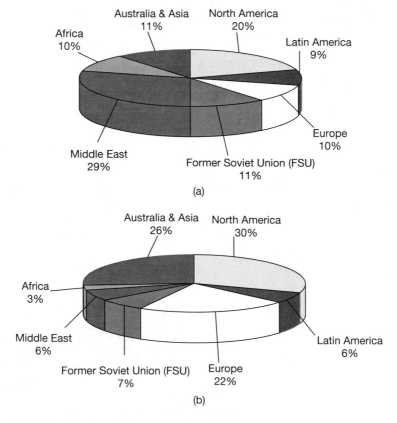

(a)

(b)

FIGURE 9.27 Percentage comparison of the countries and regions that produce oil (a) with those that consume it (b). Notice that the Middle East produces 29% of the oil and North America and Europe consume 52% of it. Australia and Asia consume an amount of oil nearly equivalent to all that produced by the Middle East. (*Source:* http://www.eia.doe.gov/.)

World oil consumption decreased for a few years following the 1973 oil embargo and in the mid-1980s, which resulted in an increase in cost. The decreases led to more use of alternative forms of energy and the necessity of more energy-efficient measures. Some countries, such as Japan and France, have gone the route of finding and developing other sources of energy. However, the United States is again relying heavily on foreign oil imports, importing 66% of its oil in 2007.

Natural gas, which is mostly methane and to a lesser extent propane, is relatively abundant and provides 20% of commercial global energy. It was a long neglected energy source and was commonly burned off when oil was extracted from the ground. It is the world's fastest growing energy source and is used as an alternative to oil. The world's total natural gas reserves are estimated at 6183 trillion cubic feet. The Former Soviet Union (FSU) has about 40% of the proven gas reserves, followed by the Middle East with 32%, and 28% is found in all other regions of the world. North America has only 6% of the reserves but produces 33% of the gas consumed (Figure 9.28). The Energy Information Agency estimates the total natural gas recoverable resource in the United States as 1191 trillion cubic feet, of which the proven reserves are about 14% of the total.

Natural gas reserves in the world may last about 60 years at current projected rates of usage. It is a relatively clean fuel when burned, because it emits only about half the amount of carbon dioxide as does the burning of coal and is virtually sulfur free, so it does not contribute heavily to acid rain problems. Because of reasonably large United States reserves compared to the nation's oil reserves, its increased use in the United States could reduce the nation's reliance on imported oil and help bridge the gap to alternative energy sources. However, it should be kept in mind that the burning of natural gas will still release the greenhouse gas CO_2 into the atmosphere.

Fossil Fuel Forecasts. Forecasting is a difficult exercise because of the many assumptions that are made, and forecasts, be they population, economic, energy, or climatic, must be viewed with caution. The same can be said for any forecast of future oil, gas, and coal production. However, one driving force throughout history, as shown in Figure 9.29 for oil, is population growth. Although other factors are involved, clearly oil and population are closely correlated. This should not be surprising, since the more people in the world, the more transportation, manufacturing, food, and almost everything required by the world's people

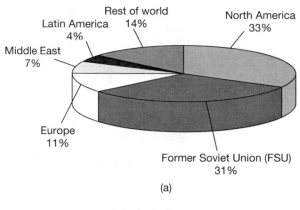

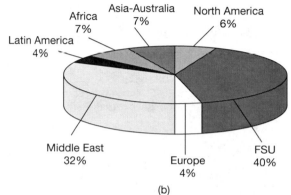

FIGURE 9.28 Percentage of natural gas production (a) and proven reserves (b) for various regions of the world. Notice that North America and the FSU produce 64% of the natural gas in the world, while the FSU and the Middle East have 72% of the proven reserves. (*Source:* Energy Information Administration, 2008; http://www.eia.doe.gov/.)

and hence the more energy needed from oil. Thus, population growth is a factor in forecasting future consumption of oil.

Oil fields come in all sizes, with half of the oil produced in the world coming from just 0.03% of all oil fields. The ages of the fields are quite variable. The 120 giant fields, defined as fields with at least 100,000 barrels of oil per day production, account for 47% of total world production of oil. Four thousand smaller fields produce 53% of the oil. The world's giant fields are aging, with the 14 largest fields having ages of 43.5 years and the 19 largest, 70 years. With aging comes a decline in production. In general, a large oil field goes into production decline after 50 years. In addition, the number of giant oil fields discovered is declining and those discovered are smaller fields. From these considerations alone, it is very likely that the oil bonanza of the past is over and we are depleting rapidly this valuable natural resource.

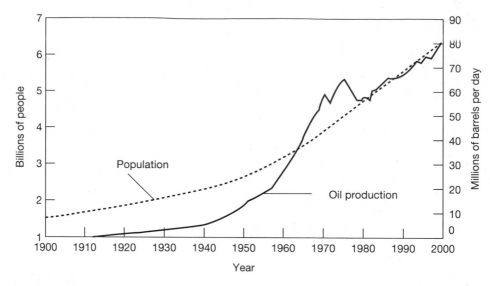

FIGURE 9.29 Relationship between world population and global oil production from 1900 to 2005. Notice the reasonable correlation between the two variables. (*Source:* Hicks and Elder, 2008; http://canada.theoildrum.com/node/2516.)

Peak Oil is a theory recognizing the fact that regional oil fields tend to show increasing production rates for some time after discovery, and reach a peak around their halfway point in production. Then the rate of production declines. The Peak Oil theory was first proposed by a geologist, M. K. Hubbert of Shell Oil Company and later the United States Geological Survey (USGS), in 1956. The peak in production is called Hubbert's Peak. Hubbert concluded that the United States would reach its oil production peak around the year 1970. In 1971, the United States began its relentless decline in oil production (Figure 9.30). This is a case of a prediction being right on target!

Using various methodologies that include, for example, synthesizing and summing the data on the historical rates of oil production from individual oil fields throughout the world, and various definitions of the type of oil that is counted, a number of analysts have attempted to predict the peak in world oil production. Figure 9.31 is one of the more recent attempts at forecasting future world oil production. According to this model, world oil production peaked in the year 2005 and steadily will decline toward the middle of the twenty-first century. Later consensus has placed the peak sometime during the years 2009–2012. The oil production of the Organization of Petroleum Exporting Countries (OPEC), mostly Persian Gulf countries and Venezuela, was predicted to cross that of non-OPEC nations in the year 2007, probably a few years later based on recent opinion. Thus, very early in the twenty-first century, OPEC nations will be the dominant source of oil to fuel the world's economies. The United States currently receives 25% of its oil from the OPEC nations. Future possible production of oil from the Arctic National Wildlife Reserve is estimated at only 2 to 5 billion barrels of oil, or at the present United States consumption rate of oil, equivalent to about one year of oil. Despite the focus on regional oil sources and the cost of oil in the early twenty-first century, the economic, strategic, and environmental implications of possible dependence on OPEC oil in the future are still little appreciated or understood in the United States and many other countries.

We may conclude this section on fossil fuel forecasting by looking briefly at the phenomenon of Peak Coal and also that of the possibility of Peak Gas. Figure 9.32 illustrates that worldwide coal production may peak by the 2020, following the global peak in oil. Notice that various regions of the world are predicted to peak at different times with, for example, the OECD countries of Europe peaking very early, China in the second decade of the twenty-first century, and the United States between 2020 and 2030. Globally liquefied natural gas (LNG) production is predicted to peak in the second decade of the twenty-first century. This will affect imports of natural gas to the United States and other countries since this is one form in which natural gas is transported from region to region. In North America, peak production of natural gas most likely occurred in 2001. This peak in production necessitated more natural gas imports from overseas, but imports remained relatively flat

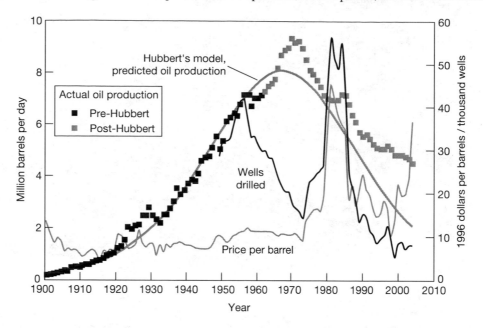

FIGURE 9.30 M. King Hubbert's 1956 model of United States oil production compared with actual oil production from 1900 to 2004. Also shown are the price per barrel of oil and the number of wells drilled during this period. Notice that in the early 1980s when the price per barrel of oil rose, the number of wells drilled increased. This did not happen in the early twenty-first century probably due in part to the inability of the oil industry to obtain sufficient oil profitably from the dwindling oil reserve in the United States.   (From "Planning for the Peak in World Oil Production" by Robert K. Kaufmann, *World Watch* magazine, January/February 2006, p. 19. Used by permission of Worldwatch Institute, www.worldwatch.org <http://www.worldwatch.org/>)

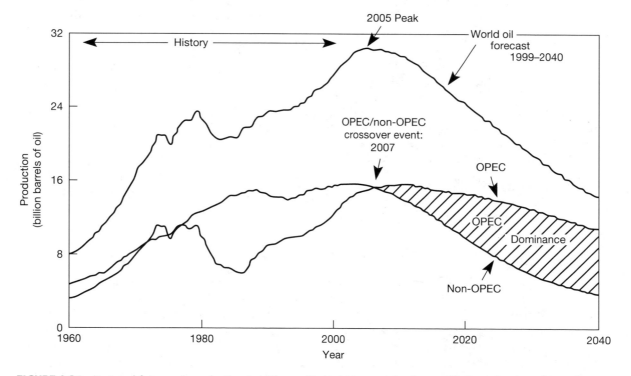

FIGURE 9.31 Past and future oil production in billions of barrels per year for the world, Organization of Petroleum Exporting Countries (OPEC), and non-OPEC countries. Notice that it was forecasted that in 2005, world oil production would peak and then fall. Also note that production from non-OPEC and OPEC nations would cross in the year 2007, after which OPEC nations would dominate world oil production.   (*Source:* Duncan, 2001.)

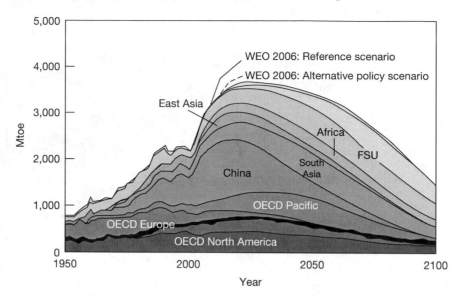

FIGURE 9.32 The possibility of global Peak Coal. Coal production in million metric tons of oil equivalent is shown from 1950 projected to 2100. The production of 909 billion tons of worldwide reserves of coal is projected to reach a peak in 2020. Also shown are two projections of the global production curve from the International Monetary Fund (IMF) World Economic Outlook (WEO). (From *Profit from the Peak* by Brian Hicks and Chris Nelder. Copyright © 2008 by Angel Publishing, LLC. Reproduced with permission of John Wiley & Sons, Inc.)

through 2005. Canada's supply of natural gas to the United States, which represents 50% of its export capacity, started to decline rapidly in 2004 and the decline is projected to continue. United States net imports of LNG from overseas were projected to rise from 0.5 trillion cubic feet in 2004 to 4.5 trillion feet in 2030 by the Energy Information Agency. Such large natural gas consumption by the United States and also Western Europe countries alone will place strong demands on natural gas reserves and although the jury is still out on when Peak Gas will arrive, it will surely come about, probably sooner than later.

Nuclear Energy Nuclear energy could be an almost inexhaustible source of energy. Currently, it is mainly produced from the splitting of the uranium atom (nuclear fission of ^{235}U) in uranium fuel resulting in the release of energy. The nuclear reactor then produces steam that drives turbines that produce electricity. The first commercial nuclear reactor was the Shippingport reactor in Pennsylvania put on line in 1957. Approximately 5% of the commercial energy of the world is derived from nuclear plants. The nuclear energy produced is mainly used for electrical power generation, accounting for 16% of the energy used in that sector worldwide. Total nuclear power generation has grown from less than 100 terrawatts (TW, 1TW = 10^{12} watts) in the 1960s to about 2550 TW in 2002 (Figure 9.33).

There were about 435 nuclear reactors operating in 25 countries in 2000. The United States, France, and Japan account for nearly 60% of the world's nuclear production (Figure 9.34). France, Belgium, Hungary, Sweden, and South Korea derive at least half of their electrical power from nuclear energy. Many people see nuclear power as a viable solution to the world energy problem because the decline and eventual depletion of fossil fuels are a certainty. However, after the Chernobyl nuclear disaster, concerns about reactor safety have lessened the public appeal for this fuel. Also, the disposal of the radioactive waste generated by this fuel source is a major problem. Despite these problems, there has been renewed interest early this century in the nuclear energy option in the United States because of Peak Oil considerations, the increased necessity to import more oil from politically unstable regions overseas, and the problem of global warming.

RENEWABLE RESOURCES Renewable energy sources (solar, geothermal, wind, and ocean, with the exception of hydropower) account for less than 1% of the total global energy usage; biomass accounts for another 14%. The development of alternative forms of energy is partially dependent on the marketplace. The relatively cheap cost of fossil fuels, as compared to the renewable sources, has slowed much research and development in the area of renewable energy. Some countries have geared their economies toward

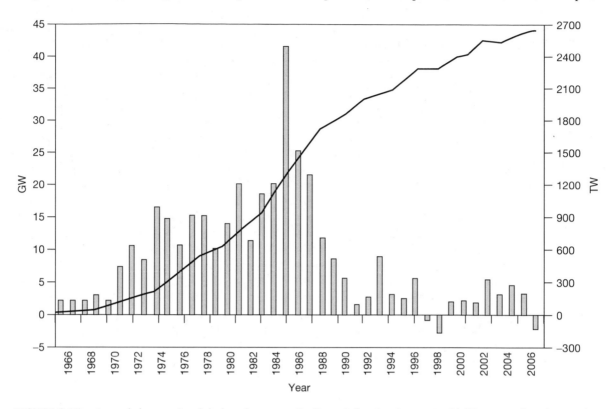

FIGURE 9.33 Annual changes in global nuclear capacity (bars, left axis, gigawatts, 1 billion watts) and annual nuclear electricity generation (line, right axis, terawatts, 1 trillion watts) from 1966 to 2006. Notice the peak in annual changes in capacity in 1985 and the general decline thereafter. This is in part due to the public's negative perception of the use of nuclear energy after the Chernobyl nuclear reactor accident in the Ukraine in 1986, which was then part of the Soviet Union.   (From *Survey of Energy Resources* World Energy Council, 2007. Used by permission.)

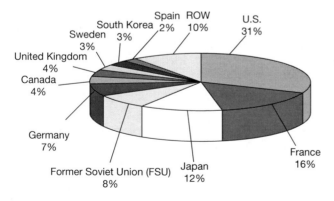

FIGURE 9.34 The world's nuclear energy production in percentage by country. Notice that 59% of the nuclear energy produced, which is commonly used for generation of electricity, is produced by the United States, France, and Japan. France supplies about 80% of its electricity from nuclear plants, Japan 30%, and the United States 20%. The United States is the world's largest producer of nuclear energy. ROW is the Rest of the World.   (*Source: BP Statistical Review of World Energy*, 2009.)

the use of alternative energy sources and are becoming more energy efficient in their current use of fossil fuel energy. With eventual depletion of fossil fuels, renewable energy sources will become more competitive on world markets, and more nations will follow their lead.

Solar power is an unlimited source of energy, which if utilized effectively could make any country energy self-reliant. The relatively cheap cost of oil and dropping of tax credits in the United States for alternative energy research resulted in less research development in solar energy in the 1980s and 1990s but this situation is changing as this book is being written, with more emphasis being placed on alternative energy sources by President Obama's administration. Photovoltaic cells can convert sunlight directly into electricity, require little or no maintenance, have no moving parts, and produce no pollution. The big drawback has been the price. However, in outlying areas or in developing countries that have an abundance of sunshine, the cost might be less than that of installing power lines. The production of photovoltaic cells has risen 265% since 1995 and in 2005 alone skyrocketed 45%, as costs have been falling.

Geothermal energy can be obtained from the upper 5 kilometers of the crust of Earth and is equal to 40 million times the energy in the world reserves of oil and gas. However, only a small part of the geothermal energy reserve is exploitable. Geothermal steam reservoirs are generally found near plate margins where hot magma intrudes close to the surface of Earth, and above oceanic hot spots like that of Hawaii. The steam produced can power electricity-generating plants. Geothermal power has been used for 50 years in Iceland and Italy. Newer plants exist in Japan, the former USSR, New Zealand, Africa, Mexico, and the Salton Sea area of California in the United States. There are a number of additional locations of geothermal steam in the United States that have not been developed.

Wind power has steadily increased in its development and use as the cost factor decreases. Denmark and California have been installing wind-energy machines since 1974. A big drawback is the unreliability of wind power; thus, it must be used with other energy sources. Ocean energy (waves, tides, and ocean thermal) is a new technology. For 20 years, France has operated the world's largest tidal power plant. In 1986, Norway constructed its first wave-powered plant.

Hydropower originates from the energy of the sun through the water cycle and provides almost 7% of commercial global energy. It is a constantly renewable energy source. To harness this power, dams must be built. The force of the water moving through the dams drives the turbines that produce electricity. The developing countries, mainly in Latin America, have the greatest potential for deriving energy from the development and use of hydropower because of their extensive river systems. Hydropower does have its drawbacks: The populations in the river basins often must be dispersed, fertile land and forests are flooded in the reservoir area, irrigation waters may become saline and sources of noxious gases, and the areas behind dams may eventually silt up and restrict water flow. China, however, is attempting to reduce these problems by building smaller dams that service only local areas. However, China is also building what will be the largest dam ever constructed in the world, the Three Gorges Dam, on the Yangtze River.

Wood, organic waste, and other biomass account for approximately 14% of total world energy.

Fuelwood is the only source of energy for nearly half the population of the world. It is not referred to as a commercial energy source because of the difficulty in assessing the amount of biomass used. Therefore, data dealing with commercial global energy consumption may be somewhat misleading because they do not necessarily take into account the energy usage of half of the world population. Biomass is burned as a fuel mainly in developing nations, where the cost of fossil fuels is too great. Fuelwood would be a recyclable fuel provided new trees were planted for those removed. However, in many areas, although considered a renewable resource, it is being depleted faster than it is being replaced. In sub-Saharan Africa, fuelwood provides 80% of total energy consumed. Its increased consumption may contribute to severe loss of forests and environmental deterioration, including desertification. It takes about 30 years to replace wood as a resource.

Summary

There are many forms of energy on Earth. We do not necessarily have an energy shortage in the world. In the long term, we only have a shortage of specific kinds of energy, namely nonrenewable fossil fuels, whose burning fills our air with carbon dioxide and other substances. Alternative forms of energy are available that have a less negative impact on our environment than the fossil fuels we now burn. Recently, we have used renewable energy on a limited basis in the form of hydro, geothermal, solar, and wind power, but because of political, economic, and social considerations, nonrenewable fossil fuels continue to be the primary energy source for the developed, industrialized nations and many developing countries of the world. This need not be the case. Figure 9.35 shows one example of how the different energy sources could change in the future, with solar energy especially occupying progressively more and more of the global energy mix of sources. In later chapters, we will explore the human-induced environmental problems of our planet and see that society's use of fossil fuels is a major cause of the problems of degradation of the environment and is a factor in the environmental issues of deforestation, aquatic pollution, acid rain, photochemical smog, plastics in the ocean, stratospheric ozone depletion, and global climatic change. Thus, correcting the energy mix in the future along the lines of Figure 9.35 would go a long way toward decreasing the environmental ills of the planet.

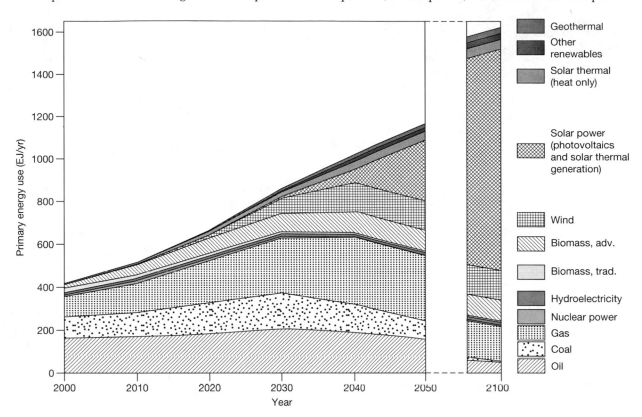

FIGURE 9.35   One possibility for transforming the global energy mix of sources going into the twenty-first century. Primary energy use by source is in exajoules per year (EJ/yr, 1×10^{18} joules). Notice the strong reliance on solar power from photovoltaics and solar thermal generation. This is a viable scenario but requires considerable political will in countries around the world and acceptance by the populaces of the countries.   (From *Survey of Energy Resources* World Energy Council, 2007. Used by permission.)

Mineral Resources

In a nonspecific sense, the term **mineral** encompasses a variety of substances taken from the Earth. Minerals include metals, such as mercury, iron, and aluminum; industrial materials, like lime, gypsum, and soda ash; construction materials, such as stone, sand, and gravel; and minerals for energy, like uranium, coal, oil, and natural gas. Specifically, the term **mineral** refers to a naturally occurring inorganic chemical compound with a particular internal structure and a continuous range of composition (see Chapter 2 and Appendix). Industry relies heavily on about 80 of the 2200 minerals that have been identified. Some minerals are used just as they are found in the ground; examples are diamonds and emeralds. Others require processing before elements, like iron, copper, silver, and tin, can be extracted into a usable form. Table 9.4 lists the world production of some selected mineral commodities.

Many of our material goods are products derived from metallic and nonmetallic elements that formed during the history of Earth. Metals are found in their native form or are contained in ore minerals from which they must be extracted. Gold, platinum, and palladium are examples of the former, and iron and manganese of the latter. Different types of mineral deposits are formed in different plate tectonic environments (Figure 9.36; see also Chapter 3). For example, manganese is found on the ocean floor as small nodules and crusts of manganese oxides. Copper may be found as copper sulfide associated with veins formed during the formation of granitic igneous rock bodies on the continents. Zinc occurs in massive sulfide deposits associated with modern and ancient midocean ridge systems.

The mining of mineral ores occurs all over the world. Copper is found associated with igneous rocks in the United States and Mexico and in copper-rich

TABLE 9.4 Estimated Early Twenty-First Century World Production of Selected Minerals

Mineral	Production (Thousand Tons)
Metals	
Pig iron	705,400
Aluminum	29,000
Copper	14,600
Manganese	11,900
Zinc	9,200
Chromium	5,930
Lead	3,370
Nickel	1,500
Tin	321
Molybdenum	139
Titanium	2,854 as concentrate and slag
Tungsten	90
Cobalt	56
Silver	20
Cadmium	19
Mercury	1.2
Gold	2.5
Platinum-group metals	0.5
Nonmetals	
Stone	1,770,000[a]
Lime	283,000
Salt	260,000
Phosphate rock	156,000
Gypsum	151,000
Industrial sand and gravel	127,000
Soda ash	44,800
Clay: kaolin	39,000
Potash	36,000

[a]United States only. Most data are from the United States Geological Survey.

Minerals Yearbooks (http://minerals.usgs.gov/minerals/pubs/commodity/myb/).

sediments of central Europe (stratiform copper deposits) and Zambia. Lead and zinc are found in sulfide ores associated with limestones in Oklahoma and Missouri, United States, and are found in the former USSR and Australia. Nickel is found associated with iron- and magnesium-rich igneous rock (mafic) bodies in Sudbury, Ontario, and Thompson Lake, Manitoba, Canada; in the tropical weathering zones developed on these igneous rock bodies in New Caledonia and Cuba; and in the former USSR. Silver is found generally as vein material that was deposited at high temperature from hydrothermal solutions in Mexico, Canada, Peru, the former USSR, the United States, and Australia. The scarce metals of platinum, palladium, rhodium, iridium, ruthenium, and osmium are mined from mafic igneous rocks of the former USSR, Republic of South Africa, Canada, and the United States. Gold is an important and scarce element mined in its elemental state from sedimentary mineral deposits of the Witwatersrand district of South Africa. Tin and tungsten occur in veins and in contact zones associated with igneous rocks. Most of the world's known tin reserves are found in a belt along the Malay Peninsula and southeast into Java and Indonesia. North America is almost devoid of tin. Half of the world's production of tungsten comes from eastern Asia.

AVAILABILITY It appears that most metals are abundant, but there are some in scarce supply. For example, iron, aluminum, chromium, titanium, manganese, and magnesium are abundant metals, whereas copper, zinc, lead, nickel, antimony, lithium, and molybdenum are considered as scarce metals. The world consumption of metals between 1930 and 1990 rose by a factor of about eight from about 300 million to 2500 million metric tons. The recession of the early 1980s caused a slight downturn in consumption, but in the late 1980s–early 1990s, consumption continued growing at a linear rate. The prospects for the future of mineral supplies involve a number of contingencies, including such factors as the rate of population growth and per capita consumption of mineral resources, our ability to find new reserves and to develop methods of extraction of metals from low-grade ore materials, our ability to develop and adopt conservation practices, and market forces. Economic forces of the marketplace play a strong role in the prospects of an adequate supply of mineral and other resources for future generations. As potential or actual shortages of mineral resources are realized, market prices rise. The price rise can lead to increased exploration and discovery of mineral resources, improvements in efficiency, possibilities for substitution, and technological innovations.

For example, the rise in prices of energy and metals in the late 1960s encouraged efficiency gains

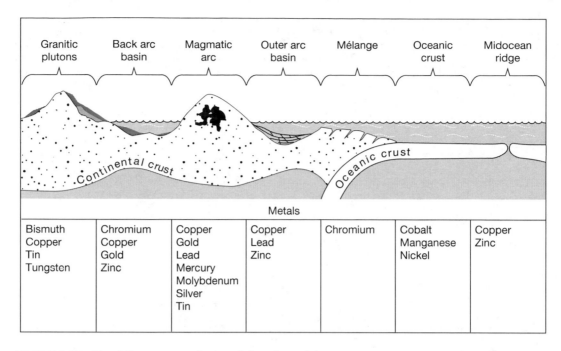

FIGURE 9.36 The different types of mineral deposits and their major metal content. The major plate tectonic environments in which the different types of deposits form are also shown.   (*Source:* Skinner and Porter, 1987.)

and substitutions that ultimately reduced some of the growth in demand. Fiber optics replaced copper in telecommunications, and aluminum recycling slowed down the demand on the original nonrenewable resource of aluminum ore. Between 1970 and 1988, the annual consumption of mineral resources as a percentage of reserves decreased for bauxite (an ore of aluminum), zinc, and tin but increased for copper, iron ore, lead, and nickel. However, as V. E. McKelvey, former director of the United States Geological Survey, pointed out, "We can say unequivocally that resources are not adequate to support consumption if it continues at exponential rates" (in Laporte, 1975). Although this statement was made in the mid-1970s, it still holds today.

The problem with our exploitation of mineral resources is not so much their present availability but the effects associated with their mining, extraction, and consumption. Ninety-five percent of the total material removed from Earth is waste. Table 9.5 shows the waste associated with the mining of mineral ores. In 1991, waste constituted nearly 85% of the total tonnage of mineral ore mined. For copper, 990 million tons of ore were mined to produce just 9 million tons of copper metal. For gold, only a trivial percent of the

gold ore actually ended up as metallic gold; the majority of the ore became waste (Table 9.5). Mining operations continue to generate substantial amounts of waste.

The mining and extraction operations and the generation of wastes lead to pollution and destruction of natural habitats. Rivers, groundwater, and soils may be contaminated with the wastes associated with the mining and extraction processes. For example, the use of mercury in gold-mining operations in the Amazon has led to contamination of waters draining the mining area. The problem of disposal of mineral wastes is difficult and probably will increase in the future with the likely necessity of mining ores containing lower percentages of the elements sought.

CONSUMPTION, USAGE, AND RECYCLING There is a growing dependence of the richer nations on foreign mineral supplies. The industrialized countries are great importers and consumers of minerals. As the world prices of metals and minerals declined erratically from 1980 to 2000, demand increased. With only 23% of the world population, the developed countries consume at least 75% of the

TABLE 9.5 Estimated 1991 Ore Production, Average Grade, and Waste Generation of Major Minerals

Mineral	Ore (Million Tons)	Average Grade (Percent)	Waste (Million Tons)
Copper	1,000	0.91	990
Gold	620	0.00033	620
Iron	906	40.0	540
Phosphate	160	9.3	140
Potash	160	17.0	130
Lead	135	2.5	130
Aluminum/bauxite	109	23.0	84
Nickel	38	2.5	37
Tin	21	1.0	21
Manganese	22	30.0	16
Tungsten	15	0.25	15
Chromium/chromite	13	30.0	9
Total	3,200		2,700

Waste figures do not include overburden. Totals do not add due to rounding.   (*Source:* Young, 1992.)

world's mineral resources. The United States alone, with about 5% of the world's population, consumes approximately 20% of the total global production of minerals. The problem of disposal of wastes in developing countries is especially critical because of the lack of strong environmental controls. Also, the economy of some of these nations is strongly tied to their mineral industries. For example, Zambia's copper mineral trade produces more than half of its national income and provides 86% of its export revenue. When copper prices fell during the economic downturn of the early 1980s, Zambia's economy went into a tailspin. This led to twice as many Zambian children dying of malnutrition in 1984 as in 1980. At least 14 developing countries receive a third or more of their revenues from mineral resources that are exported to the developed world. Thus, in these developing nations, strong conflicts exist between mineral resource development and environmental concerns.

Modern technology is continuously searching for new products usable by society. Minerals provide the substances necessary for these products. For example, aside from the use of metals in the world society, the nonmetallic minerals (Table 9.4), like stone, sand and gravel, salt, lime, and phosphate rock, are used in building and construction materials,

for road and railroad train beds, for the winter salting of roads, in the production of fertilizers, and for many other reasons. As a result, exploration for mineral resources never ceases because many areas on Earth may harbor minerals that someday may be commercially valuable. However, most of the continents of the world have been thoroughly explored for mineral resources. An exception is Antarctica. It is the least explored and exploited continent. Antarctica has been declared off limits to all countries for mineral exploitation for at least the next 50 years to preserve temporarily its pristine environment.

The reprocessing of plant scrap and postconsumer discards of material, that is recycling, is an important means of slowing extraction of the materials from sources and conserving resources. For example, large percentages of aluminum, iron and steel, lead, copper, nickel, titanium, paper, and paperboard are recycled every year. In 1992, about 44% of the requirement for copper in the United States was met by using plant and postconsumer scrap materials. The achievement in the net reduction of material through recycling involves specific information on each material to be recycled. However, recycling can and often does reduce energy and material needs per product compared with a nonrecycled product made of the same materials.

CONSERVATION

As population increases and resources become less available, the need for conservation of energy and other natural resources increases. Advances in technology are leading to ways to decrease the amount of energy expended and thus save on energy. Major improvements have been made in the energy efficiency of automobiles; the cars are lighter and the engines require less fuel. Some commercial and residential buildings are better insulated and have energy-conserving lighting (fluorescent light bulbs that use 50 to 75% less energy) and new energy-efficient appliances (refrigerators that use one-third less energy). Sweden has designed buildings that require 90% less heat than traditional buildings.

Energy intensity is the amount of energy needed to produce each unit of GNP or GDP and includes energy involved in transportation. The amount of energy needed to produce products determines the energy efficiency of a country. **Energy efficiency** is a term that is used in discussing the relative conservative or nonconservative use of energy by

various countries. Figure 9.37 shows trends in the total energy required per unit of gross domestic product for some nations of the world between 1970 and 1991. Notice that since the economic shock of the early 1970s brought about initially by the Arab embargo on oil, the energy intensities of a number of developed countries have decreased. For the 24 nations of the OECD, the use of energy rose only 20% as much as economic growth between 1973 and 1989. However, not all countries have performed the same. Greece and New Zealand, for example, have been less energy efficient for this period than Japan and the United Kingdom. Japan required 30% more energy in 1970 to produce one unit of gross domestic product than it does today, whereas in 1988 Greece required almost 40% more energy per unit of GDP than it did in 1970. Between 1949 and 2004, the United States improved its energy intensity index by 53%. Since the mid-1980s, the rate of improvement of energy intensity has tended to diminish in the developed world. In developing countries, rapid industrialization and urbanization, particularly since 1980, have led to an overall decrease in energy efficiency in

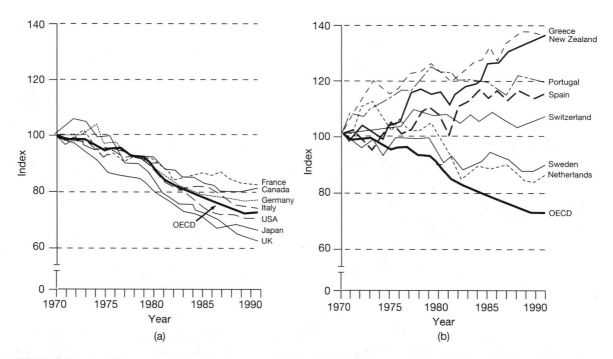

FIGURE 9.37 Trends in energy intensities between 1970 and 1991 for some countries of the world. *Energy intensity* is total energy requirement per unit of gross domestic product. The indices for the various countries and the average for the 24 countries of the Organization for Economic Cooperation and Development (OECD) are calculated by assigning a value of 100 to the 1970 ratios; (a) and (b) show comparisons of individual nations with the average for OECD. Notice that the energy efficiencies for some nations have decreased while others have increased over the 18-year period. Energy intensities for most of the world's countries declined slightly to 2000 and probably will continue to do so to 2010. (*Source:* OECD, 1991, 1994.)

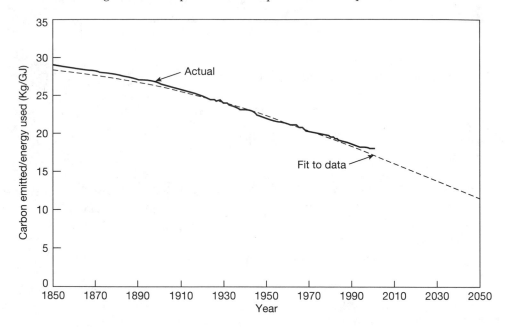

FIGURE 9.38 The decarbonization of the global primary energy sources from 1850 projected to 2050. Historically there has been declining carbon intensity of global primary energy. The ratio of the amount of carbon emitted to total energy used is given in units of kg/GJ. (Data from IIASA, BP, 1965 to 2001, and CDIAC http://cdiac.esd.ornl.gov/trends/emis/em_cont.htm; figure courtesy of Eric Gaidos.)

these countries. During the period 1973 to 1989, the use of energy in the developing countries expanded at a rate 20% faster than that of economic growth. Despite these recent trends, the International Energy Agency projects energy intensities for most of the world's countries will decrease to the year 2010.

New conservation methods for energy utilization and less energy consumption can reduce per capita energy use. Economic growth is not necessarily dependent on increased energy consumption. Indeed, the strong coupling between growth in GNP and growth in energy utilization for the developed nations of the world was broken in the 1970s, and since 1850 the world has produced energy more efficiently in terms of carbon emissions (Figure 9.38). There has been a decarbonization of the global primary energy sources over time and the trend is predicted to continue into the twenty-first century. However, even with an energy-efficient and conservation-minded populace, the continuous population growth of the world probably will necessitate an increased need for energy and other natural resources, and the demand on these resources will increase.

Concluding Remarks and the 21st Century

The human population has been relatively stable for the majority of time that humans have existed on the planet with the exception of the two rapid growth periods at the beginning of the Agricultural and Industrial Revolutions (Figure 9.39). The human race in the early twenty-first century is continuing on a course of rapid population growth, this time with the growth concentrated in the developing nations. An increase in resource consumption is a direct outcome of this population growth. Large amounts of natural resources, both energy and other material substances, are needed to sustain this growing population. For example, by the year 2010, according to the International Energy Agency, total world energy demand will be 48% greater than in 1990. The fossil fuels of coal, oil, and gas will continue to supply about 90% of commercial energy. The worldwide carbon emissions from the burning of these fossil fuels starting at close

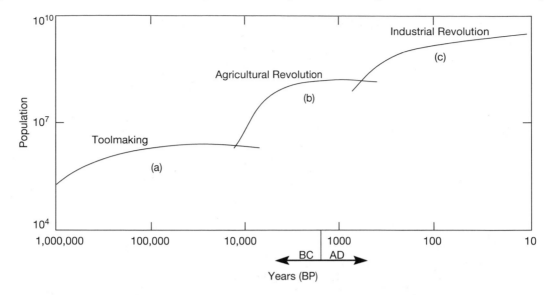

FIGURE 9.39 The growth of the world population for the past million years. This portrayal of population is logarithmic, making it possible to plot the growth of population over a longer period from 10,000 to 6 billion persons. The three rapid increases in population reflect (a) the toolmaking or Cultural Revolution, (b) the Agricultural Revolution, and (c) the Scientific–Industrial Revolution.

to zero in 1850 has reached a level of approximately 8.7 billion tons of carbon released into the atmosphere in 2008 (Figure 9.40). Between 1990 and 1999, the fossil fuel emissions of carbon grew at a rate of 0.9% per year but increased to 3.4% per year during 2000 to 2008. The 2008 carbon emissions represent an increase in emissions of 41% in just 18 years, relative to the 1990 level!

Figure 9.41 shows the relationship between per capita GNP and per capita carbon emissions from fossil fuels for 2008 for various countries of the world. The per capita worldwide emissions of

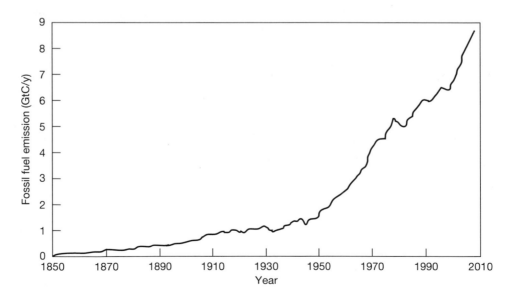

FIGURE 9.40 Emissions of carbon from fossil fuel burning and cement manufacturing (about 1% of the total) to the atmosphere from 1850 to 2008. (*Source:* Carbon Dioxide Information Analysis Center, CDIAC, cdiac.ornl.gov/, Marland et al., 2009.)

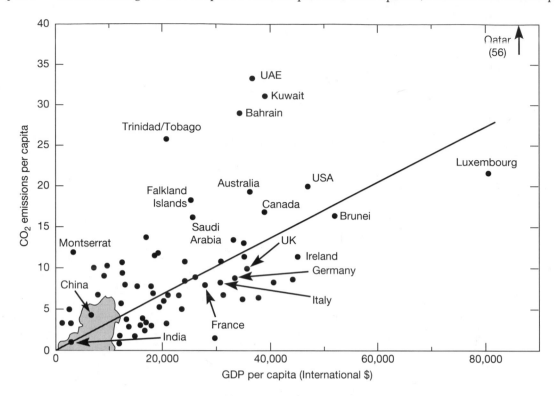

FIGURE 9.41 Per capita GNP in 1000 $US per person versus per capita CO_2 emissions in metric tons per person. Shaded area denotes a large group of countries of the developing world. Data are for 2008 for per capita GNP from the World Bank and for per capita carbon emissions for 2006 from the United States Department of Energy Carbon Dioxide Information Analysis Center (CDIAC).

carbon from fossil fuels in metric tons per person has varied from a low of 1.07 in 1983 to a high of 1.22 in 2006. In 2050, with a world population of perhaps around 9.5 billion people (Figure 9.3), if there were no change in carbon emissions per capita from that of the year 2006, the emissions of carbon to the atmosphere from coal, oil, and gas combustion would be about 11 billion tons annually, about double 1998 emissions. There are continuing large differences in per capita emissions between nations of the developed world and those of the developing world. For example, Kuwait, Singapore, and the United States have relatively high emissions per person, whereas many European countries have lower emissions per capita. This is because of reliance on other forms of energy, such as nuclear for France, and conservation and energy intensity differences among Kuwait, Singapore, the United

States, and many European countries. It can also be seen from the figure that in general most developed nations have higher emissions per capita than developing nations.

Earth has a limit to its carrying capacity for species—that is, the number of a species that can exist at any one time and utilize natural resources while maintaining a dynamic and nondeteriorating ecosphere. No one is sure what this capacity is for the human species because people have been able to extend the natural carrying capacity of Earth through the use of tools, medicines, agriculture, products of industry, and modern technology. However, the rapid growth in the human population and corresponding increased resource consumption rates are stressing natural systems and will require creative engineering to provide for the planet and its species into the future. The longer the time that exponential growth rates in population continue, the greater the likelihood of damage to the ecosphere and, most likely, to the prosperity of the world's people. Ultimately, the only answer to ensuring a continued healthy planet is to stabilize the current human population and to conserve the natural resources available on Earth. The burgeoning population and corresponding resource consumption are the two single greatest dangers to the health of the planet and to ourselves.

Study Questions

1. (a) What are the four stages of demographic transition?
   (b) What is the stage presently characterizing many of the world's poorer nations?
2. If the rate of increase of the world population is 2%, what is the doubling time of the population?
3. What variable is used to gauge the strength of a nation's economy, and what is the approximate range of this variable?
4. What variable may be used to determine how well off an individual is in a particular nation, and what is the approximate range of this variable?
5. (a) From analysis of Figure 9.4, what is the approximate percent increase in the population of the developing world vs. the developed world from 1900 to 2100?
   (b) Why is such a change in the distribution of the population important?
6. What are three problems associated with the heavy use of biomass as an energy source?
7. What are three environmental consequences related to increased reliance on coal as an energy source?
8. From Table 9.5, how much nickel metal was obtained in 1991 from the mining of 38 million tons of nickel ore?
9. (a) From Figure 9.17, what is the approximate percentage increase in GWP and GWP/capita from 1950 to 2008?
   (b) How are these two variables tied to energy consumption?
10. The following conversions hold: one petajoule of energy = 1×10^{15} joules of energy = 163,000 "United Nations" standard barrels of oil or 34,140 standard metric tons of coal. One metric ton of oil = 7.33 barrels of oil. In 2000, approximately 2.7 billion metric tons of oil equivalent of coal (Figure 9.18) were burned worldwide. How many metric tons of coal does this represent?
11. How has the use of various types of energy in the United States changed during the last 150 years?
12. Why the concern with the consumption rate of energy of developing countries relative to developed countries since the late 1970s?
13. From Figure 9.21, what are the relative percentages by the demand sectors of transportation, industrial, residential and commercial, and electric power of the total energy used in these sectors? What is the percentage contribution of the fossil fuels to the demand sectors in terms of total energy required?
14. (a) In 1990, the United States per capita energy consumption was 7822 kilograms of oil equivalent. With a population of 250 million people, what was the total energy consumed by the United States in 1990? How many barrels of oil equivalent does this represent? (One metric ton of oil = 1000 kilograms of oil = 7.33 barrels of oil.) Calculate from inspection of Figure 9.20 the United States per capita energy consumption in 2007 with a population of 306 million people.
    (b) If there were to be a national consumption tax on energy utilization in the United States based on total energy use, how much tax revenue in 2007 could have been collected at a tax rate of $2.00 per barrel oil equivalent? This rate is equivalent to about 2% of the cost of a barrel of oil in late 2007. How much did oil cost per barrel in late 2007?
15. Why could the exploitation of methane hydrates be of importance to the future of world energy supplies?
16. Oil and gas form in source rocks, mainly fine-grained sediments, and migrate into traps. How are these fluids trapped?
17. What does M. King Hubbert's model predict for United States oil production? Does the prediction fit the observations of actual oil production in the United States?
18. Why the concern on the part of the United States that OPEC production may have exceeded non-OPEC production in 2007?
19. What percentages of the world's total oil and gas resources are located in the Middle East?
20. What are four metals obtained from ores associated with granitic igneous rock bodies of the continental crust?
21. Three events in the history of the world are associated with rapid population growth. What are they?

Additional Sources of Information

Bailey, R. (Ed.), 2000, *Earth Report 2000*. McGraw-Hill, New York, 362 pp.

BP Statistical Review of World Energy, June 2009. BP.com/statisticalreview, 45 pp.

Craig, J. R., Vaughan, D. J. and Skinner, B. J., 1988, *Resources of the Earth*. Prentice Hall, Englewood Cliffs, NJ.

Hicks, B. and Nelder, C., 2008, *Profit from the Peak*. John Wiley & Sons, Hoboken, NJ, 286 pp.

International Energy Agency, 1993, *World Energy Outlook to the Year 2010*. Organization for Economic Cooperation and Development/International Energy Agency, Paris, 71 pp.

Energy Information Administration, 2009, *Annual Energy Review 2008*. www.eia.doe.gov, 407 pp.

Jacobsen, J. E., 1993, *Population Growth*. Global Change Instruction Program, University Corporation for Atmospheric Research, Boulder, CO, 32 pp.

Millennium Assessments, 2009. http://www.millenniumassessment.org/en/index.aspx.

The Millenium Development Goals Report 2008. United Nations, New York, 52 pp.

World Bank, 1994, *The World Bank Atlas 1995*. The World Bank, Washington, DC, 36 pp. and http://www.worldbank.org/.

Worldwatch Institute, 2001, *State of the World 2001*. W. W. Norton & Company, New York, 275 pp.

Worldwatch Institute, 2001, *Vital Signs 2002*. W. W. Norton & Company, New York, 192 pp.

World Resources Institute, 1996, *World Resources 1996–97*. Oxford University Press, New York, 365 pp.

World Resources Institute, 2000, *World Resources 2000–2001*. World Resources Institute, Washington, DC, 389 pp.

10 The Changing Earth Surface: Terrestrial Vegetation

Man has gone to the Moon but he does not know yet how to make a flame tree or a birdsong. Let us keep our dear countries free from irreversible mistakes which would lead us in the future to long for these same birds and trees.

PRESIDENT HOUPHOUET-BOIGNY, PRESIDENT OF CÔTE D'IVOIRE (THE IVORY COAST), A COUNTRY THAT HAS LOST 66% OF ITS FORESTS AND WOODLANDS IN THE LAST 25 YEARS

This chapter begins our exploration of human impacts on the ecosphere resulting from the factors of population growth, industrialization and utilization of resources, growth in affluence, and technological innovation. It is important to recognize that the ecosphere is an interconnected system and hence environmental impacts due to the above factors, which in themselves are interconnected, are often also intertwined. For example, the burning of fossil fuels not only fills our atmosphere with greenhouse gases leading to global warming but results in acid rain and the acidification of ocean waters.

The Earth's vegetated and nonvegetated landscape has changed naturally over time, but human activities are also dramatically changing the landscape. Land use changes include the conversion of land by the growing human population for grazing and agriculture (food production), urbanization (building and development of towns and cities), energy production (dams), and transportation systems (roads, trains, etc.). Changes in land use may result in deforestation and soil degradation, such as desertification and salinization, the loss of wetlands, and changes in the distribution and chemical and biological properties of aquatic systems. This chapter and the following discuss changing land use practices and their effects on the environment. In this chapter, the focus is on two terrestrial ecosystems. We will initially discuss the world's forests, their types, functions, changes in area, and effects and causes of change. The emphasis will then shift to a discussion of domesticated ecosystems, such as croplands. Chapter 11 deals with the terrestrial realm of soils and land and their link to aquatic systems, including lakes, streams, coastal wetlands and estuaries, and oceans, and the human impacts on these systems.

FORESTS WORLDWIDE

Global Forest Assessments

Several global assessments have been made of the state of the world's forests, and the Food and Agriculture Organization (FAO) of the United Nations has been coordinating global forest resource assessments every 5 to 10 years since 1946. None of the assessments are without problems. The first global forest assessment, the 1980 Tropical Forest Resource Assessment, was completed in 1982 by the FAO and the United Nations Environment Programme (UNEP). It was the major source of information on the state of the world's forests at that time. However, this global assessment is now outdated but does provide data against which more recent assessments can be compared. Since the 1982 assessment, a number of regional forest studies indicated major changes in the state of the world's forests. As a result, in late 1991, an interim report on the condition of the world's forests was released by the Forest Resources Assessment 1990 Project of the FAO. A complete global assessment, the 1990 Forest Assessment, was presented to the public in early 1993. In 1997, *The State of the World's Forests 1997* was published by the FAO with revised estimates of forest cover change between 1980 and 1990 and new information on global forest cover and change since 1990. In 2000, the *Global Forest Resources Assessment 2000* and in 2005, the most comprehensive report, the *Global Forest Resources Assessment 2005,* were released by the FAO. The 2005 assessment is based on reporting by 172 national correspondents and their networks of professionals comprising more than 800 people from 229 countries and areas, and encompassing three points in time: 1990, 2000, and 2005. A further assessment is forthcoming from the FAO in 2010 (see End Note). In addition, the World Resources Institute (WRI) has made some separate forest assessments primarily during the latter part of the twentieth century, and the Millennium Ecosystem Assessment (MA, 2005) has investigated how changes in ecosystem services owing to land use changes and other drivers of change interact with and influence human well-being. These various assessments are the major global databases available and represent the best worldwide synthesis of land use changes and forest information at this time. These sources do provide some feeling for the state of our forests worldwide. However, the global forest studies are still somewhat incomplete and controversial. Caution must be used in interpretation of the conclusions of the assessments, mainly because different methodologies have been used over the years and the baseline information on forested areas has changed from assessment to assessment. Thus, there is still some uncertainty in the sizes of terrestrial forest biomes and the extent of deforestation/reforestation practices.

Historical Estimates of Wooded and Forested Land

One example of the uncertainty in forest data is the historical range in estimates of wooded and forested land area in the world (Figure 10.1). The wide range in estimates is due to the changing database over the years, difficulties in obtaining measurements of areas of forested and woodland ecosystems in various regions of the world, and to some extent the lack of consistent definitions of these ecosystems. For example, the 1997 FAO assessment for 1995 uses terms for forest with different definitions for developed and developing countries. In developed countries, the term **forest** was defined as land with tree crown cover of more than 20% of the area on which there is continuous forest growing to more than 7 meters in height. The forest is able to produce wood. In developing countries, a forest was defined as an ecosystem with a minimum of 10% crown cover of trees and/or bamboo. The area so defined generally included wild flora and fauna, natural soil conditions, and was not subject to agricultural practices. These definitions are applicable to FAO Forest Resources Assessment forest cover data for 1990 and 1995 but not earlier. In addition, the 2000 FAO assessment uses 10% tree crown cover for both developed and developing countries for the definition of forest, making that assessment difficult to compare with the 1990 assessment. In the 2005 FAO report, information was sought on the current status and change over time of the following four variables: areas of "forest," "other wooded land," and "other land with tree cover"; characteristics of primary, modified natural, seminatural, protective plantation, and productive plantation forests; standing volume of wood; and carbon stocks contained in woody biomass, dead wood, litter, and forest soils. So once more there is some inconstancy with previous reports because of methodology but the 2005 report remains the most authoritative to date.

Historical estimates of wooded and forested land have varied by a factor of three over time (Figure 10.1), although estimates are converging on a value of about 5300 million hectares (one hectare = 10,000 square meters = 2.471 acres). For example, in *World Resources 1990–91*, the extent of forested and woodland area is given as 4081 million hectares for

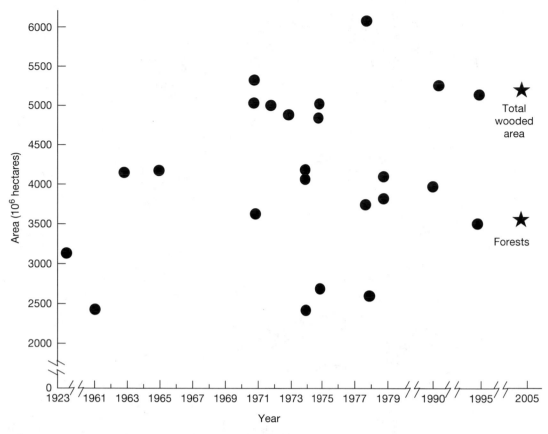

FIGURE 10.1 Historical estimates of total wooded and forested land over time. Estimates shown by the stars are the 2005 estimates of the FAO. (*Source:* Sedjo and Clawson, 1984; after FAO, 2006.)

the 1980s. Thirty percent of the forested land was considered to be open and 70% was closed (see definitions following). In the 1992–1993 edition, the total extent of forested and woodland area for the 1980s was given as 5289 million hectares, of which the natural forest constituted 67% (14% open and 53% closed). Plantation forests and other wooded areas were two additional categories that were listed in this later edition. Plantation forests constituted less than 1% and other wooded area 32% of total forested land (see World Resources Institute, 1990, 1992). The FAO in 1997 (FAO, 1997) estimated the total net wooded area of the land, which includes forests, open woodland, and brushland, as 5100 million hectares and the forests of the world as about 3450 million hectares. This estimate takes into account natural forest loss as well as growth in industrial forest plantations. In 2000, the FAO assessment was 5163 million hectares as total wooded area and 3860 million hectares as forest but the baseline regarding the definitions had changed (see above). In 2005, the FAO estimates were 5328 million hectares of total wooded area and 3952 million hectares of forests. More than 50% of the

world's forests were reported to be located in five countries: Russian Federation (22.1%), Brazil (15.9%), Canada (7.1%), United States (6.2%), and China (3.9%). Figure 10.2 illustrates the distribution of forest areas and the total above-ground woody biomass by country, and Table 10.1 shows the regional distribution of forested areas according to major subregions.

It should be kept in mind that despite recent advances in acquiring quantitative forest data, it is still difficult on a global and regional scale to assess the degree of land use changes and condition. In addition, the wide range in estimates of variables concerned with land use has led to controversy among scientists, environmentalists, politicians, and developers. Such confusion in forest and land use data can lead to problems for policymakers who ultimately must set conditions and standards for land use.

Forests of the Past

To set the stage for discussion of today's forests, we will explore briefly the evolution of forest types of the past (see also Chapter 8). The first forests on Earth

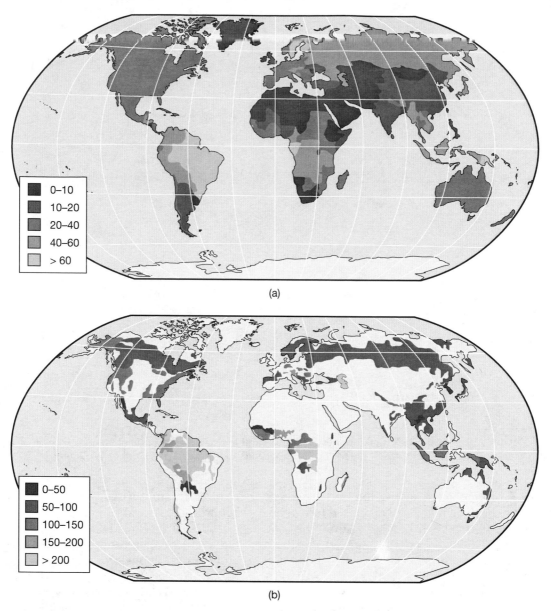

FIGURE 10.2 Proportion of forest area by country in percent of land area (a) and aboveground woody biomass by country in metric tons per hectare (b).   (Map from http://www.fao.org/forestry/en/. © FAO 2006. Reprinted by permission of the Food and Agricultural Organization of the United Nations.)

evolved during the warm Devonian Period, about 365 million years ago. The first insects and amphibians lived in these forests of huge club mosses and ferns. These plants were flowerless and seedless plants but had vessels for carrying and circulating fluids and thus were the first true vascular plants. Vast swamp forests appeared about 345 million years ago near the beginning of the Carboniferous Period, the Age of Amphibians. Large flying insects also evolved at this time, some with wingspans of more than 20 centimeters. Swamp forests, consisting of 54-meter-tall club mosses and 8-meter-tall tree ferns, covered a large

portion of the continent now referred to as North America. When these giant plants died and fell into the swamp sediments, they were eventually buried and altered. Their altered remains constitute the vast Carboniferous coal deposits of today. During the Permian Period, the swamp forests slowly disappeared as the climate cooled. Conifers (gymnosperms) gradually became the dominant form of plant life in the Permian.

The gymnosperm flora, which were uncommon in the late Paleozoic Era, were abundant during the Mesozoic. Gymnosperms are vascular plants with seeds that are not enclosed in a fruit or seed case, and

TABLE 10.1 Distribution of Forests by Subregion 2005 (FAO, 2006)

Region/Subregion	Forest Area (1 000 ha)	% of Global Forest Area
Eastern and Southern Africa	226,534	5.7
Northern Africa	131,048	3.3
Western and Central Africa	277,829	7.0
Total Africa	**635,412**	**16.1**
East Asia	244,862	6.2
South and Southeast Asia	283,127	7.2
Western and Central Asia	43,588	1.1
Total Asia	**571,577**	**14.5**
Total Europe	**1,001,394**	**25.3**
Caribbean	5,974	0.2
Central America	22,411	0.6
North America	677,464	17.1
Total North and Central America	**705,849**	**17.9**
Total Oceania	**206,254**	**5.2**
Total South America	**831,540**	**21.0**
World	**3,952,025**	**100.0**

From *Global Forest Resources Assessment, 2005,* ftp://ftp.fao.org/docrep/fao/008/A0400E/A0400E00.pdf.
© FAO 2006. Reprinted by permission of the Food and Agricultural Organization of the United Nations.

include pine, fir, spruce, and other cone-bearing trees and shrubs. These gymnosperm forests consisted of primitive conifers, ginkgoes, and the flowerless and seedless tree ferns. Amphibians, insects, and reptiles lived in these forests. The first dinosaurs, mammals, and birds evolved within them.

The first flowering plants, angiosperms, which are vascular plants and produce seeds encased in a fruit or seed case, appeared about 130 million years ago. They became the dominant type of flora during the warm Cretaceous Period, a second coal-forming period. At this time, insects flourished and were successful pollinators of the flowering plants. Today, the angiosperms are the dominant group of plants and include oak, maple, eucalyptus, and koa trees.

The climate began to cool approximately 65 million years ago at the beginning of the Cenozoic Era, and vast temperate forests spread over the North American, European, and Asian continents. Conifers and flowering plants dominated the plant life of the planet. The dinosaurs were gone and bird, insect, and mammal species began to expand.

The Earth entered the Pleistocene Epoch 2 million years ago. The cold glacial stages of the Pleistocene were times of destruction of forests of the high northern latitudes because continental glaciers 2 to 3 kilometers thick covered the land. With the last glacial retreat 10,000 years ago, the forests moved north with the melting of the ice, and today the temperate and boreal forests of the northern latitudes cover this previously glaciated area. The current distribution of forest types, shifting from oak-hemlock to pine and then spruce with increasing latitude, arose after the Pleistocene ice age.

Changes in the distribution and type of world forests have taken place relatively slowly over centuries to thousands to millions of years. The current distribution is not a permanent one because forests are sensitive to natural climatic change and other factors. However, since the Agricultural Revolution began some 10,000 years ago, human activities have had a significant impact on the distribution and types of forests of the world.

Forests of Today

Several systems are used to classify the different forest types of the world. They may be classified by the characteristics of dominant trees in a region, such as an evergreen, deciduous, or coniferous forest, or by the usefulness of the trees, such as hardwood or softwood forests. Another way of classifying the forests is as open or closed. A **closed forest,** as defined by the FAO, is "land where trees cover a higher proportion of the ground and where grass does not form a continuous layer on the ground." Closed forests covered an area of approximately 2800 million hectares worldwide in 1990; of this area, about 57% was in the temperate zone, and 43% was in the tropics. **Open forests** have trees that are more sparsely spaced among grassland and bush. It is estimated that open forests covered more than 1200 million hectares in 1990. However, the estimates of open forested areas are uncertain. Plantation farms are forested areas established artificially for industrial or nonindustrial use. Plantation forests constituted 5% (195 million hectares) of total forested land in 2000. Some forested areas are commonly referred to as "other wooded areas." Other wooded areas and plantation areas may or may not be included in estimates of open forested areas.

Another forest classification is based on the concept of an ecosystem, whereby forests with similar soil types, rainfall, and climate are grouped. One example of an ecosystem classification of forests is that of temperate boreal coniferous, temperate deciduous, temperate coniferous, woodlands, tropical rain forest, chaparral, tropical deciduous, and savanna biomes (see Figure 6.11). The vast boreal forests of the high latitudes of the Northern Hemisphere are composed mainly of evergreen coniferous trees that have adapted to extremely cold winters and a short growing season. Bogs are abundant in these forested areas. Farther south in the temperate belt, where a climate of warm summers and cold winters prevails, two main types of forest grow: the temperate deciduous, such as oak and beech, that shed their leaves in winter; and the temperate evergreen (coniferous). Woodlands, located in the temperate belt of the United States and southern Europe, are areas of lower rainfall and widely spaced trees scattered through extensive grasslands.

Temperate-zone forests are generally found in the developed countries of the world. Deciduous trees, as well as some conifers, grow in eastern North America, Europe, the former USSR, and eastern China. Temperate evergreen forests grow near coastal areas and are found along the northwest coast of North America, in Chile, New Zealand, and southeast Australia.

The world's forests that have attracted the most attention in the late twentieth century and early twenty-first century are the tropical forests of the developing countries. This is because of the current rate of destruction of these ecosystems. There are different types of tropical forests. It is often unclear in the literature whether the terminology used applies to one or more tropical forest types. For our purposes, the definitions and extent of tropical forests will be those defined by the FAO and used in the *World Resources* volumes from 1988 through 1996.

Tropical forests include open and closed forests, shrub land, and forest fallows (slash-and-burn areas). Savannas are open tropical forests, with trees scattered through extensive grasslands. They often experience long periods of drought and receive less than 10 centimeters of rainfall per year. Estimates of open tropical forest areas vary significantly.

The closed tropical forests are of two main types. The trees of the **wet tropical forests** (tropical rain forests or evergreen forests) do not shed their leaves and have a thick closed canopy. The trees of the **dry tropical seasonal forests** (moist deciduous or monsoon forests) lose their leaves in the dry season. Both wet and dry tropical forests are collectively referred to as tropical moist forests. In estimates done in 1990, approximately 33% of tropical moist forest area was dry tropical forest, and 66% was wet tropical rain forests. The remaining 1% of the closed tropical forest area was various types of deciduous and semideciduous forests.

The dry tropical forest receives about 100 to 380 centimeters of rain per year. The temperature of the environment is less constant than that of the wet rain forest. The wet season brings torrential monsoon rains, but during the 4-month dry season, the trees lose their leaves and the tree canopy opens up. This enables sunlight to penetrate to ground level, allowing for the growth of a thick underbrush.

The tropical wet rain forest, unlike the dry forest, never dies back. An equatorial evergreen tropical rain forest has a high rainfall, 380 to 960 centimeters per year, and a high average temperature of 27°C. These evergreens have a luxuriant canopy that allows little underbrush to grow below them in the understory because the canopy excludes light from reaching the forest floor (Figure 10.3). The ground is thickly covered by fungi, ferns, and other luxuriant ground cover. Shrubs and small trees less than 10 meters tall constitute the understory. One can see why it would be so difficult to determine the total amount of organic matter as dry matter or as carbon in the aboveground biomass of such a complex ecosystem.

Tropical forests lie in a belt around the globe between the Tropic of Cancer and the Tropic of

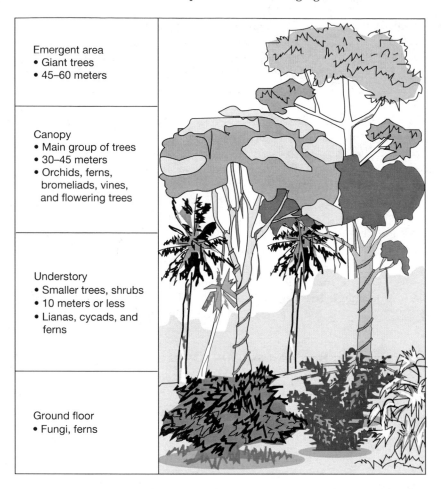

Emergent area
• Giant trees
• 45–60 meters

Canopy
• Main group of trees
• 30–45 meters
• Orchids, ferns, bromeliads, vines, and flowering trees

Understory
• Smaller trees, shrubs
• 10 meters or less
• Lianas, cycads, and ferns

Ground floor
• Fungi, ferns

FIGURE 10.3 The levels of vegetation found in a tropical rain forest.

Capricorn in areas where rainfall and temperature conditions are conducive to their growth. The savannas are located in the arid dry lands of Africa, Asia, Australia, and Latin America. Tropical moist forests are located in three major regions of the world, in the developing countries of Central and South America, Southeast Asia, and Africa (see Figure 6.11). Latin America had 57%, Southeast Asia 25%, and Africa 18% of the remaining tropical moist forests of the world in the 1990 FAO assessment. In the Americas, patches of tropical moist forests are found in southern Mexico and in the Caribbean. Earth's largest tropical moist forest system is Amazonia in South America, which is drained by the largest river in the world, the Amazon. More than 50% of the area of the Amazonian forest is in Brazil. As a result, 27% of the tropical moist forest area of the world is in that country. The second largest tropical forest region is in Southeast Asia. Indonesia has the largest forested area within this region, with 6% of the tropical forest area of the world within its borders. The third major

tropical forest region of the world is found in Africa. The Ituri forest, drained by the Zaire River and tributaries, is the second largest contiguous tropical moist forest in the world. The Republic of the Congo contains 9% of the tropical forest area of the world. In addition to these three major regions, tropical moist forests also exist on the islands of Mauritius, Madagascar, and the Seychelles.

The Impact of Human Activities on Forests

Two major problems of the world's forests related to human activities are pollution that affects the health of trees and deforestation, especially in the developing countries. Deforestation is largely driven by the world's demand for wood. Since 1950 the amount of wood consumed has increased more than 250%. In 1961, some 1000 million cubic meters of wood were consumed by the world economy, and in 1993, 1500 million cubic meters. Approximately 43% of this consumption was wood burned as fuel, mostly in developing countries. The remaining 57% was cut as timber and milled into boards, plywood, veneer, chipboard, paper, paper products, and other products, mainly in industrialized countries. Wood fuels account for more than half of biomass energy consumed in developing countries.

World consumption of paper and paperboard alone has increased more than 20 times since 1913, reaching about 354 million tons in 2005. This amounts to approximately 55 kilograms of paper and paperboard per person per year. This demand for wood and wood products is one of the major forces behind deforestation. World wood consumption is predicted to increase to 2250 million cubic meters in 2010 and then by more than 50% by 2050. Figure 10.4 shows the trends in wood removal from forests as industrial roundwood (wood material finished to either round or half-round shape and generally 2 feet to 8 feet in diameter) and wood for fuel from 1990 to 2005.

AIR POLLUTION AND ACID DEPOSITION   Air pollution and acid deposition are environmental stresses that are contributing to the degradation and dieback of forests

in areas reached by pollution generated by industrial and transportation sources. Forests in the northeastern part of the United States and those of Europe have been especially affected by these stresses. About 6 million hectares of European forests have been destroyed or severely damaged because of air pollution originating mainly from factory and transportation sources of acid gases and hydrocarbons and formation of tropospheric ozone. Forests in the developing countries of the world are also showing the effects of pollution stress. Landsat satellite imagery is being used to assess the degree of dieback of forests from various environmental stresses (http://landsat .gsfc.nasa.gov/). Air pollution and acid deposition problems are more thoroughly discussed in Chapter 12.

DEFORESTATION PRACTICES   Deforestation was a term used originally for the process of clearing forests from the land by burning or logging practices solely for agriculture or for settlement purposes. Land that was logged by clear-cutting and left to regrow—that is, afforested—was not necessarily referred to as deforested land. However, in 2000, the FAO specifically defined deforestation as the

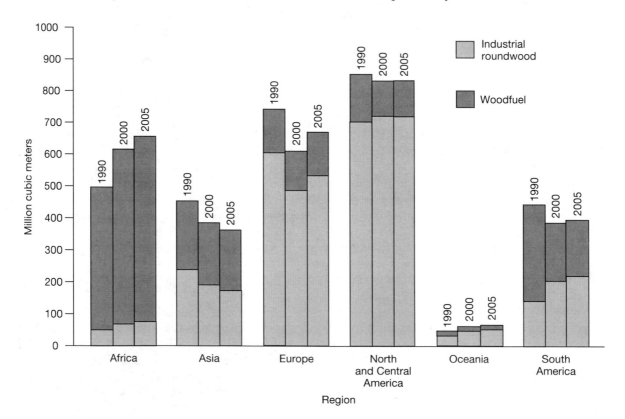

FIGURE 10.4   Trends in wood removal from the forests as industrial roundwood and wood for fuel from 1900 to 2005 in millions of cubic meters. One-third of the forests are used for these products.   (*From Global Forest Resources Assessment, 2005,* ftp://ftp.fao.org/docrep/fao/008/A0400E/A0400E00.pdf. © FAO 2006. Reprinted by permission of the Food and Agricultural Organization of the United Nations.)

conversion of forest to another land use or the long-term reduction of the forest tree canopy cover to below a 10% threshold. Remember in 1990 that the FAO definition for developing countries was less than 20% reduction in the tree canopy. **Afforestation** is now viewed as the establishment of forest plantations in areas not previously in forest, whereas **reforestation** refers to the establishment of forests through planting, seeding, or by another means after a temporary loss of forest. Thus, changing definitions of the terms related to forest practices have resulted in some confusion in the literature as to the extent of regional and global deforestation practices. This is a factor that leads to problems in estimates of the amount of forest cover and hence the amount of carbon tied up in the world's forests.

Slash-and-burn techniques (shifting cultivation) to clear land have been a practice extensively employed by humans for centuries (Figure 10.5; see http://earthobservatory.nasa.gov/IOTD/view.php?id=2786). The nutrients in the burned plants help to fertilize the soils and provide nutrients for future crops. This practice generally occurs over small areas of land, and the people move on to new patches of land when those currently farmed can no longer sustain crops. Over the years, the old ecosystem of the deforested area eventually rejuvenates itself (fallow forests), and the people return to slash and burn the same plots. In the latter part of the twentieth century, the number of people using slash-and-burn clearing

for subsistence agriculture increased. This activity has become a major cause of deforestation worldwide in the twenty-first century.

Clear-cutting of the forest is a logging practice that removes most of the vegetation along with the nutrients that are contained in the aboveground biomass. Clear-cutting today is a common practice in the tropical rain forests, for example, once done in Costa Rica, and in the temperate forests, such as those found in the states of Washington and Alaska in the United States (Figure 10.6). The practice of select-fell is employed when it is desired to retain part of the forest; hence, only selected trees in an area are cut. The rest are allowed to grow to maturity to replenish the stock.

In some cases, trees are planted in an originally deforested area; that is, the area is reforested. Some of the reforested areas are plantation forests that may contain varieties of trees other than those that originally grew there. Reforestation practices, although more common in the industrialized nations of the temperate zone than in the developing nations of the tropics, are not always employed in all clear-cut areas.

Because of the preceding practices, approximately 30% of the area of the world's forests has been lost since preagricultural times, a decrease in area from perhaps as much as 6 billion hectares to approximately 4 billion hectares reported for 2005. The developing countries, with 59% of total land area of the world, have approximately 57% of world forest area, most of which is tropical.

FIGURE 10.5 Burned farmland near Santa Fé (Veraguas Province), Panamá. (DirkvdM, Wikipedia)

(a)

(b)

FIGURE 10.6 Photograph of the deforestation practice of clear-cutting in (a) Washington, United States, and (b) Costa Rica. (Photographer, Gary Braasch.)

FOREST AREA TRENDS Figure 10.7 shows qualitatively the global distribution of original and remaining forests of the world. People throughout the world have deforested land for thousands of years. At one time original, primary, old-growth forests blanketed the European and North American continents. As the population on both of these continents grew, massive deforestation took place to provide lumber, additional land for agriculture, fuelwood, and land for mining purposes. Europe was extensively deforested of its old-growth forests. Today, the forests are almost entirely regrowth or replanted trees. In the United States, deforestation by European settlers occurred in New England in the seventeenth century and then proceeded toward the Midwest. In a single lifetime, an area the size of Europe was deforested. Approximately 85% of the primary, old-growth forests of the United States was destroyed by the original European settlers. The majority of the forests in the United States today, with the exception of the old-growth forests of the Pacific Northwest and Alaska, are regrowth or replanted trees.

Extensive deforestation continues to occur in some areas of the developed world. In the United States, about 43,000 hectares of primary old-growth forests are cut yearly by clear-cutting, and much of the clearing of forests occurs legally in national forests. The forests of the Pacific Northwest and Alaska are some of the most severely affected. A satellite photograph released in 1993 of clear-cut areas in the Gifford Pinchot National Forest in Washington State showed that clear-cutting of forest in this national forest is nearly as extensive as that in the state of Amazonas in Brazil. Despite the extensive cutting of primary forests in the United States, the country still gained forested area in the early twenty-first century because of the regeneration of previously cut forests and the development of new forest plantations.

A few thousand years ago, the tropical moist forests covered approximately 2 billion hectares, or about 14%, of the planet's land surface. People have destroyed about half of that forested area in the last 200 years. Most of this destruction occurred after World War II, with most of the intensive deforestation happening in the 1980s. In 1990, tropical moist forests occupied approximately 1.3 billion hectares, or about 6 to 7% of the planet's land surface.

The information currently available suggests that the temperate and boreal forests of Europe are stabilized and grew in area from 1990 to 2005

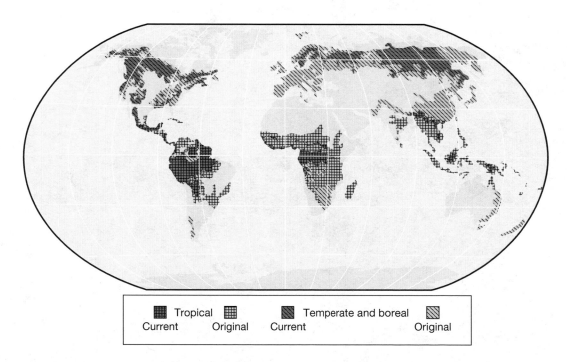

Tropical Current   Tropical Original   Temperate and boreal Current   Temperate and boreal Original

FIGURE 10.7 This map shows the approximate original distribution of temperate and tropical forest cover under current (post-glacial) climatic conditions and before significant human impacts, and the distribution of remaining forest. Approximately half the world's original forest cover has disappeared, and more original forest has been replaced by secondary tree cover, which cannot be distinguished on this map. (*Source:* Global Generalised 'Original' Forest dataset (V 1.0) and Global Generalized 'Current' Forest dataset (V 3.0) prepared at UNEP World Conservation Monitoring Centre, March 1998)

(Figure 10.8). However, this conclusion is based strongly on the changing state of the forests of the Russian Federation. These forests constitute about 80% of the European forested area and 20% of the world's forested area. North and Central America lost forest area between 1990 and 2005 mainly, but not exclusively, due to deforestation of the tropical forests of Central America. Although Asia lost forest area between 1990 and 2000, it apparently gained forest area between 2000 and 2005, due primarily to large-scale afforestation in China. South America and Africa exhibited the most significant loss in forest area, followed by Oceania. The areas lost were mainly those of tropical forests.

Figure 10.9 shows the extent of deforestation in 1978 in the Brazilian Amazon compared with that in 1988, as obtained from Landsat-4 and Landsat-5 satellite imagery. The total area affected by deforestation (including deforested, isolated forest, and adversely affected forested area) was reported as increasing from about 20.8 million to 58.8 million hectares during the decade. The total deforested area increased by nearly 200% during this time period (from about 7.8 million to 23 million hectares). It has

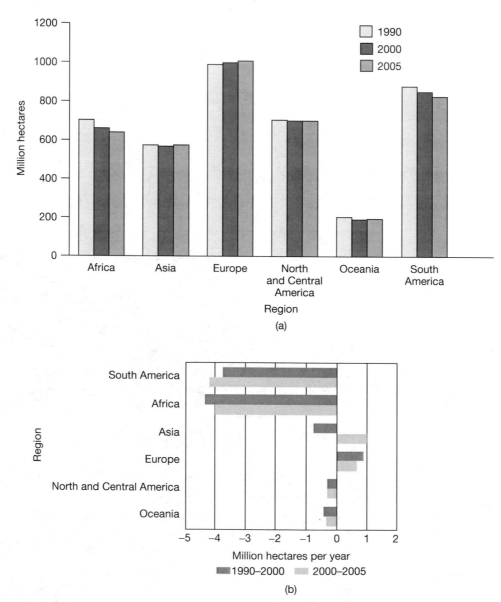

(a)

(b)

FIGURE 10.8 Global trends in forested area by region in millions of hectares (a) and annual net changes in forested area by region in millions of hectares per year from 1990 to 2005 (b). (From *Global Forest Resources Assessment, 2005*, ftp://ftp.fao .org/docrep/fao/008/A0400E/A0400E00.pdf. © FAO 2006. Reprinted by permission of the Food and Agricultural Organization of the United Nations.)

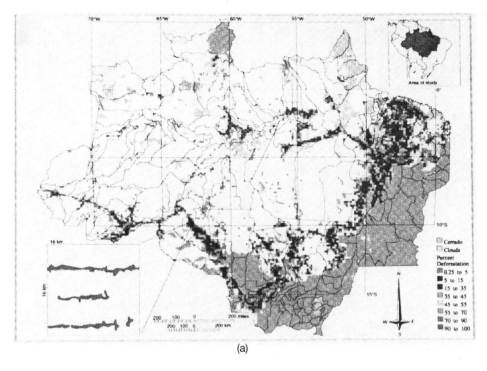

(a)

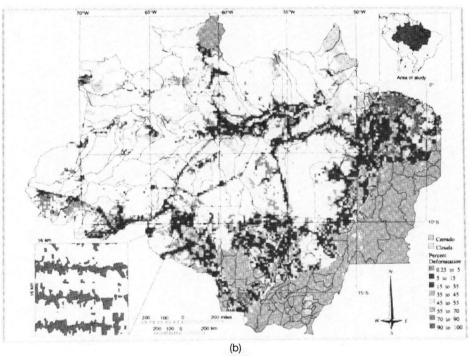

(b)

FIGURE 10.9 Comparison of the extent of deforestation in percent for the Brazilian Amazon between 1978 (a) and 1988 (b) produced from data acquired from the Landsat-4 and Landsat-5 satellites.   (Courtesy of the National Aeronautics and Space Administration, 1993.)

been estimated that the rate of deforestation in Brazil amounted to 2.6 million hectares of forested area per year for the period 1990 to 1995 (an area about the size of the states of Rhode Island and West Virginia).

Table 10.2 gives the rates of deforestation for the country from 1988 to 2009. A detailed satellite view of the continuing extent of deforestation in the Brazilian Amazon between June 17, 2002 and June 28, 2006 is

TABLE 10.2 Deforestation Rates for Brazil

Year	Deforestation [sq mi]	Deforestation [sq km]	Change [%]
1988	8,127	21,050	
1989	6,861	17,770	−16%
1990	5,301	13,730	−23%
1991	4,259	11,030	−20%
1992	5,323	13,786	25%
1993	5,751	14,896	8%
1994	5,751	14,896	0%
1995	11,220	29,059	95%
1996	7,012	18,161	−38%
1997	5,107	13,227	−27%
1998	6,712	17,383	31%
1999	6,664	17,258	−1%
2000	7,037	18,226	6%
2001	7,014	18,165	0%
2002	8,260	21,394	17%
2003	9,748	25,247	19%
2004	10,588	27,423	9%
2005	7,276	18,846	−31%
2006	5,447	14,109	−49%
2007	4,453	11,532	−18%
2008	4,621	11,968	4%
(est) 2009	<3,860	<10,000	N/A

given for the state of Mato Grosso, Brazil at http://earthobservatory.nasa.gov/IOTD/view.php?id=6811.

The amount of deforestation continuing to take place in the world and the ramifications of the deforestation practices for local and global communities and the environment are a major problem. Between 1980 and 1995, the area of the world's forests decreased by about 180 million hectares. Most of this area was lost in tropical regions. This rate amounts to as much as 33,000 hectares of tropical forests deforested daily. However, annual global deforestation rates slowed between the 1980s and 1995 but still exceeded 13 million hectares per year in 2005. At the same time, the net loss of forest area has significantly declined because of forest planting, landscape restoration, and natural expansion of forests. Planta-

tion forests alone have expanded in area throughout the world, and for most regions, forests designated for conservation have increased in area (Figure 10.10). However, the annual tropical deforestation rate was still −0.62% of forested area between the 1990s and 2005. The FAO estimates that 10.4 million hectares of tropical forest were permanently destroyed each year in the period from 2000 to 2005. Figure 10.11 illustrates regions with high net changes in forested area for the period 2000 to 2005.

This loss of forest habitat has led to poorly known rates of habitat and species loss (see below). The controversy in the United States in the early 1990s concerning the spotted owl involved not only the possible extinction of a single species, but also the destruction of an entire ecosystem, one of the

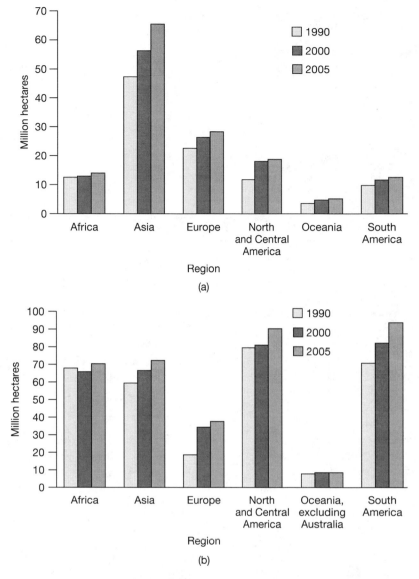

FIGURE 10.10 Worldwide changes in plantation area in millions of hectares (a) and the forested area designated for conservation in millions of hectares (b) by regions from 1990 to 2005. Eleven percent of forests are now used for conservation of biodiversity. (From *Global Forest Resources Assessment, 2005*, ftp://ftp.fao.org/docrep/fao/008/A0400E/A0400E00.pdf. © FAO 2006. Reprinted by permission of the Food and Agricultural Organization of the United Nations.)

last primary, old-growth forests in the continental United States.

Forest Ecosystems: The Effects of Change

Forest ecosystems are vital to the maintenance and health of the ecosphere. Figure 10.12 shows the resources and processes that the forests of the world provide, in other words, their ecosystem services. The forests serve as watersheds, effectively control erosion of the soils, and are essential in the carbon dioxide and oxygen balance of the atmosphere. Forest ecosystems cycle life-giving nutrients, modulate the water cycle, regulate temperature, and provide a habitat for a considerable variety and number of species on the planet. Each forest system is unique in how it functions. Temperate forests have been studied more extensively than have tropical forests, and thus, the various functions of tropical forest ecosystems are less well known. Modification of forest ecosystem functions and services, or their loss, results in major changes in the environment and has

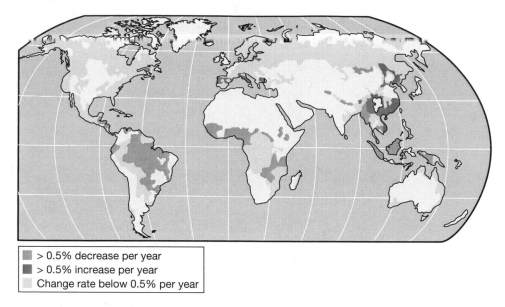

> 0.5% decrease per year
> 0.5% increase per year
Change rate below 0.5% per year

FIGURE 10.11   Net changes in forested area from 2000 to 2005 for the world.   (Map from http://www.fao.org/forestry/fra/41256/en/, © FAO 2006. Reprinted by permission of the Food and Agricultural Organization of the United Nations.)

many long-term ramifications. For example, 45% of the area of Ethiopia was covered with trees in 1900; only 1 to 3% of forested land remains today. In 1920, the island of Haiti was 60% forested; by 1987, only 2% of its forests were left. Ninety-three percent of the island of Madagascar is deforested. These countries are among the most densely populated and poorest of the world. Their forests, a major natural resource, are gone. In this section, we discuss some functions and services of forest ecosystems and the effects of their loss.

SOIL, BIOMASS, AND NUTRIENT CYCLES   In forested ecosystems, nutrients are recycled between the biomass and soil. Tropical rain forests and temperate forests differ in their manner of cycling of nutrients (Figure 10.13). Temperate forests generally have rich, thick, fertile topsoil extending to a depth of several meters or more, with a humus-rich layer one-third of a meter or more thick on top. Nutrients cycle deep into the ground with much of the available nutrient stored in soils. Clearing of temperate forests may still leave a considerable amount of nutrients in the soils,

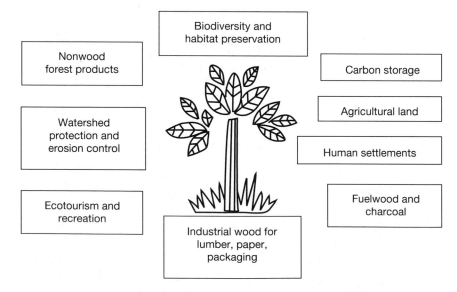

Biodiversity and habitat preservation

Nonwood forest products

Carbon storage

Watershed protection and erosion control

Agricultural land

Human settlements

Ecotourism and recreation

Fuelwood and charcoal

Industrial wood for lumber, paper, packaging

FIGURE 10.12   Schematic diagram showing the ecosystem services and resources of the forests of the world.

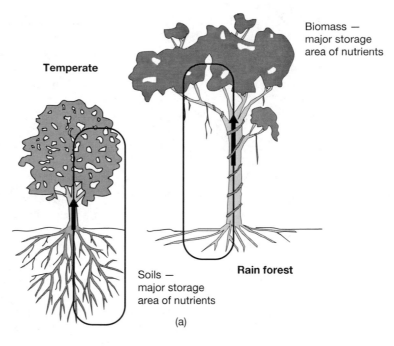

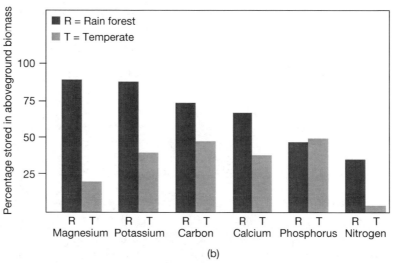

FIGURE 10.13   (a) Differences in nutrient recycling between temperate forests and tropical rain forests. (b) The percentages of nutrients locked in biomass and soil of the two systems.   (*Source:* Newman, 1990.)

resulting in fertile land for forest regrowth or growing agricultural crops.

The fertility of the soil, a mixture of rock, water, air, organic matter, and nutrients from decomposing vegetation, is partially dependent on the type of rock that underlies the soil (see Chapter 2). Most of the soils of tropical rain forests are found in areas of geologically old rocks. The weathering of these rocks under tropical climatic conditions forms a red lateritic clay that is deficient in nutrients. The tropical forest

soil on top of the clay is generally only about 5 centimeters thick, with the upper humus-rich layer averaging only 2.5 centimeters in thickness. The soils of most tropical forests are very infertile. Under these circumstances, how does the luxuriant growth of the tropical forest sustain itself?

Much of the nutrient stock of the rain forest is found in the huge canopy and not in the soils. On average a rain forest stores 90% of its potassium and magnesium, about 70% of its carbon and calcium,

50% of its phosphorus, and 35% of its nitrogen (Figure 10.13) in aboveground biomass. Furthermore, the canopy of the rain forest supports more than 66% of the plant and animal life of the biome. The luxuriant growth found in the tropics is mainly a result of nutrients cycled above the ground through the biota and not through the soils. When the biota die, decomposition is rapid, and the decomposers release nutrients directly back into the aboveground system with relatively little loss or wastes. It was once thought that the root systems of trees of the tropical forests were very shallow and spread out, but in fact the root system of some trees of the drier tropical forests may be deep, enabling water to be brought from depth to the growing vegetation at the land surface.

Because of the nutrient-poor soil underlying the tropical wet forests, very few nutrients are left in the ground after deforestation. In the Brazilian Amazon, nutrients are also transported to the rain forest in dust plumes from Africa. Regardless, the long-term use of tropical forested land worldwide for agricultural purposes, regrowth, or reforestation of the ecosystem has proven to be difficult and sometimes not practical because of nutrient deficiencies.

WATER CYCLE The amount of water that falls in the rain forest exceeds that of the temperate forests. The water cycle in the rain forest is essentially a closed system. For example, in the Amazonian rain forest, 75% of the water falling as rain evaporates or is transpired directly back to the atmosphere, only to fall again as precipitation (Figure 10.14). One acre of rain forest releases approximately 76,000 liters of water per day into the atmosphere to form clouds. These clouds precipitate the water directly back to the forest and provide the abundance of water that is characteristic of the rain forest. To evaporate this much water from the ocean requires an area 20 times larger.

The elimination of a tropical forest disrupts the regional water cycle. The loss of forests results in the loss of soil moisture, and consequently, an entirely different ecosystem may develop in the region of the original forest. In deforested areas, less water is evaporated than before deforestation, and the recycling of water between the land and atmosphere and return is lessened by as much as 75%. Thus, deforestation keeps water from returning to the atmosphere, resulting in changes in a number of characteristics of the watershed, including erosion rates, temperature, and long-term climate of the region. For example, in Madagascar the regional climate has been severely affected because of the removal of more than 90% of the island's forests. Approximately 130 tons of soil per acre are now lost annually by erosion. This value is 260 times greater than the global average of about 0.5 tons per acre. Where once vast rain forests stood, only barren, eroded land now remains (see http://earthobservatory.nasa.gov/IOTD/view.php?id=3615). Obviously, such a severe change in environmental conditions has a negative effect on the productivity of the land and the life of the people.

BIODIVERSITY Forest biomes are home to many species of plant and animal life. The rain forests are one of the oldest biomes on Earth, dating back about 60 million years. Over millions of years, the relatively constant, hot, moist climate of this ecosystem has allowed for the extreme diversity of species by providing many ecological niches. Approximately half of the living species in the world are found in the rain forest. According to Caufield (1991), the rain forest contains from 20 to 86 species of trees per acre, as compared to a temperate forest that has only 4 species per acre. The North American temperate zone forest contains 400 species of trees, whereas the Madagascar rain forest contained 2000. The tropical forest in Panama has as many plant species as found on the entire European continent. Five hundred bird species, or four times as many as that found in all the temperate forests of eastern North America, were found in a 300-square-mile forest area between Panama and Costa Rica. A typical 4-square-mile area of rain forest contains 750 species of trees, 120 species of animals, 400 species of birds, 100 species of reptiles, 60 species of amphibians, 1500 species of flowering plants, and 150 species of butterflies. Many of the rain forest species are endemic to just one area. Because of this, there are relatively few individuals of some species.

Forest plants of the world have proved to be highly beneficial to people. Many of the plants have medicinal or agricultural value. People worldwide use 15,000 plants for medicine, food, and other purposes. Fifty percent of the medicines used originate from plants. Multiple sclerosis, Parkinson's disease, glaucoma, leukemia, high blood pressure, cancer, Hodgkin's disease, and amoebic dysentery are just a few of the diseases that have been treated using compounds derived from plants. However, less than 1% of the tropical plants of the world have been looked at in terms of their potential medical benefits. In addition to the wealth of medicines derived from the forests, almost 80% of the world's food stock is derived from the tropics. Table 10.3 gives the world export of crops originating in the

FIGURE 10.14 The water cycle in the Amazonian rain forest. Of the rain that falls, 75% of it returns to the atmosphere to be available again for precipitation, and 25% runs off in river flow.

tropical forests. In 1991, the value of the export sales of these crops approached $24 billion.

Soil, plant, and animal species and their habitats are being lost at an astonishing rate as a result of deforestation. As many as 4000 to 6000 species a year may be lost. This is a rate that is 10,000 times greater than naturally occurring extinction rates of the past. As an example of species loss, Figure 10.15a shows one interpretation of species extinction estimates for the tropical closed-canopy forests of the world from 1990 projected to 2040. At the current rate of deforestation, 15 to 35% of total species present in these forests could be lost by the year 2040 because of deforestation. Habitat destruction has been responsible for 36% of the known causes of animal extinctions since 1600 (Figure 10.15b).

Loss of habitat and species could have serious ramifications in the future for the sustainability of the ecosystems of the world. The Convention on Biological Diversity, formulated at the United Nations Conference on Environment and Development (UNCED, commonly called the Earth Summit) in Rio de Janeiro in 1992, provides a broad framework of national and international obligations to preserve biodiversity in

TABLE 10.3 World Crop Exports Originating from Tropical Forest Plants in 1991

Crop	Value (Billion Dollars)
Coffee	7.6
Citrus fruit	3.8
Rubber	3.4
Banana	3.1
Palm oil	2.8
Cacao	2.1
Pineapple	0.9
Vanilla	0.1
Total	23.8

(*Source:* Durning, 1993.)

tropical forests and in other ecosystems. It recognizes the intrinsic value of biodiversity, its importance for human welfare, and the sovereign right of a country over its own biodiversity and its responsibility for conserving it. Although still contentious, it appears to be a start toward preservation and conservation of tropical forests and other ecosystem habitats and biodiversity worldwide.

GLOBAL CARBON BALANCE Through the many millennia of time, carbon has been stored in the aboveground biomass of the forests and in the organic matter of their soils. This storage has helped to maintain the carbon cycle in balance and has enabled life to exist on the planet. The total carbon locked up in land vegetation is equal to 600 billion metric tons. The organic matter of soils and litter (leaves, stems, and their comminuted remains) contains another 1500 billion tons of carbon. In contrast, the atmosphere holds 800 billion tons of carbon. Thus, if all the carbon in terrestrial organic matter resided in the atmosphere, atmospheric carbon dioxide concentration would be increased nearly threefold. Each year, about 16%, or 120 billion tons of carbon as carbon dioxide, are taken out of the atmosphere through gross photosynthesis and converted into organic matter in leaves, wood, and roots of plants. This amount is nearly balanced by a return of carbon dioxide to the atmosphere by respiration of plants, animals, and microbes.

The removal of forests in North America and Europe during the past 200 years has led to an increase in the carbon content of the atmosphere. The current cutting and burning of the forest, in particular the tropical rain forests, contribute significantly to the amount of carbon dioxide released to the atmosphere by human activities. The amount of carbon released by deforestation practices is difficult to estimate. To do so requires better knowledge than we have at present of living and dead aboveground and belowground biomass, their carbon contents, and rates of deforestation. However, attempts have been made to estimate carbon emissions for both tropical and temperate forests. Carbon emissions from tropical deforestation have grown over time and for the period 2000 to 2008, an estimate of the average amount of carbon released to the atmosphere owing to tropical deforestation was 1.4 billion tons annually (Figure 10.16). Thus, about 15% of the total anthropogenic carbon dioxide emitted to the atmosphere by human activities is due to deforestation. At least 50% of this carbon emission originated from Brazil, Indonesia, Colombia, the Ivory Coast of Africa, and Thailand.

Figure 10.17 shows an interpretation of the history of carbon release from deforestation since 1850 and several scenarios for the future. The carbon release from deforestation is compared to that due to fossil fuel burning. Up until about 1900, it appears that the carbon flux from deforestation exceeded that of fossil fuel burning. The deforestation of temperate forests was the main source of carbon from deforestation practices until about 1935. Since then, the fossil fuel carbon flux has risen exponentially and the tropical forests have become the main source of carbon to the atmosphere from deforestation practices.

Changes in land use including, for example, deforestation and reforestation, have led to an increase in the global area of cropland, pasture, and shifting cultivation by 2230 million hectares and the release of about 100 billion tons of carbon to the atmosphere during this period of time. Also, about 23 billion tons of carbon were transferred from live vegetation to wood products and dead plant material. During this period, about 1070 million hectares of forest were logged. The combined processes of logging and plant regrowth contributed another 23 billion tons of carbon to the atmosphere. Thus, the combined processes of land use, logging, and the regrowth of vegetation resulted in the release of 123 billion tons of carbon to the atmosphere between 1850 and 1990.

It can be seen in Figure 10.17 that how the global community manages the forests of the world in the future will determine the amount and direction of the carbon flux between forests and the atmosphere.

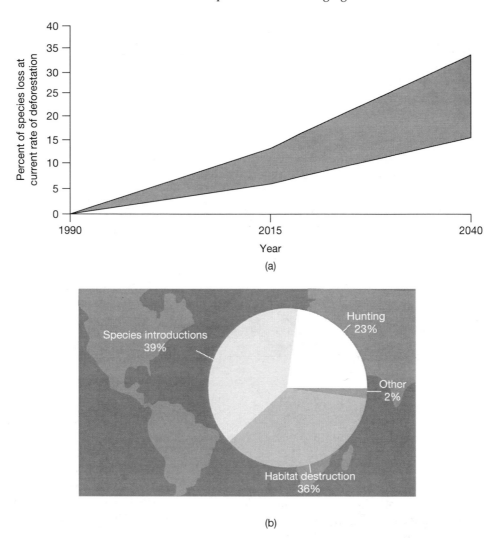

FIGURE 10.15 Species extinction estimates for tropical, closed-canopy forests from 1990 projected to 2040 as percent of species loss at the late twentieth century rate of deforestation. (After Whitmore and Sayer, 1992) and (b) the known causes of animal extinctions in percent since the year 1600. The worldwide current total loss of species is 4000 to 6000 species per year. (*Source:* World Conservation Monitoring Center, 1992; http://www.wcmc.org.uk/.)

There is both good and bad news in the early twenty-first century. The good news is that carbon stocks in North and Central America and Europe have increased for the period 1990 to 2005 (Figure 10.18). The bad news is that Africa, Asia, and South America lost considerable forest carbon, and for the world as whole, carbon stocks in forested biomass decreased by 1.1 Gt C annually during this period of time. Obviously exponential rates of deforestation will cause high carbon releases and a rapid loss of the world's forests. However, a massive reforestation program could lead to the forests of the world being a net sink of atmospheric carbon dioxide and to removal of carbon from the atmosphere. With present land use

practices, this latter scenario is unlikely to happen in the near future, but the rate of loss of carbon from the world's forests could be slowed with proper management techniques.

Carbon dioxide is accumulating in the atmosphere because of deforestation and fossil fuel burning. This increase in atmospheric carbon dioxide plus other factors have stimulated plant growth in undisturbed and regrowth forest ecosystems and led to carbon removal from the atmosphere, offsetting to some extent the large carbon releases from tropical forests. Changes in the heat balance and climate of Earth may result from emissions of trace gases, such as carbon dioxide, methane, and nitrous oxide, into

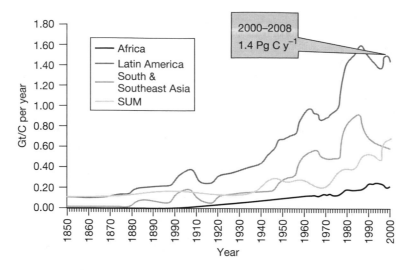

FIGURE 10.16 Historical emissions of carbon dioxide to the atmosphere from tropical deforestation practices in billions of tons of carbon (Gt C) per year for the world's tropical forests and for African, Latin American, and South and Southeast Asian tropical forests. For the period 2000 to 2008, the average emission rate was 1.4 Gt C/yr. (*Source:* R. A. Houghton, unpublished; http://www.globalcarbonproject.org/carbontrends/index.htm.)

the atmosphere from deforestation and other land use activities. This accumulation is the backbone of the problem of the enhanced greenhouse effect and global climatic change (see Chapter 14).

Causes of Deforestation

The tropical forests of the developing world are rapidly being deforested because of complex social, economic, and political issues of both developed and developing nations. The land is deforested for debt payment, for monetary profit for a few, to maintain land ownership, to provide lumber for resettlement and migrant farming, for conversion to cattle pasture and agricultural land, for fuelwood, for industrial and urban expansion, for hydroelectric dams, and for export. Table 10.4 lists the major factors involved in forest loss; some causes are discussed in more detail in the following sections.

DEBT PAYMENT The developing countries have aspirations similar to those of the industrialized world for their future development, but they are also experiencing rapid population growth. Providing electricity, medical services, housing, food, and all the necessities of the developed world for the burgeoning populations of developing countries requires money. Money has been borrowed by the developing nations from the world community, the wealthy nations of the North. Many developing countries cannot repay

the loans. The total debt of the 18 most highly indebted nations approximates $40 billion. The interest payments alone on this debt require large sums of cash. One source of revenue for the developing countries is from the sale of their natural resources, such as timber and minerals.

LAND DISTRIBUTION AND RESETTLEMENT It has been argued that overpopulation in developing countries is the cause of emigration of people into forested areas. Overpopulation is certainly a major concern, but the actual problem is a combination of overpopulation and land distribution. In the developing countries of the South, a very small percentage of the populace owns most of the land and controls most of the wealth. For example, in Latin America 7% of landowners possess more than 90% of the arable land. In Brazil, 4.5% of Brazilians own 81% of the farmland. Eighty-five percent of the people of Java are essentially landless, while one-third of the land is in the hands of 1% of the people. In El Salvador, 2000 families own 40% of the land. The list goes on.

New land is needed for the growing population of developing countries. As a result, the populace is encouraged to settle the forested areas. Roads are built, opening up the forests for settlement purposes. The landless people venture into the forests, clearing the land and searching for the possibility of a better life. Unfortunately, the land of the tropical forests is

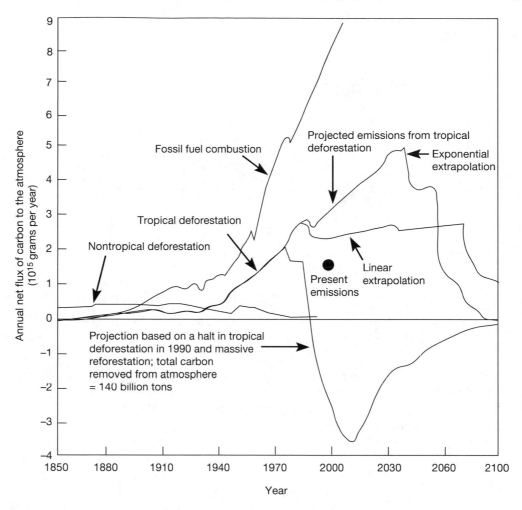

FIGURE 10.17 Emissions of carbon to the atmosphere from nontropical deforestation, tropical deforestation, and fossil fuel burning. Future emissions from tropical deforestation are model projections based on linear and exponential extrapolations of deforestation rates and on a halt in tropical deforestation in 1990 and a massive reforestation program that withdraws carbon from the atmosphere (negative values).   (*Source:* Houghton, 1995.)

very infertile, and it is difficult to make a living on the degraded soil of the deforested land.

Forest dwellers worldwide are severely affected by present practices of land use. Indigenous people generally have not been consulted when land use decisions are made. Tribal lands have been destroyed and confiscated and whole tribal cultures have disappeared. In A.D. 1500, the indigenous people of the Amazonian Basin may have numbered 6 million to 9 million people. By 1900 Brazil had only a million indigenous people. In the 1990s, the number was about 200,000. There are about 1000 indigenous tribes left throughout the world. Many are on the verge of disappearing as unified entities, in part because of loss of land and the resources provided by the land.

CONVERSION OF FORESTS TO CATTLE PASTURE AND AGRICULTURAL LAND Forest and grasslands are burned and then converted to cattle pasture and agricultural land, especially in Latin America. Pasture expansion began in Brazil in the 1960s because of construction of paved roads, lucrative financial incentives, tax credits, and land offerings to investors. Tax credits were eliminated in 1989 but expansion in usage of land for cattle grazing continues.

The grasses for cattle grazing grow well the first few years after burning. However, the tropical soil generally has little nutrient-retaining capacity and is gradually degraded over a few years. The cattle require larger and larger areas to feed because of the nutrient-poor soil, and hence, lack of sufficient and

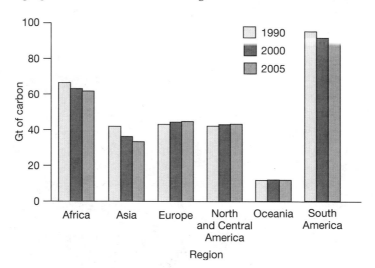

FIGURE 10.18　Changes in the carbon stocks of forest biomass in gigatons of carbon (Gt C) from 1990 to 2005 for various regions of the world.　(From *Global Forest Resources Assessment, 2005,* ftp://ftp.fao.org/docrep/fao/008/A0400E/A0400E00.pdf. © FAO 2006. Reprinted by permission of the Food and Agricultural Organization of the United Nations.)

good-quality grass. More land is deforested, and the cattle move on, leaving degraded land behind.

Forests are also cut to provide land for farming. There is concern by some that pressure is growing for the conversion of forests and other environmentally valuable land to sugarcane for the purpose of producing ethanol. However, in Brazil, the world's second largest producer of ethanol fuel and its greatest exporter, most of the sugar cane plantations are located far from the Amazonian rain forest and expansion of sugar cane area during the past 25 years has taken place mainly in the south-central part of Brazil.

INTERNATIONAL LOGGING　Despite the international outcry, logging is increasing in the developing countries. In the 1980s, Asian nations supplied 80% of tropical hardwood exports, and nearly two-thirds of all kinds of timber used for international trade. Indonesia has already allocated most of its forest to international timber corporations. Much of Indonesia may be deforested by the early twenty-first century, and the Asian/Pacific countries will supply only 10% of total tropical hardwood export, owing to the depletion of their hardwood trees.

Why is this happening? Third World nations need foreign skills and capital and often sell their natural resources in exchange. In this case, their forests are sold to politically and economically powerful foreign timber corporations. The major concerns of the corporations to date have been those of monetary profit and

how long the politically unstable countries in the Third World will continue to exist in their current political state and honor logging agreements. Also, deforestation is proceeding at a rapid rate because there is a fear of new restrictions on logging owing to the international outcry against deforestation practices.

Some timber corporations have paid little attention to the developing countries and the potentially devastating future impact for their economies of the elimination of part of their natural resource base. Forests are being depleted faster than nature can replenish them. The foreign corporations are generally allowed to do what they wish because they provide enormous investment capital and build roads for their logging needs. These roads provide access to new land for the increasing population in the developing countries. The economic benefits of logging rarely help the local community or government in the long run. Most of the money is made by timber corporations, the wealthy few of the developing nations, and the politicians or military officers who grant logging concessions.

The nation of Japan is over 60% forested, more than any other industrialized nation outside of Finland in Scandinavia. However, its importation of wood has been greater than that of any other nation. With only 7% of the world's population, Japan consumes about 50% of the total tropical timber that is cut. Approximately half of the 50 major logging companies of the world are owned by the Japanese. The World

TABLE 10.4 Major Factors in Forest Loss

Underlying Causes	Direct Causes
Population change:	Agricultural clearing of forests:
Growth	Subsistence agriculture
Density	Commercial farming
Migration	Cattle ranching
Economic growth:	Industrial logging
Rising income	
Dietary change	Infrastructure and industrial development:
Housing preference	Roads
	Dams
Poverty	Mining
	Housing
Market failures:	
Inadequate property rights	Fuelwood and charcoal production
Inappropriate valuation of forest goods and services	
Policy failures:	
Price and taxation policies	
Population resettlement programs	
Corruption	

Wildlife Fund concluded that mainly these lumber companies, with enormous capital invested in Southeast Asia, are practicing deforestation with erosional and flooding consequences for the region.

The tropical forests in Southeast Asia appear to be on the road to total extinction by the early part of the twenty-first century. However, an abundance of these forests will still exist in South America. The major concern for the Amazonian rain forest, unless forest practices change more rapidly, is continued destruction of this important ecosystem.

HYDROPOWER The history of construction and utilization of hydroelectric dams has been one of mixed results. Certainly, these dams have provided the energy necessary for the economies of the developed and developing countries. The potential for deriving energy from dams in Brazil and other countries with large river systems is great. However, dams have not had a completely positive impact on the forest environment in some countries.

Deforestation taking place concurrently with the construction of dams for hydroelectric power, irrigation, and flood control purposes also contributes to a variety of social and environmental problems. These problems include water acidity, human diseases, the displacement of indigenous people from the land, flooding of original waterways with resultant loss of prime land, and the siltation of dam reservoirs and drainage areas. The effect of erosion and resultant siltation of dam reservoirs is further discussed in Chapter 11.

The construction of dams on the waterways of tropical forests leads to extensive flooding of land behind the dams, resulting in decomposition of submerged trees in the standing water. The extent to which tropical forests decompose after flooding affects the quality of the water. If the flooded area is deep and stagnant, the water may become acidic, leading to death of aquatic life. If the area is shallow, the decaying trees release nutrients when they decompose, and the waters are an excellent habitat for many water-weeds. These waterweeds absorb the excess nutrients and grow rapidly. Other plants are crowded out, and fish that depend on them die off.

In Latin America, the contained waters behind the dams provide breeding sites for the carriers of two major diseases: malaria, passed on to people by the bite of mosquitoes, and schistosomiasis, carried by a parasitic flatworm that can penetrate a body immersed in water. The malaria-carrying mosquito generally lives high in the trees and feeds on mammals in the tree canopy. When the trees are felled, mosquitoes descend to the ground to prey on other animals, including people. Careful future planning could eliminate some of the negative consequences of dam construction that have arisen in the past.

FUELWOOD DEMAND The lack of fuelwood is a major problem for the developing countries. Two billion people in the developing countries, or 30% of the world population, depend on this dwindling resource to provide for their basic needs of heating and cooking. About half of these 2 billion people cannot find enough wood for their needs. As the population grows and the need for this natural resource increases, the supply will steadily decline, unless fuelwood is replaced by another energy source. The trees simply cannot grow fast enough to satisfy the needs of the growing population. In the early twenty-first century, approximately 1900 million cubic meters of fuelwood and charcoal were consumed worldwide; about 75% of this consumption was in Africa and Asia.

Summary

The 4 billion hectares of forested land on the planet cover 30% of the total land area. This amounts to 0.62 hectares per capita globally but the area of forest is unevenly divided, with only 10 countries accounting for two-thirds of total forested area. Primary forests accounted for 36% of total forest area in 2005 and were being lost at a rate of 6 million hectares per year from 2000 to 2005. Total forested area is decreasing due to deforestation, mainly the conversion of forests to agricultural lands, at a rate of 13 million hectares per year. But the rate of deforestation appears to be slowing in the first decade of the twenty-first century, despite strong demand for additional land for biofuel crops like sugarcane. Forest planting, landscape restoration, and the natural expansion of forests are responsible for the reduction in the *net* loss of forested area. Net change in forested area in the period 2000 to 2005 was estimated as −7.3 million hectares per year, down from −8.9 million hectares per year for the period 1990 to 2000. South America suffered the greatest regional net loss of forest area, about 4.3 million hectares per year from 2000 to 2005, followed by Africa with 4 million hectares per year.

Forests in many countries are managed by professional foresters or are under some type of governmental control and protection. About 10 million people worldwide are employed in forest conservation and management activities. Forest plantations constituted less than 5% of total forested area in 2005 but for the period 2000 to 2005 increased in area by 2.8 million hectares per year. Forests designated for conservation have increased in all regions of the world from 1990 to 2005. In the early twenty-first century, the total coverage of protected land areas was estimated at 348 million hectares, with 11% of the world's forests designated for the conservation of biological biodiversity. Some small but poorly known forested area is used for recreation and education globally. In Europe (excluding the Russian Federation), 72% of the forested area provides social services. Overall, roughly 6% of global land area is maintained in a natural state and closed to exploitation or extraction uses. However, one-third of the world's forested area is still used for the production of wood and nonwood forest products.

Forests constitute a vital sink of atmospheric carbon dioxide. The world's forests store about 280 Gt C in living biomass. Forest biomass carbon decreased in Africa, Asia, and South America during the period 1990 to 2005 and increased in other regions. Overall for the world, carbon stocks in biomass are still decreasing owing to continued deforestation and forest degradation, in part offset by expansion of forested area and by increases in the rate of growth of stock per hectare in some regions.

There are differences between developed and developing countries in forestry practices that lie in complex political, economic, and social policies. The World Bank has funded many projects in developing countries that initially were considered worthwhile from an economic and social viewpoint. In hindsight, however, some of these projects proved to be ecological and social disasters. The consequences of the construction of huge dams and roads built into the interior of the forests that opened up vast areas to settlement and deforestation were, to some extent, unforeseen. These include land erosion, loss of species diversity, carbon dioxide and other trace-gas emissions to the atmosphere, and the displacement of indigenous people. The world community and environmental groups both within and outside the countries affected are taking a different view of the forests and are attempting to establish new and better forest-management programs.

The world community's reaction to the deforestation of the tropical forests and the lack of management of these forests has been one of extreme criticism directed at the developing countries. The developing countries of the world have reacted to the world community's interference in their practices and policies. They point out that the industrialized world has already deforested their own temperate forests and currently uses most of the energy of the world, leading to the most talked-about problem of human-induced global environmental change, that of global climatic change. The developing countries, with their growing populations, high poverty levels, and high industrialization needs, feel they should be allowed to develop their resources with no interference from the developed countries (Figure 10.19).

This schism between the two "Worlds" was amply demonstrated at the United Nations Conference on Environment and Development (Earth Summit) held in Rio de Janeiro in 1992. A weakened climate convention was signed by the nations of the world, but consideration of a forest convention was put off into the future, and as of 2009, one has not been put in force.

DOMESTICATED ECOSYSTEMS: AGROSYSTEMS AS AN EXAMPLE

Natural ecosystems contain myriad organisms that interact with one another for their mutual benefit while developing ways to adapt to any infringement on their ability to survive. In general, these systems are not managed by some external force like humans. In contrast, "domesticated" ecosystems are heavily managed and controlled by humans. Forest plantations are an example of a domesticated ecosystem; agroecosystems are another. In terms of acreage, croplands and pastures are now among the largest ecosystems on Earth and occupy about 35% of the ice-free land surface (Figure 10.20). Agroecosystems provide the

FIGURE 10.19 Cartoon illustrating the perceived difference between the way in which the developed and developing countries of the world view deforestation.   (Scott Willis © 1989 Used with permission of San Jose Mercury News.)

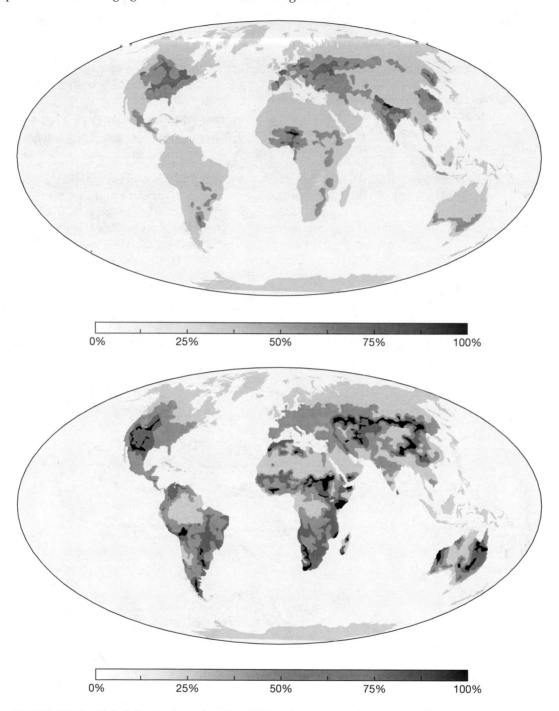

FIGURE 10.20 Global distribution of croplands (a) and pastures and rangelands (b) in percent. Approximately 35% of the world's ice-free land area was devoted to agriculture in the year 2000. These managed ecosystems are among the largest on Earth.   (*Source:* Foley et al., 2007.)

overwhelming majority of crops, livestock feed, and livestock on which humans depend. Globally, agricultural production provides most of our animal and plant protein consumption and 99% of all the calories consumed by humans. Agroecosystems also provide

most of the fiber used by humans in the form of cotton, flax, hemp, jute, and other fiber crops.

Agroecosystems survive because they are carefully tended and nurtured by humans. New food crops developed and grown in the 1950s in North

America and Western Europe produced higher yields than older varieties and led to the Green Revolution. The Green Revolution spread to Asia and Latin America in the 1960s. High-yield varieties of wheat, rice, and corn are now staple crops worldwide. The history of world grain production and grain production per capita since 1950 is shown in Figure 10.21a and b, respectively. World grain production has generally increased during the past 50 plus years. In order to maintain high crop yields to achieve this growth of world grain production, it has been necessary to apply substantial amounts of fertilizers, pesticides, and irrigation water to croplands. The application of these materials to croplands has inadvertently led to several environmental problems, and despite their application, world grain production per capita and world grain reserves (Figure 10.21b, c) have fallen erratically since the late 1980s. This is due at least in part to the rapidly increasing world population, the inability of nutrient subsidies to increase crop yields substantially, declining availability of cropland area worldwide, and soil degradation, which has slowed crop yields in some regions of the world. However, overall *global food production* has grown faster than population in recent decades mainly due to improved seeds and increased use of fertilizer and irrigation.

Fertilizers

There has been considerable concern in recent decades over the adequacy of global food supplies and the costs of food to consumers. In addition, there has been increased use of cropland in the United States, Brazil, and elsewhere for corn and sugar cane crops to use to produce ethanol for fuel (see example for the United States in Figure 10.22). These factors have spawned an increased interest in the use and role of fertilizers in agricultural and food production.

Farmers around the world have used natural organic fertilizers for centuries. People have recycled nutrients by applying animal dung and carcasses to croplands to obtain better crop yields. When the first settlers arrived on the American continent, the North American Indians showed them how to bury a fish with each maize seed planted. The decaying fish provided nutrients for the soil in which the maize grew. In medieval times, European farmers applied dung to the soil and planted nitrogen-fixing legumes, such as clover, on agricultural land to increase the productivity of the soil. By the time of the Industrial Revolution, the use of organic fertilizers was a well-established practice. However, the reasons for the

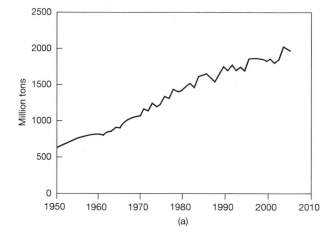

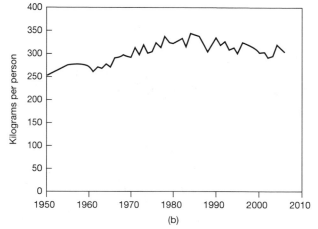

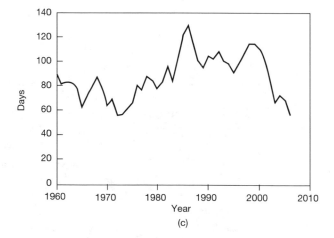

FIGURE 10.21 Trends in total world grain production and production per capita from 1950 to 2006 (a, b) and world grain reserves (stocks) from 1960 to 2006 (c). Notice the general overall decline in grain production per person since the mid-1980s despite the continuous rise in total world grain production. Reserves dropped about 54% between 1987 and 2006. (From "World Grain Stocks Fall to 57 Days of Consumption: Grain prices Starting to Rise" by Lester R. Brown, *Energy Bulletin*, June 15, 2006. Copyright © 2006 Earth Policy Institute, www.earthpolicy.org. Used by permission of Earth Policy Institute.)

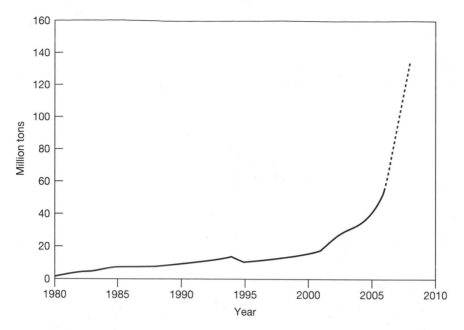

FIGURE 10.22 The use of corn for ethanol in the United States from 1980 to 2008. (*Source:* United States Department of Agriculture; after http://www .earth-policy.org/.)

effectiveness of these chemical compounds in stimulating plant productivity were not well understood. In 1840, Justus von Liebig discovered that nutrients removed from the ground by crops could be chemically replaced. This discovery eventually led to the expanded use of chemical fertilizers. Such use has been both a blessing and a curse.

Prior to the Green Revolution of the 1950s, most increases in crop yield were due to expansion of areas of cultivated land. After 1950, crop-yield increases were mainly the result of chemical fertilization of the existing land. Nitrogen, phosphorus, and potassium are the three basic plant nutrients that are generally applied to the soils. These nutrients are naturally found in organic matter. Most of the phosphorus and potassium used today as fertilizer is the product of mining of ores of these elements, and the nitrogen, principally as ammonia, is the result usually of an industrial chemical process. In this process, known as the Haber-Bosch process, atmospheric nitrogen is reacted with hydrogen derived from natural gas. This reaction of nitrogen fixation is

$$N_2 + 3H_2 = 2NH_3.$$

The product is ammonia, which is then used as a fertilizer in liquefied form or as a salt.

The nutrient compounds that constitute fertilizers are generally mixed in various proportions and then sold commercially. Written on the product labels, for example, may be the numbers 20–15–20. This nomenclature means that 20% of the bulk of the fertilizer is nitrogen, 15% is phosphorus pentoxide (P_2O_5), and 20% is potassium oxide (potash).

In 1950, about 14 million tons of chemical fertilizers were applied to the world's crops. By 1990, this figure had increased to 143 million tons (Figure 10.23). This amounted to about 60 million tons of nitrogen in 1980, with an increase to about 80 million tons or about 5 grams of nitrogen per square meter of agricultural land in 1990. During 1990 to 1995, fertilizer use dropped due to an economic downturn to 122 million tons per year but increased again to 141 million tons per year in 2000 and to 212 million tons per year in 2008. This increase was almost entirely due to nitrogen, with 136 million metric tons of nitrogenous fertilizer used in 2008. As a result of fertilizer application, crop yields have increased, as has the expanded use of energy to mine, produce, and transport these fertilizers.

In the last 40 years, global cropland production has doubled but the cropland area has increased by only 12%. The total area harvested for grain has decreased about 10%. Because of the growing world population, the area harvested per person fell by 48% between 1955 and 1995. Many of the world's croplands are being used more intensely as the opportunities for

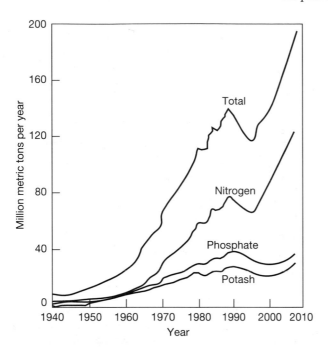

FIGURE 10.23 World total, nitrogen, phosphate, and potash fertilizer consumption in metric tons/yr from 1940 to 2008.

world population, although this production is not always distributed wisely.

To increase per capita food supplies during the last half of the twentieth century, it has proven necessary to develop new high-yield varieties of crops and to use large amounts of fertilizer and water. The per capita use of fertilizer increased during the last half of the twentieth century, decreasing significantly in the developed world in the early 1990s. However, the rate of per capita fertilizer use in the developing world continues to increase (Figure 10.24). In the mid-1980s, China surpassed the United States in both grain production and fertilizer use, and India in the year 2000 used as much fertilizer as the United States. The Asian region is becoming a major consumer of commercial fertilizers (Figure 10.25), mainly nitrogenous fertilizers, and the Far East consumes nearly 60% of this fertilizer. Thus, it is very likely that commercial fertilizer production and use will grow on into the twenty-first century. Excess nitrogen- and phosphorus-bearing fertilizers added to croplands may be leached and washed into rivers and groundwaters and transported by these flows downstream to lakes or coastal marine areas, where the nutrients may cause problems such as eutrophication (see Chapter 11).

expansion are shrinking. Some cropland regions show increased net primary production over natural NPP rates while crop NPP in other regions—particularly India, and parts of Africa, Russia, Asia Minor, South and North America—is less than natural NPP rates. However, overall, chemical fertilization, irrigation, and increased mechanization have allowed the world to continue to produce sufficient food for the expanding

Pesticides

Agricultural pests such as fungi, insects, rodents, and weeds are responsible for reducing world food production by about half. As a result, the effort to keep pests at bay is given high priority by farmers. Pesticide

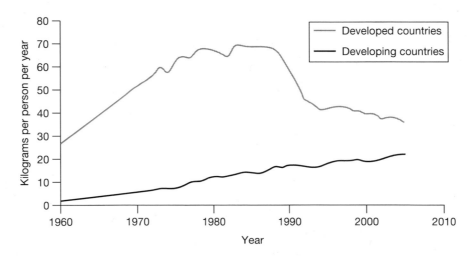

FIGURE 10.24 Per capita fertilizer use in developed and developing countries in kg/capita/yr from 1960 to 2005. Notice the sharp decline in per capita fertilizer use in 1988 for the developed world. (Data from http://faostat.fao.org/site/422/Desktop Default.aspx?PageID=422#ancor and http://esa.un.org/unpp/).

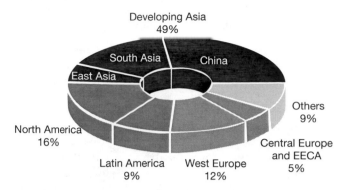

FIGURE 10.25 Regional distribution of the use of fertilizers in the world in percent averaged for the period 1998 to 2001. Notice that China, South and East Asia used nearly 50% of the fertilizer at this time and the use of fertilizer continues to grow in these regions. (*Source:* http://www.fertilizer.org/.)

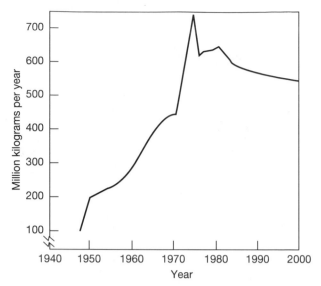

FIGURE 10.26 Annual production of commercial pesticides in the United States in kg/yr from 1948 to 2000. Notice the decline in pesticide production since the early 1970s.

chemicals, which include insecticides (for insects), herbicides (for plants), and fungicides (for fungi), have been used extensively. Different measures to control pests have been employed by people since the beginning of agriculture. In 1939, Paul Muller developed the synthetic pesticide DDT (dichlorodiphenyl-trichloroethane). This pesticide helped to prevent the destruction of food crops by pests and diseases carried by mosquitoes. DDT was the first of the synthetic organic pesticides.

Modern synthetic pesticides, especially the organophosphate insecticides, were developed from chemicals tested for use in the manufacture of nerve gas weapons during World War II. In the 1940s, industrialized countries began to use these substances for the control of agricultural pests. Subsequently, their use increased substantially. Currently, more than 33,000 commercial pesticides are in use. In 2000, 2.3 million tons of pesticides valued at about $30 billion were consumed worldwide. This represented an increase in pesticide consumption of 240% since 1970. Eighty percent of the world's pesticide production is consumed in North America, Western Europe, and the Far East. In the United States alone, 860 active-ingredient pesticide chemicals are formulated into 21,000 commercial products, and about 546 thousand tons of pesticides are produced annually (Figure 10.26).

On a global scale, most of the pesticides used in agriculture are applied to a small number of crops—corn, rice, cotton, soybeans, and wheat. Despite Rachel Carson's warnings in her book *Silent Spring* (1962) concerning the environmental effects of pesticides, many of the toxic pesticides she discussed are still in use. One important exception is DDT, which because of its adverse effects has been banned in most countries of the world. Pesticides, as with fertilizers, may enter the water cycle and be carried far from their source of application and enter biological systems. Organochlorine-based insecticides are volatile and are transported through the atmosphere as well as by water. These chemicals have been found in tree bark collected from every continent, with relatively high concentrations in the United States, Europe, the Middle East, and China, and low concentrations at high northern latitudes. Pesticides are an important source of water pollution in the United States and elsewhere. It has been estimated that, worldwide, 500,000 to 2 million pesticide poisonings occur each year. In the United States, perhaps as many as 45,000 farm and chemical workers annually suffer pesticide poisoning.

The percentage of crops lost to pests during the 50 plus years of the Green Revolution has not declined despite the large amounts of pesticides used on croplands. This has led many persons to suggest that the use of pesticides might not be the most efficient way to control pests in agroecosystems. The growing concern over the biological and ecological effects of pesticides in the environment and their efficacy in controlling pests have led to increased awareness and caution in the use of pesticides in the developed countries. However, such caution is only slowly spreading to most developing countries.

Irrigation Systems

Irrigation systems have greatly contributed to the growth in crop production. Water from rivers and dammed reservoirs and water withdrawal from groundwater systems have significantly enhanced the quality of agricultural lands. Agriculture worldwide accounts for about 65% of the total annual demand for water of 4430 cubic kilometers, or about 2880 cubic kilometers of water (see Chapter 11, Table 11.1). This is equivalent to a per capita withdrawal rate of 505 cubic meters per year. Industry accounts for most of the remaining water use (22%) worldwide, with only 7% withdrawn for domestic use. In the United States, nearly equivalent volumes of water are withdrawn annually for agricultural and industrial use. Domestic yearly use accounts for about 12% of total water withdrawal of 467 cubic kilometers, or 216 cubic meters per person. Water use is rapidly lowering groundwater tables in some regions of the world as agriculture, industry, and domestic uses compete for water supplies. Chapter 11 discusses further fresh water use and consumption.

Food Resources

Less than one-tenth of 1% of all plants and animals of the world are used by humans for food. There are approximately 75,000 edible plant species on Earth. However, people use only about 5000 species for food. Thirty crops provide 95% of our nutrition; wheat, rice, maize, potatoes, barley, sweet potatoes, and cassava are

notable (Figure 10.27). In addition, only a few domesticated animals provide all of our meat needs, such as pork, beef, poultry, and lamb. Only a few species of fish and shellfish are products of aquaculture, and a limited number of fish species are hunted for food.

Modern-day agricultural overreliance on only a few species of crops creates concern because of the potential loss of genetic resistance to disease or pest infestations of these crops. If a plant of one species becomes vulnerable to a pest or disease, other plants may be equally affected. New single or monospecies often have little resistance to pests and disease. The simplified human-maintained ecosystems remove the natural forces that keep disease and pests in check. Problems develop when the system is not managed properly, which may lead to the system becoming unbalanced and degraded. For example, the potato blight, a disease of the potato plant, affected Ireland in 1846. Potatoes were a major source of carbohydrates for the Irish, but only a few species were grown in Ireland at the time of the blight. The potato infestation caused the death of many people by starvation. Others abandoned their homeland and emigrated to the United States and Canada. However, there are more than 200 known species of potato plants, mostly available from Peru. If Ireland had had a greater number of potato species, some might have been immune to the blight.

Other examples of crop loss owing to cultivation of restricted varieties of agricultural crops abound. The USSR experienced a wheat crop loss in 1972, and citrus canker seriously affected orange and grapefruit

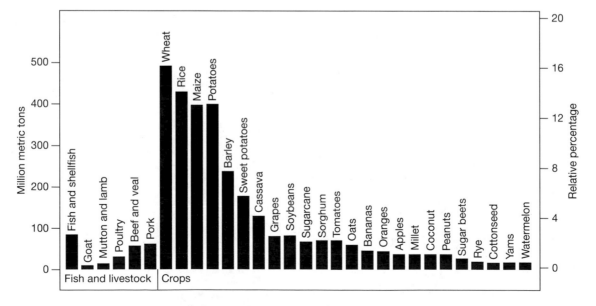

FIGURE 10.27 Production of crops, fish, and livestock. Annual production of 24 crops in 1980 was 2.5 billion metric tons. There are potentially 75,000 food crops available for consumption. However, the people of the world depend heavily on only a few crops and livestock for their diet and nutrition. (*Source:* Myers, 1984.)

trees in Florida in 1984. In 1970, the United States lost one-quarter of its corn crop to a leaf blight, and in 1980 the peanut crop of the United States, which consisted of only two varieties, was almost entirely destroyed.

Strains of wild varieties of plants can be crossed with cultured varieties to enhance growth and protect plants from diseases and pests. This crossbreeding enables monocultured plants to receive the protection that nature accords the wild varieties.

The wild varieties of our major sources of plant food come from all parts of the world (Figure 10.28).

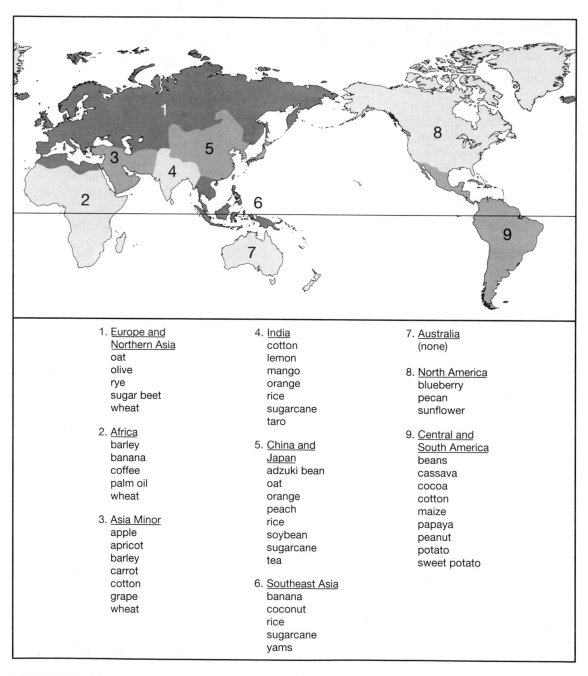

1. Europe and
   Northern Asia
   oat
   olive
   rye
   sugar beet
   wheat

2. Africa
   barley
   banana
   coffee
   palm oil
   wheat

3. Asia Minor
   apple
   apricot
   barley
   carrot
   cotton
   grape
   wheat

4. India
   cotton
   lemon
   mango
   orange
   rice
   sugarcane
   taro

5. China and
   Japan
   adzuki bean
   oat
   orange
   peach
   rice
   soybean
   sugarcane
   tea

6. Southeast Asia
   banana
   coconut
   rice
   sugarcane
   yams

7. Australia
   (none)

8. North America
   blueberry
   pecan
   sunflower

9. Central and
   South America
   beans
   cassava
   cocoa
   cotton
   maize
   papaya
   peanut
   potato
   sweet potato

FIGURE 10.28   The great genetic treasure map. This map shows the different areas around the world that hold the greatest concentration of germ plasm important to modern agriculture and world food production. Although evidence indicates that some of the crops listed originated in their respective areas, no one knows for sure exactly where most crops first got started.   (*Source:* Newman, 1990.)

The loss of these wild varieties through the destruction of their ecosystems can have an indirect effect on the future world production of food. In addition, developing countries now want monetary reimbursement for any biotechnology developed from the wild varieties that they hold in reserve. On a positive vein, in 2008 a vast underground vault, which will store millions of seeds, was opened on the remote Svalbard Islands between Norway and the North Pole. It is anticipated that the Svalbard Global Seed Vault or as called by some, the "Doomsday Vault", will be a safety net for the world's seed collections and support the long-term conservation of crop biodiversity, protecting them from a wide range of threats.

Concluding Remarks and the 21st Century

Land use changes, particularly in the latter half of the twentieth century, have become an important factor in the ecosphere of the planet. The changes have had obvious beneficial results, including increased food production, construction of shelter for human beings, increased and more efficient transportation systems, development of energy and mineral resources, and provision of land for industrial growth. At the same time, these changes have modified fluxes of gases to the atmosphere and led to deterioration of the land surface and loss of forests, habitat, and biodiversity. The increased gas fluxes owing to forest and other biomass burning are partly responsible for the changing chemistry of Earth's atmosphere and the potential of climatic change. Land use and degradation have led to enhanced riverine fluxes of nutrients and suspended sediments to freshwater and marine ecosystems and the pollution of these water bodies.

There is much controversy concerning how many mouths the planet can feed. This controversy is not new and dates back almost 200 years to the work of Thomas Malthus. One argument is that the Green Revolution has ceased or at least slowed, and it will be very difficult to feed the expanding population of the twenty-first century. One certainty is that if the world grain production per unit area per person continues to decrease into the twenty-first century, there will be need for additional fertilizer, water, and pesticides to increase grain yields and/or for new high-yield crops. The grain area supporting each person in the world fell from 0.23 hectares per person in 1950 to 0.12 hectares in 2000, less than one-sixth the size of a soccer field. It is projected to decrease to 0.08 hectares per person in 2030 (Figure 10.29a). World consumption of nitrogen fertilizer was about 10 kilograms per capita in 1970 and is projected to grow to 16 kilograms per capita in 2020. Projections of total nitrogen fertilizer consumption to the year 2020 show that the reliance of Asian countries on commercial fertilizers will grow significantly in the future (Figure 10.29b).

Continuous loss of cropland due to urbanization, groundwater depletion and diversion, soil erosion and other degradation (Chapter 11), and use of farmland for nonfood crops, such as cotton and coffee, will directly affect the ability of farmers to feed the growing world population in the twenty-first century. Also the increased use of feed to produce beef, which consumes 12.7 kg of feed per kg of weight (18 times more than that needed to produce eggs), will require more cropland or higher yields. The protein conversion efficiency for beef is only 5%, roughly three times less than that of pork, 5 times less than that of chicken, and 6 times less than of carp fish and eggs. In other words, it takes a lot more feed to produce a pound of beef than a pound of carp. The additional nutrients applied to cropland to increase grain yields and produce beef and other products relying on grain will be of concern because of the potential of leaching of the nutrients into aquatic systems and the problem of eutrophication (Chapter 11) and of emissions of oxides of nitrogen to the atmosphere from the applied nitrogen fertilizers (Chapters 12 and 14).

Another analysis contends that grain yields per hectare have increased significantly during the past few decades and into the twenty-first century in many countries around the world. It has been argued that better distribution of improved crop varieties, farm machinery, and technical knowledge may increase worldwide grain harvests by as much as 50% in the future. This predicted increase is in spite of the environmental ills of soil degradation, depletion of groundwater for irrigation, and loss of arable land. Whatever the case, even if yields can be increased and farming soils can be suitably maintained for the growth of crops, world hunger will not end. Much of the problem of curtailing world hunger is not that related to a lack of agricultural capacity but one related to the inefficient distribution of food and the low purchasing power of the Third World countries.

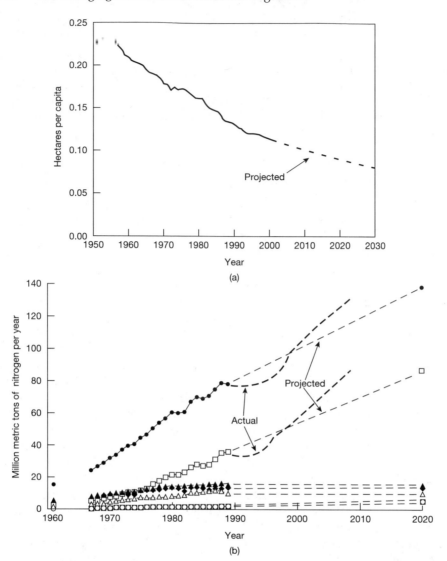

FIGURE 10.29 Some aspects of the world's croplands. (a) World grain area in hectares harvested per person from 1950 to 2000 and projected to the year 2030. (*Source:* Gardner, 1996.) (b) Total nitrogen fertilizer consumption from 1960 to 1989 and projected to the year 2020 for the world (filled circle), Asia (open square), Europe (filled triangle), North and Central America (filled diamond), former USSR (open triangle), Africa (filled square), and South America (open diamond). Notice the difference between the projections for the world and Asia between 1989 and 2008 and actual nitrogen consumption. (*Source:* Galloway et al., 1995.)

End Note

On completion of this book, State of the World's Forests 2009 was published by the Food and Agricultural Organization of the United Nations. The interested reader is referred to this document for the most up-to-date information on the world's forest resources.

Study Questions

1. The total area of forest plus woodland in one estimate is 5.3 billion hectares. Thirty-two percent of this area is listed as woodland. How many square meters of woodland are there?

2. The average specific biomass of a woodland as dry matter (DM) is 6 kilograms per square meter.
   (a) How much dry matter is present in the woodlands of the world?

(b) How much carbon does this represent? (The conversion factor for DM to carbon is 0.45.)

3. (a) If all the woodland in question 2 were burned, how many grams of carbon dioxide could be released to the atmosphere?

(b) What percentage is this flux of the present atmospheric carbon dioxide reservoir of 830 billion tons?

4. What are the two major types of higher plants of the kingdom Plantae, and how do they differ?

5. From an environmental standpoint, what are two important reasons to know the areas of forest land?

6. Where are the three major regions of tropical forests located?

7. How do forest practices in the developed and developing world differ, and why the worldwide concern with forest practices in the developing countries?

8. (a) How does nutrient cycling in the tropical rain forest differ from that in temperate forests?

(b) Why is this difference so critical to recovery of the two systems after deforestation?

9. How does deforestation affect the water cycle of tropical rain forests?

10. (a) What is one reason for the large biodiversity of the tropical rain forests?

(b) If 10 square kilometers of rain forest were deforested, how many species could be affected?

11. What are five major effects of deforestation of the tropical rain forest?

12. In Figure 10.17, it is shown that a massive reforestation program could lead to removal of 140 billion tons of carbon from the atmosphere. If all this carbon were stored as dry matter in forests with an area of 4.08 billion hectares, what would be the specific additional biomass in kg DM/m^2?

13. What are five major causes of deforestation in developing countries?

14. The worldwide deforestation rate of the tropical forests is about 0.8% per year. With a total forested area of 1.3 billion hectares, how many square meters of forest are deforested annually?

15. What major steps could be taken to lessen the rate of destruction of the world's forests and at the same time be economically practical to both developed and developing nations?

16. What three materials have been important to the production of crops with high yields?

17. If the 136 million tons of nitrogen applied to agricultural land all entered the world's rivers, what would be the change in concentration of nitrogen in rivers? (Assume all the nitrogen leaches off the land surface as nitrate [NO_3^-], and the average concentration of nitrate in the world's rivers with a flow of 3.7×10^{16} liters per year is 1 milligram per liter.)

18. What activities take place in perturbed (domesticated) ecosystems that can lead to environmental degradation?

19. How do the per capita world fertilizer use trends over time differ between the developed countries and the developing countries?

20. The use of pesticides in the early twenty-first century in the United States amounted to 546 million kg/yr. With a farm area of 922 million acres (40.8% of the land area of 2.26 billion acres of the United States) and a population of 304 million, what are the specific area and per capita application of pesticides to crops and other plants in kg/hectare and kg/person. (1 acre = 0.405 hectares.) Why is there concern with the extent of pesticide use in the United States and the world?

Additional Sources of Information

Carbon Budget 2007, 2008, http://www.globalcarbonproject.org/carbontrends/index.htm.

Caufield, C., 1991, *In the Rainforest*. University of Chicago Press, Chicago, 310 pp.

Foley, J. A., Monfreda, C., Ramankutty, N. and Zaks, D., 2007, *Our share of the planetary pie*. Proceedings of the National Academy of Sciences, v. 104(31), 12585–12586.

Food and Agriculture Organization of the United Nations, 1997, *State of the World's Forests, 1997*. Author, Rome, Italy, 201 pp.

Food and Agriculture Organization of the United Nations, 2000, *Global Forest Resources Assessment 2000*. Author, Rome, Italy, 479 pp.

Food and Agriculture Organization of the United Nations, 2006, *Global Forest Resources Assessment 2005*. Author, Rome, Italy, 320 pp.

Food and Agriculture Organization of the United Nations, 2008, *Current and World Fertilizer Trends and Outlook to 2001/12*. Author, Rome, Italy, 44 pp.

Gardner, G., 1996, *Shrinking Fields: Cropland Loss in a World of Eight Billion People*. Worldwatch Institute, Washington, DC, 56 pp.

Houghton, R. A., 1995, Land-use change and the carbon cycle. *Global Change Biology*, v.1, pp. 275–287.

Millennium Ecosystem Assessment, 2005, *Ecosystems and Human Well-being: Synthesis*. Island Press, Washington, DC, 137 pp.

Ramakrishna, K. and Woodwell, G. M., 1993, *World Forests in the Future, Their Use and Conservation*. New Haven, CT, Yale University Press, 156 pp.

World Resources Institute, 2000, *World Resources 2000–2001*. World Resources Institute, Washington, DC, 389 pp. Also editions of 1986, 1988, 1990, 1992, 1994, and 1996.

The Changing Earth Surface:
Land and Water

A continent ages quickly once we come.

ERNEST HEMINGWAY

The land, water (freshwater and marine), and atmosphere are intimately coupled. Water falls on the land surface, weathers rock minerals, and leads to erosion of rocks and transport of solid and dissolved products of soil formation by overland flow of water, by leakage into groundwater systems, or by river and stream flows. The rivers may discharge into lakes or coastal marine areas where the materials they carry are deposited. Thus, water plays a dominant role in transferring chemicals, nutrients, and solid matter about the land surface and into aquatic environments where these materials may temporarily or permanently reside.

Eroded soils transported by rivers have created the great deltas of the world, and their deposition has transformed the boundaries of the continents. Today, nutrient-rich soils are being degraded and their dispersal modified by human activities. For example, the completion of the Aswan Dam on the Nile River in 1970 has certainly been beneficial to some degree. The flooding of the river is controlled, and Egypt and the Sudan are assured of a steady supply of water each year. However, prior to the building of the dam, the yearly flooding of the Nile brought nutrient-rich, fertile soil materials to the Nile valley and delta. With the construction of the dam, there has been a decrease in the supply of nutrients to the lands of the lower Nile. These nutrients are being replaced with expensive synthetic fertilizers. Also, the Nile delta is now retreating because of the reduced amount of sediments transported by the waters of the Nile and the slow rise of sea level. Saltwater is intruding into coastal freshwater aquifers. The loss of nutrients entering the delta region also has been partially responsible for a decline in the productivity of marine waters offshore of the Nile delta. This change probably was partially responsible for the decline in Egypt's sardine catch from 18,000 metric tons in 1962 to only several hundred metric tons in 1978. The catch has subsequently risen to about 8600 metric tons.

Winds blow across the land surface and carry dust into the atmosphere. The dust may settle back down to the land surface or be transported out over the ocean where it may fall in wet or dry deposition. At the sea surface, aerosols (suspension of fine-grained solid or liquid particles or both in air) are generated by the capping of waves and the impact of raindrops. These aerosols may simply recycle back onto the sea surface in precipitation or be transported over the continents to fall in wet and dry deposition. All along the natural circulation route, the coupled system of land, water, and air is continuously influenced by human activities.

For example, North Africa loses large quantities of soil particles that are eroded by the wind from its semiarid and desert land. This airborne dust, carried by tropospheric winds, transports nutrients from the Sahara to South America (see also Chapter 1). The nutrient-rich dust aids in the fertilization of the Brazilian rain forest. Atmospheric transport of dust from North Africa has occurred for millennia, but recently an extended drought period of natural origin and human activities of cattle grazing and farming have enhanced the rate of erosion of topsoil and its transport in the atmosphere. It is estimated that between 1971 and 1981 approximately 1 million to 4 million tons of African soils were transported away from the continent by winds. These soil particles cover vast expanses of the atmosphere in the Atlantic region and can be seen as large plumes extending westward from Africa in satellite images (http://earthobservatory.nasa.gov/NaturalHazards/view.php?id=39337).

On the other side of the world in the Pacific, air samples collected at the Mauna Loa Observatory in Hawaii from 1944 to 1982 and later measurements indicate that dust particles in the atmosphere of this region originate on the Asian mainland. Similar to the Atlantic situation, this dust has been transported for thousands of years across the Pacific. However, an increase in the quantity of dust particles in the atmosphere at Hawaii has been observed to coincide with the beginning of the spring plowing season in Asia. Thus, in both the Atlantic and Pacific, it has been demonstrated that atmospheric transport of dust particles has been increased by human activities.

It is the modification of the global land–water system by people that is the subject of this chapter. We will look at the land–water system initially by examining how human activities have impacted the land and soil of the planet. The coupled system of land and water will be considered from the standpoint of problems of eutrophication and physical and chemical pollution of both freshwater and marine aquatic systems, with emphasis placed on coastal zones. Chapters 12 and 14 discuss human impacts on the atmosphere.

THE SOIL ECOSYSTEM

The land surface area of Earth is 14.9 billion hectares or approximately 29% of the exterior of the planet. Ice, rock, semidesert, and desert regions comprise about 3.4 billion hectares of this area. The remaining 11.5 billion hectares are vegetated land, with soils of variable thickness and physical characteristics forming the upper layer of the land surface.

To gain a more quantitative picture of the nature of soil, a more detailed representation of the structure and composition of soil with increasing depth than given in Figure 2.7 is shown in Figure 11.1. The soil depicted is a mature, well-developed soil on granitic bedrock in a region of moderate rainfall. The surface horizon of the soil (the O-horizon) is mainly void space filled with gases of N_2, O_2, and CO_2, and rich in organic matter derived from the vegetation living above the soil and organisms living within the soil. With increasing depth in the soil, inorganic mineral and rock fragments become more abundant and gases are reduced in content. Much of the organic matter disappears with increasing depth in the soil. Water in the soil is found mainly in the A-, B-, and C-horizons. The granitic bedrock is low in water content unless it is highly fractured, which would provide void space for water to accumulate.

Soils are an important natural resource. They not only provide an anchorage for terrestrial plants but filter water, recycle nutrients, and provide habitats for many life forms. They are perhaps the least appreciated of all our natural resources. Without soils plants could not grow, herbivores could not live, and carnivores would perish. The hydrologic cycle would not function like today nor would the exchange of gases between Earth's surface and the atmosphere. Earth would be a considerably different planet (see Soils, Chapter 2).

Soil Conditions

The Global Assessment of Soil Degradation (GLASOD), sponsored by the United Nations Environmental Programme (UNEP) (http://gcmd.nasa.gov/records/GCMD_GNV00018_171.html), has provided some preliminary baseline data on the status of world soils over the 45-year period from 1945 to 1990. This study is the first of a series of studies designed to assess the

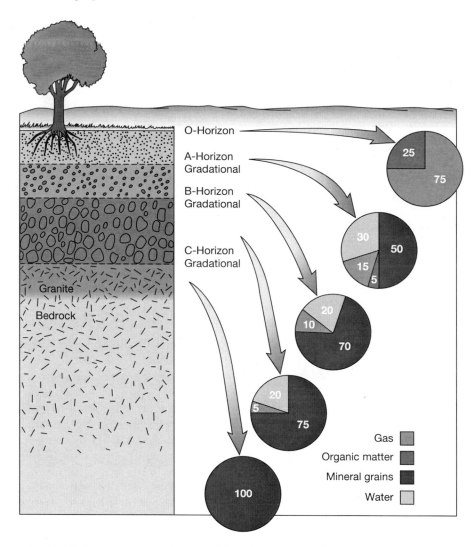

FIGURE 11.1 The structure and composition of a mature, well-developed soil on granitic bedrock. Numbers in the pie diagrams refer to percentages of various components.

quality and quantity of the soils of the world. The data, although incomplete, provide some indication of the status of the soils of the world.

Degraded soil is that which has been affected by human processes such that its current or future capacity to support human life is lowered. The extent of degradation of soil may be light, moderate, severe, or extreme. Degraded vegetated land is estimated to cover approximately 1.9 billion hectares or approximately 17% of the total vegetated land area (Figure 11.2), with approximately 9.6 billion hectares (83%) of vegetated land left undegraded. The degraded vegetated land is a result principally of human activities of the past two centuries and especially from 1945 continuing into the twenty-first century.

TYPES OF SOIL DEGRADATION DUE TO HUMAN ACTIVITIES   Land may be degraded in a number of ways by human activities (Figure 11.3). Soils may be physically removed by water and wind erosion when land is improperly cultivated or is deforested. After vegetation is removed from the land, erosion by water increases, and fertile soils are washed from the land surface. With vegetation and topsoil removed, the potential for further soil loss by wind erosion is increased. The ultimate result of these processes can be **desertification,** a term generally used to describe the encroachment of the desert onto nondesert land. Figure 11.4 is a dramatic example from Papua New Guinea of changes in erosion rates due to human activities of deforestation, cultivation,

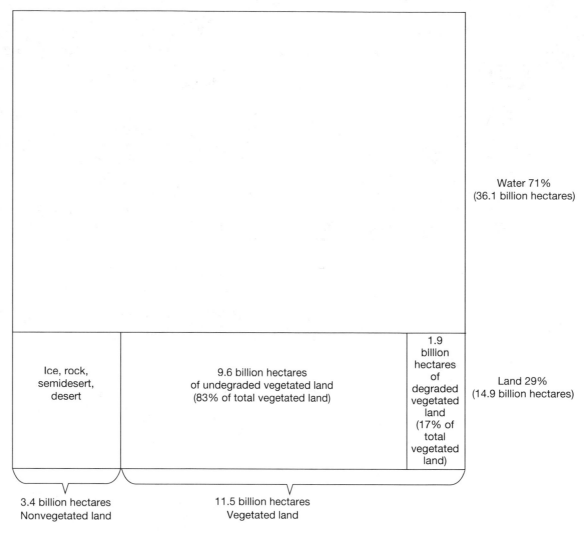

Water 71%
(36.1 billion hectares)

Ice, rock,
semidesert,
desert

9.6 billion hectares
of undegraded vegetated land
(83% of total vegetated land)

1.9 billion hectares of degraded vegetated land (17% of total vegetated land)

Land 29%
(14.9 billion hectares)

3.4 billion hectares
Nonvegetated land

11.5 billion hectares
Vegetated land

FIGURE 11.2 Percentage distribution of land types and water on Earth. Total land area of the world is 14.9 billion hectares, of which 77% is vegetated land.

and development. Erosion rates since about 9000 years before present have increased by more than 30,000%!

For many centuries, the United States has been losing soil to inland lakes and to the coastal ocean via river runoff due to natural erosion. With the development of extensive agricultural, transportation, urban, and industrial activities, soil is being lost from the surface of the country at an increasing rate. Some of this is stored behind dams, but much detritus reaches lacustrine and coastal marine environments. Figure 11.5 illustrates the amount of soil lost annually from various regions of the United States. Much of the soil lost in the eastern United States west of the Appalachian Mountains and in the Midwest enters tributaries that flow into the Mississippi River. The river then transports this material to the Mississippi delta, where it is

sedimented out or the detritus bypasses the delta to be discharged into the Gulf of Mexico.

Another type of land degradation is that of the stripping of land of its nutrients through clearing of vegetation. Continuous harvesting can result in less fertile soil, unless nutrients are returned to the ground through the application of fertilizers. However, repeated application of fertilizers may also lower the pH of the soil and result in acidification. Salinization is another form of chemical degradation, whereby salts become concentrated in the soils, and the land becomes unfit for agriculture. The excess salts decrease the ability of plant life to survive. Additional chemical pollutants may reach soils through urban and industrial wastes, pesticide use, dry and wet acid deposition, and chemical and oil spills.

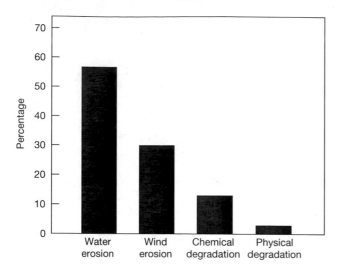

FIGURE 11.3 Types of soil degradation in percent of total degraded land. (*Source:* Oldeman et al., 1990.)

Soils can also be degraded by a set of processes known collectively as physical deterioration. The soil is not eroded or chemically changed but is degraded because of physical mechanisms, such as overcompaction by heavy machinery or trampling by grazing cattle, sheep, and goats. Waterlogging occurs when excess water is applied to the land surface or when the land is submerged because of human-induced changes in natural drainage systems. This leads to saturation of the pores of soils with water, which can physically deteriorate the land.

Figure 11.6 shows a global assessment of the status of human-induced soil degradation as of 1990. The various causes of degradation and their extent and degree, their degradation severity, vary from region to region.

CAUSES OF SOIL DEGRADATION The leading causes of soil degradation are overgrazing, deforestation (including land conversion and logging), agricultural practices, fuelwood gathering, industrialization, and industrial pollution (Figure 11.7). However, the underlying reasons for land degradation because of human activities are related to poverty, the expansion of human settlements, mismanagement, short-term profit, and lack of understanding of how Earth systems function. These are the same underlying reasons that affect the quality and quantity of the forests of the world (Chapter 10).

Grazing Practices Overgrazing by livestock can destroy vegetation, and as a result, increases the rate of erosion of the land by wind and water. There are about 15 billion domestic animals in the world. Nearly 75% are poultry; most of the rest are large grazing animals. The large animals trample and compact the ground, making it difficult for water and root systems to penetrate into the soil. Overgrazing affects 35% of all degraded vegetated land and occurs in both industrialized and developing countries. Much of United States grassland is publicly owned and is managed by the federal Bureau of Land

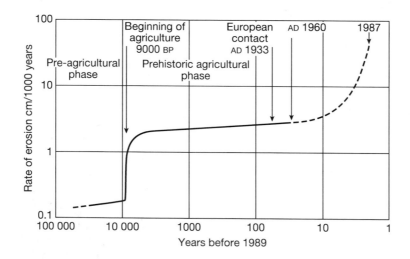

FIGURE 11.4 Rates of erosion during the Holocene for Papua New Guinea. Notice how the beginning of agriculture 9000 years before present (B.P.) and shortly after contact with Europeans led to increased erosion rates of the land surface. Since preagricultural times, erosion rates have increased more than 300 times, mainly due to human activities. (*Source:* Hughes et al., 1991.)

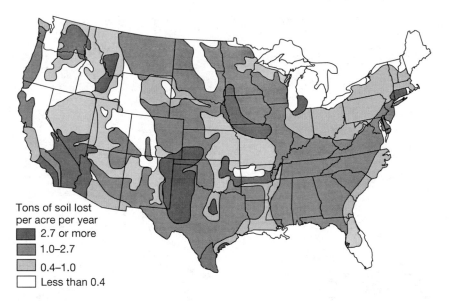

Tons of soil lost
per acre per year

- 2.7 or more
- 1.0–2.7
- 0.4–1.0
- Less than 0.4

FIGURE 11.5 Tons of topsoil lost from the contiguous United States per acre of land per year. The darker the pattern, the more the rate of loss within a region. (*Source:* http://www.nrcs.usda.gov/technical/NRI/2003/nri03eros-mrb.html.)

Management (BLM), which leases grazing rights to ranchers. This grassland is primarily grazed by cattle, often beyond a sustainable limit over the long term. As a result, more than 50% of publicly owned grassland in the United States is considered degraded. The conversion of forestland to pastureland in Brazil has also led to degradation of the land through loss of nutrients and compaction of the soil.

Overgrazing, excessive withdrawal of ground-water, or other unsuitable water-usage practices can also lead to desertification. The Dust Bowl created in the United States in the 1930s and the current ongoing desertification of the Sahel (Arabic for "border"; the land running west to east across Africa sandwiched between the Sahara desert and the tropics) are excellent examples of what can happen to improperly managed land. In the Sahel, the human population increased dramatically, as did the cattle population, in the years from 1935 to 1970. The rapid growth of these two factors has exerted excessive pressure on the land. In the early 1970s, the Sahel underwent the worst drought of the twentieth century. The drought, combined with overgrazing of the sparsely vegetated land, resulted in severe land degradation and further loss of vegetation in the Sahel. These events have led to the death of livestock and starvation for many people. This situation continues to this day (http://landsat.gsfc.nasa.gov/images/archive/e0013.html).

Deforestation   Deforestation accounts for about 30% of degraded vegetated land. Trees generally act

as a cushion, buffering the erosional effects of strong rains and downpours. During and immediately after rainfalls, water is evaporated directly back to the air from plant surfaces, is captured in the soils to be taken up and slowly released by plants, with the remainder entering streams and rivers through groundwater or by direct runoff. When forests are cleared, much of the rainfall runs directly off the land or percolates rapidly into the ground. As a result, an increase in stream flow generally follows extensive deforestation. Soil erosion and flooding become more prevalent, and streams and rivers run muddy.

One regional example of this set of processes is that of Thailand, where deforestation has caused extensive flooding and erosion and wreaked havoc both environmentally and socially through much of the country. As a result, Thailand has prohibited some logging practices (http://landsat.gsfc.nasa.gov/about/Application5.1.html). In India, flooding, a result of excessive logging in the Himalayan Mountains, has reportedly led to damage of 12 million acres of land. In the 1978 monsoon season, 66,000 villages in India were destroyed because of flooding. Although monsoonal rains are natural events, this massive damage occurred in part because of large-scale deforestation of the watersheds of Indian rivers.

Rain forests that are converted into agricultural lands are especially susceptible to soil degradation. Once tropical rain forests are cut and removed, nutrients stored in soils are easily lost and unavailable for subsequent agricultural production or forest

GLOBAL ASSESSMENT OF THE STATUS OF HUMAN-
INDUCED SOIL DEGRADATION (1990)

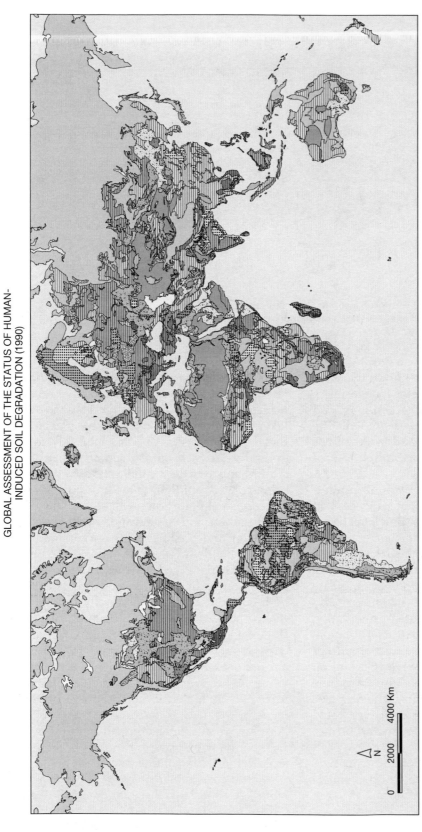

DEGRADATION SEVERITY (Extent + Degree)

Water erosion
- Loss of topsoil
- Terrain deformation/
  mass movement

Wind erosion
- Loss of topsoil
- Terrain deformation
- Overblowing

Chemical deterioration
- Loss of nutrients/
  organic matter
- Salinization/alkalinization
- Acidification
- Pollution

Physical deterioration
- Compaction/crusting
- Waterlogging
- Subsidence of organic soils

Stable terrain
- Stable under
  natural conditions
- Stable without vegetation
- Stabilized by human
  intervention

Other

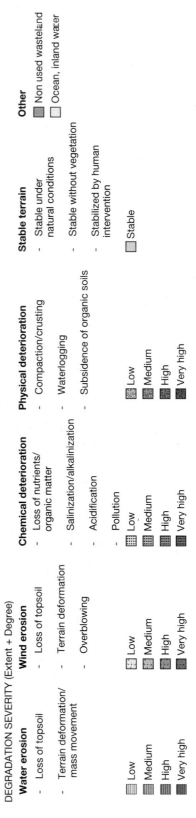

FIGURE 11.6   A global assessment of the status of human-induced soil degradation in 1990. The extent and degree of soil degradation, that is, its degradation severity, are shown for various causes of soil degradation (*Source:* http://www.isirc.org/webdocs/images/glasod_mercator1400.jpg).

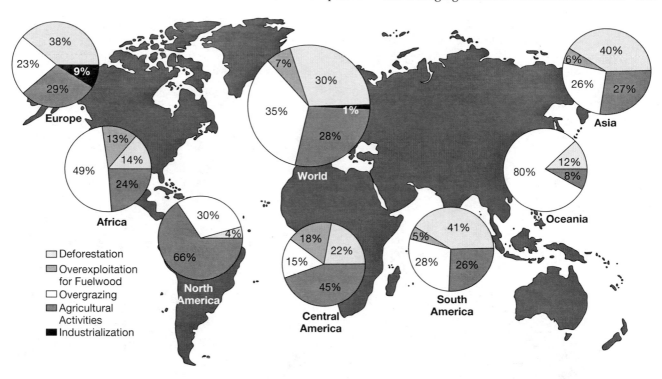

FIGURE 11.7 Causes of soil degradation for the world and selected regions of the world in percent of degraded land. (*Source:* Oldeman et al., 1990.)

regrowth. As a result, new plants find it difficult to grow in the nutrient-depleted soils. Also, the soil is generally unsuited for long-term agricultural purposes. For example, fertilization of tropical soils has been difficult because of the lateritic clay that underlies the thin soil of this region. Only 22% of the soils in the tropical areas of the world are suitable for agriculture, and many of these soils are already heavily used. Unfortunately, research to find ways to fertilize tropical forestland is limited to date because it has been far less expensive and easier to simply abandon depleted land and deforest more land for short-term agricultural use.

Agricultural Practices Of the vegetated soils that cover the planet, only about 13% (1.5 billion hectares) are naturally suitable for growing crops. Another 15% (1.7 billion hectares) of the vegetated soil area may be suitable for agriculture with proper management. The remaining vegetated soil area is too dry, cold, hot, or low in nutrients, or has soils too thin to support crop production. Most soil degradation is occurring on the available vegetated land that is currently conducive to agricultural use.

Poor agricultural practices account for about 28% of the degraded land of the world. Tilling the land across the contours of hills and valleys or tilling it parallel to the prevailing wind direction exposes soils to wind and water erosion and leads to degradation. By

the 1980s, soil erosion in the United States exceeded soil formation for about one-third of the area of cropland. Furthermore, 6 million to 7 million hectares of agricultural land in Third World countries are made unproductive each year because of erosion.

Agricultural lands are also susceptible to nutrient depletion. As a result, fertilizers are added to replace nutrients lost when the crops are removed. In addition, pesticides are frequently added to raise crop yields. These chemicals can drain off into groundwater or other waterways and contaminate them. Groundwaters are particularly susceptible to this type of chemical pollution because of their slow movement and long time of isolation from Earth's surface. Chemicals may remain in these systems for decades or longer.

Salinization of soils can occur when unsuitable irrigation techniques are used, such as irrigating with water of high salt content or providing inadequate water or drainage for water, which through evaporation might lead to an increase in the salt content of soils. For example, southern Iraq's soils are badly damaged by saline waters, and roughly a third of Iraq's cultivatable land has already been lost to salt buildup. More will be lost during the Iraq War. The reasons for this situation are complicated. They involve damming of the Euphrates River, especially upstream of Iraq in Turkey with the construction of the Ataturk Dam. Overirrigation and poor drainage of soils compound the problem. Ponds of stagnant water are left on the fields to evaporate, leaving behind a crust of halite (NaCl) and other salts. In the United States, the mighty Colorado River flowing south into Mexico becomes increasingly salty. This is because of the huge amounts of water withdrawn from the Colorado River in the United States for irrigation and other purposes. Salinization of soils is a result of the reduced flow of the Colorado and its increasing salinity through evaporation as it enters Mexico. Because of diversion of the waters of the Colorado, virtually no suspended sediment or river water enters the Gulf of California at present (Figure 11.8a; http://geologia.cicese.mx/RCdelta/animation.htm).

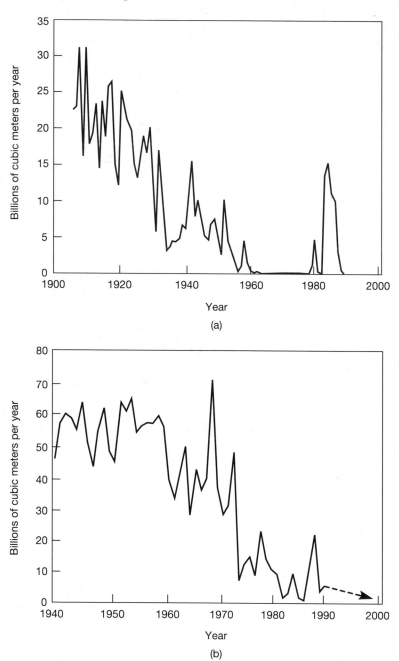

FIGURE 11.8 Effect of freshwater diversions on river flow. (a) The flow of Colorado River water in billions of cubic meters per year below the sites of major dams and diversion of water to other areas. (b) River flow in billions of cubic meters per year into the Aral Sea between 1940 and 1990. Neither system has recovered since 1990. (*Source:* Postel, 1996.)

The Colorado is not a unique example of the effect of water diversion on river flows. The flow of water into the Aral Sea lying between southern Kazakhstan and northwest Uzbekistan from the Amu Dar'ya and Syr Dar'ya Rivers diminished by a factor of about six between 1940 and 1990 because of large diversions for the irrigation of cotton (Figure 11.8b; http://landsat.gsfc.nasa.gov/images/archive/f0012.html). The sea was once the fourth-largest inland body of water in the world. Because of water diversion and subsequent reduction of river flows into the sea, the area of the sea has decreased by 50%, and its volume by three-quarters. The salinity of the sea has tripled. Because of these changes, the fish catch has dropped to zero, 20 of 24 fish species have disappeared, and contaminated water has become a problem for the populace living in the area.

An abundance of water and suitable farming methods can aid in keeping the salt content of agricultural soils low. An example of such a practice is in Lovelock, Nevada in the United States, where farmland is irrigated by shallow irrigation canals and drained by deep drainage ditches to prevent the salts from accumulating at the land surface. Unfortunately, because of water limitations, this practice is not possible in many regions.

Fuelwood Gathering Stripping land for fuelwood is responsible for 7% of the degraded land worldwide. Biomass burning is a source of energy for approximately 50% of the world population and accounts for 14% of the total energy produced worldwide each year. The removal of biomass exposes the soil to processes of deterioration, much like overgrazing and deforestation. Fuelwood gathering in the Sahel region of Africa has been particularly devastating to the soil environment, leading to a substantial increase in erosional rate, particularly by the wind. Soils may also suffer indirectly. For example, the high rate of removal of fuelwood from the land in some countries has led to scarcities in this fuel source and an increase in the rate of burning of cattle dung. The burning of the dung results in impoverishment of the soil in nutrients because these substances are not recycled back into the ground but are released to the atmosphere during the burning process and may be transported to other regions. Also, dung burning is often done in small huts, exposing the inhabitants to high concentrations of chemical pollutants in the air.

Industrial and Municipal Waste Pollutants One percent of degraded land is the result of industrial and waste-disposal activities. Radioactive wastes from nuclear power plants, hospitals, and other institutions and military operations, toxic wastes from industry, and municipal wastes from households and community activities are by-products of human civilization. All of these wastes continue to increase worldwide in both absolute and per capita terms. The wastes may be disposed of temporarily or permanently on land. Such disposal can lead to degradation of the land because of covering of originally productive land by waste material and leaching of toxic and radioactive wastes into soils and eventually into groundwater systems. Figure 11.9 shows the surface disposal sites and underground injection facilities for hazardous waste disposal in the United States. Most industrialized countries have similar sites but in the developing world, waste disposal sites are generally lacking.

The United States manages to produce a quarter of the world's waste stream. The mass of all solid wastes in the United States from agriculture, industry, mining, and municipal activities is 4.5 billion metric tons of solid waste annually, equivalent to 17 tons per person per year, or about 47 kilograms of waste per person per day. About 5% of the solid-waste stream is made up of municipal solid waste, and 10% is generated by industrial activities. The larger proportion comes from agriculture (50%) in the form of animal manure, crop residues, and other agricultural by-products and mining in the form of waste rock, dirt, and slag (36%).

Figure 11.10 shows the amount of municipal wastes generated per capita by households, businesses, and institutions in some countries of the world in the mid-1980s. As is true for many of the statistics on the environment, North America stands out in terms of waste generation. The United States generated about 178 million metric tons of municipal waste in 1990, equivalent to 712 kilograms per person. In the early twenty-first century, about 236 million tons of municipal waste or 770 kilograms per person were generated annually in the United States. About 70% of this waste was in the form of paper and cardboard and yard wastes. Plastics, food wastes, rubber, leather, textiles, and wood, glass, and metals each constituted less than 8% of the waste stream. The amount of municipal waste generation in the United States was projected to increase 40% between 1990 and the year 2010.

Industrial wastes are produced in a range of industrial processes and involve a variety of materials from different sources. Wastes are produced by the treatment of metals and plastics, by the production of biocides, by oil operations, by industrial use

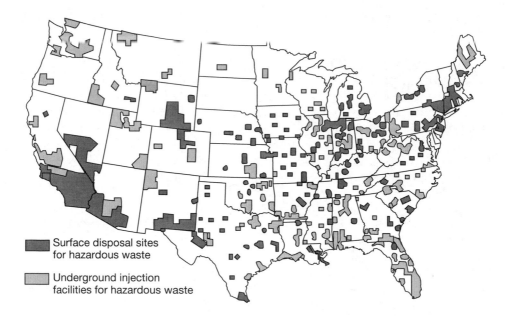

FIGURE 11.9 Surface disposal sites and underground injection facilities for hazardous waste in the contiguous United States. (*Source:* http://www.epa.gov/waste/hazard/index.htm.)

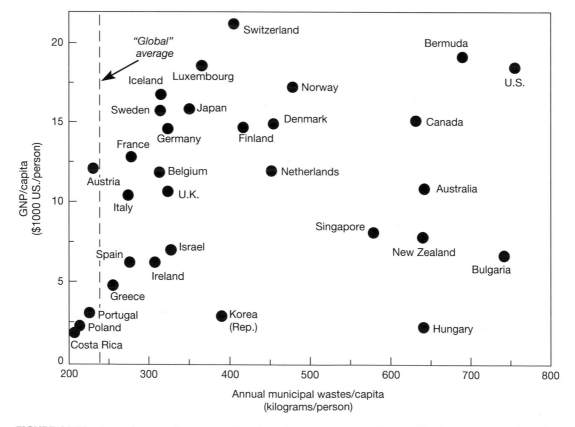

FIGURE 11.10 Annual per capita gross national product versus per capita municipal waste generation of various countries of the world in the mid-1980s. The global annual average per capita municipal waste generation was about 240 kilograms.

of PCBs (polychlorinated biphenyls), by clinical and pharmaceutical practices, and by the production and use of photographic materials, organic solvents, paints and pigments, and resins and latex. Total industrial waste of this nature generated in the United States in 1990 was at least 130 million metric tons of organic solvent waste alone, amounting to 280 kilograms per capita. In energy equivalents, in 2007 the United States produced about 111,000 terrajoules (10^{12} joules) of industrial waste, equivalent to 0.3% of total United States oil consumption.

In urban settings, municipal and industrial waste generation appears to increase with affluence. The wealthier cities of the developed world, like Los Angeles and New York, produce larger quantities of solid waste than do cities in the developing world, like Calcutta or Bangkok. For example, Washington, D.C. generated 1246 kilograms of solid waste per capita, while Saõ Paulo, Brazil, generated less than one-third that amount in the early 1990s.

In the United States, most solid waste is disposed of in legal landfills. However, in many cities of the developing world, this is not true. Only 30 to 50% of the waste is collected; the rest may be burned or dumped in unregulated landfills. The uncontrolled disposal of these urban wastes is a major cause of land, surface water, and groundwater pollution in the developing countries.

Atmospheric deposition of acidic wastes derived from industrial and transportation fossil-fuel-burning sources can lead to acidification of soils and loss of soil productivity. It should be pointed out that some scientists argue that nitrogen compounds derived from the burning of fossil fuels have actually led to fertilization of forests in Western Europe and elsewhere. However, the consensus opinion of most scientists is that a combination of tropospheric ozone, smog, and acid deposition produced by fossil fuel combustion is detrimental to terrestrial ecosystems, including soils, natural vegetation, and crops (see Chapter 12).

Summary

Water and wind erosion, chemical inputs, and physical deterioration induced by human activities can alter the land surface. These human activities include overgrazing, deforestation, agricultural practices, fuelwood gathering, and release of industrial and municipal waste pollutants. Figure 11.11 shows how human disturbances of land vegetation (Chapter 10) through a remarkable series of effects of the disturbances on the land surface affect the soil, landforms, climate, water, and fauna of the land. Notice in particular how these latter systems are interlinked so that an effect on one can be translated to another. There are a number of feedbacks built into the diagram shown in Figure 11.11. For example, the removal of vegetation from a large area, like a tropical forest, eliminates a sink for atmospheric CO_2 and can lead to increasing concentrations of CO_2 in the atmosphere. The

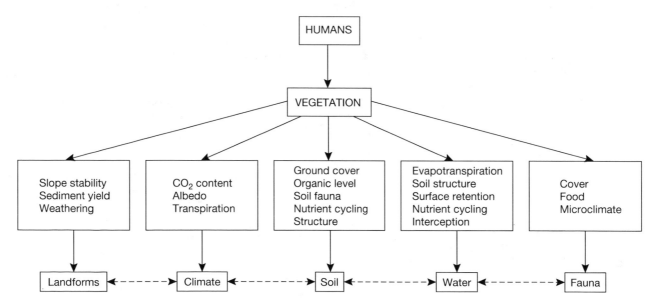

FIGURE 11.11 Diagram illustrating how human disturbances of the terrestrial vegetation and effects of the disturbance on the land surface can result in changes in surface landforms, climate, soil, water, and fauna. Notice how changes in these latter systems are interlinked; an effect on one can affect all.

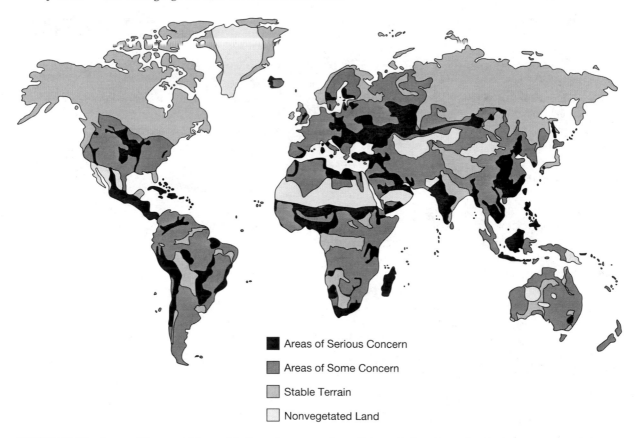

FIGURE 11.12 Areas of the world for which there is concern for soil degradation. Note that for most regions of the world, there is some to serious concern. (*Source:* International Soil Reference and Information Centre, unpublished map, Wageningen, Netherlands.)

removal of the vegetation also exposes soil containing organic matter to the atmosphere and atmospheric oxygen. This leads to increased oxidation of the soil organic matter to CO_2 and its release to the atmosphere. This is a positive feedback on the rise of CO_2 in the atmosphere first created by the removal of the tropical vegetation. An opposite series of events could occur: The growth of the forest removing CO_2 from the atmosphere can result in an increase in ground cover and soil organic matter storing more atmospheric CO_2 in organic matter, a negative feedback to the increase of CO_2 in the atmosphere.

The areas of concern for soil degradation for the world and for regions of the world are increasing in the twenty-first century (Figure 11.12). Nearly the entire surface of the land is affected by soil degradation, including the land areas of both the developed and developing countries. This is a serious concern because the soil ecosystem is important to the health of vegetated ecosystems, water and nutrient cycles, climate, and the production of crops to feed the world population.

Soil is intimately coupled to the hydrosphere and atmosphere through a number of interactions and processes (Figure 11.11). The alteration of soil conditions by human activities heavily impacts and alters the global water system. In the next section, we consider the interactions between the land surface and the water environment, emphasizing water pollution and the nature of the coupling between the land and the coastal zone of the ocean.

THE COUPLED LAND–WATER ECOSYSTEM

Water is a major component of global ecosystems. It is a major agent of erosion, an essential ingredient for vegetative growth and all other life processes, and a key factor in regulating climate. It circulates throughout the ecosphere, traveling through the atmosphere, to the land below, across the land, or into the ground as groundwater. It works its way into tributaries and rivers and eventually flows into the oceans, where evaporation returns it to the atmosphere to begin the journey again. Water is very abundant on Earth.

However, at any moment about 97% of water on Earth is in the salty oceans, 2% is locked up in the icy parts of Earth, and most of the remaining 1% is in inaccessible aquifers (see Chapter 4). As a result, people have only a minuscule fraction of the water on the planet available for their use. If all the waters of the world represented 100 liters, the amount of water available to people would be equivalent to approximately one-half teaspoon.

Freshwater Resources

Regional water scarcity is an important and growing problem. In some regions, water use exceeds that replenished each year by recharge. One-third of the world's population lives in countries with moderate to high water stress, and 80 countries, constituting 40% of the world's population, were having water shortages by the mid-1990s. By 2020 water use is expected to increase by 40% and by 2025, 1.8 billion people will be living in regions with a scarcity of water.

Freshwater is used for municipal, industrial, and agricultural purposes by the human population. Table 11.1 gives the estimated global water demand and consumption by sector. The demand and consumption of water in agriculture are equivalent to 65% of the total global water demand of 4430 cubic kilometers per year and 82% of the consumption of 2285 cubic kilometers per year. By 2020, 17% more water will be required for food production to feed the growing world population. Industrial water consumption is ranked second to agriculture. Municipal consumption is only about 2% of total consumption, whereas reservoir losses through evaporation account for the remaining 12% of consumption. The efficient use of municipal water has increased in many urban areas of the United States in recent decades. However for the United States and the world, per capita household water consumption will continue to increase into the twenty-first century. The growing southwest region of the United States is already facing critical problems in terms of future water supplies because of limited water resources and climatic change. Figure 11.13 shows one prediction of how worldwide water use in different sectors will change from 1995 to 2025. Use of water resources in irrigation practices in Asia constitutes an important percentage of the pie both now and in the future.

Intervention in the natural water cycle has had great benefits for people. Irrigation of croplands for food supplies, dams for hydroelectric power, reliability of an adequate water supply, and flood protection are all clear benefits. In addition, industrial use of water has led to economic growth and development. However, the utilization of water by the world's population has greatly impacted the water cycle. Humanity now uses the equivalent of about 25% of evapotranspiration over land and about 55% of the accessible water runoff from the continents, but the use and availability of water vary widely on regional scales. Figure 11.14 shows the renewable water resources of various regions of the world and the per capita regional withdrawals of water, with per capita water withdrawal for selected countries shown in Figure 11.15. In 1995, at least 20 countries, mostly in Africa and the Middle East, had annual renewable water resources of less than 1000 cubic meters per capita. Below this level, water supplies are generally insufficient to supply basic water needs. It is projected that in the year 2050 at least 45 countries, mainly in the developing world, will have annual renewable water resources of less than 1000 cubic meters per

TABLE 11.1 Estimated Global Water Demand and Consumption in Agriculture, Industry, Municipalities, and Due to Evaporation from Reservoirs

Sector	Estimated Demand (km³ per year)	Share of Total (%)	Estimated Consumption (km³ per year)	Share of Total (%)
Agriculture	2880	65	1870	82
Industry	975	22	90	4
Municipalities	300	7	50	2
Reservoir losses	275	6	275	12
Total	4430	100	2285	100

(*Source:* Postel et al., 1996.)

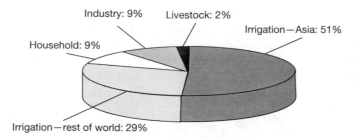

Worldwide Water Use, 1995

Industry: 9%   Livestock: 2%

Household: 9%   Irrigation—Asia: 51%

Irrigation—rest of world: 29%

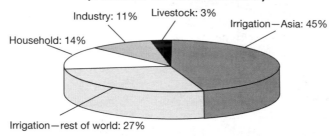

Worldwide Water Use, 2025

(Business–as–usual scenario)

Industry: 11%   Livestock: 3%

Household: 14%   Irrigation—Asia: 45%

Irrigation—rest of world: 27%

FIGURE 11.13 Worldwide water use for different sectors for the year 1995 and projected to the year 2025 under a Business as Usual (BAU) scenario of usage (From Rosegrant, M. W., Cai, X., and Cline, S. A., 2002, *World Water and Food to 2025: Dealing with Scarcity.* Washington, D.C.: International Food Policy Research Institute. Reproduced with permission from the International Food Policy Research Institute, www.ifpri.org. This paper can be found at: http://www.ifpri.org/sites/default/files/publications/water2025.pdf).

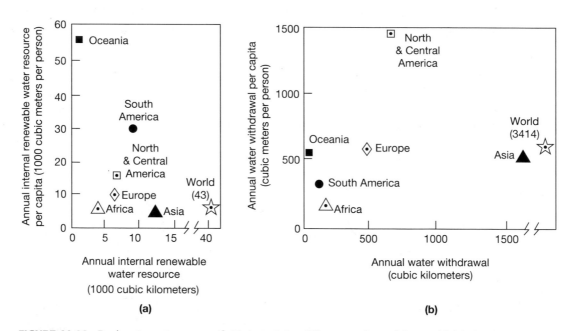

(a)

(b)

FIGURE 11.14 Freshwater resources and withdrawals for different regions of the world. (a) The 1995 annual internal renewable water resource, which includes the sum of estimates of annual runoff into rivers and recharge of groundwaters for the different regions. Notice the large water resource per capita available in Oceania, which includes Australia, New Zealand, and several of the island nations of the Pacific region. (b) Annual water withdrawals for various regions of the world. Estimates are mainly for the late 1980s and 1995. Notice the large per capita withdrawal in North America and Central America. The internal renewable water resource of the world is about 43,000 cubic kilometers per year, while annual freshwater withdrawal is 3414 cubic kilometers, equivalent to about 8% of the resource globally.   (*Source:* World Resources Institute, 2000.)

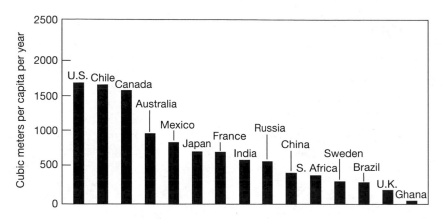

FIGURE 11.15 Water withdrawn per capita for some selected countries of the world. (*Source:* World Resources Institute, 2000.)

capita. Not surprisingly, consumption increases with affluence and availability. Some authorities are predicting that regional, if not global, water resources will become a scarcity in this century, although the Earth will never "run out" of freshwater or reach a "Peak Water" state as with oil. Water is a renewable resource, while oil is not. However, there are real limits on water use for an increasing number of regions of the world. For example, when quality water supplies are limited, food may not be produced in sufficient quantities to feed the population and economic growth may be constrained. China is an excellent example of a country facing severe and complex water problems in the twenty-first century because its water resources are over-allocated, inefficiently used, and severely polluted by human and industrial wastes. Sixteen of the twenty most seriously polluted cities in the world are found in China. It is possible that China's remarkable growth in GDP of recent decades could be slowed and its political stability weakened if it does not deal with its future water problems in a sustainable manner.

The United States currently withdraws 2.5 times more water per capita from its resource annually than Japan, 8 times more than Brazil, and 9 times more than the average for Africa. In the year 2000, about 20% of river runoff was withdrawn in Asia, and in North and Central America, about 15% was withdrawn. Human population growth, coupled with development and intensive irrigation, currently is placing a heavy stress on water resources of the Mediterranean region. Israel and Libya are already using more water annually than is recharged into their groundwater systems. The over-pumping of fresh groundwater in continental interior settings and along coasts can lead to the intrusion of salty

water from deeper saline water aquifers into a freshwater well. Figure 11.16 illustrates the situation of saltwater intrusion into a water well modifying the configuration of the interface between the freshwater phreatic zone and that of seawater origin. Notice how the interface bulges upward near the pumped well. With continuous over-pumping, the pumped water will become more and more saline and eventually nonpotable. Over-pumping and saline water intrusion are already a critical problem in the Midwest of the United States. Along with rising sea level, over-pumping has also led to the reduction in the availability of fresh water resources for island nations, like the Laura atoll region of the Republic of the Marshall Islands.

Water Pollution

Water carries natural substances and nutrients throughout the ecosphere. In the modern world, in addition to natural materials, pollutants, such as chemicals and excess eroded soils, also go along for the ride. As a result, pollutants produced in one area may contaminate areas far from their source of entry into the environment. In the atmosphere, precipitation may contain chemicals such as sulfur and nitrogen oxides from human activities. When rain falls on the soils, the water may pick up pesticides, other organic chemicals, and excess nutrients from urban, rural, and agricultural sources. These materials may then percolate into the ground to reach groundwater flows and eventually, rivers. As the river flows, the materials in it may be joined by eroded soils from deforestation and other human activities, and by sewage and industrial chemicals that have been discharged directly into the rivers. This accumulated

Natural conditions

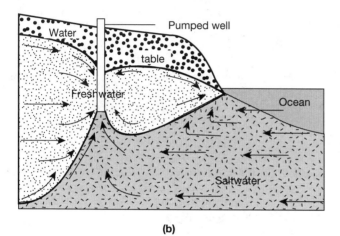

(a)

Salt-water intrusion

(b)

FIGURE 11.16 The configuration of the water table (boundary between vadose and phreatic zones; see Chapter 3) and the freshwater phreatic zone and saltwater phreatic zone interface in a coastal area under natural conditions (a) and under conditions of overpumping of a freshwater well in which saltwater intrudes the well (b). The arrows represent the flow paths of the water.

flow of substances may eventually be carried into lakes, bays, wetlands, or coastal estuaries. Coastal marine systems are particularly vulnerable to contamination by pollutants transported to them by rivers or directly discharged into them (Figure 11.17).

Water pollution is any physical or chemical change in the water from both natural and anthropogenic sources that adversely affects the organisms living in it. The distinction between point and nonpoint pollution sources is discussed in the next section, followed by a detailed look at three major types of pollution: excess nutrients leading to eutrophication, sedimentation, and toxic chemical contamination.

SOURCES The majority of humans live near freshwater or in coastal areas, with the few exceptions of people who live in deserts or polar regions. River basins in particular have traditionally supported the activities of many people, providing fertile soil for agriculture and waterways for transportation. The pollution of waterways, lakes, estuaries, wetlands, and coastal marine margins has thus been accomplished easily, and in many cases inadvertently, because of the proximity of human settlements to water supplies.

Water pollution sources are frequently categorized as either point or nonpoint sources (Figure 11.17). Roughly speaking, each contributes about half of the total amount of known pollutants to aquatic systems. **Point sources** include easily identified individual factory outlet pipes and sewage-treatment plant outlets. Nonpoint sources are by definition more difficult to pin down. Typical **nonpoint sources** include deposition of airborne pollutants derived from automobile and factory emissions and wastes from mining operations, agricultural practices, and urban lawn and street runoff and drainage. Nonpoint sources are particularly difficult to control because of their dispersed nature. Both the control and the abatement of water pollution activities are difficult measures to implement and achieve because of the many sources of pollution.

As mentioned earlier, a number of substances are involved in water pollution. Two nutrients of special concern are nitrogen and phosphorus. Excess inputs of these nutrients to aquatic systems, sometimes in conjunction with excess organic carbon and sediment inputs, can lead to the problem of eutrophication (see the following). A substantial amount of nutrient overloading of freshwater and coastal marine systems originates upstream of the affected system. For example, in the United States in the 1990s, about 68% of the nitrogen and 59% of the phosphorus delivered to the coastal zone were from upstream sources. The balance of the nitrogen entering coastal environments was from coastal point and nonpoint sources, with about equivalent amounts from each source. For phosphorus, about 35% was derived from coastal point sources and 6% from nonpoint sources.

The major sources of excess nitrogen are chemical fertilizers, sewage, runoff from feedlots, and sediments, while the sources of phosphorus are synthetic laundry detergents, fertilizers, and water-treatment chemicals. For example, small streams draining deforested, agricultural, and urbanized areas in temperate regions may contain as much as 30 times more nutrient and organic carbon than streams draining pristine forests. The total daily per capita discharge of

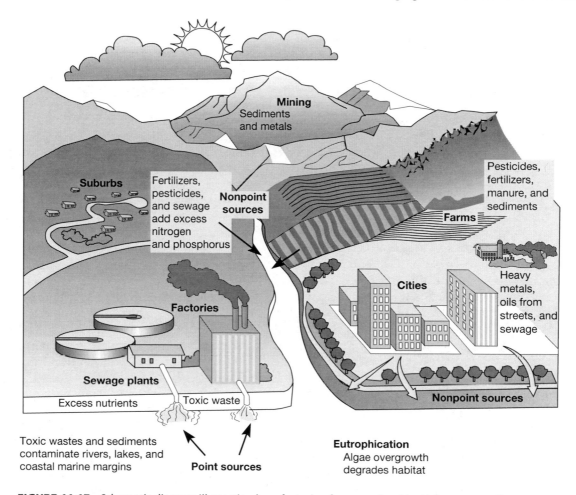

FIGURE 11.17 Schematic diagram illustrating how factories, farms, and residential areas contribute to aquatic environmental problems. Both nonpoint and point sources of aquatic pollutants are shown. This pollution is carried far downstream and impacts the coastal margin.

carbon, nitrogen, and phosphorus from domestic activity (excrement, household wastewater, washing powder) into aquatic systems in industrialized countries was 65 grams in the 1990s. For a population of 1.3 billion people, this discharge amounted to 32 million tons annually, of which 78% was organic carbon, 16% was nitrogen, and 6% phosphorus. For some Western European countries, less than 10% of the total nitrogen and phosphorus found in surface waters of these countries is of natural origin. The rest is derived from anthropogenic sources, mainly as discharges from municipal sewage plants and agriculture. In the Federal Republic of Germany, only 2% of the total phosphorus and 6.5% of the total nitrogen entering surface waters were of natural origin in the 1990s. The domestic activity discharges are very rich in phosphorus relative to living matter. This excess phosphorus and other nutrients in water can result in the eutrophication of an aquatic system.

EUTROPHICATION Eutrophication is the process of being fed too well. To understand eutrophication and the impact it has on water, it is necessary to understand how a natural aquatic system functions. A body of water generally has a clear appearance. This clarity is produced by the limitation of phytoplankton growth because of lack of the nutrients nitrogen and phosphorus. Nutrients generally do not reach aquatic ecosystems in great abundance because large amounts are recycled in the land–biota ecosystem. The nutrients that do enter bodies of water often settle near the mouth of the waterway in biological matter.

Nutrients entering aquatic systems are in dissolved and particulate organic and inorganic forms. Some portion of the particulate nutrients may enter into reactions in the aquatic system that lead to their conversion to dissolved forms. The dissolved nutrients can be taken up in the process of photosynthesis by phytoplankton and benthic plants, leading to the formation of organic matter. When the plants or organisms that fed on the plants die, some of the dead organic matter along with its nutrients accumulates in the sediments. In wetlands and at riverine entrances to lakes or coastal marine environments, the growth of plants may be increased because of the addition of nutrients by water flows into the systems. The productivity of the nearshore areas can be substantial. These areas provide habitat for many species of fish and other organisms. With time, some natural aquatic systems tend to become shallower as sediments accumulate. Eventually the sediment surface reaches the euphotic zone, and benthic plants and algae begin to grow, increasing productivity and sediment accumulation. These natural aquatic systems may change from waters exhibiting much clarity and low levels of plant production and biomass to waters that are murky, have high productivity, and are choked with plant material. At the same time, the production of excess organic matter fuels the processes of decay and respiration and can lead to substantial oxygen depletion of the waters of an aquatic system. This is a natural evolutionary path for many aquatic systems, including lakes and some coastal marine environments. Because the systems progress with time toward a eutrophic state, the set of processes associated with this evolutionary change is termed **eutrophication.**

With the addition of excess nutrients from agricultural, urban, or industrial sources, the process of eutrophication may be accelerated or initiated prematurely in an aquatic system. Effluents from sewage-treatment plants, agriculture, and lawn runoff can overload bodies of water with nutrients. The growth of vegetation as a result of the excess nutrients can be rapid. Algae blooms may develop. With a constant flow of nutrients to the body of water, algae may continue to proliferate, as may other aquatic plants. When these plants die and settle toward the bottom of the aquatic system, decomposers such as bacteria consume part of the dead organic matter. Bacteria growth is continuous with the constant availability of organic matter. Dissolved oxygen from the water is needed by some bacteria to oxidize organic matter; thus, the water may become depleted of its dissolved oxygen content. Fish and other organisms that depend on the dissolved oxygen for respiration may die of asphyxiation. Eventually an aquatic system depleted of oxygen and choked with plant material may develop. This overall set of evolutionary processes induced in a natural aquatic system by human activities is often referred to as "**cultural eutrophication.**" (Figure 11.18).

Aeration systems have been placed in many small lakes and ponds to counteract the loss of oxygen in the water. Water is continuously ejected into the air to enhance its oxygen content and to slow the process of cultural eutrophication.

The global scale of nutrient additions to water courses because of human activities is amply demonstrated in Figure 11.19. Several major rivers of the

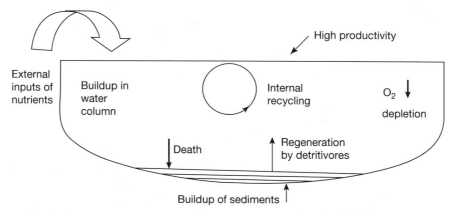

FIGURE 11.18 Schematic diagram of the processes of cultural eutrophication, the anthropogenic acceleration of eutrophication, of an aquatic system. Nutrients are introduced into the aquatic system from anthropogenic sources via rivers, groundwater, overland flow, and the atmosphere. The nutrients build up in the system and fuel biological production of organic matter. On death of the organisms, their remains sink toward the bottom of the aquatic system and on route are oxidized by bacteria in the water column or in the very shallow, upper layers of the sediment, depleting O_2 from the aquatic system and leading to anoxia. The nutrients may be regenerated in the sediments by bacterial detritivores and released back into the water fueling more productivity.

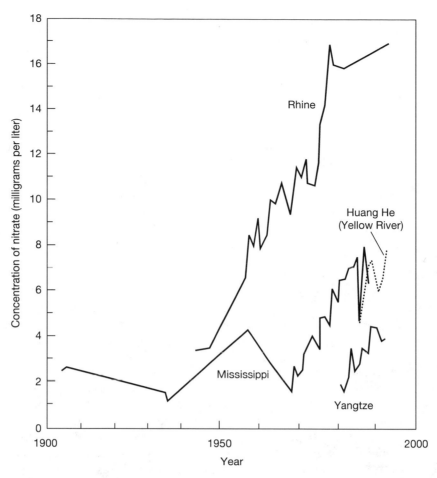

FIGURE 11.19 Historical trends in the concentration of nitrate in some of the world's major rivers. (*Source:* Smil, 1997.)

world show historical trends of increasing nitrate concentrations, including the Rhine, Huang He, Mississippi, Yangtze, and others. These increased concentrations are mainly due to the application of nitrogen fertilizers to croplands and subsequent leaching of the nitrogen and discharge of treated and untreated sewage into freshwater courses. For the Rhine, nitrate levels rose significantly between 1940 and 1970, then decreased slightly and were showing a moderate rise again in the 1990s. Over the last two decades of the twentieth century, sewage treatment decreased one form of nitrogen in the Rhine, ammonium. However, the treatment had little effect on nitrate concentrations because most of the excess nitrogen as nitrate in the river is believed to come from fertilizers used on croplands that border the river. One significant improvement in the water quality of the Rhine is the decrease in organic pollutants in the river. These pollutants include organic halogens (mainly chlorine-containing compounds), detergents,

hydrocarbons, and humic acids, which are products of the biological degradation of organic matter in sewage plants. These pollutants tend to consume oxygen and make the river water unfit for sustaining life.

Between 1969 and 1976, concentrations of organic pollutants peaked in the Rhine. Intensive international efforts have resulted in a decrease in these pollutants in Rhine River water since the early 1970s. Both the biological oxygen demand (BOD) and the chemical oxygen demand (COD) of Rhine water samples have fallen during the past few decades because of the decrease in the concentrations of the pollutants. The BOD of a water is measured by adding bacteria and nutrients to a water sample, and the consumption of oxygen, generally in milligrams of dissolved oxygen per liter of water, is measured over a period of time of typically five days. The COD of water is measured by adding concentrated sulfuric acid and dissolved chromium to a water sample and then determining the maximum oxygen consumption of the sample. Both BOD and COD establish the effects or characteristics of all the organic substances in the water. Generally, the higher the BOD and COD of a water, the greater the amount of reactive organic compounds in the water. As a result of decreasing organic pollutant inputs and concentrations, the oxygen levels of Rhine River water have increased (Figure 11.20). This is an example of at least the beginning of success in reclaiming a major polluted river system of the world.

The cultural eutrophication of Lake Erie in the United States is another excellent example of the effect of excess nutrient loading of a freshwater aquatic system, in this case mainly the nutrient phosphorus, and recovery of the system after the nutrient loading is reduced. In the 1960s, the amount of phosphorus entering Lake Erie was high, amounting to about 1000 metric tons per year, mainly from municipal and industrial point source outfalls. Phosphorus detergents were an important source of this P input. In addition, total phosphorus concentrations in the lake water of the central basin of Lake Erie were high, as were chlorophyll-a concentrations, a phytoplankton pigment and a measure of the biomass of phytoplankton in the lake and indirectly its productivity. Oxygen levels were low and were being depleted at a rate of 2–2.5 ppm O_2/month. Cyanobacteria were the dominant forms of phytoplankton in the Central Basin lake waters. The surface waters of the lake, the hypolimnion, were very turbid. Since about 1970, a set of controls focused on P inputs to the lake have been put in place with dramatic reduction in inputs to approximately 10 metric tons of P per year in the late 1990s. These inputs are mainly from nonpoint sources such as agricultural runoff. The reduction in P inputs has led to lower phosphorus and chlorophyll concentrations in lake waters, clearer waters, generally reduced depletion rates of O_2 from lake waters, a change in the phytoplankton composition from cyanobacteria in the late 1960s to diatoms and green algae, and a dramatic increase in the mayfly population in 1997. The fish populations of whitefish and walleye have recovered to some degree from their low stocks of the 1960s, primarily because of improvement in the oxygen regime of lake waters. However, native populations of blue pike, sauger, lake trout, and other species were virtually missing from the lake in the year 2000 because of this major cultural eutrophication event (see also following section on toxic chemical pollutants).

SEDIMENTS AND SEDIMENTATION Natural soil erosion and sedimentation continuously alter the morphology of the land and the shape of Earth's surface. Agriculture, deforestation, mining, and urban and industrial activities have increased the rate of transportation of sediments and the sedimentation of particulate materials in receiving waters far downstream from sediment sources. In many areas, sediment has become a major water pollutant, destroying the feeding grounds of fish, decreasing shellfish populations, filling in lakes and streams, and decreasing the light necessary for the growth of aquatic life.

Coral reef and wetland habitats and the fishing and tourism industries that depend on these habitats are particularly at risk because of increased sedimentation. Increased sediment loads are a major cause of

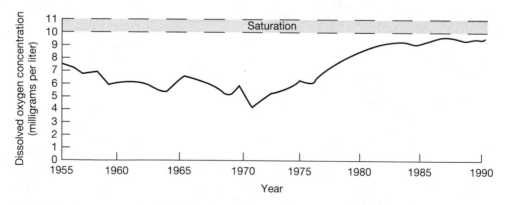

FIGURE 11.20 Historical trend in the dissolved oxygen content of the Rhine River in milligrams of dissolved oxygen per liter of river water from 1955 to 1990. In the summer of 1970, the oxygen content of the Rhine was so low that certain parts of the river had too little oxygen to sustain aquatic life. *Saturation* refers to the concentration level of oxygen in the river water at equilibrium with the atmosphere and at the temperature of the water. The Rhine River's pollution condition continues to improve in the twenty-first century. (*Source:* Malle, 1996.)

the worldwide degradation of coral reefs. The particulate materials transported by runoff into bays and lagoons may settle out onto living corals. Because many coral species inhabiting coral reefs cannot effectively rid themselves of the thin sediment layer that collects on them, they may die of suffocation. Also, the increased suspended sediment concentrations within the usually clear waters of coral reef ecosystems will increase the turbidity of the waters, preventing light penetration. This in turn can lead to decreased phytoplankton, coral, and benthic marine plant productivity.

Topsoil Erosion   Increased topsoil erosion because of human activities, particularly agriculture, is an important source of increased sediment discharge to aquatic environments. During the past few decades, it is estimated that 480 billion tons of topsoil have been lost to the world's farmers by erosion. If this estimate is correct, the rate of loss exceeds the worldwide transport of about 18 billion tons annually of suspended materials by rivers toward the ocean. Therefore, much of this topsoil loss probably represents material moving from farmlands to another terrestrial ecosystem.

Some scientists estimate that land use activities have doubled the river discharge of suspended material to the ocean during the last few centuries. If the topsoil erosion figures are correct, it is likely we will see even greater discharges of suspended sediment by rivers to coastal marine ecosystems (estuaries, bays, and marshes) in the future. However, this conclusion depends on how long it takes for the topsoil sediment derived from cropland and other sources and transported to another area of deposition on land to make its way to the ocean. Also, the rate of future dam construction will be critical because the reservoirs associated with dams are important repositories of sediment. Damming of rivers in Southeast Asia and Oceania will be particularly important in this context because the sources of nearly 75% of the suspended sediment reaching the ocean annually via rivers are in these regions.

Dams   Dams play a critical role in both the hydrologic cycle and in sediment dispersal and storage on land. Nearly 45,000 large dams are now located on the world's rivers. The construction of large multipurpose dams accelerated in the late 1940s, reaching an annual rate of more than 500 dams inaugurated annually in the 1960s. Dam construction has fallen since the 1960s but still continues at a rate of about 170 large dams per year. Globally, dams provide slightly more than 2000 billion cubic meters of reservoir volume. This volume could accommodate 280 times the total amount of sediment transported annually to the ocean by the world's rivers. It was estimated that in the early part of the 1990s, more than 13% of the global river flow to the ocean was dammed or diverted, and by the early part of the twenty-first century, this figure may exceed 20%.

In addition to affecting ecosystems as a pollutant, increased sediment loads can have significant impacts on dams and manageable waterways. The accumulation of mud and silt in standing water bodies (the process of siltation) decreases the life span of dams and interferes with traffic in waterways. For example, the Ambuklao Dam in the Philippines was built to last 56 years but its reservoir will be filled with silt in just 11 years. The Panama Canal experiences siltation problems owing to runoff from adjacent deforested land. China has considered building smaller dams serving small localized areas to cut down on potential siltation problems. However, in 1993, China decided to construct the largest dam in the world. The Three Gorges Dam is being constructed on the main stem of the mighty Yangtze River, producing a reservoir 500 kilometers long. In the process, 41,000 hectares of farmland will be destroyed, and 2 million people will be displaced. Without doubt, dam construction worldwide is currently modifying the global hydrologic cycle and river discharge of suspended solids to aquatic systems, including the ocean.

Unfortunately, even the most carefully engineered dams can create unforeseen environmental problems, some of which were mentioned in Chapter 10. Here we cite some effects of damming occurring far downstream from the diversion. The Iron Gates Dam crosses the Danube River in Russia 1000 kilometers from the mouth of the river where it enters the Black Sea. The damming of the river course has resulted in a decrease in the amount of dissolved silica that enters the Black Sea. This in turn has led to decreased dissolved silica concentrations in the surface waters of the Black Sea. Because of the decrease in silica concentrations, there has been a shift in the phytoplankton species in the Black Sea from siliceous diatoms to nonsiliceous phytoplankton. On the Huang He (Yellow River) of China, low rainfall, dam construction on the river course, and an increase in soil conservation have led to an approximately 50% decrease in both sediment and water discharge, a situation similar to the Colorado River. Competition for water from the river is increasing in the North China plain. The Chinese government plans an expensive south-to-north diversion of water from the Yangtze River into the Huang He.

In 1988, the lapse of a water agreement between India and Bangladesh, which dealt with the amount of water released from the Farakka Dam in India, resulted in a 75% reduction in Ganges River flow to Bangladesh. Once fertile farmland in the lowlands of Bangladesh became desert, and saltwater intruded into the Sunderlands, an important mangrove ecosystem and home of the Bengal tiger. Bangladesh lost annually about US$4 billion, mainly owing to poor crop production and damaged fisheries, because of lack of Ganges River water. The situation has since been resolved, although tensions between the two countries still exist. However, continued low flow downstream of the Farakka Dam will probably lead to an increase in the extraction of groundwater in the low-lying Bengal Delta, resulting in subsidence of the delta and an increase in the rise of sea level locally. The construction of the Aswan Dam in Egypt has led to a similar situation in the Nile Delta (p. 296).

TOXIC CHEMICAL POLLUTANTS Urban and industrial areas are significant contributors to water pollution. Pesticides placed on lawns and gardens and oil droppings on streets enter sewers and eventually reach lakes, streams, and rivers. Industry and mining contribute synthetic organic chemicals, such as PCBs, certain pesticides, and inorganic compounds, such as heavy metals, to the waterways. Synthetic organic compounds come from pharmaceutical manufacturing, chemical plants and petroleum refineries, and the paper and food-processing industries. Industrial sources of inorganic compounds include metal processing, use of metals in paint, plastics, and so forth, and leaching from solid-waste dumps. Toxic waste from industrial factories may be directly discharged into waterways or into the atmosphere.

Many of these chemicals are taken up by aquatic biota and concentrate in organisms at the upper end of the food chain, with relatively large amounts accumulating in fish and shellfish and in people who consume them, a process known as bioaccumulation. Some of the substances are carcinogenic and cause cancer, whereas others contribute to other illnesses either directly or by interfering with the immune system. Some chemicals are capable of mimicking estrogen in the reproductive system and can have significant effects on fertility. Heavy metals, such as lead and mercury, pose serious health problems to people in some areas.

Mercury not only enters watercourses directly but also is vented to the atmosphere from natural processes and anthropogenic activities like combustion of fossil fuels, mainly coal in utility, industrial, and residential boilers. Mercury is a toxic, persistent, and long-lived element in the atmosphere and can be transported long distances. The atmospheric mercury emissions will be deposited back on land and make their way into soils and aquatic systems. For example, mercury emitted from anthropogenic sources in China can be deposited in North America and the Arctic. Of the approximately 2270 metric tons of *total* mercury emitted to the environment in 2000 from anthropogenic sources, two thirds of the emissions were from fossil fuel combustion. Nearly 80% of total emissions of mercury to the environment came from Asia (52%), Africa (18%), and North America (9%).

In one of the most notorious examples of chemical pollution and human health effects, high levels of mercury in fish from Minamata Bay, Japan, led to an outbreak of mercury poisoning in area residents in the mid-1950s. Mercury had been introduced into the bay from discharges of a chemical plant located on a river flowing into the bay. The mercury accumulated in the food chain of the bay, particularly in edible fish. The Japanese ate the fish and developed mercury poisoning, the Mad Hatter's disease, a disease that affects the nervous system. Older Japanese living around the bay today still show evidence of the disease, and a number of people have died from it.

Table 11.2 illustrates an encapsulated history of the Minamata Bay disease incident. The Chisso factory responsible for the mercury inputs from production of acetaldehyde using mercury as a catalyst was built in 1908. The use of the mercury catalyst began in 1932. The disease was officially discovered in 1956, but the various legal and other actions dealing with the situation lasted until the year 2000. This is not an unusual situation when it comes to chemical pollutants. The introduction of the cancer-causing agent hexavalent chromium (+6) ion into groundwaters of Hinkley, California, by a utility company and the subsequent release of the movie *Erin Brockovitch* documenting the problem reawakened the public to the chemical contamination of waters in the United States. It took years to prove the source and cause of the chromium (+6) pollution incident at Hinkley, and for the injured parties to be compensated. There still remain skeptics who contend that chromium had little to do with the death and sickness of people in the area surrounding the utility plant. Table 11.3 gives the 15 top organic and inorganic hazardous substances that enter the soil, aquatic, and atmospheric environments. Their sources and potential toxic effects are also given.

TABLE 11.2 History of Events for the Minamata Bay Pollution Incident Showing the Long Time It Took Between Discovery of Minamata Disease and the Conclusion of Legal and Other Actions

Year	Event
1908	Factory built in Minamata after successful campaign by local elite
1932	Chisso begins production of acetaldehyde using mercury catalyst
Round 1	
5/1/56	Official discovery of Minamata disease
7/20/59	Kumamoto University research group announces probable cause is mercury
8/29/59	Compensation agreement between Minamata Fishing Cooperative and Chisso
10/7/59	Cat no. 400 develops Minamata disease after being fed factory waste in secret experiments at factory hospital
12/18/59	Compensation agreement between Kumamoto Prefectural Alliance of Fishing Cooperatives and Chisso
12/19/59	Cyclator completed by Chisso, waste supposedly safe
12/30/59	*Solution # I:* solatium agreement between Chisso and Minamata disease victims: $889 for deaths, annual payments of $278 for adults and $83 for children; further demands prohibited
5/31/65	Official discovery of Niigata Minamata disease
Round 2	
1/12/68	Citizens' Council for Minamata Disease Countermeasures established
9/26/68	Official government finding that Chisso's organic mercury causes Minamata disease (acetaldehyde production had ended 5/18/1968)
6/14/69	112 people (in 29 families) sue Chisso in Kumamoto District Court, other patients trust government to arbitrate compensation
10/71	New patients certified; some begin direct negotiations with Chisso, others accept government mediation
3/20/73	Victims win suit against Chisso
7/9/73	*Solution #2:* compensation agreement with Chisso by both trial group and direct negotiations group; applied to all certified patients: one-time payments of $59,000–66,000, monthly stipends of $74–221 (later adjusted for inflation), medical coverage
6/16/78	Kumamoto prefecture begins issuing bonds to finance loans to Chisso so it can continue compensation payments; Chisso owes prefecture over $2 billion by 1997
12/15/1995	*Solution #3:* Cabinet approves plan to subsidize Chisso in paying $26,000 to each eligible uncertified patient and $51,500,000 to victims' groups. Prime Minister Murayama offers government apology. Nearly all victims' groups accept and agree to drop suits (some 2,300 had been suing over certification and government responsibility)
9/1/1997	Net surrounding Minamata Bay removed after fish declared safe
3/31/2000	2,264 victims of Kumamoto Minamata disease (878 living; 1,386 deceased) have been certified and compensated under 7/9/73 agreement (out of 17,138 applicants): 10,353 victims have been compensated under 12/15/95 plan, in return for agreement never to sue or request certification as patients

(*Source:* George, 2001.)

The Great Lakes, which comprise one-fifth of the world's standing freshwater, are an excellent example of the deterioration of water quality due to both eutrophication and chemical pollution. For example, as described on p. 316, Lake Erie was once a pristine lake. Settlements have continually grown up around Lake Erie, with the lake being the repository of the waste of the adjacent communities. By the 1960s, the lake had become severely polluted by nutrient overloading and toxic chemical contamination.

TABLE 11.3 Fifteen Most Important Organic and Inorganic Hazardous Substances that Enter the Soil Aquatic, and Atmospheric Environments

Substance	Source	Toxic Effects
Lead	Lead-based paint	Neurological damage. Affects brain development in children.
	Lead additives in gasoline	Large doses affect brain and kidneys in adults and children.
Arsenic	From elevated levels in soil or water	Multiple organ systems affected. Heart and blood vessel abnormalities, liver and kidney damage, impaired nervous system function.
Metallic mercury	Air or water at contaminated sites	Permanent damage to brain, kidneys, developing fetus.
Vinyl chloride	Plastics manufacturing	Acute effects: dizziness, headache, unconsciousness, death.
	Air or water at contaminated sites	Chronic effects: liver, lung, and circulatory damage.
Benzene	Industrial exposure	Acute effects: drowsiness, headache, death at high levels.
	Glues, cleaning products, gasoline	Chronic effects: damages blood-forming tissues and immune system, also carcinogenic.
Polychlorinated biphenyls (PCBs)	Eating contaminated fish	Probable carcinogens. Acne and skin lesions.
	Industrial exposure	
Cadmium	Released during combustion	Probable carcinogen, kidney damage, lung damage, high blood pressure.
	Living near a smelter or power plant	
	Picked up in food	
Benzo[a]pyrene	Product of combustion of gasoline or other fuels. In smoke and soot	Probable carcinogen, possible birth defects.
Chloroform	Contaminated air and water	Affects central nervous system, liver, and kidneys; probable carcinogen.
	Many kinds of industrial settings	
Benzo[b]fluoranthene	Product of combustion of gasoline and other fuels	Probable carcinogen.
	Inhaled in smoke	
DDT	From food with low levels of contamination, still used as pesticide in parts of the world	Probable carcinogen, possible long-term effect on liver, possible reproductive problems.
Aroclor 1260 (a mixture of PCBs)	From food and air	Probable carcinogens.
		Acne and skin lesions.
Trichloroethylene	Used as a degreaser, evaporates into air	Dizziness, numbness, unconsciousness, death.
Aroclor 1254 (a mixture of PCBs)	From food and air	Probable carcinogens.
		Acne and skin lesions.
Chromium (+6)	From food, water, and air	Ulcers of the skin, irritation of nose and gastrointestinal tract, also affects kidney and liver.
	Originates from combustion source	

(*Source:* Enger and Smith, 2002.)

The lake literally stank and fish disappeared or were so severely contaminated that they were inedible. By the late 1970s, the lake was officially dead. At this time, new control measures on waste dumping in the lake were instituted, and the lake began a slow recovery. At present, the smell has dissipated, but the fish remain contaminated. The lake has improved but has still not regained its former pristine state. Part of the problem involves the slow leakage of phosphorus from the pore waters of the contaminated lake sediments into overlying lake waters. This leakage helps to promote continued eutrophication of parts of Lake Erie. The phosphorus originally accumulated in the sediments via enhanced organic matter deposition caused by loading of the lake waters with phosphorus from detergent and fertilizer inputs.

Water-related diseases, such as diarrheal diseases, hepatitis, and schistosomiasis, are particularly prevalent in the developing world. Rapid urbanization and industrialization are leading to greater production of chemical contaminants and exposure of the populations of these countries to waterborne chemicals and pathogens. In contrast, in the United States because of regulation of such substances, there is some indication that toxic trace elements, fecal coliform bacteria, and certain organic compounds have decreased in concentration in the water environment since 1970. However, even in the United States with its relatively strong regulatory safe drinking water standards, new chemicals, such as the widely used herbicides atrazine and alachlor, are showing up in freshwater systems at concentrations exceeding established health limits.

Coastal Zones

Because the coastal zone represents the transition between the land and ocean and because of its vulnerability to environmental degradation worldwide, the rest of this chapter is devoted to its features. First we consider human impacts on the system and then look at the past, present, and future behavior of this important ecosystem in the context of nutrient, organic matter, and sediment inputs.

The coastal zone includes the coastal plains and continental shelves, accounting for 8% of the surface area of the ocean (29×10^6 km^2) and 3.3% of the volume of the surface ocean layer to a depth of 300 m (108×10^6 km^3) (Figure 11.21). Tidal marshes, estuaries, seagrass beds, wetlands, coral reefs, mangroves, and banks are typical coastal environments. Although small in total area, organic production in coastal marine environments accounts for 10% to perhaps as much as 30% of total oceanic production. This is because of the fertilizing effect of nutrients from both riverine inputs and the upwelling of ocean waters rich in nutrients.

Estuaries, located near the mouths of rivers where freshwater meets and mixes with saltwater, comprise about 0.4% of total ocean area. However, the primary productivity of estuaries exceeds that of the continental shelves and the open ocean, amounting to

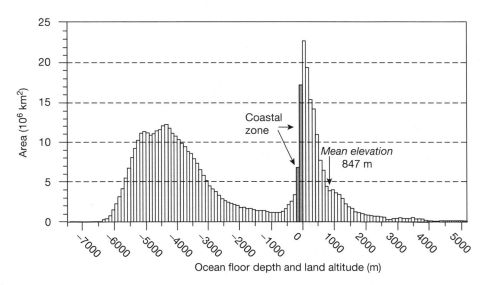

FIGURE 11.21 Global hypsometric curve showing the surface areas of the ocean floor and the land surface in 100-meter intervals. The coastal zone area between 0 and −200 m is the darker two bars. The coastal zone area is between 24×10^6 to 29×10^6 km^2. (*Source:* Mackenzie and Lerman, 2006.)

about 190 grams of organic carbon per square meter per year (see Biomass and Productivity in Chapter 6). In contrast the primary production of the continental shelves and the open ocean is about 160 and 130 g/C m²/yr, respectively. Although the productivity of waters bathing the continental shelves is not as high as in estuaries and coastal wetlands, their total net organic production is 4.3 billion tons of carbon per year, or about 10% of global oceanic production. This production supports important marine fisheries, such as that found on the Grand Banks offshore of Newfoundland, Canada.

IMPACT OF HUMAN ACTIVITIES ON THE COASTAL ZONES
The coastal zone is heavily impacted by human activities (Figure 11.17) because 40 to 60% of the global population lives within 100 kilometers of the coast, and nearly a billion people live in coastal urban centers that are now experiencing unprecedented growth. Furthermore, coastal areas are affected by pollution activities occurring far upstream in watercourses that enter these areas. As the world population grows, these impacts will increase because of the continual movement of people to urban centers located on or near the coast or within major river basins. Despite the obvious potential for damage to coastal systems, few countries have effective plans to manage coastal zone environments to protect the

ecosystem. Such management is a complex problem because it requires consideration of potentially polluting human activities occurring at point source and nonpoint source locations that are both adjacent to and far removed from the coastal environment.

According to the World Resources Institute, 34% of the world's coasts have a high potential risk of degradation (Table 11.4). Another 17% is at moderate risk. Most of the coasts threatened by development are in northern equatorial or temperate regions. Europe and Asia have 86% and 69%, respectively, of their coasts at moderate to high risk of degradation. Desert, subarctic, and arctic regions are least threatened by degradation.

Wetlands, estuaries, and coral reefs are under severe pressure from increasing population growth and from stream and river courses that deliver dissolved and solid materials of anthropogenic origin to them. The increased population leads to increased discharges of waste, suspended sediments, fertilizers, synthetic and organic chemicals, and toxic metals to coastal marine systems, leading to potential degradation of habitat and species loss. Nutrient pollution of coastal marine systems is especially important and widespread. Some recent studies indicate that up to 70% of pollution owing to excess nitrogen and phosphorus nutrient inputs to coastal environments comes from sources upstream of these environments. As

TABLE 11.4 Percentage of Coastline of Certain Regions of the World Threatened with Potential Degradation because of Development-Related Activities

Region	Percent of Coastline Under Potential Threat		
	Low	Moderate	High
Africa	49	14	38
Asia	31	17	52
North and Central America	71	12	17
South America	50	24	26
Europe	14	16	70
Former Soviet Union	64	24	12
Oceania	56	20	24
World	49	17	34

The qualitative assessment from low to high reflects increasing population density, increasing density of roads, and increasing density of pipelines in the coastal area.
(*Source:* Bryant et al., 1995.)

shown in the next section, nitrogen and phosphorus from sewage, agricultural runoff, human-induced increases in erosion rate, and other land-based sources have increased the nutrient load of rivers entering the coastal zone by 50 to 200% over the preindustrial fluxes.

It can be shown that a correlation exists between the amount of nutrients entering coastal environments and the population of people living in watersheds upstream of these environments (Figure 11.22). As the population density of a watershed increases, the nitrate and soluble, biologically reactive phosphorus fluxes

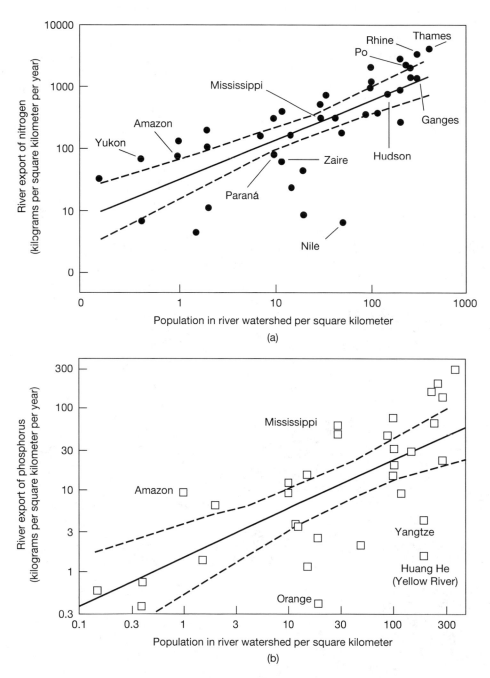

(a)

(b)

FIGURE 11.22 Relationship between the population density in a river watershed and the export of dissolved nitrogen as nitrate (NO_3^-) and phosphorus as soluble, biologically reactive phosphorus by the river to the ocean. The dark solid lines show the linear trend through the data, and the dashed lines are measures of the degree of confidence in the relationship. (a, after Cole et al., 1993) (b, after Caraco, 1995)

of the major river draining that watershed and entering a coastal environment increase. This relationship portends increased nutrient inputs into coastal environments and their enhanced eutrophication because of increasing population density unless steps are taken worldwide to reduce nutrient inputs from agricultural and other sources. The potential for enhanced eutrophication, coupled with increased inputs of pathogenic bacteria and viruses, heavy metals, and synthetic organic compounds into the coastal zone, suggests increased degradation of coastal margin ecosystems and loss of habitat and species. Because major fishing and shellfish industries are located in coastal margin environments, coastal habitat destruction and consequent species loss can have severe economic and social repercussions.

An excellent example of what can happen to a coastal environment because of excess nutrient delivery via rivers is that of the development of the Gulf of Mexico hypoxic zone (Figure 11.23). The Mississippi River and its tributaries in the United States deliver nutrients via runoff to the Gulf derived from natural sources and anthropogenic sources of fertilizer and farm wastes, sewage, industrial and urban wastes, and combustion products of nitrogen. For example, the loading of the landscape by nitrogen from nitrogenous fertilizers and by deposition of combustion nitrogen is particularly heavy in the Midwest. The excess nutrients derived from anthropogenic sources

reaching the Gulf lead to excess algal growth mainly to the west of the Mississippi River's main discharge owing to the average westward flow of nearshore, shallow-water currents in the area (Figure 11.24a). The bacterial decomposition of the dead algae sinking through the water column leads to O_2 depletion of the water and when bottom water O_2 levels reach 2 mg/L or lower, the water is considered hypoxic. Such low O_2 levels lead to the death of most benthic organisms and impact significantly the abundance of fishes and shrimp living in the area. The hypoxic zone is especially well developed in the summer because of the development of a shallow seasonal thermocline. Figure 24b shows the comparative size of the Gulf of Mexico hypoxic zone from 1986 to 2006. The United States Environmental Protection Agency (EPA) has developed an action plan to alleviate the problem of the hypoxic zone, but this is based mainly on voluntary efforts, such as farmers voluntarily using less fertilizer and using it more efficiently. This plan has the potential to reduce the 5-year running average (1996 to 2000) of the size of the hypoxic zone of 14,128 square kilometers to a 5-year running average of less than 5000 square kilometers in 2015. Hypoxic zones throughout the world are developing more frequently, mainly because of nutrient pollution. The United States and Europe apparently have the largest concentrations of these zones (Figure 11.25).

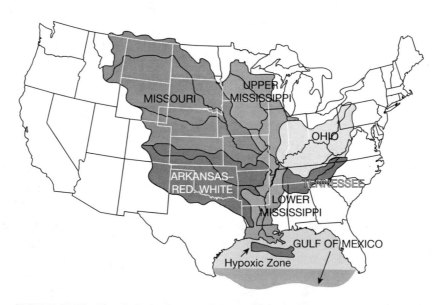

FIGURE 11.23 The drainage basins of the Mississippi River and its major tributaries. These flows transport nutrients to the Gulf of Mexico from fertilizer and farm wastes, sewage, and industrial and urban sources. The nutrients fuel high productivity in nearshore Gulf waters leading to development of the hypoxic zone (see text).

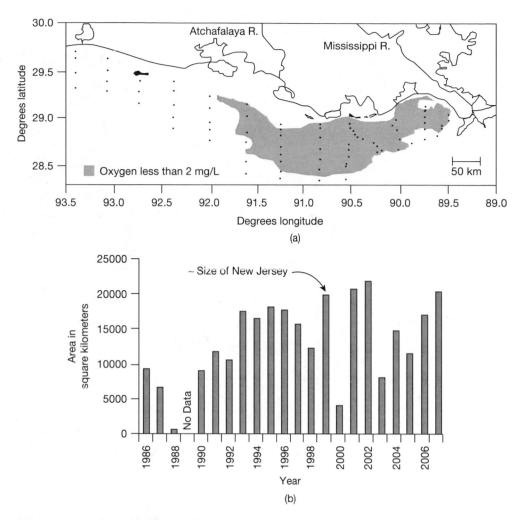

(a)

(b)

FIGURE 11.24   The Gulf of Mexico hypoxic zone. (a) The extent of bottom water hypoxia (shaded area, less than 2 mg/L dissolved O_2) during July 21–25, 1998. (b) The area of the hypoxic zone in the northern Gulf of Mexico from 1986 to 2007 in square kilometers. The size of the hypoxic zone in 1999 was about equivalent to the area of the State of New Jersey in the United States (*Source:* http://toxics.usgs.gov/hypoxia/hypoxic_zone.html).

Another impact of human activities on the coastal zone and the ocean in general is that of marine litter. Floating litter is found throughout the world's oceans and is particularly evident along coasts. Marine litter is floating plastic, wood, and other materials that have been dumped from ships or blown or washed from land. Little is known about the effects of these materials on marine populations; however, many marine organisms have died after ingesting litter or becoming entangled in it. Such litter certainly detracts from the intrinsic beauty of beaches.

A study on Inaccessible Island, a remote island in the Tristan de Cunha group in the South Atlantic Ocean, showed that litter washing up on the island increased dramatically from 1984 to 1990. The litter density in 1984 was 500 pieces per kilometer and in 1990, almost 2500 pieces per kilometer of shoreline. More recently, a large patch of marine litter, mainly high concentrations of suspended plastics, located within the eastern part of the North Pacific Gyre has attracted considerable attention. Eighty percent of the garbage in this patch is from land-based sources. The area of the patch is estimated to be about twice the size of Texas and has been called the Great Pacific Garbage Patch (also known as the Eastern Garbage Patch or Pacific Trash Vortex). Although more needs to be known about the impacts of marine litter on organisms, such evidence of rapid increase in its production, its unsightliness on beaches, and the fact marine organisms, like turtles,

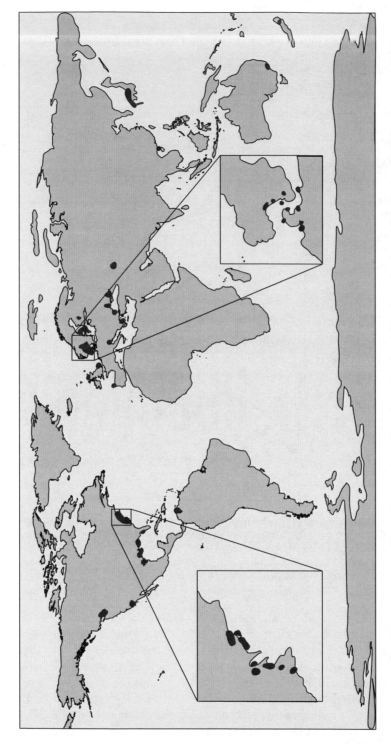

FIGURE 11.25 Global occurrence of hypoxic zones (Map from http://earthtrends.wri.org/text/coastal-marine/map-201.html. Lauretta Burke, Yumiko Kura, Ken Kassem, Carmen Reverge, Mark Spalding and Doug McAllister, 2000. *Pilot Analysis of Global Ecosystems: Coastal Ecosystems.* Reprinted by permission of World Resources Institute: Washington, Dc.).

ingest the litter argues now for strenuous efforts to curb its release.

Wetlands Because their usefulness is not readily apparent, wetlands have often been filled in for agricultural land, urban expansion, or garbage dumps. It has been estimated that wetlands in the United States once comprised approximately 70 million hectares. Less than one-third of this area remains today, although wetland losses in the United States have been slowing over time and apparently between 1998 and 2004, the United States actually gained 32,000 acres of wetland land (Figure 11.26).

In tropical countries, mangroves dominate many wetland areas. Where data exist, more than 50% of mangrove area alone has been lost since the preagricultural era. Table 11.5 gives the extent and percentage loss of mangrove ecosystems for some selected countries. Mangroves, like coral reefs, protect the shoreline from erosion and are important habitats for invertebrates and fish. Coastal wetlands also filter nutrients and prevent them from reaching coastal environments. These systems are particularly vulnerable to sea-level rise and flooding, a potential consequence of climatic change. With removal of a wetland, during a rainstorm, there are higher flood water levels and water flows than when the wetland was present.

Temperate wetland marshes are commonly characterized by an abundant growth of salt-marsh hay, the plant *Spartina*. Interestingly, one of the changes noted when a healthy estuarine salt marsh is disturbed by human activities is the rapid shift in growth from *Spartina* to the plume grass *Phragmites*. This is due to the accumulation of excess sediment and organic detritus in the marsh. The change in vegetation leads to a change in the distribution of species found in the marsh. Tolerant fish like killifish become abundant and important food fish and shellfish such as oysters and mussels may be lost because of the polluted conditions. The Hackensack Meadowlands of New Jersey are a highly disturbed salt-marsh system, now extensively covered by plume grass. Another example of a highly disturbed wetland system is that of the Florida Everglades. Chemical pollution, excess nutrient inputs, and diversion of waterways are leading to rapid deterioration of this ecosystem.

Coral Reefs Nearshore coral reefs (fringing and barrier reefs) are especially vulnerable ecosystems. They are being threatened by chemical pollutants,

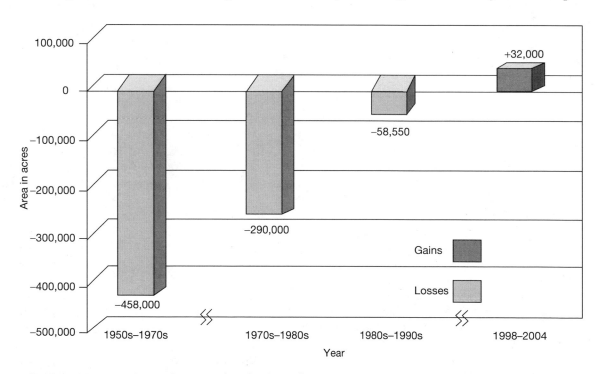

FIGURE 11.26 Annual rate of change of wetland area from the 1950s to 1970s to 1998–2004 in the contiguous United States. Much of the United States wetland area in the twentieth century was lost in the three decades after World War II. In the early twenty-first century, wetland area was actually gained. (*Source:* http://wetlandsfws.er.usgs.gov/status_trends/National_Reports/trends_2005_report.pdf_report.pdf.)

TABLE 11.5 The Areal Extent and Percentage of Loss of Area of Mangrove Ecosystems for Some Selected Countries of the World

Region and Country	Current Extent (1000s of Hectares)	Approximate Percent Lost	Period Covered
Asia			
India	100–700	50	1963–77
Peninsular Malaysia	98.3	17	1965–85
Philippines	140 +	70	1920s to circa 1990
Singapore	0.5–0.6	20–30 +	Preagricultural period to present
Thailand	196–269	25	1979–87
Viet Nam	200	50	1943 to early 1990s
Latin America			
Puerto Rico	6.5	75	Precolonial to present
Ecuador	117 +	30 +	Preagricultural to present
Guatemala	16	30 +	1965–90
Africa			
Cameroon	306	40	Preagricultural to mid-1980s
Kenya	53–62	70	Preagricultural to mid-1980s
Guinea-Bissau	237	75 +	Preagricultural to mid-1980s
Liberia	20	70	Preagricultural to mid-1980s

(*Source:* World Resources Institute, 1996.)

excess nitrogen and phosphorus inputs, siltation from erosion occurring upstream in the drainage basins that are commonly associated with these reef environments, the use of dynamite and poisons to obtain fish from reef environments, and the mining of reef materials for construction purposes. In addition, elevated sea temperatures (a consequence of a global warming brought about by an enhanced greenhouse effect) may induce bleaching in corals. Bleaching is a result of the loss of a coral's symbiotic algae (zooxanthellae) that impart color to the coral. Zooxanthellae contribute to coral nutrition, and without them the coral would eventually die. Bleaching events in recent years have been observed in Bermuda, the Caribbean, and the Pacific. The reasons for these events are controversial and are probably the result of a number of environmental stresses, including certainly warming of the waters in which the corals live. However, with global warming and a predicted average temperature increase of 3°C or more by the year 2100, and other stresses of an enhanced greenhouse (sea-level rise, lowering of the pH of surface seawater), it is very likely that an increasing number of coral ecosystems could be severely degraded in the near future.

Southeast Asia has 30% of the coral reef area of the world. Some 60 to 70% of the reefs are in poor condition because of deforestation and resultant sediment discharge to coastal waters, and because of coral mining and fishing with dynamite. In the Indian Ocean, with 24% of the world's coral reefs, 20% of the reefs have been lost because of coral mining, fishing practices that use explosives, and chemical and nutrient pollution. Caribbean reefs are threatened by deforestation and development along the coasts of the countries of this area. In particular, the reefs of countries with high rates of both population growth and poverty, like Haiti, Jamaica, and the Dominican Republic, are heavily stressed. Reefs of the Atlantic Ocean are threatened by coastal development and tourism. Because of fishing and other stresses on the Bermuda reefs, much of the coral reef area of this North Atlantic group of islands is now in a protected reserve. Pacific Ocean reef systems, with

25% of global reef area, and Middle East reefs, like those of the Red Sea, are least threatened by human activities.

On a global scale to date, 5 to 10% of coral reef ecosystems have been lost, and at current rates of destruction, 60% of these important ecosystems could be lost in the next several decades. Aside from their scenic beauty and intrinsic value, the loss of coral reefs as habitats for fish, as providers of calcareous sediment, and as barriers to beach erosion will have serious economic and social implications.

LONG-TERM PERSPECTIVE In this section, we discuss briefly how the world's coastal zone functioned prior to major human impacts on the land–water system, how it looks today, and the potential future of this important global environment.

Past Prior to extensive land use and fossil fuel burning activities of humankind, carbon, nutrient, and suspended sediment inputs via rivers to the coastal zone were virtually steady and did not change greatly on a decade to centuries time scale. Riverine calcium and bicarbonate inputs were deposited on the seafloor

in the skeletons of organisms, like foraminifera, mollusks, corals, and echinoids. The suspended sediment flux of rivers was about 10 billion tons per year and on entering the ocean settled to the seafloor. The flux of organic carbon in organic matter was 400 million tons per year, and fluxes of nitrogen and phosphorus, about 14 million and 1.4 million tons per year, respectively. The organic matter of terrestrial origin entering the coastal zone via rivers was sedimented and buried on the seafloor, was transported to the open ocean, or was decayed/respired and ultimately its carbon returned to the atmosphere as carbon dioxide gas. A geological quasi-steady state existed (Figure 11.27a).

Present In the past, the coastal zone was a net source of carbon dioxide gas to the atmosphere. It was a net heterotrophic system, consuming more organic carbon than it produced. Also, the precipitation of calcium carbonate led to the release of carbon dioxide to coastal waters and subsequently to the atmosphere. Now with the burning of fossil fuels and land use activities of deforestation, agriculture, and urbanization, the world's coastal ocean has changed (Figure 11.27b). The surface waters of the continental shelf and the

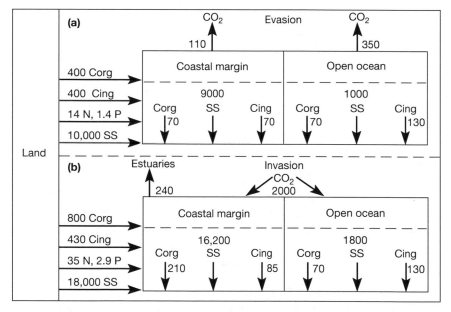

FIGURE 11.27 A portion of the biogeochemical cycles of inorganic carbon (Cing) and organic carbon (Corg), nutrient N and P, and suspended solids (SS) in the land–ocean system. (a) Geological, long-term system; (b) one possible situation today. In (b), the fluxes of organic and inorganic carbon and suspended solids to the seafloor are increased over their pristine geological values in (a). These increases are due to human activities. Notice the net heterotrophic nature of the ocean giving rise to a net flux of CO_2 to the atmosphere due to this organic state and deposition of inorganic carbon prior to human interference in the carbon cycle. Now more CO_2 enters the ocean because of the burning of fossil fuels and deforestation practices (see Chapter 14). Fluxes are in millions of tons of C, N, P, and suspended solids per year. (*Source:* Wollast and Mackenzie, 1989.)

open ocean absorb anthropogenic atmospheric CO_2. However, estuaries have very high fluxes of CO_2 to the atmosphere because of net heterotrophy and calcification. These fluxes offset the uptake of CO_2 by the shelf such that the *global* coastal ocean in the early twenty-first century is still a source of CO_2 to the atmosphere. As atmospheric CO_2 continues to rise because of the emissions of this gas to the atmosphere from human activities, the global coastal ocean will become a sink of atmospheric CO_2 later in this century, like the shelf and open ocean are at present.

The flux of suspended solids via rivers toward the oceans has nearly doubled, to 18 billion tons annually. Perhaps as much as 30% of this sediment flux is retained behind dams or stored in other reservoirs at present, reducing the sediment load to the ocean. Much of this suspended sediment comes from rivers draining Southeast Asia and Oceania, where about 30% of the coral reef cover of the world and much of the production of shallow-water calcareous sediments are found. The rate of sediment discharge is changing today because of competing practices of damming river courses that block sediment discharge to the ocean and other land use practices such as deforestation, which increase the discharge. More than 100 billion metric tons of sediment and 1 to 3 billion metric tons of carbon are now sequestered in reservoirs constructed mainly within the past 50 years. African and Asian rivers transport a greatly reduced sediment load, and Indonesian rivers deliver much more sediment to coastal areas at present than in the past. Most of the sediment that reaches the oceans settles out in the nearshore coastal zone; only 10 to 20% is transported out to the shelf and open ocean. As mentioned previously, the enhanced sediment flux stemming from human activities is responsible for siltation of freshwater and marine water bodies and is contributing to ecosystem deterioration in many regions of the world.

Organic carbon and nitrogen and phosphorus nutrient discharges via rivers to the coastal ocean may have also doubled over their pristine values (Figure 11.27). This has led to enhanced productivity and cultural eutrophication of some coastal marine environments. In these environments, it is likely that there has been a modification of the exchanges between the sea surface and the atmosphere of gases involved in biological activities, such as carbon dioxide, nitrous oxide (N_2O), and dimethylsulfide (DMS). These gases are involved in climatic change; carbon dioxide and nitrous oxide are greenhouse gases and DMS and its conversion in the atmosphere to sulfate aerosol affect the radiative properties of the atmosphere (see Chapters 7 and 13).

The Scheldt and Rhine estuaries of Western Europe, Mobile Bay, San Francisco Bay, Chesapeake Bay, Delaware Bay, Hudson Bay, Potomac Bay, and Kaneohe Bay of the United States, and the Baia de Guanahara of Brazil are excellent examples of the worldwide problem of human modification of estuarine systems and (except for San Francisco Bay) their eutrophication. San Francisco Bay has been extensively modified by human activities. Its wetlands have been diked and filled, resulting in the elimination of habitat for fish and waterfowl. Of the original 2200 square kilometers of tidal marsh, only 6% remain undiked today. Approximately 100 exotic invertebrate species, including the eastern soft-shelled clam, the Japanese littleneck oyster, and the oyster drill and shipworm pests, have been introduced to the bay, changing the composition of its aquatic communities. Freshwater flow to the estuary has been reduced by nearly one-half, leading to changes in the biogeochemical cycles of elements that in turn affect plant and animal communities. Disposal of toxic wastes has contaminated bay sediments and organisms. Many of the major changes in this system occurred decades ago. Modern important stresses involve agricultural, industrial, and domestic waste inputs and further reduction in freshwater inflow to the bay.

Many estuarine systems of the world have suffered similar changes to those of San Francisco Bay and in addition exhibit permanent or seasonal oxygen depletion in their waters. These depletions are brought about by high inputs of organic waste and nutrients of nitrogen and phosphorus, and by inflows of toxic wastes and suspended sediments. An example of a heavily modified and eutrophicated estuarine system is the Scheldt (Escaut) Estuary of Belgium and the Netherlands. Figure 11.28 shows how the dissolved oxygen in waters of this estuary is reduced nearly to zero because of consumption of oxygen via oxidation of the excess organic matter in the estuary. The excess organic matter comes from organic production brought about via nutrient loading of the estuary. The nutrients are leached from fertilizers applied to the land surface in the watershed of the Scheldt River and its tributaries. Organic substances, trace metals, hydrocarbons, and pesticides also enter the Scheldt Estuary from urban, industrial, and agricultural sources located on the shore of the estuary or in its watershed.

The inputs of materials derived from human activities to the Scheldt have not only led to eutrophication of the estuary but have modified many of the natural biogeochemical cycles of substances. Because

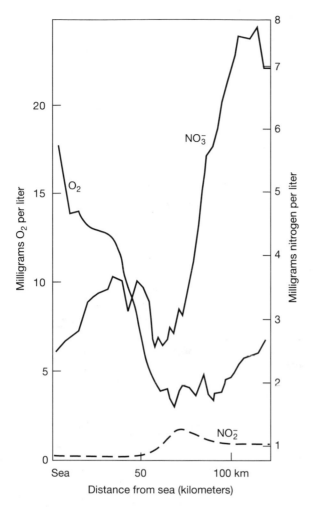

FIGURE 11.28 Distribution of oxygen (O_2), nitrate (NO_3^-), and nitrite (NO_2^-) in waters of the Scheldt Estuary. Notice the very low O_2 and high NO_3^- contents in the upper reaches of the estuary. The increase in O_2 down the estuary is due to fresh oxygenated waters entering the estuary from the North Sea. The decrease in NO_3^- and NO_2^- is due to denitrification to N_2 and N_2O gas. (*Source:* Somville, 1980.)

of the change in the oxygen content of Scheldt waters, manganese and nitrogen cycling in the estuary has changed. In the front between oxygen-depleted and oxygenated waters, manganese is rapidly deposited as manganese oxide as it changes valence state from 2^+ to 4^+, and nitrogen present as nitrate (NO_3^-) and nitrite (NO_2^-) is converted to the gases N_2 and nitrous oxide because of the process of denitrification and is lost to the atmosphere (Figure 11.28). In addition, heavy metal inputs to the estuary have increased because of industrial activities, leading to modification of biological uptake of these metals in the estuary. Furthermore, siltation of the estuary because of upstream agricultural and urbanization

activities is a recurring problem. Despite some regulations governing anthropogenic inputs, the Scheldt remains an example of a heavily polluted estuarine system, one that is not uncommon in the world. Its neighbor, the Rhine, is another example.

In light of the preceding discussions concerning the San Francisco Bay and Scheldt Estuary, it is informative to look at another land–coastal environment, that of islands. The example chosen is the state of Hawaii in the United States, located in the Subtropical Gyre of the north Pacific Ocean and the Northeast Trade Wind belt. The lower southernmost Hawaiian Islands have the vast majority of the population of Hawaii of 1.2 million persons. The island of Oahu has the highest coastal-zone population density of any coastal state in the United States. Figure 11.29 shows the flows of nitrogen through the state of Hawaii in its pristine state before extensive modification of the landscape by both the indigenous Hawaiians and subsequent immigrant populations and the situation in 1992. Notice how the number of sources of nitrogen has increased from the pristine situation to the 1990s. These new sources are mainly the application of nitrogenous fertilizers to croplands, the combustion of fossil fuels, and the importation of feed for livestock and food. Notice also that there are new sources of nitrogen to the atmosphere and increased fluxes of nitrogen to coastal bays, estuaries, and shelf/open ocean environments by river and groundwater flows owing to human activities. The total throughput of nitrogen in the Hawaiian environment in the 1990s has doubled since pristine time because of these activities. This doubling of the nitrogen flux is a similar factor increase to the doubling of the global flux of nitrogen owing to human activities. Despite the fact that nitrogen flows to the coastal zone have increased substantially because of agricultural and combustion activities in Hawaii, there are few coastal ecosystems in Hawaii that exhibit significant cultural eutrophication. This is mainly due to the fact that the residence time of excess nitrogen from human activities in most coastal waters of Hawaii is relatively short. The excess nitrogen is rapidly diluted by low-nutrient shelf and open ocean waters and transported out of coastal ecosystems to the open ocean, only a short distance seaward.

Another problem particularly affects estuaries in semiarid and arid regions of the world, like the Nile River of Egypt. In these climatic regions, the diversion of river water for agricultural use and the impoundment of sediment behind dams on rivers upstream of an estuary further complicate other human modifications of the estuary. Reduced water and sediment

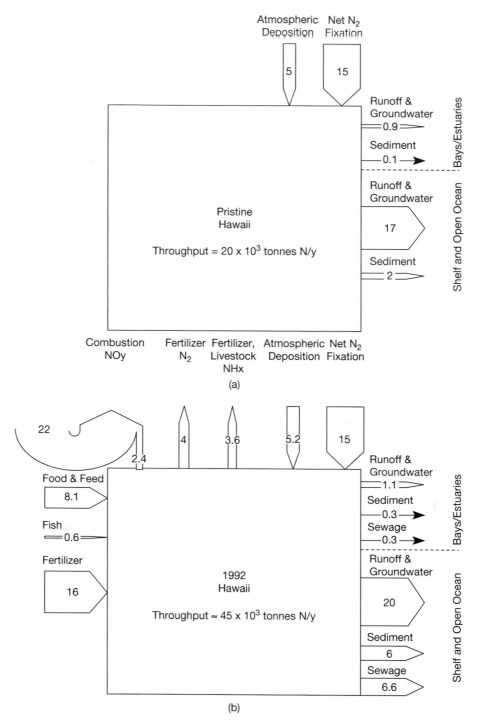

FIGURE 11.29 Flows of nitrogen in the state of Hawaii, United States, during pristine time (before extensive human interference in the system) and in 1992. Notice how the number of flows has increased since pristine times, mainly owing to human activities of fossil fuel combustion, application of nitrogenous fertilizer to croplands, and importation of food and feed for livestock. Note also the increased fluxes of nitrogen to the ocean mainly because of the human activities on land. Fluxes in 1000 metric tons N per year. (*Source:* Hoover and Mackenzie, unpublished data.)

discharge into the estuarine system results in decreased sediment deposition and nutrient inputs. This may give rise to flooding of wetlands by seawater and to decreased organic productivity.

Future   What will coastal margins look like in the future? This is a difficult question to answer, because the answer involves a number of future economic and social considerations, as well as environmental ones. However, it is clear that unless we change our agricultural and industrial practices worldwide to decrease discharges of potential pollutants, coastal margin ecosystems will continue to deteriorate. The continuous movement of populations to the coastal zone will also increase stresses on the system. Nutrient loading of estuaries, bays, and other coastal environments will lead to increases in production of organic matter in these systems; gross primary productivity will increase. Figure 11.30 shows that as gross primary productivity increases, coastal environments tend to become more heterotrophic, and the gap between production and consumption of organic carbon widens. As a result, more carbon is released from coastal waters to the atmosphere as carbon dioxide. Such a situation could lead to loss of oxygen via decay/respiration processes from an increasing number of coastal margin environments in the future and to an enhanced potential for eutrophication. Loading of these environments with chemical pollutants like trace metals, pesticides, and hydrocarbons as well as organic matter from external sources will tend to lead to even more rapid environmental deterioration of a system undergoing eutrophication.

It should be pointed out that on a global scale it is difficult to estimate the magnitude or even direction of change in the metabolism (degree of heterotrophy or autotrophy) of estuarine systems. The principal problem is that the database is simply insufficient. Figure 11.30 shows that two estuarine systems strongly affected by human activities exhibit very different trophic states. On one hand, Narragansett Bay is net autotrophic, producing more organic carbon than it consumes. This system is heavily polluted owing to large inputs of sewage-derived inorganic nutrients, for example, NO_3^-. On the other hand, Southhampton Water is strongly heterotrophic. This system is the site of a major oil refinery and receives significant inputs of industrial effluents and organic matter. Thus, the question of how the net metabolism of the coastal ocean will change in the future owing to increased stresses brought about by human activities, including climatic change, is difficult to answer. However, the suspicion is that these systems on a worldwide basis will change trophic status and become more eutrophic.

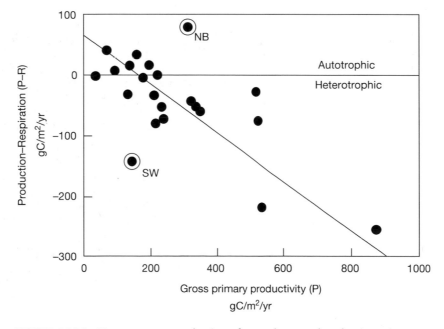

FIGURE 11.30   Net ecosystem production of coastal waters (production minus respiration) as a function of gross primary productivity. NB is Narragansett Bay and SW is Southhampton Water (see text).   (*Source:* Smith and Hollibaugh, 1993.)

Coastal margin ecosystems are being severely impacted by the activities of humankind. They are perhaps some of the most vulnerable systems on Earth, and on a worldwide basis, their future is in jeopardy. These systems are important not only for their innate beauty and recreational use, but because they are involved in the exchange of carbon dioxide gas between the ocean and atmosphere, they act as breeding grounds for many marine organisms, and they support fish and shellfish populations and the industries that depend on these populations.

Concluding Remarks and the 21st Century

Studies of natural or human-induced change in Earth's biogeochemical systems must recognize that these biogeochemical systems are complexly interwoven and interactive. This statement certainly holds true for the behavior of the coupled land–water system and its response to human activities. For example, to understand change in the coastal ocean involves considerations of this system as a linked component of the whole Earth system. The coastal ocean is influenced by the open ocean and atmosphere, and, in turn, this affects the open ocean, atmosphere, and at least nearby areas of land, if not entire continents. Materials entering the coastal zone via rivers and groundwater flow and in deposition from the atmosphere are sedimented on the seafloor, exit to the open ocean, or return to the atmosphere, perhaps to be deposited back on land. Thus, to understand the role and response of the land–water system to global change necessitates an evaluation of how this system *in a global context* has functioned in the past, how it works today, and how it will function in the future in response to human engineering and hydrological and environmental impacts.

Table 11.6 is a summary of the myriad types, sources, and effects of pollution in the coastal marine environment. Nearly three-quarters of marine pollution stems from human activities on land. Because of the many types and sources of coastal marine pollution, both the monitoring and the control of potential pollutants in coastal environments are difficult. The management of coastal ecosystems in light of these pollution problems will be a particularly difficult task for the developing countries in the twenty-first century as their economies develop and expand. If they follow the same pathway of development as the developed nations have since the Industrial Revolution, coastal ecosystems will be severely stressed, and we can anticipate continued environmental degradation of these systems worldwide.

TABLE 11.6 Type, Source, and Potential Effects of Pollution on Coastal Marine Life and the Ecosystem

Type	Primary source/cause	Effect
Nutrients	Runoff: approximately 50% from sewage; 50% from forestry, farming, and other land use. Also airborne nitrogen oxides from power plants, cars, etc.	Feed algal blooms in coastal waters. Decomposing algae depletes water of oxygen, killing other marine life. Can spur toxic algal blooms (red tides), releasing toxins that can kill fish and poison people.
Sediments	Erosion from mining, forestry, farming, and other land use; coastal dredging and mining.	Cloud water; impede photosynthesis below surface waters. Clog gills of fish. Smother and bury coastal ecosystems. Carry toxins and excess nutrients.
Pathogens	Sewage, livestock.	Contaminate coastal swimming areas and seafood, spreading cholera, typhoid, and other diseases.
Alien species	Several thousand per day transported in ballast water; also spread through canals linking bodies of water and fishery enhancement projects.	Outcompete native species and reduce biological diversity. Introduce new marine diseases. Associated with increased incidence of red tides and other algal blooms. Problem in major ports.
Persistent toxins, (PCBs, DDT, heavy metals, etc.)	Industrial discharge; wastewater from cities; pesticides from farms, forests, home use, etc.; seepage from landfills.	Poison or cause disease in coastal marine life, especially near major cities and industry. Contaminate seafood. Fat-soluble toxins that bioaccumulate in predators can cause disease and reproductive failure.

Type	Primary Source/Cause	Effect
Oil	46% from cars, heavy machinery, industry, other land-based sources; 32% from oil tanker operations and other shipping; 13% from accidents at sea; also offshore oil drilling and natural seepage.	Low-level contamination can kill larvae and cause disease in marine life. Oil slicks kill marine life, especially in coastal habitats. Tar balls from coagulated oil litter beaches and coastal habitat. Oil pollution is down 60% from 1981.
Plastics	Fishing nets; cargo and cruise ships; beach litter; wastes from plastics industry and landfills.	Discarded fishing gear continues to catch fish. Other plastic debris entangles marine life or is mistaken for food. Plastics litter beaches and coasts and may persist for 200 to 400 years.
Radioactive isotopes	Discarded nuclear submarines and military waste; atmospheric fallout; also industrial wastes.	Hot spots of radioactivity. Can enter food chain and cause disease in marine life. Concentrate in top predators and shellfish, which are eaten by people.
Thermal	Cooling water from power plants and industrial sites.	Kill off corals and other temperature-sensitive sedentary species. Displace other marine life.
Noise	Supertankers, other large vessels, and machinery.	Can be heard thousands of kilometers away under water. May stress and disrupt marine life.

(*Source:* Weber, 1993.)

Study Questions

1. What percent of the land surface is ice, rock, semidesert, and desert?
2. What are the four major types of soil degradation?
3. What are five causes of soil degradation?
4. Thirty percent of degraded vegetated land is a result of deforestation. How many square meters of land does this represent?
5. (a) If 1.7 billion hectares of land are under cultivation and during the last 20 years 480 billion tons of topsoil have been lost from agricultural land, how much topsoil has been lost per square meter?
   (b) What thickness of soil has been lost?
6. The global average annual municipal waste generation per capita is 240 kilograms per person.
   (a) What is the total waste generated annually by a human population of 6.7 billion?
   (b) How many grams of waste generation is this per square meter of vegetated land area?
   (c) Compare Canada and Germany in terms of their municipal waste efficiency.
7. The total energy consumed by the world in 2006 was 472 quadrillion Btu. Fourteen percent of this energy consumption was provided by biomass, a source of energy for 50% of the world's people.
   (a) How much world energy was provided by biomass?
   (b) How much degraded land was produced, assuming an initial biomass of 25 kg/m^2? [1 green metric ton of wood (50% moisture) $= 8 \times 10^6$ Btu].
   (c) What was the per capita amount of degraded land produced in 2006?

(d) What were the potential total emissions of carbon dioxide to the atmosphere from burning of this biomass? (Assume 50% of the biomass by weight is C.)
   (e) What was the per capita emission?
8. From Figure 11.14, calculate the percentage of annual water withdrawal per capita relative to annual internal renewable water resource per capita for each region of the world. Rank the regions according to decreasing percentage. Does this calculation tell you anything about how the countries of the different regions are treating their water resource?
9. What are three major types of water pollutants?
10. What is cultural eutrophication?
11. (a) What are some major human activities leading to destruction of coral reef ecosystems?
    (b) What is the most threatened ocean area of coral reef destruction?
12. (a) A highly soluble and very stable mercury pollutant is suddenly and then continuously dumped into a lake. This has happened at Minamata Bay in Japan, for example, where the Hg input from a chemical factory located on a river entering the bay poisoned people. The input rate for our fictitious lake is 0.16 metric tons of Hg per day. The volume of the lake is $4 \times 10^7 \text{ m}^3$ and the mean rate of flow of water through the lake is $8 \times 10^{14} \text{ m}^3/\text{day}$. There is no net evaporation or precipitation for the lake; these fluxes are balanced, and the Hg is well mixed throughout the lake. What is the residence time of water in the lake and what will be the eventual

steady-state concentration of Hg in the lake in parts per million (ppm)? (1 m^3 of water equals 1 metric ton.)

(b) Assume there is a federal regulation prohibiting this input of Hg into the lake. The company responsible is brought into court and after trial is compelled to stop the Hg pollution of the lake. The company complies quickly and stops the Hg input into the lake. Approximately how long in days will it take the lake to return to its "pristine" water quality state with respect to Hg? This time is called *renewal time (Hint:* See the box on residence time in your book.)

13. What are some changes in coastal environments during the past three centuries?

14. (a) How have fluxes of nutrients, suspended sediments, and organic matter by rivers to coastal margins changed during the past few centuries?

(b) What impacts have these increased fluxes had on coastal environments?

15. Based on your reading of this chapter, what is your opinion of the future of aquatic systems?

Additional Sources of Information

Caraco, N. F., 1995, Influence of human populations on P transfers to aquatic systems: A regional scale study using large rivers. In H. Tiessen (ed.), *Phosphorus in the Global Environment: Transfers, Cycles and Management*, SCOPE 54. John Wiley, New York, pp. 235–244.

Gleick, P. H., 2009, *The World's Water, 2008–2009: The Biennial Report on Freshwater Resources*. Island Press, Washington, DC, 402 pp.

Laws, E. A., 2000, *Aquatic Pollution: An Introductory Text*. John Wiley, New York, 639 pp.

Malle, K.-G., 1996, Cleaning up the River Rhine. *Scientific American*, January, pp. 70–75.

Millennium Ecosystem Assessment, 2005, *Ecosystems and Well-being: Synthesis*, Island Press, Washington, DC, 37 pp.

Pimentel, D. C., Harvey, P., Resosudarmo, P., Sinclair, K. et al., 1995, Environmental and economic costs of soil erosion and conservation benefits. *Science*, v. 267, pp. 1117–1123.

Postel, S., 1996, *Dividing the Waters: Food Security, Ecosystem Health, and the New Politics of Scarcity*. Worldwatch Paper 132, Worldwatch Institute, Washington, DC, 76 pp.

Postel, S. L., Daly, G. C., and Ehrlich, P. R., 1996, Human appropriation of renewable fresh water. *Science*, v. 271, pp. 785–788.

Vitousek, P. M., Aber, J. D., Howarth, R. W., Likens, G. E. et al., 1997, Human Alteration of the Global Nitrogen Cycle: Sources and Consequences. *Ecological Applications*, v. 7, pp. 737–750.

The Changing Atmosphere: Acid Deposition and Photochemical Smog

Human activity, while bringing unprecedented gains in welfare, is also endangering the ability of the Earth and its environmental systems to support future progress.

LEWIS PRESTON, PRESIDENT, WORLD BANK, 1992

The gaseous composition of the atmosphere has changed greatly during the evolution of the planet. Without much doubt, early Archean atmospheric composition was dominated by nitrogen, as is the case today. However, it is very likely that at this time, there was no oxygen in the primitive atmosphere, and carbon dioxide concentrations were about a hundred times or more greater than the levels of today or of the past several millions of years. Methane and other reduced gas concentrations were also most likely higher than present-day levels. As the planetary atmosphere evolved, oxygen levels rose and carbon dioxide and methane concentrations fell. For at least the past 600 million years (the Phanerozoic), the atmosphere of Earth has been fully oxygenated. However, Phanerozoic atmospheric composition did vary. An excellent example of compositional variation is that of carbon dioxide, which changed in concentration in an oscillatory fashion during the Phanerozoic by more than a factor of 10 (see Chapter 7).

Nitrogen, oxygen, and argon currently comprise approximately 99.9% of the gases in the planet's atmosphere, and trace gases of carbon dioxide, sulfur and nitrogen oxides, methane, and ozone account for less than 0.1% (see Chapter 4). However, atmospheric emissions of some trace gases, including CO_2, CH_4, N_2O, SO_x, NO_x, and the human-made halogenated carbon gases, like chlorofluorocarbons (CFCs) and hydrofluorocarbons (HFCs), have been increasing at a rapid rate during much of this and the past century as a result of human activities. These increased emissions have had a recognizable impact on atmospheric composition and have resulted in environmental problems such as acid deposition (a regional–continental scale problem), stratospheric ozone depletion (a global problem), and a potential warming of the planet because of the enhanced greenhouse effect (a global problem).

There are a number of atmospheric pollution issues, as shown in Table 12.1. These air pollution problems have different time and space scales (Figure 12.1). In some cases only a local area is involved, and pollutants may appear and disappear in a matter of hours. Indoor air pollution caused by smoke, nitrogen oxides, and volatile organic emissions from cooking, building materials, and machinery is an example of air pollution restricted to small, enclosed areas and usually of short duration. In other cases, pollutants may travel tens or hundreds of kilometers and remain in the atmosphere for days, weeks, or longer. On a time scale of hours to days, urban air pollution and photochemical smog, which may impact areas downwind of urban areas, are examples of air pollution that affects regions of moderate size. Acid deposition can be a regional- to continental-scale pollution

problem, while stratospheric ozone depletion and increased concentrations of greenhouse gases and global warming are global-scale problems.

In this chapter, the shorter time and space scale atmospheric pollution problems are considered. The first is the deposition of acidic materials of sulfur and nitrogen derived from human activities (acid deposition) and implications for ecosystems and human society. The second is the formation of photochemical smog and tropospheric ozone and haze. These problems have been overshadowed by all the talk about global climate change but they remain environmental issues of concern, particularly for the developing and industrializing countries. The global environmental problems (Figure 12.1) of global warming and stratospheric ozone depletion are discussed in Chapter 14.

TABLE 12.1 Important Issues of Atmospheric Pollution

Indoor air pollution	Smoke, nitrogen oxide, and/or organic pollutant emissions from cooking, building materials, and machinery may reach hazardous levels in confined atmospheres of homes and offices.
Urban air pollution	Elevated levels of sulfur oxides, nitrogen oxides, carbon monoxide, and respirable particles are often found in urban areas, with their high densities of people, industries, and vehicles. In some countries, rural areas are also affected by these pollutants.
Photochemical smog	The combination of sunlight, nitrogen oxides, and hydrocarbons results in the production of ozone and other oxidants, which may reach hazardous levels in and downwind of urban areas.
Acid deposition	Emissions of sulfur and nitrogen oxides result in the acidification of precipitation, acid deposition, and damage to receiving ecosystems.
Long-range transport of air pollutants	Many pollutant species exist long enough in the atmosphere to be transported for long distances, often across political boundaries.
Toxic trace metals and synthetic organic compounds	Many trace metal and organic pollutants emitted into the environment are persistent and toxic and enter food chains in distant terrestrial and marine ecosystems, e.g., in the Antarctic and remote oceans.
Regional and global background air pollution	Many pollutants are now found in remote areas throughout the global atmosphere, altering atmospheric composition far from source regions.
Anthropogenic perturbations to biogeochemical cycles	Anthropogenic emissions of several substances—e.g., sulfur, nitrogen, carbon, and mercury—now rival or exceed natural emissions, resulting in significant perturbations to natural biogeochemical cycles.
Tropospheric ozone	Increasing levels of tropospheric ozone in the Northern Hemisphere, as a result of increasing emissions of precursor nitrogen oxides and hydrocarbons, are raising concerns about loss in the oxidizing capacity of the atmosphere.
Accidental releases of hazardous materials	Unexpected releases of pollutants into the atmosphere, e.g., from the Chernobyl accident or Kuwait oil fires, can have widespread and long-term adverse effects on the environment and human health.

(*Source*: Whelpdale, 1991.)

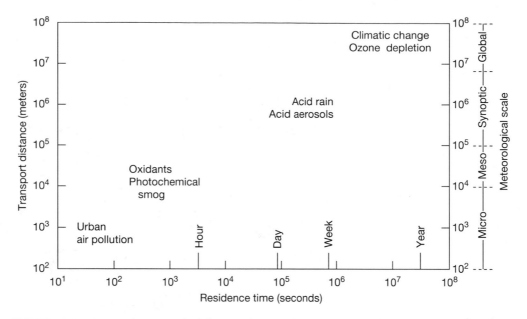

FIGURE 12.1 Time and space scales of atmospheric pollution in terms of transport distance of various atmospheric chemical species and the terms applied to the meteorological scale of transport.

ACID DEPOSITION

Acid deposition is the deposition of acids from the atmosphere in precipitation that falls as rain, sleet, snow, gases absorbed on plant surfaces, and particulates on the planetary surface (Figure 12.2). Fog and dew also can be acidic. Acid deposition is caused by two principal airborne pollutant gases in the atmosphere, sulfur oxides (SO_x) and nitrogen oxides (NO_x). The gases react chemically with moisture (H_2O), hydroxyl radical (OH^*), and sunlight to form microscopic droplets of sulfuric (H_2SO_4) and nitric (HNO_3) acids and generally less importantly, ammonium sulfate [$(NH_4)_2SO_4$] and ammonium nitrate (NH_4NO_3). A number of factors can affect these reactions, including temperature, humidity, light intensity, atmospheric transport, and the surface characteristics of atmospheric particulates. For sulfur dioxide (SO_2), the first step in conversion of this gas to sulfuric acid can be its reaction with liquid water in clouds:

$$H_2O + SO_2 \rightarrow HSO_3^- + H^+.$$

The HSO_3^- then further reacts with an oxidizing agent like hydrogen peroxide (H_2O_2) or ozone (O_3) to form sulfuric acid. The hydroxyl radical (OH^*), an atmospheric compound whose molecules have electrons that are not paired with other electrons, can also initiate the oxidation of SO_2. The hydroxyl radical also plays an important role in the oxidation of NO_x gases in the atmosphere to nitric acid. It is an oxidizing agent for many reduced gases that enter the atmosphere.

Acid rain is formed when the chemical reactions take place in the atmosphere and the acidic constituents are dissolved in rainwater. The resultant precipitation that occurs is referred to as **wet deposition. Dry deposition** takes place when sulfur- and nitrogen-bearing, fine-grained particles (aerosols) fall on Earth's surface. On the land surface, these soluble particles then react with water to form sulfuric and nitric acids. Both processes may cause acidification problems in aquatic and land systems.

Formation of Nitrogen and Sulfur Oxides

Air is composed of 78% diatomic nitrogen (N_2) gas, and at normal temperatures nitrogen does not combine with other elements via inorganic reactions. At higher temperatures, as in the exhausts of combustion engines and in the smoke stacks of some industrial plants, atmospheric nitrogen combines with oxygen to form nitrogen oxides (NO_x): nitric oxide (NO), nitrogen dioxide (NO_2), and nitrogen tetraoxide (N_2O_4). The colorless and odorless nitric oxide gas and the pungent, red-brown nitrogen dioxide gas are important components of polluted air. An example reaction is:

$$N_2 + O_2 = 2NO.$$

The important point is that in the case of anthropogenic nitrogen oxide emissions, the major source of

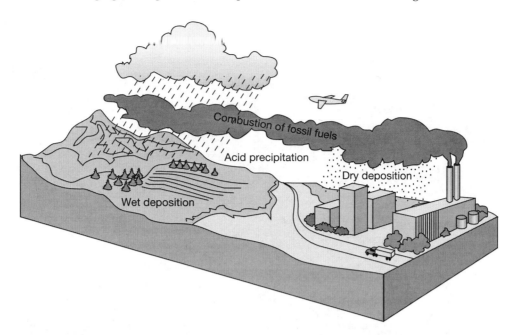

FIGURE 12.2 Diagram illustrating how acid precipitation is formed. Emissions of acid gases of sulfur and nitrogen to the atmosphere from combustion of fossil fuels in transportation and industrial sources chemically react in the atmosphere and precipitate as acid precipitation of wet and dry deposition.

the nitrogen is the diatomic nitrogen gas of the atmosphere. Lesser amounts of NO_x are derived from organic nitrogen in the fuel source.

On the contrary, in the case of sulfur oxide (SO_x) emissions, the sulfur is derived from within the fuel source, where it is found bound in organic compounds and, as in coal, contained in fine-grained sulfides of iron (e.g., FeS_2, the mineral pyrite). On burning, these sulfur-bearing compounds in fossil fuels are oxidized and released to the atmosphere as sulfur dioxide (SO_2) and other sulfur oxides. The colorless sulfur dioxide gas, with a suffocating, choking odor, can be an important compound found in polluted air. Table 12.2 summarizes the series of reactions that sulfur and nitrogen compounds undergo when gases of these elements are produced in internal combustion and furnace sources. The generalized reactions involving the emission products in the atmosphere in the formation of sulfuric and nitric acids are also given in the table.

pH Factor

The degree of acidity is an important factor in acid deposition. The acidity of aqueous solutions is usually stated in terms of pH, where pH is defined as the negative of the common logarithm of the H^+

TABLE 12.2 Steps in the Formation of Sulfuric and Nitric Acid from Source to Atmosphere

The Formation of Sulfuric Acid

Step 1. The oxidation of sulfur when coal is burned:

$$S + O_2 = SO_2$$

Step 2. Further oxidation in the atmosphere and combination with water vapor to form sulfuric acid. An example overall reaction is:

$$OH^* + SO_2(+O_2, H_2O) = H_2SO_4 + HO_2.$$

The Formation of Nitric Acid

Step 1. Nitrogen fixation in hot engines and furnaces:

$$N_2 + O_2 = 2NO$$

Step 2. Further oxidation in the atmosphere to form nitric acid. An example set of reactions is:

$$HO_2 + NO = OH^* + NO_2$$
$$OH^* + NO_2 + M = HNO_3 + M$$

where M is a third body chemical species that dissipates energy and HO_2 is the atmospheric hydroperoxyl radical.

(*Source:* Christensen, 1991.)

ion concentration (actually activity) (Figure 12.3). The hydrolysis of water may be written as

$$H_2O + H^+ + OH^-.$$

If the concentrations of H^+ and OH^- are both expressed in moles per liter, then at 25°C, the product of their concentrations is approximately equal to 10^{-14}:

$$(H^+)(OH^-) = 10^{-14}.$$

Parentheses denote concentrations. In pure water that is neutral, the concentration of H^+ equals that of OH^-; that is, $(H^+) = (OH^-) = 10^{-7}$ moles per liter. The pH of the water is 7. The scale for pH ranges from 0 to 14, with 7 being neutral (Figure 12.3). Solutions of vinegar and lemon have pH values of less than 7 and are considered acidic. Baking soda ($NaHCO_3$) and lime solutions [$Ca(OH)_2$, lime water] have pH values greater than 7 and are referred to as basic or alkaline solutions. Pure distilled water has a pH of 7.

Natural precipitation is slightly acidic because carbon dioxide and trace quantities of other gases in the atmosphere dissolve in rainwater and form a weak acid dominated by carbonic acid (H_2CO_3) with small amounts of sulfuric (H_2SO_4), nitric (HNO_3), and organic acids (e.g., formic acid, HCOOH). For CO_2, the chemical reactions involving solution of this gas in water are as follows (see Chapter 5):

$$CO_{2(gas)} + H_2O = H_2CO_3$$
$$H_2CO_3 = H^+ + HCO_3^-$$
$$HCO_3^- = H^+ + CO_3^{2-}.$$

Because the H_2CO_3 (carbonic acid) is a source of hydrogen ions, when atmospheric CO_2 is dissolved in rainwater, the water will contain an excess of H^+ ions over OH^- ions, and the water will have a pH of less than 7. A pH of about 5.6 is what one would expect for rainwater in equilibrium with the concentration of CO_2 in the clean atmosphere. Because of the other naturally occurring acid sulfur and nitrogen gases in the atmosphere, the pH of unpolluted rainwater normally registers 5.2 to 5.6 on the pH scale. Acid rain is defined as precipitation with a pH less than about 5.2 on an annual basis (Figure 12.3). Rain in many industrial areas may be up to 1000 times more acidic than natural precipitation. For example, water obtained from fog in Western Europe may have a pH of less than 3, and pH values below 2 have been measured in some very acid fogs. In early December 1982, the city of Los Angeles experienced severe acid fog conditions for a period of two days. The pH of the water in the highly acidic mist particles was 1.7!

Sweden's Dying Lakes

Acid precipitation is not a new situation. Evidence of the phenomenon has existed for more than a century in the records of the sulfate concentration of rainfall near industrialized areas, like those in Great Britain. In the mid-1800s, Robert Angus Smith conducted studies on acid rain in urban areas of England. In the 1940s, a network of stations was set up in Sweden and eventually throughout Europe to sample precipitation and determine its composition. In 1955, acid precipitation was recorded in the English countryside far from the urban sources of the SO_x and NO_x emissions responsible for the acidity. In the early 1950s, Swedish scientists began to observe pH values lower than normal (more acidic, less than 5.2) in precipitation over their country. Initially they were not sure of the cause of these abnormally low values or of

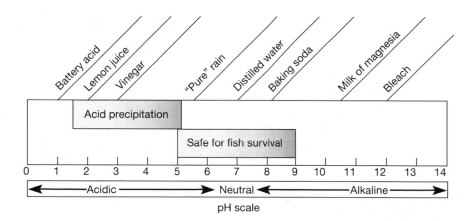

FIGURE 12.3 The pH scale with pH values of some common solutions and the range of pH of acid precipitation. A pH of 7 is considered neutral on this scale. Solutions with lower values are acidic; those with higher values are alkaline.

the ecological and other effects that might occur because of the high acidity of the rainwater. Nearly a decade later, monitoring of the pH of United States precipitation also revealed low pH values in the eastern part of the country. By the late 1960s, fish populations in lakes in the Adirondack Mountains of New York State were found to be declining rapidly. This decline was initially thought to be caused by local pollution from nearby sources. Nearly concurrently, salamander populations were observed to be declining rapidly in the northeastern United States. Also, at about the same time, declining fish populations were observed in lakes of Sweden and Norway and were a major concern to the local fishing industry.

Scientists in Sweden, the United States, Canada, and England conducted a series of research programs designed to ascertain the source of the problem. The longest-term experimental study of acid deposition in the United States has been conducted in the United States Forest Service Hubbard Brook Experimental Forest in the White Mountains of New Hampshire. Between 1964 and 1974, the average annual pH values of acid precipitation in this area ranged from 4.0 to 4.2, and the total annual hydrogen ion input to the forested ecosystem had increased by 36% (Figure 12.4). In Canada the Experimental Lakes Area program in southern Ontario began artificial lake acidification experiments in the mid-1970s. This program was designed to assess the effects of acid deposition on lake organisms. A large study of acid precipitation

was begun in 1980 in North America under the auspices of the National Acid Precipitation Assessment Program (NAPAP). This 10-year research program resulted from the United States Acid Precipitation Act of 1980. In 1990, the program's 27 technical reports and three-volume integrated assessment were released (NAPAP, 1990). In 1990, at the end of the initial 10 years, NAPAP was reauthorized as an open-ended program under the United States Clean Air Act Amendments. On the basis of these studies and others, it has been clearly demonstrated that the acidity of lakes and soils increases when rainwater that is low in pH (acid rain) falls onto watersheds or directly into lakes.

The Swedish studies concluded that acid deposition and the resultant changes in the pH of lake waters were responsible for the killing of fish and plants in lakes in Scandinavia. Sulfur and nitrogen, which were responsible for the acidification of rainwater, were arriving in Sweden and Norway in airborne pollution generated from the combustion of fossil fuels, both coal and oil, in the industrialized nations of Europe. Sweden's dying lakes brought international attention to the problem of acid deposition. Sweden has more than 85,000 lakes that are greater than one hectare in size. Of these, 14,000 are acidified by air pollution and 4,000 are severely acidified and unable to support lake organisms that are sensitive to acidic conditions. Acid-sensitive fish, insects, and other species are absent from nearly

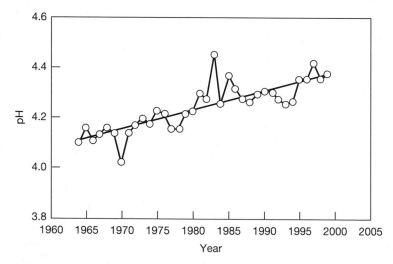

FIGURE 12.4 Annual values of pH weighted by volume of precipitation at Hubbard Brook, New Hampshire, United States. The Hubbard Brook experimental ecosystem has the longest continuous record of precipitation chemistry in North America. The recent increases in pH are due mainly to a decrease in anthropogenic emissions of sulfur dioxide to the atmosphere in the eastern United States. (From Hubbard Brook Ecosystem Study, http://www.hubbardbrook.org/overview/research_activities.htm. Data provided by Gene E. Likens through research funded by the NSF (including LTER and LTREB programs) and The Andrew W. Mellon Foundation. Used by permission of Gene E. Likens.)

40% of Sweden's rivers and streams. The number of acidified lakes would be on the order of 17,500 if liming (addition of calcium carbonate to the lake waters to buffer its pH) had not been carried out to restore the pH of many of Sweden's waters in recent years. Figure 12.5 illustrates some examples of Swedish lakes acidified from acid deposition in the late 1960s. During the period 1990 to 2004, 160 lakes in Sweden and Norway exhibited clear signs of recovery from acidification and higher pH due to the reduction in European emissions of acid gases to the atmosphere.

Similarly, pollution from industrial and transportation sources in the northeastern and midwestern areas of the United States and southeastern Canada has been transported over large areas of eastern North America. By the 1970s, in the Adirondack Mountains, more than 50% of the lakes above 600 meters had a pH of less than 5, and 90% of the lakes contained no fish species. Thus, in both Europe and North America, natural fluxes are still being overwhelmed by sulfur and nitrogen oxides released from fossil fuel burning electrical-generating and other industrial plants, as well as from vehicles. Subsequent enhanced acidification of precipitation has been shown to have

a major impact on the environment of lakes, soils, and possibly forests. Since the 1960s, lake acidification has been recognized in a number of countries (Table 12.3). Acid deposition is a major pollution problem of regional or continental dimensions that crosses political boundaries. However, for North America the integrated assessment report of 1990 concluded that the effects of acid deposition on ecosystems were probably less than previously thought.

As a matter of historical interest, prior to the problem of acid precipitation in Sweden, there is evidence for changes in lake acidity because of deglaciation and the introduction of agriculture to the region. The newly deglaciated land of Sweden 10,000 years ago was rich in alkaline cations like calcium. The pH of Swedish lakes at that time was close to neutral. However, in the time interval prior to the introduction of agriculture to the region, the alkaline cations were leached from the soils, and acids built up in the region because of the establishment of new forests and concomitant production of organic acids by the decomposing forest litter in the soils. Thus, the flux of alkaline cations to Swedish lakes declined as the flux of organic acids increased. This led to an overall

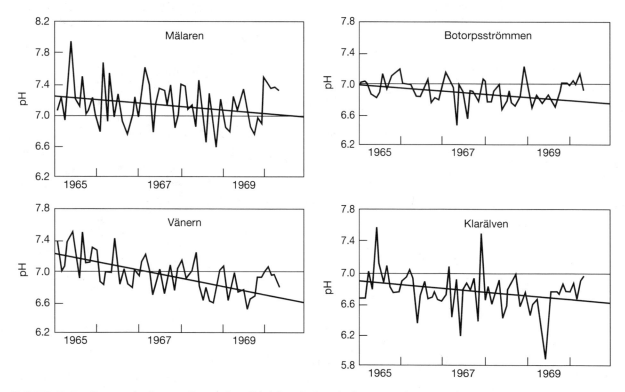

FIGURE 12.5 Changes in the pH of some Swedish lakes during the late 1960s because of acid deposition. The overall trends (solid black linear line) in pH are downward and indicate a factor of 1.5 to 3 increase in the hydrogen ion concentrations of these lakes. In more recent years, a number of Swedish lakes have begun to recover from acidification due to a reduction in European emissions of acid gases to the atmosphere, especially sulfur dioxide. (*Source:* Smil, 1997.)

TABLE 12.3 Evidence for Acidified Lakes from Various Countries of the World

Country	Evidence
Canada	More than 14,000 lakes strongly acidified, and 150,000 in the east (one in seven) suffering biological damage.
Finland	Survey of 1000 lakes indicates that those with a low acid-neutralizing capacity are distributed across the country; 8% of these lakes have no neutralizing capacity; most strongly acidified ones are located in southern Finland.
Norway	Fish eliminated in waters covering 13,000 square kilometers and otherwise affected in waters over an additional 20,000 square kilometers.
Sweden	About 14,000 lakes unable to support sensitive aquatic life and 2200 nearly lifeless.
United Kingdom	Some acidified lakes in southwestern Scotland, western Wales, and the Lake District.
United States	About 1000 acidified lakes and 3000 marginally acidic ones, according to the Environmental Defense Fund; a 1984 government study found 552 strongly acidic lakes and 964 marginally acidic ones.

(*Source:* French, 1990.)

decline in the pH of Swedish lakes to below 6. With the establishment of agricultural activities of conversion of forests to farmland, burning of vegetation, and plowing about 1000 years ago, the pH of some lakes rose gradually to above 6. The more recent acidification of Swedish lakes to below 5 in some instances has been a rapid and widespread phenomenon. This acidification is mainly due to the deposition of acidic constituents of sulfur and nitrogen derived from distant sources related to human activities outside Sweden. It is anticipated that as the regulatory controls on emissions of these constituents throughout Europe are made effective and maintained, emissions will continue to fall, and the pH of Swedish lakes will slowly rise. Liming with $CaCO_3$ of low pH dead lakes can also help to counteract the acidification of the lakes.

Sources of Sulfur and Nitrogen Oxides

Sulfur and nitrogen oxides in the atmosphere are derived from both natural and anthropogenic sources. The major natural sources of oxidized sulfur are volcanic eruptions, sea aerosol, and oxidation of reduced sulfur gases in the atmosphere. For NO_x the major natural sources are soils, lightning, and oxidation of atmospheric ammonia. Natural forest fires are minor sources of both sulfur and nitrogen oxides. The amount of sulfur and nitrogen oxides in the environment had been in a quasi-balanced state prior to major human intervention in the biogeochemical cycles of sulfur and nitrogen. Today the global emissions of sulfur gases to the atmosphere from the

burning of fossil fuels and other human activities exceed those of natural sources. Most of the global anthropogenic sulfur emissions (70%) are derived from electric power plants that burn coal, while most of the nitrogen emissions come from motor vehicle sources and some from electric power plants. Until recently the industrialized nations of the world generally have had the most severe acid deposition problems. The source of the problem has usually been the heavily industrialized areas in these countries. In the twenty-first century, the problem of acid deposition is shifting to the developing and industrializing countries, like China and India.

Figure 12.6 is a simplified diagram of the global biogeochemical cycle of the oxidized chemical species of sulfur involving transfers between Earth's surface and the atmosphere (see, for example, http://asd-www .larc.nasa.gov/biomass_burn/globe_impact.html). In the early twenty-first century, human activities, mainly the combustion of coal and oil and metal smelting, led to an annual release of nearly 62 million tons of sulfur to the atmosphere as sulfur dioxide gas, falling from a peak of 74–80 million tons in the late 1970s. These emissions are equivalent to about 50% of the total annual emissions of oxidized sulfur to the atmosphere from both natural and anthropogenic sources (excluding SO_2 formation from reduced sulfur gases in the atmosphere). Because of these emissions and reactions involving them in the atmosphere, of the approximately 187 million tons of oxidized sulfur falling on Earth's surface yearly in rain and dry deposition, about 35% is derived from human activities.

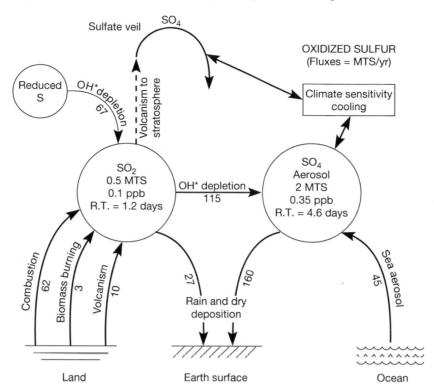

FIGURE 12.6 Earth surface–atmosphere global biogeochemical cycle of oxidized sulfur species. Sulfur dioxide (SO_2) released from the land surface to the atmosphere reacts with hydroxyl radical to form sulfate (SO_4) aerosol. This aerosol, plus that generated at the sea surface, is removed in rain and dry deposition. Both in the troposphere and stratosphere, sulfate aerosol directly or indirectly potentially exerts a cooling influence on the climate of the planet. Values on arrows are fluxes in units of million tons of sulfur per year (MTS/yr). The masses in the sulfur dioxide and sulfate aerosol reservoirs are units of million tons of sulfur (MTS). R.T. = residence time. (*Source:* Mackenzie, 2003.)

A similar situation holds true for the global biogeochemical cycle of NO_x gases (Figure 12.7; see, for example, http://www.eoearth.org/article/Air_pollution_emissions). The total global emissions of NO_x from Earth's surface, principally the land, to the atmosphere in the early twenty-first century amounted to about 47 million tons of nitrogen per year. Of these emissions, nearly 75% was derived from combustion and biomass burning sources. Of the 59 million tons of oxidized nitrogen chemical species deposited on Earth's surface annually in wet and dry deposition, nearly 60% was of anthropogenic origin.

On a regional scale, anthropogenic fluxes can overwhelm natural fluxes of SO_x and NO_x to the atmosphere. European annual emissions of sulfur and nitrogen oxide gases to the atmosphere from anthropogenic sources are about 16 and 5 million metric tons, respectively. For the eastern United States, anthropogenic emissions are about 12 million and 4 million tons annually, respectively, of sulfur and nitrogen oxides. For both gases, natural emissions of these gases in both regions are trivial. Consequently, the deposition of sulfur and nitrogen in wet and dry deposition in both Europe and the eastern United States is dominated by materials derived from industrial and transportation sources in these regions. In the eastern United States, of the total anthropogenic nitrogen and sulfur emissions, nearly 50% is deposited in this region. The rest is transported by atmospheric circulation to the North Atlantic Ocean and to eastern Canada. In Europe, more than 60% of the sulfur emissions of a country is transported by winds to other countries and deposited. More than 80% of the sulfur deposited on the land and waters of Norway, Sweden, Austria, and Switzerland is derived from the emissions of other nations. These observations illustrate well the regional- and subcontinental-scale dimensions of the acid deposition problem.

The tall smoke stacks that are used by industry to limit local pollution compound the problem of acid

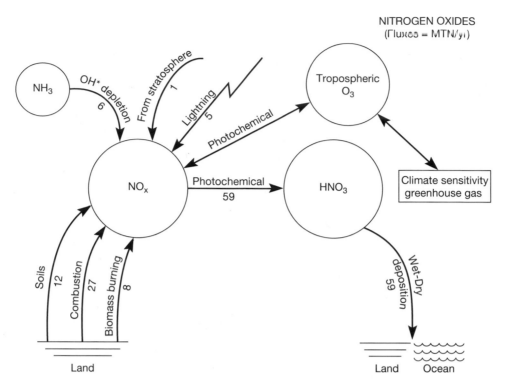

FIGURE 12.7   Earth surface–atmosphere global biogeochemical cycle of nitrogen oxides (NO_x). Nitrogen oxides react in the atmosphere, generating the tropospheric greenhouse gas ozone; are converted by photochemical reactions to nitric acid (HNO_3) and other minor constituents; and are removed in wet and dry deposition. Values on arrows are fluxes in units of million tons of nitrogen per year (MTN/yr).   (*Source:* Mackenzie, 2003.)

deposition (http://www.worldofstock.com/closeups/BIN1397.php). The sulfur and nitrogen oxide gases released at the greater heights are transported by wind and air currents far from their original sources before they are deposited on a distant surface of the planet. An example of this regional dispersion of pollutants is the eastern part of the United States. Plants with tall stacks in the upper Mississippi and Ohio river valleys of the industrial Midwest produce sulfur- and nitrogen-containing gaseous pollutants that are transported eastward to the New England states, New York State, and Canada. Since 1950 the proportion of emissions of SO_x from smoke stacks taller than 50 meters has more than doubled in the United States (Figure 12.8).

Sensitivity to Acid Deposition

Fish and other aquatic organisms, forests, crops, and soils are highly sensitive to pH change. Human-made structures of metal, concrete, and limestone may also be damaged by changes in the pH of the precipitation falling upon them.

AQUATIC ECOSYSTEMS   Figure 12.9 illustrates the pH range of the environment that is suitable for fish and other aquatic organisms. The pH of acid precipitation is commonly below that considered tolerable by a variety of aquatic organisms. Certain species of fish, salamanders, crayfish, frogs, snails, freshwater clams, and mayflies are particularly sensitive to acidification of their environment.

Lakes may become devoid of most of their aquatic organisms when lake water pH values drop because of acid deposition. These acidification situations have developed in hundreds of lakes in regions (Europe, the northeastern United States, and parts of southeastern Canada) that are located in the path of airborne gases that cause acid precipitation. Figure 12.10 shows maps of average annual pH values of rainfall over part of eastern North America between 1955 and 1990. Notice in particular that the average pH of rain between 1955 and the mid- to late-1980s decreased. There are no isopleths (lines of equal pH value) of average rain pH in 1985 and 1990 that are above 5. Values of pH above 5 are seen on the 1955–1956 and 1972–1973 maps of rain pH. In the map for 1990,

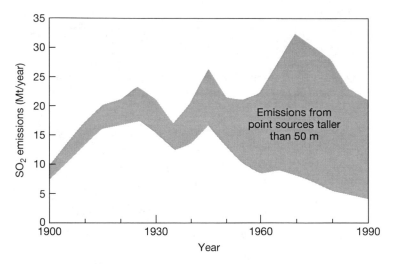

FIGURE 12.8 United States annual emissions of SO_2 from point sources from 1900 to 1990. Notice how the emissions from point sources with chimneys taller than 50 meters grew over time as bigger and bigger smoke stacks were built. Since 1950, the proportion of SO_2 emissions from tall chimneys has more than doubled. Some of the taller chimneys are now being destroyed and made smaller to decrease the spatial scale of emission transport. (*Source:* Smil, 1997.)

there is some suggestion that the area of the region of very low pH values has decreased. However, the latest data available for the United States as a whole (Figure 12.11a) still show pH values below 5 for much of the eastern one-third of the United States, although the acidity of rain has been falling in New York state since the late 1980s (Figure 12.11b) and at the Hubbard Brook Ecosystem Study site in central New Hampshire, United States (Figure 12.4). The recent increases in pH are largely a result of the decreases in SO_x emissions in the Midwest and eastern United States because of regulatory measures.

Thus, because rainfall is the major source of water for aquatic systems in eastern North America, low pH rain falling or running off into these systems can lower their pH. Furthermore, the dry deposition of SO_2 is approximately equal to that of sulfate and is an important acid deposition pathway in addition to rainfall.

Headwater streams, high-altitude lakes, and small lakes are especially vulnerable to pH change. The pH-sensitive organisms (Figure 12.9) found in the waters or living on the bottom of these aquatic systems and found in the watersheds surrounding them may not survive under the more acidic conditions. Also, because of the lower pH of the rainfall, metals and nutrients may be leached from the rocks and soils of the watershed at a faster rate than before the rainfall became acidified by anthropogenic sources of sulfur and nitrogen. Such changes can

affect the productivity of both aquatic and terrestrial ecosystems. For example, the increased leaching of aluminum from soils because of acid deposition can lead to increased fluxes of aluminum to aquatic systems. On entering a lake, the aluminum may precipitate out on the gills of fish and impede the transfer of oxygen across the gill membrane. As a result, the fish may die. This process and that of acidification are thought to be the principal causes of fish morbidity and mortality in freshwater aquatic systems affected by acid deposition.

Interestingly, although emission abatement programs have been in effect in the United States since the early 1970s, there has been a less than anticipated effect on the low pH values of rainfall in the eastern United States (see Figures 12.4 and 12.11). This is despite the fact that the United States emissions of both SO_x and NO_x slowed in the 1980s and early 1990s and fallout of acidic components of sulfur decreased. Part of the reason for this situation may be related to the fact that beginning with the Air Pollution Control Act of 1955, particulate emissions to the atmosphere from anthropogenic sources in the United States began to fall. This reduced the burden of basic particulates in the atmosphere, which could help to neutralize the acidic components of sulfur and nitrogen. Thus, it has taken longer for the reductions in emissions to have a large impact on the pH of rainfall because the atmosphere was partly cleansed of the anthropogenic basic particulates as emissions of the

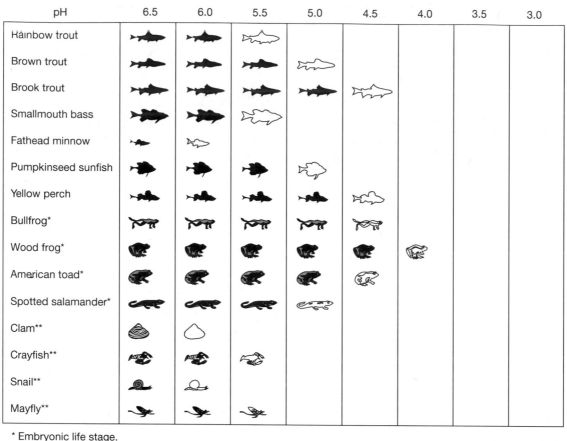

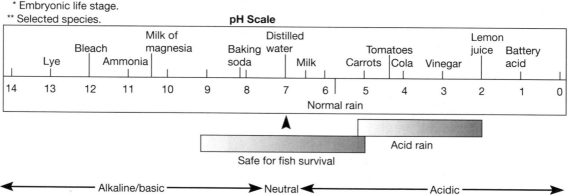

FIGURE 12.9 The effect of pH change on some organisms. The pH scale, the pH range of water considered safe for fish to survive, and the pH range of acid rain are shown for comparison. The approximate pH at which an organism dies with extended exposure to water with this pH or below is shown by an organism pattern in white. (*Source:* Christensen, 1991.)

acidic components were being reduced (see following section on emissions and atmospheric dust).

The pattern of acidification observed in eastern North America is similar to that of other regions of the world in that regions near or downwind of anthropogenic SO_x and NO_x emissions exhibit acidified precipitation. Acid deposition is also the cause of damage to aquatic and other ecosystems in these regions.

SOIL AND FOREST ECOSYSTEMS On the land, natural decay of organic matter in soils produces organic and inorganic acids. These acids are involved in rock-weathering processes leading to neutralization of the acids. Thus, in general, soils do not become overly acidic and consequently healthy plants can grow. However, the massive deforestation of the temperate forests in Europe and North America in the nineteenth

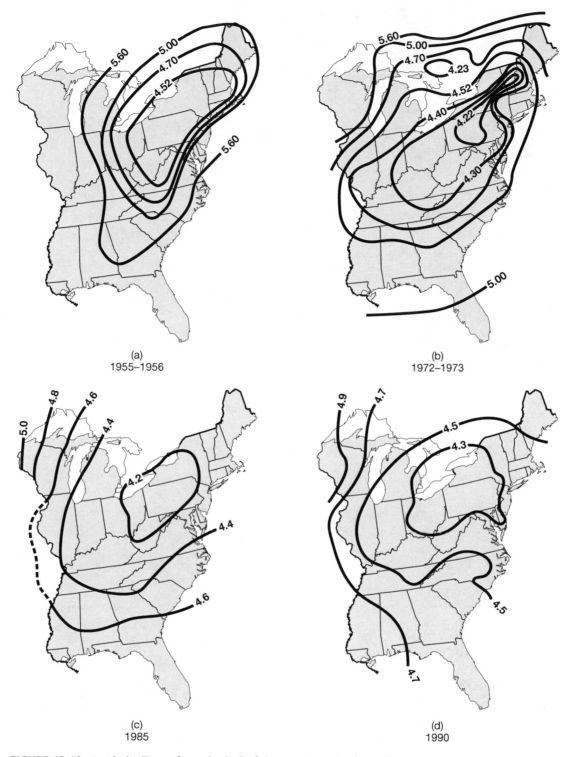

FIGURE 12.10 Isopleths (lines of equal value) of the average annual pH of precipitation over Eastern North America for different years. Notice the broadening of the region of pH values lower than 5 from 1955 to 1985 and the apparent decrease in the area of low pH values in 1990, probably resulting from the implementation of SO_x emission controls and conversion to cleaner-burning fuels.   (*Source:* Laws, 2000; Smil, 1997.)

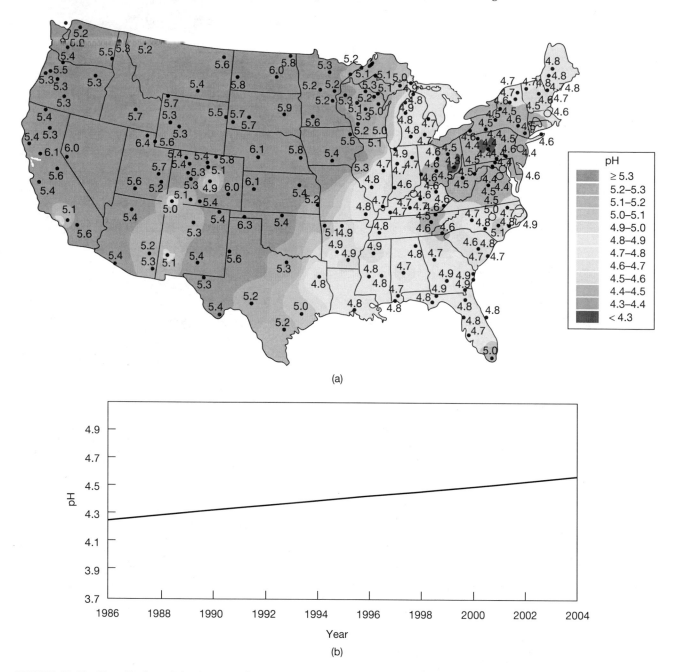

(a)

(b)

FIGURE 12.11 The pH of precipitation over the contiguous United States in the early twenty-first century (a) and the trend line of pH for acid deposition in New York state, United States from 1986 to 2004 (b). Even in the early twenty-first century, the pH of much of the precipitation over the eastern United States is still below 5 despite reductions in the emissions of sulfur dioxide to the atmosphere from anthropogenic sources. However, the acidity of rain in New York state has decreased over time due mainly to emission reductions.   (*Source:* NADP, 2008.)

and early part of the twentieth centuries, coupled with the recovery of large tracts of this deforested land, have led to excessive soil acidity in some locations because nitrous oxide gas is emitted from exposed soils when forests are cleared. The bacterial production and emission of this gas in the soil result in the release of hydrogen ions and consequently a lower pH of the soil. Thus, prior to potential acidification of soils from acid deposition, in some regions of Western Europe and eastern North America, it is possible that the soils were already somewhat more acidic than normal because of deforestation.

It is likely that acid deposition in some areas is playing a role in the dieback of acid-sensitive forests. **Dieback** is the gradual dying of plant shoots, starting at their tips. The reasons for the dieback are not fully understood, but circumstantial evidence points to stress on the trees induced by air pollutants, acid rain, and acidified soils. Acid deposition acidifies the soil. Sulfur and nitrogen oxide gases and ozone in the atmosphere surrounding plants are absorbed on the leaves or needles of trees. These gases, especially ozone, may impair photosynthesis. One evidence of ozone damage shows up in the leaves of trees as lesions (http://www.mobot.org/gardeninghelp/images/Pests/Pest763.jpg). This impairment may have implications for global warming (see Chapter 14) because impaired photosynthetic ability reduces primary production and results in less CO_2 being removed from the atmosphere.

Nitrogen saturation on land occurs when supplies of ammonium and nitrate are in excess of the total combined plant and microbial demand. The term is applied to ecosystems where the biota are unable to utilize all of the N that is added to the system, either through N fixation, atmospheric N inputs, or other sources. Nitrogen saturation frequently leads to chronic acidification of streams and other freshwater bodies. Nitrogen saturation can also result in significant decreases in soil fertility and nutrient deficiencies. As nitrate and sulfate anions move through the soil, cations (most often Ca and Mg) are removed from the exchange sites on soil minerals to maintain charge neutrality of the soil solution. These nutrients are then lost from the soil. Areas of nitrogen-saturated soils of forests and acidic surface waters in the United States are shown in Figure 12.12, exhibiting some correlation to the regions in the United States of waters and watersheds sensitive to acid deposition (Figure 12.13).

The mobilization and leaching of toxic metals from soils and rocks may also play some role in damaging the forest trees. Soils may be degraded by acidic components, leading to an accumulation of heavy metals, a loss of nutrients, and mobilization of toxic aluminum. This combination of stresses produced by air pollution and changes in the character of soil is especially harmful to forested ecosystems. As a result, the trees probably lose their resistance to insect infestations and droughts because of their weakened

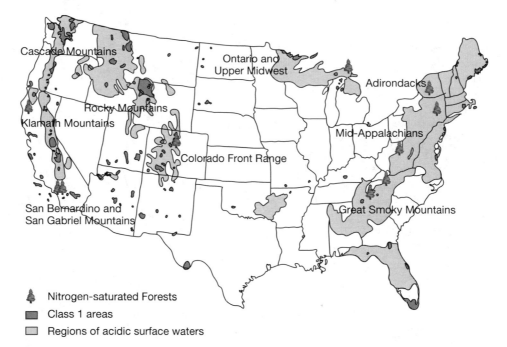

FIGURE 12.12 Map of the contiguous United States showing the forests that are nitrogen-saturated and the regions of acidic surface waters. Class 1 areas are those regions in which visibility is protected more stringently than under the United States ambient air quality standards and include national parks, wilderness areas, monuments, and other areas of special national or cultural significance in the United States. Many of these regions show signs of continental-scale haze conditions. (*Source:* EPA, 2006.)

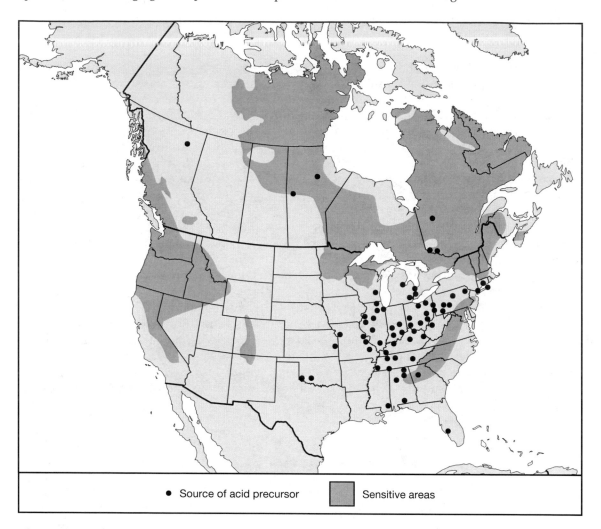

• Source of acid precursor   Sensitive areas

FIGURE 12.13 The major sources of acidic materials in North America, and the regions of the continent considered sensitive to acid deposition. Notice the large number of sources in the American Midwest, a region not particularly sensitive to acid deposition.   (*Source:* Environment Canada, 1981.)

state. Damaged forests have been observed in Europe, the United States, and Canada. Although there is considerable scientific debate, it appears that some of this forest damage can be linked to air pollution.

An excellent regional example of the relationship between air pollution and forest damage is that of Europe. In central Europe, trees covering more than 5 million hectares in the mid-1980s showed signs of injury stemming from air pollution. As an example, consider the forest damage that was reported in a survey for West Germany in 1982–1983. Nationwide the survey showed that 75% of the fir trees were affected by air pollution in 1983, up from 60% in 1982. The damage to spruce, the most important species to the West German forest products industry, had risen from 9 to 41% between 1982 and 1983. The portion of the total forest affected was 34% in 1983, more than

four times the area in 1982. Table 12.4 gives the 1988 figures for forest damage in Europe. The estimated area damaged is nearly 50 million hectares, or 35% of the forested land. Notice that forest damage in West Germany had increased to 52%, up 18% from the 1983 estimate. Documented forest damage in North America has been less dramatic and is mostly restricted to isolated high-altitude environments. In the developing world, scientists in several nations, including China, are reporting evidence of forest damage caused by air pollution.

NEUTRALIZATION AND ACID-SENSITIVE AREAS Different rivers, lakes, and soils are not equally affected by or sensitive to acid deposition (Figure 12.13). Natural water bodies contain a variety of dissolved and particulate substances that buffer the pH. These

TABLE 12.4 Forest Damage Estimated for Europe in 1988

Country or Area	Total Forest Area (1000 Hectares)	Estimated Area Damaged (1000 Hectares)	Share of Total (Percent)
Czechoslovakia	4,578	3,250	71
Greece	2,034	1,302	64
United Kingdom	2,200	1,408	64
Estonia	1,795	933	52
West Germany	7,360	3,827	52
Tuscany, Italy	150	77	51
Liechtenstein	8	4	50
Norway	5,925	2,963	50
Denmark	466	228	49
Poland	8,654	4,240	49
Netherlands	311	149	48
Flanders, Belgium	115	53	46
East Germany	2,955	1,300	44
Bulgaria	3,627	1,560	43
Switzerland	1,186	510	43
Luxembourg	88	37	42
Finland	20,059	7,823	39
Sweden	23,700	9,243	39
Wallonia, Belgium	248	87	35
Yugoslavia	4,889	1,564	32
Spain	11,792	3,656	31
Ireland	334	100	30
Austria	3,754	1,089	29
France	14,440	3,321	23
Hungary	1,637	360	22
Lithuania	1,810	380	21
Bolzano, Italy	307	61	20
Portugal	3,060	122	4
Other	13,474	n.a.	n.a.
Total	140,956	49,647	35

(*Source:* French, 1990.)

substances tend to neutralize acids and bases that are added to the water body and to maintain the pH at a certain value. Although many freshwater bodies have pH values close to neutral, the natural pH of these systems varies from one region to another depending on several factors, including the composition of the underlying bedrock, terrain features, the quality of the soil, the extent and type of plant cover, and the microbial life present in the environment. The extent of the impact of acid deposition on freshwater depends on the buffering capacity of the system, which is mainly determined by the amount of bicarbonate (HCO_3^-) and carbonate (CO_3^{2-}) ion in the water. The **buffering capacity** of the water is a measure of the incremental change in pH of the water due to an incremental addition of acid. A convenient indirect measure of the buffering capacity of water has been the **acid neutralizing capacity** (ANC), the difference between the concentrations of strong bases and strong acids in the water. For lake waters in the northeastern United States, an ANC above about 100 microequivalents per liter has protected these waters from rapid acidification by acid deposition. In other words, the buffering capacity of these waters is high, and acids added to them are continuously neutralized.

If soils and the underlying rock are rich in calcium carbonate (limestone), then the deposited acids are partially neutralized. Limestone rock can help to buffer the pH of soil, river, and lake waters, leading to a delay in the acidification of these aquatic systems. In essence what happens is that the acid readily dissolves the calcium carbonate ($CaCO_3$) and a hydrogen ion of the acid is locked up with the carbon from the limestone as dissolved bicarbonate (HCO_3^-) that is,

$$CaCO_3 + H_2SO_4 \rightarrow Ca^{2+} + SO_4^{2-} + HCO_3^- + H^+.$$

This process neutralizes the acidic deposition. However, lake, river, and soil waters associated with areas underlain by rocks containing little calcium carbonate, such as regions of the United States Adirondacks and the Fennoscandinavian region of Europe, are more susceptible to acidification stemming from acid deposition. In these regions, crystalline rocks rich in silicate minerals are abundant. These minerals are not as effective a buffering agent as is calcium carbonate. Thus, acid deposition falling on crystalline terrane is only slightly neutralized, and waters draining or located in these terranes are more susceptible to acidification. Such susceptible regions have been identified in North America and Europe. In North America, the acid-sensitive regions are the Canadian

Shield and parts of western and eastern North America that are underlain by crystalline rock (Figure 12.10). A large number of sources of acid pollutants to the atmosphere in the United States are located in areas of the American Midwest that are not particularly susceptible to acid deposition.

In 1977, Sweden, a country located on the crystalline rock terrane of Fennoscandinavia, embarked on a program to neutralize the acid in its lakes by adding lime to the lakes. This process simulates the natural process involving the reaction of acid constituents with limestone. The operation has been expensive and is only a temporary cure. A single application of lime may help to neutralize an acidic lake for only a few years unless inputs of acid constituents are halted. This procedure is analogous to taking bicarbonate of soda stomach tablets (antacid) to neutralize a sour stomach.

BUILDING AND STRUCTURE SENSITIVITY In addition to the effects on the natural environment, acid deposition may have a deleterious effect on buildings and statues because acids, as shown previously, are capable of dissolving marble and limestone. Evidence of this effect are the blackened and etched statues and buildings in many cities of the world, particularly the antiquities of Europe. Even age-old statues and artifacts of ancient civilizations are currently being damaged by acid deposition. The disintegration of these ruins by exposure to the modern-day atmosphere has proceeded faster in recent decades than in the previous thousands of years of exposure. For example, the Sistine Chapel, the Acropolis in Greece, and the ruins of the ancient civilizations of Mexico and Egypt are deteriorating prematurely. To determine the processes and rates of deterioration of marble owing to acid deposition, marble tombstones have been analyzed. Because the tombstones are well dated, the progressive degradation of the marble can be evaluated.

The effects of acid deposition are not limited to stone buildings, statues, and other structures. Acid deposition also affects modern metallic structures because the rate of corrosion of metal under acidic conditions increases with the frequency and intensity of acid deposition events.

Most of the material damage caused by acid deposition occurs in urban areas because of the large number of motor vehicles and metallic, painted, limestone, and concrete surfaces. Because of the decrease in emissions in North American and European cities and elsewhere, corrosion and deterioration problems have decreased during the last two decades. In some European cities, like Brussels, Belgium and Athens,

Greece, stone and concrete churches and other buildings are being resurfaced by removal of the surface scale produced by acid deposition. Thus, estimates of costs attributed to acid deposition damage have also decreased. However, annual global material damage costs owing to the composite effects of acid deposition and air pollutants probably still total on the order of billions of dollars each year.

Emission Abatement Measures

As awareness increased concerning the potential hazards to human health of sulfur and nitrogen emissions from fossil fuel burning, emission abatement policies and standards were implemented by regulatory agencies in the United States, Canada, and Europe. These regulations have reduced or slowed the total emissions of sulfur and nitrogen to the atmosphere in North America and in a number of European countries.

SULFUR OXIDE EMISSIONS Sulfur emissions are considered to be the greatest health hazard. Some countries have replaced the burning of high-sulfur fossil fuels with low-sulfur fuels. The use of primarily low-sulfur coal has had important economic consequences for communities in the United States in areas such as Ohio, Illinois, Virginia, and West Virginia, where high-sulfur coal is mined. In these coal-mining areas of the United States, mines have closed and unemployment is high. Figure 12.14b shows the SO_x emissions by source sector in the United States for 2002. The use of fossil fuels to generate electricity is the major source of SO_x to the atmosphere of the United States. To reduce the impact of this source, scrubbers that remove sulfur dioxide gases from smoke stacks were required to be placed on all fossil fuel burning electrical-generating and other industrial plants built after 1969 in the United States. In addition, in some plants, fossil fuels are burned in contact with fluidized beds of lime (CaO) or finely ground limestone that removes sulfur dioxide from the flue gases. As a result of these measures, sulfur dioxide emissions declined in the United States by about 53% between 1970 and 2002 (Figure 12.14a). Other countries have instituted similar pollution abatement measures. For example, Japan's emissions of sulfur dioxide declined by 39% between 1973 and 1984 owing to abatement programs. The European Economic Community has adopted a protocol for a 30% reduction in emissions of SO_2. Figure 12.15 illustrates the regional trends in anthropogenic sulfur emissions between 1970 and the early twenty-first

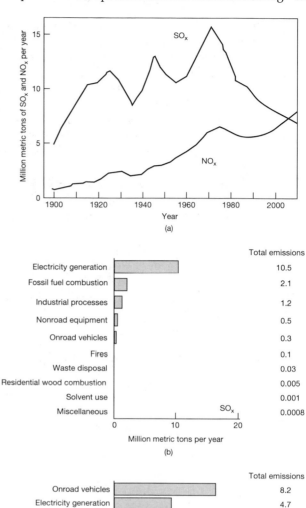

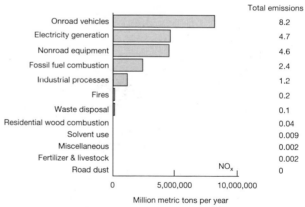

FIGURE 12.14 Annual United States anthropogenic emissions of sulfur and nitrogen to the atmosphere as SO_x and NO_x between 1900 and 2002 and projected to the year 2010 (a). The emissions come primarily from fossil fuel burned in the United States in various sources. Notice the leveling off of sulfur and nitrogen emissions in the 1980s following the decline in emissions since 1970, one year after the first United States Clean Air Act. Also, note the declines in sulfur emissions following the slowdown in economic activity immediately after each world war. Also shown are the anthropogenic emissions in million metric tons per year by sector of SO_x and NO_x to the atmosphere in the United States (b, c). (After Gschwandtner et al., 1988; Kohout et al., 1990; Air Quality Committee, 1994; (*Source:* http://www .cleanairnet.org/lac/1471/articles-58235_pdf.pdf.)

century. Dramatic declines are observed in sulfur emissions for Eastern and Western Europe and North America but emissions for this period of time have been growing in Asia and South America. However, there are signs that the growth rate is slowing or perhaps even reversing in these regions.

Emission abatement measures have had a considerable effect on the deposition of sulfate in wet precipitation over the eastern United States and parts of Canada. The maps in Figure 12.16 show that wet sulfate deposition has been consistently highest in eastern North America, particularly around the lower Great Lakes. In 1990, a considerable area of the eastern United States and southeastern Canada had annual deposition rates of greater than 20 kg per hectare of sulfate. The map patterns illustrate that significant reductions occurred in wet sulfate deposition in both the eastern United States and much of eastern Canada from 1990 to 2005. By 2005, the region receiving more than 30 kg per hectare per year of wet sulfate deposition had essentially disappeared. The wet sulfate deposition reductions are considered to be directly related to decreases in SO_2 emissions in both Canada and the United States. For the period 1980 to 2002, total SO_2 emissions decreased about 38% in the United States. A strong correlation exists between the annual total SO_2 emissions to the atmosphere for eastern North America and the total annual wet deposition of sulfate over the region. The decrease in the sulfate wet deposition fluxes implies a decrease in the amount of hydrogen ion deposited over eastern North America and probably accounts for the apparent decrease in the acidity of rain at Hubbard Brook and in New York state as seen in Figures 12.4 and 12.11b, respectively. However, rainfall pH values are still in the range of 4.2 to 4.8 in much of the northeast United States (Figure 12.11a).

NITROGEN OXIDE EMISSIONS United States government regulations have been more relaxed for nitrogen oxide emissions because these compounds have not been considered to be as great a health hazard as sulfur oxide emissions. Nitrogen oxide emissions have been more difficult to estimate than sulfur dioxide emissions, but it is known that they have risen considerably in the United States because of the increased use of motor vehicles (Figure 12.14a, c). Nitrogen oxide emissions have grown relative to sulfur emissions in their contribution to acid deposition in the last two decades. The installation of catalytic converters on cars is capable of cutting nitrogen oxide emissions by 76%. This device has led to some curtailment of nitrogen oxide emissions into the atmosphere.

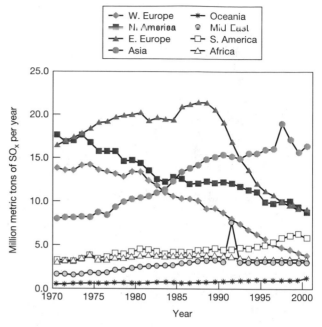

FIGURE 12.15 Trends in SO_x emissions by region from 1970 to the early twenty-first century. Notice especially the overall general trend patterns of increases in emissions from Asia and South America as national economies improve and the regions use increasing supplies of fossil fuels and the declines for Eastern Europe, North America, and Western Europe. For the former region, much of this decline is due to the breakup of the former Soviet Union and consequent effects on the economies of countries in this region; for the latter regions, the declines are mainly due to implementation of regulations on anthropogenic emissions of acid gases to the atmosphere. (*Source:* Stern, 2006.)

Interestingly, Japan's standards for automobile nitrogen oxide emissions are far stricter than those imposed on the automobile industry in the United States to date.

In contrast to SO_2 emissions, NO_x emissions in the United States have continued to rise from 1900 to 2002, except for the period of the Great Depression years and the severe recession in the early 1980s. The map patterns of wet nitrate deposition (Figure 12.17) show a similar southwest-to-northeast axis of high deposition and reduction in the depositional rate from 1990 to 2005 as exhibited by wet sulfate depositional patterns. The region of greater than 15 kg per hectare of annual wet nitrate deposition disappeared during the period of 2000 to 2005. One can conclude from this and the situation with respect to sulfate deposition in the United States and Canada that emission abatement measures play a major role in reducing inputs of acid constituents to aquatic and land ecosystems.

EMISSIONS AND ATMOSPHERIC DUST An interesting link occurs between the phenomenon of acid deposition and atmospheric dust particles. Atmospheric

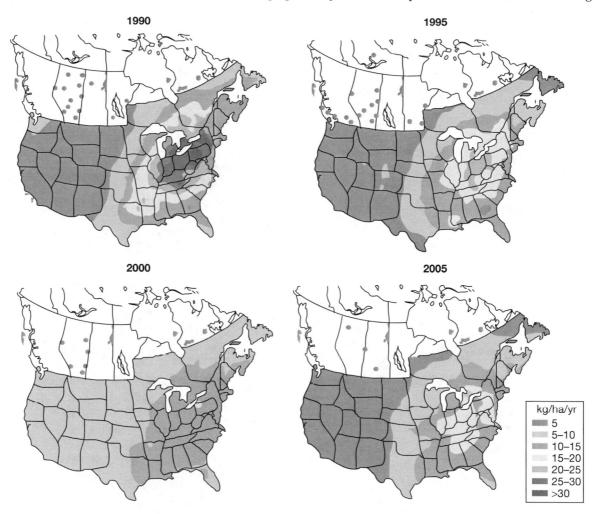

FIGURE 12.16 Annual sulfate wet deposition for the years 1990, 1995, 2000, and 2005 in kilograms of sulfur per hectare per year for the United States and parts of Canada. From 1990 to the early twenty-first century, deposition fluxes of sulfate in the eastern part of the United States and Canada have decreased mainly due to the implementation of regulations on anthropogenic sulfate emissions in these countries. Few areas showed deposition fluxes of greater than 30 kg/ha/yr in 2005. (From *Canada-United States Air Quality Agreement, Progress Report 2008*. © Her Majesty The Queen in Right of Canada, Environment Canada, 2008. Reproduced with the permission of the Minister of Public Works and Government Services Canada.)

dust particles originate from several natural and anthropogenic sources. Fossil fuel combustion; industrial activities, such as cement manufacturing, mining operations, and metal processing; agricultural practices, such as the plowing of fields; and traffic on unpaved roads generate atmospheric dust. Forest fires and wind erosion are the two most important natural sources of particles, but both phenomena can be influenced by human activities. Dust particles can contain significant amounts of calcium and magnesium carbonate minerals that act as bases when dissolved in water. In the air, the dust particles can neutralize the acidic constituents present in acid precipitation, and upon deposition on land, they serve as a source of fine-grained and hence very reactive base

materials in soils that may serve to neutralize acid rain. In the soil, the particles can neutralize acids by dissolving and by exchange of cations located on the particle surfaces with hydrogen ions in the water percolating through the soils. Thus, the hydrogen ions can be sequestered on the surfaces of the particles while the base cations of calcium and magnesium are released to the soil water. Dust particles also serve as a source of the essential nutrients of calcium, magnesium, potassium, and sodium for the growth of plants.

Regulations to limit dust emissions to the atmosphere from various human activities in the United States (Figure 12.18) and some European countries have led to a decrease of dust in the atmosphere of these regions concomitant with the reductions in

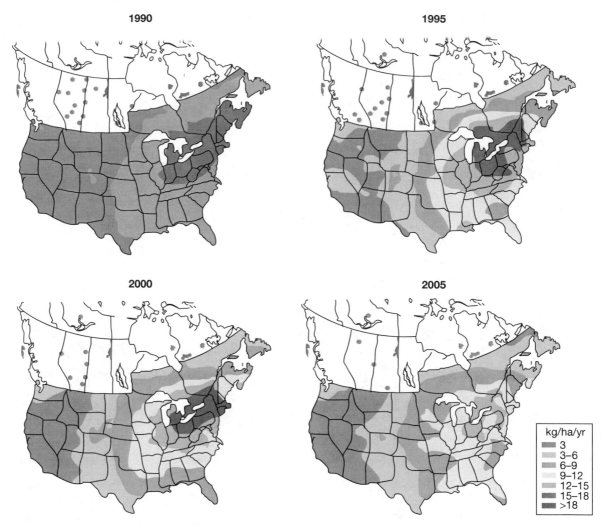

FIGURE 12.17 Annual nitrate wet deposition for the years 1990, 1995, 2000, and 2005 in kilograms of nitrate per hectare per year for the United States and parts of Canada. From 1990 to the early twenty-first century, deposition fluxes of nitrate in the eastern part of the United States and Canada have decreased mainly due to the implementation of regulations on anthropogenic nitrate emissions in these countries. No areas showed deposition fluxes of greater than 15 kg/ha/yr in 2005. Notice the similarity in the pattern of nitrate deposition fluxes to that of sulfate in Figure 12.16. (From *Canada-United States Air Quality Agreement, Progress Report 2008.* © Her Majesty The Queen in Right of Canada, Environment Canada, 2008. Reproduced with the permission of the Minister of Public Works and Government Services Canada.)

acidic sulfur emissions (see example for the United States in Figure 12.27, p. 367). Thus, the environmental benefits of reductions in acidic emissions have been partially offset by a decline in the bases associated with dust in the atmosphere and falling on soils. In other words, there was less basic dust around in the year 2008 than there was in 1970 in the United States and parts of Europe and hence a reduction in the ability of the atmosphere and soils to neutralize acid precipitation. This may be one reason that, despite substantial reductions in sulfur emissions to the atmosphere from human activities in the United States since

1970, the expected moderately strong increase in the pH of rainfall over this period of time has not been observed.

GLOBAL EMISSIONS Figure 12.19a shows the global annual emissions of sulfur and nitrogen oxides to the atmosphere mainly from the burning of fossil fuels from 1860 to the year 2002. Information on the total emissions by region and the sources of those emissions are available for sulfur (Figure 12.19b, c). North America and Europe account for about 30% of the global sulfur emissions. The burning of coal is

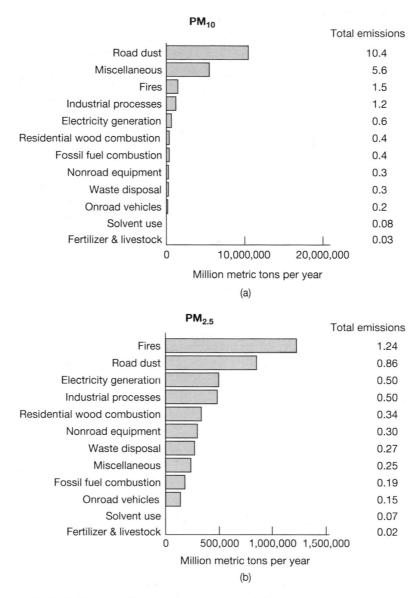

FIGURE 12.18   United States emissions of particulates by source sector to the atmosphere in 2002. Emissions are in millions of metric tons per year. PM_{10} (a) is for emissions of particles smaller than or equal to 10 micrometers in diameter. $PM_{2.5}$ (b) is for fine-grained emissions smaller than or equal to 2.5 micrometers in diameter. These fine particles can be especially dangerous to health.   (*Source:* http://www.epa.gov.)

responsible for about 50% of the global emissions of sulfur to the atmosphere. As a means of comparison through time, the total 1980 emissions of sulfur and nitrogen oxides from different regions of the world and projections for the future are shown in Figures 12.20 and 12.21. In contrast to the trends in the United States and some European countries, the global picture is one of steadily increasing emissions

on into the twenty-first century. In the future, emissions from sources in Europe and North America will probably not increase greatly (Figure 12.21), if at all. However, in the former USSR and in Eastern Europe, high-sulfur coal is still used because it is plentiful and cheap. China also has tremendous coal reserves and plans to double consumption into the twenty-first century. The emissions of sulfur

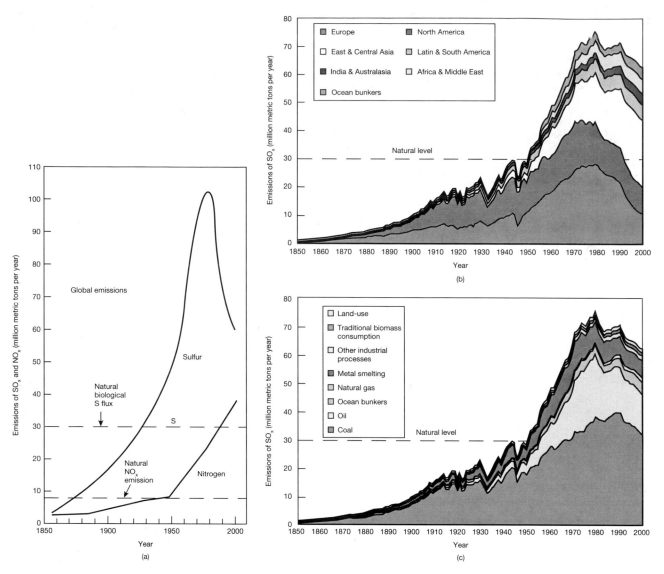

FIGURE 12.19 Annual anthropogenic global emissions of sulfur and nitrogen as SO$_x$ and NO$_x$ to the atmosphere from 1860 to the year 2002 (a) and total anthropogenic sulfur emissions from different regions of the world (b) and from different source sectors (c). Estimates of the natural *biological* sulfur and nitrogen fluxes to the atmosphere are given for comparison (dashed line). (*Source:* Dignon and Hameed, 1989; Hameed and Dignon, 1992; Smil, 1997; Smith et al., 2004, http://www.pnl.gov.)

and nitrogen oxides by the developing countries in Asia, Africa, and South America are very likely to grow substantially because of their increasing populations and demands for energy, industrialization, and a higher standard of living. With a steadily increasing population and increasing per capita sulfur and nitrogen emissions, global emissions in the year 2020 could approach 210 million and 55 million tons per year of sulfur and nitrogen, respectively. These estimates are about three times greater than 1990 emissions of sulfur and two times greater than 1990 emissions of nitrogen to

the atmosphere from anthropogenic sources. Even with no increase in the per capita global production of sulfur and nitrogen oxides, global emissions of these acid constituents will most likely increase in the future because of population growth (Figure 12.21).

Because of the projected changes in the geographical distribution of sulfur and nitrogen emissions to the atmosphere in the next century, the geographical distribution of the concentrations of oxides of sulfur and nitrogen in the atmosphere and deposition rates of these acidic substances will

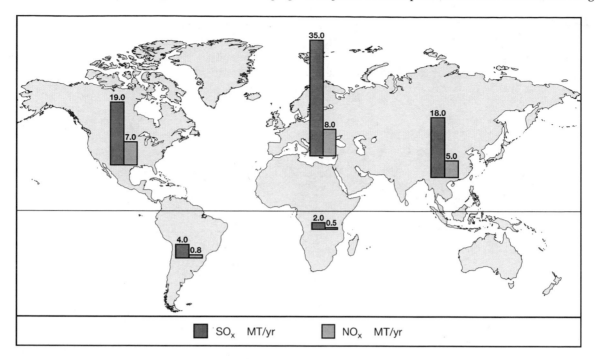

FIGURE 12.20 The emissions of sulfur and nitrogen oxide gases into the atmosphere from five regions of the world for 1980. The heights of the bars give the relative emissions in millions of tons of sulfur and nitrogen per year (MT/yr). (*Source:* Galloway, 1989.)

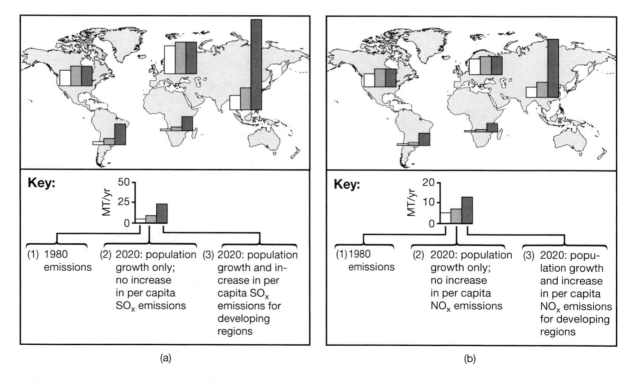

FIGURE 12.21 The emissions of (a) sulfur and (b) nitrogen gases to the atmosphere from different regions of the world. The emissions are principally from the burning of fossil fuels. The bars on the graphs show (1) the emissions in 1980 and (2,3) the projected emissions for 2020. The projections are for two different situations: (2) The world population continues to grow into the future, but there is no increase in sulfur or nitrogen emissions per capita, and (3) both population and sulfur and nitrogen emissions per capita increase to the year 2020. The heights of the bars give the relative emissions in millions of tons of sulfur and nitrogen per year (MT/yr). (*Source:* Galloway, 1989.)

change substantially in the twenty-first century. Figures 12.22 and 12.23 show a comparison between the years 1990 projected to 2050 of the atmospheric deposition rates of sulfur and from 1860 projected to 2050 for the rates of total nitrogen deposition as obtained from global models of the sulfur and nitrogen biogeochemical cycles. There is large spatial variability in these rates for both years, with most of the sulfur and nitrogen deposited in the Northern Hemisphere in the early 1990s. This is also the region

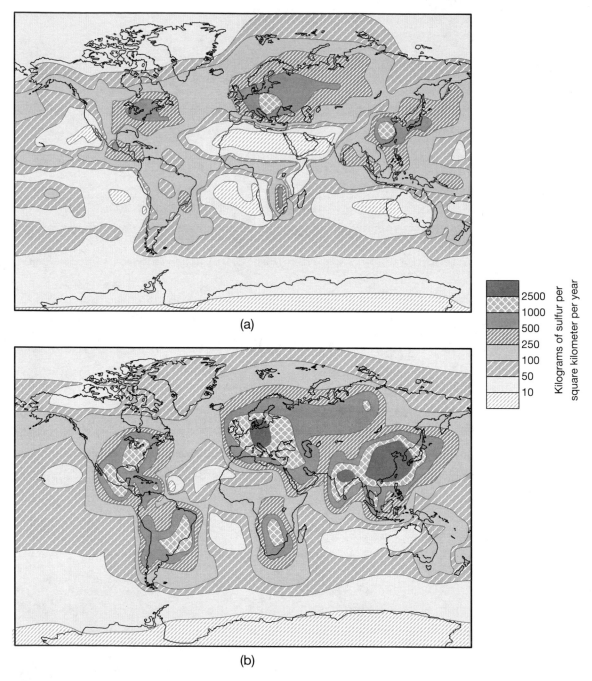

(a)

(b)

FIGURE 12.22 Comparison of the magnitude of atmospheric sulfur deposition for the years 1990 (a) and 2050 (b). Notice the large increases in both spatial extent and intensity of sulfur deposition in both hemispheres and the increase in importance of Asia, Africa, and South America as sites of sulfur deposition between 1990 and 2050. The values on the diagrams are in units of kilograms of sulfur deposited per square kilometer per year. (*Source:* Rodhe et al., 1995.)

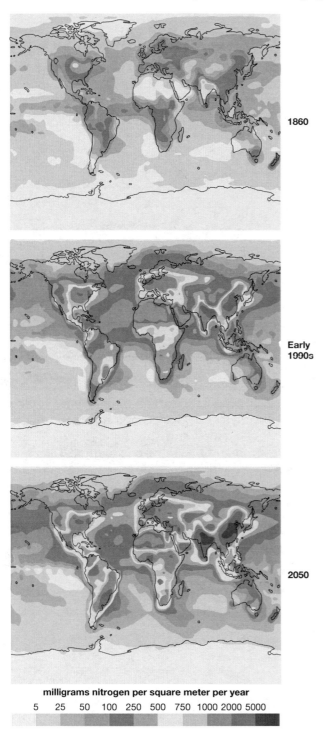

1860

Early 1990s

2050

milligrams nitrogen per square meter per year

5   25   50   100   250   500   750   1000 2000 5000

FIGURE 12.23 An estimate of the total reactive nitrogen deposition from the atmosphere in 1860, in the 1990s, and projected to the year 2050. Notice how the pattern of nitrogen deposition shifts to the developing countries of South and Southeast Asia and South America into the twenty-first century and deposition rates increase in the developed countries of the Northern Hemisphere. Deposition rates are given in milligrams of nitrogen per square meter per year. (*Source:* Galoway et al., 2004.)

of highest concentrations of sulfate aerosol in the atmosphere and the present source of most of the anthropogenic sulfur and nitrogen emissions to the atmosphere (Figure 12.20). In 2050 there are projected large increases in both spatial extent and intensity of sulfur and nitrogen deposition relative to 1990. In addition, large increases in sulfur and nitrogen deposition rates are projected to occur in eastern and southern Asia and in parts of Africa and South America. This change in the pattern of sulfur and nitrogen deposition fluxes between 1990 and 2050 reflects the projected important future increases in emissions of sulfur and nitrogen to the atmosphere from Asia, Africa, and South America (compare Figures 12.20 and 12.21). Projected increased oxidized sulfur and nitrogen deposition fluxes in the twenty-first century in southern and eastern Asia portend for these regions a greater risk of acid deposition problems in the future. The future projected increased atmospheric burden of sulfate aerosol and its regional distribution also have important climatic implications (see Chapter 14).

It is technically possible to reduce both sulfur and nitrogen oxide emissions to levels that have essentially little negative effect on the environment. Among other measures, political pressure and monetary incentives are particularly needed to meet this end. Emissions goals and a cap and trade policy have been very successful in reducing emissions. Table 12.5 shows the emissions reduction goals for the United States and Canada for sulfur as an example of the steps various countries are taking to deal with the problem of acid deposition. In 2007, the United States reduced its annual SO_x emissions from the electrical power sector to below its national emission cap for sulfur of 8.95 million tons and exceeded its NO_x emissions goal of reducing these emissions to 2 million tons below the projected annual emission levels for the year 2000 of 8.1 million tons. Canada has also been successful in reducing its SO_x emissions by more than 55% since 1980 and has surpassed its NO_x reduction target. The cap and trade policy with respect to SO_x emissions in the United States has been an environmental success story: Emissions have been reduced, wet sulfate and nitrate deposition has decreased, and in some regions, precipitation has become less acidic. At the same time, despite initial business, industry, and governmental hesitancy to support this policy because of projected costs, the cost of the effort to reduce SO_x emissions has remained essentially constant from 1994 to 2002 (Figure 12.24).

TABLE 12.5 Emission Reduction Goals for Canada and the United States

Canada	United States
SO$_2$ emissions reduction in 7 easternmost provinces to 2.07 million tons by 1994	SO$_2$ emissions reduction of 10 million tons from 1980 levels by the year 2000
Maintenance of 2.07 million tons annual cap for eastern Canada through December 1999	National cap of 5.6 million tons for industrial source emissions beginning in 1995
Permanent national cap on SO$_2$ emissions of 2.88 million tons by the year 2000	Permanent national cap of 8.95 million tons for electric utilities by the year 2010

(*Source:* Air Quality Committee, 1994.)

PHOTOCHEMICAL SMOG

Photochemical smog refers to the mix of natural atmospheric chemicals with anthropogenic emissions derived mainly from fossil fuel burning produced in the presence of solar radiation. The chemical reactions in this type of smog are aided by radiant energy from the sun, hence the term **photochemical.** The chemical mixtures produced by these reactions lead to the generation of reddish, yellow-brown, and gray hazes in the sky and can be detrimental to living organisms and ecosystems. Carbon monoxide (CO), sulfur and nitrogen oxides, lead, toxic hydrocarbons (organic compounds of hydrogen and carbon) such as benzene and toluene, and particulates can be major contributors to smog. Ozone is also a major product of the reactions involved in the generation of smog. In contrast to the protective layer of ozone that is found in the stratosphere, excess ozone in the lower troposphere can have a deleterious effect on plants and animals. Table 12.6 shows examples of the complex of substances in urban air leading to the development of smog. Vehicles used in transportation are an important source of the carbon monoxide, hydrocarbons, NO$_x$, SO$_x$, and particulates found in urban, as well as rural, air pollution. Industrial sources contribute substantial amounts of SO$_x$ and particulates in some cities.

Smog often occurs in many cities of the world where meteorological conditions lead to development of a temperature inversion (Figure 12.25). During normal atmospheric conditions, temperature decreases with increasing height above ground level and gases and aerosols, including pollutants, rise and are well mixed and often transported out of the source region and dispersed by winds. In a temperature inversion, warm, less dense air overlies colder

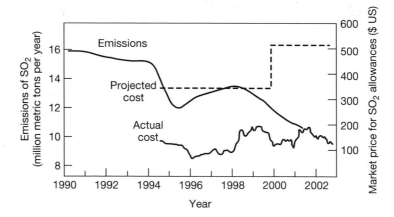

FIGURE 12.24 Pattern of declining SO$_2$ emissions in the United States from 1990 to 2002, and projected and actual costs of implementing clean air regulations to curb emissions. Notice that the projected cost estimates were much higher than the actual costs, and the actual costs have not changed greatly over time.

TABLE 12.6 The Complex of Substances Present in Urban Air Derived from Transportation Sources in Various Cities

Region	Year	Total Pollutants from all Sources (Thousand Metric Tons)	Percent Attributable to Road Transport					
			CO	HC	NO_x	SO_x	Particulates	Total
Mexico City	1987	5,027	99	89	64	2	9	80
São Paulo	1981	3,150	96	83	89	26	24	86
	1987	2,110	94	76	89	59	22	86
Ankara	1980	690	77	73	44	3	2	57
Manila	1987	500	93	82	73	12	60	71
Kuala Lumpur	1987	435	97	95	46	1	46	79
Seoul	1983	—	15	40	60	7	35	35
Hong Kong	1987	219	—	—	75	—	44	—
Athens	1976	394	97	81	51	6	18	59
Gothenburg	1980	124	96	89	70	2	50	78
London	1978	1,200	97	94	65	5	46	86
Los Angeles	1976	4,698	99	61	71	12	—	88
	1982	3,391	99	50	64	21	—	87
Munich	1974/5	213	82	96	69	12	56	73
Osaka	1982	141	100	17	60	43	24	59
Phoenix	1986	1,240	87	64	77	91	1	28

(*Source:* Faiz et al., 1990.)

and denser air, forming a lid that traps air pollutants below it. Typically, an inversion occurs when a warm air mass collides with a cold air mass (see Chapter 4). The warm air mass front overrides the cold air mass front and produces the inversion in temperature. In some cases, like in the United States cities of Los Angeles and Denver, long periods of calm weather conditions and light winds that trap pollutants against nearby mountain ranges exacerbate conditions produced by an inversion.

Even moderately populated nonindustrial areas, like rural areas and grasslands, are not devoid of photochemical smogs. The burning of savanna grasslands and sugarcane areas can produce smog because of the release of NO_x and reactive hydrocarbons that are necessary for smog formation. The fact that this burning takes place in tropical and subtropical climatic regions complicates the situation because of the intense sunlight that promotes smog formation. Ozone levels during such burnings may be several times normal background levels. Figure 12.26 shows the strong correlation between the concentration of ozone in the troposphere and biomass burning in the Southern Hemisphere. Because biomass burning in Africa, South America, and Southeast Asia is a seasonal phenomenon, tropospheric ozone concentrations vary seasonally, with highest levels in the austral (Southern Hemisphere) springtime. During intense biomass burning periods in Africa, plumes of smoke and aerosol can be seen by satellite imagery moving westward across the South Atlantic Ocean from the African continent (http://visibleearth.nasa.gov/view_rec.php?id=18611).

Urban Pollution

Examples of pollution from the burning of fossil fuel date as far back as 1306, when King Edward I initially banned the burning of coal by local London craftsmen. The Industrial Revolution led to intense burning

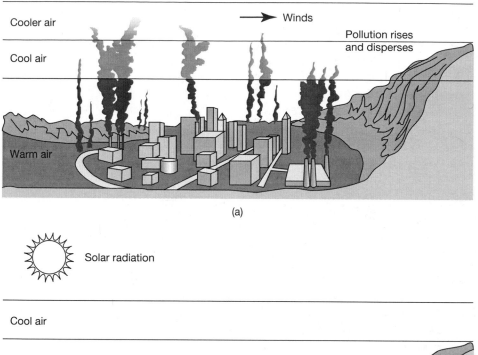

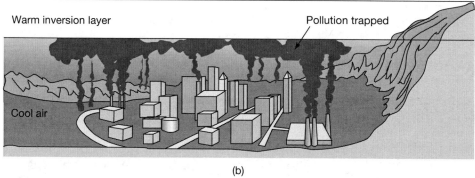

FIGURE 12.25 (a) Normal vertical pattern of temperature in the air above a city and (b) the pattern developed during a temperature inversion. In the normal situation, pollution is dispersed by winds and turbulence in the air, but during inversions, pollution is trapped below the inversion. (*Source:* Chiras, 1988.)

of fossil fuels and a rise in anthropogenic sulfur and nitrogen oxide emissions. By 1950, many countries were using high-sulfur coals and oils as sources of energy. Pollution generated by the burning of these fossil fuels was held responsible for the deaths of people during pollution incidents that occurred, for example, in the city of Donora, Pennsylvania, in 1948; the killer fog in London in 1952 that resulted in 2500 to 4000 deaths; and in New York City in the 1960s.

The witch's brew of urban pollution differs from city to city. Sulfur dioxide and particulate-matter concentration levels in urban areas can vary greatly. In Beijing, Shanghai, and Shenyang, China; Mexico City, Mexico; Rio de Janeiro, Brazil; and Seoul, Korea, from 1988 to 1992, the average annual concentration of sulfur dioxide in the air of these cities was in the range of 100 to 200 micrograms (millionths of grams) per cubic meter. In contrast, in Auckland, New Zealand, and Melbourne, Australia, for the same period, average

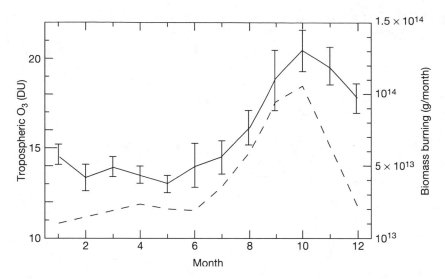

FIGURE 12.26 Monthly tropospheric ozone concentrations (solid line) averaged for the period from 1979 to 1992 compared with biomass burning (dashed line) in the latitude belt of 20° to 30°S. The concentration of ozone is given in Dobson units (DU), where one DU is equivalent to the amount of ozone present in a column of air 1/100 of a millimeter thick at 0°C and one atmosphere of total pressure (see Chapter 14). Biomass burning is in units of grams of dry matter burned per month (to convert to carbon, multiply by 0.45). (*Source:* Jiang and Yung, 1996.)

annual concentrations were less than 10 micrograms per cubic meter. The World Health Organization (WHO) guideline for sulfur dioxide is between 40 and 60 micrograms sulfur dioxide per cubic meter of air. Concentrations above this range are considered unhealthy. Between 1980 and 1984, 23 of 54 cities studied had annual average sulfur dioxide concentration levels above this level during one or more years.

Particulate-matter concentrations in cities are also very variable, ranging from about 10 to 20 to more than 1000 micrograms per cubic meter. Cities like Calcutta and New Delhi, India; Shenyang and Xian, China; Cairo, Egypt; Mexico City, Mexico; and Jakarta, Indonesia have particularly high concentrations. The WHO guideline for suspended particulate matter in the air is between 60 and 90 micrograms per cubic meter of air. Between 1980 and 1984, 29 of 41 cities studied had average annual particulate matter levels in their air exceeding this range during one or more years. In the United States, older cities like New York and St. Louis are gray-air cities characterized by smog generated in moist air mainly from sulfur oxides and particulates. Relatively nonindustrialized cities like Denver and Los Angeles are brown-air cities where ozone is an important component of the smog.

Clean air regulations in the United States and western European countries are leading to a reduction in the burdens of particulate matter in the air above these countries. Figure 12.27 illustrates the sources of the very fine-grained particulate matter ($PM_{2.5}$) and trend in their emissions with time for the United States. Notice the 45% reduction in total

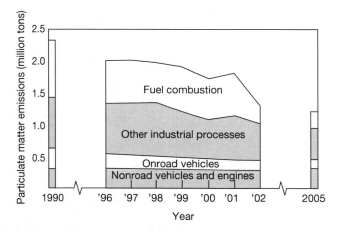

FIGURE 12.27 Anthropogenic particulate $PM_{2.5}$ emissions by source sector in the United States from 1990 to 2005. Notice the overall decline in emissions of particulates to the atmosphere of the United States over time. (*Source:* EPA, 2009, http://www.epa.gov.)

particulate emissions from 1990 to 2005, leading to cleaner air for the United States.

Tropospheric Ozone

Tropospheric ozone is a major component of urban smog. In the troposphere, besides being an atmospheric pollutant, ozone is also a greenhouse gas and can affect the radiation balance of the planet. Ozone is a product of a complex series of chemical reactions in the presence of sunlight that involves nitrogen oxides (NO_x) and organic compounds that are volatile (commonly referred to as VOCs, volatile organic compounds, including hydrocarbons, halogen-containing chemical compounds, alcohols, ethers, etc.). Methane is the most important VOC on a global scale.

The formation of ozone in the troposphere involves a natural cycle in which shortwave ultraviolet radiation breaks the chemical bond joining nitrogen and oxygen in nitrogen dioxide (NO_2) (Figure 12.28). This process results in the generation of nitric oxide (NO) and a free oxygen atom. The free

oxygen atom then joins with an oxygen molecule (O_2) to make an ozone molecule. The cycle regenerates itself in that the ozone molecule reacts with nitric oxide to produce an ordinary oxygen molecule and nitrogen dioxide. The nitrogen dioxide generated starts the cycle over again. In this fashion, ozone is produced and destroyed in equal amounts. There is no net increase or decrease in the concentration of ozone in the atmosphere.

The ozone cycle can be disrupted by the presence of VOCs that are chemically reactive in sunlight (Figure 12.28). These VOCs can be derived from natural sources or from human activities. The disruption of the ordinary cycle allows ozone to accumulate in the atmosphere. What happens? The highly reactive VOCs are able to convert nitric oxide directly to nitrogen dioxide without destroying an ozone molecule in the process. Thus, the ozone is not broken down by the nitric oxide and is free to accumulate in the atmosphere.

Because the ozone cycle is strongly linked to sunlight, ozone tends to accumulate in the atmosphere

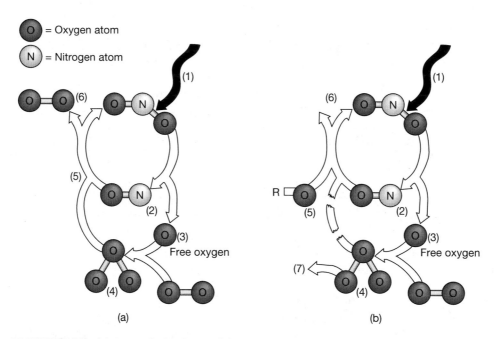

FIGURE 12.28 (a) Ozone formation and destruction in the natural environment and (b) disruption of the ozone cycle and accumulation of ozone in an environment containing volatile organic compounds. In (a), nitrogen dioxide is broken down by ultraviolet light (1) to nitric oxide (2) and a free oxygen atom (3). The free oxygen unites with an oxygen molecule to generate ozone (4). Ozone then reacts with nitric oxide (5) to generate an oxygen molecule and nitrogen dioxide (6). The cycle begins again. In (b), reactive volatile organic compounds (5) convert nitric oxide to nitrogen dioxide (6), and there is no destruction of ozone in the process. Ozone accumulates in the atmosphere (7). R is reactive volatile organic compounds (VOCs). (*Source:* EPRI, 1988.)

during the day, especially during the long-lit hours of the summer. Weather conditions and the ratio of VOCs to nitrogen oxides in the atmosphere determine to a great extent the amount of ozone formed. The precursor VOCs and nitrogen oxides and ozone can be carried by winds to areas remote (1000 to 1500 kilometers) from their sources and remain in the atmosphere for days. High springtime ozone levels are even found over the higher latitudes of Canada and the Arctic, a result of long-range transport of air pollution into the Arctic in the winter and spring and conversion of some of the pollution to ozone when more sunlight reaches the Arctic in the springtime (http://earthobservatory.nasa.gov/IOTD/view.php?id=3062). Thus, the oxidants ozone and nitrogen oxides, along with the chemically reduced VOC compounds, are atmospheric pollutants that can affect ecosystems on a regional scale.

Sources of VOCs and Nitrogen Oxides

VOCs and NO_x are emitted to the atmosphere from both natural sources and human activities. Natural sources of nitrogen oxides include forest fires, decaying organic matter in soils, and lightning. VOCs are emitted naturally from forest fires, growing terrestrial vegetation, and the ocean and from anthropogenic sources. In the United States, 90% of the VOCs derived from human activities come from road vehicles and nonroad equipment, solvent use,

fires, and the residential wood combustion sector of the economy (Figure 12.29). Vehicles, electricity, nonroad equipment, fossil fuel combustion, and industrial processes account for 99% of the anthropogenic emissions of nitrogen oxides in the United States (Figure 12.14c).

There is a strong seasonal pattern in the global emissions of VOCs to the atmosphere reflecting seasonal changes in the fluxes of VOCs from both natural sources and those related to industry, transportation, and fuel combustion. For example, some VOC emissions from terrestrial vegetation are high during the Northern Hemisphere summer and low during the winter, reflecting the seasonal pattern in vegetation growth. The austral summer is a time of important emissions of some VOCs to the atmosphere from oceanic regions in the middle latitudes of the Southern Hemisphere, reflecting increased marine productivity at this time of year. Increased combustion of fossil fuels during the winters in both hemispheres increases the emissions of some VOCs, as well as NO_x, to the atmosphere.

A portion of the emitted VOCs is nonmethane hydrocarbons (NMHCs), that is, compounds of hydrogen and carbon without a methane (CH_4) component, including ethene (C_2H_4), propene (C_3H_6), terpene ($C_{10}H_{16}$), and isoprene (C_5H_8). Figure 12.30 shows the global biogeochemical cycle of NMHCs involving the transfer of these gaseous VOCs between Earth's surface and the atmosphere. Notice that the

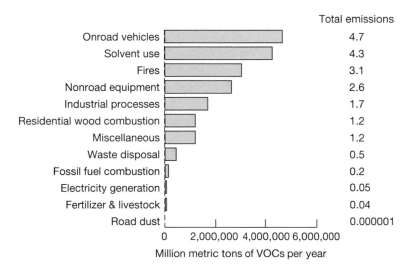

FIGURE 12.29 United States emissions of volatile organic compounds (VOCs) in millions of metric tons per year according to the source sector. (*Source:* EPA, 2009, http://www.epa.gov.)

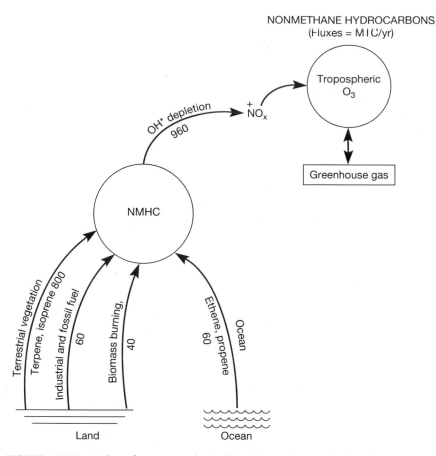

FIGURE 12.30 Earth surface–atmosphere global biogeochemical cycle of nonmethane hydrocarbons (NMHCs). The NMHCs react in the atmosphere with hydroxyl radical in the presence of nitrogen oxides to form tropospheric ozone, an atmospheric pollutant, and a greenhouse gas. Some products of this reaction are removed as organic and inorganic substances in wet and dry deposition. The values on the arrows are fluxes in units of million tons of carbon per year (MTC/yr). (*Source:* Mackenzie, 2003.)

NMHCs react with atmospheric hydroxyl radical (OH^*) in the presence of elevated levels of NO_x in a complex series of reactions that lead to the formation of peroxy $(H_2C_3O_2)$ and odd-hydrogen radicals (e.g., CH_2O and CH_3OOH) and eventually to ozone. In the process hydroxyl radical concentrations are decreased. The blue haze of the Smoky Mountains of the southeast United States has been attributed in part to the NMHC terpene emitted from the forest trees in this area.

Because transportation is such an important source of both VOCs and nitrogen oxides throughout the world in urban centers, many cities of the world have high ozone levels. In Mexico City, the one-hour ozone peak standard of 0.11 parts per million, which is not to be exceeded more than once daily, is surpassed 300 days a year. In Los Angeles in 1988, ozone levels exceeded the United States federal standard on 172 days. Table 12.7 shows the summer ozone levels for several European countries, which can be 100 to 200% higher than natural levels, with concentrations in urban centers even higher. In summer 2009, 226 of 1072 stations across Europe exceeded the European Union standard of 180 micrograms ozone per cubic meter (Figure 12.31). Even in the United States with its strict air quality standards, 345 United States counties violated the ozone standard of 0.075 ppm in 2008, resulting in 900 to

TABLE 12.7 Summer Ozone Levels for Selected European Countries

Country	Upper Daily Average (Micrograms Per Cubic Meter)	Peak	Increase of Daily Average Over Natural Levels (%)
Netherlands	80–130	500	180
Germany	100–150	400–500	200
United Kingdom	90–165	200–500	210
Belgium	—	300	—
France	70–120	—	160
Norway	—	200–300	—

(*Source:* Postel, 1984.)

1100 more deaths, 5600 more hospital visits, and US$8 billion in costs.

Effects of Air Pollution on Health

The buildup of ozone, sulfur dioxide, carbon monoxide, and other pollutants in the trapped air of an inversion can be detrimental to human health. Table 12.8 gives the health effects of major air pollutants on humans. In the case of ozone, ozone levels that exceed standards, which in the United States are set by the United States Environmental Protection Agency (EPA), can cause respiratory problems, such as emphysema and asthma. The elderly and highly sensitive people with preexisting pulmonary problems are especially vulnerable. In terms of vegetation and ecosystems, it is known that ozone can impair the process of photosynthesis and lead to reduced plant growth and a decrease in agricultural crop yield. Prolonged exposure to ozone may also weaken the defense mechanisms of plants and leave them more vulnerable to insect infestations and drought. The cost of destruction of crops and lower yields of crops because of photochemical smog in California alone is about $15 billion annually.

Controlling Tropospheric Ozone

Control strategies aimed at reducing the amount of ozone in the lower atmosphere will to a certain extent depend on the region or nation affected. This is because the principal precursors of ozone, VOCs and nitrogen oxides, come from a variety of sources. VOCs escape in vapor form from chemical plants, refineries, gas stations, and motor vehicles. There are emission control devices, such as catalytic converters and vapor recovery systems on cars, that can be used to trap, recycle, or transform these potentially harmful vapors. The VOCs that escape to the atmosphere in the use of products such as paints, solvents, and pesticides are more difficult to control. The combustion of coal, oil, and gas is the main source of nitrogen oxides to the atmosphere from human activities. Low-nitrogen oxide burners on power generating plants and engines that burn a lean mix of fuel can help to reduce the emissions of nitrogen oxides. Because transportation is a major source of both VOCs and NO_x, the best methods of controlling their emissions are improvement of vehicular design and the use of alternative fuels. However, the exact control methods probably will differ from region to region because the effectiveness of controls depends on the relative proportions of VOCs and NO_x emitted, the daily hours of sunlight, and other conditions. Certainly, the development of transportation and industrial practices that do not rely so heavily on fossil fuel combustion would go a long way in ameliorating tropospheric ozone accumulation and air pollution in general. Figure 12.32 illustrates the reduction of VOC emissions to the atmosphere in the United States because of regulatory measures from 1980 projected to 2010. No such decrease in VOC emissions for Canada is apparent in the data.

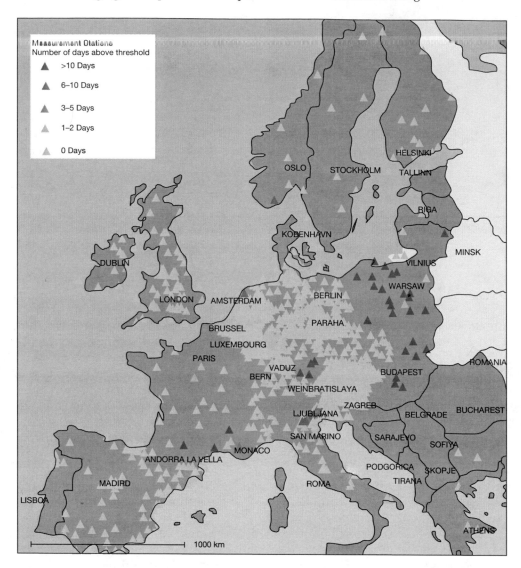

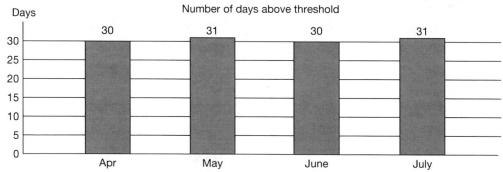

FIGURE 12.31 Number of days in Europe when ozone levels exceeded certain concentration levels. Only 226 of 1072 monitoring stations across Europe exceeded the European information threshold of greater than 180 micrograms per cubic meter in summer 2009. (*Source:* http://www .eea.europa.eu.)

TABLE 12.8 Health Effects of Major Air Pollutants

Sulfur dioxide	Respiratory irritation, shortness of breath, impaired pulmonary function, increased susceptibility to infection, illness in the lower respiratory tract (particularly in children), chronic lung disease, and pulmonary fibrosis. Increased toxicity in combination with other pollutants.
Respirable particulate matter	Irritation, altered immune defense, systemic toxicity, decreased pulmonary function, and stress on the heart. Acts in combination with SO_2; effects depend on the chemical and biological properties of the individual particles.
Oxides of nitrogen	Eye and nasal irritation, respiratory tract disease, lung damage, decreased pulmonary function, and stress on heart.
Carbon monoxide	Interferes with oxygen uptake into the blood (chronic anoxia). Can result in heart and brain damage, impaired perception, asphyxiation, or, in low doses, in weakness, fatigue, headaches, and nausea.
Lead	Kidney disease and neurological impairments. Primarily affects children. Lead in the environment has decreased dramatically since regulations restricting its use as an antiknock agent in gasoline and use in paint were put in effect.
Photochemical oxidants (e.g., ozone)	Decreased pulmonary function, heart stress or failure, emphysema, fibrosis, and aging of lung and respiratory tissue.

(*Source:* Whelpdale, 1991.)

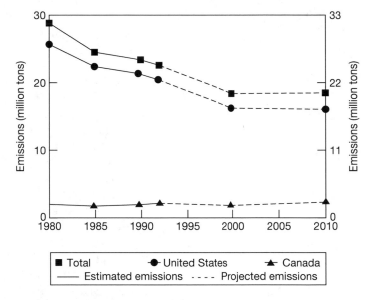

FIGURE 12.32 Emissions of volatile organic compounds (VOCs) from sources in the United States and Canada between 1980 and projected to 2010. Notice the decline in United States emissions but the nearly constant emissions of VOCs from Canadian sources. VOCs are mainly derived from industrial processes and transportation sources. (*Source:* Air Quality Committee, 1994.)

Concluding Remarks and the 21st Century

Air pollution and acid deposition in general are problems that affect regional and continent-size systems as well as local areas. Transportation of pollutants on the mesoscale level (Figure 12.1) at distances of tens to thousands of kilometers from sources has led to atmospheric chemistry changes for most of the industrialized world and portends such changes for the developing nations. In these latter nations, the current picture is mainly one of local or subregional-scale air pollution (Table 12.9). However, as these nations continue to industrialize, it is very likely that smog and haze will become more widely distributed. Figure 12.33 shows a model calculation of the seasonal distribution of surface ozone over China, South, and Southeast Asia. Notice the strong plume of ozone pollution extending out over the western Pacific Ocean. If maximum feasible reductions legislation is not met, air quality problems will be severe over Southeast and East Asia in the twenty-first century. This situation will certainly occur if the industrializing nations continue to rely on fossil fuels for transportation and industrial practices. This statement is partially supported by consideration of the present condition of the atmosphere over the continental United States.

The United States relied heavily on fossil fuels during the last century to sustain its economic development and continues to do so. As a result, atmospheric hazes of local and regional extent have been identified over most of the continental United States. These hazes are so extensive that visibility is impaired even in national parks in the United States about 90% of the time. Visibility in the eastern United States is 10 times lower than that of the western states. Fine particulates and gases such as fly ash (a carbon-rich product of combustion), sulfur and nitrogen oxides, and hydrocarbons form the major components of atmospheric haze. In the United States, sulfur dioxide emissions from urban and industrial sources are the principal precursor of the sulfate aerosol that contributes to atmospheric haze, except in the northwestern United States. In this region, organic aerosols composed of hydrocarbons and their breakdown products are the dominant cause of haze.

As an example of how the haze distribution in the air over the United States has changed over time, Figure 12.34 shows the average annual standard range of visibility in the United States for 2000 to 2004. The standard visual range is defined as the farthest distance a large dark object can be seen during daylight hours. The visual range under naturally occurring conditions without pollution in the United States is typically from 75 to 150 km in the eastern United States and 200 to 300 kilometers in the western United States. From 1948 through the mid-1980s, a trend of decreasing visibility, particularly over the eastern United States in the spring and summer months, was clearly evident. Notice in Figure 12.34 that even in the early twenty-first century that most of the United States has standard visual ranges less than that estimated for natural conditions. The eastern part of the country is especially hazy and the pattern of distribution of the haze is similar to the patterns for the anthropogenic sources of acid gases to the atmosphere, wet deposition of sulfate and nitrate, and lower pH values of rainfall. Sulfate and organic aerosols produced by human activities, in particular the burning of fossil fuels, are primarily responsible for these hazy conditions. Sailors on ships at sea hundreds of kilometers off the East Coast of the United States can see the reddish-brown haze that overlies the continent. Indeed, this haze is responsible for the brilliant sunsets often observed by the sailors.

Since the early 1970s, emissions of the air pollutants of sulfur dioxide, particulates, carbon monoxide,

TABLE 12.9 Examples of Air Pollution in Some Indian Cities. World Health Organization Standards for Annual Mean Concentrations of NO_x, SO_x, and SPM (Suspended Particulate Matter) are 150 (24 hrs.), 40–60, and 60–90 Micrograms Per Cubic Meter of Air, Respectively.

City	(Micrograms Per Cubic Meter)		
	NO_x	SO_x	SPM
Bombay			
(Jan. 1991 average)	115	59	579
Madras			
(Apr.–Dec. 1991 maximum)	85	228	490
New Delhi			
(1991 summer average)	10–15	15–20	350–500
(1989 maximum)	31	42	1,062
Calcutta			
(1989 mean)	109	88	554
(1989 maximum)	363	230	1,093
Bangalore			
(1989 mean)	23	27	141
(1989 maximum)	70	104	478
Ahmedabad			
(1989 mean)	50	25	232
(1989 maximum)	188	123	514

(*Source: The Hindu Survey of the Environment*, 1992.)

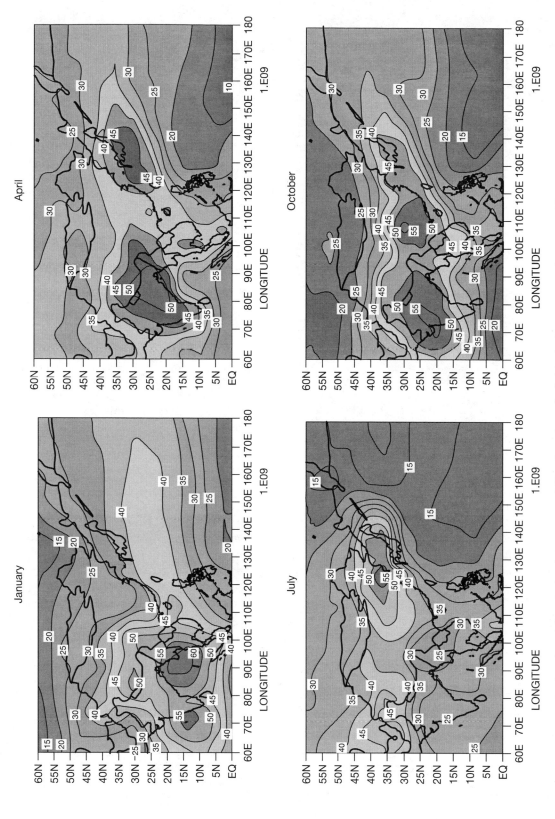

FIGURE 12.33 Simulated model ozone levels in parts per billion by volume (ppbv) in the lower troposphere across China, South, and Southeast Asia in January, April, July, and October of 1996. Notice the distinct plume of pollution extending out across the western Pacific Ocean. (*Source:* Liu et al., 2007.)

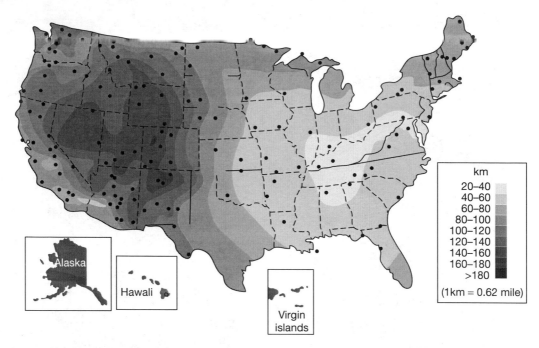

FIGURE 12.34 Annual average standard visual range in kilometers in the contiguous United States for the period of 2000 to 2004. (From *Canada-United States Air Quality Agreement, Progress Report 2008.* © Her Majesty The Queen in Right of Canada, Environment Canada, 2008. Reproduced with the permission of the Minister of Public Works and Government Services Canada.)

and, most dramatically, the trace element lead (because of the elimination of leaded gasoline) in the United States have fallen. However, it is worth pointing out in terms of pollutant trace element emissions to the atmosphere that global mercury (Hg) emissions primarily from coal combustion were about 3000 metric tons per year in the 1970s. In the mid-1990s, there was a peak in Hg emissions because of the rapid industrial development in Asia, Africa, and South America at this time using coal as a major energy resource. Mercury emissions have since stabilized at about 2200 metric tons per year but continue to be an environmental concern. North America accounts for about 9% and Asia for about 52% of current global emissions of Hg to the environment (see Chapter 11).

Nitrogen oxide emissions have remained nearly constant in the United States, although they are projected to increase in the early part of the twenty-first century. Tropospheric ozone is now the most widespread air pollutant in the United States. Because of the decrease in air pollutant emissions, the air quality over the conterminous United States has improved from the late 1980s into the twenty-first century. Continental hazes have become less intense and visibility has improved, but haze produced by pollution is still prevalent over most of the country. As the industrializing countries of the world continue to develop and to rely on fossil fuels as a major energy source, large-scale continental hazes are likely to develop in these countries and to follow a path of development similar to that of the United States.

Study Questions

1. How do the time and space scales of stratospheric ozone depletion and urban air pollution compare?
2. What is acid deposition, and what atmospheric constituents related to human activities are involved in acid deposition?
3. Knowing the hydroxyl radical is involved in the oxidation of SO_x and NO_x and other reduced gases in the atmosphere derived from human activities, what concern

might you have for the fate of the hydroxyl radical in the atmosphere?
4. (a) What are the major anthropogenic sources of SO_x and NO_x to the atmosphere?
   (b) What is the major difference in the character of these source materials?
5. Define pH. What is the concentration of hydrogen ions in a solution with a pH of 5.6?

6. What observations led a large group of scientists in the 1950s and 1960s to become concerned about the problem of acid deposition?

7. Why are freshwater aquatic systems located on crystalline rock terrane more sensitive to acidification from acid deposition than systems found in limestone areas?

8. The global anthropogenic emissions of sulfur to the atmosphere from combustion of fossil fuels in 2000 were 62 million tons per year.

   (a) When will the projected population double at an exponential growth rate of 1.7% per year, and what will be the population?

   (b) If the rate of growth of these emissions in the future were simply related to population growth, what will be the anthropogenic sulfur emissions to the atmosphere at that time?

   (c) Why might the emissions actually be higher than calculated?

9. (a) What are the major chemical components of photochemical smog?

   (b) Why are human-induced smogs termed *photochemical*?

10. What role do temperature inversions play in the development of photochemical smogs in urban areas?

11. What are the major natural and anthropogenic sources of VOCs and NO_x to the atmosphere?

12. Explain how the natural cycle of ozone in the troposphere is disrupted by anthropogenic emissions of VOCs and NO_x.

13. What are some effects of tropospheric ozone accumulation on human health and ecosystems?

14. Referring to Figure 12.6, calculate the percentage of total rain and dry deposition of sulfur on Earth's surface that is due to the anthropogenic activities of fossil fuel combustion and biomass burning. What is the principal process that acts as a sink for sulfur dioxide gas in the atmosphere? What would you expect to be the climatic response to increasing concentrations of sulfate aerosol in the atmosphere because of human activities, and why?

15. Referring to Figure 12.7, of the total flux of NO_x to the atmosphere, what percentage is from human activities? What is the principal type of reaction that destroys NO_x in the atmosphere? Write the reaction for the destruction of NO_2 and its subsequent removal as HNO_3.

16. SO_2, a major atmospheric pollutant in photochemical smog, has an average concentration of 0.2 ppbv (parts per billion by volume) in remote air. How many moles of SO_2 are there in the background atmosphere?

17. What is the residence time of SO_2 in the remote atmosphere, assuming a natural flux of 1×10^8 metric tons of S per year into the atmosphere? What does the value of this residence time tell you qualitatively about how far from a fossil fuel combustion source (e.g., coal-fired electrical generating plant) SO_2 released into the atmosphere may travel?

18. Why is there a potential for greater regional and subcontinental development of smog and haze in the industrializing and developing countries during this century?

Additional Sources of Information

Air Quality Committee, 1994, *United States–Canada Air Quality Agreement: 1994 Progress Report.* United States–Canada Air Quality Committee, International Joint Commission, Washington, DC, 64 pp.

Elsom, D., 1996, *Smog Alert: Managing Urban Air Quality.* Earthscan Publications, London, 226 pp.

Environmental Protection Agency (EPA), various dates, http://www.epa.gov.

French, H., 1990, *Clearing the Air: A Global Agenda.* Worldwatch Paper 94, Worldwatch Institute, Washington, DC, 54 pp.

Galloway, J. N., F. J. Dentener, D. G. Capone, E. W. Boyer, R. W. Howarth, S. P. Seitzinger, G. P. Asner, C. C. Cleveland, P. A. Green, E. A. Holland, D. M. Karl, A. F. Michaels, J. H. Porter, A. R. Townsend, and C. J. Vorosmarty, 2004, Nitrogen cycles: past, present, and future. Biogeochemistry, v. 70, pp. 153–226.

Gauss, M., Ellingsen, K., Isaksen, I. S. A., Dentener, F. J., Stevenson, D. S., Amann, M., and Cofala, J., 2007, Changes in nitrogen dioxide and ozone over Southeast and East Asia between years 2000 and 2030 with fixed meteorology. Terrestral, Atmosphere, and Ocean Sciences, v. 3, 475–492.

Hedin, L. O. and Likens, G. E., 1996, Atmospheric dust and acid rain. *Scientific American,* v. 275, December, pp. 88–92.

Lents, J. M. and Kelly, W. J., 1993, Clearing the air in Los Angeles. *Scientific American,* v. 269, October, pp. 32–39.

Liu, L., Sundet, J. K., Liu, Y., Berntsen, T., and Isaksen, I. S. A., 2007, A study of tropospheric ozone over China with a 3-D global CTM model. *Terrestrial, Atmosphere, and Ocean Sciences,* v. 18, pp. 515–545.

Mohnen, V. A., 1988, The challenge of acid rain. *Scientific American,* v. 259, August, pp. 30–38.

National Atmospheric Deposition Program (NADP), http://nadp.sws.uiuc.edu/.

Pacyna, J. M., 2008, Global anthropogenic emissions of mercury to the atmosphere. *Encyclopedia of Earth,* http://www.eoearth.org/, 7 pp.

Postel, S., 1984, *Air Pollution, Acid Rain, and the Future of Forests.* Worldwatch Paper 58, Worldwatch Institute, Washington, DC, 54 pp.

Smith, S. J., Andres, R., Conception, E. and Lurz, J., 2004, *Historical Sulfur Dioxide Emissions 1850–2000: Methods and Results.* Pacific Northwest National Laboratory, National Technical Information Service, United States Department of Commerce, Springfield, VA, 14 pp.

Stern, D. I., 2006, Reversal of the trend in global anthropogenic sulfur emissions. *Global Environmental Change,* v. 16, pp. 207–220.

CHAPTER

13

The Changing Ecosphere: Pleistocene and Holocene Environmental Change

The development of these huge ice sheets must have led to the destruction of all organic life at the Earth's surface. The ground of Europe, previously covered with tropical vegetation and inhabited by herds of great elephants, enormous hippopotami, and gigantic carnivora, became suddenly buried under a vast expanse of ice covering plains, lakes, sea, and plateaus alike.

LOUIS AGASSIZ, 1840

The external environment of the Earth's ecosphere has changed dramatically since the planet's creation. In Chapter 8 we discussed in broad terms the changes in the surface environment of the planet since its beginning. In this and the following chapters, more recent changes in the Earth's ecosphere are discussed, particularly concentrating on the links between changes in atmospheric composition and the physical climatic system and their effects and interactions with the hydrosphere, biosphere, cryosphere, and lithosphere. Although the subject matter of this chapter deals mainly with the natural environment of the past, returning to the theme of Chapter 8, the material is discussed at this stage to provide necessary background information for the major global environmental atmospheric issues of global warming and stratospheric ozone depletion considered in Chapter 14.

Modern technological advances in scientific instrumentation and enhanced scientific research are providing information and new insights into potential causes of climatic change during the geological past and are offering predictions for the future. Many factors govern changes in climate: the composition and concentration of gases and aerosols in the atmosphere; the amount of solar radiation reaching Earth from the sun; the effect of orbital parameters on the radiation budget of Earth; the shape and location of continents and oceans; the atmosphere/ocean circulation patterns; the interaction of biogeochemical cycles in ecosystems such as lakes, tundra, deserts, forests, and grasslands; the reflectivity (albedo) of different parts of the surface of Earth; catastrophic events of nature such as volcanic explosions and impacts of meteorites; and more recently, the human volcano, the development of the technosphere. These are the major factors that have influenced our climate throughout time (Figure 13.1).

This chapter begins with a brief discussion of some ways in which the climate of the past is determined. Factors influencing climate are then discussed. An encapsulated review of Mesozoic and Cenozoic climatic change follows with emphasis on Pleistocene and Holocene environmental conditions (see

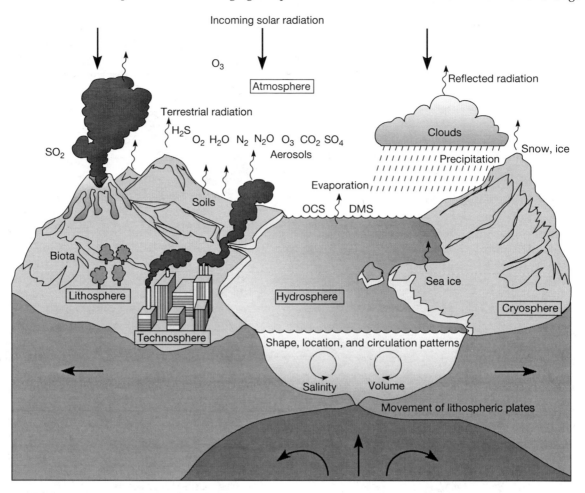

FIGURE 13.1   Earth's climate system and factors governing the world's climate. The system consists of the interactive subsystems of the atmosphere, cryosphere, oceans, land, and biota and is driven by solar energy. Some incoming energy is reflected back to space by clouds, snow, ice, and atmospheric aerosols derived from human activities and natural sources. Surface relief and the relative positions of continents and oceans are controlled by plate tectonic motions. Gases from fossil fuel burning and other sources affect atmospheric composition and hence climate.   (*Source:* Skinner and Porter, 1995.)

Chapter 8 for more information on the long-term environmental and evolutionary history of the Mesozoic and Cenozoic).

RECORDS OF CLIMATE

There is a wealth of data on past climates. Proxies of long-term temperature records are found in ice cores and deep-sea sediment cores that have been collected for many years. By extracting air from the bubbles trapped in ice cores and examining the isotopic composition of the ice water and particles trapped in the ice over time, the composition and chemical changes in atmospheric gases, as well as temperatures, ocean chemistry, and winds, of the planet can be determined during Pleistocene and Holocene times. A few long ice cores bottoming out at several kilometers in depth and a number of shorter cores have been drilled and extracted from continental ice sheets. Ice cores with the longest record of change are from Vostok (Vostok Station) and Dome C (Concordia Station), Antarctica, and the summit of the Greenland ice cap. The composite of the Vostok and Dome C ice core from Antarctica contains climatic and atmospheric compositional records extending back in time for 800,000 years and includes eight glacial cycles. The record from Vostok alone includes four ice ages and five warm interglacial periods. Analyses of the cored ice and its meltwater and air bubbles in the ice have provided a wealth of information on the history of the global environment. Analyses of ice cores from low-latitude mountain ice caps are also providing data, particularly for climatic and environmental changes of regional scale.

Cores of sediments and sedimentary rocks contain fossils of microscopic plants, animals, and pollen that provide clues to climate and environment (see Chapter 2). Plant pollen found in these cores is easily identified and can indicate the amount and kinds of vegetation that inhabited an area in the past and its climate. Historical records of pollen found in lake sediments have been used to determine the timing of human migration into an area and the consequent landscape and climatic effects on the vegetation and temperature of the region. In addition, the oxygen and carbon isotope composition of microscopic plants and animals that are found in sediment cores provide information on seawater temperature and composition and atmospheric composition.

Analyses of tree rings also provide information on past climates (Figure 13.2a), as well as analyses of pack-rat middens, which can be used to obtain environmental data for the past 40,000 years. Even fossilized beetle species have been used as indicators of past climate (Figure 13.2b). Also, ship's logs, records from the trading posts of the Hudson Bay Company during the seventeenth to nineteenth centuries, and documents from past civilizations, in particular, the climatic records from China dating back more than two millennia (Figure 13.3), have provided a great deal of data on more modern climates. Table 13.1 lists some sources of paleoclimatic data and their characteristics. In the modern world, satellites, weather balloons, ships at sea, and land-based stations are employed to record temperature and other climatic variables to enable day-by-day weather forecasting and long-term climatic predictions.

FACTORS INFLUENCING CLIMATE

In this section the major factors mentioned previously influencing global and regional climate change and climate variability are considered. Some of these factors were touched upon in Chapter 4. The greenhouse effect and the absorption of solar energy by stratospheric ozone are atmospheric phenomena that are discussed in more detail in following chapters

in the context of the enhanced greenhouse effect and global warming and stratospheric ozone depletion. These two environmental issues have become worldwide the biggest environmental problems of scientific interest and public concern and policy in recent decades.

Fluctuations in Solar Energy

The sun's energy is known to fluctuate periodically. Humans have looked at the sun for centuries through telescopes and observed changes in the number of sunspots (dark circular regions) visible on the sun's surface. This nearly 500-year record of human observations of the sun's surface indicates that there is an approximately 11-year cycle in the number of sunspots visible on the surface of the sun (Figure 13.4). In addition, when sunspots are abundant, it has been observed that there are strong solar emissions from bright rings (faculae) around the darker sunspots and from the sun's polar regions. At quiet times on the sun's surface, these emissions appear less intense. Thus, the sunspot record and the appearance of the sun suggest that the strength of the sun varies over an 11-year cycle and has done so for centuries. This conclusion has been confirmed by satellite observations of the amount of radiation received at the top of the Earth's atmosphere during three complete sunspot cycles. Since 1978, when the first satellite measurements of solar radiation arriving at the top of the Earth's atmosphere were made, there appears to be a correlation between the record of sunspots and that of solar radiation (Figure 13.5). Solar radiation varies by about 2.5 W/m^2 from 1366 W/m^2, or 0.18% over the 11-year sunspot cycles. Such a change in incoming solar radiation could affect a change in Earth's temperature by about plus or minus 0.1°C in response to the 11-year variation in the intensity of the sun's radiation. During the Sporer and Maunder periods of sunspot minima (p. 406), the sun's strength might have been as much as 0.25% weaker than it is now. Recently it has been observed that the vigor of sunspots from 1992 to 2009 in terms of magnetic strength and area has greatly diminished. This could portend a long-term omen of sunspot decline as observed for the Maunder Minimum period (Figure 13.4). However, as of this writing the jury is still out regarding this conclusion since other indicators of the solar activity cycle suggest a return to sunspot vigor soon.

The linkages between changes in the number of sunspots, the sun's strength, and the climate of the Earth are still somewhat uncertain at a decadal to

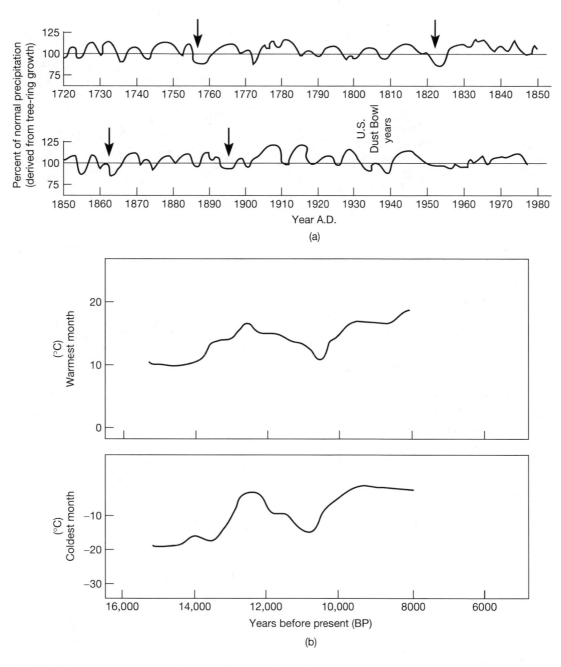

FIGURE 13.2 Two examples of climate indicators. (a) The history of drought in the Great Plains region of the United States. The graph shows tree ring growth, which is related to precipitation, relative to normal growth. The Dust Bowl years of the 1930s were similar to those of the 1750s, 1820s, 1860s, and 1890s (arrows). However, the years 1934, 1938, and 1939 were among the 10 driest years since 1700. (b) Fossil beetle record of temperatures in degrees centigrade for warmest and coldest months from 8000 to 15,000 years before present. Notice that the warming and cooling trends cover several degrees or more of temperature change.   (*Source:* (a) Stockton et al., 1985, and (b) Atkinson et al., 1987.)

century time scale. There are probably feedback mechanisms of both a positive and negative nature that respond to the initial perturbation of climate brought about by a change in the sun's strength and the amount of radiation the Earth receives. It is of interest to note in terms of the discussion on global warming in the following chapters that a significant correlation existed between the Northern Hemisphere land temperature

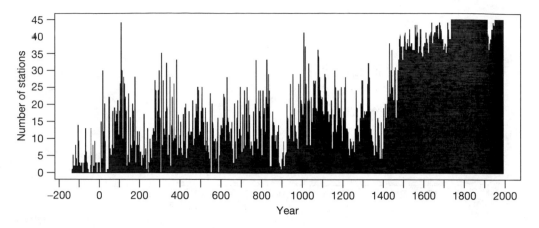

FIGURE 13.3 The two-millennia record of the number of stations in China that recorded data for drought, floods, timing of bird migrations, flowering of plants, distribution of species, abnormal weather, or natural disasters.

record of the last 100 years and the length of the sunspot cycle (Figure 13.6). The lower temperatures of the late 1800s to early 1900s coincided with a relatively long length of the sunspot cycle (about 11.5 years). From about 1910, the length of the sunspot cycle decreased erratically to about 9.7 years in 1989, and Northern Hemisphere land air temperatures increased irregularly about 0.5°C. This matching of the sunspot and temperature record is close, but a causal relationship among sunspot activity, solar irradiance, and temperature has been difficult to establish. However, most models of climate now incorporate a link between temperature and solar variability. The model studies show that solar variability does explain some of the trends observed in the temperature record of the past 100 years. However, since about 1990 the solar cycle has

not dominated the long-term variation in the Northern Hemisphere land-air temperature record. There is no indication of a systematic trend in the level of solar activity and the most recent warming of the planet.

Orbital Parameters

The amount of radiation received from the sun, as well as its distribution on Earth's surface, varies according to the relative position between the sun and Earth. These natural variations in orbital parameters impact the climate of the planet. They appear to set the initial conditions for the cooler and warmer periods of glacial-interglacial stages.

Milutin Milankovitch (1920) first proposed the theory that changes in climatic cycles of glacial-interglacial

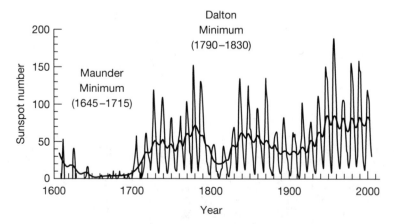

FIGURE 13.4 The 400-year record of the number of sunspots showing the Maunder and Dalton minima in sunspot number. Another sunspot minimum, the Sporer Minimum, occurred from 1460 to 1550. The 11-year cycle in sunspot activity is readily apparent in the figure. The generalized pattern of average sunspot number per cycle is shown as a black line. (*Source:* http://www.ossfoundation.us/projects/environment/global_warming/solar.)

TABLE 13.1 Some Paleoclimatic Data Sources and Their Characteristics

Data Source	Variable Measured	Potential Geographical Coverage	Period Open to Study (Years B.P.)	Climate Inference
Ocean sediments (cores, accumulation rate of <2 cm/1000 years)	Isotopic composition of planktonic fossils; benthic fossils; mineralogic composition	Global ocean	1,000,000+	Sea-surface temperature, global ice volume; bottom temperature and bottom-water flux; bottom water chemistry
Ancient soils	Soil type	Lower and midlatitudes	1,000,000	Temperature, precipitation, drainage
Marine shorelines	Coastal features, reef growth	Stable coasts, oceanic islands	400,000	Sea level, ice volume
Ocean sediments (common deep-sea cores, 2–5 cm/1000 years)	Ash and sand accumulation	Global ocean (outside red clay areas)	200,000	Wind direction
Ocean sediments (common deep-sea cores, 2–5 cm/1000 years)	Fossil plankton composition	Global ocean (outside red clay areas)	200,000	Sea-surface temperature, surface salinity, sea-ice extent
Ocean sediments (common deep-sea cores, 2–5 cm/1000 years)	Isotopic composition of planktonic fossils; benthic fossils; mineralogic composition	Global ocean (above $CaCO_3$ compensation level)	200,000	Surface temperature, global ice volume; bottom temperature and bottom-water flux; bottom-water chemistry
Layered ice cores	Oxygen-isotope concentration (long cores)	Antarctica; Greenland	100,000+	Temperature
Closed-basin lakes	Lake level	Lower and midlatitudes	50,000	Evaporation, runoff, precipitation, temperature
Mountain glaciers	Terminal positions	45°S to 70°N	50,000	Extent of mountain glaciers
Ice sheets	Terminal positions	Midlatitudes High latitudes	25,000 to 1,000,000	Area of ice sheets
Bog or lake sediments	Pollen type and concentration; mineralogic composition	50°S to 70°N	10,000+ to 200,000	Temperature, precipitation, soil moisture
Ocean sediments (rare cores, >10 cm/1000 years)	Isotopic composition of planktonic fossils; benthic fossils; mineralogic composition	Along continental margins	10,000+	Surface temperature, global ice volume; bottom temperature and bottom-water flux; bottom water chemistry
Layered ice cores	Oxygen-isotope concentration, thickness (short cores)	Antarctica; Greenland	10,000+	Temperature, accumulation
Layered lake sediments	Pollen type and concentration (annually layered core)	Midlatitude continents	10,000+	Temperature, precipitation, soil moisture
Tree rings	Ring width anomaly, density, isotopic composition	Midlatitude and high-latitude continents	1,000 to 8,000	Temperature, runoff, precipitation, soil moisture
Written records	Phenology, weather logs, sailing logs, etc.	Global	1,000+	Varied
Archaeological records	Varied	Global	10,000+	Varied

(*Source:* Kutzbach, 1975.)

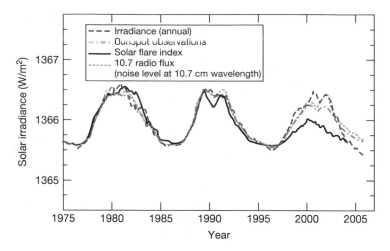

FIGURE 13.5 Correlation between solar irradiance in watts per square meter and sunspot observations. Also shown are the trends in the index of the number of solar flares and the noise level associated with the flares at 10.7 cm wavelength. (*Source:* http://en.wikipedia.org/wiki/Solar_variation.)

periods were initiated by both the amount and distribution of radiation received from the sun. He believed that Earth should be cooler or warmer depending on its relationship to the sun. Milankovitch argued that the different periods of major glaciations were initiated by changes in Earth's orbital parameters of eccentricity, axial tilt (obliquity), and time of perihelion (precession) (Figure 13.7). These three factors change the amount and more importantly the distribution of the sun's energy over the planetary surface at regular time intervals. In particular, every 100,000 years or so, the orbital parameters vary in such a way as to reduce the amount of energy received at midlatitudes in the Northern Hemisphere. This change is thought to lead to the start of an ice age. Major interglaciations occur when the three factors line up so as to give the Northern Hemisphere the greatest amount of summer isolation

Milankovitch's theory has been verified to some extent by the environmental data acquired from the Vostok, Concordia Dome C [European Project for Ice Coring in Antarctica (EPICA)] and Taylor Dome, Antarctica, ice cores and those drilled at the summit of the Greenland ice cap. The temperature records from the ice cores exhibit a strong 100,000-year cyclicity, as predicted by the Milankovitch theory. The records of temperature, atmospheric carbon dioxide, methane, and nitrous oxide roughly track each other in the data from the more than three-kilometer-long composite ice core records from the Vostok, Dome C, and Taylor Dome, Antarctica, drill sites for the past 800,000 years (Figure 13.8). This 800,000-year late Quaternary record includes the present interglaciation plus eight previous

ones and nine glaciations. Carbon dioxide during this whole interval oscillated in approximately 100,000-year cycles by about 125 ppmv from 172 to 300 ppmv, and methane by about 350 ppbv from 350 to 700 ppbv. Temperatures oscillated over a range of approximately 12°C. Figure 13.9 is a detailed view of the correlation between CO_2 concentrations and temperature variations as recorded in the ice at the Vostok drill site. The strong correlation between the two variables is obvious. However, in general, changes in CO_2 lag changes in temperature by $600 +/- 400$ years.

Notice the changes in Figure 13.8 in the amplitudes of the carbon dioxide and temperature variations from about 450,000 to 800,000 years before present (b.p.) and the lower CO_2 values from 650,000 to 800,000 years b.p. These changes are likely related to some reorganization of the carbon reservoir of the global ocean and perhaps changes in weathering rates on land. The long-term trend in the temperature record as obtained from the Greenland Ice Core Project (GRIP) ice core generally follows that of Vostok since the last interglaciation, the period of time for which the ice cores overlap in age. In addition, dust in the Vostok ice core also varies in an approximately 100,000-year cycle. High concentrations of dust particles in the Vostok ice core correlate with glacial stages and low concentrations with interglaciations (Figure 13.8).

The 100,000-year cycle observed in the ice cores is also seen in the oxygen isotopic composition ($\delta^{18}O$) of foraminifera collected from deep-sea calcareous oozes. As discussed in Chapter 2, the $\delta^{18}O$ record can be used as a proxy for ocean water surface temperature

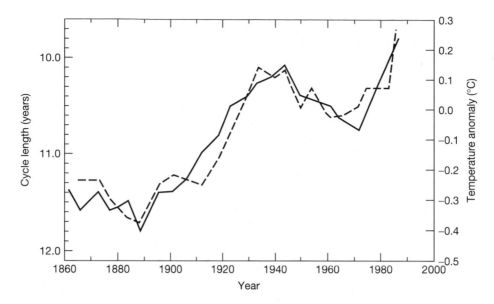

FIGURE 13.6 Correlation between the variation of the sunspot cycle length (solid dark line, lefthand scale) and Northern Hemisphere land temperature anomalies (dashed line, righthand scale) from 1861 to 1989. The temperature anomalies are the deviations in temperatures relative to the period of 1951 to 1980. (*Source:* Friis-Christensen and Lassen, 1991.)

extending back nearly 800,000 years. In addition, the sedimentary record of major glacial advances and recessions found in Pleistocene glacial deposits of North America, Europe, and Asia also exhibits a 100,000-year cyclicity. Thus, the 100,000-year pattern is seen not only in ice core environmental parameters but also in other records from both the terrestrial and marine realms.

Despite the good agreement between the historical pattern of climate records and the 100,000-year eccentricity solar insolation changes as predicted by the theory of Milankovitch, there are problems with the theory itself. The geological record of ice ages also shows periodicities of 19,000 and 23,000 years that can be attributable to the precessional changes and the roughly 41,000-year periodicity owing to obliquity changes (Figures 13.8 and 9). However, during the past 800,000 years, the strongest signals in the climate records have a frequency of 100,000 years. This coincides with the frequency of the eccentricity cycle, but the variations in eccentricity are thought to be the weakest of the orbital effects in terms of their influence on solar radiation.

Various hypotheses have been advanced to explain the puzzle. One of the most recent, interesting, and still debatable involves the relationship between the inclination of the Earth's orbital plane and interplanetary dust. The idea is that changes in orbital inclination should have a 100,000-year periodicity and

could modulate the accretion of interplanetary dust particles to the planet. This in turn could affect the amount of solar radiation reaching Earth's surface and play a role in development of the cycles of glaciation. More likely is the contention that the weak changes in the eccentricity forcing initiate a series of amplifiers to the weak forcing, such as feedbacks in the biogeochemical cycles of the natural greenhouse gases, continental and sea salt aerosol loading of the atmosphere, sea ice coverage, water vapor content of the atmosphere, cloud cover, ice sheets and their geographical extent, changes in the storage rates of inorganic and organic carbon in the ocean, changes in the size of the coastal ocean, and changes in weathering on land. These same amplifiers of the 100,000-year Milankovitch forcing also could be applicable to the 19,000, 23,000 and 41,000-year cycles. In addition, reorganization of the ocean's thermohaline circulation and fluctuations in solar activity also acted as direct forcings on Earth's climate during glacial-interglacial stages.

Furthermore, in the Hot Houses of the Phanerozoic when carbon dioxide levels and hence temperatures were high, there appear to have been no large, continental-scale glaciations of the magnitude of the Pleistocene Epoch. Thus, an extended period of lower carbon dioxide levels and hence colder conditions in an Ice House seem to aid in the development of large continental-scale glaciations. Table 13.2 gives one interpretation of some changes in a sequence of events

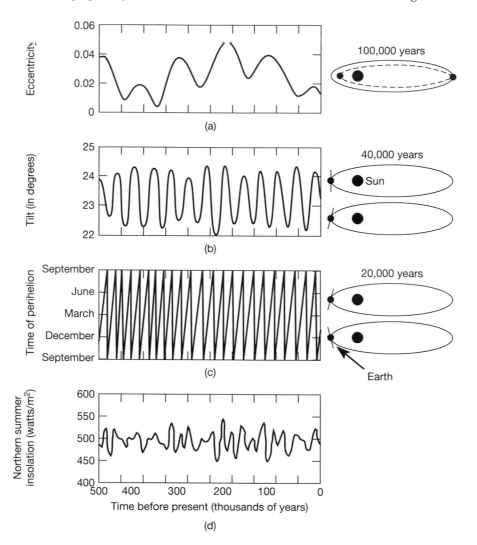

FIGURE 13.7 The Milankovitch theory of climatic change during the Pleistocene. The onset of ice ages is due to variations in three orbital parameters of Earth. (a) The *eccentricity* is the degree to which Earth's orbit departs from a circle. Times of maximum eccentricity are separated by roughly 100,000 years. (b) The *tilt angle* is the angle between Earth's axis and a line perpendicular to the plane of the orbit of the planet. (c) The *time of perihelion* involves the tilt of Earth's axis at its closest approach to the sun. The cycles of tilt and time of perihelion are roughly 40,000 and 20,000 years, respectively. (d) The calculated amount of sunlight received at 60° to 70° north latitude during the summer (summer insolation, July), based on the cycles of variation of Earth's orbital parameters. One watt = 0.0569 British thermal units (Btu) per minute = 14.28 calories per minute.   (*Source:* Covey, 1984.)

that led to the development of a glacial stage in which Milankovitch forcing (changes in orbital parameters) plays a key role in initiating the sequence of events.

The causes of glacial-interglacial environmental changes are not well understood. Variations in orbital parameters of Earth mentioned previously appear to be an initial forcing that determines the onset of ice ages. However, other factors are also involved in climate and environmental change during the glacial and interglacial stages of the past 800,000 years. For example, there have been at least 11 hypotheses put forth for the changes observed in atmospheric carbon dioxide for the past 420,000 years alone. To one degree or another, they involve changes in ocean conveyor belt circulation patterns, changes in the carbonate system in the ocean, and changes in nutrient distributions in the oceans, or some combination of all these. In general the atmospheric and climatic oscillations of the

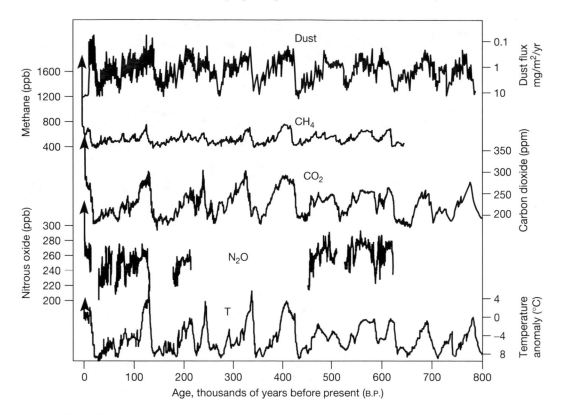

FIGURE 13.8 The trends in atmospheric CO_2, CH_4, N_2O, temperature, and dust as recorded in a composite of the Vostok, Dome C, and Taylor Dome ice core records from Antarctica. (*Source:* Petit et al., 1999; Siegenthaler et al., 2005; Luthi et al., 2008.)

glacial-interglacial stages represent complex interactions between terrestrial and marine ecosystems, changes in solar radiation and ocean circulation, and greenhouse gas concentrations. The understanding of these changes in ice core properties is a major area of research today. There is still much to be learned about the behavior of glacial-interglacial cycles, but if we cannot understand this most recent phenomenon of climatic change, then our chances of understanding the climate of the future are diminished.

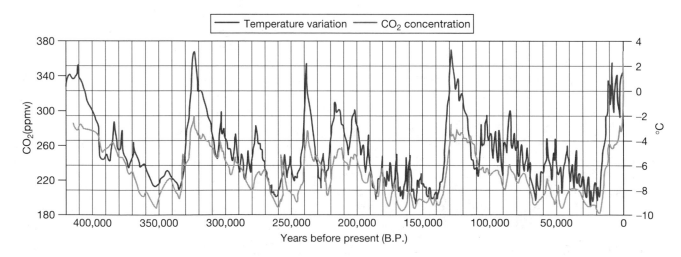

FIGURE 13.9 Correlation between temperatures and carbon dioxide concentrations for the Vostok ice core. In general, changes in temperature precede changes in carbon dioxide content by several hundreds of years. (*Source:* Petit et al., 1999; Luthi et al., 2008.)

TABLE 13.2 Possible Series of Some Events Leading to Development of a Glacial Stage in Which Changes in Orbital Parameters Initiate the Sequence of Events

1. Start with the relatively low atmospheric carbon dioxide and temperature of this Ice House, which began about 35 million years ago.

2. Milankovitch forcing leads to less insolation (less direct solar and sky radiation) at Northern Hemisphere intermediate latitudes, giving rise to relatively warm winters and cool summers. Snow accumulates and is converted to ice in continental glaciers. There is a slight overall cooling of the Earth's surface due to a change in planetary albedo.

3. Sea level falls. There is more nutrient input to the ocean from erosion of fresh, organic-rich, continental margin sediments. There is more efficient utilization of nutrients in the ocean. Greater surface biological productivity draws down atmospheric carbon dioxide. Positive feedback leads to more cooling.

4. Oceanic conveyor belt is disturbed and less heat is transported to the northern Atlantic. The planet is further cooled. The conveyor belt circulation can turn on and off on short time scales, perhaps the decadal to century time scale.

5. Continental glaciers advance over North America and Europe, reaching thicknesses of more than 2 kilometers. Sea level falls about 120 meters.

6. Glaciers cover the tundra and higher-latitude land ecosystems, and the flux of methane from boreal zones and tropical wetlands and that of nitrous oxide from terrestrial ecosystems decreases. This results in lower atmospheric methane and nitrous oxide concentrations; temperature falls—another positive feedback.

7. Stronger winds of the glacial stage promote more atmospheric dust transport and deposition and may be a source of iron for increasing glacial oceanic productivity.

8. The enhanced organic production of the glacial stage gives rise to higher dimethylsulfide emissions from the oceans and enhanced deposition of non–sea-salt sulfate. This may be a positive feedback to cooling because of the link between emissions of dimethylsulfide, cloud condensation nuclei, sulfate aerosol, and cooling (pp. 390–391).

9. Orbital forcing relaxes, and the planet rapidly enters an interglacial stage following to some extent a reversal of the changes described in 2–8 above.

The ice cores provide temperature, atmospheric composition, and other environmental data that give an impression of the state of the planet prior to major human modification of the ecosphere. It is apparent from comparison of the ice core data on atmospheric trace gas composition with that of the present atmosphere that present atmospheric trace gas composition is significantly different from that characteristic of the past 800,000 years of Earth history. This difference is mainly a result of the development of human agricultural, industrial, and transportation activities since the Industrial Revolution, resulting in the emissions of trace gases and particulate materials to the atmosphere.

Planetary Albedo

The reflectivity or albedo of Earth affects the heat budget of the planet. Aerosols, clouds, ice, water, land, and plant surfaces all contribute to this effect (see Chapter 4).

AEROSOLS Aerosols are very small airborne particles, micrometers in diameter in size. The term has commonly been used to describe the fine spray from aerosol cans that dispense deodorants and other items such as hair sprays. Natural aerosols come from volcanoes, wildfires, windblown dust of soils, land and ocean emissions of biologically produced gases, and sea-salt spray. The size and distribution of the aerosols determine whether the surface temperature of Earth is increased or decreased. In general, the larger the number of aerosol particles, the greater the amount of heat radiated back to space. Aerosols generally have contributed to Earth's radiation budget by producing a cooling effect.

Plate tectonic processes operate on a global scale, have evolved over billions of years, and continuously exert control over many present-day planetary processes, such as volcanic eruptions (see Chapters 3 and 8). Volcanic eruptions produce lava, particles, and gases like sulfur dioxide (SO_2). The erupted sulfur dioxide gas may be converted to sulfate (SO_4)-bearing aerosols in the atmosphere. The aerosol produced is similar to that generated from fossil fuel burning emissions of sulfur dioxide. Volcanic eruptions and subsequent aerosol production can affect the transparency of the atmosphere to light and hence modulate both the regional and global climate. Aerosols reaching the stratosphere may remain there for several years. Volcanic activities have been cited as a probable cause of the extremes in cold-weather conditions of the Little Ice Age.

The eruption of Tambora in Indonesia in 1815 near the end of the Little Ice Age is an example of volcanism that affected the global climate by rapidly cooling it. The dust and aerosol that spewed from the volcano entered the stratosphere and reduced the amount of solar radiation reaching Earth. The reduction in solar radiation led to a cooling of the planet. It is estimated that the immediate effects of the eruption and the starvation that followed the event killed 92,000 people. Longer-term climatic effects were felt worldwide. In 1816, because of late snows and continuous rain, Europe experienced a year without a summer. The cool conditions continued in Europe and elsewhere for three years, resulting in massive crop failures. Crop failure brought on starvation and a near-collapse of society in Europe.

The eruptions in 1600 B.C. of Santorini, an island north of Crete, of Hekla 3 in Iceland between 1150 and 1136 B.C., of Mt. Etna in Sicily in 42 B.C. , and Krakatoa (Krakatau) in 1883 are all linked with climatic changes and catastrophic effects on human societies. The explosion of Krakatoa was equivalent to 200 megatons of TNT, about 13,000 times greater than the yield of the bomb that devastated Hiroshima, Japan. The eruption resulted in the deaths of 36,417 people, mainly because of the tsunamis that resulted from and followed the eruption. The eruption of Mt. Pinatubo in the Philippines in 1991 ejected 10 billion metric tons of magma and 20 million metric tons of SO_2. The resulting aerosols generated a global haze of sulfuric acid temporarily cooling the planet about 0.5°C. The cooling lasted for several years (Figure 13.10, http://www.gsfc.nasa.gov/gsfc/service/gallery/fact_sheets/earthsci/eos/volcanoes.pdf). The eruption had important unforeseen consequences, including an unprecedented ozone hole

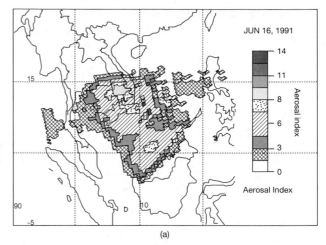

(a)

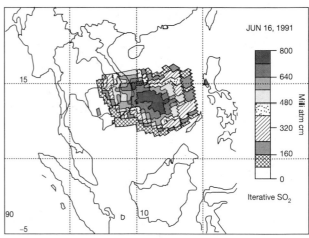

(b)

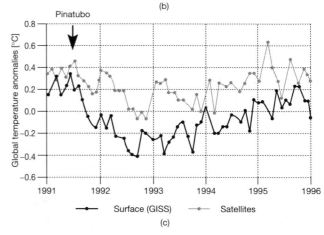

(c)

FIGURE 13.10 The eruption of Mt. Pinatubo in the Philippines in 1991. (a) The aerosol index is a measure of the amount of aerosol in the plume from the eruption. (b) The iterative SO_2 is a measure of the amount of SO_2 in the plume. Increasing values of both indices imply increasing atmospheric burdens of these materials. Data obtained from the NIMBUS-7 spacecraft Total Ozone Mapping Spectrometer (TOMS). (c) The global temperature record at the surface and from satellites for 1991 to 1995 showing the dramatic effect of the eruption on global temperatures. (*Source:* http://toms.umbc.edu/.)

above Antarctica in 1993 and a reduction in the rate of growth of atmospheric carbon dioxide in the early 1990s.

Biological emissions of sulfur gases from the land and ocean surface also lead to creation of aerosols in the atmosphere and hence affect climate. These emissions are dominated by the reduced form of sulfur in the gases dimethylsulfide (DMS), hydrogen sulfide (H_2S), and carbonyl sulfide (OCS) (see Chapter 7). The reduced sulfur-containing biogenic gases can be oxidized to sulfur dioxide gas in the atmosphere and then to sulfate aerosol. The biogenic sulfur gases released from the ocean and the land surface differ in composition. Dimethylsulfide and carbonyl sulfide characterize emissions from the ocean surface, whereas hydrogen sulfide and dimethylsulfide are the predominant forms of sulfur released from decaying terrestrial vegetation.

Carbonyl sulfide is the most abundant gaseous sulfur compound in the atmosphere of remote regions of the ocean. It is not very reactive and therefore resides in the atmosphere a moderately long time. Its lifetime is about 30 years. Consequently, although it is derived from biological reactions at the surface of Earth, it can be mixed by winds and air currents into the stratosphere. In the stratosphere, carbonyl sulfide is destroyed by ultraviolet radiation of wavelength 250 nanometers (250×10^{-9} meters) and atoms of oxygen. It is converted to sulfur dioxide gas and on to sulfate aerosol. Emissions of this reduced sulfur gas account for about half of the sulfate aerosol found in the sulfate veil, or Junge layer, in the stratosphere.

In the troposphere, dimethysulfide that has entered the atmosphere from an aquatic system is oxidized rapidly on the time scale of one day. It reacts with hydroxyl radical to produce sulfur dioxide. The sulfur dioxide is then oxidized to sulfate aerosol, which is removed from the atmosphere to Earth's surface in rain. The connections of dimethylsulfide and carbonyl sulfide to climate are as follows:

1. The major source of cloud condensation nuclei (CCN) in the remote atmosphere above the ocean is dimethysulfide escaping from the sea surface (Figure 13.11). The amount of DMS gas leaving the surface of the ocean is related to phytoplankton production and bacterial degradation of organic products of production. It has been suggested that if the planet warmed, phytoplankton production would increase, leading to more emissions of DMS to the atmosphere. The increased dimethylsulfide flux would lead to increased production of sulfate aerosol and hence cloud condensation nuclei. The increased atmospheric burden of sulfate aerosol and the greater

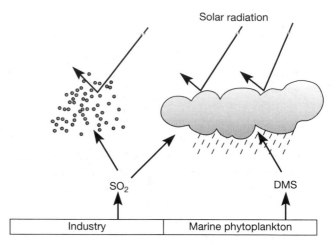

FIGURE 13.11 Diagram illustrating the similarities and differences between industrially derived and naturally produced sulfur gases. Sulfur dioxide gas produced by the burning of fossil fuels may lead to development of atmospheric haze or to formation of sulfate aerosol, which provides surfaces for cloud formation. Dimethylsulfide gas, a product of marine phytoplankton, on emission from the ocean may give rise to sulfate aerosol, which acts as cloud condensation nuclei, and formation of clouds over remote oceanic areas. Both processes may lead to an increase in planetary albedo and cooling.

degree of cloudiness would reflect solar radiation and lead to a cooling of the troposphere. The acronym of this hypothesis, CLAW, is derived from the first letters of the authors' surnames: Charlson, Lovelock, Andreae, and Warren (Figure 13.12). If the CLAW hypothesis is true, then if Earth were to warm, say, because of an enhanced greenhouse effect, the initial warming could result in a cooling effect. This effect would counteract the initial temperature rise. Such a moderating effect on an initial disturbance is a negative feedback.

One problem with the CLAW hypothesis is that in the atmosphere, DMS reacts to form either sulfur dioxide or methane sulfonic acid (MSA) that is converted to aerosols and can act as cloud condensation nuclei. Some of the aerosol is deposited back to Earth's surface and can be incorporated in glacial ice. From the Vostok ice core data, it has been found that the record of MSA in the ice core varies with temperature. It appears that MSA was more abundant in the atmosphere over the Southern Ocean during glaciations than during interglaciations. This is just the opposite of what one would expect if the CLAW hypothesis were correct.

2. Any change in the emissions of carbonyl sulfide to the atmosphere on a time scale of decades can result in a change in the sulfate burden of the

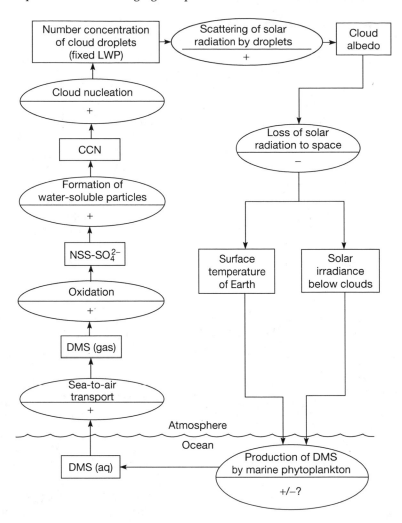

FIGURE 13.12 Conceptual diagram of the CLAW hypothesis. The rectangular boxes are quantities that are measurable; the ovals represent processes that link the quantities. The plus and minus signs in the ovals indicate the effect of a positive change in the quantity in a preceding rectangle on that in the succeeding rectangle. DMS is dimethylsulfide; NSS-SO₄, non–sea-salt sulfate; CCN, cloud condensation nuclei; and LWP, liquid water path. For the CLAW hypothesis to work as described in the text, the sign of the cloud albedo rectangle would have to be positive.   (*Source:* Charleson et al., 1987.)

stratospheric sulfate veil. An increased flux would give rise to an increased burden and cooling of the atmosphere because of reflection of solar radiation back to space. The opposite would occur if carbonyl sulfide emissions decreased.

Industrial activities, such as fossil fuel burning and resultant emissions of sulfur dioxide, can lead to formation of sulfate aerosols that contribute to the tropospheric sulfate-aerosol burden. These aerosols potentially could have a cooling effect on climate (Figure 13.11). It has been argued that anthropogenic emissions of sulfur dioxide to the atmosphere have increased the level of sulfate aerosols mainly in the Northern Hemisphere. This enhanced sulfate-aerosol burden probably helped cool the planet during the twentieth century. The cooling effect offset part of the expected temperature increase owing to anthropogenic emissions of greenhouse gases to the atmosphere. In particular, the slight overall decrease in global temperatures observed between the late 1930s and early 1970s has been in part attributed to this phenomenon. The effects of sulfate aerosols derived from human activities on recent and future climate are discussed more fully later in Chapter 14.

ICE, WATER, LAND, AND PLANT SURFACES Differences in the surface characteristics of the planet affect the amount of radiation reflected from Earth's surface back to space, the planetary albedo. Ice and clean snow surfaces have high albedos. When sunlight strikes ice on the planet, most of the light and heat energy is reflected. Assuming other factors influencing climate remain constant, the greater the area covered by continental ice sheets, the more energy reflected back to space and the cooler the planet. Deserts and other nonvegetated areas have albedos less than that of ice and reflect about half the radiation that they receive. The oceans have a low albedo and absorb most of the radiation that they receive and are warmed in the process. Land and aquatic plants absorb almost all the radiation reaching them and use the sun's radiation in the process of photosynthesis. If other climatic factors remain constant, increases in the areas of the ocean and of land vegetation will lead to a decrease in planetary albedo and warming of Earth. The present areal distribution of oceans, ice and snow, deserts, and terrestrial vegetation helps to maintain the present mode of climate. However, these areas can change with warming or cooling of the planet. Consequently, global warming caused by an enhanced greenhouse effect can change the natural delicate balance of ecosystem areas and thus the planetary albedo. This change in albedo will in turn act as a feedback in the system and affect the climate of the planet.

CLIMATE AND ENVIRONMENT OF THE MORE RECENT PAST

Throughout much of its history, Earth has generally been a warm planet, much warmer than today. For at least 2.5 billion years of Earth history, sedimentary rocks and their fossil record show evidence of these warmer times and also of shorter colder periods. The cold periods of Ice Houses (low levels of carbon dioxide) last several tens of millions of years before the planet returns to its more prevalent Hot House ("greenhouse") state (high levels of carbon dioxide) (see Chapter 8). There have been four reasonably well-documented Ice Houses. Ice Houses occurred about 2.5 billion, 700 million, and 300 million years ago, and the last Ice House began 35 million years ago. Within these Ice Houses, evidence of continental-scale glaciations is found in the sedimentary rock record. Ice ages are those periods of time when large areas of the continents were covered with thick glaciers.

The Mesozoic Era, which followed the Ice House about 300 million years ago, was characterized by high global mean temperatures (at times about 8 to 10°C warmer than today), high carbon dioxide levels, and an absence of ice sheets and deserts. This warm, generally moist climate was followed by a general, but somewhat erratic, cooling during the Cenozoic Era. The Paleocene–early Eocene was relatively warm. Temperatures began falling gradually in the mid-Eocene, then fell sharply about 35 million years ago at the Eocene–Oligocene boundary. This event marks the beginning of the last Ice House. Since that time, the overall climate of Earth has continued to cool erratically (Figure 13.13a). The average late pre-industrial surface temperature of Earth was about 15°C.

Pleistocene Epoch

The Pleistocene Epoch is within the latter part of the latest Ice House and is a time of waxing and waning of continental-scale glaciers. The cold periods within the Pleistocene are glacial stages and are times of advance of continental glaciers, falling sea level, and low atmospheric carbon dioxide concentration. Interglacial stages are periods between glacial stages, when the ice retreated from the continents, the climate was warmer, and atmospheric carbon dioxide and sea level were higher. The planet is currently in an interglacial stage, the Holocene Epoch (e.g., Figure 13.8).

Major glacial stages are longer in duration than interglacials. The former last about 80,000 years, and the latter have durations of about 20,000 years. Recovery from the maximum of a glacial stage and development of a major interglacial usually take about 10,000 years. The transition from an interglacial into a fully developed glacial period is not as abrupt as the glacial to interglacial transition. During both glacial-interglacial and interglacial-glacial transitions, ecosystems changed as the great continental glaciers advanced and retreated across the planetary surface. However, the transitions between the extremes of environments of glacial and interglacial stages were generally gradual enough to allow many ecosystems to adapt to the changing physical environment.

During the last million years, there have been approximately 10 major and 40 minor periods of these glacial cold cycles, interspersed with warmer interglacial times (Figure 13.13b). These glacial and interglacial stages are documented in the oxygen isotope record of deep-sea sediments and for the late Pleistocene in the temperature and atmospheric composition record derived from ice cores drilled into the

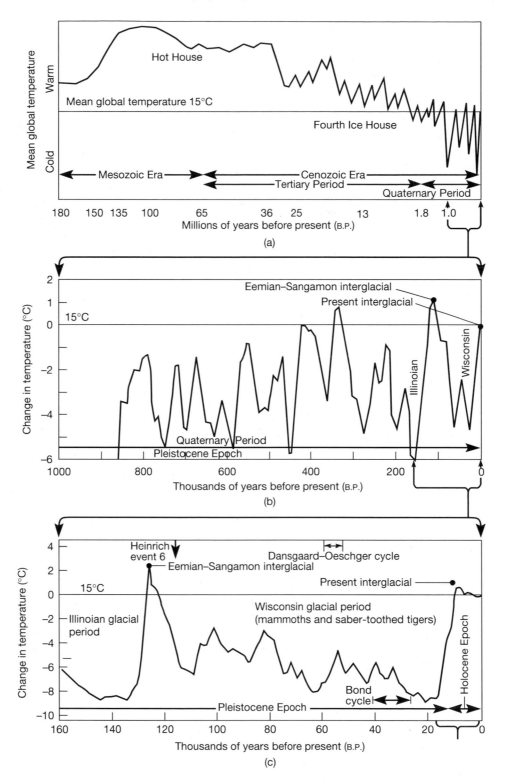

FIGURE 13.13 Changes in the temperature of Earth over time. (a) The temperature record of Earth during the past 180 million years; (b) an expanded representation of the last 1 million years; and (c) an expanded view of the last 160,000 years. (*Source:* UCAR/OIES, 1991.)

Antarctic and Greenland ice caps (Figure 13.8). One of the warmer interglacials (named Riss-Würm in Europe and Sangamon in North America), the "Eemian," peaked about 125,000 years ago. In the Eemian, the climate was warmer and more humid than today. Atmospheric carbon dioxide and methane concentrations were higher than during glacial stages, reaching levels of 300 ppmv (parts per million by volume) and 700 ppbv (parts per billion by volume), respectively. The average global temperature was at least 1°C warmer than that of late preindustrial time. Lions and elephants roamed the land of Cornwall, England. Hippopotami lived along the Thames River of England and along the Rhine River of Europe. Much of the huge ice sheet of West Antarctica had melted away, leaving only ice sheets on Greenland and eastern Antarctica. Sea level was higher than today because of less ice locked up in the cryosphere. This high sea level is recorded by rocks deposited at this time 6 to 8 meters above present-day sea level on the islands of Barbados, Bermuda, Papua New Guinea, and Hawaii.

Ice-core data from Greenland originally suggested that the Eemian may not have been a period of stable climate, at least not in the North Atlantic. Data from the GISP2 3029-meter ice core drilled at the summit of the Greenland ice cap show sharp fluctuations in climate during Eemian time. The data indicate that the warmth of the Eemian interglacial was interrupted several times by cold spells lasting about 70 to 750 years. Temperatures in the North Atlantic may have dropped by as much as 10°C, sometimes over periods of time as short as 10 to 30 years. However, subsequent analysis of a second deep ice core obtained from Greenland (the GRIP ice core) does not corroborate the findings of the GISP2 core. It appears that the ice drilled at depth in the GISP2 ice core is deformed because of ice flow. Nevertheless, both GRIP ice-core data and pollen data from laminated lake sediments of Eemian age from Europe indicate the possibility of greater climatic instability during the Eemian than the Holocene. The possibility of greater instability of the warm climate of the Eemian than that of the Holocene is of concern in terms of the problem of the enhanced greenhouse effect. Changing the composition of our present interglacial atmosphere because of human activities might force the climate system into an unsteady state, giving rise to natural climatic fluctuations of large magnitude.

The Eemian interglacial was followed by the last glacial stage (named Würm in Europe and Wisconsin in North America), which reached its maximum coldness about 18,000 years ago (Figure 13.13c). The Wisconsin glaciation is characterized by irregular shifts between relatively warm and cold periods in Earth's climate. The warm periods, or interstadials, lasted 500 to 2000 years and had a common pattern of abrupt warming and gradual cooling. The warmer interstadials had temperatures that may have approached the minima in temperatures of the preceding Eemian interglacial stage. At the maximum of the Wisconsin glaciation, atmospheric carbon dioxide, methane, and nitrous oxide concentrations had reached minimum values of 180 ppmv, 350 ppbv, and 185 ppbv, respectively, and sea level was 120 meters lower than its present-day level. This is the ice age of cave people, mammoths, and saber-toothed tigers with which many people are familiar.

Figure 13.14 contrasts the extent of ice coverage during the Last Glacial Maximum (LGM) and vegetation types (biomes; see Chapter 6) with present ice coverage and biomes. During the glacial maximum, ice covered about 41 million square kilometers of Earth's surface, or about 28% of the present-day land area. The ice reached an average thickness of about 2 kilometers. Present-day ice covers about 15 million square kilometers of Earth's surface, or about 10% of the modern land area. The additional ice present on Earth during the Last Glacial Maximum was about 49 million cubic kilometers, equivalent to 44.4 million cubic kilometers of water and accounting for the 120-meter lower sea level of the glacial maximum.

During the Last Glacial Maximum, Northern Hemisphere temperate biomes had shrunk in area, while the tundra biome of eastern Siberia was larger in area (Figure 13.14). In addition, the tropical rainforests of Brazil, Africa, and Southeast Asia were probably smaller in area than today. Furthermore, tundra was present in southern South America rather than the grasslands that currently occupy the region, and the Sahara desert of North Africa was larger in area at the climax of the last glaciation.

At the climax of the Wisconsin ice age, the North Atlantic Ocean was frozen from Greenland to Great Britain and to the south and covered with pack ice and drifting and melting icebergs. Cold southward-flowing surface currents dominated much of the high-latitude North Atlantic Ocean. The warm waters of the Gulf Stream were restricted from flowing to the high northern latitudes, except for a weak flow to the north in the eastern part of the North

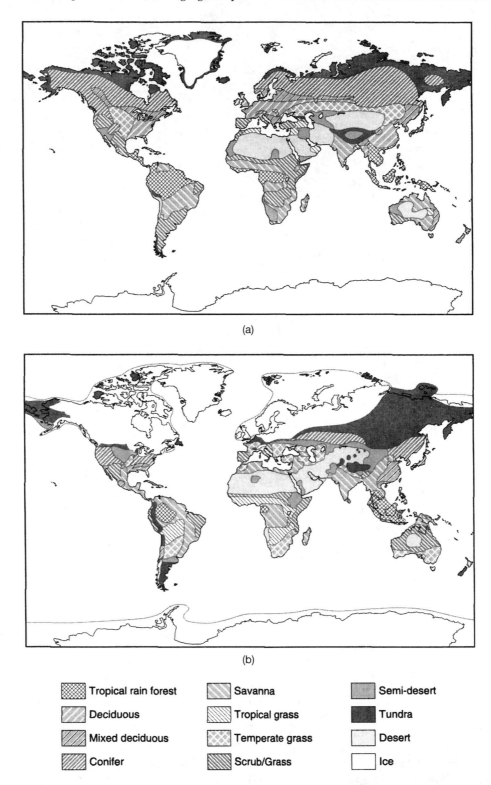

(a)

(b)

Tropical rain forest	Savanna	Semi-desert
Deciduous	Tropical grass	Tundra
Mixed deciduous	Temperate grass	Desert
Conifer	Scrub/Grass	Ice

FIGURE 13.14 Model calculations of the extent of ice coverage (white areas) and the distribution of vegetation types (biomes) (a) for the modern world and (b) for the last glacial maximum. (*Source:* Crowley, 1995.)

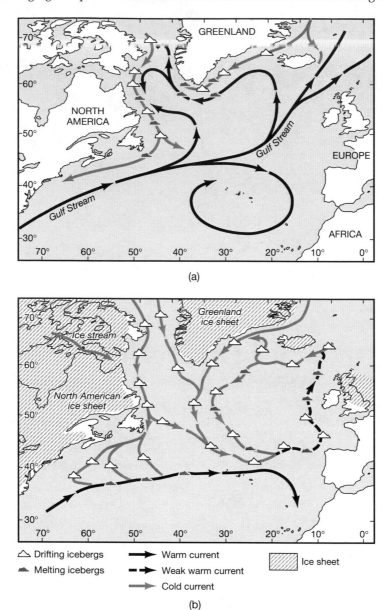

FIGURE 13.15 Schematic diagram illustrating the direction of flow of surface ocean currents and drifting and melting icebergs at (a) present and (b) during the last ice age. The stream of ice that drained the Hudson Bay portion of the Laurentide ice sheet of North America, as well as the extent of the Greenland and North American ice sheets, and the ice covering Great Britain are also shown. Surges in the ice stream from the Laurentide ice sheet might have been responsible for Heinrich events. (*Source:* Bond et al., 1992.)

Atlantic (Figure 13.15). Instead, the Gulf Stream flowed directly across from Florida to the Azores, warming only Africa and not Western Europe as it does today (see Chapter 5). Drifting and melting icebergs extended farther south into the North Atlantic Ocean than they do at present. Evidence for this is found in Pleistocene sediments of the North Atlantic. As the icebergs melted, the sand and gravel incorporated in the ice were released, sank through the water column, and were deposited on the seafloor. These deposits of ice-rafted debris record the geographical extent of the drifting and melting icebergs of the Pleistocene in the same way that modern glacial debris on the North Atlantic seafloor does.

EVENTS, CYCLES, AND OCEAN CIRCULATION In 1988 Hartmut Heinrich recognized six sedimentary layers rich in ice-rafted debris derived from Canada, and poor in foraminifera shells in cores from the Dreizack seamounts of the eastern North Atlantic. Since then, other cores have been collected from the North Atlantic with identical records. The layers are now known as Heinrich events 1 to 6. For example, Heinrich event number 6 marks the transition from the warm northern Atlantic region of the Eemian interglacial to the cold conditions prevailing during the last glacial stage of the Wisconsin (Figure 13.13c). Heinrich event 1 marks the beginning of the termination of the last glacial stage (Figure 13.16a).

Figure 13.17 shows an example of the sedimentary evidence for these millennial-scale Heinrich events. The events are the result of climate change in the North Atlantic region. The core depicted in Figure 13.17 shows the distribution with age of detrital, ice-rafted material and planktonic foraminifera shells and was obtained in the Irminger Basin of the high-latitude North Atlantic Ocean. The region where the core was taken received ice-rafted debris mainly brought in by icebergs that originated in the Norwegian–Greenland Sea. The Heinrich events are evident in the core by the increases in the number of detrital grains presumably resulting from melting icebergs. In this case, the foraminifera content of the sediment also increases with the detrital content. This correlation probably reflects the retreat of sea ice synchronously with periods of iceberg discharges and exposure of surface seawater suitable for planktonic foraminifera production. At midlatitudes an inverse relationship is usually observed between the planktonic foraminifera content of the sediment and ice-rafted detrital grains.

Prior to the recognition of Heinrich events, it was known from Greenland ice-core records that the northern Atlantic region experienced repeated large and abrupt changes in climate. Two more recent deep cores from Greenland, GRIP and GISP2, confirm these rapid climatic changes. The Wisconsin glacial portion of the Greenland ice-core record contains a series of climatic cycles averaging several thousands of years in duration. These cycles have been called the Dansgaard-Oeschger cycles. Each cycle is characterized by a period of gradual cooling terminated by an abrupt warming event (see Figure 13.13c). The bundling of several progressively colder Dansgaard-Oeschger cycles into a package gives rise to so-called Bond cycles (Figure 13.13c). The most intense cold phase of a package of several Dansgaard-Oeschger cycles (a Bond cycle) is terminated by the deposition of ice-rafted debris, a Heinrich event. Heinrich events occurred when the North Atlantic was at its coldest and represent great armadas of melting icebergs that flooded the northern Atlantic several times during the last great ice age (Figure 13.15). The massive iceberg discharges were probably a result of the collapse of the Laurentide ice sheet covering eastern North America. Although somewhat more speculative, it is likely that following the collapse and subsequent retreat of the ice sheet, meltwater discharge to the ocean decreased. This in turn enhanced the thermohaline circulation of the ocean (see Chapter 5), giving rise to increased transport of heat from the tropics to the high latitudes of the Atlantic region and abrupt climatic warming, the start of another Dansgaard-Oeschger cycle.

Conversely, when the ice sheet advanced, a lid of relatively fresh water from glacial meltwaters covered the northern Atlantic. The cap of freshwater disrupted the thermohaline circulation of the ocean, cooling the northern Atlantic region. Figure 13.18 contrasts the differences between the conveyor belt circulation patterns of the glacial and modern Atlantic Ocean to provide an impression of how this circulation might have changed between the last glacial stage and today. The lower sea-surface temperatures of high-latitude North Atlantic waters (Figure 13.19a), the formation of sea ice, the discharge and melting of icebergs (Figure 13.15), and resulting changes in sea-surface salinity (Figure 13.19b) could lead to a different pattern in the distribution of North Atlantic surface seawater density and hence the geographical area(s) of formation of deep water and its magnitude and flow direction. The problem is complex, and Figure 13.18 is but one interpretation of the situation. Whatever the case, such a modification of the glacial ocean conveyor belt circulation would have shifted the supply of heat from the higher North Atlantic to lower latitudes (Figure 13.18), and northern Europe would have been colder than today. Changes in the conveyor belt circulation pattern of the ocean are of considerable concern in the modern environmental problem of global warming (Chapter 14).

Evidence is accumulating that the various cycles mentioned previously have global climatic signatures, and their expression is not simply restricted to the northern Atlantic. However, the mechanism that drives the various cycles and their interactions is a matter of debate. The resolution of the debate will only come about after the dynamics of the coupled ocean-atmosphere-cryosphere system is more fully understood and there is substantially more paleoclimatic information from around the world.

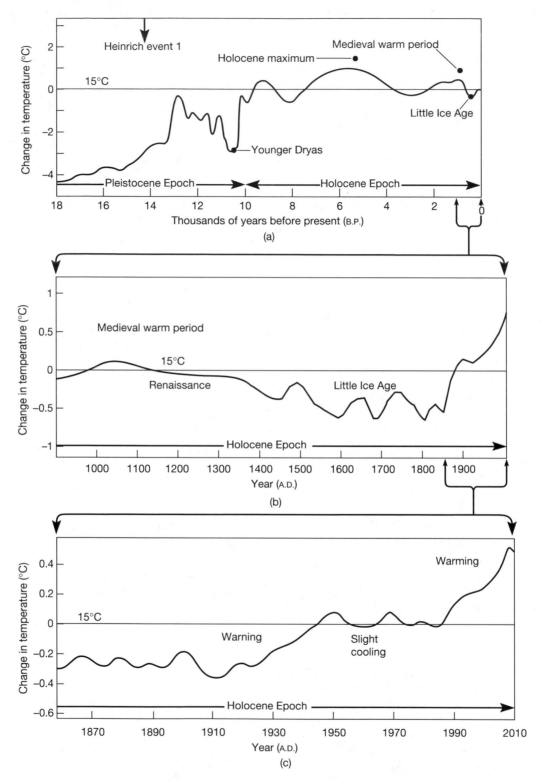

FIGURE 13.16 (a) Temperature history of Earth during the last 18,000 years; (b) and (c) are expanded versions of the period A.D. 1000 to 2001 and A.D. 1870 to 2008, respectively. Notice the change in scale for temperature in (a), (b), and (c). (*Source:* UCAR/OIES, 1991 and http://data.giss.nasa.gov/gistemp/graphs/Fig.A2.lrg.gif.)

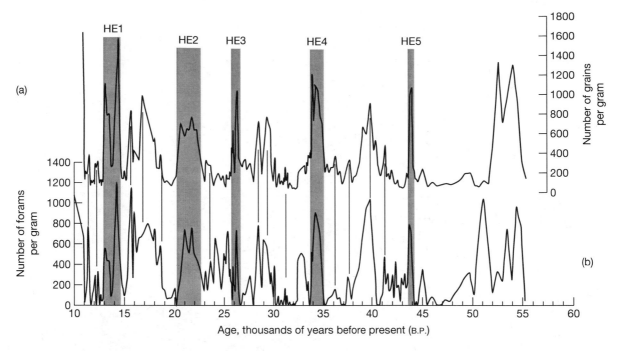

FIGURE 13.17 Variations in the number of ice-rafted detrital grains (a) and planktonic foraminifera shells (b) per gram of sediment in a core collected from the Irminger Basin in the high-latitude North Atlantic Ocean. The ice-rafted detritus came from icebergs originating north of the Irminger Basin in the Norwegian–Greenland Sea area. HE represents a Heinrich event. (*Source:* Duplessy, 2001.)

In addition to the differences mentioned previously between the glacial world of 18,000 years ago and the modern, there were others. Although somewhat controversial, and there were certainly ocean-to-ocean differences, it is likely that the pH of the glacial deep ocean was higher than that of the modern ocean, perhaps as much as 0.3 pH units. Glacial deep-ocean water was more alkaline than the deep water of the modern ocean. This would imply more carbonate ion (CO_3^{2-}) dissolved in the glacial deep ocean than in the modern ocean. Because of this chemical difference, calcium carbonate shells formed in the shallow waters of the ocean by plankton would have dissolved at a greater depth in the glacial ocean than they do in the modern ocean. Ocean productivity was higher and nutrient distributions differed from those of the modern ocean in the Last Glacial Maximum ocean. In addition, ice albedo was higher because of the large areal extent of the continental ice sheets. Snowlines in the North and South American Cordillera were lower than present-day snowlines. The atmospheric dust burden was much larger during glacial times. It is also likely that the planet was windier during glacial time, and the water vapor content of the glacial

atmosphere lower than that of the modern atmosphere. Finally, the concentrations of the greenhouse gases CO_2, CH_4, and N_2O were at historic lows. As mentioned previously, CO_2 concentration was at a level of 180 ppmv compared to 280 ppmv in the late pre-industrial atmosphere. Methane concentration was about 350 ppbv 18,000 years ago compared to 700 ppbv in late preindustrial time. Nitrous oxide was about 185 ppbv compared to 270 ppbv. The environment of the glacial world was vastly different from that of the modern.

THE INITIAL RECOVERY FROM THE LAST GLACIAL MAXIMUM At the end of the Last Glacial Maximum, global mean temperatures were about 5 to 7°C below the temperatures of the late pre-industrial world. The Earth entered a warming trend at the end of the Last Glacial Maximum. The warming has not been continuous and has been interrupted by several cool intervals and several intervals even warmer than the climate characterizing the beginning of the Industrial Revolution of 1850 (Figure 13.16). Thus, an important point to keep in mind is that the warming trend of the past 18,000 years has not been continuous and monotonic and, in addition, is predominantly the

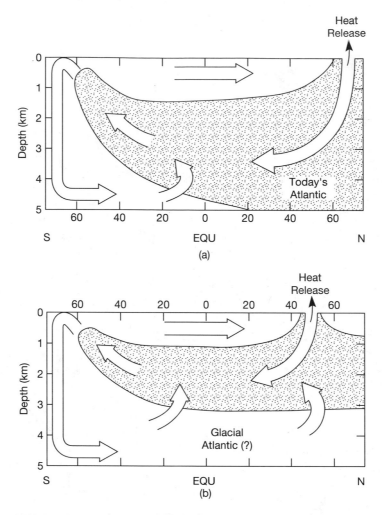

FIGURE 13.18 Schematic diagrams illustrating the deep-water conveyor belt circulation (see Chapter 5) of the modern (a) and glacial (b) Atlantic Ocean. Notice in particular that in the glacial ocean as opposed to that of today, the circulation apparently did not penetrate as deeply into the ocean's interior and that the circulation formed at more southerly latitudes. The large arrows represent the general transport paths of water and the stippled pattern the deep water that originates in the North Atlantic. (*Source:* Broecker, 1993.)

result of variations in natural factors that drive climatic change.

The major warming of the Earth began between 10,000 to 15,000 years ago (Figure 13.16). The low summer solar insolation levels (Figure 13.20) of the Last Glacial Maximum rose to a maximum about 10,000 years ago as winter insolation levels dropped. The ice sheets began to melt and, as meltwater was returned to the ocean, sea level started to rise (Figure 13.21). Once exposed, continental shelves began to be flooded, and

the large volumes of meltwater produced by the melting and receding continental glaciers eroded and hence modified the preexisting glacial landscape. The climate ameliorated, and precipitation and temperature patterns changed. Soils developed on the landscape once covered by ice, new vegetation patterns began to emerge, and biomes migrated (compare Figure 13.14a and b). The drainage patterns of rivers changed and new river systems emerged. The distribution of foraminifera adapted to cold polar waters found in

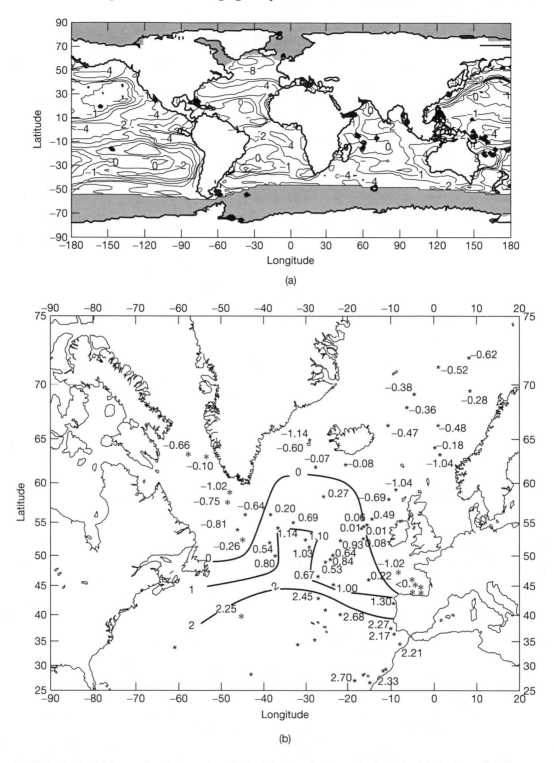

FIGURE 13.19 (a) Sea-surface temperature (SST) differences between the Last Glacial Maximum (LGM) ocean 18,000 years ago (*Source:* CLIMAP, 1981) and the modern conditions for the Northern Hemisphere summer and (b) the mean sea-surface salinity (SSS) for the North Atlantic Ocean Last Glacial Maximum (LGM) expressed as salinity minus salinity of 35. (*Source:* Duplessy et al., 1991.)

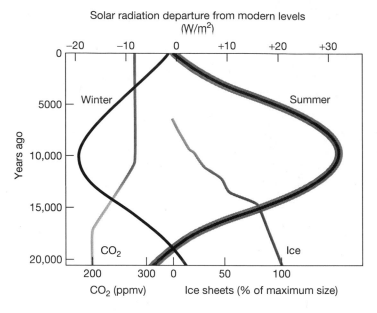

FIGURE 13.20 Schematic diagram of summer and winter solar radiation (insolation) in W/m² departure from modern levels, atmospheric CO₂ concentration in ppmv, and the extent of the continental ice sheets in percent of maximum size since the Last Glacial Maximum to late pre-industrial time. (*Source:* Kutzbach, 1998.)

high-latitude North Atlantic sediment cores indicates that the cold polar waters of the North Atlantic moved northward as the climate ameliorated (Figure 13.22a, b). In addition, pollen analyses of lake sediments in northern Europe (Figure 13.22c) show a general increase in the density of scrub plants such as juniper followed by willow and later in some areas by birch woodland as the climate warmed. By 11,000 years ago, most of Scotland was deglaciated. Summer temperatures in England were nearly 10°C warmer than at the Last Glacial Maximum (Figure 13.22d). The North American ice sheets, by 13,000 years ago, had migrated out of most of the United States and their margins were mainly in Canada. Then a major and abrupt cooling event took place about 13,000 years ago. Early evidence suggested the event was mainly restricted to the high-latitude North Atlantic and Europe, but now evidence for it can be found far from the North Atlantic. This reversal in climate is called the Younger Dryas event, named after the Dryas flower that is found currently only in arctic and alpine regions. The Younger Dryas event brought an end to the more benign climatic conditions that were beginning to develop after the Last Glacial Maximum.

As the Younger Dryas proceeded, cold polar waters moved southward in the North Atlantic (Figure 13.22a), and the climate of Europe cooled (Figure 13.22d). A new ice sheet developed over parts

of Scotland, and valley glaciers advanced in Europe. Cold-tolerant vegetation established itself once more in Europe (Figure 13.22c). The rates of accumulation of snow in the Greenland ice sheet increased as temperatures fell, and the atmosphere of the North Atlantic became dustier because of the colder, drier, and windier climate of the Younger Dryas (Figure 13.23). In addition, herblike vegetation replaced forests in the Pacific Northwest of the United States, North Africa became more arid and lake levels in the region declined, northwestern Pacific Ocean surface waters cooled, and mountain glaciers advanced in New Zealand.

Figure 13.23 shows that whatever the cause of the Younger Dryas event, the event began rapidly and ended abruptly. The debate still goes on as to what caused the Younger Dryas event. One possibility that has received a lot of publicity is that proposed by geochemist Wally Broecker of Lamont-Doherty Observatory of Columbia University. Broecker suggests that the path of meltwater flow from the North American ice sheet changed during the Younger Dryas event. The glacial meltwater flow was rerouted from delivery to the Gulf of Mexico by the Mississippi River prior to the Younger Dryas event to the North Atlantic Ocean through the St. Lawrence seaway at the beginning of the event. This pulse of low-salinity water to the North Atlantic led to a rearrangement of ocean circulation

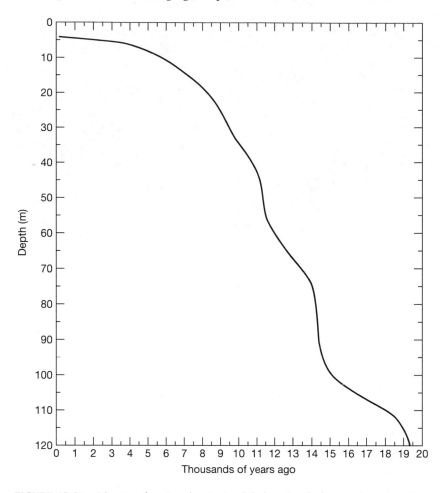

FIGURE 13.21 Diagram showing the rise in global sea level after the last glaciation as obtained from radioactive dating of corals submerged off the island of Barbados. The rise is due to the return of meltwater from the continental ice sheets to the ocean. (*Source:* Fairbanks, 1989; Bard et al., 1990a,b.)

temporarily decreasing or cutting off the rate of deep-water formation in the high-latitude North Atlantic Ocean (the conveyor belt circulation, Chapter 5). Because the formation of deep water at high latitude gives off heat from the surface waters of the ocean to the atmosphere to warm the northern Atlantic region, the decrease in deep-water flow cooled the North Atlantic and surrounding regions. At the termination of the Younger Dryas event, most of the flow to the east to the North Atlantic ceased and presumably the conveyor belt circulation turned back on. The climate began to warm again. This is but one possibility for the cause of the Younger Dryas event focusing on a mechanism that operates in the North Atlantic region. It is also possible that the cause of the Younger Dryas event is more complex and may be the result of a set of processes that operate at a larger and even global scale.

Holocene Epoch

The end of the Younger Dryas terminated the Pleistocene Epoch about 10,000 years ago, and since that time Earth has been in an interglacial stage known as the Holocene (Figure 13.16a; see also Chapter 8). Summer insolation levels during this time span have dropped toward the present while winter insolation levels have risen (Figure 13.20). Although remnants of the great North American Laurentide ice sheet persisted until about 7000 years before present (B.P.), in general, after the Younger Dryas event, the climate continued to warm, although erratically. It appears that the climate overall was warmer in the early Holocene from approximately 10,000 to 4500 years before present than for the past 4500 years. By 6000 years ago, the low atmospheric CO_2, CH_4, and N_2O levels of the Last Glacial Maximum had risen to their full interglacial

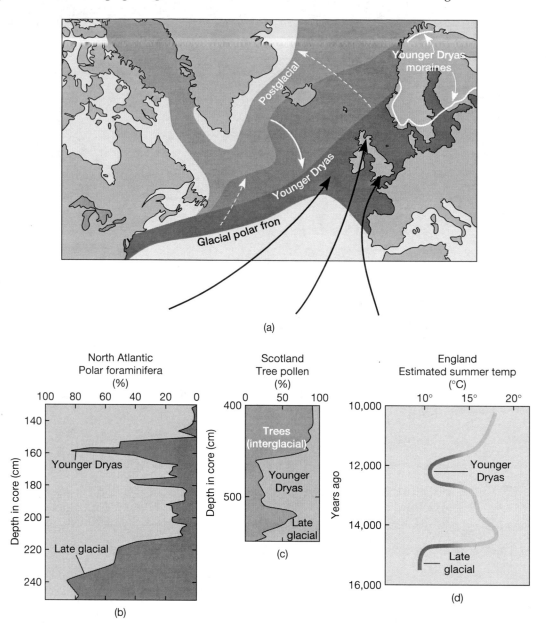

FIGURE 13.22 Evidences for the Younger Dryas cold event in the North Atlantic region. (a) Changes in the position of cold polar water as seen in the evidence obtained from cold-water foraminifera distributions in sediment cores (b); arctic vegetation trends as obtained from pollen records (c); and estimated temperatures for England as obtained from fossil insect populations in Britain (d). (*Source:* Ruddiman, 2001.)

levels of 280 ppmv, 700 ppbv, and 275 ppbv, respectively. The great continental ice sheets were fully melted, except for Greenland and Antarctica, and the summers in the north polar latitudes had reached their maximum warmth. Low-latitude vegetation biomes were displaced northward in eastern North America and eastern Europe, and vertical displacements of mountain vegetation zones and/or glaciers occurred in western North America, the European Alps, and New Guinea. Sea level at least in the Pacific Basin was

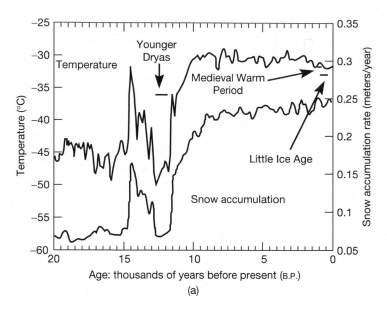

(a)

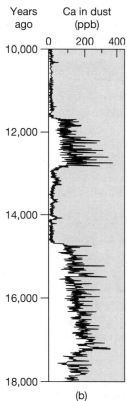

(b)

FIGURE 13.23   Ice core temperatures and rates of accumulation of snow (a) and concentrations of calcium in windblown dust (b) in the ice of the Greenland ice sheet before, during, and after the Younger Dryas event. Note how during the cool Younger Dryas period, ice accumulation rates, and windblown dust levels return to late glacial values.   (*Source:* Ruddiman, 2001 and http://www.ncdc.noaa.gov/paleo/abrupt/data_glacial.html.)

higher than at present. This period of time, which was warmer and wetter than the modern interval, has been called by some scientists the hypsithermal or the Holocene Climatic Optimum and may represent global mean temperatures more than 1°C above those of the late pre-industrial era (Figure 13.16a).

Since about 6000 years ago, summer insolation levels (Figure 13.20) have been decreasing at high latitudes, actually falling below levels that led to the melting of the ice sheets. Thus, the Earth at high northern latitudes during the past several thousand years has been experiencing a renewed natural cooling. This cooling, interrupted by the several-hundred-year-long Medieval Climatic Optimum (Figure 13.16b), culminated in the Little Ice Age of 1350 to 1850 A.D. and perhaps continued into the early twentieth century as the planet recovered from the extreme cold events of the Little Ice Age (Figure 13.16b). During the Little Ice Age, global mean temperatures were nearly 0.5°C lower than those of the latter part of the nineteenth century. Mountain glaciers reached their maximum extent during the seventeenth and eighteenth centuries, and large-scale advances of mountain glaciers occurred in the Alps of Switzerland and Austria and the mountains of Norway in the fourteenth and fifteenth centuries. As mentioned previously, the year 1816 was so cold in Europe that it has been called the year without a summer. The settlements that had been established during the Medieval Climatic Optimum by Nordic people in Greenland were abandoned during the Little Ice Age because of the inhospitable climate. There were crop failures in Europe because of the colder winters and shorter growing seasons, and people suffered greatly. The cause of the Little Ice Age, and especially the sharp extreme cold events, are still controversial. The cooling may have been the culmination of the slow cooling of the past 6000 years or it could represent some millennial-scale oscillation in climate. There were three distinct periods of decreased sunspot activity known as the Sporer sunspot minimum from 1460 to 1550, the Maunder sunspot minimum from 1645 to 1715, and the Dalton sunspot minimum from 1790 to 1830 during the Little Ice Age. These periods of decreased sunspot activity correlate with some of the colder periods of the Little Ice Age. The decrease in sunspot activity could have been accompanied by a decrease in solar radiation and hence a cooling of the planet. Some of the cold, sharp excursions during the Little Ice Age were probably also the result of volcanic explosions generating atmospheric sulfate aerosol that cooled the planet. The eruption of Tambora in Indonesia is an example of such an eruption that cooled the planet near the end of the Little Ice Age.

Among other changes, the longer-term renewed cooling of the past several thousand years is evidenced in the (1) advance of mountain glaciers to lower elevations in the Northern Hemisphere, (2) decrease in summer melting of ice and advance of ice caps on arctic islands, (3) increase in the frequency of sea ice off of Greenland, (4) lower temperatures of North Atlantic Ocean waters in some regions, and (5) southward shift in the boundary between the high-latitude tundra biome and the lower-latitude spruce forest biome in northern Canada.

From the end of the nineteenth century to the present day, the planet has been experiencing a general warming trend at the surface, with minor reversals of temperatures, including the period from about 1940 to the early 1970s (Figure 13.16c). This temperature record and its causes are addressed in Chapter 14.

Broadly speaking, for the last 18,000 years, the global climate warmed several degrees, and sea level rose about 120 meters. Atmospheric carbon dioxide concentrations climbed from 180 to 280 ppmv. Atmospheric methane concentration rose 350 ppbv and nitrous oxide concentration 90 ppbv. In 2008, the concentration of atmospheric carbon dioxide was about 387 ppmv, that of methane was 1785 ppbv, and that of nitrous oxide was 322 ppbv. These increases are due mainly to fossil fuel and biomass burning and agricultural and industrial activities of human society.

Concluding Remarks and the 21st Century

The purpose of this chapter to some extent is to set the stage for the global atmospheric environmental issues of global warming and stratospheric ozone depletion discussed in following chapters. In many ways our planet has left the environmental domain that defined the Earth system for the 800,000 years before the Industrial Revolution. A brief review of the preceding chapters in this book and further reading in Chapters 14 and 15 well document this conclusion, as does Table 13.3. The table lists some changes in Earth's surface environment during late Pleistocene glacial-interglacial stages. It also shows the global environmental conditions of the present with present rates of change and questionable qualitative projections into the near future. The present conditions are a result, to a significant degree, of inputs of materials

TABLE 13.3 Some Historical and Present Global Environmental Conditions of the Earth's Surface Environment and Recent and Future Changes in These Conditions: (y) Indicates a Quantity Not Accurately Known at Present

Component	Conditions			Change in Rates	
	Glacial	Interglacial	Present	Early Twenty-First Century[a]	Future[a]
CO_2 concentration	180 ppmv	280 ppmv	387 ppmv	↑ 1.93 ppmv/yr	↑
CH_4 concentration	0.35 ppmv	0.7 ppmv	1.78 ppmv	↑4.9 ppbv/yr until the late 1990s and a recent flattening in the rate of rise	↑
N_2O concentration	185 ppbv	275 ppbv	322 ppbv	↑0.8 ppbv/yr and rising linearly	↑
CFC-11 concentration	0	0	242 pptv	↓ Slightly negative	↓
CFC-12 concentration	0	0	530 pptv	↓ Slightly negative	↓
SO_2 emissions	—	—	$62 \times 10^9 \, k \, y^{-1}$	↓	↑?
DMS emissions	$60 - 400 \times 10^9 \, kg \, y^{-1}$	$40 \times 10^9 \, kg \, y^{-1}$	$40 \times 10^9 \, kg \, y^{-1}$	↑?	↑?
Temperature	10–11°C	15°C	15.8 °C	↑	↑
Mineral aerosol flux	$5 - 10 \times$ (y)	(y)	$2 \times 10^{12} \, kg \, y^{-1}$	↑	↑
Land runoff (H_2O) flux	$2 \times 10^{16} \, kg \, y^{-1}$	$3.7 \times 10^{16} \, kg \, y^{-1}$	$3.7 \times 10^{16} \, kg \, y^{-1}$	$\downarrow - 0.1\% \, y^{-1}$	↑[b]
Particulate erosion products in runoff flux	$\sim 1 \times 10^{13} \, kg \, y^{-1}$	$\sim 7 \times 10^{12} \, kg \, y^{-1}$	$1.5 \times 10^{13} \, kg \, y^{-1}$	↓	↑[b]
Dissolved salts in runoff flux	—	—	$4 \times 10^{12} \, kg \, y^{-1}$	↑	↑
N_{total} riverine flux	—	$1.4 \times 10^{10} \, kg \, N \, y^{-1}$	$3.5 \times 10^{10} \, kg \, N \, y^{-1}$	↑	↑
P_{total} riverine flux	—	$1.4 \times 10^9 \, kg \, P \, y^{-1}$	$3 \times 10^9 \, kg \, P \, y^{-1}$	↑	↑
$C_{organic}$ riverine flux	—	$4 \times 10^{11} \, kg \, C \, y^{-1}$	$8 \times 10^{11} \, kg \, C \, y^{-1}$	↑	↑
Total marine net primary production	>(y)	(y)	$3.8 \times 10^{13} \, kg \, y^{-1}$	↑	↑

[a]↑increase; ↓decrease; pptv = parts per trillion by volume.

[b]Damming will decrease runoff and particulate erosion flux; temperature increase will increase runoff.

(*Source:* Mackenzie et al., 1991.)

from human activities into the natural system. Some of these inputs represent materials that have circulated through the ecosphere naturally and whose fluxes have been modified by human activities (e.g., riverine nitrogen flux). Others represent synthetic compounds that have been manufactured by humans (e.g., chorofluorocarbons, CFCs). The inputs from human activities have modified the natural chemical composition of the atmosphere; changed the direction of transfer of materials from one major Earth surface reservoir to another; modified the stream fluxes of nitrogen, phosphorus, and suspended particulate matter to aquatic systems; introduced chemical compounds to the ecosphere that naturally do not occur there; and affected the climate system of the planet. In addition, humans have directly changed the water cycle by their use of water and the construction of dams, impounding both water and sediments in the reservoirs associated with dams on major river courses. There is no doubt that humans have become a "geologic force" in the ecosphere and will continue to be such in the 21st century.

One example of a climate variable that apparently has undergone modification because of this geologic force is that of the more recent surface temperature history of the Earth. As preparation for discussion of the global climatic issues of the following chapters, we consider here the temperature history of the surface environment during the past millennium. It is possible to reconstruct the Northern Hemisphere surface temperature record for the past 1300 years from data obtained from thermometer measurements and proxies of temperature based on tree rings, corals, and ice-core records of temperature. Figure 13.24 shows

several renditions of attempts to reconstruct this temperature record. There are large uncertainties in this record, particularly for the period prior to A.D. 1400. In addition, there are problems in agreement between the satellite record of tropospheric temperatures and the surface thermometer record since 1978 (see Chapter 14). However, there appears to be a modest cooling of the planet for most of the past millennium after the Medieval Warm Period, culminating in the Little Ice Age. The twentieth-century warming of the Northern Hemisphere as a whole appears to be large, abrupt, and significant in comparison with the variability exhibited in past centuries and is counter to the millennial-scale direction of temperature change. Why this apparently abrupt change? Is it a real trend? What are the causes of the change? These questions and others related to global warming and stratospheric ozone depletion are discussed in the following chapters.

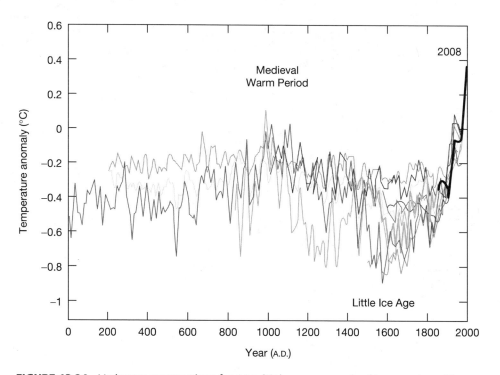

FIGURE 13.24 Various reconstructions from multiple proxy records of temperature, like those found in tree rings, ice cores, and corals, of the Northern Hemisphere surface temperatures for the past 2,000 years. The instrumental thermometer record shown as the dark black line extends from 1850 to 2008. (From *Climate Change 2007: The Physical Science Basis. Contribution of Working Group I to the Fourth Assessment Report of the Intergovernmental Panel on Climate Change.* © Intergovernmental Panel on Climate Change 2007. Published by Cambridge University Press. Used by permission of Intergovernmental Panel on Climate Change.)

Study Questions

1. Describe briefly temperature change during the Cenozoic Era.
2. What types of data records are used to provide information on historical climatic change?
3. What is the relationship between the sunspot cycle and the temperature record of the last 100 years?
4. Describe the three orbital parameters that affect climate on the time scale of Pleistocene glacial-interglacial change.
5. What are aerosols, and how do they affect climate?
6. What is the source of dimethysulfide and carbonyl sulfide gas to the atmosphere, and what role do these gases play in climate?
7. How did the tundra and conifer biomes change from the Last Glacial Maximum (LGM) to present time?
8. What are two sedimentary evidences for Heinrich events?
9. How might the conveyor belt circulation pattern of the ocean have changed between the LGM and the modern ocean?
10. How did the water vapor content and winds of the LGM differ from those of modern times?
11. How has summer and winter solar insolation changed since the LGM?
12. What is the Younger Dryas event?
13. What are three evidences for the Younger Dryas cold oscillation?
14. Since the hypsithermal of about 6000 years ago, what has been the general millennial-scale change in temperature of the Northern Hemisphere? What are two important climatic periods during this time, and what did they represent for climate and people?
15. What is the CLAW hypothesis?

Additional Sources of Information

Broecker, W. S., 2004, *The Role of the Ocean in Climate Yesterday, Today, and Tomorrow.* Eldigio Press, Palisades, New York, 176 pp.

Charlson, R. J. and Wigley, T. M. L., 1994, Sulfate aerosol and climate change. *Scientific American,* February, pp. 48–57.

Crowley, T. J. and North, G. R., 1991, *Paleoclimatology.* Oxford University Press, New York, 349 pp.

Delmas, R. J., 1992, Environmental information from ice cores. *Reviews of Geophysics,* v. 30, pp. 1–21.

Foukal, P. V., 1990, The variable sun. *Scientific American,* February, pp. 34–41.

Imbrie, J. and Imbrie, K. P., 1979, *Ice Ages: Solving the Mystery.* Macmillan, London, 229 pp.

IPCC, 2007, *Climate Change 2007: The Physical Science Basis. Contribution of Working Group 1 to the Fourth Assessment Report of the Intergovernmental Panel on Climate Change* [Solomon, S., D. Qin, M. Manning, Z. Chen, M. Marquis, K. B. Averyt, M. Tignor, and H. L. Miller (eds,)], Cambridge University Press, United Kingdom and New York, 996 pp.

Ruddiman, W. F., 2001, *Earth's Climate: Past and Future.* W. H. Freeman and Company, New York, 465 pp.

14 The Changing Atmosphere: Global Warming and Stratospheric Ozone Depletion

Humankind is now carrying out a large-scale geophysical experiment of a kind which could not have happened in the past nor be repeated in the future. Within a few centuries we are returning to the air and oceans the concentrated organic carbon stored over millions of years.

R. REVELLE AND H. SEUSS, 1957

This chapter is concerned with the two Earth system phenomena that recently have had a substantial amount of scientific research devoted to them and have become a part of the everyday vocabulary of the press and the general public. They are the topics of global warming and stratospheric ozone depletion.

Relevant to the topic of global warming is the fact that we as a human society are tinkering with Earth's energy budget. This tinkering may affect the climate of the planet. Human activities are essentially dirtying the atmospheric window that permits the transfer of radiation between the planetary surface and space. The current concern about the greenhouse effect and climatic change stems from the amounts of greenhouse gases that are being released into the atmosphere from the burning of fossil fuels (coal, oil, and gas), deforestation, agricultural and industrial practices, production and release of human-made chlorofluorocarbons (CFCs do not occur naturally in the environment), and other activities of humankind. Accumulation of these heat-absorbing greenhouse gases in the atmosphere could result in an enhanced greenhouse effect and consequent global warming induced by human activities. The concern is that an enhanced greenhouse effect may raise the temperature of Earth above that experienced by the planet in the last several hundred thousand years; a super interglacial stage may occur (Figure 14.1). If Earth were left to its own recourse, it is likely that the planet would reenter the natural cooling trend of a glacial stage during the next few thousand years. As succinctly pointed out by Wallace S. Broecker of Lamont-Doherty Observatory of Columbia University in the United States, we are toying with an angry beast of global warming due to anthropogenic emissions of CO_2 and other greenhouse gages to the atmosphere and do not well understand the beast's reactions (Figure 14.2).

The effects that the accelerating rates of increase of greenhouse gases in the atmosphere will have on our climate is a topic of much debate and uncertainty because of the many variables that are involved in evaluating the problem. This debate has to some extent culminated in the contrary conclusions reached

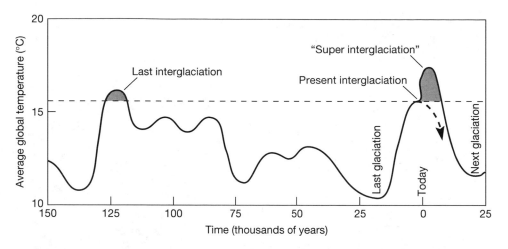

FIGURE 14.1 Earth's climate during the last 150,000 years and an interpretation of its future. The natural course of climatic change would involve a cooling, reaching a glacial maximum about 23,000 years from now (dashed line). With an enhanced greenhouse effect, because of the human-induced emissions of carbon dioxide and other greenhouse gases to the atmosphere, a continued warming may lead to a super interglacial stage in the next couple of hundred years. The temperature may be higher than that of the last interglacial stage. Consequently, the decline in temperature leading to a future glaciation could be delayed by several thousand years. (*Source:* Imbrie and Imbrie, 1979.)

in two influential reports. The Intergovernmental Panel on Climate Change (IPCC) report *Climate Change 2007: The Physical Science Basis* (IPCC, 2007) concluded that "Most of the observed increase in global average temperatures since the mid-20th century is *very likely* due to the observed increase in anthropogenic greenhouse gas concentrations" (IPCC, p. 10). To some extent the conclusions of the IPCC have been popularized by former Vice President of the United States Al Gore in his book *An Inconvenient Truth* (2006). On the contrary, another less well documented report entitled *Nature, Not Human Activity, Rules the Climate* (Singer, 2008) concluded ". . . that the human greenhouse contribution to current warming is insignificant" (p. 27). The IPCC shared the Nobel Peace Prize in 2007 with former Vice President Al Gore "for their efforts to build up and disseminate greater knowledge about man-made climate change, and to lay the foundations for the measures that are needed to counteract such change." In light of this global climate debate, we will explore several questions concerning global warming in this chapter and the following. What will happen as humans continue to put an excess of greenhouse gases into the atmosphere? Will Earth warm further? What will be the degree of warming? What will be the effects of climatic change on ecosystems? What are the social, political, and economic consequences of an enhanced greenhouse and global warming of the planet?

The second topic considered in this chapter, that of stratospheric ozone depletion, the "hole in the sky," is not independent of the problems of global warming. Because the warming of the stratosphere is due mainly to the absorption of solar radiation by stratospheric ozone, providing a warm lid to the troposphere below (see Chapter 4), and because CFCs are both greenhouse gases in the lower atmosphere and are responsible for depletion of the stratospheric ozone layer, it is appropriate to consider global warming and stratospheric ozone depletion concurrently.

On October 11, 1995, the Royal Swedish Academy of Sciences in Stockholm, Sweden, awarded the Nobel Prize in Chemistry to three atmospheric scientists for their role in determining the chemical processes that affect stratospheric ozone formation and destruction and thus the concentration of O_3 in the ozone layer. Paul Crutzen of the Max Planck Institute for Chemistry in Mainz, Germany; Mario Molina of the Massachusetts Institute of Technology in Cambridge, Massachusetts; and F. Sherwood Rowland of the University of California at Irvine shared the $1 million prize. In 1970, Crutzen showed that nitrogen oxides catalyze reactions involving the destruction of ozone. Five years later, Molina and Rowland hypothesized that chlorofluorocarbons and other chlorinated hydrocarbons are a threat to the integrity of the stratospheric ozone layer. The hypothesis was initially controversial, but the discovery of

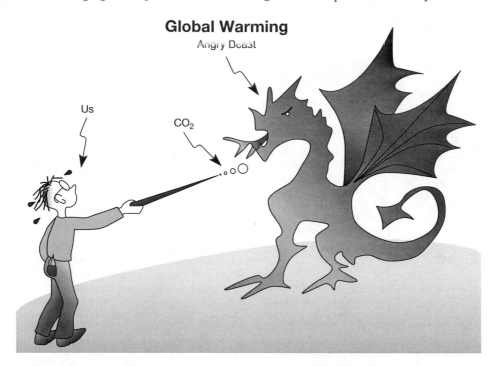

FIGURE 14.2 The angry climate beast of global warming. As human society puts CO_2 and other greenhouse gases into the atmosphere, the response of the beast to the changes in atmospheric composition may surprise us. (From *Fossil Fuel CO$_2$ and the Angry Climate Beast* by W. Broecker, 2004. Illustration used by permission of Patricia Catanzaro.)

the "hole in the sky," the ozone hole, above Antarctica in 1985 supported the hypothesis. Later, Crutzen and his colleagues demonstrated that the severe depletion of ozone in the Antarctic stratosphere during the austral spring is due to particles within polar stratospheric clouds (PSCs) (see Chapter 4) that serve as sites for chemical reactions to take place that deplete ozone. In this chapter, we explore the global environmental issue of stratospheric ozone depletion, in part to understand the background for the award of the Nobel Prize in Chemistry to these three atmospheric scientists.

THE GREENHOUSE EFFECT AND CLIMATIC CHANGE

The natural greenhouse effect was explained initially in Chapter 4. To set the stage for the rest of this chapter, further discussion of the phenomenon is provided here. The greenhouse effect is an important phenomenon that regulates the global climate of the planet. However, the term *greenhouse* is actually a misnomer. A greenhouse lets sunlight in and the sunlight warms the interior of the greenhouse because the air within absorbs infrared radiation. The manner in which the greenhouse is constructed prevents

convective heat loss and the greenhouse stays warm. To understand how the global natural greenhouse functions, we will first look to the major source of energy for our planet, the sun.

The sun's surface temperature is about 5480°C. This stellar body radiates light energy into space (see Box 4–1: Energy). The temperature of Earth is determined and maintained by the balance between the flux of incoming solar radiation and the amount of outgoing infrared radiation radiated back to space from Earth (see Figure 4.3 and the more quantitative Earth radiation balance in Figure 14.3). The distribution of radiation at the top of the atmosphere depends on the geometry of the globe, its rotation, and its elliptical orbit around the sun. On entering Earth's atmosphere, solar radiation is absorbed and scattered. The absorbed radiation is added to the heat budget of the planet, while the scattered radiation is returned to space or continues its passage through the atmosphere, where it is further scattered or absorbed. Atmospheric gases, aerosols, and clouds are responsible for the scattering and absorption of solar radiation.

Some of the light energy penetrating the transparent atmosphere of the planet is absorbed in the stratosphere by ozone, and warming takes place. A

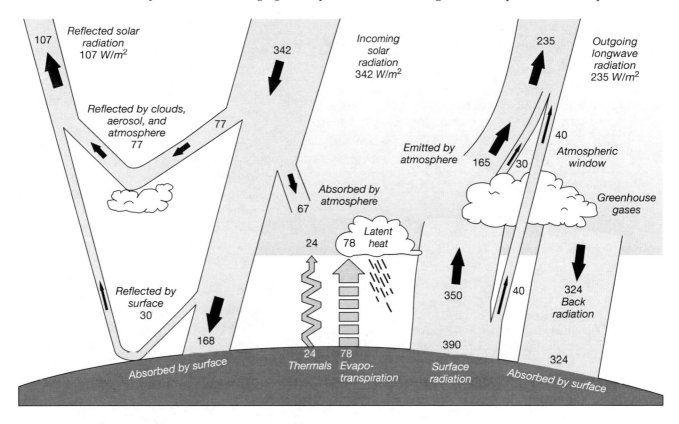

FIGURE 14.3 Radiation balance of Earth in units of W/m². Notice that 235 W/m² of incoming short wave solar radiation of 342 W/m² are absorbed by the atmosphere and the surface of Earth. Rising thermals, evapotranspiration, and surface radiation return 492 W/m² of long wave heat radiation to the atmosphere, some of which escapes Earth's atmosphere but due to the greenhouse gases in the atmosphere, much is back-radiated to warm the planet.

portion of the remaining solar energy is reflected back to space by clouds that form in the stratosphere and troposphere. Without clouds Earth would reflect only about 10% of the incoming solar radiation. With clouds nearly 30% of the radiation is reflected. Thus, clouds globally exert a cooling influence on the planet, amounting to −13 W/m² to −20 W/m². However, the immediate effect of clouds on the radiation budget depends on the type of clouds, their location, and their elevation. Clouds both reflect solar radiation and trap infrared radiation emitted from Earth's surface below. The entrapment has a heating effect and complements the warming effect of greenhouse gases. Thus, clouds can either help to warm or help to cool the planet. On one hand, low-level stratus and mid-level storm clouds tend to cool more than they warm. On the other hand, cold, icy, high-altitude cirrus clouds tend to be net warmers (see Chapter 4 and, later in this chapter, the section on water vapor and clouds).

Atmospheric aerosols (from volcanoes, sea salt, and windblown dust particles) and sulfates in the air

generally reflect energy back to space and tend to cool the atmosphere. When the sun's radiation impacts the surface of the planet, it is partially reflected back to space. The amount reflected depends on the albedo of ice, land, ocean, and vegetation surfaces. Solar energy that is not reflected is absorbed by the surface of Earth.

The incoming shortwave radiation from the sun only partially warms the atmosphere. Most of the warming is due to the absorbed solar energy that is radiated back from the surface of the planet in the form of long-wave infrared (heat) radiation. The greenhouse gases of water vapor, carbon dioxide, methane, nitrous oxide, and other gases that reside in the atmosphere absorb the infrared radiation. As a result, the atmosphere is warmed (see Figure 4.2). The trapping of heat by greenhouse gases is the process that warms our atmosphere and makes the planet habitable for life. This is the **"natural" greenhouse effect.** Without a greenhouse effect, the planet would be about 33°C cooler than it is, or −18°C.

Figure 14.3 illustrates that incoming solar radiation delivers to the top of Earth's atmosphere about

342 watts per square meter (W/m^2) of energy. About 235 W/m^2 penetrate the atmosphere to drive the climate system, with 67 W/m^2 absorbed by the atmosphere and 168 W/m^2 absorbed by the planetary surface. Rising thermals add 24 W/m^2 and evapotranspiration through latent heat released in the condensation of water vapor to form clouds and rain adds another 78 W/m^2 to the amount of heat absorbed by the atmosphere. The naturally occurring greenhouse gases of H_2O, CO_2, CH_4, N_2O, and tropospheric O_3 trap heat energy from the Earth's surface approximately equivalent to 90% of the 390 W/m^2 of the infrared energy emitted as surface radiation from Earth, or 350 W/m^2, in the natural greenhouse effect (see Figures 4.2, 4.3, and 14.3). Thus, the greenhouse gases absorb the long wave surface Earth radiation and reradiate most of it (324 W/m^2) back toward the planetary surface, warming the Earth's surface and lower atmosphere.

The combination of these warming and cooling components of the atmosphere balance Earth's radiation budget (Figure 14.3). What comes in must go out for the temperature to remain relatively constant and to maintain the system at steady state. If one of the preceding climatic factors were altered, then an adjustment in the temperature and the rest of the climate system to another state would occur.

Earth has remained comfortable and suitable for life in *recent* geologic time because the global average near-surface temperature of the Earth of about 15°C has not varied greatly. Early hominids were able to survive the frigid temperatures of glacial stages and the warmth of the interglacials and to evolve under the changing climatic conditions. In more recent times, the Holocene Climatic Optimum, the Medieval Warm Period, and the Little Ice Age (see Chapter 13) were times of relatively small natural changes in temperature. Human populations were affected on a regional scale, particularly in the temperate and polar regions of the Northern Hemisphere, by cold snaps and crop failures during the Little Ice Age, and by drought, but more equable temperatures and longer growing seasons, during the Holocene Climatic Optimum and the Medieval Warm Period. The presence of greenhouse gases in our atmosphere has kept the natural temperature swings of the recent past relatively small due to the dynamic balance in the many interconnected cycles of the lithosphere, hydrosphere, atmosphere, cryosphere, and biosphere that have evolved throughout the history of the planet.

Enhanced Greenhouse Effect

If the reader were studying this book in the middle of the twentieth century, the emphasis on global climatic change might have been on that of global cooling and the possibility of another ice age. Indeed even some college and high school natural science materials emphasized this possibility at this time, and some scientists suggested that climatic cooling would be the near-term climatic future of the planet. Much of the reasoning for this assertion was based on the fact that the observational global mean surface temperature record obtained from thermometers showed little directional change from about 1940 to 1970, except for perhaps a slight decrease in temperature over that time period of approximately 0.1°C per decade. However, ever since the work of the Swedish scientist Svante Arrhenius in 1896, there has been concern that the buildup of carbon dioxide in the atmosphere could lead to global warming of the planet. Arrhenius's calculations showed that a doubling of the concentration of CO_2 in the atmosphere could lead to an increase in the global mean surface temperature of 5 to 6°C, an estimate that is remarkably close to our present understanding. In 1940, G. S. Callendar was the first twentieth century scientist to calculate the degree of warming that the increasing carbon dioxide concentrations coming from the combustion of fossil fuels could have on the Earth.

Another reason for the emphasis on global cooling in the middle of the twentieth century was the fact that major interglacial stages appear to occur on a 100,000-year frequency (see Chapter 13). The Earth is currently in an interglaciation; the last one occurred about 100,000 years ago. This led some scientists to suggest that the cool period of the mid-twentieth century was the beginning of a future glacial stage. What has changed this perspective to a significant extent is the observational thermometer record of land and sea surface global mean temperatures since the early 1970s. The temperatures have risen relatively rapidly, albeit irregularly, for more than 30 years. This observation—along with the continuously increasing concentrations of greenhouse gases in the atmosphere derived from human activities and the fact that the rate of increase of atmospheric CO_2 at Mauna Loa, Hawaii, essentially parallels the trend of emissions of CO_2 to the atmosphere from fossil fuel burning (Figure 14.4)—has led a large group of scientists to claim that the planet is now in the midst of an enhanced greenhouse effect.

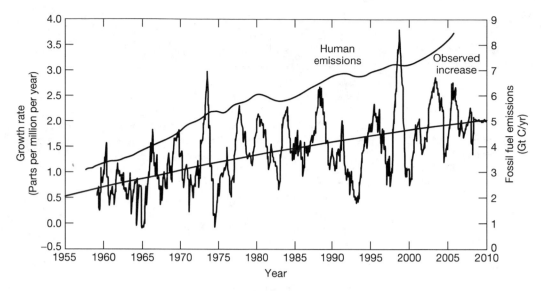

FIGURE 14.4   Correlation between fossil fuel emissions of CO_2 to the atmosphere and growth rate of rising atmospheric CO_2 concentrations from 1960 to 2008. This correlation has led many individuals to conclude that the rising atmospheric CO_2 concentrations are due in part to the combustion of coal, oil, and gas by human society, leading to emissions of CO_2 to the atmosphere.   (*Source:* NOAA, 2008 and Carbon Dioxide Information Analysis Center, CDIAC, cdiac.ornl.gov/, after Boden et al., 2009.)

The greenhouse gases added to the atmosphere by human activities add to the natural greenhouse effect by means of a direct radiative forcing; that is, a forcing that occurs even without any of the feedbacks that might occur in the Earth's climate system (see Box 14–1). Although there is some controversy about the magnitude of the cumulative effect of all the greenhouse gases added to the atmosphere by human activities over the past couple of centuries, the trapping effect of these gases in the year 2005 appears to be equivalent to about 2.99 W/m² or 1% of the total amount of incoming solar radiation that is not scattered and reflected back to space. Thus, the 2.99 W/m² represent a 1.7% addition to the amount of energy absorbed by Earth's surface. The total additional energy is referred to as the **enhanced greenhouse effect.** Now let us consider the sources of the greenhouse gases for the atmosphere and the recent changes in the concentrations of these gases in the atmosphere.

Atmospheric Greenhouse Gases

Natural atmospheric greenhouse gases comprise less than 0.1% of the constituents of air. However, they are necessary gases involved, for example, as water vapor in evapotranspiration and condensation, as carbon dioxide in photosynthesis and respiration, as methane in the formation of CH_4 hydrates, as

nitrous oxide in the destruction of stratospheric ozone, and as tropospheric ozone acting as an oxidizing agent of carbon monoxide and hydrocarbons in the atmosphere. Very importantly these gases also maintain an equable climate on Earth, and their rising concentrations along with those of the synthetic chlorofluorocarbons create the enhanced greenhouse effect and the problem of global warming.

CARBON DIOXIDE (CO_2)   The natural carbon cycle involves the cycling of carbon through the reservoirs of the lithosphere, hydrosphere, atmosphere, and biosphere over a wide range of time and space scales. This cycling has gone on throughout the history of the planet, and variations in fluxes in the cycle are responsible for changes in atmospheric CO_2 concentrations. Figures 7.3 and 7.4 show the complex of processes that involve the transfers of carbon in the exogenic cycle at various time scales. Figure 14.5 illustrates the many processes involving exchange of carbon as carbon dioxide between the surface of the Earth and its atmosphere, and the anthropogenic fluxes due to fossil fuel combustion and land-use (mainly deforestation) changes and uptake by the oceans and the land for the middle part of the first decade of the twenty-first century. Some important aspects of this diagram are discussed in the following sections.

BOX 14.1

Radiative Transfer, Forcing, and Global Warming Potential (GWP)

Radiative transfer is a mathematical way of describing the exchange of energy between the Earth and the atmosphere. The energy transferred or transmitted by electromagnetic radiation occurs as discrete quanta called **photons.** The rate of transfer is called the **radiant flux,** which usually has units of joules per second, (J/sec) or watts (W). The sun's radiant flux received by the Earth each second is about 3.9×10^{26} watts. If we divide the radiant flux by the area through which the energy is passing, we have a term known as **irradiance.** The units for irradiance are watts per meter squared (W/m^2).

The **radiative forcing** at the surface of the Earth can be defined as the change in net irradiance in watts per square meter at the tropopause after allowing for stratospheric temperatures to readjust to equilibrium, but with surface and tropospheric temperatures held fixed at their unperturbed values. The concept of radiative forcing is tied to climate model calculations that indicate an approximately linear relationship between the global mean radiative forcing at the tropopause and the equilibrium global mean surface temperature change. The relationship seems to be unaffected by the nature of the forcing, be it changes in greenhouse gas concentrations or solar radiation. The change in the global mean radiative forcing may be expressed as

$$\Delta F = \Delta T_s / \lambda$$

where ΔF is the change in mean radiative forcing in watts per square meter (W/m^2), ΔT_s is the change in global mean surface temperature in kelvin (K), and λ is a parameter that represents the equilibrium sensitivity of the climate to change; that is, the equilibrium change in average surface air temperature following a unit change in radiative forcing. The sensitivity has units of $K/(W/m^2)$. This parameter can be defined for a black body Earth or for an Earth with an albedo (gray Earth) or for one like the modern realistic Earth in which λ depends on several processes, including feedback mechanisms associated with water vapor and cloud formation and the reflectivity of polar ice (albedo). Instead of the above definition of λ, the global **climate sensitivity** can also be expressed as the temperature change ΔT_{x2}, following a doubling of the atmospheric CO_2 concentration. Such an increase in concentration is equivalent to a radiative forcing of about $3.8 \ W/m^2$. Thus, $\Delta T_{x2} = 3.8 \ W/m^2 \lambda$. The value of λ is difficult to determine, but it is probably in the range of 0.3 to 1.4 K per watt per square meter ($K/W/m^2$). The 2007 IPCC report (IPCC, 2007) gives the equilibrium climate sensitivity for the modern Earth (ΔT_{x2}) in the range of 2°C to 4.5°C with a best estimate of about 3°C, based on global circulation model (GCM) calculations and observations, but the actual value and its range are still debated. However, it is unlikely that λ is less than 1.5 °C. Whatever the case, the actual value is critical to any discussion of future climatic changes. Figure 14.55 (discussed later) shows the radiative forcing for major factors (mechanisms) involved in climatic change on the time scale of concern for global warming. It is interesting to note that the eruption of Mt. Pinatubo in the Philippines in 1991 produced a negative radiative forcing of $-4 \ W/m^2$, that is, a cooling that lasted for a few years. The overall radiative forcing from the accumulation of greenhouse gases in the atmosphere is only $+2.99 \ W/m^2$.

The **Global Warming Potential** (GWP, Table 14.5) is tied in concept to that of the mean radiative forcing in that the GWP is a simple measure of the radiative effects arising from the emissions of greenhouse gases to the atmosphere. The GWP is the cumulative radiative forcing now and at times in the future relative to the reference gas CO_2. The GWP provides a measure of the relative radiative effects of the emissions of various greenhouse trace gases to the atmosphere. The GWP is a function of the radiative forcing due to a unit increase in the concentration of the trace gas in the atmosphere, the concentration of the trace gas remaining at a certain time after its release, and the number of years over which the GWP calculation is performed. The GWP is typically used to contrast different greenhouse gases relative to CO_2. When calculating the GWP of a specific greenhouse gas in which the reference gas is CO_2 and the time horizon is 100 years, the GWP of the gas is mathematically:

$$\text{GWP}_{\text{GAS}} = \text{AGWP}_{(\text{GAS})} / \text{AGWP}_{(\text{CO2})} = \int_0^{100 \text{ years}} a_{\text{GAS}} \times [\text{GAS}_{(t)}] dt / \int_0^{100 \text{ years}} a_{\text{CO2}} \times [\text{CO}_{2(t)}] dt$$

where AGWP_{GAS} and AGWP_{CO2} are the absolute global warming potentials, which are the integrated forcings of pulse emissions of the greenhouse gas and CO_2, respectively, over 100 years and a is the radiative efficiencies of the gas and CO_2 in $W/m^2/kg_{\text{GAS}}$ or $W/m^2/ppmv$.

On short time scales, volcanoes, plants, animals, natural forest and grass fires, and decaying organic matter contribute carbon dioxide to the atmosphere. Photosynthesis takes carbon dioxide from the atmosphere and stores it as carbon in the tissues of plants on land and in the ocean. Medium-term cycling involving geologic processes stores carbon in organic matter and its alteration products of kerogen, coal, oil, and gas. This carbon is weathered to carbon dioxide on exposure at Earth's surface. Long-term cycling

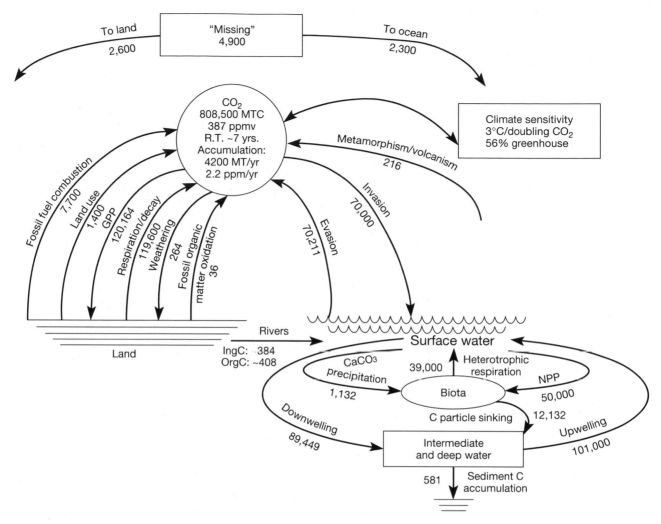

FIGURE 14.5 Non–steady state global biogeochemical cycle of carbon dioxide as carbon for the early twenty-first century. The major processes affecting transfer of carbon as carbon dioxide between the surface of Earth and its atmosphere are shown, as well as the rates of transfer (fluxes). Fluxes are in millions of metric tons of carbon (MTC) per year. The reservoir size of carbon dioxide is in millions of tons of carbon. Residence time (R.T.) is calculated with respect to gross primary production (GPP) on land. The lifetime of CO_2 is on the order of 50 to 100 years but molecules of the gas from anthropogenic activities can remain in the atmosphere from thousands to a hundred thousand of years. Fifty-six percent of the enhanced greenhouse effect over the past two centuries is due to carbon dioxide. A doubling of atmospheric CO_2 concentration could lead to a 3°C increase in temperature. (*Source:* Mackenzie, 1995, 1997, with additional information from Houghton et al., 1996; Watson et al., 2000; IPCC, 2007.)

involving the weathering of rocks removes carbon dioxide from the atmosphere, and volcanic-metamorphic processes restore it. The concentration of carbon dioxide in the atmosphere has evolved and fluctuated during geologic time in response to changes in these natural fluxes of carbon in the exogenic cycle (see Chapter 7).

Carbon Dioxide on Earth, Mars, and Venus Earth, Mars, and Venus were formed from the same cosmos at about the same time. Earth and Venus are of similar size, but Mars is only one-tenth of the mass of

Earth. Of these three planets, Earth alone provided an environment that is habitable for life (see Chapter 8). Venus today is thought to have evolved along the path of a runaway greenhouse effect. It is believed that at one time water might have been present on Venus and thus the planet had the potential for life. However, today there is no efficient mechanism on Venus to extract carbon dioxide from the atmosphere (no plants or fossil organic matter to store carbon and no deposition of carbon in sediments). As a result, carbon dioxide is concentrated in the atmosphere and not in the rock reservoirs (Figure 14.6). Because

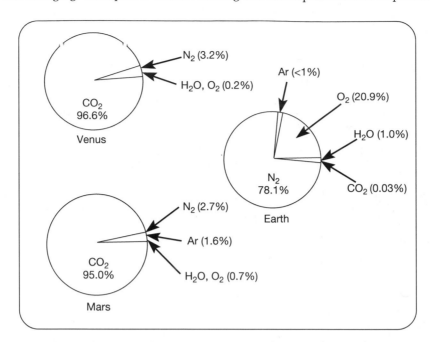

FIGURE 14.6 Comparison of the major atmospheric gas composition of Mars, Earth, and Venus. (*Source:* Rambler et al., 1989.)

of the abundance of carbon dioxide in its atmosphere, the average temperature of Venus is about 460°C. At this temperature all water vapor has long since evaporated, and the potential for life has disappeared.

On Mars it is a different story (Figure 14.6). There is evidence of ancient river beds. It is thought that the surface of Mars might once have been warm enough to have surface water and a substantial amount of atmospheric carbon dioxide. However, something went amiss. It is likely that there are huge amounts of carbon stored in the form of carbonate rocks along with frozen water in the rocks of Mars. Why did Mars evolve this way? Perhaps there was no way to recycle carbon back into the atmosphere to keep this small planet warm as the planet lost internal heat and plate tectonic motions ceased. Thus, with no carbon dioxide gas to act as a greenhouse gas and warm the atmosphere, Mars became a frozen planet with an average temperature of −50°C.

Only the planet Earth provided the right combination of a position in the solar system relative to the sun and atmospheric gases to produce and maintain a greenhouse effect that enables life to exist on the planet (Figure 14.6). It is because carbon is stored in plant and animal tissue and sequestered in the rocks of the planet (as calcium carbonate and organic matter) and not in the atmosphere and because of plate tectonic processes recycling this carbon that life is possible on Earth.

Anthropogenic Carbon Dioxide Atmospheric carbon dioxide concentrations have varied from about 180 ppmv in recent glacial periods to 300 ppmv in interglacial periods. The warm Eemian-Sangamon interglacial of 125,000 years ago had an atmosphere containing 280 to 300 ppmv carbon dioxide, while the atmosphere at the climax of the Wisconsin ice age contained 180 to 190 ppmv carbon dioxide. Atmospheric carbon dioxide increased to 280 ppmv as the planet warmed during the last 18,000 years prior to the Industrial Revolution.

In 1958, Charles Keeling of Scripps Institution of Oceanography in the United States began a continuous sampling of carbon dioxide in the atmosphere at the Mauna Loa Observatory located at about 3400 meters elevation on the slopes of the mountain and above the atmospheric boundary layer on the island of Hawaii. Atmospheric carbon dioxide has risen continuously from that time to today, as documented by the Mauna Loa carbon dioxide measurements and measurements from other areas (Figure 14.7a). In the year 2009, the concentration of carbon dioxide in the atmosphere was approaching 387 ppmv. The annual growth rate of CO_2 in the atmosphere above Mauna Loa is shown in Figure 14.7b. Notice the pronounced fluctuations in the growth rate. These fluctuations are the result of a complex set of processes involving the oceans and terrestrial biosphere, volcanism, and changes in the anthropogenic

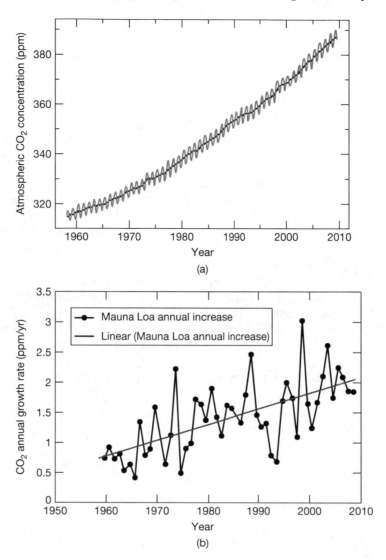

FIGURE 14.7 Record of measurements of atmospheric carbon dioxide between 1958 and 2008 at Mauna Loa Observatory, Hawaii. (a) Long-term trend (black line) and the oscillations in carbon dioxide which are seasonal cycles superimposed on the long-term increase caused by fossil fuel emissions and deforestation. The cycles represent the "breathing" of the Northern Hemisphere land biosphere—in summer carbon dioxide levels drop because of photosynthesis and in winter they rise because of respiration. (b) Annual growth rate of atmospheric CO_2 at Mauna Loa. The yearly averages are shown as well as the long-term trend line. Notice that the growth rate has been rising irregularly at a rate of 0.025 ppmv per year.   [*Source:* National Oceanic and Atmospheric Administration (NOAA) Climate Monitoring and Diagnostics Laboratory (CMDL), Carbon Cycle Greenhouse Gases, www.cmdl.noaa.gov/ccgg/figures/figures.html.)

emissions of CO_2 to the atmosphere. For example, during an El Niño event, the equatorial surface waters of the Pacific warm and there is decreased upwelling of cold, CO_2- and nutrient-rich waters in the equatorial region and along the coast of South America. Less

CO_2 is taken up by the ocean because of decreased phytoplankton productivity, and the warming of the surface waters, which cannot hold as much CO_2 as colder waters (this is the so-called *soda-fizz* effect). The regional net sea-to-air CO_2 flux decreases. Thus, during some, but not all, El Niño events, the annual growth rate of CO_2 in the atmosphere increases.

Within the last 300 years, carbon dioxide concentrations have risen by 107 ppmv, or about 38% (Figure 14.8). Most of this increase is due to the fossil fuel burning and deforestation activities of human society (see the temporal pattern of changes in these fluxes to the atmosphere in Figures 9.40 and 10.16). An amount of carbon equivalent to roughly half of the total carbon emitted to the atmosphere by fossil fuel burning has remained there. The rest has gone elsewhere, mainly into the oceans and the terrestrial environment of land plants and humus.

Fossil fuel burning is the major source of anthropogenic carbon dioxide for the atmosphere. Our use and reliance on this source of energy will impact the composition of the atmosphere into the twenty-first century and beyond. The burning of fossil fuels and cement manufacturing (about 2.5% of the total) in 2008 released about 8.7 billion tons of carbon as carbon dioxide into the atmosphere. Deforestation, a practice that has gone on at least since the beginning of slash-and-burn agriculture about

10,000 years ago, and other land use practices also release carbon dioxide into the atmosphere. It is estimated that for the year 2008, 1.2 billion tons of carbon per year were vented into the atmosphere as a result of deforestation practices, but the value of this flux is not well constrained. Carbon is released to the atmosphere because of biomass burning associated with the practice of deforestation. Carbon is no longer stored in the tree biomass and the organic matter of litter and soils. In addition, agricultural crops or pasturelands store far less carbon than do the temperate or tropical forests they replace. The disruption of the forest soils by plowing allows carbon, once held as organic carbon in root systems of trees and in soil, to be oxidized and to escape into the air as carbon dioxide. Even lumber derived from forest trees decays with time and releases carbon dioxide to the atmosphere.

The rate of rise of atmospheric CO_2 during the 1970s was 1.3 pppv/yr; for the 1980s, 1.6 ppmv/yr; and for the 1990s, 1.5 ppmv/yr. Following the eruption of Mt. Pinatubo in 1991, the rate of rise of atmospheric carbon dioxide slowed dramatically. By 1992, the growth rate had decreased by an amount

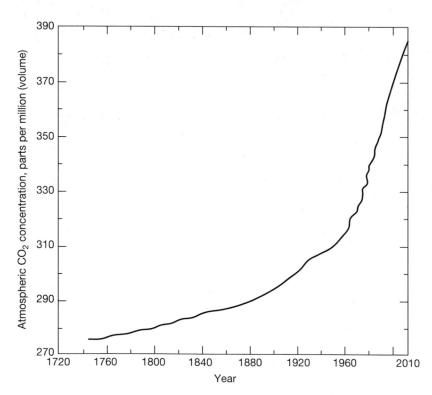

FIGURE 14.8 Atmospheric carbon dioxide concentrations over the last approximately 300 years from ice-core data and atmospheric measurements at Mauna Loa Observatory, Hawaii. [*Source:* Siegenthaler and Oeschger, 1987; Boden et al., 1991; Halpert and Ropelewski, 1993; National Oceanic and Atmospheric Administration (NOAA) Climate Monitoring and Diagnostics Laboratory (CMDL), Carbon Cycle Greenhouse Gases, www.cmdl.noaa.gov/ccgg/figures/figures.html.]

unprecedented in the 50-year record of atmospheric carbon dioxide recorded at the Mauna Loa Observatory in Hawaii (about 0.4 ppm per year). It is still not clear what processes caused the change. The favored candidates are the cooling effect of Mt. Pinatubo's sulfate aerosol plume that may have affected the balance between photosynthesis and respiration on land or cooled the ocean sufficiently to increase the uptake of carbon dioxide. Another mechanism is fertilization of ocean waters by iron from the volcanic eruption deposited on the sea surface. The iron acting as a nutrient might have stimulated plant production in the ocean, leading to removal of carbon dioxide from the atmosphere. However, in the first decade of the twenty-first century, the rate of rise of atmospheric CO_2 had increased to 2.0 pppm/yr and in 2008, the rate of rise was 1.79 ppmv/yr. (Figure 14.7b).

System Imbalance The modern carbon system of Earth was difficult to balance in terms of inputs and outputs in the late twentieth century. There was a problem as to where all of the carbon that had been emitted to the atmosphere by human activities was going. There has been some resolution of the problem in the past decade. Of the average flux of 9.1 billion tons of carbon that were released annually into the atmosphere for the period 2000 to 2007 by human activities, 4.2 billion tons per year remained in the atmosphere, leaving about 4.9 billion tons per year unaccounted for (Figure 14.5; Table 14.1). Some scientists have referred to this latter carbon as the "missing carbon" (Figure 14.5).

The surface waters of the oceans have proven to be a major sink of this "missing" anthropogenic carbon dioxide through absorption of the gas, although the strength of this sink varies yearly and its magnitude is difficult to determine accurately. However, such determinations have been made. Figure 14.9a shows the depth distribution of the anthropogenic CO_2 that has been absorbed by surface seawater and mixed to deeper depths in the ocean for oceanographic cruise sections in the Atlantic, Pacific, and Indian oceans based on observational data. Also shown in the figure is the total uptake of anthropogenic CO_2 integrated over the depth of the water column for the world's oceans (Figure 14.9b). Based on these data and modeling, the total uptake of CO_2 from 1800 to 1994 was estimated to be $118 +/- 19$ gigatons of carbon, accounting for 48% of the total fossil fuel and cement manufacturing CO_2 emissions during this period.

For the early twenty-first century, estimates vary as to how much CO_2 is being absorbed by the

oceans, but on the order of 2.3 billion tons of carbon per year appears to be a reasonable estimate (Figure 14.5). However, the oceans are a limited repository of carbon on a decadal to centurial time scale because of the slow mixing of surface water with deep water, Thus, it is very likely that some anthropogenic atmospheric carbon dioxide has been stored on land, especially since about 1950 (Figures 14.5 and 14.11, Table 14.1). This carbon probably resides in the enhanced carbon uptake related to the regrowth of forests in the Northern Hemisphere since the vast deforestation of these forests in the eighteenth through early twentieth centuries, for example, in Europe and North America. In addition, the forests of the Northern Hemisphere in particular have been stimulated by the additional carbon dioxide in the atmosphere, by the global warming of the past century, and by nitrogen (and perhaps phosphorus) added to the landscape by human activities. This fertilization of the forests appears to have led to increased productivity and terrestrial uptake of atmospheric carbon dioxide. Some investigators contend that the carbon being stored on land is in the undisturbed tropical forests of the world.

One experimental piece of evidence for the fertilization effect comes from chamber studies of the rates of photosynthesis of various plants under changing conditions of ambient CO_2 levels and temperature. Many of these studies have involved agricultural plants, such as cabbage, potato, soybean, cotton, and lemon trees, but some have dealt with nonagricultural plants. Figure 14.10 shows the effect of temperature and carbon dioxide levels on C_3 and C_4 plants. C_3 plants fix carbon through photosynthesis via the Calvin cycle that is employed by the majority (95% of known plant species) of higher plants, algae, and autotrophic bacteria. In the Calvin cycle of photosynthesis, the initial organic product synthesized is a 3-carbon chain organic acid. The process involves the enzymatic carboxylation (introduction of a carboxyl group, COOH, into a compound) of CO_2 by ribulose-1.5-bisphosphate (RuBP) carboxylase. The C_4 pathway of photosynthesis involves the initial formation of a 4-carbon chain organic acid. The C_4 plants are all angiosperms and comprise about 1% of all plant species. The C_4 plants evolved an anatomy and biochemistry that led to the concentration of CO_2 into the bundle sheath cells of leaves and subsequent assimilation of the carbon by the C_3 pathway. Because CO_2 concentrations are already elevated in C_4 leaves, these plants should respond less than C_3 plants to elevated CO_2 concentrations (see also Chapter 6, Box 2: The Photosynthetic

TABLE 14.1 Comparison of Various Estimates of the Sources and Sinks of Anthropogenic Carbon Dioxide[a] Over Time. Compare with Figure 14.5 for 2007 Figures. Fluxes are in Millions of Metric Tons of Carbon Per Year (MTC/yr)

	1980s			1990s			2000–2005	2000–2007
	TOTEM[b]	IPCC SAR[c]	IPCC TAR[d]	TOTEM[e]	IPCC TAR	SRLU-LUCF[f]	IPCC AR42[g]	Canadell et al.[h]
Atmospheric increase	3140	3300 ± 100	3300 ± 100	4800	3200 ± 100	3300 ± 100	4100 ± 100	4200
Fossil fuel/ cement emissions	5380	5500 ± 300	5400 ± 300	7000	6400 ± 400	6300 ± 400	7200 ± 300	7500
Net ocean-atmosphere	−1625	−2000 ± 500	−1900 ± 600	−2300	−1700 ± 500	−2300 ± 500	−2200 ± 500	−2300
Net land-atmosphere	−610	−200 ± 600	−200 ± 700	−220	−1400 ± 700	−700 ± 600	−900 ± 600	−1100
Land use change	1525	1600 ± 1000	1700 (600–2500)	1800	—	1600 ± 800	1600 (500–2700)	1500
Terrestrial uptake	−2135	−1800 ± 1600	−1900 (−3800–300)	1700	—	−2300 ± 1300	−2600 (−4300 to −900)	−2600

[a] Positive values are fluxes to the atmosphere; negative values are uptake fluxes from the atmosphere.

[b] Terrestrial Ocean Atmosphere Ecosystem Model.

[c] IPCC Second Assessment Report (Houghton et al., 1996).

[d] IPCC Third Assessment Report (Houghton and Yihui, 2001).

[e] Mackenzie, et al., 2001; flux estimates for 1999.

[f] IPCC Special Report on Land Use, Land-Use Change, and Forestry (SRLU-LUCF) (Watson et al., 2000).

[g] IPCC Fourth Assessment Report.

[h] Canadell et al., 2008 (http://www.globalcarbonproject.org/carbontrends/index.htm).

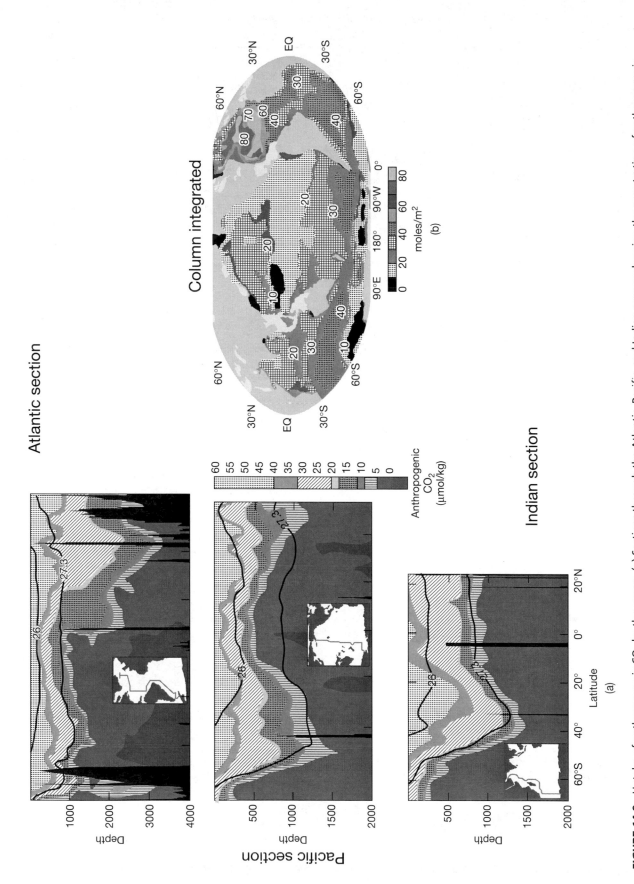

FIGURE 14.9 Uptake of anthropogenic CO_2 by the ocean. (a) Sections through the Atlantic, Pacific, and Indian oceans showing the penetration of anthropogenic CO_2 to depth in the oceans as concentrations in micromoles per kilogram (μmol/kg) of CO_2. Black lines represent salinity contours. (b) The amount of anthropogenic CO_2 held by the ocean shown as CO_2 integrated over ocean depth in moles of CO_2 per square meter of ocean surface (moles/m²). (*Source:* Sabine et al., 2004.)

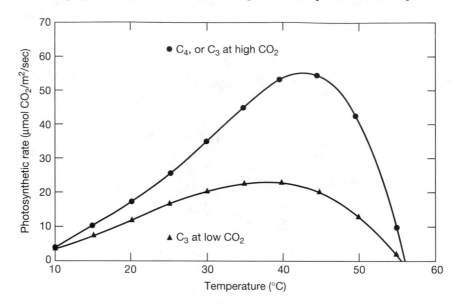

FIGURE 14.10 Photosynthetic rate versus temperature for the leaves of C_3 and C_4 plants. Notice in particular that the photosynthetic rates of the abundant C_3 plants are enhanced by increasing temperature and CO_2 levels. One micromol (μmol) equals 12 micrograms of CO_2. (*Source:* Allen and Amthor, 1995.)

Process). As shown in Figure 14.10, CO_2 enrichment in combination with higher temperatures enhances net photosynthesis in C_3 plants. Thus, the combination of rising atmospheric CO_2 concentration and temperature during the past century could have led to enhanced photosynthesis in C_3 plants. If the enhanced photosynthetic rates translated into an increase in the storage of carbon in plants, this fertilization process could act as a negative feedback on atmospheric CO_2 increase. It has been shown that the average growth response of forest plants to CO_2 enrichment alone is about 1.32. This is equivalent to a 32% increase in the total plant dry mass induced by elevated CO_2 concentrations. Excess nutrients added to the land surface from atmospheric deposition of anthropogenic nitrogen and from the application of nitrogenous and phosphorus fertilizers to croplands would also help to fertilize the terrestrial biomass, as well as any increase in precipitation.

The enhanced storage of carbon in forests also qualifies as a sink of anthropogenic carbon dioxide and was equivalent to about 2.6 billion tons of carbon per year in the early twenty-first century (Figure 14.5, Table 14.1). Thus, although land use practices of deforestation; conversion of forests to grasslands, pasture, and croplands; and urbanization lead to the release of carbon to the atmosphere from the land, it is also the case that some forests and their soils are storing carbon because of the fertilization effect.

Anthropogenic atmospheric carbon dioxide may also reside on land as organic carbon in the sediments of reservoirs, lakes, and in floodplains.

The problem of terrestrial uptake of anthropogenic atmospheric carbon dioxide is not resolved. The processes and fluxes related to this potential sink are still not well known. This is also true for the sources and fluxes due to land use changes (Figure 14.5, Table 14.1). However, the flux estimates are becoming more and more refined over time and the resolution of the CO_2 land-use source and fertilization sink problem is probably not too far in the future.

In summary, Figure 14.11 shows the modeled temporal trend from 1850 to 2008 for the anthropogenic fluxes of CO_2 to the atmosphere and the sinks of these emissions. For this period, the cumulative amounts of carbon released to the atmosphere by fossil fuel burning and land-use activities (mainly deforestation) were, respectively, about 340 billion metric tons and 210 billion metric tons. The atmosphere, ocean, and land gained, respectively, about 240, 150, and 160 billion tons of carbon, although during this whole period of time the land underwent an overall net loss of carbon from soils and plant biomass.

NITROUS OXIDE (N_2O) Nitrous oxide should not be confused with nitric oxide (NO) or nitrogen dioxide (NO_2). Neither NO nor NO_2 is a greenhouse gas,

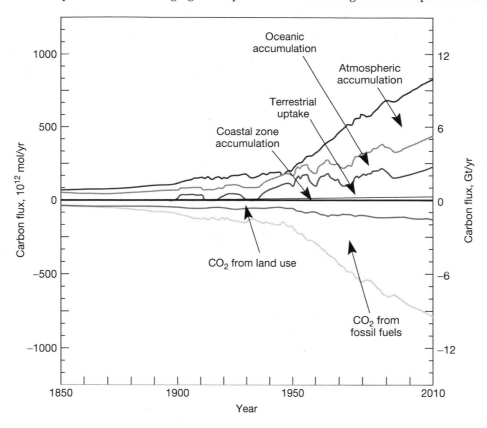

FIGURE 14.11   Sources and sinks of anthropogenic CO_2. The magnitudes of the sources are shown below the zero line and the magnitudes of the sinks above the line in units of 10^{12} mol C/yr and gigatons (Gt) of carbon per year. One mole of carbon = 12 grams. One gigaton = 1 billion tons = 10^{15} grams.   (Courtesy of Bouchez, Lerman, and Mackenzie, 2009.)

although both are involved in the formation of tropospheric ozone, which is a greenhouse gas. There are several natural and anthropogenic sources of nitrous oxide for the atmosphere. These fluxes have been difficult to quantify. Figure 14.12 shows one interpretation of the global biogeochemical cycle of nitrous oxide involving transfers between Earth's surface and the atmosphere. Nitrous oxide is a naturally occurring gas that is produced by microbial activity in soils and the ocean (see Chapter 7). After production, the nitrous oxide gas is vented to the atmosphere. The anthropogenic global nitrous oxide flux to the atmosphere is due to the combustion of fossil fuels, biomass burning, industrial processes like the production of nitric acid, the application of fertilizers to agricultural crops, and other activities of human society. Some of the nitrogen used in the fertilizers of urea and ammonium nitrate is converted by bacteria in the soils to nitrous oxide and then released to the atmosphere. It is estimated that as much as 4 million tons of nitrogen as nitrous oxide

may be added to the atmosphere annually because of agricultural activities.

The nitrous oxide concentration in the atmosphere has increased nearly linearly since the late 1970s owing to these releases (Figure 14.13). Its concentration in the late pre-industrial atmosphere was 275 ppbv and in 2008 it was 322 ppbv, an increase of about 17% owing mainly to human activities. Nitrous oxide contributes to the enhanced greenhouse effect in the same way as carbon dioxide. It captures reradiated infrared radiation from Earth's surface and warms the surface and lower troposphere. Nitrous oxide is responsible for about 5% of the enhanced greenhouse effect. The gas is chemically inert in the troposphere (its lifetime is about 120 years) and eventually makes its way into the stratosphere, where it goes through a chain of photochemical reactions that decompose nitrous oxide and lead to the destruction of stratospheric ozone (O_3) (Figure 14.12). Nitrous oxide has become the largest stratospheric ozone-depleting substance emitted through human activities, and is

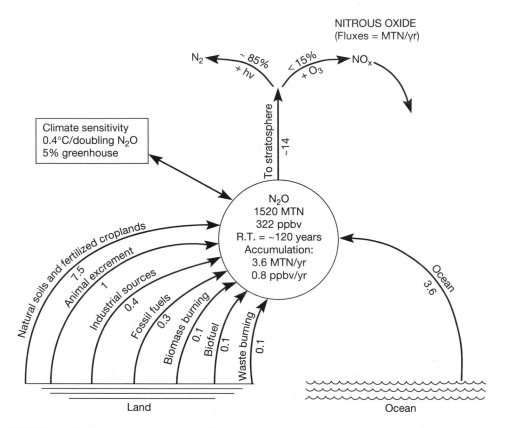

FIGURE 14.12 Non–steady state global biogeochemical cycle of nitrous oxide in the early twenty-first century. The major processes and fluxes involved in transfer of nitrogen as nitrous oxide between the surface of Earth and its atmosphere are shown. Fluxes are in millions of tons of nitrogen (MTN) per year, and the reservoir size of nitrous oxide is in millions of tons of nitrogen. Five percent of the enhanced greenhouse effect over the past couple of centuries is due to nitrous oxide. A doubling of atmospheric N_2O concentration could lead to a 0.4°C increase in temperature. (*Source:* Mackenzie, 1995, 1997 with additional information from Houghton et al., 1996, 2001; IPCC, 2007.)

expected to remain the largest throughout the twenty-first century.

METHANE (CH_4) Methane is a very effective greenhouse gas with a relatively short residence time in the atmosphere of about eight years. Its atmospheric concentration is much less than that of carbon dioxide, but it is 21 times more effective than carbon dioxide at trapping infrared radiation. Sixteen percent of the enhanced greenhouse effect is attributed to methane gas. Another 2% of the enhanced greenhouse effect is due to conversion of methane in the stratosphere to water vapor. The global biogeochemical cycle of methane is shown in Figure 14.14. Fluxes in the cycle are poorly known, and therefore the cycle is difficult to balance. It is estimated that as much as 60% or more of the total flux of methane to the atmosphere is from activities related to human society. These include emissions from enteric fermentation

processes in domesticated livestock, from cultivated rice paddies, from fossil fuel and biomass burning, and from landfills. Methane is derived from the decay of plant material by anaerobic bacteria in the stomachs of cows and sheep, the guts of termites, and from oxygen-depleted, wet soils of swamps and rice paddies. Belching and flatulating cattle produce about one-half pound of methane per cow per day. Termites rely on the bacteria in their stomachs for digestion and produce methane as a by-product of their digestive processes. Rice paddies transfer methane from their anaerobic soils to the air through hollow tubes in rice plants. It is estimated that the increase in the number of domesticated cattle and cultivated rice paddies may contribute as much as 50% of the methane released to the atmosphere. Burning of wood and vegetation also releases methane to the atmosphere. Smaller amounts are released in the mining of coal, oil, and gas.

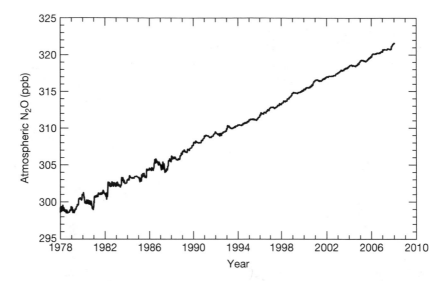

FIGURE 14.13 Atmospheric nitrous oxide levels from 1978 to 2008.   (*Source:* Hielman, 1989; Boden et al., 1991; Prinn et al., 1990; Houghton et al., 1996, 2001; http://www .esrl.noaa.gov/gmd/aggi/.)

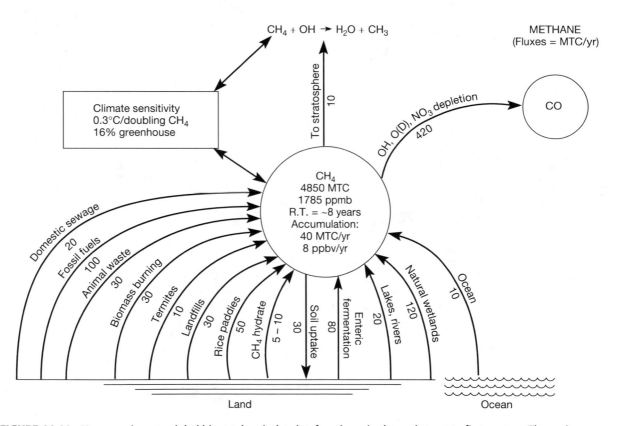

FIGURE 14.14   Non–steady state global biogeochemical cycle of methane in the early twenty-first century. The major processes and fluxes involving transfer of carbon as methane between the surface of Earth and its atmosphere are shown. Fluxes are in millions of tons of carbon per year, and the reservoir size of methane is in millions of tons of carbon. Sixteen percent of the enhanced greenhouse effect within the past couple of centuries is due to methane. A doubling of atmospheric CH_4 concentration could lead to a 0.3°C increase in temperature   (*Source:* Mackenzie, 1995, 1997; with additional information from Houghton et al., 1996, 2001; IPCC, 2007.)

Methane has significantly increased in concentration in the atmosphere because of human activities (Figure 14.15a). In the last 200 years, atmospheric methane has more than doubled in concentration. Its concentration in the atmosphere was 730 ppbv in late pre-industrial time and has risen to 1785 ppbv in 2008. This is an increase of 145%, mainly due to anthropogenic activities. Methane was accumulating in the atmosphere at a rate of about 34 million tons of carbon per year, or at a little less than 1% annually, in the late 1970s. The rate of increase of methane in the atmosphere has slowed in recent decades. Methane was growing at 20 ppbv per year in the late 1970s and at 13.3 ppbv per year in 1983, but its rate of growth

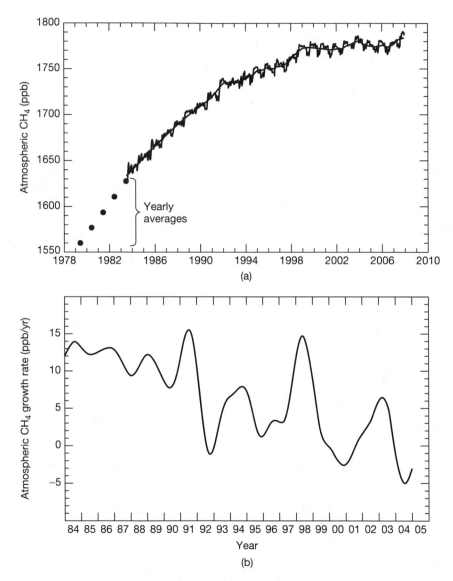

(a)

(b)

FIGURE 14.15 Atmospheric methane levels from 1978 to 2008. (a) The long-term trend (black line) and the oscillations which are related to ENSO events, changes in seasonal wetland emissions of CH_4, and quasi-biennial oscillations in meteorological conditions. The rate of rise in the concentration of methane has decreased in recent years. (b) Atmospheric growth rate of methane in ppbv per year showing the general overall reduction in growth rate since the mid-1980s. (*Source:* Hielman, 1989; Boden et al., 1991; Halpert and Ropelewski, 1993; Houghton et al., 1996; National Oceanic and Atmospheric Administration (NOAA) Climate Monitoring and Diagnostics Laboratory (CMDL), Carbon Cycle Greenhouse Gases, www.cmdl.noaa.gov/ccgg/figures/figures.html.)

slowed to 9.5 ppbv in 1990, and around the middle of 1992, methane concentrations briefly stopped growing. In the early twenty-first century, there has been a near-zero growth rate, ranging from an annual 0.3 % increase to a 0.2% decrease (Figure 14.15b). The reasons for these changes in growth rate are not completely known. Perhaps the total flux of methane to the atmosphere has decreased or the rate of oxidation of methane in the atmosphere to carbon monoxide has increased. Although present opinion is divided, it appears that the slowing in the growth rate of atmospheric methane is in part due to decreased fluxes from rice cultivation and domesticated livestock. Recent efforts to reduce natural gas leaks of methane, particularly in the former USSR, might have had an effect on the rate of growth of atmospheric methane, as did the 1991 eruption of Mt. Pinatubo. Whatever the case, since a great deal of the methane flux to the atmosphere can be attributable to human activities, the slowing of the rate of rise, although it may not continue, implies that controlling human emissions could have a significant impact on future concentrations because of the short lifetime of methane in the atmosphere.

Methane is stored in cold environments of peat bogs in tundra biomes and in methane hydrates (frozen methane-ice compounds, clathrates) found in permafrost regions and in sediments beneath the sea of the margins of the continents. At the temperature and pressure conditions under which the methane hydrates are found, they are generally stable (Figure 14.16). The amount of methane sequestered as methane hydrates is difficult to estimate but is on the order of the equivalent of 1000 billion to 100,000 billion tons of carbon. In Arctic permafrost regions alone, the amount of carbon in methane of gas hydrates is estimated at about 8 billion to 18,000 billion tons. If Earth's climate were to continue to warm, it is possible that the methane hydrates, particularly in the Arctic where the warming is projected to be greatest, would be destabilized (Figure 14.16). Methane would then be released to the atmosphere from the resultant breakdown of these materials. This event would be a positive feedback to any warming resulting from an enhanced greenhouse effect. The positive feedback strengthens the initial climatic disturbance brought about by the enhanced greenhouse effect. There is already evidence of melting permafrost, due to Arctic warming, and increased emissions of CH_4 to the atmosphere. However, it is likely that the positive feedback on the global climate of methane hydrate destabilization in Arctic marine sediments on a decadal to century time scale will be minimal, unless Arctic temperatures rise quite dramatically.

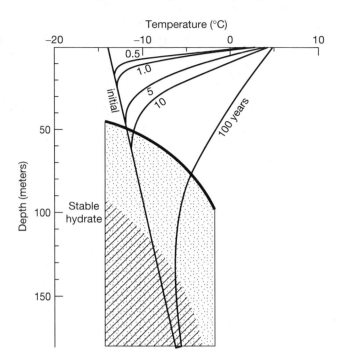

FIGURE 14.16 A temperature–depth diagram illustrating the potential of gas release from CH_4 hydrate. The initial temperature distribution with depth is shown as a straight line. The stippled area shows the depth–temperature relations within which hydrate under lithostatic pressure is stable (see Chapter 3); the top of the stability region lies deeper under hydrostatic pressure (cross-hatched area). The diagram illustrates the effect of a significant warming from $-14°C$ to $+5°C$, and the number of years for the warming pulse to reach the hydrate zone and destabilize it, releasing gas to the ocean–atmosphere system. (*Source:* Nisbet, 1990.)

HALOCARBONS Halocarbons are carbon-based compounds that contain chlorine, fluorine, bromine, or iodine. Eleven percent of the enhanced greenhouse effect is attributed to halocarbons. The compounds that contain only carbon, chlorine, and fluorine are called chlorofluorocarbons and include the freons CFC-11 (CCl_3F) and CFC-12 (CCl_2F_2). The CFCs, halons (bromine-containing CFCs), and hydrofluorocarbons (HFCs), as well as methylchloroform (CH_3CCl_3) and carbon tetrachloride (CCl_4), are exclusively of industrial origin and did not exist prior to 80 years ago. The halocarbons are responsible for the environmental problem of stratospheric ozone depletion (see below), and several of the compounds are strong greenhouse gases. The most publicized of these compounds are the CFCs that were used as coolants in refrigeration and air conditioners, as propellants in spray cans and other devices, in the manufacturing of plastic foam used in cups and similar products, and as solvents for industrial purposes. The CFCs are far less abundant in the atmosphere than carbon dioxide, but they are 10,000 times more powerful as greenhouse gases and are long-lived atmospheric gases. CFC-11 remains in the atmosphere for 45 years, and CFC-12 has a 100-year life span. Figure 14.17 shows the annual global releases of CFC-11 and CFC-12 to the atmosphere from 1960 to 2005. Since 1960, 8.8 million metric tons and 13.2 million metric tons of CFC-11 and CFC-12, respectively, have been emitted into the atmosphere. Most of the emissions came from the temperate counties of the Northern Hemisphere.

Because of the releases of CFC-11 and CFC-12 to the atmosphere from industrial activities, the concentrations of these synthetic gases in the atmosphere were steadily increasing until the 1990s (Figure 14.18a). However, since the Montreal Protocol of 1987 (see below) and its further amendments restricting the production of chlorofluorocarbons, the rate of increase in the tropospheric concentrations of CFC-11 and CFC-12 has declined dramatically. CFC-11 in 1995 was actually decreasing at a rate of -0.6 pptv (parts per trillion by volume) per year, down from a maximum growth rate of 12 pptv per year in the mid 1980s. In the same year, CFC-12 had a yearly growth rate of 5.9 pptv, down from a maximum rate of 22 pptv annually in 1986–1987. It appears that CFC-11 reached its peak atmospheric concentration in 1994 of about 276 pptv, and CFC-12 concentrations have also begun to decline (Figure 14.18a). Figure 14.18b illustrates the dramatic reduction in carbon tetrachloride (CCl_4) and methychloroform (CH_3CCl_3) atmospheric concentrations since the early 1990s due to the restriction in the uses of these gases resulting from the implementation of the Montreal Protocol. Carbon tetrachloride has declined at a nearly steady rate of 1% per year since the mid-1990s as its use as a feedstock for producing CFCs decreased, and methylchloroform has declined nearly exponentially since 1998, consistent with small emissions of this gas from its use as an industrial solvent. The halocarbons are discussed further in the section on stratospheric ozone depletion.

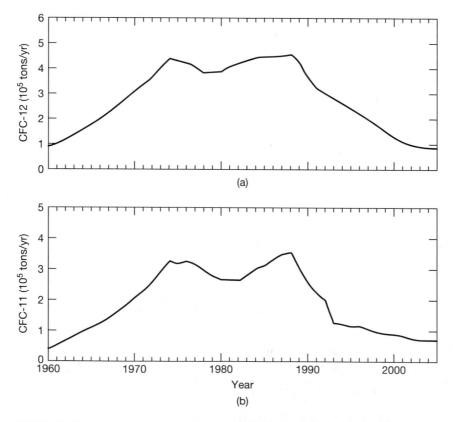

FIGURE 14.17 Annual global emissions of the halocarbons CFC-12 (a) and CFC-11 (b) to the atmosphere from 1972 to 2005, principally from their use as refrigerants and agents for the blowing of foam. Notice the decline in emissions following enactment of the Montreal Protocol in 1987. (*Source:* WMO, 2007.)

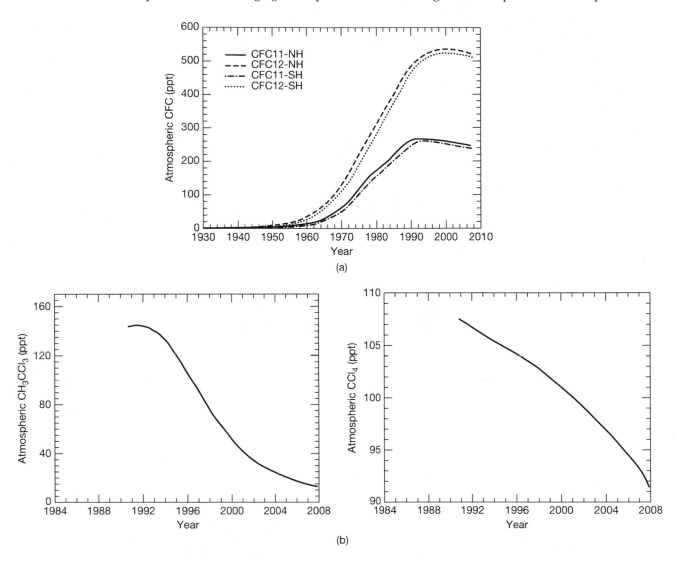

FIGURE 14.18 Concentrations of halocarbons in the atmosphere. (a) Concentrations of chlorofluorocarbon-11 and chlorofluorocarbon-12 in the atmosphere from 1940 to 2008. Notice the higher concentrations in the Northern Hemisphere (NH)—the source of major emissions of CFCs to the atmosphere from human activities before the Montreal Protocol—relative to the Southern Hemisphere (SH), and the decrease in the rate of growth of the two gases in the atmosphere since the late 1980s. (b) Concentrations of methyl chloroform (CH_3CCl_3) and carbon tetrachloride (CCl_4) in the atmosphere from 1990 to 2008. Notice the rapidly declining concentrations due to regulations limiting these emissions.   [*Source:* National Oceanic and Atmospheric Administration (NOAA) Climate Monitoring and Diagnostics Laboratory (CMDL), Carbon Cycle Greenhouse Gases, www.cmdl.noaa.gov/ccgg/figures/figures.html.]

TROPOSPHERIC OZONE (O_3)   The natural sources of tropospheric ozone are electrical storms, decomposing plants, and forest fires. As discussed in Chapter 12, tropospheric ozone, as distinguished from stratospheric ozone, is produced in the lower troposphere as a result of complex chemical reactions involving precursor nitrogen oxides and volatile organic compounds (VOCs) released from natural soil microbial processes and as terpenes from trees and from the burning of fossil fuels and biomass. Thus, tropospheric ozone is generally found at higher concentrations near the surface of Earth. Changes in tropospheric ozone concentrations are variable, both spatially and vertically, in the atmosphere. In the Northern Hemisphere, there is evidence for increases in tropospheric ozone concentrations since 1900. Model calculations and observations suggest that tropospheric ozone concentrations may have increased about 25 ppbv since late pre-industrial times, representing a doubling of concentration in the Northern Hemisphere.

In the Southern Hemisphere, the data are insufficient to determine any trend in tropospheric ozone concentrations. An exception is the South Pole, where a decrease in ozone concentration has been observed since the mid-1980s. Ozone levels are as low as 20 ppbv in the austral spring and summer. Twelve percent of the enhanced greenhouse effect is attributed to tropospheric ozone.

There are strong seasonal variations in tropospheric ozone (Figure 14.19; see http://www.jpl.nasa.gov/index.cfm for information on ozone measurements using instruments on board the Aura satellite). During the summertime in the Northern Hemisphere, on average, tropospheric ozone levels are relatively high while in the Southern Hemisphere, the concentrations are low (Figure 14.19a). The contrary is true in the fall months. In general during the winter months, global tropospheric ozone levels are low worldwide (Figure 14.19b). The pattern of seasonal change is due to a complex interplay between stratospheric and tropospheric exchange of ozone, enhanced convectional processes in the atmosphere of the Pacific region, lightning in both Africa and South America, industrial pollution in the extra-tropical regions, particularly in the Northern Hemisphere, and biomass burning in the tropics. In the summer

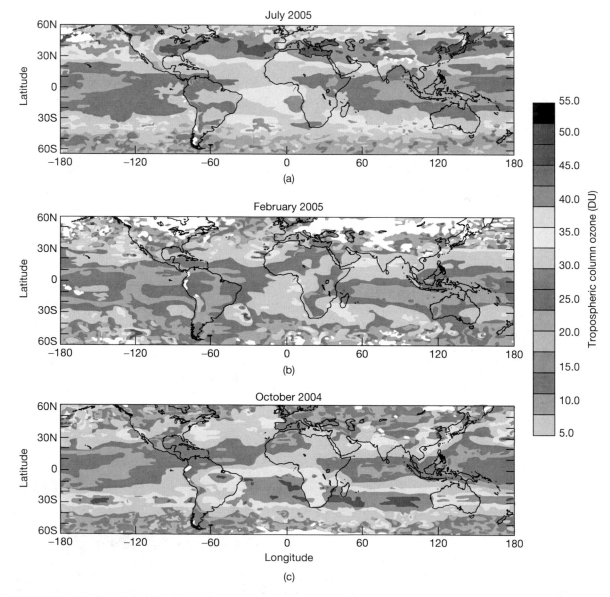

FIGURE 14.19 Global distribution of the concentrations of tropospheric ozone in Dobson Units (DU) in the summer (a), winter (b), and fall (c). (*Source:* Ziemke, et al., 2006.)

months, the enhancement of ozone concentrations found in the atmosphere of the subtropics and mid-latitudes of the Northern Hemisphere is due to photochemical production of ozone from both anthropogenic pollution and biogenic volatile organic compounds (VOCs) and NO_x gases. The high values of ozone in the austral spring extending from the equator to 30 to 40° south latitude (Figure 14.19c) along all longitudes are in part a result of biomass burning in the Southern Hemisphere at this time. The tropical Pacific and Antarctic regions have substantially lower ozone levels due to low levels of ozone-producing sources and in the case of the Antarctic, only a small influence of exchange of air between the troposphere and stratosphere.

Within many large urban cities and their vicinities, tropospheric ozone is a pollutant that is capable of forming a thick smog. Increases in tropospheric ozone have the potential to increase temperature directly because it is a greenhouse gas. Furthermore, because ozone is toxic, it may decrease primary production and potentially lead to increasing atmospheric carbon dioxide levels because of less uptake of carbon in the biota.

WATER VAPOR (H_2O) AND CLOUDS Water vapor in the atmosphere is the principal greenhouse gas. Because the amount of water vapor in the atmosphere is dependent on the temperature of the planet, any initial warming of the planet will probably lead to more water vapor in the atmosphere. The increased water vapor would absorb more infrared radiation reradiated from the planetary surface and thus lead to further warming. This is another example of a positive feedback. However, it has been pointed out that there are ways in which water vapor in the troposphere can produce a negative feedback directly by increasing CO_2 concentrations, producing a drier upper troposphere, or indirectly though reduced cirrus cloud coverage.

Water droplets form in the atmosphere by condensation of water on aerosol particles. These particles may absorb or reflect radiant energy. The amount of water vapor in part determines the types and distribution of clouds that form in the atmosphere. To predict the effects of increasing concentrations of greenhouse gases in the atmosphere on temperature and other climatic variables, general circulation models (GCMs) and other types of models have been used. The GCMs are very complex computer representations of the atmosphere (AGCM), ocean (OGCM), or atmosphere–ocean system (AOGCM) that are used in the modeling of global change. These models are key components of Global Climate Models that include

sea-ice and land-surface components and other factors. The effects of clouds on the radiant energy budget of the planet in these models represent a major source of uncertainty in attempting to predict future climatic change.

Clouds regulate the radiative heating of the planet (see Chapter 4). They reflect a significant part of the incoming solar radiation. Clouds also absorb long-wave, infrared radiation emitted by a warm Earth. At the cold tops of clouds, energy is emitted to space. In 1984 the Earth Radiation Budget Experiment (ERBE) was launched. This experiment involves a system of three satellites employed to provide data on incoming and outgoing radiation. One result of this experiment so far is the demonstration that clouds have a net cooling effect on Earth. On a global scale, clouds reduce the amount of radiative heating of the planet by -13.2 to -20 watts per square meter. One watt is equal to 14.3 calories of heat energy per minute. The energy associated with cloud cooling is a large number when compared to the $+2.99$ watts per square meter attributed to the modern increases in atmospheric greenhouse gases. It is also large compared to the radiative heating that could arise from a doubling of atmospheric carbon dioxide concentrations over late pre-industrial levels during the twenty-first century. This value is about $+4$ watts per square meter. Thus, a small change in the types and distribution of clouds may have a large effect on the radiation budget compared to the direct effect of changing greenhouse gas composition owing to human activities.

In a world initially warmed by greenhouse gases released from human activities, it is difficult to predict what will happen to the types and distribution of clouds. Too little is known about the complex, small spatial and temporal scale physical processes of cloud formation and, in particular, how they will respond to a warming Earth. Also, these complex processes are difficult to simulate in the global circulation models and are parameterized and dealt with empirically, that is, using recipes for how the average characteristics of clouds vary with local temperature and water vapor content. Clouds may act as a positive or negative feedback in a future Earth warmed by the enhanced greenhouse effect. This ambiguity accounts in part for the range of *best* estimates of the average temperature increase predicted from the global circulation models for various greenhouse gas emissions scenarios and atmospheric concentrations for the year 2100. The range is $+1.8$ to $+4.2$°C.

Before concluding this section, it is infomative to consider briefly some further problems related to clouds. One is the connection between the galactic cosmic ray flux, cloud formation, and climate (Figure 14.20). Galactic cosmic rays are highly energetic particles, mainly protons and alpha particles that enter the solar system from outside. The magnetic fields of the Earth and sun affect the path of the particles. The galactic cosmic ray flux to Earth depends on variations in solar activity and the intensity of the flux is modulated by the strength of the solar wind and its magnetic field effect on the cosmic rays. It has been argued that variations in solar activity are responsible for varying solar wind strength. A stronger wind reduces the flux of cosmic rays reaching the Earth, and a weaker wind does the opposite. Since cosmic rays are responsible for ionization of particles in the troposphere, an increase in solar activity could result in a reduction in tropospheric ionization rate, and in formation of cloud condensation nuclei, and hence a reduction in cloud cover. If the Earth's blanket of low altitude clouds, which have a net cooling effect on climate, is reduced, there could be a net warming effect. Presumably a decrease in solar activity would lead to a reverse pattern of change and cooling of the planet. Some in the scientific community contend that solar wind variablity is the primary cause of climate change on a decadal time scale and not rising atmospheric greenhouse gas concentrations. This conclusion has been strongly criticized.

The second problem has to do with the relative effects of water vapor and cloud formation on the climate. There is some observational evidence supporting the conclusion that atmospheric water vapor has been increasing along the lines predicted by calculations from GCMs. This *positive* water vapor feedback produces roughly twice the global temperature rise of that obtained from an anthropogenically enhanced greenhouse gas effect alone. However, water is actually involved in two feedback processes: the first through water vapor and the second through cloud formation. Roy W. Spencer, a Principal Research Scientist at the University of Alabama in Huntsville, Alabama, finds that cloud cover changes produce a *negative* feedback on the climate system; that is, they tend to cool the climate. When he adds the two water feedbacks together, Spencer finds that the total feedback from warming-induced changes of water in the atmosphere is *negative;* that is, the effect of cloud formation dominates the water vapor effect. Although this conclusion is not generally accepted at the moment, it does point to the fact that we need to know both the effect of water vapor and that of cloud cover on the climate system to deduce its behavior in the future.

World Distribution of Human Sources of Greenhouse Gases

Human sources of greenhouse gases that may affect our world environment come from fossil fuel burning, rice paddy cultivation, deforestation, agricultural practices, biomass burning, fertilizer consumption, and use of synthetic materials such as CFCs. The gases from these sources continue to increase in concentration in the atmosphere today. Table 14.2 summarizes some characteristics of atmospheric gases that have sources related to human activities. Notice

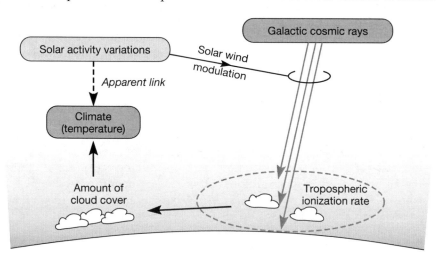

FIGURE 14.20 The possible linkages between the flux of galactic cosmic rays, variations in solar activity, and the development of cloud cover that affects the climate of Earth. (From "Cosmic Rays and Climate" by Nir J. Shaviv http://www.sciencebits .com/CosmicRaysClimate. © Nir J. Shaviv. Used by permission of the author.)

TABLE 14.2 Atmospheric Trace Gases that have Sources Related to Human Activities and are of Significances to Global Environmental Change

	Carbon Dioxide CO_2	Methane CH_4	Nitrous Oxide N_2O	Chlorofluoro-carbons CFCs	Tropospheric Ozone O_3	Carbon Monoxide CO	Water Vapor H_2O
Greenhouse role	Heating	Heating	Heating	Heating	Heating	None	Heats in air; cools in clouds
Effect on stratospheric ozone layer	Can increase or decrease	Can increase or decrease	Can increase or decrease	Decrease	None	None	Decrease
Principal anthropogenic sources	Fossil fuels; deforestation	Rice culture; cattle; fossil fuels; biomass burning	Fertilizer; land use conversion	Refrigerants aerosols; industrial processes	Hydrocarbons (with NO_x); biomass burning	Fossil fuels; biomass burning; deforestation	Land conversion; irrigation
Principal natural sources	Respiration balanced in nature	Wetlands	Soils; tropical forests	None	Hydrocarbons	Hydrocarbon oxidation	Evapotranspiration
Atmospheric lifetime	50 to 200 years	8 years	120 years	60 to 100 years	Weeks to months	Months	Days
Present atmospheric concentration in parts per billion by volume at surface	387,000	1785	322	CFC-11: ~0.240 CFC-12: ~0.520	20 to 40	100	3000–6000 in stratosphere; ~10,000 in troposphere
Preindustrial concentration (1750 to 1800) at surface	280,000	790	288	0	10	40 to 80	Unknown
Annual rate of increase (early twenty-first century)	0.57%	0.45%	0.25%	CFCs: decreasing	0.5 to 2.0%	0.7 to 1.0%	Poorly known but probably increasing
Relative contribution to the anthropogenic greenhouse effect (total forcing, including tropospheric ozone = 2.99 W/m^2)	56%	16%	5%	11%	12%	None	2.3% due to oxidation of CH_4 in the stratosphere[1]

[1]Tropospheric water vapor positive feedback amounts to +1.8 W/m^2.

(*Source:* Graedel and Crutzen, 1993; UCAR/OIES, 1991.)

that for the greenhouse gases, carbon dioxide is responsible for nearly 60% of the enhanced greenhouse effect. Methane, nitrous oxide, chlorofluorocarbons, and tropospheric ozone are the culprits for the remainder of the effect.

The sources of greenhouse gases are distributed throughout the world in both developed and developing countries. In 2008, global emissions of carbon dioxide, the principal greenhouse gas, amounted to nearly 10 billion tons of carbon per year. This estimate includes emissions from land use activities. Eighty-two percent of these emissions was related to the industrial emissions of burning of fossil fuels, cement manufacturing, and flaring of natural gas (Figure 14.21a). Industrial emissions in 2009 are predicted to drop by 2.6%, the largest decline in 40 years, because of the global financial crisis of 2008 and 2009. The global per capita industrial emissions were 1.22 metric tons of carbon (4.47 metric tons of CO_2) annually in 2008 (Figure 14.21b). In 2006, the United

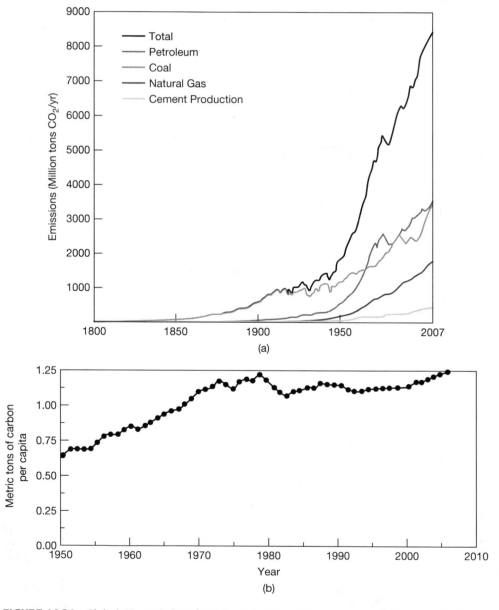

FIGURE 14.21 Global CO_2 emissions from fossil fuel burning, cement production, and flaring of gas from 1800 to 2007 (a) and global per capita CO_2 emission estimates from 1950 to 2007 (b). In 2008, the per capita emission estimate was 1.22 metric tons of carbon. (*Source:* http://cdiac.esd.ornl.gov/trends/emis/glo.htm.)

States industrial emissions of carbon dioxide per capita were 19 metric tons of CO_2 per person, ranking ninth in the world, following Qatar (56.2 tons per capita), United Arab Emirates (32.8 tons per capita), Kuwait (31.2 tons per capita), Bahrain (28.8 tons per capita), Trinidad and Tobago (25.3 tons per capita), Luxembourg (24.5 tons per capita), Netherlands Antilles (22.8 tons per capita), and Aruba (22.3 tons

per capita). Six countries—mainland China, United States, Russia, India, Japan, and Germany—were responsible for about 58% of the total industrial emissions to the atmosphere in 2006 (Figure 14.22a; Table 14.3). The United States, China, Russian Federation, Germany, and Japan from 1960 to 2005 have contributed slightly more than half of the world total industrial emissions of the atmosphere (Figure 14.22b).

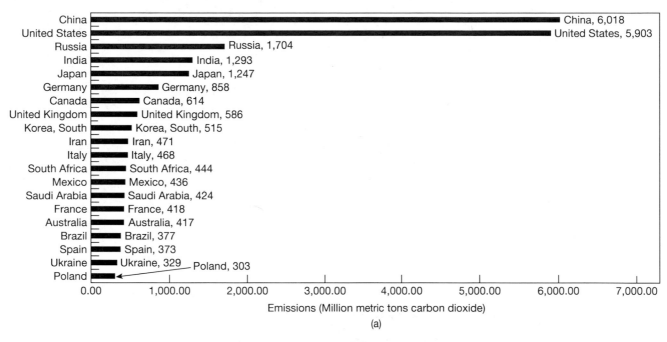

(a)

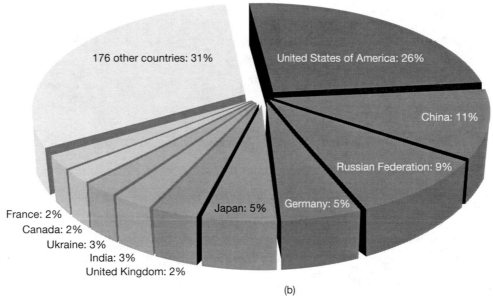

(b)

FIGURE 14.22 Twenty top countries in CO_2 emissions in 2006 (a). Cumulative CO_2 emissions from 1960 to 2005 in percentage of world total emissions (b). (From Union of Concerned Scientists, http://www.ucsusa.org/global_warming/science_and_impacts/science/graph-showing-each-countrys.html. Copyright © 2009 by Union of Concerned Scientists. Used by permission.)

TABLE 14.3 Ranking of Countries in 2006 in Terms of Industrial CO_2 Emissions in Thousands of Metric Tons of CO_2. Also Shown are CO_2 Emissions Per Capita.

Rank	Country	Annual CO_2 Emissions (in Thousands of Metric Tons)	Percentage of Total Emissions	CO_2 Emissions Per Capita (Metric Tons Per Person)
—	World	28,431,741	100.0	4.34
1	China	6,103,493	21.5	4.6
2	United States	5,752,289	20.2	19
3	Russia	1,564,669	5.5	10.9
4	India	1,510,351	5.3	1.3
5	Japan	1,293,409	4.6	10.1
6	Germany	805,090	2.8	9.7
7	United Kingdom	568,520	2.0	9.4
8	Canada	544,680	1.9	16.7
9	South Korea	475,248	1.7	9.9
10	Italy	474,148	1.7	8.1
11	Iran	466,976	1.6	6.6
12	Mexico	436,150	1.6	4.1
13	South Africa	414,649	1.5	8.6
14	France	383,148	1.4	6.2
15	Saudi Arabia	381,564	1.3	15.8
16	Australia	372,013	1.3	18.1
17	Brazil	352,524	1.2	1.9
18	Spain	352,235	1.2	8
19	Indonesia	333,483	1.2	1.5
20	Ukraine	319,158	1.1	6.9
21	Poland	318,219	1.1	8.3
22	Thailand	272,521	1.0	4.3
23	Turkey	269,452	1.0	3.6
24	Kazakhstan	193,508	0.7	12.6
25	Malaysia	187,865	0.7	7.2
26	Argentina	173,536	0.6	4.4
27	Venezuela	171,593	0.6	6.3
28	Netherlands	168,513	0.6	10.3
29	Egypt	166,800	0.6	2.2
30	Pakistan	142,659	0.5	0.9
31	United Arab Emirates	139,553	0.5	32.8

Rank	Country	Annual CO_2 Emissions (in Thousands of Metric Tons)	Percentage of Total Emissions	CO_2 Emissions Per Capita (Metric Tons Per Person)
32	Algeria	132,715	0.5	4
33	Uzbekistan	115,672	0.4	4.3
34	Czech Republic	116,991	0.4	11.3
35	Belgium	107,199	0.4	10.3
36	Vietnam	106,132	0.4	1.2
37	Romania	98,490	0.4	4.6
38	Nigeria	97,262	0.3	0.7
39	Greece	96,382	0.3	8.7
40	Iraq	92,572	0.3	3.2
41	Kuwait	86,599	0.3	31.2
42	North Korea	79,111	0.3	3.6
43	Austria	71,834	0.3	8.6
44	Israel	70,440	0.3	10.3
45	Belarus	68,849	0.2	7.1
46	Syria	68,460	0.2	3.5
47	Philippines	68,328	0.2	0.8
48	Finland	66,693	0.2	12.7
49	Colombia	63,422	0.2	1.4
50	Chile	60,100	0.2	3.7
51	Portugal	60,001	0.2	5.7
52	Hungary	57,644	0.2	5.7
53	Singapore	56,217	0.2	12.8
54	Libya	55,495	0.2	9.2
55	Denmark	53,944	0.2	9.9
56	Serbia and Montenegro	53,266	0.2	5.1
57	Sweden	50,875	0.2	5.6
58	Bulgaria	48,085	0.2	6.3
59	Qatar	46,193	0.2	56.2
60	Morocco	45,316	0.2	1.5

Source: Carbon Dioxide Information Analysis Center (CDIAC, 2007)

In contrast, carbon dioxide emissions from Brazil and several other tropical developing countries are mainly due to the massive deforestation that is occurring in the tropical forests of these countries (see Chapter 10; http://www.edf.org/documents/ 4930_tropicaldeforestation_and_climatechange.pdf). Deforestation also releases significant quantities of methane and nitrous oxide to the atmosphere. On average for the early twenty-first century, 1.5 billion tons of carbon per year were emitted to the atmosphere because of changes in tropical land use (Figure 10.16). This CO_2 flux was accompanied by 275 million tons of CH_4 and 5 million tons of N_2O in the year 2000. The greenhouse gas emissions from deforestation may account for nearly a quarter of the enhanced greenhouse effect.

Tropical South and Central America, tropical Asia, and tropical Africa were the source of 40%, 40%, and 20%, respectively, of the tropical forest deforestation CO_2 emissions, involving 13 million hectares of deforestation annually. Carbon dioxide emissions from deforestation for some tropical countries are the principal source of anthropogenic CO_2 emissions to the atmosphere. For example, in the early twenty-first century, about 60% of the total CO_2 emissions for Brazil was due to deforestation. About 70% of the total carbon dioxide emissions from the Democratic Republic of the Congo was due to deforestation and 80% from Indonesia was from destruction of peat lands and deforestation. However, keep in mind that carbon dioxide emissions data related to land use activities are less well constrained than those for fossil fuel burning and cement manufacturing.

Although not well quantified, annual global anthropogenic methane emissions in the early twenty-first century were equivalent to about 340 million tons of carbon (Figure 14.14). Fossil fuel mining, processing, and use contributed about 29% of global emissions, followed by livestock flatulation and belching, 24%, rice cultivation, 15%, landfills, animal waste, and biomass burning, about 9% each, and domestic sewage 6%. About 52% of the global emissions of methane came from Asia, principally from sources related to rice cultivation, livestock, and coal mining. The rice fields of China, India, Indonesia, Thailand, Viet Nam, Bangladesh, Myanmar, and the Philippines, in that order, are important sources of methane to the atmosphere. The United States was responsible for approximately 9% of global methane emissions in 2007, with about 65% coming from enteric fermentation, landfills, and natural gas systems. China, India, and the Russian Federation were responsible for 17%, 12%, and 6%, respectively, of emissions. These latter four countries contributed nearly one-half of the global anthropogenic methane emissions to the atmosphere in the early twenty-first century.

Anthropogenic nitrous oxide emissions to the atmosphere are even more difficult to quantify than methane emissions (Figure 14.12). The anthropogenic nitrous oxide flux for the early twenty-first century was equivalent to about 3 million metric tons of nitrogen per year or higher. North American emissions to the atmosphere may have been as high as 4.3 million metric tons of N_2O. If so, it is likely that global emissions were higher than estimated in Figure 14.12.

The global emissions of CFCs to the atmosphere are shown in Figure 14.17. They are decreasing rapidly because of the regulations on emissions proposed by the Montreal Protocol and further amendments (see below).

Future emissions of greenhouse gases to the atmosphere from human activities depend on the course of economic development and the trade-off between development and the concern of many people for the environment (see Chapter 15). There are several economic development scenarios used by the Intergovernmental Panel on Climate Change (1996, 2001, 2007) and other sources for the future of greenhouse gas emissions and the concentrations of these gases in the atmosphere in the twenty-first century. There is little doubt that the emissions of carbon dioxide, methane, and nitrous oxide will continue to increase in the twenty-first century. Many developing countries need and want what the industrialized world has obtained during the last two centuries of growth. The burning of fossil fuels and clearing of land for agriculture, livestock, and urban development are currently part of the development plans of most developing countries. How the industrialized world manages its emissions of greenhouse gases in the future and participates in the economic and environmental development of developing and industrializing nations will influence the regional and global rates of emission of greenhouse gases into the atmosphere in this century. So far, the only strongly regulated gases are the halocarbons, and the CO_2 emission targets set in the Kyoto Protocol, which was initially adopted for use in 1997 and entered into force in 2005 without United States participation, in general have not been met.

Climate and Ecologic Consequences of an Enhanced Greenhouse Effect

Atmospheric carbon dioxide, methane, nitrous oxide, tropospheric ozone, chlorofluorocarbon concentrations, cloud reflectivity, water vapor content, aerosol type and distribution, the sun's radiance, and other factors have a profound influence on the climate. These factors act in concert in the warming and cooling of Earth. Air, land, water, and life are all affected by climatic change. The effects or consequences of climatic change because of natural or human-induced factors are open to much debate, speculation, and uncertainty, but also much is known about the physical climate system and its interactions with the ecosphere. There presently are sufficient scientific data available that shed light on the *potential* climatic and ecologic consequences resulting from increasing concentrations of carbon dioxide and other greenhouse gases in the atmosphere and an enhancement of the greenhouse effect. Some of the consequences were mentioned previously. Our knowledge of the effects comes from modeling the climate system on high-speed computers, from historical records of climatic change, and from study of the modern climate system and its interactions with the ecosphere.

TEMPERATURE AND PRECIPITATION CHANGES The most obvious consequence of an enhanced greenhouse effect is that of increasing temperature. The global average late pre-industrial air surface temperature of Earth, excluding Antarctica, was about 15°C. There are several different historical records of global temperature, including thermometer records for surface air temperatures of land areas, sea-surface temperatures, combined land-surface air and sea-surface temperatures (SSTs), and shorter records of temperatures as recorded by instrumentation on balloons and satellites. Records are available for both the Northern and Southern Hemispheres. In addition, several agencies develop global temperature records, including the Meteorological Office Hadley Centre in England, the NASA Goddard Institute for Space Studies (GISS) and the National Climatic Data Center of NOAA in the United States. Satellite temperature measurements have been obtained from measurements of radiance in various wavelength bands using Microwave Sounding Unit (MSU) instruments since 1978 and Advanced Micrometer Sounding Unit (AMSU) instruments since 1998 flown on satellites. The radiance measurements are then mathematically inverted to get temperature. In 2004 the last MSU instrument degraded in data quality. Several groups including NASA GISS, Remote Sensing Systems (http://remss.com/), and scientists at the University of Alabama at Huntsville analyze the satellite data.

Estimates of temperature variations near the Earth's surface are obtained from thermometer readings taken daily at thousands of stations on land and by thousands of ships at sea. In general, the number of measurement stations and the spatial coverage of the globe have increased during the past two centuries. Hence there is a built-in sampling bias in the temperature data used to calculate a time series of global mean temperatures because of changing coverage. However, dating back to the late nineteenth century, data coverage is thought to be sufficiently

dense to make a calculation of the temporal trend in global mean temperature. The records of global mean temperature change for both the lower atmosphere and ocean show an increase for the past century. Figure 14.23 shows two time series of global average surface temperatures from combined land-surface and sea-surface measurements from the nineteenth century to 2008. The global increase is not steady, but irregular. In actual fact, the global mean temperature of the surface in some analyses (Figure 14.23b) exhibits a very slight cooling trend between the early 1940s and the early 1970s, when anthropogenic greenhouse gas emissions and the carbon dioxide content of the atmosphere were rising rapidly. Recent evidence suggests the abrupt drop in the temperature anomaly in the mid-1940s may be due to how scientists measured sea surface temperatures of the ocean.

The time series is an area-weighted average of the surface air temperature over land and the temperature of the water at the surface of the ocean. Stations over land exhibiting an urban heat effect have been eliminated from the time series analyses. The urban heat effect biases the temperature record toward higher temperatures since urban areas usually show higher temperatures than their surroundings because of industrial, transportation, and residential activities, presence of buildings and concreted and asphalted roads, and congregations of people that add heat to the urban environment. In the scientific literature on climate change, temperatures are commonly shown in terms of departures from the local climatic mean temperature for a specified reference period. The departures of temperature from the climatic means are referred to as temperature anomalies. In Figure 14.23a the anomalies are calculated as differences from the global mean temperature for 1961 to 1990 and in Figure 14.23b, from 1951 to 1980.

The time series of global mean temperatures displays interannual fluctuations that are primarily related to El Niño/La Niña events and volcanic eruptions. For example, the warming related to the El Niño event of 1997–1998 is readily apparent in the global mean temperature time series, as is the atmospheric cooling due to the eruption of Mt. Pinatubo in 1991. Decadal and centurial changes are also recognizable in the temperature record. For example, the generally cyclical nature of the temperature record in the late nineteenth century into the early decades of the twentieth century is likely due to some extent to recovery of the planet from the colder temperatures of the Little Ice Age. The Little Ice Age cold episode ended in 1850 and probably was a result, along with short-lived volcanic explosions, of the Maunder

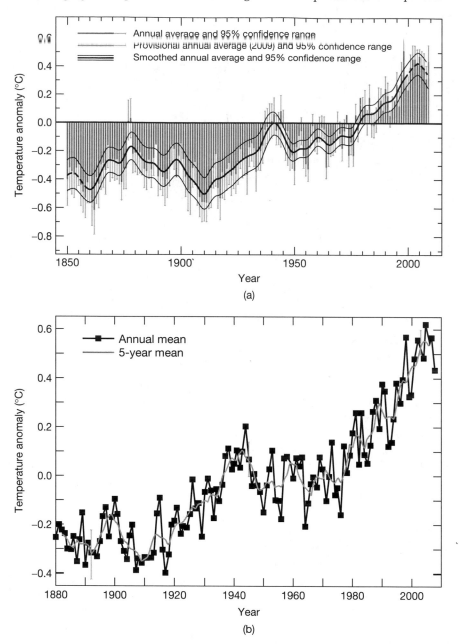

FIGURE 14.23 Global combined land-surface air and sea-surface temperature annual anomalies in degrees centigrade as obtained by thermometer measurements. (a) Data from the Met Office Hadley Centre for 1850 to 2008 shown as deviations from the 1961 to 1990 global mean temperature. (http://hadobs.metoffice .com/hadcrut3/diagnostics/global/nh+sh/.) (b) Data from the Goddard Institute for Space Studies (GISS) for the period 1880 to 2008 shown as deviations from the 1951 to 1980 global mean temperature. (http://data.giss.nasa.gov/gistemp/2008/.)

minimum in solar activity. Although the long-term trend in solar irradiance since the Little Ice Age remains uncertain, the trend seems to be slightly upward, probably accounting for some of the rise in temperature since the Little Ice Age into the twentieth century. The direct radiative forcing of the sun, however, is weak compared to the forcing of greenhouse ages and tropospheric sulfate aerosols (see Figure 14.55) on climate for the past 150 years, unless it is modified by the solar wind's influence on the solar flux (see above), a controversial assertion.

The globally averaged time series is not necessarily representative of local or regional conditions. For example, Canada and Siberia have warmed much more rapidly for the past 20 years than the global average, and parts of the North Pacific and southern oceans actually show regions of cooling in the first decade of the twenty-first century (Figure 14.24). Nevertheless, a global warming during the past century or so appears to be a robust feature of the time-series record. The global warming for 1850 to 2008 has been approximately 0.8°C. For the past 30 years or so, there has been a surface warming of about 0.42 to 0.55°C, equivalent to approximately 0.14 to 0.18°C per decade. The decade of 1998 to 2007, following the El Niño event in 1997 to 1998, which led to a greater positive temperature anomaly, was the warmest on record. In the NASA GISS record (Figure 14.23b), the 14 warmest years on record have all occurred since 1990. The year 2007 tied with 1998 as the second warmest year in a century, exhibiting a 0.57°C departure from the 1961–1990 global average in the NASA GISS record. This is despite the fact that solar irradiance was at a minimum and the equatorial Pacific Ocean was experiencing a cool phase of its natural ENSO cycle. The lower temperature anomaly of 2008 is due the imprint of cooling from a La Niña event on the global temperature rise. As the Pacific Ocean returns to normal ENSO conditions and enters the new 2009 phase of El Niño, it is likely that 2009 will have a more positive temperature anomaly than 2008 (Figure 14.23a).

The global surface warming of the planet should also be apparent in the lower and mid-troposphere. Temperatures at altitude can be obtained in a number of ways, including the use of radiosondes and satellites. For the former, balloon-borne instrument packages that include thermometers are released daily or twice daily from observing stations throughout the world for the monitoring of climate. The radiosonde unit measures various atmospheric parameters and transmits data to a receiver at a radio frequency of 403 MHZ or 1680 MHZ. For the latter, instruments mentioned above onboard satellites make measurements of microwave radiation emitted by oxygen gas in the atmosphere of the lower to mid troposphere. The microwave radiation is an indicator of the temperature of the air. The radiosonde network has been in operation since the late 1940s, and since the mid-1960s has been suitable for estimating global temperature changes. The continuous MSU temperature time series began in 1978.

Figure 14.25 shows the time-series record of global mean temperature anomalies, relative to the period mean of 1979 to 1997, of the lower stratosphere, mid to upper troposphere, and lower troposphere, as obtained from radiosonde and satellite observations, and the surface record from thermometer measurements from three governmental agencies for the period 1958 to 2005. The major features apparent in the time-series record of the lower stratosphere are the overall downward trend in temperature and the warming events following volcanic eruptions produced by sulfate aerosols injected into the stratosphere. By absorbing both solar and terrestrial radiation, the volcanically produced sulfate aerosol layer heats the stratosphere, as the lower atmosphere cools from reflection of incoming solar radiation by the bright sulfate aerosols injected into it (Figure 14.25). The overall downward trend in stratospheric temperatures is due to emissions of ozone-depleting CFCs mixing into the stratosphere, increased carbon dioxide concentrations, and changes in the water vapor content of the stratosphere, leading to overall cooling of the lower stratosphere. In contrast, the overall temperature trends for the mid to upper troposphere, the lower troposphere, and the surface have been upward for this period of time. The general trend of increasing temperature is irregular, reflecting to some extent the effects of volcanic explosions and ENSO events on the temperature record (Figure 14.25).

There has been considerable debate about the apparent disparity between the surface thermometer record of temperatures and the radiosonde/satellite records of tropospheric temperatures. However, it now appears that the warming rates obtained from balloon-borne and satellite measurements of temperatures are rather similar to the surface thermometer measurements (Figures 14.25 and 14.26), although there are still some who question this conclusion. In addition, cooling of the lower stratosphere and warming of the troposphere appear to be responsible for an increase in the height of the tropopause by 200 meters during 1979 to 2001. Such an observation provides support for the conclusion that the surface of the planet has been warming globally.

The Intergovernmental Panel on Climate Change (IPCC) concluded in 2007 that warming of the climate system is unequivocal based on observations of global average air and sea temperatures, rising sea levels, and widespread melting of snow and ice. The IPCC report also concluded that there is *very high confidence* in the conclusion that the global average net effect since 1750 of human activities on climate has been one of warming. The net overall radiative forcing of the several warming and

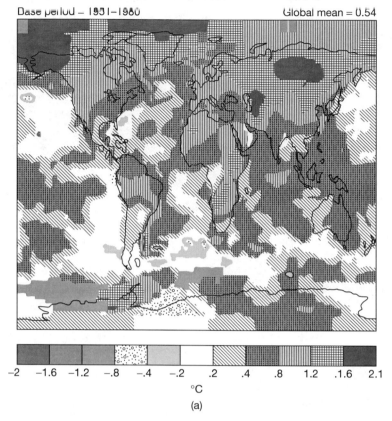

2001–2005 Mean surface temperature anomaly (°C)

Base period – 1951–1980   Global mean = 0.54

−2	−1.6	−1.2	−.8	−.4	−.2	.2	.4	.8	1.2	.1.6	2.1

°C

(a)

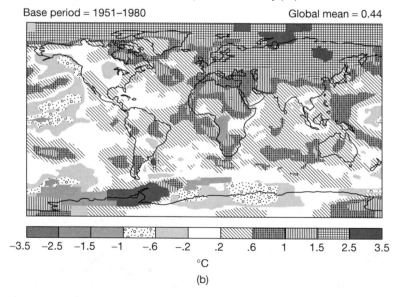

2008 Surface temperature anomaly (°C)

Base period = 1951–1980   Global mean = 0.44

−3.5	−2.5	−1.5	−1	−.6	−.2	.2	.6	1	1.5	2.5	3.5

°C

(b)

FIGURE 14.24 Mean surface temperature anomaly for 2001 to 2005 (a) and for 2008 (b) calculated relative to the global mean temperature of 1951 to 1980. The *global* mean anomaly for 2001 to 2005 was 0.54°C and that for 2008 was 0.44°C. (*Source:* http://data.giss.nasa.gov/gistemp/2008/.)

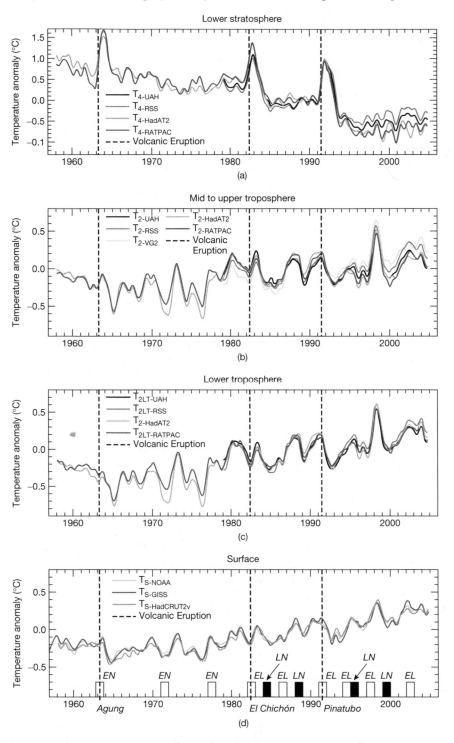

FIGURE 14.25 Global mean surface and upper-air temperature anomalies as departures from the 1979 to 1997 average temperature. (a) Lower stratosphere (T4 MSU satellite analysis); (b) mid- to upper troposphere (T2 MSU satellite analysis); (c) lower troposphere [T2 MSU satellite analysis and United Kingdom Met Office (UKMO) HadAT2 and NOAA RATPAC radiosonde observations]; and (d) surface thermometer record [NOAA, NASA/GISS, and UKMO/CRU (HadCRUT2v)]. El Niño (warming), La Niña (cooling), and volcanic cooling episodes are shown below the time-series records. (*Source:* http://www.ncdc.noaa.gov/img/climate/globalwarming/ar4-fig-3-17.gif.)

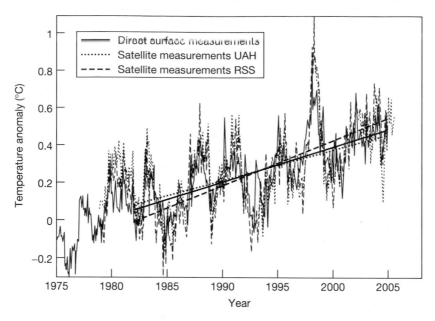

FIGURE 14.26 Comparison of surface temperature measurements (——) and satellite temperature measurements by University of Alabama at Huntsville (UAH) (.......) and Remote Sensing Systems (RSS) (– – –) from 1975 to 2005. Notice the close similarity in the linear least square fit lines for the three time trends. (http://en.wikipedia.org/wiki/Satellite_temperature_measurements.)

cooling factors involved in this overall warming trend since 1750 was +1.66 (+0.6 to +2.4) W/m², with +2.99 W/m² radiative forcing attributed to increased greenhouse gas concentrations in the atmosphere due to human activities. (Figure 14.55). A few investigators argue against this conclusion and contend that most of the modern warming is due to natural causes.

The future course of temperature for the planet depends on a number of factors. During at least the past 150 years, five greenhouse gases have accumulated in our atmosphere because of human activities, resulting in today's enhanced greenhouse effect: carbon dioxide (56% of the relative contribution to the enhanced greenhouse effect), methane (16%), tropospheric ozone (12%), chlorofluorocarbons (11%), and nitrous oxide (5%) (Table 14.2). Also, anthropogenic SO_x emissions have resulted in changes in the sulfate aerosol burden of the atmosphere and acted as a cooling agent on surface temperatures of the planet. Perhaps the most important factor influencing the path of a human-induced global warming of the planet is the future rate of anthropogenic emissions of the greenhouse gases to the atmosphere. These rates and the SO_x emissions rate will depend on future population growth and present and future policies developed by the nations of the world relative to their use of fossil fuels and their deforestation, industrial,

agricultural, and transportation practices. Another important factor involves the behavior of the coupled climate–ecosphere system. This behavior includes the complex chain of processes in the atmosphere and other components of the physical climate system and how they respond to future anthropogenic greenhouse gas and SO_x emissions and physical and biogechemical feedbacks. Different groups including the IPCC have developed a number of future emissions scenarios. In *The Special Report on Emissions Scenarios of the IPCC* in 2000, there are 40 different scenarios (Table 14.4).

Figure 14.27a shows one scenario for the future of carbon dioxide emissions to the atmosphere from fossil fuel burning. This projection is based on a so-called BAU (business as usual) scenario (IS92a) of the IPCC, and is a reasonable scenario for the world, if the world's nations do little to modify their behavior in terms of energy utilization in the twenty-first century; that is, if they continue to rely heavily on the use of coal, oil, and gas to drive their economic development. This BAU scenario projects carbon dioxide fossil fuel emissions equivalent to approximately 20 billion tons of carbon per year by the year 2100. This emission value is roughly three times the fossil fuel CO_2 emissions at the beginning of the twenty-first century. Other economic and policy scenarios (e.g., *The Special Report on Emissions Scenarios of the*

TABLE 14.4 The Special Report on Emissions Scenarios Classification of Families or Story Lines. There are 40 Total Scenarios Under the Various Families

A1 storyline and family

The A1 scenarios are of a more integrated world. The A1 family of scenarios is characterized by:

Rapid economic growth.
A global population that reaches 9 billion in 2050 and then gradually declines.
The quick spread of new and efficient technologies.
A convergent world; income and way of life converge between regions. Extensive social and cultural interactions worldwide.

There are subsets to the A1 family based on their technological emphasis:

A1FI: An emphasis on fossil-fuels.
A1B: A balanced emphasis on all energy sources.
A1T: Emphasis on nonfossil energy sources.

A2 storyline and family

The A2 scenarios are of a more divided world. The A2 family of scenarios is characterized by:

A world of independently operating, self-reliant nations.
Continuously increasing population.
Regionally oriented economic development.
Slower and more fragmented technological changes and improvements to per capita income.

B1 storyline and scenario family

The B1 scenarios are of a world more integrated, and more ecologically friendly. The B1 scenarios are characterized by:

Rapid economic growth as in A1, but with rapid changes towards a service and information economy.
Population rising to 9 billion in 2050 and then declining as in A1.
Reductions in material intensity and the introduction of clean and resource efficient technologies.
An emphasis on global solutions to economic, social, and environmental stability.

B2 storyline and scenario family

The B2 scenarios are of a world more divided, but more ecologically friendly. The B2 scenarios are characterized by:

Continuously increasing population, but at a slower rate than in A2.
Emphasis on local rather than global solutions to economic, social, and environmental stability.
Intermediate levels of economic development.
Less rapid and more fragmented technological change than in A1 and B1.

(*Source:* IPCC, 2000).

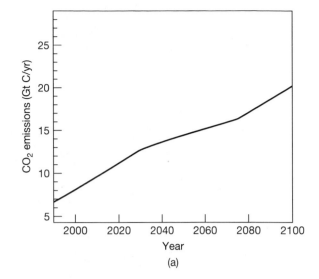

(a)

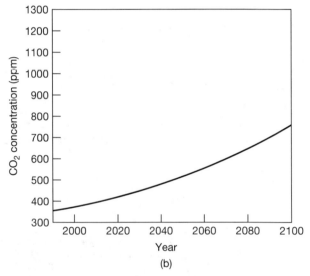

(b)

FIGURE 14.27 (a) Future carbon dioxide emissions from fossil fuel burning to the atmosphere based on the Intergovernmental Panel on Climate Change (IPCC) BAU (Business as Usual) scenario (IS92a), and (b) concentrations of atmospheric CO_2 based on that scenario. The range of emissions calculated from various emission scenarios in the year 2100 is from nearly 30 to less than 5 billion tons of carbon per year and that of atmospheric CO_2 concentrations from 540 to 960 ppmv. (*Source:* Houghton et al., 1996, 2001.)

IPCC, 2000) project a range of CO_2 emissions by the end of the twenty-first century of nearly 30 to below 5 billion tons of carbon per year. The latter would require major changes in the sources of energy for driving economic development of the world's nations and in demography and technology. Based on various emission scenarios, the range of possible atmospheric CO_2 concentrations that are estimated by the use of various types of models in the year 2100 is approximately 540 to 960 ppmv. Figure 14.27b shows the time course of atmospheric CO_2 concentrations, rising to 750 ppmv in 2100, based on the BAU IS92a emissions scenario and on models that attempt to describe the behavior of the biogeochemical cycle of carbon and climate through time. The carbon-cycle models vary in the way they are structured and the extent to which they describe the various processes controlling the behavior of the carbon cycle. Thus, both the emission scenario as well the formulation of physical and biogeochemical processes and feedbacks adopted in a global model affect the projected atmospheric CO_2 concentrations and SO_x loadings of the atmosphere in the future, and hence temperature.

Based on several future greenhouse gas emission scenarios and using multimodel global averages of surface warming relative to the 1980 to 1999 period, the 2007 *Fourth Assessment Report* of the IPCC indicates that the average surface air temperature increase relative to the year 2000 for the end of the twenty-first century owing to a global warming of the planet will be about 1.4 to 4.2°C, with a best guess of around 2.5°C, as shown in Figure 14.28. Figure 14.28 also shows the trends of warming if the atmosphere were kept at a constant composition from 2000 to 2100 and if climatic forcing from anthropogenic activities were to be held constant

from 2100 to 2300. In both cases, the reduction in climate forcing leads to some stabilization of temperature but still small increases in temperature over time. This is not surprising since depending on how much CO_2 is put into the atmosphere by human activities, traces of this greenhouse gas alone will remain for thousands, even hundreds of thousands of years in the atmosphere!

The full range in estimates of the global mean temperature increase in 2100 from the various models in the 2001 *Third Assessment Report* of the IPCC was from 1.4 to 5.8°C, and in the previous IPCC projections of 1990 and 1996, the range in estimates was 2 to 5°C and 0.9 to 3.5°C, respectively. Considering the uncertainties in predicting future temperatures from models, these various estimates of the range in temperatures are in reasonably good agreement. The differences in estimates are due to several factors. The 1990 temperature estimates were not based on models that contained sulfur emissions from anthropogenic sources and the cooling effect of anthropogenic sulfate aerosols in them (see Chapter 7 and the following), but the 1996, 2001, and 2007 estimates did include sulfur emissions from anthropogenic sources and their conversion to sulfate aerosols. The 2001 and 2007 temperature estimates also included the effects on climate of aerosols from biomass burning and black carbon (soot) from combustion of

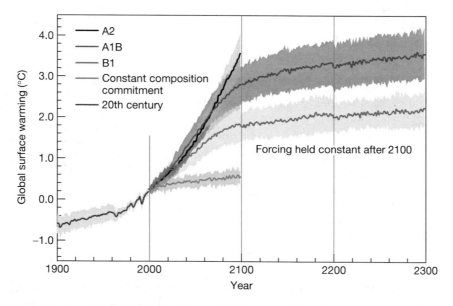

FIGURE 14.28 Ranges in estimates of changes in future global mean surface air temperatures as based on various emissions scenarios for the future (A2, A1B, and B1) and models involving climate response to the additional greenhouse gases and behavior of the atmosphere-ocean-land system. The range in temperature projections by the year 2100 is 1.4°C to 4.2°C. Also shown are projections until the year 2300 for emissions scenarios A1B and B1, a commitment in the year 2000 to no change in atmospheric composition, and the global mean temperature record since 1900 to 2000. (From *Climate Change 2007 The Physical Science Basis. Contribution of Working Group I to the Fourth Assessment Report of the Intergovernmental Panel on Climate Change.* © Intergovernmental Panel on Climate Change 2007. Published by Cambridge University Press. Used by permission of Intergovernmental Panel on Climate Change.)

fossil fuel. The 2001 and 2007 estimates differ in range in part due to the development of a more complete range of models and climate-carbon feedbacks. The 2007 IPCC *Fourth Assessment Report* estimates are probably the best because of substantial improvements in modeling and knowledge of processes involved in the physical-biogeochemical system. Figure 14.29 shows how well several model projections of global mean warming agree with the observational temperature record from 1990 to 2008, with projections of the models to 2025. The reader should keep in mind that the temperature predictions for the future of the twenty-first century are based principally on modeling of the climate system but the fact that the assemble of multi-models appears to do well in fitting the observed temperature records from the early twentieth century to 2000 both globally and regionally (Figure 14.30) lends some confidence to the model calculations for the future. In addition, the range of model calculations of the temperature anomalies better fit the observational global and regional record of temperatures if the models employ climate forcing in them that includes both anthropogenic and natural forcings (Figure 14.30), lending credence to the conclusions that the twentieth century's temperature change was not simply due to natural forcings, particularly since about 1970.

If the planet continues to warm at the preceding estimated rates, the rate of temperature increase projected for the twenty-first century will be considerably larger than the observed temperature change during the twentieth century (Figure 14.23) and very likely to be without precedent during at least the last 10,000 years (Figures 13.16 and 13.24). In addition, the temperature change will be most dramatic for the high latitudes of the Northern Hemisphere (Figure 14.31). This great warming is connected in part to a reduction in the snow and sea-ice cover of the high latitudes of the Northern Hemisphere. For the Southern Ocean, the warming is less dramatic because of uptake of heat by the large area of ocean in this region. Ocean heat uptake also results in a minimum of warming in the North Atlantic Ocean. In general, the land warms relatively rapidly everywhere. However, over industrialized areas, because of the relative abundance of sulfate aerosols and their cooling effect, the warming is somewhat moderated. Figure 14.32 illustrates modeling results of regional projections by continent of temperature change from the early twentieth century to 2050. The continental projections suggest that North America will warm most and South America and Australia least.

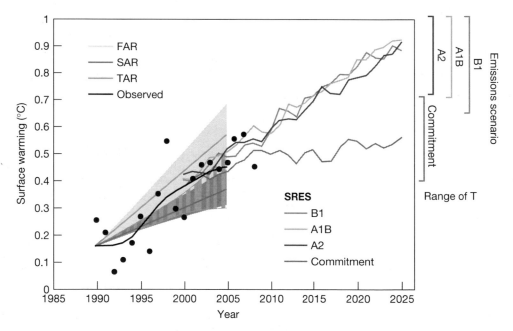

FIGURE 14.29 Comparison of the First (FAR), Second (SAR), and Third (TAR) Assessment reports of the IPCC and the more recent emissions scenarios [A2, A1B, B1, and Commitment (no change in atmospheric composition)] of the Special Report on Emissions Scenarios (SRES) of the IPCC with the global annual mean temperatures (black dots) from 1990 to 2008. Projections from the SRES scenarios and their range for temperatures until 2025 are also shown. (From *Climate Change 2007 The Physical Science Basis. Contribution of Working Group I to the Fourth Assessment Report of the Intergovernmental Panel on Climate Change.* © Intergovernmental Panel on Climate Change 2007. Published by Cambridge University Press. Used by permission of Intergovernmental Panel on Climate Change.)

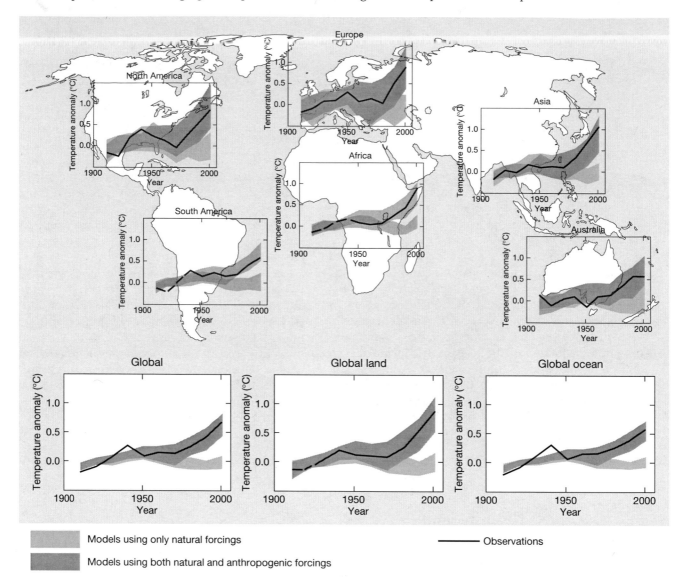

FIGURE 14.30 Comparison of an assemble of multimodel calculations of global, global land, global ocean, and regional trends in temperature employing only natural forcings and those using both natural and anthropogenic forcings with the observational record of the trend in mean temperatures from the early twentieth century to 2000 (black line). Notice that the observational temperature trends are best fit by models that employ both natural and anthropogenic forcings. (From *Climate Change 2007 The Physical Science Basis. Contribution of Working Group I to the Fourth Assessment Report of the Intergovernmental Panel on Climate Change*. © Intergovernmental Panel on Climate Change 2007. Published by Cambridge University Press. Used by permission of Intergovernmental Panel on Climate Change.)

A change in the temperature of the Earth's atmosphere and surface will affect the hydrological cycle. The water content of the global atmosphere depends on the average temperature of the planet. On a warmer Earth, the atmosphere would be expected to hold more water vapor. Because atmospheric water vapor is a strong greenhouse gas, an initial warming of Earth owing to an enhanced greenhouse effect will probably increase the water vapor content of the atmosphere. Therefore, any initial warming of the planet will be reinforced by the presence of more water vapor, a positive feedback to global warming. There is some evidence that the water vapor content of the troposphere has increased over time. It should be kept in mind that there are some scientists who suggest the water vapor feedback could be negative (see above).

On a warmer Earth, the amount and distribution of evaporation and precipitation will change, and hence patterns of soil moisture content and runoff from the land (Figures 14.33 and 14.34). The pattern of

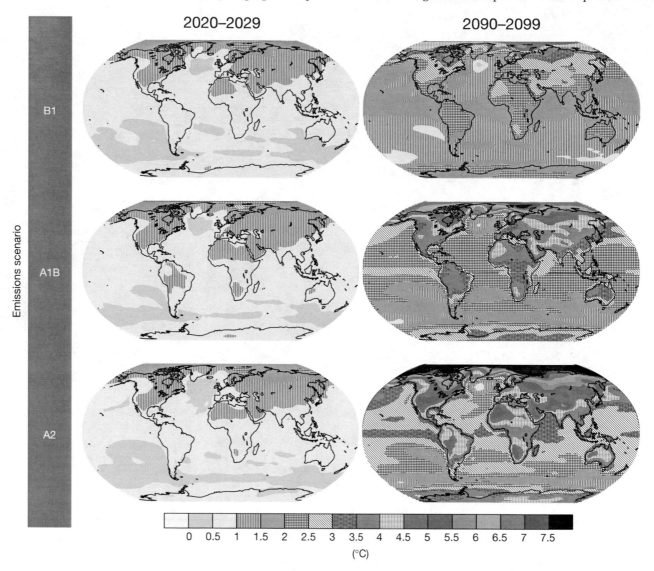

FIGURE 14.31 Geographical distribution of the differences in temperatures for the globe averaged for the period between 2020 to 2029 and that averaged for 2090 to 2099 based on model calculations. Notice the distinct pattern of warmer temperatures in the high northern latitudes. The calculations are done for three different SRES scenarios: B1, A1B, and A2. (From *Climate Change 2007 The Physical Science Basis. Contribution of Working Group I to the Fourth Assessment Report of the Intergovernmental Panel on Climate Change.* © Intergovernmental Panel on Climate Change 2007. Published by Cambridge University Press. Used by permission of Intergovernmental Panel on Climate Change.)

change is difficult to predict. However, it is a certainty that some regions will become drier and some regions will become wetter. One interpretation for a warmer Earth is that there will be increases in precipitation throughout the year at high latitudes of both hemispheres. Increases might also be observed in much of the equatorial region. The mid-latitudes, except for South and Southeast Asia, would tend to dry out. Evaporation will decrease significantly for the southern oceans and the region south of Greenland, and parts of the southwestern United States, Central and South America, northern and southern Africa, the

Middle East, and western Australia. These changes in precipitation and evaporation will lead to changes in soil moisture content and river runoff across the globe (Figures 14.33 and 14.34). Some of the impacts of future climate change on the availability of freshwater for some regions of the world are shown in Figure 14.34.

The possibility of changing patterns of soil moisture are of particular concern to people living in the semiarid, developing regions of the world, where rainfall is currently scarce and soils are already deteriorated. It is also of concern to the world community because regional changes in the balance between

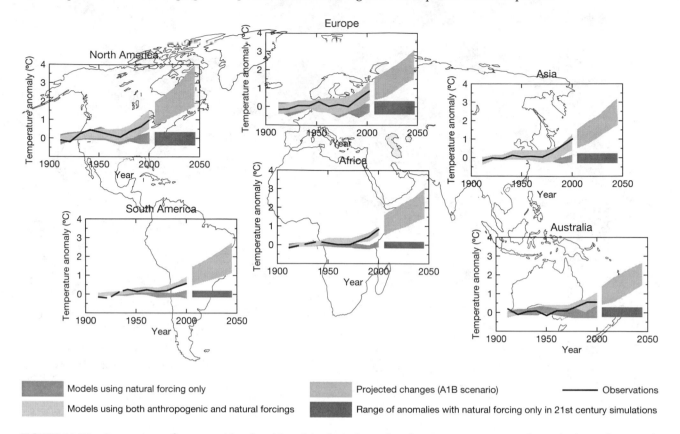

FIGURE 14.32 Comparison of an assemble of multimodel calculations of regional temperature trends employing only natural forcings and those using both natural and anthropogenic forcings with the observational records of the regional trends in mean temperatures from the early twentieth century to 2000 (black line). Notice that the regional observational temperature trends are best fit by models that employ both natural and anthropogenic forcings. Temperature projections until the year 2050 are also shown using the A1B emission scenario and for the case in which the natural forcings of the twentieth century are extended into the future. (From *Climate Change 2007 The Physical Science Basis. Contribution of Working Group I to the Fourth Assessment Report of the Intergovernmental Panel on Climate Change.* © Intergovernmental Panel on Climate Change 2007. Published by Cambridge University Press. Used by permission of Intergovernmental Panel on Climate Change.)

precipitation and evaporation and consequent effects on soil moisture could have a strong impact on agriculture and food production as some areas of the world dry out and others become wetter (Figure 14.35). Farmers may need to shift crops or relocate agricultural lands, and new varieties of drought-resistant crops will require development. The need for irrigation may increase and put stress on water resources in some regions, such as the Middle East. In addition, continuous rising atmospheric CO_2 concentrations could be beneficial and lead to increased crop growth for some regions of the world because of the fertilization effect. Crop yields could be further aided by rising temperatures in some regions, like the grain belt of the Ukraine.

OCEAN RESPONSE TO RISING ATMOSPHERIC CO_2 AND CLIMATE CHANGE The ocean and the atmosphere play important linked roles in climate and climate change. The planet absorbs radiation unequally, with most solar radiation being absorbed at the equator. This heat is transported by tropospheric winds and by ocean currents distributing the heat around the globe, as these flows circulate from the equator to the polar areas. Forty percent of the total heat transported to the polar regions is through the oceans. Interestingly, the Intertropical Convergence Zone (ITCZ; see Chapters 4 and 5), the equatorial region of convergent northeast and southeast trade winds and strong air-sea interactions, rising air and moisture, and persistent bands of showers and heavy thunderstorms, appears to have moved northward over the past 300 years at a rate of just less than one mile per year. This overall movement is most likely a response to warming of the planet since the Little Ice Age. Continuous movement northward of the ITCZ, as modern warming continues, implies less rain for parts of central Africa, the Amazon of Brazil, the Galapagos Islands, and Pacific island nations near the

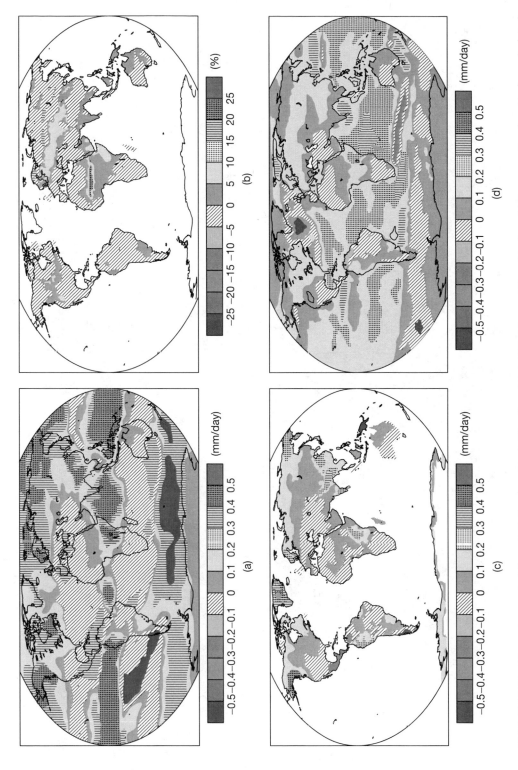

FIGURE 14.33 Mean changes in (a) precipitation (mm/day), (b) soil moisture (%), (c) runoff (mm/day), and (d) evaporation (mm/day) as calculated from an assemble of multimodels. The changes represent the annual means using the SRES A1B emissions scenario for the period 2080 to 2099 relative to 1980 to 1999. (From *Climate Change 2007 The Physical Science Basis. Contribution of Working Group I to the Fourth Assessment Report of the Intergovernmental Panel on Climate Change.* © Intergovernmental Panel on Climate Change 2007. Published by Cambridge University Press. Used by permission of Intergovernmental Panel on Climate Change.)

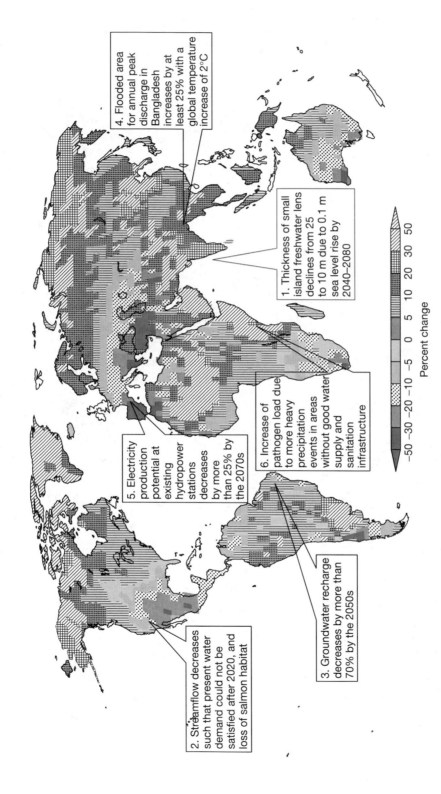

FIGURE 14.34 Some potential impacts of future climate changes on freshwater systems. The background pattern is that of Figure 14.33c for the mean changes in runoff in percent for the period 2080 to 2099 relative to 1980 to 1999. (*From Climate Change 2007: Impacts, Adaptation and Vulnerability. Contribution of Working Group II to the Fourth Assessment Report of the Intergovernmental Panel on Climate Change.* © Intergovernmental Panel on Climate Change 2007. Published by Cambridge University Press. Used by permission of Intergovernmental Panel on Climate Change.)

The following labels appear on the map:

1. Thickness of small island freshwater lens declines from 25 to 10 m due to 0.1 m sea level rise by 2040–2080

2. Streamflow decreases such that present water demand could not be satisfied after 2020, and loss of salmon habitat

3. Groundwater recharge decreases by more than 70% by the 2050s

4. Flooded area for annual peak discharge in Bangladesh increases by at least 25% with a global temperature increase of 2°C

5. Electricity production potential at existing hydropower stations decreases by more than 25% by the 2070s

6. Increase of pathogen load due to more heavy precipitation events in areas without good water supply and sanitation infrastructure

Percent change

−50 −30 −20 −10 −5 0 5 10 20 30 50

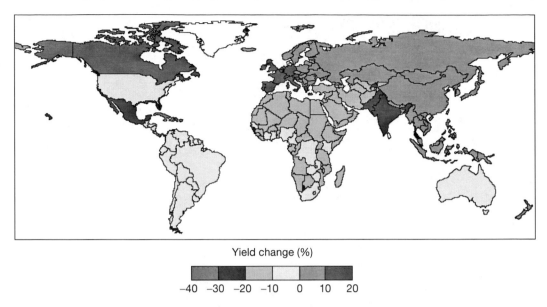

Yield change (%)

−40 −30 −20 −10 0 10 20

FIGURE 14.35 An example model calculation illustrating the projected changes in agricultural crop yields in percent change from the present day to the middle of the twenty-first century. The geographical picture of crop yield changes is shown in different patterns. The results of models of this type not only depend on projected changes in precipitation and soil moisture but also on the fertilization effects of atmospheric CO_2. (*Source:* Parry et al., 1997.)

equator. Farther northward of these regions, rainfall would be more intense because of the northward movement of the ITCZ.

Ocean Temperature and Salinity Changes The ocean's heat content and temperature have changed over time (Figure 14.36). As the climate has warmed, much of the heat has been transferred to the ocean, accounting for its increase in heat content and the rise in global sea surface temperatures (SSTs). The ocean's heat capacity is 1000 times larger than that of the atmosphere, and therefore as the planet has warmed since 1961, the change in the ocean content of heat from 0 to 3000 meters in depth has amounted to 14.2×10^{22} joules, or about 90%, of the change in the total additional energy content of the Earth of 15.9×10^{22} joules. Approximately 90% of the additional heat of the ocean is stored in the upper 300 meters of the ocean, accounting for about 95% of the thermosteric (thermal expansion of the ocean) rise in sea level. Since the mid-1980s, the heat content of the ocean has risen about 5×10^{22} joules per decade (Figure 14.36a). Global sea surface temperatures for the past century or so have risen at a rate of approximately 0.65°C per century (Figure 14.36b).

A reasonable amount of recent observational evidence shows that both the northern and southern high-latitude regions of the ocean exhibit a freshening of surface waters, perhaps reflecting higher precipitation over the regions and also other factors like higher

runoff, ice melting, changes in ocean advection patterns, and in the Meridional Overturning Circulation of the Atlantic (see below). In the near-surface waters of the more evaporative regions of the ocean, the subtropical gyres of both hemispheres, salinity appears to be increasing in the upper layers of the water column. These changes are consistent with changes in the hydrological cycle due to global warming predicted by some model calculations. However, the complexity of the temporal changes in salinity, temperature, and hence density of ocean water can be appreciated from data collected at the Hawaii Ocean Time-series (HOTs) hydrostation ALOHA (A Long-term Oligotrophic Habitat Assessment) shown in Figure 14.37. The figure illustrates the variability of the above seawater parameters down to a depth of 5000 meters. For the period 1988 to 2008 there has been an overall warming and salinity increase of the surface waters, as might be anticipated, but a trend toward less salty and cooler waters is observed at about 250 meters depth. This is followed by mostly increasing temperatures and slightly increasing salinities with depth to the bottom of the profile at about 5000 meters. The deeper water changes in temperature, salinity, and density are pre-industrial.

Meridional Overturning Circulation (MOC) Changes Deep ocean water formation occurs in high latitudes and depends on the balance of salinity and

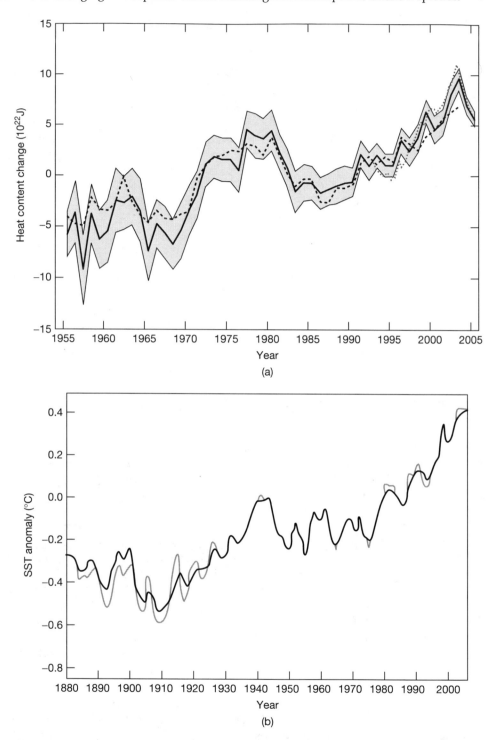

FIGURE 14.36 Trends in ocean heat content (a) and sea surface temperature (SST) (b). The time series for global annual heat content in 10^{22} joules is for the 0 to approximately 700-meter depth range of the ocean. The various curves represent estimates by three groups: black line with gray shading for the 90% confidence limits, Levitus et al., 2005; --- curve, Ishii et al., 2006; and ..... curve, Willis et al., 2004. (In IPCC, 2007.) The time series for sea surface ocean temperatures is generalized and exhibits an average trend of ~ +0.65°C per century. (*Source:* Smith et al., 2005.)

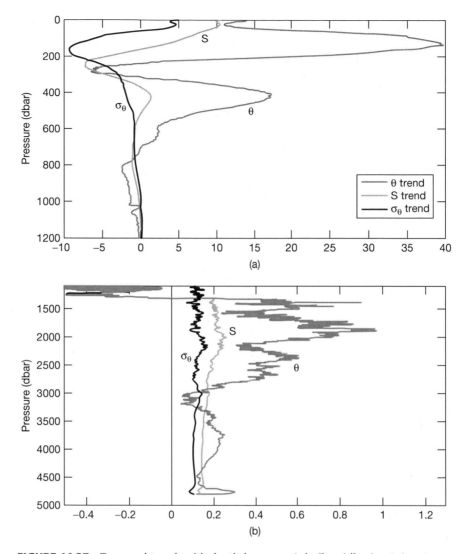

FIGURE 14.37   Temporal trends with depth (pressure, 1 decibar (dbar) = ~1 meter column of seawater = ~0.1 atmosphere) in temperature (θ), salinity (S), and density (σ_θ) at the Hawaii Ocean Time-series (HOTs) hydrostation ALOHA (A Long-Term Oligotrophic Habitat Assessment), located approximately 100 kilometers north of Oahu, Hawaii in the North Pacific Subtropical Gyre [(a), shallow trends; (b), deeper trends]. The profiles represent the composite of data for 209 hydrostations and 300 CTD casts (Conductivity, Temperature, Depth measurements) from 1988 to 2008. Deviations to the right of the 0 line mean increases in the magnitude of a variable over this period of time, and to the left, decreases. Notice the overall surface warming of the waters and the trend toward less salty and cooler waters at about 250 meters depth. The warming for the deep water (b) is pre-industrial. σ_θ trend: $\times 10^{-3}$ kg/m^3/yr; θ trend: milli°C/yr; S trend: $\times 10^{-3}$/yr. Variability in water parameters such as observed at HOTs makes it difficult to arrive at estimates of global mean SSTs.   (Courtesy of Roger Lucas, University of Hawaii.)

temperature (see Chapters 5 and 13). This balance could be upset by the observed warmer temperatures in the polar regions, leading to changes in circulation patterns in the atmosphere and oceans. The Atlantic Meridional Overturning Circulation (MOC), part of the conveyor belt circulation pattern of the ocean,

could be upset by global warming, resulting in less heat released to the atmosphere in the high-latitude North Atlantic.

Figure 14.38 illustrates one model result for the effect of global warming on the Atlantic MOC pattern of the ocean. The figure shows the effect of increases

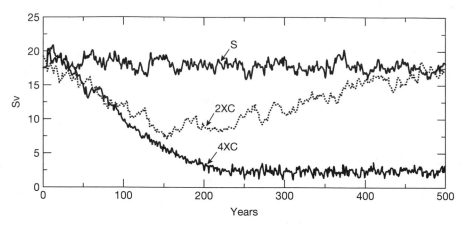

FIGURE 14.38 The results of a model calculation showing the effects of a global warming of the Earth on the intensity of the Meridional Overturning Circulation (MOC) of the ocean. The model runs start with an initial intensity of the circulation of 18 Sv (18 million cubic meters of water per second). S is a run in which the CO_2 of the atmosphere remains constant. The 2XC and 4XC scenarios are for conditions where the planetary surface warms due to increased atmospheric CO_2 concentrations twice and four times recent levels. (*Source:* Manabe and Stouffer, 1994.)

in atmospheric CO_2 to two and four times modern levels and consequent high-latitude surface temperature and precipitation increases on the intensity of the downwelling circulation. Although there is a range of estimates, downwelling in the high-latitude North Atlantic now is equivalent to about 18 Sverdups (Sv) of water per second (18 million cubic meters per second). With a global warming that could lead to higher temperatures and freshening of high-latitude North Atlantic waters because of increases in high-latitude precipitation, and to a deepening of the oceanic thermocline at lower latitudes, the intensity of the MOC could be reduced.

Figure 14.38 shows that warming could lead to a decrease in the intensity (strength) of the circulation of about 50 to 20% of the modern intensity of 18 Sv within a couple of centuries. The range given by simulations from 19 models for changes in the MOC to the year 2200 are from indistinguishable from the natural variability of the circulation to more than a 50% reduction relative to the 1960 to 1990 mean of the circulation intensity. No model simulation has shown an increase in the intensity of the MOC during the twenty-first century. The reduction in MOC intensity due to increasing greenhouse gases represents a negative feedback for warming in and around the North Atlantic since less heat would be transported from low to high latitudes, and high latitude North Atlantic SSTs would be cooler than in a situation where the MOC remained unchanged. Weakening of the MOC would not

necessarily result in a cooler Europe as might be expected because the warming over Europe due to increasing greenhouse gas concentrations would override any cooling due to a reduction in the strength of the MOC. A reduction of the MOC could perhaps result in changes in the frequency and intensity of hurricanes, as well as in global and regional wind and precipitation patterns over the ocean and continent. There is already some evidence that the Gulf Stream, the upper limb of the MOC, slowed during 1957 to 2004.

The behavior of the MOC is one example of the potential for a *climate surprise* in which there can be an abrupt transition or a temporary or permanent transition to a different state of the climate system. Figure 14.39 shows that as freshwater inflow to the North Atlantic increases due to increased precipitation and runoff, the strength of the MOC would weaken and as the freshening continues, there could be a rapid change in climate, with the abrupt cooling of Europe. This scenario assumes that the warming over Europe due to increased greenhouse gas concentrations does not override the cooling due to the decrease in the intensity of the MOC. Notice that the opposite might occur if we started out in a world with large amounts of freshwater flow into the Atlantic that decreased with time. Despite all the work to date, the behavior of the MOC as we enter the warmer world of the twenty-first century is still not well known and abrupt changes in its strength could be a climate surprise of the future.

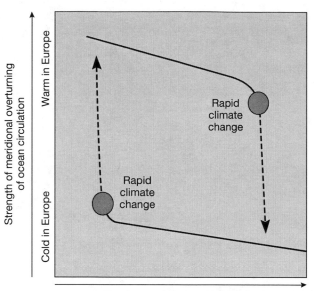

Strength of meridional overturning of ocean circulation — Warm in Europe / Cold in Europe

Rapid climate change

Rapid climate change

Freshwater inflow to North Atlantic

FIGURE 14.39 A possible climate surprise. As surface water warms and freshwater continues to flow into the North Atlantic at increasing rates due to global warming, the strength (intensity) of the meridional overturning circulation (MOC) could weaken and lead to rapid climate change. There would be less heat transported southward by ocean currents in the western North Atlantic and Europe could cool. This scenario assumes that overall global warming does not override the lessened heat transport and cause temperatures in Europe to rise. The reverse situation is shown for reduced freshwater flow.   (Modified from concept of W. S. Broecker.)

Sea Level Changes Continuous sea-level rise is likely to be another result of a warmer world. During the Pleistocene Epoch, sea level rose and fell as the climate changed periodically and continental glaciers waxed and waned. Sea level has risen continuously since the Last Glacial Maximum as continental ice caps have been shrinking. The rate of overall rise has been slower since about 6000 years ago (Figure 13.21). During the past 130 years or so, reconstructions of sea level from tide gauge and satellite altimetry measurements indicate that global mean sea level has risen on average about +20 centimeters, or approximately +1.5 millimeters per year (Figure 14.40). During the period of 1961 to 2003, the average rate of global mean sea-level rise was estimated from tide gauge data to be 1.8 +/− 0.5 millimeters per year and for the period 1993 to 2003 from TOPEX/Poseidon satellite data to be 3.1 +/− 0.77 millimeters per year. The faster rate obtained for the 1993 to 2003 may reflect decadal variability or acceleration in the rate of sea-level rise in recent decades and a longer-term time trend.

Table 14.5a shows one interpretation of the factors involved in the rise and fall of sea level during the twentieth century, the combination of which leads to an estimated range of sea-level rise during this period of +12 to +20 centimeters. Estimates of sea level and the sources of the rise are given in Table 14.5b for 1961 to 2003. The question is, what portion of this rise is due to an *enhanced* greenhouse effect, and what portion is natural? No one knows for sure. In the future, a continuous warming of the planet and resultant thermal expansion of the oceans, melting of mountain glaciers and polar ice sheets, and increased groundwater discharge to the ocean will lead to a continuous, and perhaps accelerating, rise in sea level.

Various estimates have been made of the future changes in global mean sea level that might be expected from an enhanced greenhouse effect and global warming of the planet. The extent of the rise is difficult to estimate. Sea-level rise estimates generally have been revised downward during the past two decades as we learn more about the behavior of the coupled land-ocean-atmosphere-cryosphere system. In the early 1980s, it was predicted that the probable sea-level rise by the year 2100 owing to global warming would be about 3 meters, with a large range of uncertainty of about 50 centimeters to nearly 350 centimeters. The estimates of the Intergovernmental Panel on Climate Change (IPCC, 1990, 1996, 2001, 2007) depend on the various scenarios adopted for emissions of greenhouse gases to the atmosphere in the twenty-first century and models of the response of the coupled land-ocean-atmosphere-cryosphere system to those emissions and temperature change. In the 1990 IPCC assessment, sea level was projected to rise at 6 centimeters per decade on into the twenty-first century, with an uncertainty range of 3 to 10 centimeters per decade. The projected best-estimate rise in sea level was 20 centimeters in 2030 and 65 centimeters in 2100. In the 1996 assessment, the best estimate in sea-level rise was set at about 49 centimeters by the year 2100, with a range of uncertainty of 20 to 86 centimeters. In the 2001 and 2007 assessments, the 2100 projections are, respectively, between 9 and 88 centimeters, with a central "average" value of 48 centimeters, and 20 to 50 centimeters with no overall average value (Figure 14.40). Whatever the case, local sea-level changes over the globe will not be uniform because of ocean density and circulation changes due to warming of the planet (Figure 14.41). Sea level is likely to rise most in the high northern latitudes and actually be lower in the high southern

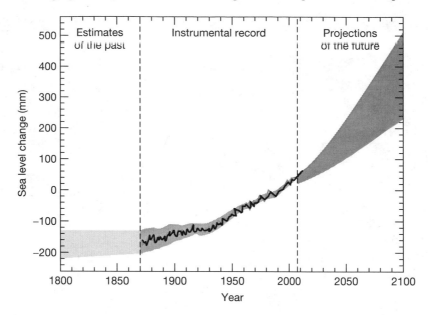

FIGURE 14.40 Sea level rise from 1800 to 2000 based on proxy records, instrumental tide gauge records, and more recently, satellite estimations of sea level with projections to the year 2100 from model calculations using the SRES A1B scenario. The 2100 projected range in sea level estimates is between 22 centimeters and 50 centimeters above the sea level of 2000. The projections do not take into account any significant discharge of water from melting of the continental ice sheets. (From *Climate Change 2007 The Physical Science Basis. Contribution of Working Group I to the Fourth Assessment Report of the Intergovernmental Panel on Climate Change.* © Intergovernmental Panel on Climate Change 2007. Published by Cambridge University Press. Used by permission of Intergovernmental Panel on Climate Change.)

latitudes *relative to the global rise* in sea level at the end of the twenty-first century.

One problem with the above estimates of sea-level projections for the twenty-first century is that the current climate models assume that the great

TABLE 14.5A Factors Involved in the Rise and Fall of Sea Level in the Twentieth Century in Centimeters

	Best Estimate	Possible Range		
Thermal expansion				
Surface water	+5	+1.5	to	+7
Deep water	?	?		
Melting of land ice				
Mountain glaciers	+3	+2	to	+4.6
Antarctic ice	?	−10	to	+13
Greenland ice	+2.5	+2.3	to	+2.5
All factors	+10.5	−4	to	+27
Observed sea-level rise	+15	+12	to	+20

(*Source:* Ruddiman, 2001.)

continental ice sheets will melt slowly in response to increasing temperature. However, ice sheets fracture as they melt and allow water to penetrate rapidly downward toward the base of the sheet with the result that an ice sheet might surge and break up. The rate of ice loss for the entire Greenland ice sheet has more than tripled in the twenty-first century (Figure 14.42) and there has been rapid breakup and loss of sea ice for the Wilkins and Larsen ice shelves in Antarctica and mass loss of ice in West Antarctica. These recent findings portend that the global rise in sea level at the end of this century could be as much as one meter or more. However, recent research employing constraints on future sea-level rise based on reconstruction of past sea-level fluctuations for the past 22,000 years and IPCC projections for warming in the twenty-first century predict 7 to 82 centimeters of sea-level rise by the year 2100. We still have a way to go to predict accurately future sea level change.

A sea-level rise of the magnitude of only 0.5 meters in 2100 will lead to inundation of low-lying areas around the world. One of the magnitude of one meter will certainly threaten the large populations living near the low-lying deltaic regions of the world, particularly populations of the Nile, Ganges-Brahmaputra, Mekong, Mississippi, Godavari, and

TABLE 14.5B Sources of Sea Level Rise and Magnitudes of Rise in Modern Time

	Sea Level Rise (mm/yr)			
	1961–2003		1993–2003	
Sources of Sea Level Rise	Observed	Modeled	Observed	Modeled
Thermal expansion	0.42 ± 0.12	0.5 ± 0.2	1.6 ± 0.5	1.5 ± 0.7
Glaciers and ice caps	0.50 ± 0.18	0.5 ± 0.2	0.77 ± 0.22	0.7 ± 0.3
Greenland Ice Sheet	0.05 ± 0.12		0.21 ± 0.07	
Antarctic Ice Sheet	0.14 ± 0.41		0.21 ± 0.35	
Sum of individual climate contributions to sea level rise	1.1 ± 0.5	1.2 ± 0.5	2.8 ± 0.7	2.6 ± 0.8
Observed total sea level rise	1.8 ± 0.5 (tide gauges)		3.1 ± 0.7 (satellite altimeter)	
Difference (Observed total minus the sum of observed climate contributions)	0.7 ± 0.7		0.3 ± 1.0	

(From *Climate Change 2007: The Physical Science Basis. Contribution of Working Group I to the Fourth Assessment Report of the Intergovernmental Panel on Climate Change.* © Intergovernmental Panel on Climate Change 2007. Published by Cambridge University Press. Used by permission of Intergovernmental Panel on Climate Change.)

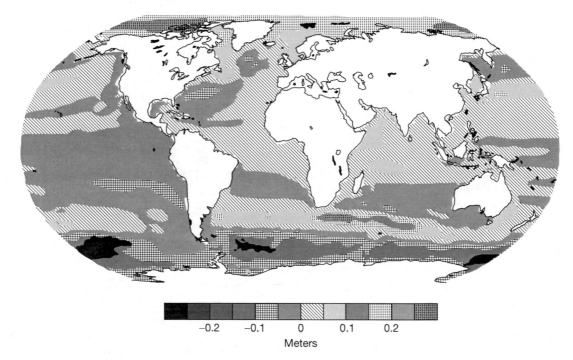

-0.2   -0.1   0   0.1   0.2
Meters

FIGURE 14.41 Sea level change in meters during the twenty-first century due to ocean density and circulation changes relative to the global average change in sea level (see Figure 14.40). Positive values indicate greater local changes in sea level than the global change, and negative values, the opposite. The results are for an ensemble of multimodels and were calculated as the differences between averages for 2080 to 2099 and 1980 to 1999 using the SRES A1B scenario. Notice in particular the higher than global average sea level for the high northern latitudes and the lower than global for the southern oceans. (From *Climate Change 2007 The Physical Science Basis. Contribution of Working Group I to the Fourth Assessment Report of the Intergovernmental Panel on Climate Change.* © Intergovernmental Panel on Climate Change 2007. Published by Cambridge University Press. Used by permission of Intergovernmental Panel on Climate Change.)

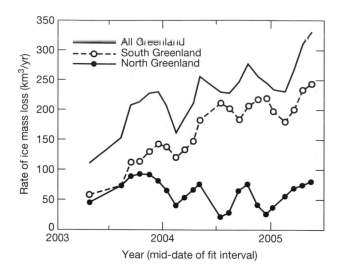

FIGURE 14.42 Ice mass loss from the Greenland ice sheet. The loss has been equivalent to 258 +/− 36 km²/yr or a global sea level rise of 0.5 +/− 0.1 mm/yr from April 2002 to April 2006. Notice that most of the increased rate of loss is from South Greenland, while North Greenland loss shows no definite trend. (*Source:* Velicogna and Wahr, 2006.)

Yangtze (Changjiang) deltaic areas. Twenty percent of the world's population lives on lands that could be inundated or dramatically changed by a sea-level rise of one meter. The areas most severely affected are likely to be the wetlands associated with river deltas and the low-lying lands of oceanic islands. An example of the former is that of Bangladesh, a country that is located in the great Bengal delta, with 50% of its land at an elevation of less than 5 meters. An extreme example of the latter is that of the Republic of the Maldives, a nation of 1190 small islands, rising only 2 meters above sea level. The possibility of more frequent and larger storm surges along coastal areas because of global warming worsens the situation for low-lying areas undergoing flooding owing to sea-level rise. Sea-level rise will also lead to changes in groundwater reservoirs associated with continental coastal areas and oceanic islands. For the Laura area on the western end of Majuro Atoll in the Marshall Islands, a one-meter rise in sea level could result in loss of nearly 30% of the land area of the island. The volume of the freshwater resource in the groundwater lens could be reduced by 50%.

It is not likely that sea-level rise will be rapid enough to drown coral reefs whose organisms must live near the sea surface. The vertical growth rates of reefs are rapid enough to keep up with sea-level rise resulting from climatic change. However, other factors related to climatic change, such as an increase in the acidity of ocean water because of the solution of anthropogenic carbon dioxide gas into the water, and an increase in temperature, will affect the growth rates of coral reef organisms.

CO₂ Exchange and Ocean Acidiffication The 1971 paper *Carbon dioxide—man's unseen artifact* by W. S. Broecker, Y.-H. Li, and T.-H. Peng was one of the first modern recognitions of the fact that carbon dioxide added to the atmosphere by fossil fuel combustion will lead to absorption of this gas by the ocean and the acidification, or lowering of the pH, of seawater. This was followed by the 1973 paper *Atmospheric carbon dioxide and radiocarbon in the natural carbon cycle: Changes for A.D. 1700 to 2070 as deduced from a geochemical model* by R. B. Bacastow and C. D. Keeling in which the authors recognized that as anthropogenic CO_2 is added to the atmosphere the "ocean surface water will become progressively more acid." Despite this early recognition of the problem, research concerning ocean acidification is only in its infancy, with a surge in research activities in the twenty-first century; thus, it is difficult to predict how individual species' responses to acidification will transfer through marine food webs and community structures and what the overall impact will be on marine ecosystems. However, it is reasonably well documented that prior to extensive emissions of anthropogenic CO_2 to the atmsphere, both the global coastal ocean and the open ocean were net sources of CO_2 to the atmosphere because of net heteroptropy and calcification (see Chapter 11, and Figure 14.43). With rising atmospheric CO_2 concentrations, the direction of the sea-to-air CO_2 exchange flux reversed, and the ocean began to absorb CO_2 in the late nineteenth century. The global coastal ocean at the time of the writing of this book was also tending toward being a net sink of atmospheric CO_2 (Figure 14.43). By the year 2050, the global ocean will be absorbing approximately 4 billion tons of carbon per year, or about 30% of the 14 billion tons of CO_2 emitted by combustion of fossil fuels under the BAU IS92a scenario of emissions.

Because the oceans are absorbing part of the anthropogenic emissons of CO_2 to the atmosphere, the content of dissolved inorganic carbon and the acidity of seawater will increase and the carbonate ion concentration and hence the carbonate saturation state of seawater will decrease (see Chapter 5, Box 5–1 for explanation of the CO_2–carbonic acid–carbonate system of seawater). The increasing acidity of the ocean has been referred to as the "other CO_2 problem." As can be seen in Figure 14.44, these changes are already occurring in parts of the Northern Hemisphere

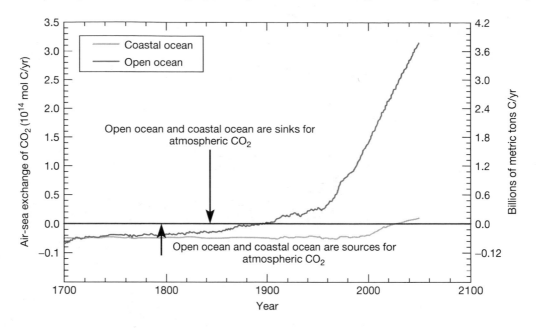

FIGURE 14.43 Oceanic uptake of anthropogenic CO_2 in 10^{12} mol C/yr and Gt C/yr (1 mol C = 12 grams) from 1700 to 2050 based on modeling. The global ocean prior to extensive fossil fuel and deforestation emissions of CO_2 to the atmosphere was a source of CO_2 to the atmosphere due to net heterotrophy and calcification. With rising atmospheric CO_2 concentrations, the open ocean became a net sink of CO_2 at about the beginning of the twenty-first century, but the coastal ocean is only now becoming a net sink.   (Courtesy of Julien Bouchez, Abraham Lerman, and Fred T. Mackenzie, 2009.)

subtropical oceans with surface water pH falling, and CO_2 and DIC rising, but what of the future? Figure 14.45 shows the results of one model of changes in surface seawater pH and saturation state from 1900 projected to 2100 assuming several different scenarios of emissions. The future trend in pH and saturation state depends very much on what happens to anthropogenic CO_2 emissions in the future. Under the BAU IS92a emissions scenario of the IPCC, the pH of surface seawater in the year 2100 would be about 7.85, a 4% reduction in pH from its late nineteenth century value of 8.18, representing a 114% increase in hydrogen ion concentration. The carbonate ion content of surface seawater would be reduced by 43%, resulting in a similar magnitude of reduction in the aragonite and calcite saturation state. Why the concern with the falling pH of surface seawater?

Changes in the pH and temperature of seawater can affect marine organisms and communities of organisms from simple bacteria to higher trophic level organisms. As an example of what might be anticipated, experimental evidence overwhemingly demonstrates that as the carbonate saturation state of seawater decreases, the rate of calcification of some marine calcifiers, like corals, coralline algae, and the pelagic Coccolithophoridae, will decrease (Figure 14.46a). In addition,

some marine calcifiers live under ambient seawater temperature conditions close to their lethal threshold (Figure 14.46b). Thus, it would be anticipated that with decreasing seawater pH and warmer sea surface temperatures in the future, the optimal conditions for coral reef growth might change for the worst and corals and coralline algae, the major framework builders of reefs, would have difficulty calcifying at present rates. In addition, because of the warmer waters, more coral bleaching events might be observed in the future, resulting in coral morbidity and mortality. Furthermore as the carbonate saturation state continues to fall, certain high-latitude regions of the surface oceans will go undersaturated with respect to aragonite and a range of compositions of calcite containing magnesium (the magnesian calcite minerals). Tropical waters will also go undersaturated with respect to a range of magnesian calcite compositions. Those organisms having shells and tests of these minerals will face conditions in which they will have extreme difficulty in calcifying and if the waters are sufficiently undersaturated, it is likely they could not calcify. Their shells and tests would be subject to dissolution.

Along with other stresses, the above seawater chemistry and temperature changes could lead to further degradation of coral reef ecosystems worldwide.

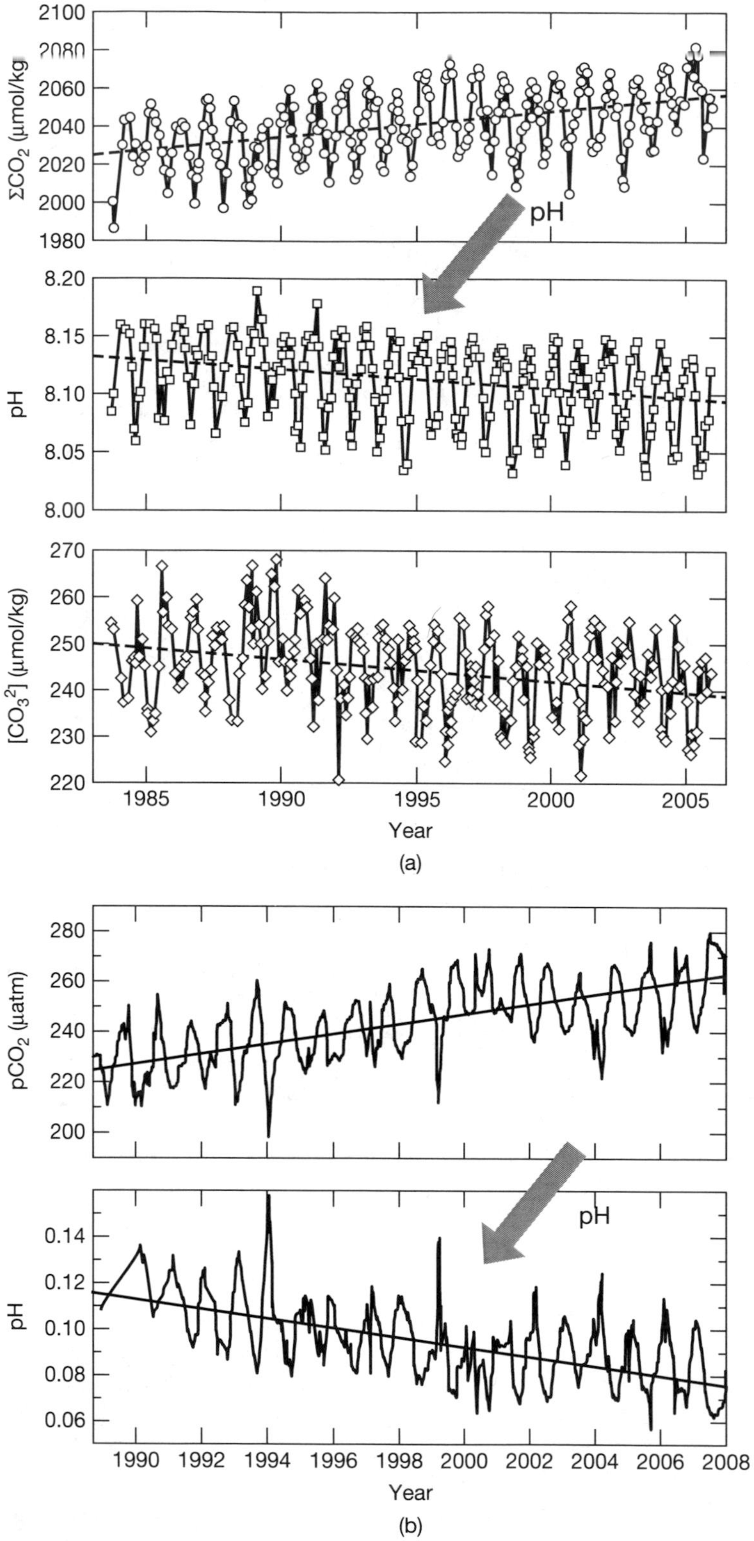

FIGURE 14.44 Time series of measurements of pH and dissolved inorganic carbon parameters for the surface waters of (a) Bermuda Atlantic Time Series (BATS) hydrostation in the North Atlantic and (b) Hawaii Ocean Time Series (HOT) hydrostation in the North Pacific. Notice the declining pH values at both stations. (*Source:* http://bats.bios.edu; http://hahana.soest.edu/hot/hot_jgofs.html.)

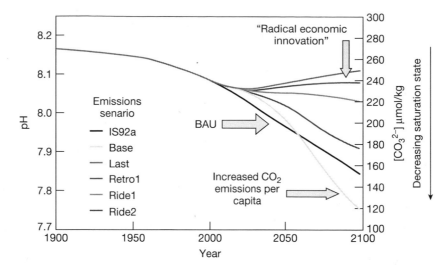

FIGURE 14.45 Time series of surface-ocean water pH values and carbonate ion (CO_3^{2-}) concentrations based on modeling from 1900 to 2100. The carbonate ion concentration is a measure of the carbonate saturation state of seawater: The lower the concentration, the less saturated the water. Time series results are shown for various CO_2 emissions scenarios, including the IPCC IS92a scenario and a variety of other scenarios that range from those in which there are radical economic innovations to reduce emissions to one in which there is increased CO_2 emissions per capita with growing world population and continued use of fossil fuels as the primary energy source. Notice that to avoid change in the future pH of surface seawater requires reductions in anthropogenic CO_2 emissions. (Courtesy of Wolf D. Grossman, Fred T. Mackenzie, and Andreas Andresson.)

For the Pacific and Caribbean regions, changes in surface seawater pH, carbonate saturation state, and temperature are likely to result in a reduction in the late nineteenth century area available for optimal reef growth to a state in the middle of this century where the environmental conditions are only marginal for reef growth (Figure 14.47). Furthermore, it has been predicted that when and if atmospheric CO_2 levels reach 560 ppm or perhaps somewhat higher, coral reef carbonate production will be less than overall dissolution and coral reefs will cease to grow and start to dissolve. Declining coral calcification rates have already been observed on the Great Barrier Reef of Australia. Although the causes of the decline are not altogether clear, the decline is likely due to declining carbonate saturation state and increasing temperature stress. The world is faced with the possibility that one of its most beautiful, diverse, and economically important marine ecosystems, coral reefs, may decline for generations to come. Continuing acidification of ocean waters can only be slowed or prevented by reductions in anthropogenic emissions of CO_2 to the atmosphere.

If the impacts of ocean acidification on calcifying organisms withstand the test of further research

work and time, then in the face of rising sea level, sea-surface warming, and lower seawater carbonate saturation state—plus the additional human-induced stresses of population growth and increased nutrient, toxic chemical, and sediment discharges to the coastal marine environment (see Chapter 11)—further degradation of coral reef and other carbonate ecosystems ecosystems could ensue. Coral reefs supply sediment to nearshore beaches and protect such areas from erosion. The degradation of these systems could allow storm surges to reach farther inland and increase the potential of inundation of low-lying areas. A combination of sea-level rise and these other stresses may make it difficult to live in many low-lying areas, particularly those of the Pacific island nations.

There are a number of other potential effects on the ocean system stemming from global warming. In the ocean, there is the possibility of changes in the rate of upwelling. Such a change could affect the rate of biological productivity in the ocean. At present, the oceans take up about 50 billion tons of carbon annually in net primary production. Global warming could result in an increase or decrease in organic production, depending on whether the upwelling of nutrient- and

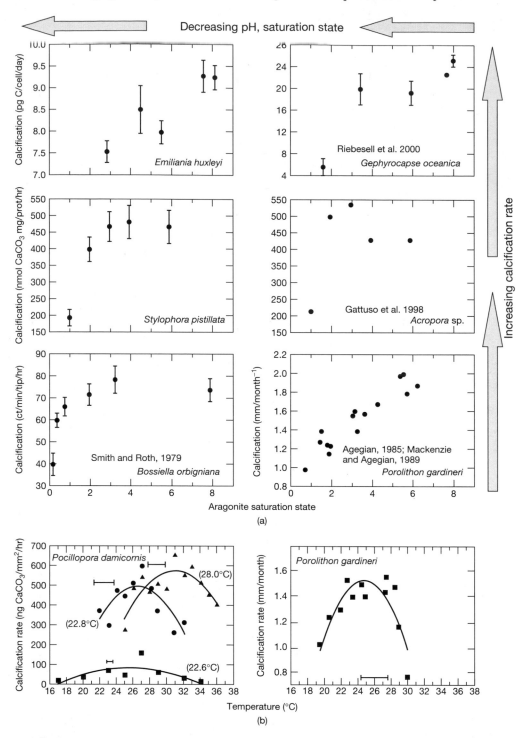

FIGURE 14.46 Experimental results for the effects of pH (carbonate saturation state) on the calcification rates of Coccolithophoridae, corals, and coralline algae (a) and temperature on the calcification rates of the coral *Pocillopora damicornis* and the coralline algal *Porolithon gardineri* (b). As pH declines, the rates of calcification of the three organism groups decline. The negative parabolic temperature-calcification rate relationships indicate that these calcifying organisms will at first calcify faster as temperature increases, but at higher temperatures the rate will slow. The ambient mean temperature at which the corals used in the experiments live is given by the temperatures in parentheses next to the curves and the range of temperatures by the horizontal bars. *Porolithon* generally lives in waters of 24°C to 28°C. (*Source:* Kleypas et al., 2006; Andersson, Mackenzie, and Lerman, 2005.)

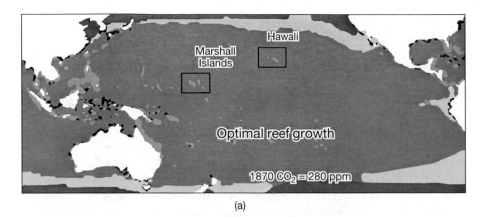

(a)

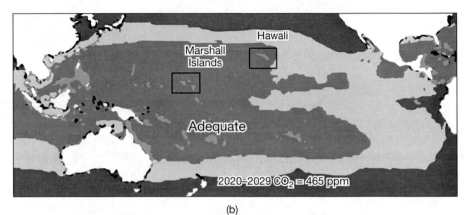

(b)

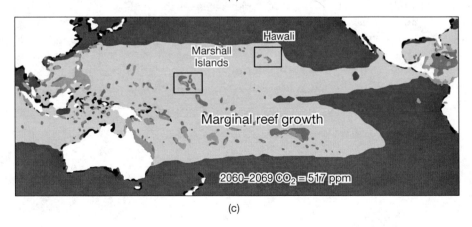

(c)

FIGURE 14.47 Evolution from the late nineteenth century to the mid twenty-first century of the development of Pacific Ocean and Caribbean seawater conditions marginal for reef growth due to changes in surface seawater carbonate saturation state and temperature. (*Source:* Guinotte et al., 2003.)

carbon-rich waters is increased or decreased. However, such changes probably would not greatly affect atmospheric carbon dioxide concentrations and hence global warming. Upwelling waters are not only enriched in nutrients, but in dissolved carbon (see Chapter 5). The dissolved carbon and nutrients are present in the waters in the proportions necessary for the production of organic matter. Thus, no carbon needs to be extracted from the atmosphere for organic production fertilized by upwelling nutrients of nitrogen and phosphorus. In fact, many upwelling regions are sources of CO_2 to the atmosphere. Global warming is most likely to result in a reduction in this source of CO_2 to the atmosphere.

A change in the upwelling rate of waters along coasts could have a significant impact on fisheries located in these regions. As mentioned previously, the warming of waters offshore of Peru because of the El Niño–Southern Oscillation events can lead to decreases in the rate of coastal upwelling. This has led to important declines in the fish catch along the Peruvian coast during these events. On one hand, global warming could increase the number of times that upwelling slows in areas like the Peruvian coastal margin. On the other hand, the possible increase in the intensity of winds along the western margins of continents associated with a global warming could enhance upwelling.

It is possible that global warming will also lead to changes in the structure of marine communities and their geographical distribution. For example, some organisms are very sensitive to the temperature of the waters in which they live. Plant and animal plankton are a good example. Certain species of plankton are associated with waters of a specific temperature range. If the geographical distribution of sea-surface temperatures changes because of global warming, one might expect to see a change in the spatial distribution of plankton and perhaps their species composition. Indeed, the North Pacific region of relatively low productivity and nutrient-deficient surface waters has already expanded to the east as the surface ocean has warmed, with poorly known consequences for pelagic fishes like tuna.

Because respiration and decay also are significantly affected by temperature change, it is likely that global warming will lead to increases in emissions of biologically produced gases from the sea surface. Some of these gases are greenhouse gases or, when vented to the atmosphere, react to produce greenhouse gases. Other gases have the potential in the atmosphere to generate aerosols that directly or through processes leading to cloud formation reflect radiant energy back to space and cool the planet. In addition, it has been reported that the oxygen content of that part of the thermocline (about 100 to 1000 meter in depth) that is ventilated by oxygen from the atmosphere by sinking of surface waters to depth has decreased in several ocean basins. The causes of this decline are not well known but may be related to changes in biological activity in the oceans, reflecting the changing balance between organic matter production and respiration, and/or may reflect a reduction in the ventilation of the thermocline due to warming of the surface waters of the ocean.

CRYOSPHERE RESPONSE TO CLIMATIC CHANGE More modern changes in the cryosphere and those anticipated for the future based on model calculations were touched upon above in terms of sea level rise. A few more observations are given here. The observed decade-averaged changes in snow cover for the Northern Hemisphere for the months of March through April from 1920 to 2009 are shown in Figure 14.48. Since about 1980, there appears to have been a more rapid reduction in the area of snow cover, which is correlated with 40°N to 60°N April temperatures, the higher the temperature, the less the

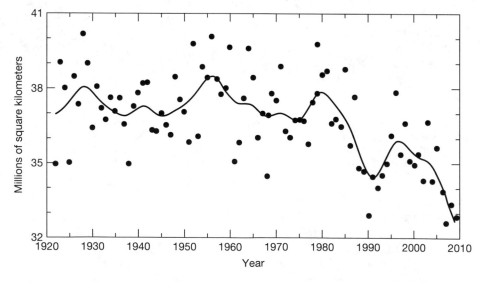

FIGURE 14.48 Trend in average snow-covered area in millions of square kilometers for the Northern Hemisphere from the early 1920s to 2009. (*Source:* http://www.natice.noaa.gov/ims/.)

snow cover. Just prior to 1980, global temperatures began to increase more rapidly, particularly in the high-latitude Northern Hemisphere. This may account for the abrupt change in the pattern of the area of snow cover starting in the late 1970s.

Another rather dramatic cryosphere response to warming of the planet has been the record of changes in the extent of sea ice in the Arctic (Figure 14.49a). There is some evidence that ice cover in the Arctic Ocean Basin has thinned more than 45% since 1950. More extensive recent data on the extent of sea ice thinning, reported as the area of ocean with at least 15% ice cover, have been available since 1979. Between 1979 and 2005, there was a loss of more than 1.6 million square kilometers of Arctic sea ice, equivalent to approximately 60,000 square kilometers per year or a 7% per decade rate of loss (Figure 14.50). In 2007 the Northwest Passage became open to ships

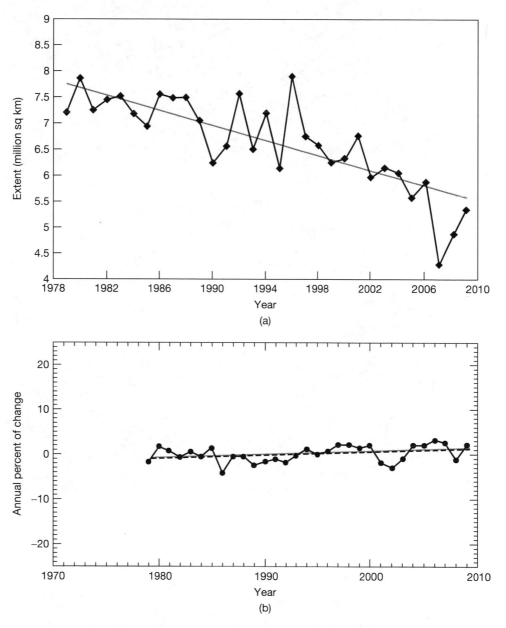

FIGURE 14.49 Annual summer minimum Arctic sea ice coverage in millions of square kilometers from 1979 to 2009 (a) and Antarctic sea ice extent anomalies calculated as percent change relative to the 1979 to 2000 mean of 18.7 million square kilometers (b). Linear best fit lines are also shown. While Arctic sea ice extent is generally shrinking, Antarctic sea ice extent appears to be increasing slightly despite recent breakups of ice shelves. (*Source:* http://nsidc.org/index.html.)

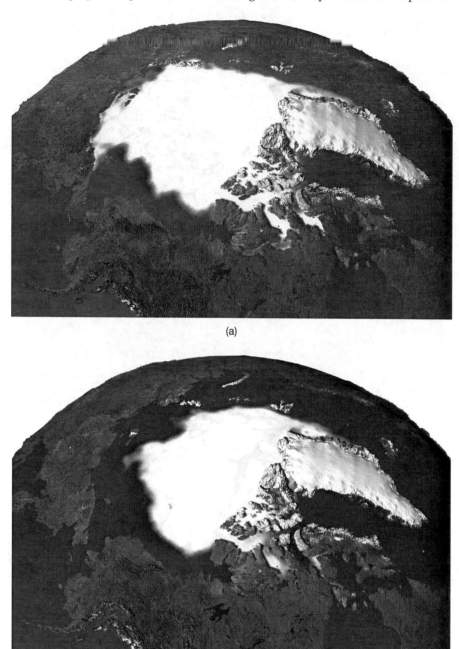

(a)

(b)

FIGURE 14.50 Satellite images of the minimum extent of Arctic sea ice in 1979 (a) and 2005 (b). (NASA, http://www.nasa.gov/home/index.html.)

without the need of an icebreaker and in September 2008 both the Northwest Passage and the Northeast Passage along Russia's Arctic coast were navigable. In 2008, the pattern of sea-ice loss reversed somewhat. It is not surprising that the extent of sea ice increased in 2008 since the global temperature anomaly for 2008 was a little lower than in preceding years due to an extended La Niña event (Figure 14.23). In 2100 the ensemble mean sea ice extent from model runs is predicted to be 3 million square kilometers, 63% decrease in area, but recent decadal loss is exceeding model prediction.

Antarctic sea ice has actually increased in area at an average rate of 0.7% per year from 1979 to June

2009 (Figure 14.49b), although there has been breakup and loss of sea ice from Antarctic ice shelves (see above). The increased growth could be the effect of changing weather patterns due to the ozone hole above Antarctica delaying the impact of global warming on the climate of the continent. Whatever the case, the sea-ice areal trends in the Arctic and Antarctic at the moment seem to be opposite of one another.

The retreat of Arctic sea ice has led countries fronting the Arctic Ocean to begin to assert their claims to sea floor under the Arctic in anticipation of exploiting the Arctic's resources, particularly for oil exploration. In August 2007, Russia laid claim to a large area of sea floor beneath the Arctic Ocean by planting a seawater-resistant titanium flag at a depth of 4200 meters below sea level at the North Pole. This is an excellent example of how climate change can impact the economic and political decisions of a nation.

LAND RESPONSE TO RISING ATMOSPHERIC CO_2 AND CLIMATE CHANGE The terrestrial biosphere response to climatic change is an uncertainty in the picture of a warming Earth caused by an enhanced greenhouse. The processes of both photosynthesis and plant and bacterial heterotrophic respiration increase with increasing temperature. At higher temperatures, respiration is generally more sensitive to a temperature increase than photosynthesis. Terrestrial ecosystems take up about 120 billion metric tons of carbon in gross primary production each year and release nearly the same amount in respiration and decay. Therefore, any change in these rates on a warming Earth could have an important effect on the carbon dioxide content of the atmosphere. One idea that has been reasonably well confirmed is that an increase in the growth rates of plants may occur with more carbon dioxide in the atmosphere (see Figure 14.10). It is known from experimental simulations that certain agricultural crop plants and a few perennials when subjected to increased carbon dioxide levels will increase their photosynthetic and growth rates. Increased atmospheric carbon dioxide may increase plant growth in a number of ways. These include stimulation of photosynthesis; depression of respiration; relief from water, low light, and nutrient stresses; and delay of senescence that prolongs the growing season. If some of this carbon taken up in increased growth were stored in plants, litter, or organic matter in soils, this storage might act as a sink of carbon dioxide produced by human activities. This storage represents a negative feedback of the terrestrial biosphere on global warming.

Also, increased carbon storage in organic matter may be aided by the increase in nutrients being added to the land surface because of the application of fertilizers to croplands and the atmospheric deposition of anthropogenic nitrogen (see Chapter 11). The fertilization or "greening" of Earth's surface potentially represents storage of anthropogenic carbon dioxide and as such is also a negative feedback of the terrestrial biosphere on global warming. There is much rapidly accumulating evidence that at least from the 1980s into the twenty-first century, this fertilization of Earth's surface was an important sink of anthropogenic CO_2 (Figure 14.5).

The future of this fertilization phenomenon is uncertain. With enhanced photosynthesis might come increased respiration, returning carbon dioxide to the atmosphere and canceling the stimulation effect on plants. As temperatures warm, respiration of carbon in litter and soils may increase and result in a period of relatively rapid and significant release of carbon dioxide, methane, and other biologically produced gases to the atmosphere. This release represents a positive feedback on an initial warming of Earth. This feedback is especially important in high-latitude tundra areas where much methane is stored in the organic soils of the region. Its release in a warming world would represent a significant flux of the greenhouse gas methane to the atmosphere. Figure 14.51 illustrates the distribution of permanently frozen, organic-rich ground (permafrost) in North America. The figure shows those areas of permafrost that are unstable, metastable, and stable with respect to a temperature change. A warming of the air of the high latitudes of North America would lead to the melting of substantial areas of permafrost. The melting could release significant quantities of carbon gases to the atmosphere acting as a positive feedback on warming of the Earth. As mentioned above, such melting has already been observed in Canada and Russia.

Table 14.6 shows various model results for the sensitivity of net primary production (NPP) to rising atmospheric CO_2 concentrations and climate change and the sensitivity of heterotrophic respiration to climate change. The interpretation for an ensemble of model results is that with a doubling of atmospheric CO_2, the NPP of the world's vegetation would increase by 48% +/− 20%; however, for a one degree Celsius temperature increase, the NPP would decrease by −1.3% +/− 2.6%, and the specific heterotrophic respiration rate would increase by 6.2% +/− 2.7%. The overall result of the combined CO_2 fertilization and climate perturbation gives rise to a decrease in the organic carbon storage in terrestrial vegetation and soils of −79% +/− 45% GTC/yr per one degree Celsius rise in temperature; that is, the

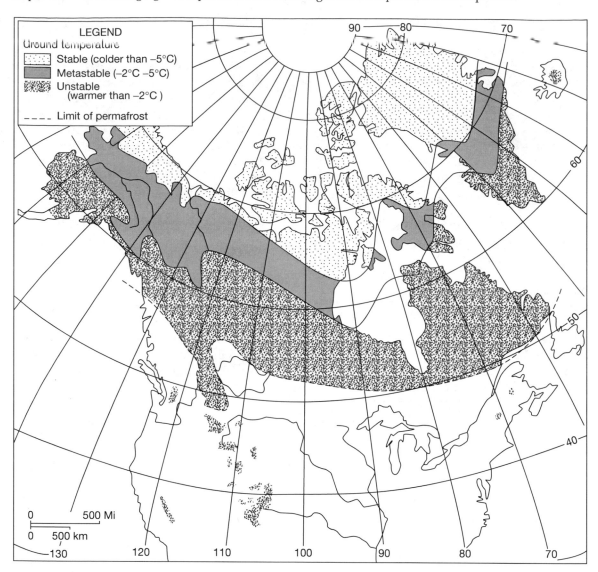

FIGURE 14.51 The distribution of permanently frozen, organic-rich ground (permafrost) in North America. The patterned areas show those regions that are stable, metastable, and unstable with respect to a global warming. For example, a warming of the air of the high northern latitudes of North America of only 2°C could result in the melting of a substantial area of permafrost. (*Source:* Harris, 1986.)

organic carbon sink on land would weaken, a positive feedback to rising atmospheric CO_2 and global warming (Table 14.7). It is also the case that the storage of carbon in the ocean will weaken over time by about $-30 +/- 15$ Gt C per degree Celsius rise in temperature, another positive feedback (Table 14.7). For certain regions of the ocean, the strength of the oceanic sink of anthropogenic CO_2 may be already weakening.

With rapid changes in temperature and precipitation, the geographical distribution of vegetation and animal habitats will change. There has already been voiced concern for polar bears and penguins

and their ability to adjust to a warmer world. The changes in vegetation ecosystems and animals associated with them will take place in part due to the changes in soil moisture patterns on a warming Earth (Figure 14.33b), which are the result of changes in the hydrologic cycle brought about by a warming planet. The distribution of arid and semi-arid lands and agricultural lands will change, as well as the water balances of whole regions. The potential change in soil moisture patterns also can be responsible for changes in the distribution of microbial and insect life and pests. Thus, soil moisture changes are one of the causes of changes in the

TABLE 14.6 Sensitivity of Net Primary Production (NPP) to CO_2 and Temperature and Sensitivity of Heterotrophic Respiration to Temperature Change. Results from Different Models are Shown with the Mean and Standard Deviation of Model Runs

Model	Sensitivity of Vegetation NPP to CO_2: % Change for a CO_2 Doubling	Sensitivity of Vegetation NPP to Climate: % Change for a 1°C Increase	Sensitivity of Specific Heterotrophic Respiration Rate to Climate: % Change for a 1°C Increase
A. HadCM3LC	57	−5.8	10.2
B. IPSL-CM2C	50	−4.5	2.3
C. MPI-M	76	−4.0	2.8
D. LLNL	73	−0.4	7.0
E. NCAR CSM- 1	34	0.8	6.2
F. FRCGC	21	1.2	7.2
G. UVic-2.7	47	−2.3	6.5
H. UMC	12	−1.6	4.8
I. BERN-CC	46	1.2	8.7
J. CLIMBER2-LPJ	44	1.9	9.4
K. IPSL-CM4-LOOP	64	−0.3	2.9
Mean	48	−1.3	6.2
Standard deviation	±20	±2.6	±2.7

(*Source:* IPCC, 2007a.)

exchange of microbially produced biogenic gases between the land surface and the atmosphere on a warming Earth. Also, some plant and animal species will increase in abundance and others will decrease. Terrestrial ecosystems will change in structure and their geographical distribution will change along with species habitats (Figure 14.52). Some ecosystems may be displaced to higher latitudes or altitudes.

For example, the tundra region of the high northern latitudes is likely to shrink in area while the boreal forest moves north in its extent. The sparsely vegetated region of Asia is projected to spread in area, as is the semiarid region of the southwest United States. However, large ecosystems like forests may not be able to migrate quickly enough to keep up with environmental changes. In addition, on a warmer Earth, it is likely that terrestrial plant species would be more impoverished, with bushes and shrubs replacing trees in some regions. Figure 14.53 shows the current and projected range of sugar maple trees in eastern North America. These trees are valued as a sugar maple source and for their intrinsic

beauty, particularly during the autumn, when the leaves turn brilliant shades of red and orange. As the climate shifts because of global warming, the climatic zone of sugar maple might shift northward into Canada. This shift could result in economic harm to the sugar maple and tourist industries of the New England states of the United States Both industries help make New England famous.

Concluding Comments

It is widely accepted that a natural planetary greenhouse effect exists, that greenhouse gases are involved in climatic changes of the past, and that the concentrations of greenhouse gases in the atmosphere are increasing at a rapid rate owing to human activities. What are debated is the modern enhanced greenhouse effect and its effects on the ecosphere. Figure 14.54 is a schematic, composite representation of the changes in global mean temperature and physical and biological systems reported for 1970 to 2004. The dots on the diagram represent areas for which there have been reported observational changes in snow, ice, and frozen ground, hydrology,

TABLE 14.7 Impact of Carbon Cycle Feedbacks from Several Models with the Mean and Standard Deviation of Model Results. Column 2 Shows the Impact of Climate Change on Atmospheric CO_2 Concentration in 2100, and Column 3 the Amplification of the Atmospheric Increase—the Climate–Carbon Cycle Feedback Factor. The Other Columns Show the Sensitivities of Various Feedbacks to CO_2 and Climate

Model	Impact of Climate Change on the CO_2 Concentration by 2100 (ppm)	Climate–Carbon Feedback Factor	Transient Climate Sensitivity to Doubling CO_2 (°C)	Land Carbon Storage Sensitivity to CO_2 (GtC/ppm)	Ocean Carbon Storage Sensitivity to CO_2 (GtC/ppm)	Land Carbon Storage Sensitivity to Climate (GtC/°C)	Ocean Carbon Storage Sensitivity to Climate (GtC/°C)
A. HadCM3LC	224	1.44	2.3	1.3	0.9	−175	−24
B. IPSL-CM2C	74	1.18	2.3	1.6	1.6	−97	−30
C. MPI-M	83	1.18	2.6	1.4	1.1	−64	−22
D. LLNL	51	1.13	2.5	2.5	0.9	−81	−14
E. NCAR CSM-1	20	1.04	1.2	1.1	0.9	−24	−17
F. FRCGC	128	1.26	2.3	1.4	1.2	−111	−47
G. Uvic-2.7	129	1.25	2.3	1.2	1.1	−97	−43
H. UMD	98	1.17	2.0	0.2	1.5	−36	−50
I. BERN-CC	65	1.15	1.5	1.6	1.3	−104	−38
J. CLIMBER2-LPJ	59	1.11	1.9	1.2	0.9	−64	−22
K. IPSL-CM4-LOOP	32	1.07	2.7	1.2	1.1	−19	−17
Mean	**87**	**1.18**	**2.1**	**1.4**	**1.1**	**−79**	**−30**
Standard deviation	±57	±0.11	±0.4	±0.5	±0.3	±45	±15

(*Source:* IPCC, 2007a.)

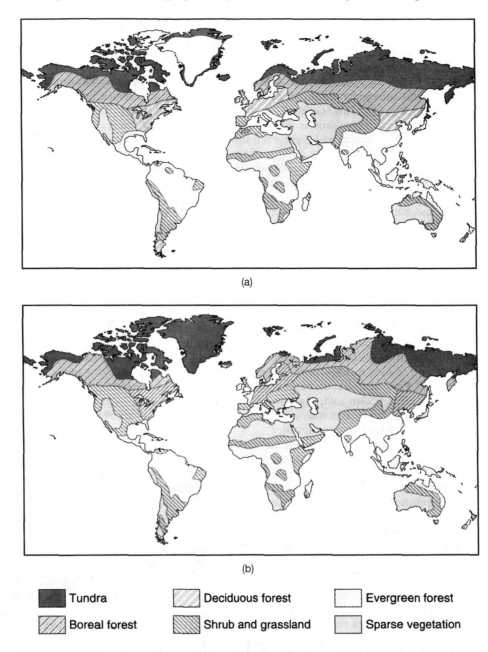

(a)

(b)

Tundra   Deciduous forest   Evergreen forest

Boreal forest   Shrub and grassland   Sparse vegetation

FIGURE 14.52 Model calculation of the distribution of existing vegetation (a) and the distribution of vegetation in a greenhouse world (b). Notice in particular the predicted spread of the boreal forest northward, the shrinkage of the area of tundra, and the spread of the area of sparse vegetation in Asia in a warmer world. (*Source:* Prentice and Sykes, 1995.)

and coastal processes, and terrestrial, marine, and freshwater ecosystems. Despite the fact that there is more land area and greater population in the Northern Hemisphere, variables that could affect the observational record, it is very evident that temperate and polar regions of the Northern Hemisphere, which have experienced greater temperature changes during 1970 to 2004, exhibit most of the

areas of reported changes in physical and biological systems. This could be a fortuitous result, but it strongly suggests that the increases in global and regional mean temperatures are playing an important role in modification of processes at the Earth's surface.

Most discussion of the future effects of the accumulation of greenhouse gases in the atmosphere is

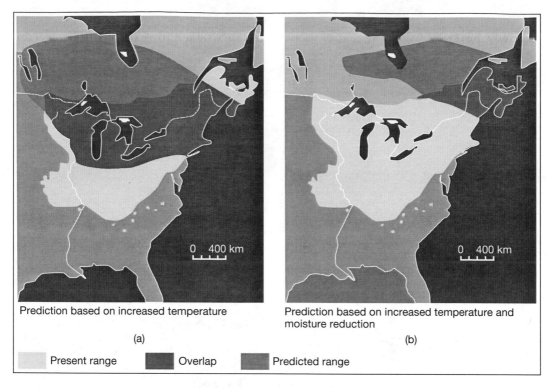

Prediction based on increased temperature

(a)

Prediction based on increased temperature and moisture reduction

(b)

Present range   Overlap   Predicted range

FIGURE 14.53 Current and projected ranges of sugar maple trees in eastern North America. The two predictions are based on increases in temperature only (a) and on increased temperature and reduction in soil mosture (b). (*Source:* Office of Science and Technology Policy, 1997.)

based on general circulation model results. These results are not necessarily in agreement with observations of global or regional historical environmental records. The observational records show trends of a moderate increase in temperature in all regions of the troposphere; increases in ocean heat content, sea surface temperatures, and acidity; a moderate increase in sea level; increases in atmospheric water vapor, precipitation, and cloudiness; and a moderate decrease in stratospheric temperatures in recent decades. There is substantial evidence for recession of mountain glaciers and an increase in the altitude of snowlines in mountain ranges, melting of the Greenland ice cap, and decreases in the area of Northern Hemisphere snow cover and Arctic sea ice. There is also evidence as warmer isotherms have moved farther north and south of the equator with global warming that vegetation tolerant of warmer conditions is slowly moving poleward accompanied by insects and bird populations adapted to the warmer conditions. Most observations are in accord with the predictions of the global circulation models but some are of a lesser magnitude. It had been anticipated from climate models that a change in the diurnal temperature range might continually occur

with global warming. This has not been observed to date and day- and night-time temperatures have risen at about the same rate, although the changes are highly variable from region to region.

For the United States, which has reasonably good climatic records for this and the past century, there are some trends in climatic variables in accord with projections for the effects of global warming on various regions. However, the trends probably require a longer time series to substantiate the pattern of changes. For example, the conterminous United States appears to be getting slightly warmer, cloudier, and wetter over recent decades. Still, since the early 1900s, there is no strong evidence for an increase in the frequency of hurricanes or severe hurricanes making landfall in the United States, as might be expected from some model scenarios of global warming. In general changes in tropical storm and hurricane frequency and intensity are masked by large natural variability. The behavior of hurricanes in a warmer world is still not resolved, although if sea surface temperatures play a dominant role in hurricane formation and duration, as some theories of hurricane formation suggest, one might anticipate an increase in at least frequency.

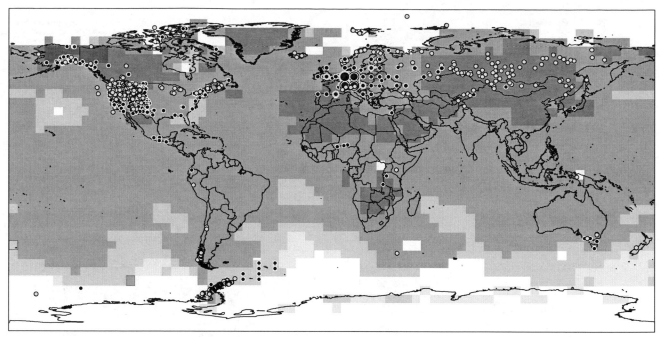

○ Physical systems (snow, ice, and frozen ground; hydrology; coastal processes)

⊛ Biological systems (terrestrial, marine, and freshwater)

Temperature change °C
1970–2004

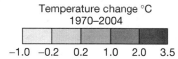

−1.0 −0.2 0.2 1.0 2.0 3.5

FIGURE 14.54 Locations of significant changes in physical and biological systems consistent with warming as a possible cause of the changes superimposed on the pattern of temperature change from 1970 to 2004. The correlation of the locations with significant physical and biological changes with regional areas of significant warming points to a linkage between the two.   (From *Climate Change 2007: Impacts, Adaptation and Vulnerability. Contribution of Working Group II to the Fourth Assessment Report of the Intergovernmental Panel on Climate Change.* © Intergovernmental Panel on Climate Change 2007. Published by Cambridge University Press. Used by permission of Intergovernmental Panel on Climate Change.)

How the climate system will react to the rapid change in atmospheric composition is not entirely predictable at the moment, but there exists general consensus among many scientists that global warming owing to human activities is occurring. Some scientists still dispute this claim, thus leaving the public and policy makers in a quandary as to who is right and subject to certain biases in the scientific community. The future of climatic change does not simply depend on the ability of changes in greenhouse gases to induce a change in climate but on a number of other factors that exert a control on climate (Figure 14.55) and are discussed in this book. One important factor that has been recognized, particularly since the early 1990s, is of particular note: the link between anthropogenic emissions of oxidized sulfur (Figure 12.19) to the atmosphere and climate. Emissions of SO_2 to the atmosphere have led to an increased mass (burden) of atmospheric sulfate

aerosol, particularly in the Northern Hemisphere (see Chapter 12). The enhanced sulfate aerosol burden of the atmosphere is surely responsible for part of the discrepancy between the observed temperature record of the past century and that predicted on the basis of radiative forcing due only to the increases in greenhouse gases in the atmosphere. When the cooling effect of the sulfate aerosols is introduced into models of climate change, the calculated temperatures fit better the global mean surface temperature record of the past century. The overall effect of atmospheric aerosols resulting from emissions from the combustion of fossil fuels, biomass burning, and other human activities is a direct climate forcing of about −0.5 watts per square meter (W/m^2) and perhaps more than −1.2 W/m^2, if the indirect effects of aerosols on climate are included (Figure 14.56; e.g., cloud formation and effect on planetary albedo). The inclusion of sulfate aerosols in climate models attempting to predict future

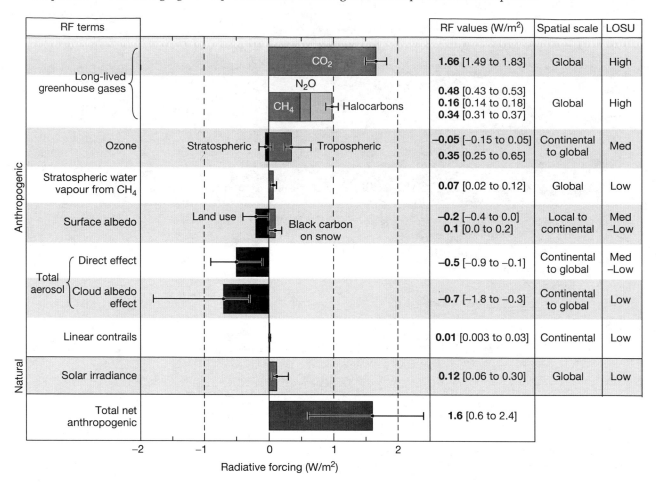

RF terms					RF values (W/m^2)	Spatial scale	LOSU

FIGURE 14.55 The global mean radiative forcing values in W/m^2 of the climate system owing to changes in the concentrations of greenhouse gases and aerosols in the atmosphere, stratospheric ozone depletion, natural changes in solar radiation, and other factors for 2005, relative to 1750. Radiative forcing in W/m^2 is a measure of the ability of a variable to induce a change in climate. Bars to the right of the zero line denote a positive forcing and hence a warming influence, whereas those to the left denote a negative forcing and hence a cooling influence. The I-beams associated with each bar show the uncertainty range of the estimates. The degree of scientific understanding of the influence of each variable on climate is shown in the column to the far right. The spatial scale of the forcing is also shown. A net positive radiative forcing of about 1.6 W/m^2, indicating a warming influence, is the best estimate of the global mean radiative forcing during the past 255 years.   (From *Climate Change 2007 The Physical Science Basis. Contribution of Working Group I to the Fourth Assessment Report of the Intergovernmental Panel on Climate Change.* © Intergovernmental Panel on Climate Change 2007. Published by Cambridge University Press. Used by permission of Intergovernmental Panel on Climate Change.)

climate change significantly lowers the rate of temperature increase during the twenty-first century and the regional patterns of global climatic change (Figure 14.56).

In addition, natural variations of solar radiation determine to some extent the temperature record of the planet in the past and are important in climate models attempting to calculate temperatures of the future. The overall radiative forcing over the past couple of centuries due to variations in solar irradiance is about 0.12 +/− 0.6 to 0.30 W/m^2 (Figure 14.55). However,

although highly questionable, one should keep in mind the possible natural effect on the temperature of the planet due to the solar wind effect on the galactic cosmic ray flux to the Earth's atmosphere mentioned above. Regardless of this phenomenon and other natural forcings, the conclusion of many scientists is that the global mean surface temperature record of the past century, particularly since 1970 and into the twenty-first century, shows strong evidence of an anthropogenic influence, but this conclusion is not without its detractors!

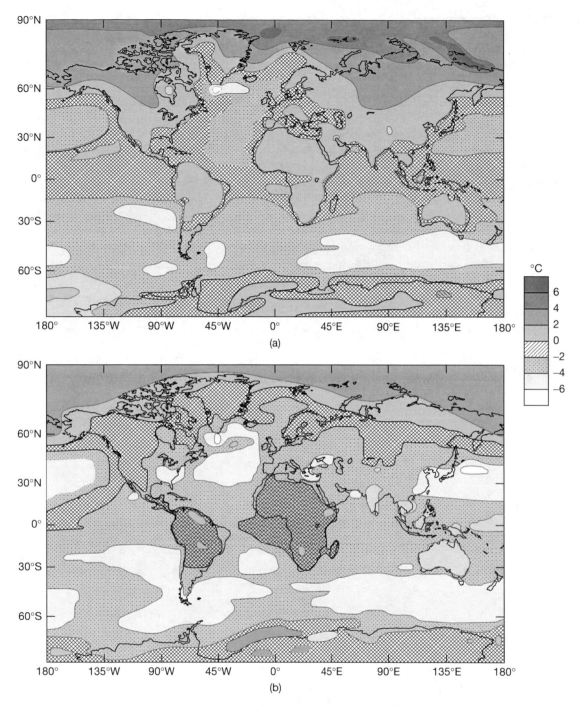

FIGURE 14.56 Examples of model calculations of annual mean temperature changes in degrees Celsius from 1795 to 2050. In (a) only the effect of increasing greenhouse gases in the atmosphere on the temperature change is considered, whereas in (b) the calculated temperature response is to both increasing greenhouse gases and aerosols in the atmosphere. Notice how the atmospheric aerosols, which are concentrated principally over continents and in the Northern Hemisphere, tend to cool the planet, mainly over continental regions.   (*Source:* Kattenberg et al., 1996.)

STRATOSPHERIC OZONE AND THE HOLE IN THE SKY

The Importance of Ozone

Ozone (O_3) is an important trace gas that is found in low and variable concentrations from the surface of the planet to a height of some 60 kilometers. The gas accounts for a trace to 0.0008% of all the gases in the atmosphere. About 10% of the gas, tropospheric ozone, resides in the lower atmosphere. Approximately 90% of the ozone in the atmosphere is found in the stratosphere in a layer from 19 to 48 kilometers above the surface of the planet in tropical regions and at lower altitudes in polar regions. Peak concentrations of ozone occur at about 30 kilometers in the tropical stratosphere and at about 20 kilometers in the polar stratosphere. This layer of ozone gas is naturally produced and constitutes the protective stratospheric ozone layer (see the following). There is strong evidence that ozone is being depleted from the stratosphere by the release of gases resulting from human activities and their chemical reactions in the stratosphere (Figure 14.57).

Stratospheric ozone serves two major functions that maintain our planet as suitable for life. Ozone gas in the stratosphere forms a thin protective layer around Earth that filters out 99% of the ultraviolet radiation (UV) reaching the planetary atmosphere from the sun. Ozone thus acts as a shield and protects life on Earth from destructive UV radiation. Ultraviolet light comprises 2% of the solar radiation reaching Earth's atmosphere. The wavelength of this energy spans the range of 1 to 400 nanometers (1 nanometer = 1 billionth of a meter) and is invisible to humans (see Box 4–1: Energy). Ultraviolet radiation from 320 to 400 nanometers is known as UV-A and generally passes through the ozone layer. It is relatively harmless radiation and actually facilitates the repair of damaged photosynthetic mechanisms. The lower range of these wavelengths, known as UV-C (200 to 280 nm), is lethal to

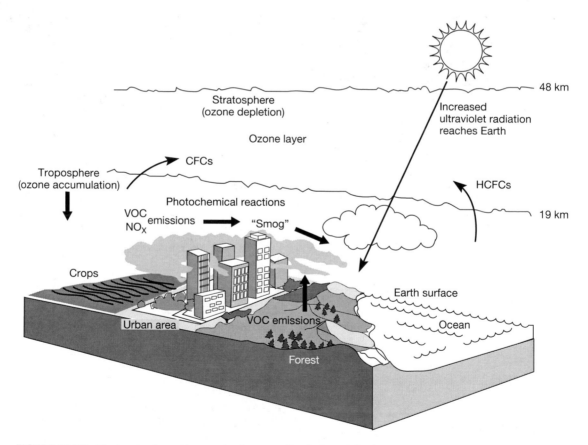

FIGURE 14.57 Stratospheric and tropospheric ozone. Depletion of the former can lead to increased ultraviolet radiation reaching Earth's surface and human health and ecosystem risks. Increase in the latter at near-ground atmospheric levels helps to create smog (see Chapter 12), a human health and ecosystem hazard. Also, tropospheric ozone is a greenhouse gas. It is possible that increased UV radiation reaching Earth's surface because of stratospheric ozone depletion could increase the production of photochemical smog near Earth's surface.

life, but this radiation is nearly totally absorbed by ozone in the atmosphere. UV-B radiation of wavelength 280 to 320 nanometers is also lethal to life even at very low intensities. Stratospheric ozone absorbs most, but not all, UV-B radiation. UV-B radiation that does reach the planet's surface can cause eye damage, skin cancer, and sunburns; negatively affect the body's immune system; reduce plant growth; and damage aquatic life. UV-B radiation affects specific plant structures like chloroplast membranes, substances like DNA, and enzymatic activity associated with processes like photosynthesis.

Ozone also plays an important role in regulating planetary temperature. In the stratosphere, ozone absorbs incoming UV radiation and warming occurs. This warm region caps the climate system of Earth below in the troposphere and helps maintain the relatively equable temperatures of the planetary surface. Ozone is an excellent example of a resource owned and shared globally.

The Measurement of Ozone

It is relatively easy to detect ozone in the atmosphere. However, it has proved difficult to make sufficiently precise and numerous measurements with extensive geographical coverage to determine changes in concentrations of only a few percent per decade. Stratospheric ozone concentrations are measured using both ground-based and satellite instrumentation. Many ground-based ozone measurement stations are distributed over the globe (e.g., http://www.esrl.noaa.gov/gmd/ozwv/). Four are in Antarctica, in the region of the now well-known ozone hole. The ground-based stations record by absorption spectroscopy the total amount of ozone overhead. This method of analysis can use the sun, the moon, or a star as a source of light. The light is reduced in intensity when it passes through Earth's atmosphere. The amount of reduction depends on the total number of ozone molecules in the stratosphere overhead. The light is recorded on an absorption spectrometer, leading to a measurement of the total amount of ozone in the column of air above the station. Balloons and rockets can also be released from the ground and are used to obtain the distribution of the concentration of ozone with height above the planetary surface, giving a profile of ozone.

Most recently, instruments for detecting ozone launched onboard satellites have led to near-global coverage of ozone measurements and contributed significantly to our knowledge of the behavior of this gas. For example, the Total Ozone Mapping Spectrometer

(TOMS) was flown on the *Nimbus 7* spacecraft from 1978 until early May 1993 (http://jwocky.gsfc.nasa.gov/n7toms/n7sat.html). A second TOMS was launched on the Russian *Meteor 3* satellite in August 1991. This instrument worked until December 1994. The instruments aboard these satellites mapped total ozone in a column of air over much of the globe. The Ozone Monitoring Instrument (OMI) aboard the National Atmospheric and Aeronautics Administration (NASA) Aura spacecraft (http://aura.gsfc.nasa.gov/instruments/omi.html) continues to record atmospheric ozone using solar backscatter ultraviolet spectrometers. These spectrometers measure ozone by observing the amount of solar radiation scattered back from the atmosphere at 12 wavelengths between 255 and 340 nanometers. Appropriate choice of wavelengths at which ozone absorbs and does not absorb radiation allows for the calculation of both the total column and the vertical concentration profile of ozone in the atmosphere. Figure 14.58 shows one method for determining total ozone in a column of air using the ratio of the amount of UV-B radiation reflected from Earth's atmosphere in comparison to that emitted from the sun and received directly by satellites.

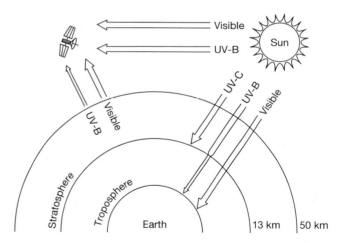

FIGURE 14.58 The fate of UV-A, UV-B, UV-C, and visible solar radiation in the stratosphere and troposphere can be determined from satellite-based measurements. The physical interactions of UV-A radiation (here included with visible radiation) are similar to those of visible radiation. Instruments in satellites measure the ratio of long- and shortwave radiation emanating directly from the sun and the ratio of long- and shortwave radiation reflected back into space from the atmosphere. The increased differences between these two ratios indicate decreasing thickness of the ozone layer. More UV-B radiation is reflected back into space where the ozone layer is thickest. (*Source:* Behrenfeld and Chapman, 1991.)

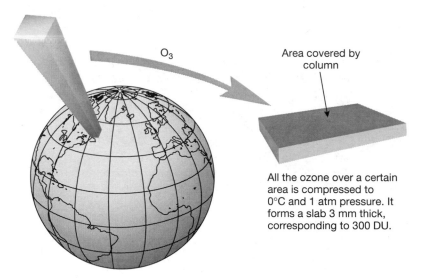

FIGURE 14.59 Schematic diagram showing a visualization of the meaning of the Dobson Unit as a concentration measure of atmospheric ozone.

The standard for measurement of the concentration of atmospheric ozone is referred to as the *Dobson unit*, named after G. M. B. Dobson, who, along with his colleagues in the late 1920s, presented the first measurements of the distribution and variability of the ozone layer. They showed that stratospheric winds play a strong role in moving ozone around the globe. One Dobson unit is equivalent to the amount of ozone present in a column of air 1/100 of a millimeter thick at 0°C and 1 atmosphere of total pressure. The average thickness of ozone around the planet is only 3.2 millimeters under these conditions. This amounts to 320 Dobson units, which is roughly equivalent to 8.6×10^{18} molecules of ozone per square centimeter of Earth's surface (Figure 14.59). Ozone concentrations are also given in millipascals, where 1 pascal (Pa) $= 10^{-5}$ bar $= 9.8692 \times 10^{-6}$ atmosphere of pressure.

The Ozone Balance

Ozone formation is a photochemical process. The amount of ozone produced in the stratosphere is almost entirely dependent on the amount of solar radiation received from the sun. The upper atmosphere is dry and cold so that the production of hydroxyl radical from water is not important and thus it does not play as strong a role in the chemistry of the stratosphere as it does in the troposphere (see Chapters 4 and 12). Chemical reactions at these high altitudes generally involve atomic oxygen.

Ozone is formed in the stratosphere when solar radiation of wavelength 180 to 240 nanometers is absorbed by diatomic oxygen molecules (O_2) (Figure 14.60). This absorption of solar radiation splits the oxygen molecule into two, single oxygen atoms (O). The two oxygen atoms independently recombine with two other oxygen molecules to form

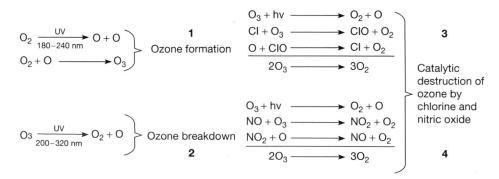

FIGURE 14.60 (1) Production and (2) destruction of ozone in the atmosphere. There may be a net loss in ozone over time due to reactions involving anthropogenic compounds of (3) chlorine (bromine and fluorine), (4) nitrogen, and hydrogen, which act as catalysts and increase the rate of destruction of ozone. (*Source:* Clarke, 1987.)

two separate ozone (O_3) molecules. When ultraviolet light of wavelength 200 to 320 nanometers strikes an ozone molecule, the molecule is destroyed by being split into an oxygen molecule and a single oxygen atom. In the process, heat is given off, which warms the stratosphere. If the process ended there, ozone would eventually be depleted, allowing UV light to penetrate to the planetary surface. However, the oxygen molecule reunites with an oxygen atom and produces ozone again (Figure 14.60; see also Chapter 4).

In addition to the breakdown of ozone by ultraviolet light, the atmosphere contains small amounts of natural gases that cause the destruction of ozone molecules (Figure 14.60). These natural gases are normally in a state of dynamic balance. The natural atmospheric gases include the oxides of nitrogen, chlorine, and hydrogen. These natural gases help to maintain stratospheric ozone in balance. However, compounds of these elements also have sources related to human activities. Chlorine in chlorofluorocarbons (CFCs) and bromine in halons are particularly important.

The combination of production and destruction of ozone as a result of UV radiation and natural trace gases in the atmosphere was in a dynamic equilibrium prior to human interference in the system. This resulted in a relatively stable amount of ozone in the stratosphere, because ozone was produced as fast as it was destroyed, resulting in the balanced system. Thus, we had a planet protected from UV radiation, as well as with relatively stable stratospheric temperatures increasing with altitude.

The Ozone-Depleting Chemicals

Trace gases released at Earth's surface to the atmosphere are emitted from natural and human sources. The natural release of these gases has been part of the system that has maintained ozone in balance in the atmosphere. However, elements of hydrogen, nitrogen, chlorine, bromine, and fluorine are found in compounds that are produced for, or as by-products of, industrial and agricultural practices. These elements are combined with one another and other elements to form gaseous molecules such as nitrous oxide, methyl chloride, methane, halons, and chlorofluorocarbons. Some of these trace gases linger in the atmosphere for many years and consequently are transported into the stratosphere. There they may speed up the breakdown of stratospheric ozone (Table 14.8).

CFCs have historically been used as propellants in spray cans, in the production of Styrofoam products, in cleaning solvents, and as a coolant in refrigerators and air conditioners. CFCs are the primary anthropogenic ozone-depleting chemicals. Once CFCs

reach the stratosphere they decompose in the presence of intense ultraviolet light. As a result, chlorine is produced. Chlorine has a great appetite for ozone. A single molecule of chlorine can devour thousands of molecules of ozone. Because of the long lifetime of CFCs in the atmosphere, the potential for further destruction of stratospheric ozone is highly possible. Figure 14.61 illustrates how CFCs destroy stratospheric ozone.

Halons (halocarbons, e.g., $CBrF_3$ and $CBrClF_2$), methane, and nitrogen are also culprits. Halons are used in fire extinguishers and are potential ozone-depleting synthetic chemicals. Methane is a trace gas that is released during rumination processes in cattle, from cultivated rice paddies, and from burning fossil fuels and biomass. Nitrogen in the form of nitrogen oxides is released into the stratosphere from the exhausts of high-flying aircraft. In addition, nitrogen-bearing fertilizers, used extensively in agriculture, release nitrous oxide into the atmosphere during the process of denitrification of the fertilizers. Despite the reductions in ozone-depleting synthetic chemical emissions, there is still considerable worldwide concern for the observation that ozone in the stratosphere is still below concentration levels observed prior to anthropogenic emissions of ozone-depleting trace gases.

History of Ozone Depletion

Concern about the destruction of the ozone layer began in the early 1970s with the proposal to use high-altitude supersonic airplanes for civilian and military purposes. The aircraft were to be flown in the lower stratosphere where ozone-destroying nitrogen oxides could be released in their exhaust fumes. However, the civilian fleets of aircraft did not materialize. Also, the early predictions of large ozone loss because of the potential use of these aircraft were later shown to be excessive. In the mid-1970s, attention was switched from supersonic aircraft to the aerosol spray can. A propellant is mixed in these cans to expel its contents. The propellants were the CFCs. When first developed, CFCs were considered environmentally safe. However, the effects of these compounds on stratospheric ozone are devastating.

By the late 1970s and early 1980s, the United States, Canada, Norway, Denmark, and Sweden had begun to ban the use of CFCs in aerosol cans in response to scientific studies showing that these chemicals could deplete ozone in the stratosphere. However, CFCs had become useful in many industrial products such as Styrofoam cups, insulation and padding, refrigerators, air conditioners, and solvents for cleaning

TABLE 14.8 Trace Gases Involved in Radiative Forcing (Warming) and those Relevant to Radiative Forcing and Ozone Depletion. Lifetime is a Measure of the Average Age of the Gas in the Atmosphere

Gas	Formula	Abundance (Year 1750)	Abundance (Year 2008)	Radiative Forcing (W/m²)	Lifetime (Years)	GWP[a]
		Gases relevant to radiative forcing only				
CO_2		278 ppmv	387 ppmv	1.66	50–200	1
CH_4		700 ppbv	1,785 ppbv	0.48	10	35
N_2O		270 ppbv	322 ppbv	0.16	120	298
CF_4		40 pptv	~80 pptv	0.003	50,000	7,390
C_2F_6		0	~3 pptv	0.001	10,000	12,200
SF_6		0	~5,8 pptv	0.002	3,200	22,800
HFC-23	CHF_3	0	~18 pptv	0.003	260	14,800
HFC-134a	CH_2FCF_3	0	~33 pptv	0.0055	14	1,430
HFC-152a	CH_3CHF_2	0	~2 to 5 pptv	0.0004	1.4	124
		Gases relevant to radiative forcing and ozone depletion				
CFC-11	CCl_3F	0	~251 pptv	0.063	50	4,750
CFC- 12	CCl_2F_2	0'	~540 pptv	0.17	102	10,900
CFC-13	$CClF_3$	0	~4 pptv	0.02	400	14,400
CFC-113	$C_2Cl_3F_3$	0	~78 pptv	0.024	85	6,130
CFC-114	$C_2Cl_2F_4$	0	~15 pptv	0.005	300	10,000
CFC-115	C_2ClF_5	0	~8 pptv	0.001	400	7,370
CCl4		0	~93 pptv	0.012	42	1,400
CH_3CCl_3		0	~4 pptv	0.004	7	146
HCFC-22	$CHClF_2$	0	~168 pptv	0.033	13	1,810
HCFC-141b	CH_3CCl_2F	0	~18 pptv	0.0025	9	725
HCFC-142b	CH_3CClF_2	0	~16 pptv	0.0031	19	2,310
Halon-1211	$CBeClF_2$	0	~4.4 pptv	0.001	20	1,890
Halon-1301	$CBrF_3$	0	~3.2 pptv	0.001	65	5,400

[a]GWP is the direct Global Warming Potential of a gas. The GWP is a measure of the radiative forcing of a gas relative to CO_2 with a value of 1. The value depends on several factors, including the atmospheric concentration of the gas, its radiative properties, its lifetime, and the number of years over which the calculation of GWP is performed. In the table, the time horizon is 100 years. (*Source:* Houghton et al., 2001; WMO, 2007; and IPCC, 2007a.)

electrical components. Consumption of CFCs presently continues but at much reduced rates because of implementation of the Montreal Protocol (see the following). Other substances have replaced these chemicals in industrial and other practices (Table 14.8). There still exists a black market in sales and distribution of CFCs.

In 1985, scientists from the British Antarctic Survey reported that the ozone layer had thinned from 1977 to 1985 over the Antarctic during the polar springtime in September and October; the "hole in the sky" was discovered. In 1987, the National Aeronautical and Space Administration (NASA) aircraft ER-2 flew south into the Antarctic polar vortex during development of the ozone hole above Antarctica. Instrumentation onboard the aircraft monitored ozone and chlorine monoxide concentrations during the flight. As the aircraft crossed the southern polar vortex boundary at 65°S, O_3 concentrations fell and ClO concentrations rose sharply (Figure 14.62). This was one of the first firm evidences of the fact that

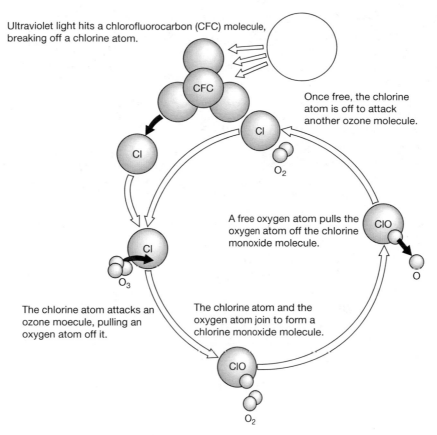

FIGURE 14.61 The chemical pathway for destruction of ozone in the stratosphere by chlorine atoms derived from the breakdown of CFCs in the atmosphere in which ClO is a precursor for the production of free chlorine.

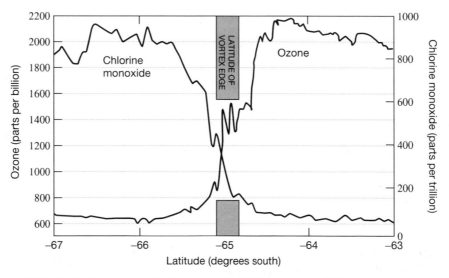

FIGURE 14.62 Changes in chlorine monoxide (ClO) and ozone (O_3) concentrations measured by the National Aeronautical and Space Administration (NASA)'s ER-2 aircraft as it flew south into the Antarctic polar vortex on September 22, 1987, during development of the ozone hole above Antarctica. This is one of the first firm evidences of the fact that human activities are responsible for development of the hole in the sky. Notice how O_3 concentrations fell and ClO concentrations rose sharply as the aircraft crossed the votex boundary at 65°S (left to right in the figure). (*Source:* Reid, 2000.)

human activities are responsible for the development of the hole in the sky. In response to the new discovery, an international treaty, the Montreal Protocol, designed to reduce the emissions of ozone-depleting chemicals to the atmosphere, was signed in 1987 by 24 nations. This step was followed by the signing of an expanded treaty, the London Agreement, by an additional 36 nations in 1990, and strengthened three times since in the agreements of Copenhagen 1992, Montreal 1997, and Beijing 1999.

GLOBAL OZONE   The greatest thinning in the ozone layer is over the Antarctic; however, ground-based instrumentation and later instrumentation on board satellites have recorded long-term changes in global mean column ozone between 60°S to 60°N from 1964 (~295DU) to 2006 (~285DU). Since the early 1980s, there was an irregular decline in total ozone concentrations in this latitudinal region until the year 1994, when ozone concentrations began to display an upswing followed by a recent decline. A strong global anomaly was seen for the year 1993, when the minimum of 273 DU was about 10 DU below the preceding and following years (Figure 14.63). This anomaly was caused principally by the eruption of Mt. Pinatubo in 1991, which injected a large amount of sulfur dioxide gas directly into the lower stratosphere at altitudes as high as 30 kilometers. Within about a month, the gas was oxidized to sulfuric acid, which was rapidly converted into a sulfate aerosol cloud that spread nearly worldwide. This sulfate aerosol cloud led to stratospheric ozone depletion both because of aerosol heating effects and by providing particle surfaces that accelerated the chemical reactions that destroy ozone. Since 1994, global ozone levels at 60°S to 60°N have begun to stabilize somewhat.

THE HOLE IN THE SKY   The Antarctic ozone hole has reappeared every austral spring from 1977 to 2008. As an example, Figure 14.64a shows a profile of the concentration of ozone above the South Pole during the development of the Antarctic ozone hole in 1992 and 1993. In this profile, nearly total destruction of ozone is seen between the altitudes of 14 and 19 kilometers in the austral springtime of 1993.

The ozone hole is caused by a complex set of conditions, including atmospheric circulation patterns over the Antarctic polar region and the superfrigid air in the region that causes ice crystals to form in high polar stratospheric clouds (PSCs). Winds circulate unimpeded, generally in an east to west circular vortex over the Antarctic region. CFCs are trapped in this polar vortex and accumulate throughout the year. The

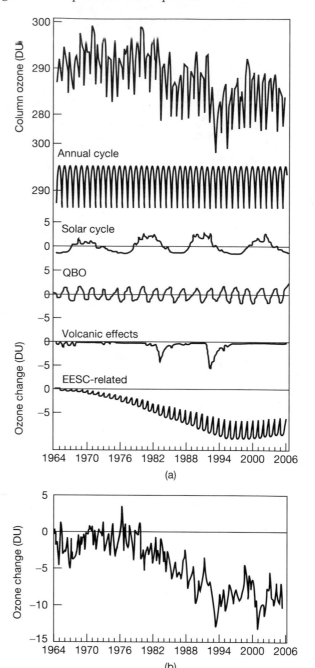

FIGURE 14.63   Total column ozone variations for 60°S to 60°N from ground-based data (top panel) and the changes in individual components that contribute to the variations, including the annual cycle, solar cycle, Quasi Biennial Oscillation (QBO) of the atmospheric system, volcanism, and Equivalent Effective Stratospheric Chlorine (ESSC) (bottom panels) (a), and the area-weighted total ozone deviations estimated from ground-based data adjusted for solar, volcanic, and QBO effects from 1964 to 2006. The result in (b) indicates that most of the depletion of stratospheric ozone is due to the manufactured synthetic ozone-depleting chemicals in the atmosphere.   (WMO (World Meteorological Organization) *Scientific Assessment of Ozone Depletion: 2006*, Global Ozone Research and Monitoring Project – Report No. 50, 527 pp. Geneva, 2007. Used by permission.)

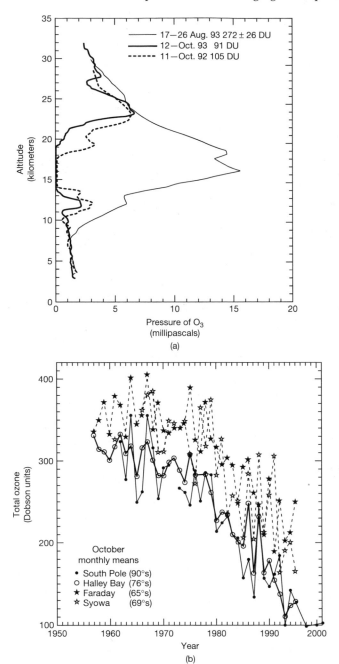

FIGURE 14.64 (a) Comparison of the ozone profile above the South Pole in 1993 before ozone depletion with the profiles observed when total ozone reached a minimum in the austral springtime of 1992 and 1993. 1 millipascal equals 0.00000001 bar. (b) Total ozone above four ozone measurement stations in Antarctica from 1957 to 2001. (*Source:* World Meteorological Organization, 1994; www.cmdl.noaa.gov/ozwv/ozson.des/spo/spotal.html, 2002.)

CFCs break down into chlorine monoxide, and when the sun rises after the long winter's night, the sunlight initiates the destruction of ozone by this chemical (Figures 14.61 and 14.62).

The decrease in total column ozone over Antarctica, measured during the austral springtime, has been about 65% since 1977. The instruments at the ground-based ozone measurement stations in Antarctica show a strong depletion in total column ozone since 1977 at the several monitoring stations of South Pole, Halley Bay, Faraday, and Syowa (Figure 14.64b). Since 1982 the minimum in Antarctic concentrations from 60°S to 90°S decreased almost linearly until 1985 and then less regularly to 1998, and has wobbled up and down since that time, with minimum ozone concentrations observed in 1992, 1998, and 2007 (Figure 14.65a). The area of the ozone hole (size of the "hole in the sky") generally increased from the early 1980s to 2001, with the rate of increase slowing about 1994 and then exhibiting an oscillatory pattern of small increases and small to moderate decreases (Figure 14.65b). Since 1980, as there has been overall growth of the size of the ozone hole, ozone hole size decreased relative to the overall trend (in 1989, 2003, and 2005). Figure 14.66a illustrates total column ozone minimum concentrations from 40°S to 90°S for the more recent years of 2003, 2006, and 2007 as compared with the range for 1990 to 2001. A similar diagram is shown in Figure 14.66b for the minimum in the ozone hole area for these years.

The depletion of Antarctic stratospheric ozone and presumably enhanced UV-B radiation reaching Earth's surface may be causing a measurable loss in biological productivity of surface waters in the ice-edge zone of the southern oceans. The change, however, is much smaller than the yearly variations in biological productivity in the region. So far, there has been no convincing evidence of any acute effects of enhanced UV-B exposure of humans and animals in the southern lands of Chile and Argentina. There also remains the possibility that a decrease in biological productivity in the ocean stemming from ozone depletion could lead to changes in the emissions of biologically produced trace gases, such as dimethyl-sulfide, to the atmosphere.

In the Arctic region, the thinning of the ozone layer has been less evident until recently. However, significant ozone depletions in the Arctic stratosphere are also taking place because of reactive halogen gases. Winds traveling over land and oceans in the Arctic region result in an uneven heating of the atmosphere and prevent a smooth flow of air. This process makes it difficult for a well-defined polar vortex to develop. Furthermore, the Arctic stratosphere is not as cold as that of the Antarctic, and thus there is less chance to form polar stratospheric clouds (PSCs) with ice crystals, surfaces that help to catalyze the destruction of ozone. In addition, the Northern

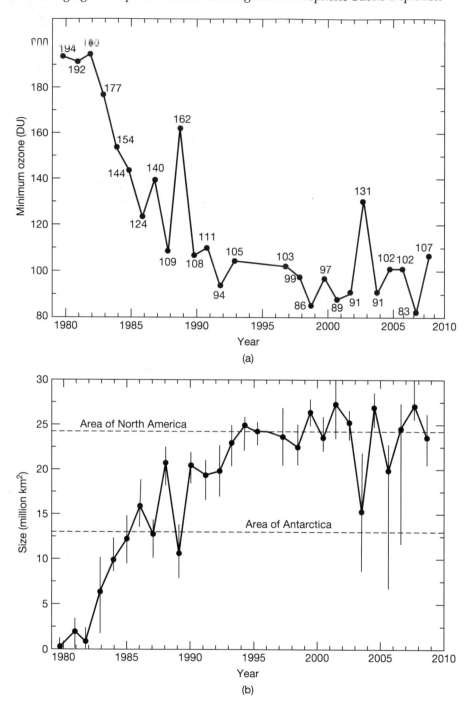

FIGURE 14.65 The Antarctic "hole in the sky" minimum ozone concentrations (a) and the average areas of the stratospheric ozone hole above Antarctica (b). The area of the ozone hole is defined as that area where stratospheric ozone levels drop below 220 Dobson units (DU). (*Source:* http://toms.gsfc.nasa.gov/eptoms/dataqual/ozone_v8.html.)

and Southern Hemisphere differ in how ozone-rich air is transported into the polar region from the lower latitudes in that the poleward and downward transport of ozone-rich air is stronger in the Northern Hemisphere during fall and winter than in the Southern Hemisphere. Ozone levels are thus much higher

in the northern polar region relatively to the southern polar region at the beginning of each winter season (Figure 14.67).

Generally in the Northern Hemisphere, changes in atmospheric circulation patterns drive the year-to-year changes in Arctic stratospheric temperatures.

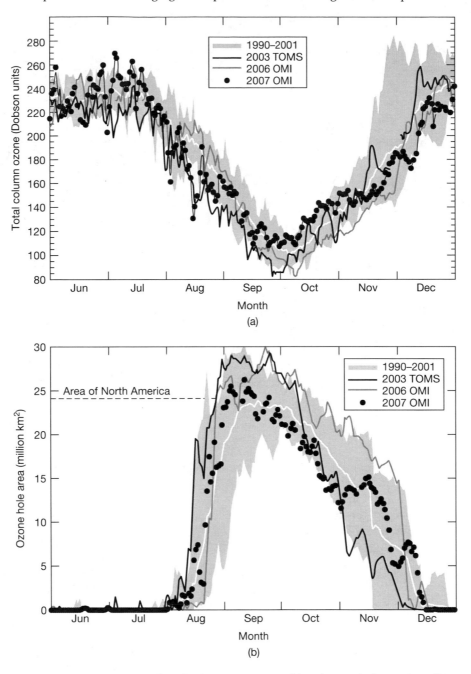

FIGURE 14.66 Comparison of total column ozone in DU (a) and ozone hole area in millions of square kilometers (b) from 40°S to 90°S for 2003, 2006, and 2007 with the period of 1990 to 2001. The area of the ozone hole is defined as that area where stratospheric ozone levels drop below 220 DU.   (*Source:* http://toms.gsfc.nasa.gov/eptoms/dataqual/ozone_v8.html.)

The changes in temperature are strongly linked to the changes in stratospheric ozone concentrations (Figure 14.68). Times of colder temperatures result in greater development of PSCs, whose ice crystals act as sites for ozone destruction. In late March and April 1997, total column ozone over the Arctic fell to unusually low levels. During March 1997, total column ozone values as recorded by instrumentation on TOMS were 40% below the mean level for 1979 to 1982. The minimum total column ozone of 219 Dobson units was recorded on March 24. In addition, total column ozone values in March 1996 were 24% below the mean. Figure 14.67 contrasts the average total ozone in the Arctic and Antarctic polar regions

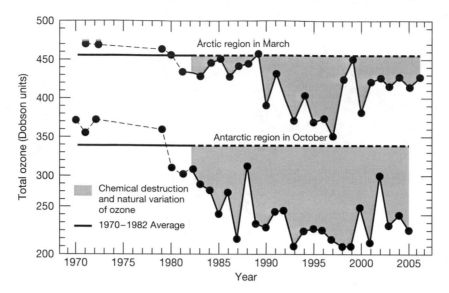

FIGURE 14.67 Trends in average total ozone in the polar regions of the Northern and the Southern Hemisphere from 1970 to 2005. Averages are for 63 to 90° north and south latitudes. The shaded areas represent both chemical destruction of ozone by ozone-depleting chemicals and natural variations in ozone. Notice the generally higher ozone levels in the Arctic region than in the Antarctic and the higher ozone levels in both the Arctic and Antarctic regions prior to the important chemical destruction of ozone. (WMO (World Meteorological Organization) *Scientific Assessment of Ozone Depletion: 2006,* Global Ozone Research and Monitoring Project - Report No. 50, 527 pp. Geneva, 2007. Used by permission.)

from 1970 to 2005. The ozone changes are due to a combination of chemical destruction and natural variations. Meteorlogical conditions influence the year-to–year changes in depletion, particularly in the Arctic because of the sharp fluctuations in temperature. Nevertheless, most of the depletion over the Antarctic and usually most of the depletion over the Arctic is due to chemical destruction of ozone by the reactive CFCs and other halogenated gases. Such depletions could put more of the human population at risk to UV-B exposure because the Arctic region is closer to large population centers than is the Antarctic.

Recent Findings Concerning Ozone Depletion

Since 1981, laboratory investigations, atmospheric observations, and theoretical and modeling studies have led to a much deeper understanding of both human-induced and natural chemical changes in the atmosphere and their causes and relationship to stratospheric ozone depletion. In this section, a summary of the major scientific findings and observations as published in the Scientific Assessment of Ozone Depletion [World Meteorological Organization (WMO), 1994, 1998, 2007] is given. This synopsis will provide the reader with a feeling for the state of

the issue of stratospheric ozone depletion in the early twenty-first century.

- Because of the Montreal Protocol and its amendments, the atmospheric growth rates of several major ozone-depleting chemicals have slowed. Table 14.9 gives the CFC and HCFC phaseout schedules of the 1987 Montreal Protocol and the further amendments to the protocol. Because of the regulations imposed on production of halogenated gases, the data of recent years show that the growth rates of CFC-11, CFC-12, halon-1301, and halon-1211 are slowing down. The abundances of carbon tetrachloride (CCl_4), methyl chloroform (CH_3CCl_3), and methyl bromide (CH_3Br) are actually decreasing at a significant rate (Figure 14.69). Total chlorine and bromine loadings from halocarbons, measured as EECl, equivalent effective chlorine, in the troposphere peaked in 1994 and had fallen about 10% from the peak by the first decade of the twenty-first century (Figure 14.70). Figure 14.71a shows how the Montreal Protocol regulations and further amendments will affect the future abundance of chlorine from halocarbons in the atmosphere.

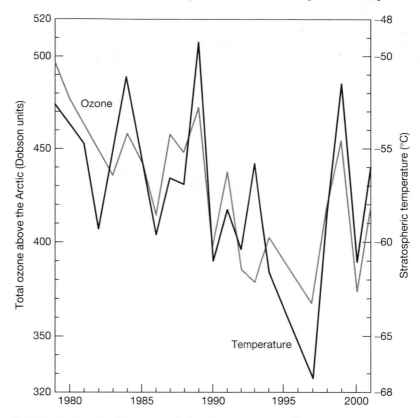

FIGURE 14.68 Significant correlation between stratospheric temperature in °C and total column ozone in DU above the Arctic from 1979 to 2001. Colder temperatures result in the formation of more polar stratospheric clouds (PSCs) and lower ozone levels. Changes in atmospheric circulation drive the annual temperature changes. Notice the overall decline in temperatures and Arctic ozone levels from 1979 to 2001. (www.giss.nasa.gov/)

TABLE 14.9A CFC Phaseout Schedules: Allowed Production and Consumption for Developed Countries (Percent of Baseline)

	1987 Original Montreal Protocol	1990 London Montreal Protocol	1992 Copenhagen Montreal Protocol	1990 United States Clean Air Act Amendments	1994 European Community Schedule
1990	100				
1991	100	100		85	
1992	100	100		80	
1993	80	80		75	
1994	80	80	25	25	50
1995	80	50	25	25	15
1996	80	50	0	0	0
1997	80	15			
1998	80	15			
1999	50	15			
2000	50	0			

TABLE 14.9B Comparison of the Montreal Protocol and United States Phaseout Schedules for HCFCs

Montreal Protocol		United States	
Year to be Implemented	% Reduction in Consumption and Production, Using the Cap as a Baseline	Year to be Implemented	Implementation of HCFC Phaseout through Clean Air Act Regulations
2004	35.0%	2003	No production and no importing of HCFC-141b
2010	75.0%	2010	No production and no importing of HCFC-142b and HCFC-22, except for the use in equipment manufactured before 1/1/2010 (so no production or importing for NEW equipment that uses these refrigerants)
2015	90.0%	2015	No production and no importing of any HCFCs, except for use as refrigerants in equipment manufactured before 1/1/2020
2020	99.5%	2020	No production and no importing of HCFC-142b and HCFC-22
2030	100.0%	2030	No production and no importing of any HCFCs

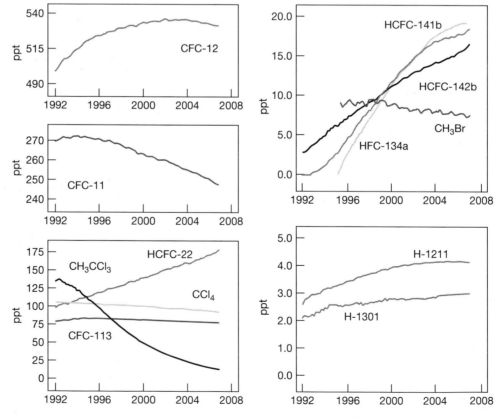

FIGURE 14.69 Recent trends in the chlorofluorocarbons and other halogenated gas concentrations in parts per trillion (ppt) in the atmosphere from 1992 to 2007. (*Source:* http://gcmd.nasa.gov/index.html.)

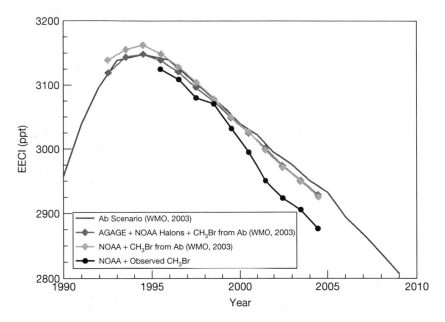

FIGURE 14.70 Trends in equivalent effective chlorine (EECl) in the troposphere in parts per trillion (ppt) based on several analyses. Notice the significant decline since 1994. Estimates are obtained from the concentrations of the different halogenated compounds in the troposphere and the number of chlorine atoms they contain. Notice that chlorine abundance in the troposphere is declining, reflecting the restrictions on emissions of halocarbons to the atmosphere imposed by the Montreal Protocol of 1987 and later amendments and transport to the stratosphere.   (WMO (World Meteorological Organization) *Scientific Assessment of Ozone Depletion: 2006,* Global Ozone Research and Monitoring Project – Report No. 50, 527 pp. Geneva, 2007. Used by permission.)

Also shown is how the adoption of the Copenhagen Amendment to the Montreal Protocol in 1992 will significantly reduce the number of projected cases of skin cancer in the future due to ozone depletion (Figure 14.71b).

- The atmospheric abundance of several of the substitutes for CFCs are still increasing: HFC (hydrofluorocarbon)-134a, HCFC (hydrochlorofluorocarbon)-22, HCFC-141b, and HCFC-142b (Figure 14.69). These substitutes are also capable of stratospheric ozone depletion, but to a much lesser extent than the CFCs. The production and importation of these gases will be completely phased out by the year 2030.

- Record low global ozone levels were observed in 1992 and 1993 because of the eruption of Mt. Pinatubo and perhaps other phenomena such as the extended El Niño of 1991 to 1994. Post-1994 global ozone levels have been almost constant since recovery from the 1991 Mt. Pinatubo eruption, except for a drop in 2000 to 2001. The eruption of Mt. Pinatubo greatly perturbed the atmosphere, altering global temperatures, the concentrations and distribution of global ozone, and Antarctic ozone hole abundance levels.

- Ozone abundances in the extrapolar regions of 60°S to 60°N have not declined further in recent years. The decline in ozone observed in the 1990s has not continued. The essentially unchanged column ozone levels over the past decade are related to the near-constancy of stratospheric ozone-depleting gas concentrations during this period.

- Springtime polar ozone depletions continue to be severe, especially under exceptionally cold stratospheric winter conditions. Variations in meteorological conditions over both north and south polar regions have played a stronger role in the observed variability of ozone levels over both poles in recent years. The ozone holes above the Antarctic in 1998, 2000, 2002, 2004, and 2007 were exceptionally large, and those of 1998, 2001, 2002, 2004, and 2007 exceptionally intense, with ozone levels reaching lows of about 90 DU. A substantial Antarctic ozone hole will continue to develop above Antarctica on into the twenty-first century because chlorine and bromine abundances will approach their pre-ozone hole levels very slowly as the atmosphere is cleansed of these chemicals.

- Cooling of the lower stratosphere (Figure 14.25a) has slowed in recent years from its overall average decline of 0.6°C/decade during 1979 to 2004.

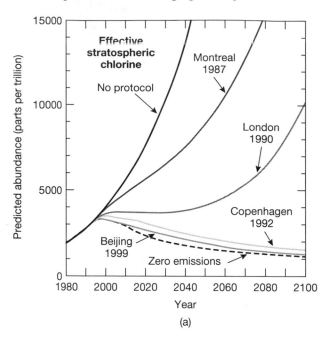

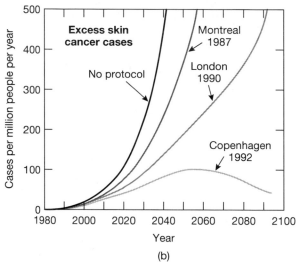

(b)

FIGURE 14.71 The effect of the Montreal Protocol and other international agreements on ozone-depleting halogenated gases from 1980 to 2100 in terms of the predicted abundance of effective stratospheric chlorine in parts per trillion (a). Without the Montreal Protocol and amendments, the abundance of effective stratospheric chlorine would have reached 15,000 parts per trillion in the year 2040. Total chlorine from halocarbons in the troposphere peaked in concentration in the mid-1990s and is now declining. Total bromine and fluorine are also decreasing. The abundance of stratospheric chlorine and bromine in the stratosphere peaked in 2000. In (b) is shown the number of excess cases of skin cancer per million people per year related to the situation with no ozone-depleting gas regulations at all and for the Montreal Protocol and its amendments. (WMO (World Meteorological Organization) *Scientific Assessment of Ozone Depletion: 2006,* Global Ozone Research and Monitoring Project – Report No. 50, 527 pp. Geneva, 2007. Used by permission.)

Ozone loss is the main reason for the cooling. Larger cooling trends are seen in the upper stratosphere, presumably a result of both decreases in ozone abundance and increases in the concentrations of the well-mixed greenhouse gases.

• The link between a decrease in stratospheric ozone and an increase in UV radiation has been strengthened by parallel observations of the two parameters. Large increases in UV radiation are observed in Antarctica and the southern part of South America during the period of development of the seasonal ozone hole. In addition, elevated UV levels were observed during 1992 and 1993 at mid- to high latitudes at the time of the unusually low ozone levels during those years. Furthermore, between 1979 and 1992, UV exposure at the ground level apparently has increased within certain latitudinal belts, as would be anticipated from declining total ozone levels during this period. Figure 14.72 shows the strong correlation between increases in erythemal (sunburning) ultraviolet (UV-B) radiation at the surface and ozone decreases at various locations. Because ozone is an effective absorber of ultraviolet radiation, decrease in ozone leads to an increase in UV-B radiation received at the surface.

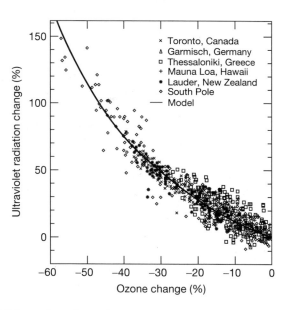

FIGURE 14.72 Relationship between erythemal (sunburning) ultraviolet radiation (UV-B) and ozone measured at several locations at ground level. The solid line is a model calculation that fits the observations reasonably well. (WMO (World Meteorological Organization) *Scientific Assessment of Ozone Depletion: 2006,* Global Ozone Research and Monitoring Project – Report No. 50, 527 pp. Geneva, 2007. Used by Permission.)

- A complex relationship exists between stratospheric and tropospheric ozone and greenhouse warming. The lower stratosphere has cooled 0.6°C per decade since 1979, mainly a result of ozone depletion. The total radiative forcing on the planet of the cooling effect due to the depletion of stratospheric ozone is −0.05 (−0.15 to +0.05) W/m². Ozone depletion appears to lead to a cooling effect that offsets a percent or two of global greenhouse warming. On the other hand, increases in tropospheric ozone abundance levels lead to a warming effect, equivalent to a total radiative forcing of +0.35 (+0.25 to +0.65) W/m². Tropospheric ozone is responsible for about 12% of the enhanced greenhouse effect. Ozone in one region

of our atmosphere cools the planet and in another warms the planet! This situation demonstrates the inherent complexity of problems of global climate change.

- Unambiguous detection of the beginning of recovery of the ozone layer will be years after the maximum loading of the stratosphere by ozone-depleting chemicals. Assuming full compliance with the Montreal Protocol and its amendments and adjustments, the stratospheric abundance of halogenated ozone-depleting compounds is expected to return to an equivalent level of about 2 ppm chlorine (the unperturbed state) by about 2050. After this, in the decades to follow, global and Antarctic ozone levels will start to recover slowly. Figure 14.73

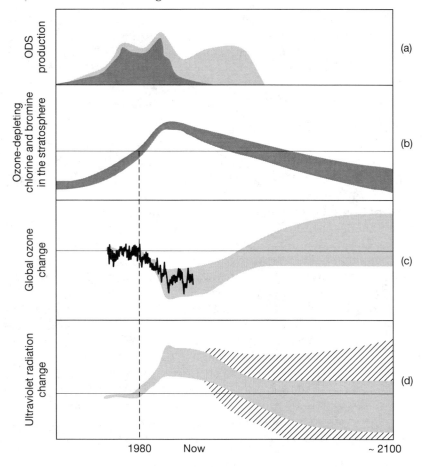

FIGURE 14.73 Schematic diagram illustrating (a) the production of ozone-depleting chemicals (black area, CFCs; gray area, HCFCs); (b) the abundance of ozone-depleting chlorine and bromine in the stratosphere; (c) the change in global ozone levels (black line is observations); and (d) the change in ultraviolet erythemal radiation (the gray shaded area is in response to the ozone changes shown in (c). The hatched area shows what might occur due to future climate-related changes in cloud type and distribution and aerosols. (WMO (World Meteorological Organization) *Scientific Assessment of Ozone Depletion: 2006,* Global Ozone Research and Monitoring Project – Report No. 50, 527 pp. Geneva, 2007. Used by permission.)

summarizes the past, present, and future of the production of the ozone-depleting chemicals of CFCs and HCFS and the like, the effective abundance of ozone-depleting chlorine and bromine in the atmosphere, the total ozone change from 60°S to 60°N, and the estimated change with time in sun-burning (erythemal) irradiance at high sun levels. Although emissions of ozone-depleting chemicals will be vanishing by the middle of the twenty-first century, and ozone-depleting bromine and chlorine will be reduced to—or nearly so—below pre-1980 levels, total global ozone and ultraviolet flux may or may not reach pre-1980 levels by the middle of the century. The large uncertainties in the projections of the latter two variables depend on future climatic conditions, and future changes in clouds and atmospheric aerosols.

Concluding Remarks and the 21st Century

Currently there is a great deal of consensus, but some disagreement, in the scientific community concerning the environmental problems of an enhanced greenhouse effect and stratospheric ozone depletion. Many scientists argue that the effects of climatic change because of human-induced perturbations of the global system will have far-reaching implications for the increasing human population. Past natural changes in the climate have resulted in displacement of ecosystems, evidenced, for example, by the Pleistocene glacial retreat that led to a change in location and demise of certain species of animals and plants, and by the more modern climatic episodes of the Holocene Climatic Optimum, Medieval Warm Period, and the Little Ice Age that led to variable patterns of regional environmental changes. The modern rates of environmental changes could potentially happen at a rate faster than that previously recorded in the history of the planet because of human activities. For example, the recent changes in the ozone composition of the stratosphere, certainly in part the result of emissions of CFCs to the atmosphere from refrigerants and other sources, have been relatively rapid. Also, the more controversial topic of warming of the atmosphere owing to an enhanced greenhouse effect caused by human activities will lead to changing ocean and atmospheric circulation patterns, thermal expansion of ocean waters, melting of glaciers, and rising sea level, and cause variations in weather patterns throughout the world on a time scale of less than a century.

Climatic change could in turn affect natural ecosystems, agriculture, and human patterns of settlement and development. Interestingly, not all of the countries of the world would necessarily find the global warming of an enhanced greenhouse undesirable. Parts of the cold interior of Russia could become extensive farmlands. The ice-locked ports of northern Russia, Europe, and Canada could be navigated year round. Growing seasons might be extended in certain regions of the world, like the Ukraine. However, the breadbasket of the United States could move northward into the plains of Canada, the desert areas around the equator could increase in size and regions of drought spread, and the island nations of the Pacific and countries with low-lying deltaic and wetland areas could be flooded. Some individuals are arguing that the world should try to maintain the warming of the Earth below a 2°C increase in global mean temperature above pre-industrial levels to prevent significant disruptions of ecosystems and human society.

Other scientists argue that the global temperature increase of 0.8°C during the past 100 years is due to factors other than the enhanced greenhouse effect and that any future warming due to this effect will be moderate. They also argue that human-made chlorofluorocarbons are not a significant and immediate threat to the ozone layer. They maintain that the Antarctic ozone hole is mainly a physical phenomenon, a result of atmospheric circulation patterns and the intensity of formation of polar stratospheric clouds.

This disagreement among scientists is leading to some ambivalence on the part of world policymakers as to what actions should be taken with regard to these global environmental problems (see Chapter 15). As happens many times in science, the "truth" concerning global climatic change probably lies somewhere between the extreme positions taken by the advocates. However, there is little doubt that the consequences of global climatic change for humans and ecosystems are a real concern to the world community. The implementation of the Montreal Protocol and later amendments regulating emissions of gases that can destroy the stratospheric ozone layer and the discussions by representatives from many of the world's nations to find ways to reduce emissions of fossil fuel CO_2 to the atmosphere (see Chapter 15, Kyoto Protocol) are indicative of steps the world community is willing to take to deal with the global atmospheric environmental problems. The stratospheric ozone depletion issue demonstrates unequivocally how regulations on harmful anthropogenic emissions to the atmosphere can mitigate and/or prevent damage to the environment globally. The problem of global warming is also fundamentally one of emissions and their regulation.

Study Questions

1. Briefly explain the natural greenhouse effect and how human activities are modifying this phenomenon.
2. Why are clouds so important in the modeling of climatic change?
3. What is the most important greenhouse gas, and what four major greenhouse gases are being emitted to the atmosphere by human activities?
4. Briefly explain how the concentration of carbon dioxide in the atmosphere has changed since the climax of the last (Wisconsin) glacial stage.
5. What are the three major sinks of anthropogenic CO_2?
6. How does N_2O affect climate and stratospheric O_3?
7. Why would decreasing anthropogenic emissions of CH_4 to the atmosphere clear the air of excess CH_4 relatively quickly?
8. Briefly discuss how temperature and precipitation might change because of an enhanced greenhouse effect brought about by emissions of greenhouse gases to the atmosphere from human activities.
9. Besides temperature and precipitation changes, what are some other consequences of an enhanced greenhouse effect?
10. Explain the CO_2 fertilization effect.
11. What is the MOC and how might it be affected by global warming?
12. What is the problem of ocean acidification? How can it be resolved?
13. How is stratospheric ozone produced and destroyed?
14. Describe the process by which stratospheric ozone is destroyed by chlorofluorocarbons.
15. What are the two important functions of stratospheric ozone that aid in maintaining the environment of Earth as suitable for life?
16. Chlorofluorocarbons are heavier than air. How then can they get to the stratosphere?
17. What is the source of most of the chlorine in the modern stratosphere?
18. Why does it appear that changes in the sun's energy and volcanic explosions are not responsible for the long-term decline in stratospheric ozone?
19. Does ozone depletion cause climatic change?
20. The United States had per capita CO_2 emissions to the atmosphere from the burning of fossil fuels in 2007 of 1.22 metric tons of carbon per person. If the world population reached 8 to 10 billion people by 2100, what would be the range of global fossil fuel CO_2 emissions in 2100, if the world per capita emissions were that of the United States in 2007? Does the answer to this question concern you in any way? If so, why?
21. In your opinion, how does the global community prevent, slow, or abet global climatic change owing to human activities?

Additional Sources of Information

Archer, D., 2008, *The Long Thaw: How Humans Are Changing the Next 100,000 Years of Earth's Climate.* Princeton University Press, Princeton, NJ, 196 pp.

Balling, R. C., Jr., 1992, *The Heated Debate: Greenhouse Predictions versus Climate Reality.* Pacific Research Institute for Public Policy, San Francisco, 195 pp.

Friedman, T. L., 2008, *Hot, Flat, and Crowded.* Farrar, Straus and Giroux, New York, 438 pp.

Graedel, T. E. and Crutzen, P. J., 1993, *Atmospheric Change: An Earth System Perspective.* W. H. Freeman, New York, 446 pp.

Holland, H. D. and Petersen, U., 1995, *Living Dangerously: The Earth, Its Resources, and the Environment.* Princeton University Press, Princeton, NJ, 490 pp.

Houghton, J., 1994, *Global Warming: The Complete Briefing.* Cambridge University Press, Cambridge, UK, 251 pp.

Houghton, J. T., Meira Filho, L. G., Callander, B. A., Harris, N., Kattenberg, A., and Maskell, K. (eds.), 1996, *Climate Change: The Science of Climate Change.* Cambridge University Press, Cambridge, UK, 572 pp.

Houghton, J. T., Ding, Y., Griggs, D. J., Noguer, M., van der Linden, P. J., Dai, X., Maskell, K., and Johnson, C. A. (eds.), 2001, *Climate Change 2001: The Scientific Basis.* Cambridge University Press, Cambridge, UK, 881 pp.

IPCC (Intergovernmental Panel on Climate Change), 2007a, *Climate Change 2007: The Physical Science Basis.* Solomon, S., D. Qin, M. Manning, Z. Chen, M. Marquis, K. B. Averyt, M. Tignor, and Miller, H. L. (eds.), Cambridge University Press, Cambridge, UK and New York, 996 pp.

IPCC (Intergovernmental Panel on Climate Change), 2007b, *Climate Change 2007: Impacts, Adaptation, and*

Vulnerability. Parry, M. L., O. F. Canziani, J. P. Palutikof, P. J. van der Linden, and C. E. Hanson (eds.), Cambridge University Press, Cambridge, UK and New York, 976 pp.

IPCC (Intergovernmental Panel on Climate Change), 2007c, *Climate Change 2007: Mitigation of Climate Change.* Metz, B., O. R. Davidson, P. R. Bosch, R. Dave, and L. A. Meyer (eds.), Cambridge University Press, Cambridge, UK and New York, 851 pp.

Jones, L. (ed.), 1997, *Global Warming: The Science and Politics.* The Fraser Institute, Vancouver, BC, 180 pp.

Karl, T. R., Nicholls, N., and Gregory, J., 1997, The coming climate. *Scientific American,* May, pp. 78–83.

Kasting, J. F., 1998, The carbon cycle, climate, and the long-term effects of fossil fuel burning. In *Consequences,* Saginaw Valley State University, University Center, MI, pp. 15–27.

Kundzewicz, Z.W., L.J., Mata, N.W. Arnell, P. Döll, P. Kabat, B. Jiménez, K.A. Miller, T. Oki, Z. Sen and Shiklomanov, I.A., 2007, Fresh water resources and their management. *Climate Change 2007: Impacts, Adaptation and Vulnerability. Contribution of Working Group II to the Fourth Assessment Report of the Intergovernmental Panel on Climate Change,* Parry, M.L., O.F. Canziani, J.P. Palutikof, P.J. van der Linden and C.E. Hanson (eds.), Cambridge University Press, Cambridge, UK, 173–210.

National Research Council, 2000, *Reconciling Observations of Global Temperature Change.* National Academy Press, Washington, DC, 85 pp.

Reid, S. J., 2000, *Ozone and Climate Change: A Beginner's Guide.* Gordon and Breach Science Publishers, Amsterdam, The Netherlands, 210 pp.

Rosenzweig, C. and Hillel, D., 1998, *Climate Change and the Global Harvest.* Oxford University Press, New York, 334 pp.

Singer, S. F. (ed.), 2008, *Nature, Not Human Activity, Rules the Climate: Summary for Policymakers of the Report of the Nongovernmental International Panel on Climate Change.* The Heartland Institute, Chicago, Illinois, 40 pp.

WMO (World Meteorological Organization), 1994, *Scientific Assessment of Ozone Depletion: 1994.* WMO, Global Ozone Research and Monitoring Project Report No. 37, Geneva, Switzerland, 369 pp.

WMO (World Meteorological Organization), 1998, *Scientific Assessment of Ozone Depletion: 1998, Executive Summary.* WMO, Global Ozone Research and Monitoring Project, Report No. 44, Geneva, Switzerland, 43 pp.

WMO (World Meteorological Organization), 2007, *Scientific Assessment of Ozone Depletion: 2006.* WMO, Global Ozone Research and Monitoring Project, Report No. 50, Geneva, Switzerland, 572 pp.

15 Human Dimensions of Global Environmental Change in the Twenty-First Century

If we do not change the direction we are going, we will end up where we are heading.

CHINESE PROVERB

In this book, it has been demonstrated that natural environmental change is a consequence of living on Earth. The effects of natural and human-induced physiochemical changes for ecosystems and the human species were discussed in detail in preceding chapters. It was shown that within the background of generally slower natural environmental change, human activities are modifying the global environment of the planet at a rapidly accelerating pace and have become a "geologic force" in the ecosphere. It should be recognized that global warming as a contemporary environmental issue has received much attention, but one should not lose track of the fact that there are a number of environmental issues—like photochemical smog, acid deposition, and cultural eutrophication—that may be more severe environmental problems at the local, regional, and subcontinental scales than warming of the planet. In addition, it is the composite of environmental issues and their socioeconomic and political overtones playing out at various time and space scales that hinder steps toward a sustainable environment and development. With this in mind, in this chapter, the book is concluded with a closer look at some of the human dimensions questions (economic, social, political, and ethical) of global environmental change and the question of the practicality of a sustainable environment and development on into the twenty-first century.

It should be kept in mind that there are uncertainties in the science of global change, particularly in predictions of future environmental changes and their consequences, but these uncertainties are diminishing. The uncertainties have at times led to a strong diversity of opinion as to what actions individual nations and the world community should take in light of such uncertainty. The very question of management of the environment or sustainable development touches on the fundamental cultural values of individuals, regions, and nations.

WHAT WE KNOW AND FUTURE CONCERNS

As demonstrated in previous chapters, there is no doubt that the global environment of the Earth has changed in the past. Change is more the nature of the planet than constancy. During the geologic past, continents have broken apart to drift across climatic zones. They have rejoined to form great mountain ranges. Ocean basins have been created and then have been destroyed by subduction. The planet has been warmer and colder than at present. Continental ice sheets have advanced and retreated. Meteorites with an impact equivalent to about 10,000 megatons of TNT strike the Earth about every several million years, and initiate changes in the physical environment, extinction rates, and biodiversity. Atmospheric and oceanic composition and circulation patterns have changed over time. The sun's energy output has varied, changing the amount of radiant energy reaching Earth's surface. River water discharge and dissolved and suspended loads have not been constant over geologic time. Sea level has risen and fallen. Ecosystems and species have come and gone and migrated with climatic change and for other natural causes. All aspects of the environment of the surface of Earth have changed during geologic time.

The physical and chemical changes in the environment have led to and constrained biological evolution. In turn, the evolution of species and ecosystems has interacted with the physiochemical system of the planet to produce the surface environment of the Earth of the past and as we know it today. The biota and environment are a product of coevolution, although the properties of the environment also exhibit cyclical characteristics. *Gaia* is a term applied to this interactive set of organic and inorganic processes that regulate the natural environment of Earth, and Earth system science or geophysiology is the study of this interactive system.

In Chapter 13 changes in Earth's surface environment owing to natural causes during late Pleistocene glacial-interglacial stages and the Holocene and the rates of change were documented. Major human evolution took place during the time of Pleistocene environmental change. Recent information on rates of natural change shows that dramatic global change can be rapid, on the time scale of several decades to centuries. Thus, it is very unlikely that the surface environment of the planet will not change in the future.

Table 13.3 lists historical changes over the past century in some surface environmental parameters of the planet to compare with glacial-interglacial change in these parameters. It can be seen that these many parameters have changed more rapidly during the last several centuries than in more recent geologic time. Aside from historical data, Table 13.3 also gives qualitatively the direction of change of environmental parameters in the late twentieth century and projections for the early part of the twenty-first century. Although the interactions between human activities and the total Earth surface system and its physical environment that harbors life are not well understood, these interactions are substantial, may be cumulative, and in many cases are accelerating. For all the environmental parameters of Table 13.3, there is still much to be learned about their future course and the effects they have on the environment worldwide.

Human-Induced Global Environmental Change

If the nature of the Earth is one of change, why is there concern with global environmental change because of human activities? It is mainly a matter of the rates of change. Human-induced global environmental change is a consequence of direct and indirect *rapid* modification of the environment by human activities, such as urbanization, transportation, cultivation of crops, and industrialization. The distribution and rates of growth of the human population and the demand for economic growth with concomitant utilization of resources are forces that are acting as agents of global environmental change (see Chapter 1, IPAT equation).

The growth of the human population, for some scientists and policymakers, is the most important factor involved in environmental change. In 1993 the first summit of the world scientific academies was held in New Delhi, India, followed by a meeting in Cairo, Egypt in 1994, to discuss global population growth. Fifty-six of the academies endorsed a 15-page statement calling for zero population growth within the lifetime of children at that time. Furthermore, they described such a scenario as necessary to deal successfully with world social, economic, and environmental problems. However, there was not universal agreement on the document. Scientists from African nations, Ireland, and the Vatican rejected the statement outright. Japan and Argentina declared that the matter was outside their competence, and the academies of Spain, Georgia, Armenia, and Italy did not sign the agreement for various reasons.

Human activities stemming from population growth and the demand for economic growth include fossil fuel and biomass combustion, land use change, agricultural practices, and halocarbon and other synthetic chemical production and release. All these activities result in emissions of chemicals to soil and aquatic systems and to the atmosphere. Furthermore, fossil fuel burning depletes a nonrenewable resource. Biomass burning, as currently done in many countries, depletes a potential renewable resource. Land use changes lead to the conversion of forests to croplands and rangelands, of wetlands to agricultural and urban uses, and the loss of habitat and biological diversity. Agricultural activities lead to the release of synthetic pesticides and nitrogenous and phosphorus-bearing fertilizers to the environment. Synthetic chemical production and use in industrial practices result in the venting of these chemicals or their by-products to aquatic systems and to the atmosphere. The production and utilization of chlorofluorocarbons and other halocarbons are promoting growth of the atmospheric concentrations of greenhouse gases and stratospheric ozone-depleting chemicals. Although it is established that human population growth and resource use are strongly coupled and are driving forces for global environmental change, the question of what to do about these forces is largely political in nature. A summary of global environmental change problems owing to human activities discussed in this book is given in Table 15.1.

THE BASIS OF CONCERN Future concerns for the environment are many. The environment is still under heavy attack by human activities. To support this conclusion, some of the statements made in preceding chapters are emphasized in the following paragraphs. Humans have used, co-opted, or foregone—that is, affected in one way or another—an amount of organic matter equivalent to as much as 40% of terrestrial primary production and several percent of marine net primary production. Habitat destruction, biological impoverishment, loss of biological diversity, and extinction of plants and animals are taking place at a significant, but difficult to define, rate (see Figure 6.19). The world population now uses an amount of water equivalent to about 25% of evapotranspiration over land and 55% of the accessible water runoff from the continents. Only 20% of the world's major rivers have pristine water quality.

About 8.7 billion tons of carbon, or about 1.22 tons per person per year, as carbon dioxide were vented to the atmosphere in 2008 by fossil fuel burning alone. Deforestation of tropical forests is proceeding at a rate of about 0.8% of forested area annually, accounting

---

TABLE 15.1 Some Problems of Global Environmental Change Owing to Human Activities[a]

- Climatic changes from anthropogenic inputs to the atmosphere of CO_2 and other greenhouse gases, and SO_2 and its fate
- Disruptions in biogeochemical cycles of C, N, P, S, trace metals, and other elements
- SO_x and NO_x emissions and acid deposition
- NO_x and VOC emissions and development of photochemical smog and tropospheric ozone
- Emissions of halocarbons and alterations in the stratospheric ozone layer and associated effects on ultraviolet radiation
- Increasing rates of tropical deforestation and other large-scale destruction of habitat, with potential effects on climate
- Disappearance of biotic diversity through explosive rates of species extinctions
- The global consequences of distribution and application of chemicals potentially harmful to the biota, e.g., pesticides
- Cultural eutrophication from agricultural runoff and municipal and industrial sewage disposal
- Exploitation of natural resources and consequent waste disposal and chemical pollution problems
- Water quality and usage
- Waste disposal: municipal, toxic chemical, and radioactive

[a] Population growth at 1.2–2.2% per year in the last 50 years is a factor common to all these problems.

---

for 15 to 20% of the total carbon dioxide emitted to the atmosphere by human activities. Human activities resulted in the release to the atmosphere of about 340 million tons of methane and 3 million tons or more annually of nitrous oxide in the early twenty-first century.

In addition, in the early twenty-first century, a total of 60 to 70 million tons of sulfur were being emitted into the atmosphere because of human activities. This mobilization of sulfur is equivalent to 30% of the sulfur rained out of the atmosphere each year. The total annual mobilization of nitrogen by human activities is 165 million tons, equivalent to about 2.8 times the amount of nitrogen falling in rain each year and five times the total annual river discharge of nitrogen to the ocean. The area of total forest damaged by acid-forming sulfur and nitrogen compounds and other air pollutants in Europe was on the rise in the late twentieth century, amounting at that time to 22% of forested area.

Furthermore, 20 to 25 million tons of phosphorus are used by humans each year, principally in agriculture, equivalent to more than five times the annual river discharge of dissolved phosphorus to the ocean. Total annual sediment discharge toward the ocean is 20 billion tons, representing a doubling of the pristine sediment discharge to the ocean. About 45,000 dams are now located on the world's rivers and streams and trap a significant amount of the riverine sediment heading toward the ocean. The reservoirs behind these dams, if filled to the top with sediment, would contain about 280 times the amount of sediment discharged by rivers to the ocean each year. In the last 45 years, almost one-third of the world's arable land has been lost by erosion, contamination, and for other reasons related to human activities. Cropland continues to be lost at a rate of 10 million hectares annually.

Moreover, global annual production of synthetic organic chemicals and materials reached about 600 million tons per year in 2000. CFC-11 and CFC-12 production, despite significant decline in their rates, still amounted to about 75,000 tons annually in 2005. Yearly fertilizer use, although falling 24% between 1989 and 1995, grew to 31.3 kilograms per capita, or 210 million tons worldwide in 2008. The worldwide use of active pesticide ingredients amounted to 2.43 million tons in 2001. In the United States alone, 561,000 tons, or 2.0 kilograms per person, of insecticides, herbicides, and fungicides were used annually. Trace-metal emissions from industrial, combustion, and other sources to the land, aquatic, and atmospheric environments were still important fluxes in the biogeochemical cycles of the metals. For example, the neurotoxin mercury has nearly become a worldwide pollutant and arsenic has contaminated major water supplies. World water supplies became scarcer and more polluted in the 1990s and on into the twenty-first century in some regions of the world. Twenty-six countries, mainly in the Middle East and Africa, had less than 1000 cubic meters of water per capita per year. In many areas, regional freshwater supplies contain significant amounts of metals, nutrient fertilizers, pesticides, and other synthetic organic chemicals derived from human activities. These and other human interferences in the environment continue and require policy decisions at both the national and international levels to slow emissions and moderate the effects of human activities on the environment.

NEAR-FUTURE OF THE ENVIRONMENT For the near future, it is likely that the degradation of the environment on a global basis will continue, if for no other reason than that the key socioeconomic factors of global population, world gross domestic product (GDP), and energy consumption mainly from fossil fuels will continue to show growth into the early part of the twenty-first century (Figure 15.1a). The recent recession of 2008–2009 will slow growth somewhat in GDP and emissions of greenhouse gases to the atmosphere. Nevertheless, as growth picks up without changes in national and international policies regarding emissions of environmentally damaging materials to the ecosphere, we will continue to see environmental degradation of natural systems with socioeconomic and political ramifications.

Growth in world GDP (Figure 15.1a) does not necessitate environmental degradation, but if such growth is fueled by the burning of coal, oil, and gas, then emissions of carbon dioxide, nitrogen and sulfur oxides, hydrocarbons, particulates, and trace metals to the environment will increase. Carbon emissions alone could reach 10 billion metric tons per year in 2015 (Figure 15.1b) and go as high as 20 billion tons per year, if future CO_2 concentrations were planned to stabilize at levels approximating 1000 ppmv (Figure 15.6).

The need for food for the growing population will necessitate increased use of fertilizers and pesticides to increase yields unless new crop varieties are developed and a more conservative use of fertilizers and pesticides is employed, particularly in the developing world. Increased fertilizer use will lead to increasing problems of eutrophication. In addition, the loss of cropland to other uses, such as urbanization, further strains our ability to feed the growing population. Industrial, agricultural, and domestic use of water will increase as the global population grows, placing strains on the water resources of certain regions because of consumption and pollution (see Chapter 11). The important point is that if the world continues on the same course of industrial development as that followed by the presently industrialized nations of the world during the last two centuries, then the near future of the planet is one of continuous environmental degradation, with the locus of environmental problems shifting somewhat to the developing world but the problem of global warming remaining a concern for all nations.

ECOSYSTEM SERVICES AS A MEASURE OF THE PLANET'S WELL-BEING Humans benefit from a multitude of resources and processes that ecosystems harbor. These are collectively known as **ecosystem services.** There has been a great deal of attention paid to ecosystem services as a measure of the well-being of the planet and to the interactions between humans and services in recent years. Ecosystem services can be grouped into

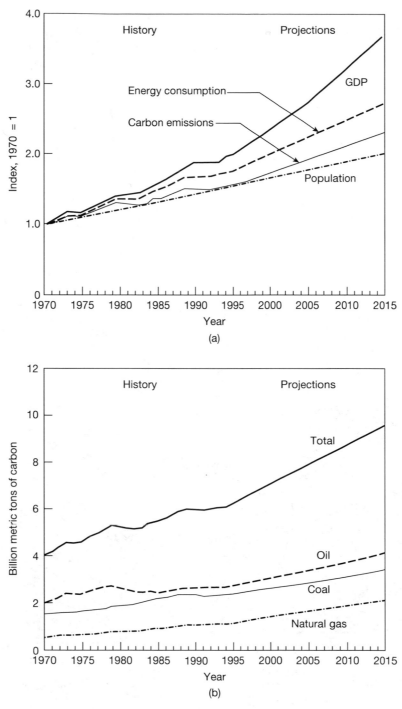

FIGURE 15.1 History and projections of (a) the key socioeconomic factors of world gross domestic product (GDP), energy consumption, and population, and (b) total and individual fuel sources of carbon emissions to the atmosphere from burning of fossil fuels. In (a), 1970 values are assigned a value of 1. (*Source:* EIA, 1997.)

four broad categories: *provisioning*, such as a system that provides food and water; *regulating*, such as the services of controlling climate and water supplies, and sequestration of high loads of nutrients in wetlands; *supporting*, such as the role of nutrient cycles and crop pollination in maintaining a healthy ecosystem; and *cultural*, the recreational and spiritual benefits of an ecosystem. Economic values have been assigned by individuals and in academic and practical studies to provide a more quantitative assessment of the value of the services provided by an ecosystem for planetary or human health.

The Millennium Ecosystem Assessment *Ecosystems and Well-Being: Synthesis of 2005* was commissioned by United Nations Secretary General Kofi Annan in 2000 to assess the consequences of ecosystem changes for overall human well-being and for reduction in poverty. The assessment also attempted to establish a scientific basis for actions required to strengthen conservation measures and sustainable use of ecosystems and their contributions to human well-being. Figure 15.2 shows the strength of the linkages between the different categories of

ecosystem services and the constituents of human well-being. The intensities of linkages vary, as do the potentials for remediation by various socioeconomic factors. There is a complex array of indirect drivers of demographic, economic, sociopolitical, science and technology, and cultural and religious (ethical) beliefs that can affect the direct drivers of change (Figure 15.3), such as changes in land use and cover, climate change, etc. The direct drivers can directly affect human well-being and poverty reduction or can do so indirectly through the effects of the drivers on the ecosystem service factors. There are also feedbacks from the human systems to the indirect drivers of change (Figure 15.3). It can easily be seen from Figures 15.2 and 15.3 that there is a complex set of

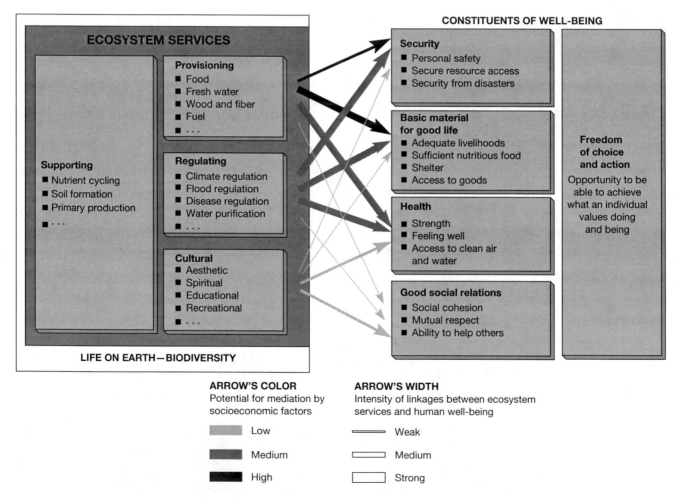

FIGURE 15.2 Relationships between supporting, provisioning, regulating, and cultural ecosystem services that support life on Earth and biodiversity to the constituents of well-being of security, basic material for a good life, health, good social service, and freedom of choice and action. The color and width of the arrows represent, respectively, the potential for mediation by socioeconomic factors and the intensity of the linkages between ecosystem services and human well-being. (*Source:* Millennium Ecosystem Assessment, 2005.)

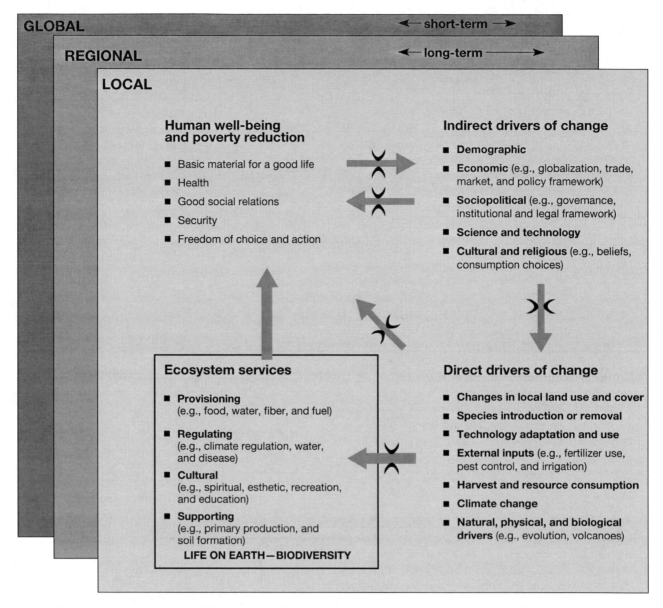

FIGURE 15.3 Conceptual framework of interactions between biodiversity, ecosystem services, human well-being, and the indirect and direct drivers of change. These interactions can involve time scales of short- to long-term and can involve spatial scales of local, regional, and global. Strategies and interventions (arrows with parabolas) can be applied at several points in this framework to enhance human well-being and conserve and preserve ecosystems. (*Source:* Millennium Ecosystem Assessment, 2005.)

interactions between biodiversity, ecosystem services, human well-being, and the sociopolitical and economic, and natural and human-induced agents of environmental changes. The interactions between the various factors can take place at various time and space scales. It is this complexity that makes it so difficult to evaluate the condition of the planet and its human population in the future and to develop strategies and interventions that can enhance human well-being and converse ecosystems. However, attempts

are being made by the world community to address these challenges.

Reactions to Global Environmental Change Issues

The world community has reacted to the problems of global environmental change. In some cases, such as the decisions in the latter part of the twentieth century involving chlorofluorocarbon production and the reductions in SO_x and NO_x emissions to the atmosphere for some countries, the actions have been reasonably aggressive. In other cases, such as those of global warming and cultural eutrophication, the development of policy is more complicated and slower to achieve. Much has been said in a negative sense concerning our ability to develop policy to deal with problems of global environmental change. In the following, some examples of actions and policies that are benefitting the environment are presented.

GLOBAL WORLD CONFERENCES ON ENVIRONMENT AND DEVELOPMENT

The Brundtland Commission   In June 1972 the United Nations Conference on the Human Environment was held in Stockholm, Sweden. At the conference, a declaration known as the Stockholm Declaration set out the principles for various environmental issues, such as human rights, natural resource management, pollution prevention, and the relationship between the environment and development. The conference led to the formation of the United Nations Environment Programme (UNEP). Then in 1983 the World Commission on Environment and Development, better known as the Brundtland Commission named after its chair, Gro Harlem Brundtland of Norway, was convened. The report of the Commission *Our Common Future* was published in 1987 by Oxford University Press. It was in this report that the most commonly used definition of sustainable development was stated: "Sustainable development is development that meets the needs of the present without compromising the ability of future generations to meet their own needs. It contains within it two key concepts: the concept of 'needs', in particular the essential needs of the world's poor, to which overriding priority should be given; and the idea of limitations imposed by the state of technology and social organization on the environment's ability to meet present and future needs."

The Earth Summit   In 1992 the United Nations Conference on Environment and Development (UNCED, also known as the Earth Summit) was held in Rio de Janeiro, Brazil, 20 years after the Stockholm Conference on the Human Environment. At the conference Agenda 21 was adopted by 172 nations representing 98% of the world's population. This document is a program of action for the implementation of the principles by which people should conduct themselves in relation to each other and the environment, as stated in the Earth Charter of UNCED. Agenda 21 is a non–legally binding statement of goals and objectives as well as a list of strategies and actions that governments and other actors should take to meet these objectives. The bold goal of Agenda 21 is to halt and reverse environmental damage to the planet and to promote environmentally sound and sustainable development in all nations on Earth. The six major themes of Agenda 21 are the quality of life on Earth; the efficient use of natural resources on the planet; the protection of our global commons of atmosphere and oceans; the management of human settlements; chemicals and the management of human and industrial waste; and sustainable economic growth.

At the Earth Summit, the terms of three intergovernmental conventions were discussed. Framework conventions on climate change and biodiversity were signed by most nations at Rio. A forestry statement was drafted but not adopted by the Earth Summit governmental representatives. By the end of 1993, the Framework Convention on Climate Change had received its fiftieth ratification. Enforcement of the convention began on March 21, 1994. The Convention on Biodiversity received its thirtieth ratification and entered into force on December 30, 1993.

The WSSD   In 2002 the World Summit on Sustainable Development (WSSD) was held in Johannesburg, South Africa. Since this meeting was held 10 years after the first Earth Summit in 1992, it was nicknamed "Rio+10". The slow progress of work on the environment is evidenced by the fact that the WSSD took place 30 years after the United Nations Conference on the Human Environment held in Stockholm, Sweden in June 1972. The WSSD was a follow-up to the Agenda 21 document and a check on its implementation. The summit focused on the five key areas of water, energy, health, agriculture, and biodiversity. There were nonbinding agreements reached on water, forests, and AIDS, but not on renewable energy or agricultural subsidies. Poverty was identified as the major impediment to sustainable development. Many nations bemoaned the lack of United States leadership at the Summit.

The Montreal Protocol   In 1990 and later in 1992 and 1999, the nations ascribing to the Montreal Protocol of 1987 revised their schedule of phasing out production and use of chlorofluorocarbons (CFCs), halons, methyl chloroform, and carbon tetrachloride. They also agreed to place new controls on hydrochlorofluorocarbons (HCFCs) and methyl bromide. The global deadlines for phaseout included cutting CFCs to 25% by 1995 and completely phasing out these synthetic chemicals by 1996. Halons were to be phased out in 1994, methyl chloroform and carbon tetrachloride in 1996, and caps begun on methyl bromide and HCFCs in 1995 and 1996, respectively. However, it will not be until 2030 that HCFC production will cease. The European Community of nations had set even more stringent deadlines for CFCs and carbon tetrachloride. Figure 15.4 emphasizes the statements made in Chapter 14 on stratospheric ozone depletion showing how the atmosphere will clear itself of the synthetic halogen gases in the future because of adoption of regulations of the Montreal Protocol and its amendments.

The Kyoto Protocol In December 1997, a historic meeting was held in the ancient and traditional city of Kyoto, Japan. International negotiations that had begun in the late 1980s and culminated in the United Nations Framework Convention on Climate Change in Rio de Janeiro in 1992 had led to the meeting in Kyoto. The various parties to the Framework Convention had met in Berlin in 1995 and at that meeting made the decision to aim for a set of agreements (a protocol) at the third meeting of the parties in Kyoto in 1997. The Kyoto Protocol was ratified in late 2004 when Russia became a signatory and took effect on February 16, 2005 with 141 nations

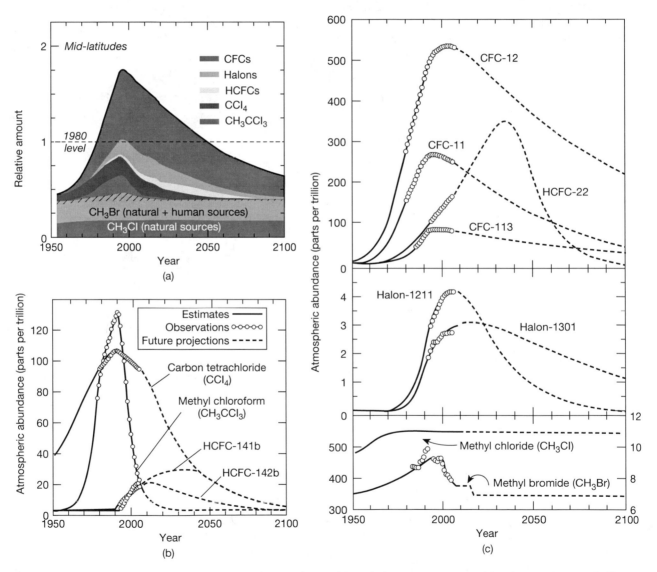

FIGURE 15.4 Historical and expected future abundances of atmospheric halogen source gases. (a) Relative amount of effective stratospheric chlorine, that is, chlorine able to destroy stratospheric O_3, related to CFCs, halons, HCFCs, CCl_4, and CH_3CCl_3; (b), (c) temporal history of the atmospheric abundance of the CFCs, HCFCs, and other compounds dangerous to the strato-spheric O_3 layer. (WMO (World Meteorological Organization) *Scientific Assessment of Ozone Depletion: 2006*, Global Ozone Research and Monitoring Project – Report No. 50, 527 pp. Geneva, 2007. Used by permission.)

signed on, but the United States refused to sign the protocol. The developing nations are not included within the protocol, despite the fact that greenhouse gas emissions from these countries could exceed those of the developed, industrialized world by about 2015.

The Kyoto Protocol requires that the 39 industrialized nations involved in the protocol reduce their greenhouse gas emissions by a specified annual percentage of their 1990 emissions by 2008–2012. The average overall reduction is targeted to be approximately 5%. However, the percentages vary from country to country: For example, for the United States, it was to be 7%; for the European Union countries, 8%; for Japan, 6%; for New Zealand, no change; and Australia and Iceland would be allowed to increase emissions 8% and 10%, respectively. The emissions of six gases are included in the protocol: CO_2, CH_4, N_2O, hydrofluorocarbons (HFCs), perfluorocarbons (PFCs), and sulfur hexafluoride (SF_6). Each country's limit is based on a radiation-weighted sum for all six gases. This weighting considers the degree of influence of each gas on the radiative balance of the atmosphere and takes into account the abundance of each gas and its individual global warming potential. Besides target levels for emissions, the protocol allows trading of greenhouse gas emissions credits and allows offset of emissions by development of sinks for greenhouse gases. For example, if a country can demonstrate that carbon is being stored in its forests because of planting trees, this carbon storage will be entered negatively into a country's net emissions, counting as a reduction and credit against emissions.

In the United States, the Kyoto Protocol has been met with strong resistance over the years on the part of most of the petroleum and coal industries, other lobbying groups, and by the Bush administration. In addition, the public has not been overly supportive of the agreement. The opposite can be said for many of the European and other nations, although the nations signed on to the Kyoto Protocol have a mixed track record of making their Kyoto emissions target goals (Table 15.2). The adverse reactions of policymakers and the public alike in the United States to some extent stem from the opinions of those individuals, both inside and outside the scientific community, who believe global warming is not a proven fact; if warming is occurring, much of the warming is due to natural causes; and the fact that the developing countries' emissions are not regulated in the protocol. Also, if the Kyoto Protocol were fully implemented as presently formulated,

TABLE 15.2 Carbon Reductions by those Countries Responsible for 98% of the Carbon Emissions from the Developed World. The Percentage Reductions are those Targeted in the Kyoto Protocol, the Emission Changes between 1990 and 2007, and the Pledges the Countries are Currently Making for the Copenhagen Meeting in December 2009. Russia's Strong Percentage Reduction between 1990 and 2007 in Part Reflects the Strong Downturn in its Economy since the Demise of the Soviet Union.

Country	Kyoto target	1990 to 2007	2020 pledges
Australia	+8%	+30%	+13 to −11%
Belarus	None	−38%	−5 to −10%
Canada	−6%	+26.2%	−3%
Ukraine	0%	−52.9%	−20%
EU; 15 members	−8%	−4.3%	−20 to −30%[a]
Japan	−6%	+8.2%	−9 to −25%
New Zealand	0%	+21.2%	−10 to −20%
Norway	+1%	+10.8%	−30%
Russia	0%	−33.9%	−10 to −15%
United States	None	+16.8%	0 to −23%

[a] Includes 27 member states of the European Union (EU). Data analyzed by the World Resources Institute, 2009.

some economists believe that there are strong reasons to conclude that it would have a negative impact on economic growth.

Since the meeting in Kyoto, there have been several meetings of the parties to evaluate progress of the Kyoto protocol and to formulate an agenda for the discussion of a new agreement to be framed in Copenhagen in December 2009. The most important meeting was that held in Bali in December 2007 for 12 days and attended by 191 counties and more than 10,000 people. The "Bali Roadmap" was produced, which is a plan for the processes involved to reach an agreement protocol at the future Copenhagen meeting. The actual agreement will be hammered out at the Copenhagen meeting in December 2009. As of this writing, the various nations of the world are putting together their strategies and proposals for consideration and debate at the Copenhagen meeting. Some countries are aiming at setting more stringent targets than those reached at Kyoto to reduce greenhouse gas emissions; others want strong emphasis on sinks to be used to offset emissions; others are proposing the strong use of cap and tráde policy; still others are proposing carbon capture and sequestration to reduce emissions; and several countries will not at the moment commit to any percentage reduction in emissions by any set date. The commitments toward reductions of emissions by 2020 so far fall below the 25 to 40% reductions (Table 15.2) needed by the developed nations alone to stabilize atmospheric CO_2 concentrations at 450 ppmv. Even at this level, the planet could still warm by more than +2°C above the late pre-industrial global mean temperature by mid–twenty-first century.

Whatever the case, the information shown in Figure 15.5 and similar scenarios developed in the IPCC 2007 reports will guide to some extent the final agreements reached in Copenhagen (see End Note on Copenhagen Conference). Figure 15.5 shows examples of model calculations of carbon dioxide emission scenarios for the future necessary to achieve goals for stabilization of atmospheric CO_2 at concentrations of 350 to 750 ppmv. To stabilize at 350 ppmv would require drastic measures to control emissions of CO_2 to the atmosphere from fossil fuel burning and land use practices. Stabilization at 750 ppmv would allow emissions to grow to about 14 billion tons of carbon per year in the middle of the twenty-first century and then decline to 4.5 billion tons of carbon per year in 2300. However, this scenario implies that by about the end of the twenty-second century, the concentration of CO_2 in the atmosphere would reach 750 ppmv, about 170% above the preindustrial level of 280 ppmv.

It is obvious that the future course of anthropogenic emissions will play a dominant role in determining the atmospheric concentrations of CO_2 on into the next century and beyond. Thus, the management of these emissions is a critical factor in the rate of global warming in the future. There are newer and slightly different scenarios than the ones in Figure 15.5—e.g., from IPCC, 2007—but the conclusions remain very much the same. One of the most recent projections (IPCC, 2007) shows the stabilization levels of atmospheric CO_2 as ranges and the expected equilibrium average temperature ranges for each range of stabilization level (Figure 15.6). Once more, the conclusion is obvious: The sooner the world community reduces CO_2 emissions to achieve lower stabilization levels of atmospheric CO_2, the less the global temperature increase from 2000 into the future. A +2°C global mean rise in temperature above the late pre-industrial temperature is considered highly threatening to society and the ecosphere by some scientists.

EXAMPLES OF POSITIVE ENVIRONMENTAL ACTIONS

For several decades in the industrialized world, a schism existed between environmentalists and business and industry. The environmentalists argued for cleanup and protection of the environment at any cost. Corporate leaders feared that burgeoning environmental regulations would limit economic growth. Most pollution legislation in the United States and Europe involved the government specifying the technology an industry or factory must use to control pollution or setting emission standards regardless of pollution source or relative costs to each source. In the late 1980s, there were indications of a change and of more cooperation between environmentalists and industry. Attempts to use the power of the market to protect the environment were set into play. Perhaps one of the best examples in the United States is the 1990 Amendments to the Clean Air Act. This bill establishes a market for tradable emissions allowances. In 1995 each power utility was granted a certain number of permits to release an amount of sulfur dioxide to the atmosphere that is based on the historical average release of the power industry. Utilities that release higher than average emissions were granted fewer permits than those that release less. The allotment of permits for each utility gradually decreased through the 1990s. By the early twenty-first century, sulfur dioxide emissions in the United States were reduced to about one-half of 1990 emissions. A unique aspect of this bill is that each utility may comply with the emissions regulations in different ways. It may develop new technology to decrease emissions; it may switch to cleaner fuels; it may invest in energy conservation; and it may purchase extra permits from utilities producing fewer emissions. In addition, there was no undue burden on the profit margins of the power utility companies, despite the outcry to the contrary when the regulations were first being suggested and then augmented (see Chapter 12).

Externalities   The Clean Air Act is not only an example of cooperation between environmentalists and industry, but it also puts a price on environmental pollution that had not been previously borne by the polluters. These indirect costs are termed **externalities,** because they are outside of the usual methods used to assess pricing. Society in general bears these costs in the form of higher income taxes, health care premiums, food prices, and so forth, and the industry, factory, or enterprise does not bear the costs. An example is the cost of swampland along the coast of Louisiana, currently priced at roughly $200 to $400 an acre. If one includes the value of the land for recreational activities, storm

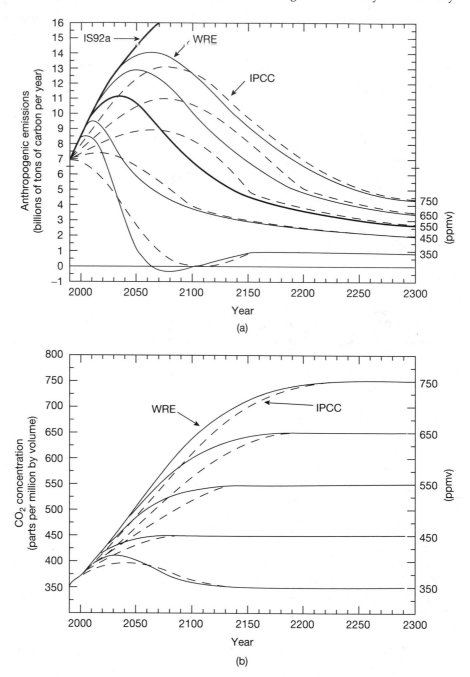

FIGURE 15.5 Examples of model calculations of future carbon dioxide emissions to the atmosphere necessary to achieve certain levels of stabilization of atmospheric CO_2 concentrations. (a) The emission scenarios starting in 2000 necessary to achieve prescribed stabilization levels of CO_2 in 2300; (b) the projected atmospheric concentrations resulting from the different emission scenarios from 2000 to 2300. The trend labeled IS92a is for a Business as Usual scenario of the Intergovernmental Panel on Climate Change. The dotted lines are scenarios developed by the IPCC (Houghton et al., 1996, 2001), and the solid lines are calculations of Wigley et al. (1996). Notice the difference in emission scenarios between the two groups leading to similar levels of stabilization of atmospheric CO_2 levels.

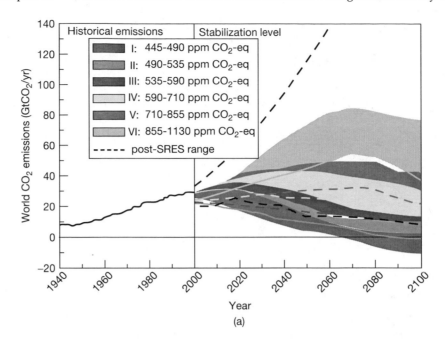

(a)

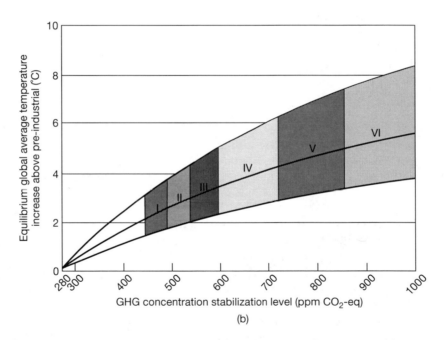

(b)

FIGURE 15.6 World carbon dioxide emissions in billion tons of CO_2 per year (a) and equilibrium temperature increases (b) for a range of stabilization levels of CO_2. Six different scenarios of stabilization ranges associated with a range in CO_2 emissions (I-VI) are shown in (a), as is the recent baseline range (black, dashed lines) published since the Special Report on Emissions Scenarios of the IPCC in 2000. In (b) the equilibrium global mean temperature is given in °C above the pre-industrial level; the best estimate climate sensitivity of 3°C is shown as a black line; an upper bound of temperature using a 4.5°C climate sensitivity is shown by the upper bounding line, and a likely lower bound with a climate sensitivity of 2°C is shown as the lower bounding line. (*Assessment Report Climate Change 2007: Synthesis Report. Contribution of Working Groups I, II, and III to the Fourth Assessment Report of the Intergovernmental Panel on Climate Change.* © Intergovernmental Panel on Climate Change 2007. Published by IPCC, Geneva, Switzerland. Used by permission of Intergovernmental Panel on Climate Change)

protection, trapping, and as a fisheries resource, the real cost may be between $2,400 and $17,000 per acre. Another example is the governmental fee charged cattlemen in the United States to use public lands for grazing cattle and sheep. The cost of land degradation and soil erosion (the environmental externalities), a result of overgrazing, is not included in setting the fee.

It is difficult to put a price on all the externalities associated with a resource. On overgrazed or logged land, how does one determine the costs of the lost recreational opportunities, increased erosion, and reduction in biodiversity? In the future, we are likely to see more attempts to include costs associated with environmental externalities in the costs of doing business and in the prices paid for a product. Such pricing would give the consumer a better idea of the actual environmental costs associated with producing a product and enable the consumer to make informed decisions.

Green Technologies There are other signs throughout the world that the global environment will receive more attention in the twenty-first century. Green technologies—that is, technologies designed to save the environment as well as produce profits—are expanding rapidly in Western European countries, Japan, and more recently in the United States and Canada. China and India, rapidly growing developing countries, also recognize the value of green technologies. An example is the wind energy business, which is booming in the early twenty-first century. In 2006, the world wind power generating capacity was about 74,223 megawatts and grew 3746% between 1990 and 2006! In 1980 the capacity was only 10,000 kilowatts. Germany, Spain, and the United States in that order lead the world in total capacity at present. There are more than 16,000 wind turbines in the United States alone, with an electrical generating capacity of about 11,600 megawatts. European wind capacity rose 19% in 2006, reaching more than 48,000 megawatts, or enough to meet more than 3% of the European Union's electricity demand. The largest wind farm in Europe is a group of 150 windmills near Gibraltar that produce 15,000 kilowatts of electricity. Asia is also experiencing strong growth in wind energy generating capacity, adding nearly 3680 megawatts of wind energy in 2006.

Other "green" energy sources have shown strong growth in recent decades. World shipments of photovoltaic cells that convert sunlight directly into electricity have risen from a 1971 energy equivalency of 100 kilowatts to 7 megawatts in 1980 to about 2500 megawatts in 2006. The cumulative production of photovoltaic cells rose from 19 megawatts in 1980 to about 8600 megawatts in 2006. The world geothermal electrical generating capacity rose from 200,000 kilowatts in 1950 to about 10,000 megawatts in 2008. Geothermal plants presently have the capacity to generate globally about 10 gigawatts of electricity (0.3% of global electricity demand) and another 28 gigawatts of geothermal heating to heat directly homes, buildings, and other facilities without conversion to electricity. About half the world's geothermal electrical generating capacity is in the United States. The Philippines and Mexico rank second and third in terms of generating electricity from geothermal energy.

The global hydroelectric generating capacity increased from 45 million kilowatts in 1950 to 777 million kilowatts in 2006. This represents a 16-fold increase in capacity in a little more than 50 years. Developing countries generate 37% of hydroelectric power. The United States and Canada produce the most hydroelectricity, each generating about 13% of the world's total capacity. There is some opposition to further strong development of hydroelectricity because of the environmental problems associated with dam building and the displacement of people by flooding of the land.

Worldwide sales of compact fluorescent bulbs (lamps) (CFLs) increased 900% between 1988 and 2000. The total number of CFLs in use globally nearly doubled between 2001 and 2003 alone, growing from an estimated 1.8 billion lamps to 3.5 billion lamps. These bulbs have four times the efficiency and ten times the lifetime of standard incandescent bulbs. The fluorescent lamps cut the demand for electricity and can decrease the total energy used by a country. As a precautionary thought, modern CFLs contain mercury, a neurotoxin. Because of this and other factors, there is a strong industrial movement on to switch to light-emitting diodes over time but currently they are more expensive.

World production of biofuels rose 28% in 2006 and annual production in 2006 was 44 billion liters of fuel. In 1975 ethanol production was 556 million liters and in 2006, 38,200 million liters, equivalent to a 6770% increase in 31 years. Biodiesel production amounted to 11 million liters in 1991 and 6153 million liters in 2006, an astounding percentage increase of 55,836% in just 15 years! The United States and Brazil combined produce nearly 90% of the world ethanol fuel with the United States passing Brazil in production in 2005. In 2006, Germany, the United States, France, and Italy were the top four producers of biodiesel in that order. There is considerable concern about the socioeconomic and environmental impacts of biofuel crops because the growing demands for the crops that are feeder for the fuels have pushed up the price of a wide range of agricultural commodities in 2006, including sugar, corn, soybeans, rapeseed, and palm oil. This in turn cut the profit margins of the biofuel producers. In addition, the planting of crops mainly for feedstock for biofuels has reduced the amounts of these crops available for use as food. There are potential environmental costs of biofuels such as loss of forest area and biodiversity, pressure to increase yields by using more nutrients, pesticides, and water to produce feeder crops, and destruction of farmland. Because of this, fuels made from United States corn, Brazilian soy, and Malaysian palm oil may be worse overall than fossil fuels in terms of net greenhouse gas emissions and degradation of ecosystems. The best alternative feeder stocks include residual products, such as recycled cooking oil and ethanol from grass or wood, for biofuel production.

Alternative energy sources are also receiving more research and development attention. The establishment of the National Renewable Energy Laboratory (NREL) in 1991 bolstered the lagging alternative energy industry in the United States. The NREL consists of more than 60 laboratories and includes centers that deal with evaluating photovoltaic materials, creating fuel-producing organisms through genetic engineering, conducting wind and photovoltaic testing, designing and constructing a solar furnace to process materials and destroy wastes, and developing ethanol from corn to serve as an energy source.

There has been an enhanced program of research and development involved with the production of cleaner cars, including Japanese automakers' hybrids and the building of electric vehicles by Chrysler and Ford. In 1993 President Bill Clinton and the chief executives of Chrysler, Ford, and General Motors began a new government–industry initiative to develop more fuel-efficient cars and at the same time adhere to strict standards of air pollution, safety, performance, and comfort. President Bush in 2002 promoted the use of hybrid vehicles using conventional fuel and an electric motor/battery. The United States Energy Security and Independence Act of 2007 imposed a standard of 35 miles per gallon on United States car manufacturers by 2020. The Obama administration has proposed that automakers meet this standard by 2016.

Efforts are under way to develop farming and forestry techniques that will be less damaging to the environment. Farmers in the United States and other parts of the world are cutting back on their use of chemical fertilizers and tilling the land in a more conservative fashion to prevent soil erosion. A more diverse selection of crops is being employed in crop rotation programs. Organic fertilizers are becoming more popular. Computerized tractors are playing an important role in modern farming methods. Geographical position is determined using a Geographical Positioning System (GPS), and crop production and fertilizer usage are monitored in real time. Organic farming methodologies have increased dramatically across the United States and Europe with less use of water, fertilizers, and pesticides and better quality of animal and plant products. However, despite these actions, many streams and rivers in the United States and worldwide are still loaded with nutrients, pesticides, and other materials leached off farmlands, leading to increasing observations of anoxia developing in coastal marine ecosystems fed by land drainage (see Chapter 11).

Although efforts to save the world's primary temperate forests are too late, there are programs in place that might protect the tropical forests from the same fate. Low-intensity harvesting, total preservation of tropical forest tracts, preplanned logging methodologies, strip clear cutting where isolated 30- to 40-meter-long swaths are cut in the forest and intervening forest tracts are left intact, use of forest resources for commodities other than lumber, and restoration of deforested land are some examples. However, tropical forests and the remaining temperate forests still remain highly threatened ecosystems (see Chapter 10).

Citizens have become more proactive and governments more responsive to urban air pollution. Air quality has considerably improved over the United States because of the regulations put in effect with the Air Quality Act of 1967 and later amendments. However, Southern California's air quality is still the worst in the United States. Unhealthful levels of air pollution are reached on half of the days each year. The United States federal standards for healthful air are violated for ozone, carbon monoxide, nitrogen oxides, and fine-grained particulates. Yet between 1970 and 1990, ozone concentrations of air in the Los Angeles area dropped from 580 ppb to 330 ppb, although the population and number of motor vehicles increased by 43 and 66%, respectively. The number of days per year that ozone levels exceed 200 to 350 ppb decreased from about 120 to 45, and the area of the Los Angeles basin covered by high ozone levels shrank considerably. Air quality in Los Angeles has improved in the last two decades because of promotion of the use of cleaner fuels, purification of engine exhaust through the use of catalytic converters, reduction in the amount of exhaust gases escaping via the pistons of cars, and the encouragement of car pooling and use of public transportation. At the time of this writing, the United States 1990 Clean Air Act is in force, and even stricter air quality standards are under consideration. This is but one example of improvements in air quality of urban areas in the industrialized world. For example, western European air quality has also improved greatly during the past decade or so. However, these improvements are not representative of many of the major cities of the world. In particular, eastern European cities, those of the former Soviet Union, and many urban areas of Southeast and East Asia and Central and South America have air quality problems (see Chapter 12).

Further Actions Bearing on the Environment   Some other positive measures bearing on the environment include the following:

1. Energy efficiency centers have been established in several cities of the world, including a joint United States–Chinese project, the Beijing Energy Efficiency Center, which opened in late 1993. China now leads the world in terms of total carbon dioxide emissions to the atmosphere, but its per capita emissions are still low compared to the United States.
2. The Global Environmental Facility (GEF) was established to handle funds donated by the nations for environmental projects, particularly in the developing world. It includes 70 governments.
3. Many nations have already issued reports on strategies and measures dealing with climatic change and established plans to stabilize carbon dioxide emissions. The measures to accomplish stabilization vary from country to country and are currently the subject of much debate within these countries. For example, the European Union of nations had originally decided on a unionwide carbon/energy tax to slow emissions and tax the polluters. The decision to tax is now left up to the individual European Union countries. In general, several national strategies and measures to reduce greenhouse gas emissions have in common a timetable of returning to 1990 levels by the year 2010.

However, as of this writing, there is little likelihood that will happen, and many nations have missed their Kyoto Protocol agreements.

4. The United Nations Commission on Sustainable Development was established in 1993. This body monitors the implementation of Agenda 21. In particular it analyzes and evaluates reports from all relevant organizations, programs, and institutions of the United Nations system, as well as information provided by governments and nongovernmental organizations. The commission reviews the progress of the industrialized nations toward the United Nation's target of providing 0.7% of their gross national product for developmental assistance.
5. A President's Council on Sustainable Development was established in the United States in 1993 and was in effect through June 1999. In response to the recommendations of Agenda 21, the council was to develop specific recommendations for a national strategy for sustainable development to be implemented by both the private and public sectors. Issues considered include sustainable agriculture and land use, efficient energy and transport systems, environmentally sound products like energy-efficient lightbulbs and refrigerators, environmental education, environmental justice, and the establishment of environmental priorities. However, this council is now defunct and relatively little effort was made, or is being made, to implement sustainable development strategies within the United States federal government. While committed individuals are working within some specific agencies, there is no individual or organizational entity at the helm to steer the executive branch or congress, or any charted course by which to steer. It appears there has been little if any strategic thinking since 1999 within the United States about how the United States as a whole can and should achieve sustainable development. Obviously the United States and other nations should develop and implement a strategy for achieving sustainable development.
6. Measures are being taken to clean up some of the major rivers emptying into coastal marine areas and to reduce direct pollutant discharges into coastal environments. The Rhine and Scheldt river-estuarine systems of Western Europe, and Chesapeake Bay and San Francisco Bay of the United States are examples of areas

where pollution abatement measures are underway. However, coastal marine environments, forests, and coral reefs (see Chapters 11 and 14) still remain the most threatened ecosystems in the world.

AN EXAMPLE OF GLOBAL ENVIRONMENTAL POLICY

The policy decisions involved with a question of global environmental change perhaps may be illustrated by considering the recommendations of the United States National Academy of Science panelists in the book *Policy Implications of Greenhouse Warming* published in 1992 (National Research Council, 1992). Global warming is a sufficient potential threat to the United States and the world to justify further forms of action now. However, what shall be the course of action so that it does not stymie economic growth or produce undue hardship on peoples of the United States and the world? The policy recommendations for the United States fall into several categories: (1) reduce or offset emissions of greenhouse gases; (2) enhance adaptation to greenhouse warming; (3) improve knowledge for future decisions; (4) evaluate geoengineering options; and (5) exercise international leadership. These recommendations to one extent or another are also applicable to the considerations of other nations facing the global warming threat.

The National Research Council Recommendations

1. Continue the phaseout of CFCs and other halocarbon production. The United States is a party to the 1987 Montreal Protocol to the Vienna Convention and its amendments, which required total phaseout of CFCs, halons, and carbon tetrachloride in 1996 in industrialized countries and by 2010 in developing countries.

2. Strive to attain the goal of full social-cost pricing of energy. The pricing of energy production and use should include the full costs of associated environmental problems (the environmental externalities). It is clear that such a policy will be difficult to design or implement and is probably best coordinated internationally.

3. Enhance energy conservation and efficiency to reduce the emissions of greenhouse gases. This step would also decrease emissions of sulfur and nitrogen oxides implicated in the acid precipitation problem. The Framework Convention

on Climate Change of the 1992 Earth Summit argued for stabilization of atmospheric greenhouse gas concentrations at levels that would prevent dangerous anthropogenic interference with the climate system. The 1990 Intergovernmental Panel on Climatic Change (IPCC) report *Climate Change: The IPCC Scientific Assessment* (Houghton et al., 1990) calculated that to stabilize greenhouse gas concentrations at 1990 levels would require immediate reductions in emissions from human activities of the long-lived gases of CO_2, N_2O, and CFCs of more than 60%, and a 15 to 20% reduction in methane emissions. If such emission reductions occurred quickly, the economic systems of the world's nations would be severely stressed. Thus, many developed nations are striving to reduce greenhouse gas emissions over a protracted period, as they increase conservation and efficiency measures and research alternative fuel sources.

4. The potential of greenhouse warming should be an important factor in planning the future energy supply mix of the United States. Alternative energy supply technologies are unable at the moment or in the near future to supply sufficient electrical energy for the country. These energy sources require increased research and development. According to the NRC report, nuclear power is the most technically feasible alternative to fossil fuels for generating electricity. However, in the aftermath of Chernobyl and the inability of the United States to come to grips with its nuclear waste disposal problems, growth of the United States nuclear industry is at a standstill, but there recently has been increased activity in the industry.

5. Reduce global deforestation rates, explore a moderate reforestation program in the United States, and support international reforestation efforts. This effort would also alleviate other environmental problems such as loss of species habitat and diversity, biological impoverishment, increased soil erosion and altered water runoff, and alterations in regional weather patterns.

6. In terms of adaptations to global warming, the recommendations addressed agriculture, water systems, long-lived structures, and preservation of biodiversity. Specifically, it was suggested to (a) maintain basic, applied, and experimental agricultural research to enable farmers and commerce to deal with global warming and ensure

an ample food supply; (b) make the water supply less vulnerable to changing patterns of weather and precipitation by increasing efficiency of use and by better management of existing supply systems; (c) plan margins of safety for long-lived structures such as bridges, dams, and levees, taking into consideration the possibility of climatic change; and (d) take actions to slow losses in biodiversity, such as controlling and managing wild species to avoid overexploitation.

7. The NRC report also identified areas to improve knowledge for future decisions involving global environmental change. They included (a) the collection and dissemination of data records of the evolving climate and of data necessary to improve climate models; (b) improvement of weather forecasts over weeks and seasons to ease any future adaptations to climatic change; (c) continued investigation of critical mechanisms that play a significant role in the response of climate to changing atmospheric gas concentrations, for example, the role of clouds and aerosols in the radiative balance of the planet; (d) field research on the response of entire ecosystems to determine how CO_2 enrichment alters species composition and changes net primary production, and how warming affects biological diversity and the potential of biological impoverishment; and (e) strengthening of research on the social and economic aspects of global warming, including an improved understanding of the costs involved in the mitigation of greenhouse gas emissions, studies of the impacts of and adaptation to climatic change, an improved understanding of the social and economic processes leading to greenhouse gas emissions, a comprehensive analysis of the policy options and strategies related to climatic change, and an improvement in the database for understanding the economic and environmental trends relating to global change.

8. The concluding sections of the recommendations argue for research and development projects aimed at improving our understanding of geoengineering options to offset global warming and for the United States to exercise international leadership in addressing responses to global warming. Under the latter, it was pointed out that controlling population growth has the potential to make significant contributions to raising world living standards as well as easing global environmental problems like that of greenhouse warming.

Factors Involved in Policy Decisions

The recommendations of the NRC give a sense of the range of topics considered in questions of policy involved with an issue of global environmental change. The policy decisions depend not only on the perceived science but also on economic, social, and political factors. It is especially difficult to define policy when the science of global change is controversial and debatable. However, the very nature of science is controversy and uncertainty. When decisions of policy are made, this fact needs to be kept in mind. The scientist arrives at a hypothesis based on theoretical considerations and experimental evidence. The hypothesis is usually formulated as a null hypothesis (see Chapter 1), which is tested by further experimental and observational data in an attempt to falsify the hypothesis. The scientist recognizes that her/his hypothesis will be tested and debated by the scientific community. The policymaker must realize that the testing and debate may go on for years before any resolution is reached. Science generally proceeds by consensus. In the time needed to reach consensus, policy must be formulated within the framework of scientific uncertainty. The uncertainty necessitates that scientific progress reports dealing with the environment, including those of the IPCC on climate change, should not be the basis of laws that cannot be modified. It also necessitates an understanding by the policymaker and the public of the basic science, processes, and impacts of global environmental change. It is the responsibility of the scientist to produce and explain preliminary findings concerning global change and to provide the public with accurate and objective scientific information. It is the responsibility of policymakers and the public to be scientifically literate so they can understand to some extent the scientific arguments.

Herein lies a fundamental problem within our educational system in which the teaching of science and mathematics in the United States has fallen behind many countries in the developed world. Furthermore, many developing countries have a considerable way to go before science and mathematics teaching reaches a level that enables the citizens of these countries to develop the wherewithal to understand the environmental issues, such as cultural eutrophication, photochemical smog, stratospheric ozone depletion, and climate change, that plague the planet.

The irregular rise in global surface temperatures since the mid-1970s is an excellent example of scientific uncertainty. Some scientists argue that the temperature increase is due to the enhanced greenhouse effect; others argue that the temperature record is too short to conclude that it is due to a human-induced global warming. Yet policy is currently being formulated with respect to greenhouse warming based on a scientific consensus contending that with continued releases of greenhouse gases to the atmosphere by human activities, a global warming with all its ramifications is here or is very likely in the near future. Indeed, the 1996 IPCC report (Houghton et al., 1996) concluded that "the balance of evidence suggests discernible human influence on global climate" (p. 4), and in the 2001 report (Houghton et al., 2001) that "all (model) simulations with greenhouse gases and sulphate aerosols that have been used in detection studies have found that a significant anthropogenic contribution is required to account for surface and tropospheric (temperature) trends over at least the last 30 years. Evidence of a human influence on climate is obtained over a substantially wider range of detection techniques" (p. 57). The 2007 IPCC report concludes that there is " . . . very high confidence that the global average net effect of human activities since 1750 has been one of warming . . . " (p. 3). However, the report (see Chapter 14) *Nature, Not Human Activity, Rules the Climate* (Singer, 2008) contends that "Most of modern warming is due to natural causes" (p. 11). Other global environmental issues to some extent involve the necessity of policy decisions within the context of continuing scientific controversy and debate. In other words, one can argue that caution in the context of uncertainty should be exerted in terms of the environmental issue. This is the **precautionary principle,** which is a moral and political principle that states if an action or policy might cause severe or irreversible harm to the public or the environment, in the absence of scientific certainty that harm would not take place as a result of the action, the burden of proof lies with those who advocate taking the action; in this case the continuous emissions of greenhouse gases to the atmosphere from human activities. The principle implies that there is a moral and political responsibility to protect the world from human-induced climate change.

APPROACHES TO GLOBAL ENVIRONMENTAL COOPERATION

The previous section discussed an example of the types of policy considerations that are encountered when dealing with a problem of global environmental change, in this case, policy recommendations dealing with the issue of global warming for the United States by the United States National Academy of Science. These policy considerations have some applicability to the nations of the world as a whole as to items they might consider in terms of formulating policy dealing with climate change. However, formulation of global policies concerning environmental problems is more difficult and requires some form of cooperation among regions and nations. There are several approaches that could be envisioned and are to some extent being used. These include, among others, business as usual, global partnership, and global governance.

Business as Usual

Business as usual implies continued incremental changes in policy framework or institutional structure at the global level. Environmental issues in this situation are dealt with on a case-by-case basis. Examples of such an approach are the Montreal Protocol and the Framework Convention on Climate Change (FCCC). The former has been very effective in reducing halocarbon emissions to the atmosphere and rendering the depletion of stratospheric ozone a less dangerous phenomenon. Implementation of the Montreal Protocol and amendments is still continuously evaluated by the various party nations involved. Implementation of the FCCC is found in the 1997 Kyoto Protocol and amendments since then. There appeared to be genuine consensus by many nations signing the Kyoto agreement that they would go ahead and try to meet the reduction targets set by the protocol, but the degree of success meeting reduction schedules so far has been anything but satisfactory (Table 15.2).

Global Partnership

The global partnership approach entails major shifts in the policies of key industrialized and developing countries and a concerted effect on the part of these countries to collaborate extensively on sustainable development. The Global Environmental Fund to some extent is an example of this approach. A global partnership between the industrialized and developing countries has to deal with several important issues. These include interests and demands of the developing countries and certain commitments sought by the industrialized nations from the developing world. The former include (1) lessening or ending the net capital drain from developing countries to industrialized nations by increasing financial flows to the less developed nations and reducing their debt burden; (2) increasing the access of developing countries to markets for manufactured goods within the industrialized countries; (3) providing provisions to the developing world for access to energy-efficient and other advanced technologies; and (4) requirements for the slowing of wasteful high per capita consumption of the highly industrialized nations. The latter include (1) a stronger commitment on the part of the developing world to the sustainable management of their forests and other resources; (2) the use of appropriate energy mixes by developing countries to help in contributing to the stabilization or reduction of atmospheric greenhouse gas emissions; (3) a greater emphasis by the developing countries on slowing population growth; (4) greater accountability by the developing countries for the financial assistance given them by the developed world; and (5) greater participation by the populace in developing countries in developmental and environmental decision making.

Global Governance

The final approach to environmental cooperation, that of global governance, is certainly the most difficult. It involves far-reaching restructuring at the global level to stem the tide of environmental degradation and natural resource depletion. An example would be a global environmental legislative body with the power to impose regulations on nation-states. Such a body could conceivably reside within the United Nations. In 1989, 24 heads of state meeting at The Hague in the Netherlands actually declared support for such a concept but at the moment, there is little enthusiasm for this approach by the nations of the world.

Because of the global or worldwide nature of the major environmental problems, cooperation among regions and states must increase in the future if we are to stem the tide of environmental degradation. Such cooperation has started at the grassroots level with education and opportunities for people to manage their local environmental resources and to influence environmental policy decisions through their political institutions. Global cooperation will entail not only studies of the science of an environmental problem, but economic and social impacts of the resolution of the problem. Individual and group ethics will also play a role, and the ultimate policies will be defined in the political arena.

THE QUESTION OF SUSTAINABILITY

In recent years, a great deal of discussion has centered on the concept of sustainability and its meaning for the environment and economic development. If we are to achieve sustainability from an environmental and human perspective, it will be necessary to ensure that the needs of the present generation and the human activities arising from these needs do not compromise the ability of future generations to meet their own needs. It is obvious that to achieve this goal will necessitate sustainable use of renewable and nonrenewable resources. This in itself is a difficult proposition because there are many instances of overexploitation of resources throughout the world. For example, consider cultivated systems that already encompass at least 24% of the terrestrial surface but will continue to increase in area as more terrestrial biomes are converted into other uses into the twenty-first century, mainly cultivated systems; and consider also the decline in the mean trophic level of fish harvested, which has been occurring for 50 years mainly because of overfishing—the Atlantic cod fish stocks off of the east coast of the United States and Newfoundland collapsed in 1992, 45% of the assessed fishery resource of the United States has been classified as overutilized, and 59% of the 78 fish stocks of European waters are considered overutilized. Also, it is difficult to determine what is meant by sustainable use of a resource. What is the time scale of sustainability of a nonrenewable resource? How do we manage a renewable resource so that it is sustainable?

As succinctly stated by Kai N. Lee (Lee, 1993), formerly professor and director of the Center for Environmental Studies at Williams College and now Program Officer with the Conservation and Science program at the David & Lucile Packard Foundation, and still applicable today, sustainable development is more than a problem of policy and administration. It is more than a union of ecological science with development of a political strategy. It is likely that any approach to sustainability will lead to a redefinition of the role of the human species in the natural order, and that this redefined role in itself will alter the form of how we do things. Sustainability will probably necessitate a new global ethics. A society can be affluent by having much or wanting little. If we cannot continue to grow forever as a global community following the industrial pathway of the past 150 years, then at some time, we must ask whether affluence means having much or wanting little. Whatever the case, the path to sustainability is not clear. Lee proposes three conditions that must be satisfied by any path to sustainable development. These are the equity, legacy, and continuity conditions.

Equity Condition

The equity condition states that the concept of a sustainable world is difficult unless sustainability can become feasible for a majority of the world population. Many of the developing countries of the world have been affected by colonialism. In general, they are poor, and their natural resources are even today being rapidly exploited. Their poverty levels are now so high that to return to a traditional way of life is virtually impossible. They must develop within the bounds of the global marketplace.

At present, wealth, income, and power are unevenly distributed throughout the world. Lee concludes that in order to preserve the biological heritage of the weaker and poorer nations, and thus achieve a sustainable environment, there will need to be either substantial redistribution of wealth and power among nations or the formation of some sort of dependency relationships among nations that are stable over many lifetimes of ecological systems. The objective probably cannot be one of equality of wealth and power but an equality of economic conditions that is sustainable. Among other factors, equality of conditions will depend on cultural values.

Legacy Condition

The legacy condition implies a linkage between present and future generations. What we do today in terms of the environment and economic growth links our fate with that of future generations. This legacy includes the economic growth of the past three centuries. Growth was the result of scientific and technological advances during this period and was accompanied by environmental changes unprecedented in most of the geologic record of change.

A question arises as to whether growth will continue. With finite resources, it is difficult to see how economic growth can expand forever without reaching some limit. Many economists debate this idea feverishly, arguing that we will always find substitutes and a way around depleting resources. Finding new resources and discovering new kinds of resources through investment in research and development can postpone the limit. However, the assumption of continued economic growth makes sustainable use of natural resources very difficult to reconcile with that growth. Because rates of return for investors are determined by the marketplace and almost always are not zero, the monetary value of a resource in the future, such as a forest, is lower than it is at present. This is because an investor can take a smaller sum of money and invest it and make an equivalent or greater amount of money to that obtained in the future. The difference between the future amount and the lesser amount today is the discount rate. The lesser the discount rate, the greater are the profits over the long term. The consequence of such investing has important implications for preservation or sustainable development or resources. K. N. Lee concludes that "if resources are traded in markets, the value of conserving them for ecologically significant lengths of time is set by markets, not biology; usually, biological conservation turns out to be worth very little" (Lee, 1993, p. 192).

There are arguments that offset this conclusion. The first is that technological progress can make once valueless parts of nature worth something, and these can displace that which was being exploited. The second is that the ownership of property can lead to husbandry of resources. The third is that cultural values can limit the extent to which resources are treated as commodities. The fourth is governmental controls that regulate economic decisions.

The greatest barrier to the legacy condition is the assumption that economic growth will continue. There is no analytical foundation to this belief, but it engenders behavior that makes the assumption inevitably incorrect. To meet the equity condition requires a change in behavior and actions of the world's peoples for a sufficiently long period of time to modify the path of economic activity.

Continuity Condition

The continuity condition recognizes the fact that 25% of the world has achieved a remarkable level of wealth and material advantage through economic growth and relative political stability. The strong correlation between global industrial carbon dioxide emissions and global gross domestic product for much of the twentieth century is one indication that economic growth has become the major force of global change in the industrialized world. The inequity in economic growth between the industrialized countries and the developing world has led to overconsumption of resources and materials in the former and underconsumption and mass poverty in the latter. The historical record of wealth creation in rich nations through economic growth has left much of the world in an impoverished state, one reason for higher population growth rates in the poorer countries.

The hope to extend prosperity to the developing world is the continuity condition. It is very likely that the extension of prosperity will necessitate technology transfer and monetary programs involving the rich and poor nations of the world. It is to the benefit of the world community to pursue this course. To ensure a path of sustainable development in both the developed and developing worlds is a form of insurance. It protects against wars of redistribution of resources and the formation of spheres of influence that continue on the pathway of unsustainable practices. In the longer run, it gives to the developing world some sense of sharing in part of the wealth produced by economic growth in the industrialized nations fueled significantly by the resources of the developing world. Also, it provides markets for goods and services of the developed world and stabilizes political situations in the developing world.

Concluding Comment

Sustainable development will be a difficult state to attain for the world community. To meet the three conditions of equity, legacy, and continuity will require political decisions on a scale rarely seen to date. Sustainable development is a direction toward which the world community moves, not a goal in itself but more like the concepts of freedom and justice. There seems to be little recourse other than the movement toward sustainable development for the planet. The views of Earth from space show the global nature of the continents, atmosphere, and oceans and the extent to which humanity and its activities are distributed over the globe. The ecological footprint of global human society, a measure of the amount of land and sea area needed to produce resources, absorb wastes, and provide space for infrastructure like roads, factories and businesses, and homes is increasing each year. We now require 1.25 Earths to supply all our resource needs. At the current consumption levels of the higher income countries, the world could only sustainably support 1.75 billion people, not the 6.7 billion on Earth today and the projected 9.5 billion in 2050.

Human activities both contribute to and are affected by global environmental change. Humans are part of the ecosphere, and as members we cannot continue to plunder the very system that maintains us. We must husband, nurture, and respect that system and cultivate a world in which we can live together in a mutually beneficial way.

Concluding Remarks and the 21st Century

Global environmental change is inevitable simply because natural change will occur and a consequence of human activities is the production of wasteful energy (entropy), some of which is pollution, which leads to environmental change. However, the phenomena of environmental change are complex and difficult to predict. Witness the problems associated with a potential global warming because of the enhanced greenhouse effect. The scientific community still debates the extent, timing, and degree of the warming. Will the warming be gradual so humans and ecosystems can adjust, or will it be rapid and thus potentially more disrupting? Will a doubling of atmospheric CO_2 lead to a mean global temperature increase of 1 or 5°C? Will the climatic change result in rising sea levels and flooding of low-lying coastal areas and atoll islands? Will the interior of continents

tend to dry out as a result of the warming and decreased soil moisture levels? One could go on and list many more questions concerning the science of a human-induced global warming. However, despite all the scientific uncertainties which over time are becoming less and less global warming must be viewed as a threat to world society. The threat justifies policy actions by the world community. For many of the environmental problems discussed in this book, one can reach a similar conclusion.

In Chapter 9 (Figure 9.35), it was shown that it is conceivable that the global mix of energy sources could be transformed as we move into the twenty-first century with increasing reliance on environmentally benign solar power from photovoltaic and solar thermal generation. This would take a large amount of political will from all nations and capital resources to transform our energy base from fossil fuel to mainly solar. Other energy scenarios to wean us from coal, oil, and gas rely more on efficiency and conservation initially, carbon capture and sequestration of flue gases or direct capture of CO_2 from the air chemically, and storage in subsurface reservoirs, with the use of solar coming on line late in the twenty-first century. Figure 15.7 is one proposal put forth as an overall strategy to cut emissions of CO_2 to the atmosphere. The proposal promotes adopting a series of strategic stabilization wedges. Each carbon-cutting wedge would reduce carbon emissions by a billion metric tons in 2057. The major set of strategies to achieve these reductions includes efficiency and conservation, carbon capture and storage, use of low-carbon fuels, increased use of renewable energy, and increase of storage of organic carbon in croplands and forests.

A major strategy for dealing with CO_2 emissions that is receiving a great deal of attention recently is that of carbon capture and sequestration (storage) (Figure 15.7). This includes both the capture of CO_2 flue gases and the direct removal of CO_2 from air, its transport to storage sites, and the sequestration of the CO_2 in the ocean, in deep continental sedimentary basins or large sediementary plies of sediment (accretionary wedges) beneath the oceans, or in permeable basaltic rocks. Currently there are at least 14 CO_2 capture and storage site projects planned or proposed and as of mid-2008, 65 projects have been anounced worldwide. One problem with sequestration of CO_2 is that to reduce significantly global emissions of CO_2, huge volumes of CO_2 must be sequestered and this limits the options for where the CO_2 can be stored. The amount of CO_2 to be stored annually is equivalent to a blanket of CO_2 spread across Manhattan in New York City covering the city to a depth of three quarters of the way up the Empire State Building (Broecker and Kunzig, 2008)!

An interesting example of an attempt to resolve this problem is that of CarbFix. CarbFix is a field-scale project beginning in 2009 at an injection site adjacent to a new geothermal power plant at Hellisheidi, Iceland. The concept is to capture as much as 30,000 metric tons of CO_2 from the power plant and inject the gas dissolved in water at a partial pressure of 25 bars at a depth of 400 to 700 meters and at a temperature of about $30°C$ into subsurface basalts. It is anticipated that reactions between the basalt and the CO_2-charged water will lead to the formation of carbonate minerals like calcite and dawsonite $[(NaAl(CO_3)(OH)_2]$, thus trapping the carbon in solid mineral phases.

Other solutions involve introducing the CO_2 into the ocean as a sinking or rising plume depending on the depth of injection, dispersing it from a ship at sea to an appropriate depth in the ocean, or creating a lake of CO_2 on the sea floor. Deep injection into offshore sedimentary accretionary wedges could even involve injection into the pile of sediment at a depth at which methane hydrate is found. At the right depth, the CO_2 could react with the methane hydrate and become a CO_2 ice (CO_2 hydrate), liberating methane which could be captured at the surface and used as a fuel.

In deep continental sedimentary formations, because of its solubility characteristics, the fresher the water into which the CO_2 is injected, the more CO_2 that will be dissolved in the water. The CO_2 could be trapped in structural traps (see Chapter 8), by dissolving in the subsurface water, by reactions with the primary minerals of the sedimentary reservoir and containment in newly formed solid mineral phases, and by capillary trapping (also called residual-phase trapping). The latter involves trapping the CO_2 primarily after the injection of the CO_2 into the subsurface ceases and water begins to enter the CO_2 plume created by injection of the gas.

Carbon dioxide could also be scrubbed directly from air with huge scubbers that suck the gas out of the atmosphere. The capture towers may contain sodium hydroxide, or a proprietary chemical compound or material, that interacts with the air flowing through the tower and bonds the CO_2 to the sodium hydroxide separating it from the air that passes through the tower. The captured CO_2 would then have to be stored and sequestered.

There is still much scientific research to be done and many questions to be answered in terms of the processes involved in CO_2 capture, storage, and sequestration and the economic implications of the methodologies proposed. However, with the slowness at which the nations of the world appear to be willing to reduce their use of fossil fuels, carbon dioxide sequestration may be one important solution to this global problem.

The proposal in Figure 15.7 is not without its detractors but is a set of strategies that to some extent will be used if the world wants to avoid the major impacts of a warming of the planet on its water availability, ecosystem health, food productivity, coastal flooding and wetland loss, and human health shown in Figure 15.8.

In the early twenty-first century, the peoples of the world are continuing to enter an era that is global in a dual sense. The whole Earth is recognized as a

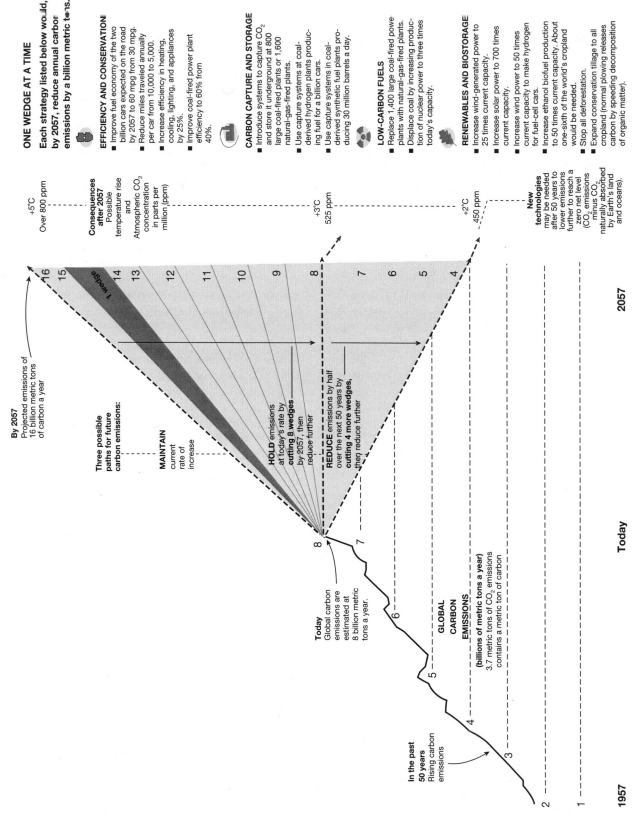

ONE WEDGE AT A TIME

Each strategy listed below would, by 2057, reduce annual carbon emissions by a billion metric tons.

EFFICIENCY AND CONSERVATION

■ Improve fuel economy of the two billion cars expected on the road by 2057 to 60 mpg from 30 mpg.

■ Reduce miles traveled annually per car from 10,000 to 5,000.

■ Increase efficiency in heating, cooling, lighting, and appliances by 25%.

■ Improve coal-fired power plant efficiency to 60% from 40%.

CARBON CAPTURE AND STORAGE

■ Introduce systems to capture CO_2 and store it underground at 800 large coal-fired plants or 1,600 natural-gas-fired plants.

■ Use capture systems at coal-derived hydrogen plants producing fuel for a billion cars.

■ Use capture systems in coal-derived synthetic fuel plants producing 30 million barrels a day.

LOW-CARBON FUELS

■ Replace 1,400 large coal-fired power plants with natural-gas-fired plants.

■ Displace coal by increasing production of nuclear power to three times today's capacity.

RENEWABLES AND BIOSTORAGE

■ Increase wind-generated power to 25 times current capacity.

■ Increase solar power to 700 times current capacity.

■ Increase wind power to 50 times current capacity to make hydrogen for fuel-cell cars.

■ Increase ethanol biofuel production to 50 times current capacity. About one-sixth of the world's cropland would be needed.

■ Stop all deforestation.

■ Expand conservation tillage to all cropland (normal plowing releases carbon by speeding decomposition of organic matter).

+5°C Over 800 ppm

Consequences after 2057
Possible temperature rise and Atmospheric CO_2 concentration in parts per million (ppm)

+3°C
525 ppm

+2°C
450 ppm

New technologies may be needed after 50 years to lower emissions further to reach a zero net level (CO_2 emissions minus CO_2 naturally absorbed by Earth's land and oceans).

By 2057
Projected emissions of 16 billion metric tons of carbon a year

16
15
1 wedge
14
13
12
11
10
9
8
7
6
5
4

Three possible paths for future carbon emissions:

MAINTAIN current rate of increase

HOLD emissions at today's rate by **cutting 8 wedges** by 2057, then reduce further

REDUCE emissions by half over the next 50 years by **cutting 4 more wedges,** then reduce further

Today
Global carbon emissions are estimated at 8 billion metric tons a year.

8
7
6
5
4
3
2

GLOBAL CARBON EMISSIONS
(billions of metric tons a year)
3.7 metric tons of CO_2 emissions contains a metric ton of carbon

In the past 50 years Rising carbon emissions

1957

Today

2057

FIGURE 15.7 A strategy employing a number of measures to reduce global carbon emissions one wedge at a time, where one wedge equals a billion metric tons by year 2057. (*Source:* Pacald and Socolow, 2004.)

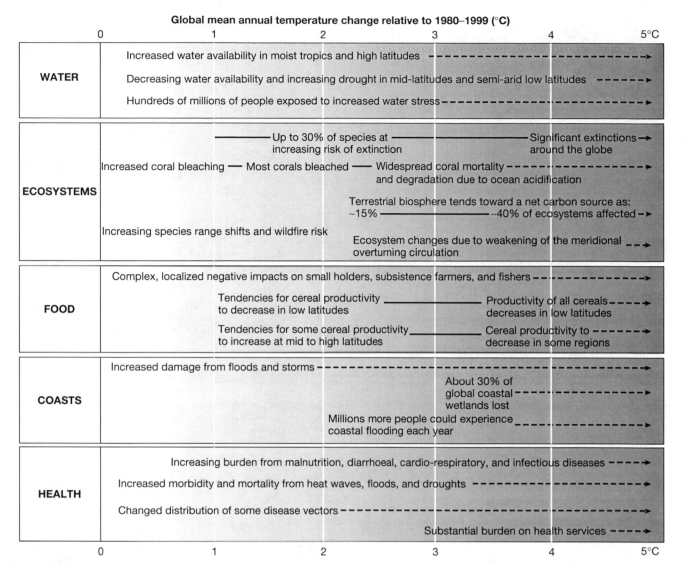

Global mean annual temperature change relative to 1980–1999 (°C)

FIGURE 15.8 Key impacts on water, ecosystems, food, coastlines, and human health as a function of global mean temperature increase relative to 1980–1999 in °C.   (From *Climate Change 2007: Impacts, Adaptation and Vulnerability. Contribution of Working Group II to the Fourth Assessment Report of the Intergovernmental Panel on Climate Change.* © Intergovernmental Panel on Climate Change 2007. Published by Cambridge University Press. Used by permission of Intergovernmental Panel on Climate Change.)

single ecosystem, and the nations of the world are moving to internationalize politically and economically in a process known as globalization. It is very likely that globalization will continue to be an important feature of the twenty-first century. Emphasis will be on sustainable patterns of development in terms of the environment and the economy. The industrialized countries must take the lead in making the transition to sustainability in their own societies, but it is also likely that to attain a secure and sustainable future for all of humankind, these countries, their corporations, and individual citizens will need to invest capital and technology in the nations of the developing world. The commitment by Ted Turner of

$1 billion to the United Nations in 1997 might be an example of such an investment by a private citizen. The former United States President William Clinton's Global Initiative program dealing with global challenges of food and energy and The Carter Center founded by the former United States President Jimmy Carter to advance human rights and alleviate unnecessary human suffering by waging peace and fighting disease are two other examples of prominent individuals becoming involved in global issues impacting the environment. However, events like the September 11, 2001, destruction of the World Trade Center in New York City, which had global repercussions, and the recent Iraq and Afghanistan wars, will

FIGURE 15.9 The insidious side of putting off actions to combat a slow and gradual warming of the planet. We will soon reach a point where human health and infrastructures are threatened, as are the world's ecosystems, and there could be very significant costs to combat the changes to the ecosphere. (Toles ©2002 The Washington Post. Reprinted with permission of UNIVERSAL UCLICK. All rights reserved.)

slow any movements toward a sustainable future for humankind.

The Earth Summit in Rio de Janeiro, Brazil, in 1992 and the 1993 population summit of the world's scientific academies in New Delhi, India, defined the environmental problems of the next century and initiated for the first time political debate of a global dimension. The signing of the Climate and Biodiversity Conventions by a large number of the world's nations and the introduction of a statement on forests at the Earth Summit, and the endorsement by 56 scientific academies of a statement calling for zero population growth, established precedents for future worldwide communication on economic development consistent with a sustainable environmental future. The Montreal Protocol and the Kyoto Protocol also set precedents for cooperation.

New academic disciplines have been established dealing with issues of global change, including ecological economics, global environmental politics, and the subject of this book, Earth system science. Learning has been recognized as of strategic importance in developing responses to global environmental change. All of these measures and actions argue for increased awareness of our obligations to exercise environmental stewardship of the planet and protection of the Earth for our children and their children. Currently we are like frogs in the pot of water slowing warming as the planet warms and undergoes environmental degradation from other causes. We will soon boil ourselves and pay the consequences of putting off actions capable of leading to sustainabilty of the ecosphere and to development of a sustainable nature (Figure 15.9).

End Note

The December 2009 United Nations Copenhagen Conference on Climate set no targets for future emissions of greenhouse gases to the atmosphere. The conference ended with an accord involving three major elements: 1. Nations agreed to cooperate to keep temperatures from rising 3.6°F above the preindustrial level; 2. Richer nations agreed to finance poorer nations $10 billion annually for three years to deal with drought and develop cleaner energy, with a goal of $100 billion annually by 2020; and 3. Developing nations will report every two years on their voluntary actions to reduce carbon emissions. Although some industrial and developing countries, including China, India, Brazil, and South Africa, had committed to reducing carbon emissions prior to the Copenhagen meeting, no legally binding international agreement was reached at the conference on emission reductions. Right now carbon emissions reductions are at the discretion of individual nations or are prescribed for nations that signed the Kyoto Protocol.

Study Questions

1. In the natural system, there have always been flows of materials throughout the environment of the planet. Why the concern about human emissions of materials to the environment?
2. Contrast the Montreal Protocol and the Kyoto Protocol in terms of measures of success and failure.
3. How do the changes in the indirect drivers of population, technology, and lifestyle lead to changes that affect ecosystem services and human well-being?
4. What are the strong linkages between ecosystem services and human well-being? What linkage between ecosystem services and the constituents of well-being has the greatest potential for mediation by socioeconomic factors?
5. From Figure 15.5, how high could anthropogenic emissions of CO_2 rise to achieve stabilization of atmospheric CO_2 at 750 ppmv in 2300? What would be the CO_2 emissions and the atmospheric CO_2 content in the year 2100 under this scenario?
6. Figure 15.7 illustrates one set of strategies for reducing anthropogenic CO_2 emissions to the atmosphere in anticipation of keeping future temperature changes minimal. What do you feel are the more easily attainable goals of this global carbon emission proposal? What do you see as drawbacks to or weaknesses of the whole concept of stabilization wedges?
7. What are the three conditions of sustainability mentioned by Kai N. Lee? How do they differ?
8. Pick one global environmental change issue and write an essay about its cause, its effects on humans and ecosystems, strategies to deal with the problem, and your assessment of whether the world community is likely to succeed in the near future in correcting the problem.

Additional Sources of Information

Bongaarts, J., 1994, Can the growing population feed itself? *Scientific American*, March 1994, pp. 36–42.

Broecker, W. S. and Kunzig, R., 2008, *Fixing Climate.* Hill and Wang, New York, NY, 253 pp.

Bruce, J., Lee, H., and Haites, E., 1996, *Climate Change 1995—Economic and Social Dimensions of Climate Change.* Contribution of Working Group III to the Second Assessment Report of the Intergovernmental Panel on Climate Change, Cambridge University Press, Cambridge, UK, 448 pp.

Flavin, C., 2008, *Low-Carbon Energy: A Roadmap.* Worldwatch Institute, Report 178, Washington, DC, 49 pp.

Gleick, P. H., 2001, Making every drop count. *Scientific American*, February 2001, pp. 40–45.

Kates, R. W., 1994, Sustaining life on Earth. *Scientific American*, October 1994, pp. 114–122.

Kareiva, P. and Marvier, M., 2007, Conservation for the people. *Scientific American*, October 2007, pp. 50–57.

Lee, K. N., 1993, *Compass and Gyroscope: Integrating Science and Politics for the Environment.* Island Press, London, 243 pp.

Millennium Ecosystem Assessment, 2005, *Ecosystems and Human Well-being: Synthesis.* Island Press, Washington DC, 137 pp.

Porter, G. and Brown, J. W., 1991, *Global Environmental Politics.* Westview Press, Boulder, CO, 208 pp.

Readings from Scientific American, 1990, Managing Planet Earth. W. H. Freeman, New York, 146 pp.

Renner, M., Sweeney, S., and Kubit, J., 2008, *Green Jobs: Working for the People and the Environment.* Worldwatch Institute, Report 177, Washington, DC, 60 pp.

White, R. M., 1996, Climate science and national interests. *Issues in Science and Technology,* v. 13, pp. 33–38.

Worldwatch Institute, 2001, *State of the World 2000.* W. W. Norton & Company, New York, 275 pp.

Worldwatch Institute, 2001, *Vital Signs 2001: The Trends That Are Shaping Our Future.* W. W. Norton & Company, New York, 192 pp.

Worldwatch Institute, 2007, *Vital Signs 2007–2008.* W. W. Norton and Company, New York, 166 pp.

Worldwatch Institute, 2009, *State of the World: Into a Warmer World.* W. W. Norton and Company, New York, 262 pp.

World Resources Institute, 2000, *World Resources 2000–2001.* World Resources Institute, Washington, DC, 389 pp.

APPENDIX: MINERALS

This appendix is designed to provide a brief discussion of major mineral groups for readers with a background in geology or the minimum required amount of basic mineral information for others. Table A.1 lists some important mineral types and their characteristics, including their chemical composition, crystal system, formula weight, and density. All crystalline minerals can be placed in one of six crystal systems depending on the patterns of arrangement of atoms that are repeated in three dimensions. Within each of the crystal systems there are specific crystal classes with distinctive symmetries. Thirty-two such classes are distributed among the six systems. The six crystal systems and their lines of reference—their crystallographic axes and angular relationship to each other—are shown in Figure A.1.

ELEMENTS

Many minerals are simply crystalline grains of the element. Among them are iron, nickel, silicon, sulfur, copper, silver, gold, arsenic, bismuth, and platinum. "Native" iron, of great importance in the deeper portions of Earth, has only a few occurrences in rocks exposed at the surface, as at Ivigtut, Greenland. Native silicon has never been observed, but may be an important alloying element in Earth's core.

OXIDES

Some of the most important oxide minerals are quartz (SiO_2), hematite (Fe_2O_3), corundum (Al_2O_3), ilmenite ($FeTiO_3$), and spinel (which includes a wide variety of compositions in a characteristic structure). The spinel structure is shown in Figure A.2. The large oxygens are in a cubic arrangement such that the interstices can be occupied by cations of two distinct size groups. Neither of the two positions need be filled with a single cation. Small cations, such as Al^{3+} or Fe^{3+}, fit into one position and somewhat larger ones, such as Fe^{2+} or Mg^{2+}, fit into the other. Representative chemical formulas are $MgAl_2O_4$ (spinel proper), $Fe^{2+}Fe_2^{3+}O_4$ (magnetite), and $Mg^{2+}Fe_2^{3+}O_4$ (magnesioferrite). Spinels probably are important minerals in the mantle. Their compact structure and high density, combined with their versatility in accepting a variety of cations, make them natural candidates as materials for the mantle. In the crust they are not a major constituent of rocks, probably because there is enough SiO_2 there to permit formation of silicates rather than oxides.

The most important oxide mineral in the crust is quartz. The low-temperature modification (alpha-quartz) is by far the most abundant in the crust and in fact makes up about 10 to 15 weight % of crustal rocks. Quartz is probably scarce or absent in the lower continental crust and in the oceanic crust, which, nevertheless, contain about 40 to 50 weight % of SiO_2 combined with other elements. Nearly pure quartz rocks are formed by depositions from wind, waves, and streams. Quartz is rather unreactive at Earth-surface temperatures; it does not dissolve or precipitate easily. Hematite is rare in the Earth as a whole, but is concentrated locally in the crust as nearly pure masses that are exploited commercially. Goethite ($FeOOH$) is an important mineral of soils.

TABLE A.1 Some important minerals

Mineral	Composition	Crystal system	Formula weight	Density (g/cm³)	Remarks
Elements					
1. Iron	Fe	Cubic	55.85	7.9	Density of solid iron about 12 under conditions of Earth's core
2. Nickel	Ni	Cubic	58.71	8.9	Probably makes up several % of core
3. Gold	Au	Cubic	197.0	19.3	Alloys commonly with Ag
4. Silver	Ag	Cubic	107.9	10.5	Too dense for core at core pressures
5. Platinum	Pt	Cubic	195.1	21.5	Densest of all minerals
6. Copper	Cu	Cubic	63.54	9.0	Too dense for core at core pressures
Oxides					
7. Quartz	SiO_2	Hexagonal	60.09	2.65	Most important oxide mineral in crust; chemically and physically durable at low T
8. Hematite	Fe_2O_3	Hexagonal	159.70	5.3	Insoluble under most surface conditions
9. Spinel	$Me^{2+}Me_2^{3+}O_4$	Cubic			The close-packed cubic structure and high density, even at surface conditions, make the spinels candidates for important mantle minerals
"The" spinel	$MgAl_2O_4$	Cubic	142.3	3.58	
Hercynite	$FeAl_2O_4$	Cubic	173.8	4.26	
Magnetite	$FeFe_2O_4$	Cubic	231.6	5.21	
Trevorite	$NiFe_2O_4$	Cubic	234.4	5.37	
10. Ilmenite	$FeTiO_3$	Hexagonal	151.8	4.79	
11. Corundum	Al_2O_3	Hexagonal	102.0	3.99	Chiefly in metamorphic rocks
12. Gibbsite	$Al_2O_3 \cdot 3H_2O$	Monoclinic	156.0	2.44	Product of extreme weathering of aluminosilicates
Silicates					
13. Olivine	$Me_2^{2+}SiO_4$				Orthorhombic olivines exhibit all compositions from pure forsterite to pure fayalite; other divalent cations also substitute; e.g., Ni^{2+}, Mn^{2+}
Forsterite	Mg_2SiO_4	Orthorhombic	140.7	3.21	
Fayalite	Fe_2SiO_4	Orthorhombic	203.8	4.39	
14. Pyroxene	$Me^{2+}SiO_3$				Important rock-formers, both in low-silica rocks of crust and in mantle
Enstatite	$MgSiO_3$	Orthorombic	100.4	3.12	
Wollastonite*	$CaSiO_3$	Triclinic	116.1	2.91	
Diopside	$CaMg(SiO_3)_2$	Monoclinic	216.6	3.28	
Hedenbergite	$CaFe(SiO_3)_2$	Monoclinic	248.1	3.55	

(Continued)

TABLE A.1 Continued

Mineral	Composition	Crystal system	Formula weight	Density (g/cm³)	Remarks
15. Amphibole					Note presence of $Si_4O_{11}^{-6}$ groups and OH^-; especially important in sheared crustal rocks; composition highly variable; Al commonly replaces some of the Si
Tremolite	$Ca_2Mg_5Si_8O_{22}(OH)_2$	Monoclinic	812.5	2.98	
Iron tremolite	$Ca_2Fe_5Si_8O_{22}(OH)_2$	Monoclinic	970.1	3.40	
Glaucophane	$Na_2Mg_3Al_2Si_8O_{22}(OH)_2$	Monoclinic	783.6	2.91	
Aluminosilicates					
Framework silicates					
16. Feldspar					Most important rock-forming minerals of Earth's crust; plagioclase exhibits all compositions from albite to anorthite
Orthoclase (K-spar)	$KAlSi_3O_8$	Monoclinic	278.4	2.55	
Plagioclase					
Albite (Na-spar)	$NaAlSi_3O_8$	Triclinic	262.2	2.62	
Anorthite	$CaAl_2Si_2O_8$	Triclinic	278.2	2.76	
17. Zeolite					Zeolites comprise a very large group of hydrated aluminosilicates; cations generally almost completely replaceable
Analcime	$NaAlSi_2O_6 \cdot H_2O$	Cubic	220.2	2.26	
Laumontite	$CaAl_2Si_4O_{12} \cdot 4H_2O$	Monoclinic	470.4	2.2–2.3	
Phillipsite	$KAlSi_2O_6 \cdot 2H_2O$	Monoclinic	254.3	2.2	
Layer silicates					
18. Chlorite	$Mg_5Al_2Si_3O_{10}(OH)_8$	Monoclinic	555.7	2.6–2.8	Most important minerals of sedimentary rocks; those given here are representative of a much larger group; generally not stable at high T and P
19. Montmorillonite	$Na_{0.33}Al_{2.33}Si_{3.67}O_{10}(OH)_2$	Monoclinic	257.5	2–3	
	$Ca_{0.17}Al_{2.33}Si_{3.67}O_{10}(OH)_2$	Monoclinic	256.6	Variable	
20. Muscovite (mica)	$KAl_3Si_3O_{10}(OH)_2$	Monoclinic	398.3	2.8	
21. Kaolinite	$Al_2Si_2O_5(OH)_4$	Monoclinic	258.1	2.65	
22. Garnet	$Me_3^{2+}Me_2^{3+}Si_3O_{12}$	Cubic			Probably important upper-mantle and lower-crust minerals
Almandine	$Fe_3Al_2Si_3O_{12}$	Cubic	443.7	4.25	
Grossular	$Ca_3Al_2Si_3O_{12}$	Cubic	396.4	3.53	
Pyrope	$Mg_3Al_2Si_3O_{12}$	Cubic	349.1	3.51	
Spessartine	$Mn_3Al_2Si_3O_{12}$	Cubic	441.0	4.18	
Carbonates					
23. Calcite	$CaCO_3$	Hexagonal	100.1	2.71	Chief minerals of limestones; aragonite is high P form but is common at low P and T

(Continued)

Mineral	Composition	Crystal system	Formula weight	Density (g/cm³)	Remarks
24. Aragonite	$CaCO_3$	Orthorhombic	100.1	2.93	
25. Dolomite	$CaMg(CO_3)_2$	Hexagonal	184.4	2.87	
26. Magnesite	$MgCO_3$	Hexagonal	84.3	3.01	
Sulfates					
27. Gypsum	$CaSO_4 \cdot 2H_2O$	Monoclinic	172.1	2.32	Major products, with NaCl, from evaporation of seawater
28. Anhydrite	$CaSO_4$	Orthorhombic	136.2	2.96	
Sulfides					
29. Pyrite	FeS_2	Cubic	120.1	5.02	Much of the sulfur in sedimentary rocks occurs in pyrite
Halides					
30. Halite	NaCl	Cubic	58.45	2.16	Major mineral in basins of evaporation

*Pyroxenoid.

(After Garrels and Mackenzie, 1971)

SILICATES

The important silicate minerals belong to three groups: the olivines, the pyroxenes, and the amphiboles. In the olivines, individual SiO_4^{4-} tetrahedra are linked by various cations. The general mineral formula is $Me_2^{2+}SiO_4$. The pyroxenes are made up of chains of SiO_3^{2-} groups, whereas the amphiboles have a double-chain structure of repeating $Si_4O_{11}^{6-}$ groups and characteristically have OH in their structure (Figure A.3). A great variety of cations are found balancing the negative charge of the silicate groups. Compositional variation is further complicated by substitution for Si, most importantly by aluminum. The ratio Si/O increases in the sequence olivine, pyroxene, amphibole.

Most investigators believe the mantle is largely composed of olivine and pyroxene material. In the crust these minerals occur abundantly in low-silica rocks, such as the lavas found in the ocean basins. The olivines form a continuous solid solution series from Mg_2SiO_4 to Fe_2SiO_4. The amphiboles are most characteristic of relatively shallow rocks that have been squeezed and distorted in the presence of H_2O. Amphiboles may be important species in the upper mantle.

ALUMINOSILICATES

The aluminosilicates are not clearly separable from the silicates, but it is convenient to distinguish the important feldspar and zeolite groups, which are highly aluminous, from the olivines, pyroxenes, and amphiboles, in which aluminum is commonly present but is not an essential structural element.

The feldspars are divided into the potassium feldspars, $KAlSi_3O_8$, and the plagioclase feldspars, which form a continuous solid solution form $NaAlSi_3O_8$ to $CaAl_2Si_2O_8$. Potassium feldspar may contain several percent sodium substituting for potassium. Potassium enters the plagioclase structure to an even more limited extent. The feldspars are the most important mineral group in the crust, making up more than 50% of many rocks. They are probably unimportant in the mantle and core.

The zeolite group contains about 25 distinct species. Many resemble hydrated feldspars from the compositional viewpoint. A major characteristic of the zeolites is their structure, in which $(Al,Si)O_4$ tetrahedra are linked together in such a way as to leave spaces in the crystal structure. Consequently, whereas the cations of feldspars are tightly enclosed and cannot be removed without disrupting the

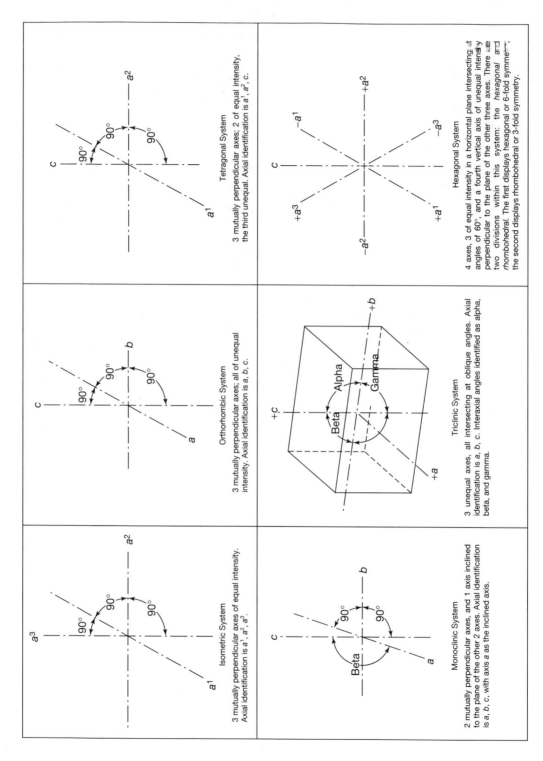

FIGURE A.1 Crystallographic axes and interaxial angular relationship of the six crystal systems.

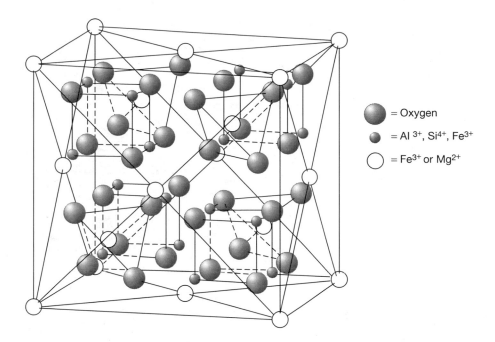

= Oxygen

= Al $^{3+}$, Si^{4+}, Fe^{3+}

= Fe^{3+} or Mg^{2+}

FIGURE A.2   Structure of spinel.

structure, in the zeolites one cation can replace another without disturbing the basic structure. Zeolites are not very important as rock-forming minerals, but they are representative "upper-crust" or surface environment species and are found abundantly in rocks of basaltic composition. Zeolites are also an important mineral of some deep-sea sediments.

LAYERED ALUMINOSILICATES

There is a large group of hydrated aluminosilicates with a sheet structure. Mica is a well-known example. The micas are made up of stacks of sheets built from Al-O octahedral and Si-O tetrahedral units, with various cations between the layers. Many show exchange properties like the zeolites

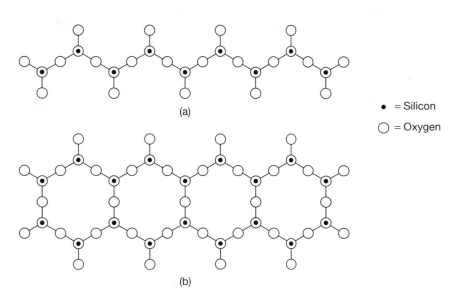

(a)

= Silicon

= Oxygen

(b)

FIGURE A.3   Structural framework of pyroxenes (a) and amphiboles (b).

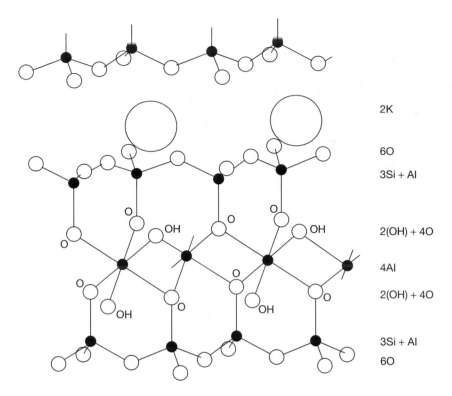

2K

6O

3Si + Al

2(OH) + 4O

4Al

2(OH) + 4O

3Si + Al

6O

FIGURE A.4 Structure of muscovite mica.

in that the interlayer cations can migrate in and out between the layers. Figure A.4 illustrates the structure of mica. Soils and sediments tend to be dominated by clay minerals, layer silicates characteristically, with a grain size on the order of 1 micrometer. The layer silicates are typical of the surface environment of Earth, and in general they are products of the weathering of other minerals. Some, such as mica and chlorite, exist as important rock-forming species in the upper crust, but at high temperatures and pressures, more compact structures are formed and the OH-groups in the layered aluminosilicate structure are expelled.

CARBONATES

The carbonate minerals calcite ($CaCO_3$), its dimorph aragonite, and dolomite [$CaMg(CO_3)_2$] make up about 10% of the sedimentary rocks of the upper crust. The structures of calcite and dolomite are shown in Figure A.5. Magnesite ($MgCO_3$) is relatively rare as a rock-forming mineral. In general, the carbonate minerals are products of surface processes, and $CaCO_3$, with or without some Mg, is precipitated by organisms to form shells and other types of skeletons. Carbonatites—rocks made up of calcium carbonate formed at many hundreds of degrees centigrade—are rare.

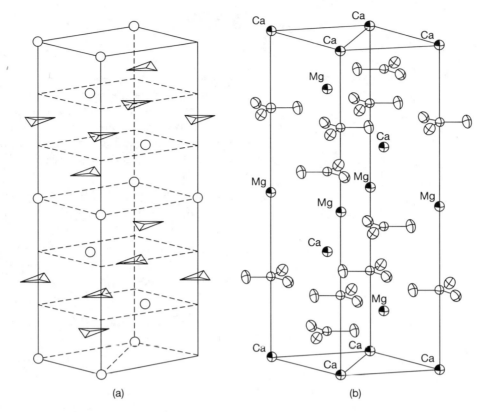

FIGURE A.5   Structural framework of (a) calcite (CaCO$_3$) and (b) dolomite [CaMg(CO$_3$)$_2$]. In (a), the opencircles are calcium (Ca) atoms and the triangles are carbonate groups (CO$_3$). In (b), the small circles represent calcium (Ca) and magnesium (Mg) atoms and the triangular shapes are carbonate groups (CO$_3$). The C axes of both structures are vertical.

SULFATES

Only two sulfates are of major importance—anhydrite (CaSO$_4$) and the dihydrate, gypsum (CaSO$_4 \cdot$ 2H$_2$O). Both occur abundantly where seawater has undergone extensive evaporation.

SULFIDES

There are many sulfide minerals, and they have great economic importance, yielding much of the copper, lead, zinc, and other metals used in industry.

However, except for pyrite (FeS$_2$), they form an insignificant fraction of Earth materials. Pyrite is formed abundantly in the surface environment in areas where bottom sediments are oxygen-deficient. It is a minor constituent of crustal rocks, but it or pyrrhotite (FeS) may well be important in the lower mantle.

GLOSSARY

abiotic of, or characterized by, the absence of life or living organisms.

achondrite a stony meteorite that does not contain spheroidal granules of olivine or pyroxene (chondrules).

acid typically a water-soluble, sour compound capable of reacting with a base to form a salt; a hydrogen-containing molecule or ion able to give up a proton to a base or to accept an unshared pair of electrons from a base.

acid deposition the fallout of acidic substances primarily of nitrogen and sulfur from the atmosphere onto Earth's surface.

acid rain rainfall with a pH of generally less than 5 on an annual basis because of the presence of acidic substances of nitrogen and sulfur dissolved in the rainwater.

adiabatic pertaining to the relationship between the pressure and volume of a gas or other fluid when the substance is compressed or expanded without giving off or receiving heat from its surroundings.

advection the movement of a fluid as an integral entity. The flow of the Gulf Stream along the eastern margin of North America is an advective flow, as is the movement of water in a river.

aerobic with oxygen; requiring the presence of free oxygen for life, or an environment that contains oxygen.

aerosol the suspension of very fine, generally micrometer-size, solid and liquid particles in the atmosphere.

air mass an extensive portion of the atmosphere in which temperature, humidity, and hydrostatic stability are relatively uniform in the horizontal direction.

albedo the amount of incident radiation that is reflected by a surface and thus does not contribute to the heating of the surface. The albedo of the entire Earth is approximately 30%. The albedo of clean snow is about 90% and that of water about 10%.

alkalinity a measure of the ability of a solution to neutralize acids and is equivalent to the stoichiometric sum of the bases in the solution. Practically, its measurement involves titration of the solution with usually hydrochloric acid to the equivalence point of carbonate ion or bicarbonate ion; that is, the point at which these ions are completely converted to dissolved carbon dioxide. Common natural chemical species that contribute to the total alkalinity of a solution are carbonate, bicarbonate, borate, hydroxide, phosphate, silicate, dissolved ammonia, sulfide, and some organic acids. A special case of alkalinity is that of carbonate alkalinity involving only bicarbonate, carbonate, hydrogen, and hydroxyl ions.

anaerobic without oxygen; an organism that does not require oxygen to carry on its metabolism, or an environment without oxygen.

angiosperm a flowering plant that produces seeds encased in a fruit or seed case, like oak, maple, and eucalyptus trees.

anoxic without oxygen.

antarctic bottom water (ABW) dense, cold water mass that forms around the Antarctic continent because of extreme chilling, sinks, and flows along the seafloor at a depth of about 4,000 meters.

antarctic intermediate water (AIW) water mass that forms in the Southern Ocean in the vicinity of sea ice by extreme chilling, sinks, and flows northward at a depth of about 1,000 to 2,000 meters.

anthracite coal a hard, natural coal generally of high luster and containing little volatile matter.

anthropogenic of, relating to, or influenced by the impact of humans on nature.

anthropogenic CO_2 increase steadily increasing concentrations of CO_2 in the atmosphere over the past two centuries.

anticyclone a fairly recognizable center of high pressure and associated wind pattern.

aquatic of, or pertaining to, water; growing or living in or frequenting water.

archaebacteria kingdom of bacteria represented by the extreme halophiles and thermophiles and methanogenic bacteria.

asthenosphere the dynamic subdivision of Earth between the lithosphere and mesosphere.

Atlantic Multidecadal Oscillation (AMO) a mode of multidecadal climate variability in the North Atlantic Ocean which has its principle expression in changing sea surface temperatures on a quasi-cycle of 70 years (based on limited temporal data). There appears to be a link between warm phases of the

AMO and the number of tropical storms that mature into major hurricanes.

atmosphere the gaseous envelope surrounding Earth; a reservoir in Earth's surface system.

atom the smallest component of an element having the chemical properties of the element; the atom consists of a nucleus containing combinations of neutrons and protons and one or more electrons bound to the nucleus by electrical attraction.

autotroph an organism that requires carbon dioxide as a source of carbon and a simple inorganic nitrogen compound for synthesis of organic matter.

autotrophic system an environment in which the difference between gross photosynthesis and gross respiration is positive. In such a terrestrial or aquatic environment, the net transfer of carbon dioxide is into the system.

autotrophy the biochemical pathway by which an organism utilizes carbon dioxide as a source of carbon and simple nutrient compounds for synthesis of organic matter.

bacteria one-celled organisms having a spherical, spiral, or rod shape belonging to the kingdom Monera.

basalt rock a dark, dense igneous rock of a lava flow or intrusion into Earth's crust composed principally of the minerals labradorite (a plagioclase) and pyroxene.

base typically a water-soluble and acrid or brackish-tasting compound capable of reacting with an acid to form a salt; molecules or ions able to take up a proton from an acid or to give up an unshared pair of electrons to an acid.

benthic of, relating to, or occurring at the bottom of a body of water.

benthic organisms organisms that inhabit the surface of, or live within, the sediments at the bottom of a body of water.

biodiversity the type and variety of living organisms; biological diversity.

bioessential elements (nutrients) those elements required by virtually all living organisms. The major elements are oxygen, carbon, nitrogen, phosphorus, sulfur, potassium, magnesium, and calcium. Minor or trace quantities of iron, manganese, copper, zinc, boron, silicon, molybdenum, chlorine, vanadium, cobalt, and sodium are also required by organisms.

biogeochemical cycle representation of biological, geological, and chemical processes that involve the movement of an element or compound about Earth's surface.

biogeochemical system the interactive system of biogeochemical processes and cycles of elements and compounds.

biogeochemistry the discipline that links various aspects of biology, geology, and chemistry to investigate the surface environment of Earth.

biological production see *productivity*; the rate of production of organic matter by producer organisms. For example, the rate may be given as grams of carbon per year for a forest ecosystem.

biological productivity rate of production per unit area of organic matter by producer organisms. For example, the rate may be given as grams of carbon per square meter per year for a marine grass community. There are several types of productivity. **Gross primary productivity** (GPP) refers to the total amount of plant material produced by photosynthesis in a defined area in an interval of time. **Net primary productivity** (NPP) is the net amount of plant material produced per unit area per unit of time and is the difference between GPP and cell respiration. **Net ecosystem productivity** (NEP) is the difference between GPP and cell respiration plus heterotrophic processes of decay.

biological pump the set of processes by which organic carbon is exported from the surface ocean to the deep sea.

biomass the amount of living matter in a unit area or volume of habitat. For example, the total biomass of the world's tropical rainforests is 42 kilograms of dry matter per square meter of forest, or a total of 420 billion tons of dry matter (equivalent to approximately 190 billion tons of carbon).

biome a major ecological community type: e.g., grasslands, coniferous forests, tundra, etc.

biosphere the living and dead organic components of the planet.

biota living organisms, such as animals, plants, fungi, etc.

birth rate the ratio between births and individuals in a specified population and time, commonly expressed as the number of live births per hundred or per thousand capita yearly.

bituminous coal a soft nonlustrous coal that contains considerable volatile matter.

blue-green algae a widely distributed group of predominately photosynthetic prokaryotic organisms of the kingdom Monera that occur singularly or in colonies; also called *cyanobacteria*.

boreal relating to or growing in northern or mountainous parts of the Northern Hemisphere.

box model a representation of a system that includes reservoirs (boxes), processes, and fluxes of materials.

Btu (British thermal unit) the quantity of heat necessary to raise the temperature of 1 pound of water from 60°F to 61°F at a constant pressure of 1 atmosphere.

C-3 pathway a photosynthetic metabolic pathway involving the initial synthesis of a three-ring organic acid by which most trees and shrubs (95% of all land vegetation) remove CO_2 from the air.

C-4 pathway a photosynthetic metabolic pathway involving the initial synthesis of a four-ring organic acid by which about 5% of all land plants extract CO_2 from the air.

calorie the quantity of heat necessary to raise the temperature of 1 gram of water by 1 °C from an initial standard temperature at 1 atmosphere pressure.

Carbon capture and sequestration (storage) (CCS)— the capturing of CO_2 emissions from fossil fuel combustion or the direct removal of CO_2 from the atmosphere and the storage and sequestration of the carbon in the ocean, continental subsurface geologic formations, or in basaltic rocks.

carbon isotopes isotopes of carbon with different atomic masses; ^{13}C and ^{12}C can be used to distinguish the sources of different kinds of carbon, and ^{14}C and ^{12}C can be used to determine the elapsed time since formation of a carbon-containing compound.

carbonaceous chondrite a chondrite meteorite containing important amounts of organic matter.

carrying capacity the population of organisms that an area will support without undergoing deterioration.

Celsius a temperature scale for which pure water freezes at 0° and boils at 100°.

chemical reaction a chemical change or transformation in which a substance decomposes, combines with other substances, or interchanges constituents with other substances.

chemical weathering the dissolving or other alteration of minerals and rocks in the presence of water and acids like carbonic acid to yield dissolved constituents and solid reaction products.

chemolithoautotrophs non-photosynthetic autotrophic microorganisms consisting entirely of the bacteria. These organisms do not use light and their energy sources for manufacturing of food are a variety of inorganic metabolites absorbed from the environment. The metabolites are often combined with oxygen in the cell resulting in energy release and a variety of inorganic products. Water and carbon dioxide are the inorganic raw materials used in subsequent food manufacture. This process of manufacturing food is termed chemosynthesis. The best-known chemoautotrophic organisms are the sulfur, iron, nitrifying, hydrogen, and methane bacteria.

chlorofluorocarbons (CFCs) synthetic organic compounds composed of methyl (CH_3) groups, chlorine, and fluorine.

chondrite a stony meteorite containing spheroidal grains of olivine and pyroxene (chondrules) in a matrix of fine-grained olivine, pyroxene, and nickel-iron with or without glassy material. The chondrites constitute more than 80% of meteorite falls.

clathrate a form of matter in which a substance, like methane, is locked in a structure of ice.

CLIMAP (Climatic Mapping and Prediction Project) a large group of scientists in the 1970s and 1980s that first attempted to obtain a picture of the ice-age Earth.

climate the characteristic, long-term environmental conditions of temperature, precipitation, winds, etc., in a region or for the globe presently and in the past (paleoclimate).

climate forcing the ability of a variable, like the concentration of a greenhouse gas in the atmosphere, to induce a change in climate.

closed system a system that does not exchange mass with its surroundings.

cloud condensation nuclei airborne particles of very small size, generally less than a micrometer in diameter, that serve as sites on which liquid cloud droplets condense when an air mass is supersaturated with water vapor. The particles are commonly composed of water-soluble material.

coccolith the micrometer-size plate of a Coccolithophoridae.

Coccolithophoridae a family of planktonic algae that build a skeleton of micrometer-size disc-shaped plates of calcite ($CaCO_3$).

compartment a subdivision of a box model in which the flow in or out of a substance is not known, although exchanges of mass may be suspected between the subdivision and its surroundings.

compound a pure substance composed of two or more elements whose composition is constant, such as water (H_2O) and common table salt (NaCl).

compressional wave the type of seismic wave that involves particle motion of alternating compression and expansion in the direction of propagation of the wave; also termed *P wave*. Sound waves are compressional waves.

concentration the fraction of the total of a substance made up of one component. For example, seawater contains 400 parts per million by weight of calcium. Concentration is also expressed in moles per liter or kilogram, in weight percent, in parts per thousand by weight (‰), or in parts per million (ppm) or billion (ppb), and so forth, by weight or volume.

consumer an organism that requires complex organic compounds for food and obtains them by preying on other organisms or by eating particles of organic matter.

continental drift lateral movement of continents resulting from the motion of the lithospheric plates.

continental ice sheet mass of ice on the order of several kilometers in thickness covering the continent or a portion thereof and moving independently of the underlying topography.

continental shelf the shallow, submarine plain of varying width that forms the border of a continent and terminates seaward at the continental slope.

convection the rising of a heated, warm fluid (air, water, ductile rock of the Earth's interior) accompanied by the sinking of colder, denser fluid elsewhere.

convection cell a distinct volume of circulating fluid that is heated from below and cooled from above. Such cells are thought to operate in the asthenosphere of Earth.

core the central portion of Earth enriched in iron and nickel and having a radius of about 3471 km. The innermost portion of the core is in a solid state, whereas the outer portion is molten.

Coriolis effect the apparent deflection of a body in motion with respect to Earth as seen by an observer on Earth. It is caused by the rotation of Earth and appears as a deflection to the right in the Northern Hemisphere and to the left in the Southern Hemisphere.

coupling a feature or characteristic of a system implying that information from one part of the system is provided to, and influences the behavior of, other parts. The biogeochemical cycles of the elements necessary for life are coupled through processes that are essential for life—e.g., photosynthesis and respiration.

crust Earth's outer layer, enriched in silicon, sodium, and potassium and having a thickness of 35 km beneath the continents and 10 km beneath the oceans.

cryosphere the icy part of Earth; consists of continental and mountain glaciers, ice sheets, and ice shelves; a reservoir in Earth's surface system.

crystalline of, or pertaining to, the nature of a crystal; having a regular molecular structure.

cultural eutrophication the various processes leading to overnourishment of an aquatic system in nutrients, rapid plant growth and death, and oxygen deficiency brought about by human-derived nutrient and carbon inputs to the aquatic system.

Curie temperature the temperature above which a material cannot be magnetized.

cyanobacteria the blue-green algae, a widely distributed group of predominantly photosynthetic prokaryotic organisms of the kingdom Monera that occur singularly or in colonies.

cyclone a fairly recognizable center of low pressure and associated wind pattern.

Dansgaard-Oeschger cycles oscillations in properties at intervals of 2,000 to 7,000 years in Greenland ice records, such as those of dust and isotopes of oxygen.

darcy a standard unit of measurement of permeability, equivalent to the passage of 1 cubic centimeter of fluid of 1 centipoise viscosity flowing under a pressure differential of 1 atmosphere in 1 second through a porous medium having a cross-sectional area of 1 square centimeter and a length of 1 centimeter. Because permeabilities are so low in natural materials of soils, sediments, and rocks, the unit of measure is usually the millidarcy, which is equivalent to 1/1,000 of a darcy.

decay the oxidative process of conversion of organic tissue to simpler organic and inorganic compounds. The oxidizing agent may be diatomic oxygen (O_2), nitrate (NO_3^-), or other chemical compounds.

decomposer various organisms such as bacteria and fungi that feed on and break down dead protoplasm and thus return the constituents of organic substances to the biogeochemical cycles.

deforestation the set of processes of clearing a forest.

denitrification the conversion, principally by bacteria, of compounds of nitrogen in soils and aquatic systems to nitrogen gas (N_2) and nitrous oxide gas (N_2O) and the eventual release of these gases to the atmosphere.

desertification the encroachment of the desert onto vegetated land.

diagenesis the collection of physical, chemical, and biological processes that operate on a sediment after deposition.

diatom planktonic and benthic freshwater and marine algae that commonly use silicon to build a skeleton of opal.

diffusion the generally slow movement of a substance through a solid, gaseous, or liquid medium from a region of higher concentration to one of lower concentration. The movement of ions through a membrane separating saltwater from freshwater is an example. **Eddy diffusion** is the movement of packets of air and water in the atmosphere and in an aquatic environment, respectively.

dinoflagellate a planktonic protozoan in the aquatic environment that may be soft-bodied or secrete a skeleton of opal.

disproportionation a chemical reaction involving the conversion of a chemical compound into product substances without the intervention of another reactant compound. For example: $2CH_2O \Rightarrow CH_4 + CO_2$.

dissolution the solution of solid substances such as minerals in rock or salt in water that leads to the production of ions dissolved in an aqueous solution.

Dissolved inorganic carbon (DIC) the summation of the concentrations of carbonate ion (CO_3^{2-}), bicarbonate ion (HCO_3^-), and aqueous CO_2 in an aqueous solution.

Dobson unit a measure of the concentration of ozone (O_3) in the atmosphere. The unit is named after G. M. B. Dobson, one of the first scientists to investigate atmospheric ozone. One Dobson unit is defined to be 0.01 mm thickness of ozone at standard temperature and pressure (STP, 0 °C and 1 atmosphere pressure)

dry deposition the deposition of material from the atmosphere onto Earth's surface in the form of solid particles. Such particles also may be "washed out" of the atmosphere by rain.

Earth radiation the long-wave, infrared (heat) radiation emitted by Earth's surface to space.

Earth system science that branch of knowledge or study dealing with Earth as a whole; study of the sum of processes operating in the atmosphere, hydrosphere, biosphere, cryosphere, and lithosphere and the interactions among these components.

eccentricity the extent to which the orbit of the Earth about the sun departs from a perfect circle

ecosphere the system that includes the biosphere and its interactions with the physical systems of planet Earth.

ecosystem the complex of a community or group of communities and the environment that functions as an ecological unit in nature.

eddy viscosity a measure of the internal friction that develops during turbulent flow in fluids when parcels of fluid, rather than individual molecules, are exchanged between one part of the moving fluid and another.

Ekman layer the upper part of the ocean where the water is influenced by friction induced by a wind blowing across the water surface.

El Niño an interannual climatic phenomenon initiated in the equatorial Pacific and accompanied by warm sea-surface temperatures off the West Coast of South America. El Niño recurs every 3 to 7 years and has global climatic consequences.

El Niño Southern-Oscillation (ENSO) a natural climatic phenomenon involving periodic changes in the ocean and atmosphere in the tropical Pacific and with teleconnections to other parts of the world; EN is the abbreviation for El Niño and represents the changes in the ocean, SO is the abbreviation for Southern Oscillation and represents the changes in the atmosphere. El Niño is the warm phase of the oscillation when the atmospheric pressure at Tahiti is relatively low compared to Darwin, Australia [the Southern Oscillation Index (SOI) is negative]. The opposite temperature and pressure conditions characterize a La Niña phase.

electromagnetic spectrum the entire range of radiation. The wavelengths (distance between adjacent peaks) of the electromagnetic waves within the spectrum range from kilometers for radio waves to billionths of a meter (nanometers) for X-rays.

element one of a class of substances that cannot be separated into simpler substances by chemical means, such as sodium (Na), gold (Au), and magnesium (Mg).

energy the capacity to do work; any source of usable power such as fossil fuel and solar power.

energy efficiency the efficiency by which a country uses its energy.

energy intensity the amount of energy needed to produce each unit of GNP or GDP.

enhanced greenhouse the additional greenhouse effect produced by the accumulation of greenhouse gases in the atmosphere derived from human activities.

entropy a scientific measure of the degree of disorder in a system. The greater the disorder the greater is the entropy of the system. The second law of thermodynamics states that entropy is always increasing.

entropy production a scientific measure of the amount of entropy change. Most natural systems produce wasteful energy, that is, entropy.

equilibrium a stable balanced system in which all influences are countered by others.

equivalent CO_2 radiative forcing changes due to changes in all greenhouse gas concentrations expressed in terms of the equivalent change in radiative forcing of atmospheric CO_2 alone.

erosion the processes by which the surface of Earth is worn away by the action of water, wind, glacial ice, etc.

estuary an arm of the ocean where freshwater from the land mixes with seawater.

eukaryote an organism having as its fundamental structural unit a cell that contains specialized organelles in the cytoplasm, a membrane-bound nucleus, and a system of cell division by mitosis or meiosis.

euphotic zone the upper lighted zone of the ocean or a lake in which most of the productivity of plants occurs. In the ocean, the euphotic zone extends from the surface to a depth where the light intensity is reduced to about 0.1–1.0% of that available at the surface. The depth of the euphotic zone depends on season and latitude.

eutrophication the various processes leading to overnourishment of an aquatic system in nutrients, rapid plant growth and death, and oxygen consumption and deficiency in the system. These processes occur naturally in some aquatic systems but may be speeded up by additions of nutrients from human activities (e.g., fertilizer application) to the systems. Human-induced eutrophication is often called *cultural eutrophication*.

evaporation the physical process by which water is converted from the aqueous state to the vapor state and is transported into the atmosphere from an aquatic environment.

evapotranspiration the combined processes of evaporation and transpiration.

evolution the pattern of development and change in a variable from one state to another. **Biological evolution** describes the pattern of emergence, development, and extinction of organic species through geologic time.

excess volatiles those components found in sedimentary rocks, atmosphere, and hydrosphere that have been derived from the degassing of Earth's interior. These are chemical constituents that cannot be derived simply from the weathering of igneous rocks.

exogenic system Earth's outer sphere, which includes the atmosphere, hydrosphere, biosphere, cryosphere, and shallow lithosphere.

exosphere the highest region of the atmosphere in which the density of air is very low and the probability of a fast-moving air molecule to escape Earth's atmosphere is high.

Fahrenheit a temperature scale for which pure water freezes at 32° and boils at 212°.

fault a break in a body of rock with dislocation along the plane of the fracture.

feedback a process or mechanism that generates a feedback loop in which some fraction of the output is returned or "fed back" to the input. Feedback loops may act in such a way as to stabilize (negative feedback) or destabilize (positive feedback) a system undergoing a perturbation. These feedback loops exist in both the biogeochemical cycles and the climate system.

fermentation the bacterial process of conversion of sugars to carbon dioxide.

fertility rate the average number of children women will have in their lifetime.

fertilization effect the increased growth rate of plants caused by adding CO_2 to the atmosphere, increasing temperature, and additions of nutrient N and P to the landscape. Often CO_2 is the only factor considered in the effect.

flocculation the process by which very fine-grained colloidal particles are held together in clotlike masses or are loosely aggregated into small lumps or granules and settle out of suspension.

flux the rate of transfer of a material; a parameter used in box models.

foraminifer any chiefly marine protozoan of the sarcodinian order Foraminifera. Also called *foram*. These microscopic creatures typically have a linear, spiral, or concentric shell perforated by small holes or pores through which pseudopodia extend. Their shells are commonly calcium carbonate in composition.

forcing function or driver a parameter that controls the behavior of a system. Such a function often makes the behavior of the system regular and predictable.

front the contact at Earth's surface between two different air masses, commonly cold and warm.

galaxy a large system of stars held together by mutual gravitational attraction and isolated from similar systems by vast regions of space.

general circulation model (GCM) a simulation of the large-scale, or general, wind and ocean systems on Earth to calculate climate and its changes using large computers.

geologic time scale a calendar of the history of Earth. The time scale is divided into variable-length time units of eon, era, period, and epoch.

geothermal of, or relating to, the heat of Earth's interior.

glacial stage a relatively cold period of time recognized during the Pleistocene Epoch during which much of the Northern Hemisphere was covered by great ice sheets.

glacier a large mass of ice formed from the accumulation and transformation of snow to ice over many years. Glaciers move down valleys as valley glaciers or move outward from centers of accumulation as continental glaciers because of gravity.

global environmental change the sum of transformations or modifications of the environment of Earth brought about both by natural and human-induced causes.

Global warming potential (GWP) a measure of how much a given mass of a greenhouse gas will contribute to global warming. The GWP is usually referenced to CO_2 whose GWP by definition is set equal to 1. The GWP of a greenhouse gas depends on several factors, including the absorption of infrared radiation by the gas, the location of the absorbing wavelengths of the gas, and the lifetime of the gas in the atmosphere. Thus gases with large infrared absorption characteristics and a long lifetime have high GWPs, like the CFCs.

gravitational force the natural force of attraction exerted by a celestial body, such as Earth, upon objects at or near its surface, tending to draw them toward the center of the body.

gravity anomaly the difference between the measured force of gravity on a given mass of material at a particular gravity station and the additive effects of corrections, such as the distance of the station from the center of Earth, the centrifugal force owing to Earth's rotation, and the local irregularities of the distribution of mass because of topography.

greenhouse effect the phenomenon of the warming of Earth's atmosphere and surface by the atmospheric greenhouse gases. These gases absorb and reradiate long-wave Earth radiation, making the global temperature of the planet reasonable. Without the natural greenhouse effect, the planet would be about 32° cooler than its global mean annual temperature of 14°C, that is, −18°C. Because of inputs to the atmosphere of greenhouse gases from human activities, these gases are increasing in concentration in the atmosphere. This may lead to an *enhanced greenhouse effect* and warming of the planet.

greenhouse gas an atmospheric gas that absorbs and radiates energy in the infrared part of the electromagnetic spectrum. Such gases include water vapor, carbon dioxide, methane, nitrous oxide, tropospheric ozone, and the synthetic chlorofluorocarbon gases. These gases warm the atmosphere and Earth's surface below. This phenomenon is frequently referred to as the *greenhouse effect* or *warming*.

gross domestic product (GDP) the market value of the total national output of a country's goods and services.

gross national product (GNP) the total output of goods and services by residents and nonresidents of a country regardless of the manner in which the output is allocated to domestic and foreign claims.

groundwater the water beneath the ground, largely formed by the seepage of surface water downward.

gymnosperm plants with seeds that are not enclosed in a fruit or seed case, including pine, fir, spruce, and other cone-bearing trees and shrubs.

gyre a spinning large cell of water in the subtropical latitudes of major ocean basins.

habitat the place where a plant or animal normally lives and grows.

hadal of, or relating to, the parts of the ocean below a depth of 6,000 meters.

Hadley cell the atmospheric vertical circulation pattern starting at the tropics with warm, moist air rising, flowing to the subtropics at altitude, and sinking at about 30° latitude, and returning as the surface flow of trade winds toward the tropics.

half-life the time required for half of the number of atoms of a radioactive element to decay to a daughter isotope.

Heinrich event a period of rapid flow of icebergs from the margins of ice sheets. The icebergs melt at lower latitudes leading to the deposition of ice-rafted debris on the seafloor.

heterotroph an organism requiring complex organic compounds of carbon for metabolic synthesis.

heterotrophic system an environment in which the difference between gross photosynthesis and gross respiration is negative. In such a terrestrial or aquatic environment, the net transfer of carbon dioxide is out of the system.

heterotrophy biochemical pathway in which organic substrates are used by organisms to make organic matter.

high pressure cell (a high) in the center of a high, the air is converging, descending, and as it does so, there is a temporary build-up of air before the air flow diverges at the surface and flows outward. The highest pressure is found at the center of the high and decreases outward from the high pressure cell's core. The balance between pressure forces and the Coriolis force causes the wind in a high to deviate to the right in the Northern Hemisphere in a clockwise manner (called anticyclonic flow). Highs are characterized mainly by precipitation-free, clear skies.

Holocene climatic optimum (hypsithermal)—warm period in Earth history between about 9,000 to 5,000 years B.P. Most of the warming occurred in the higher latitudes of the Northern Hemisphere with up to 4 °C of warming near the North Pole. During this time, the Arctic had substantially less sea ice and the current desert regions of Central Asia were extensively forested because of the warmer conditions and higher rainfall.

Hot House an extended period of geologic time during which Earth was warm.

hot spot a portion of the mantle below the crust that is molten and supplies magma to the surface of Earth. A hot spot is thought to be immobile.

humus dark organic material in the upper layers of soils produced by the decomposition of vegetable and animal matter. Humus is essential to the fertility of Earth.

hydrocarbon an organic compound containing only carbon and hydrogen found in, for example, petroleum, natural gas, and coal.

hydrolysis a chemical reaction involving water.

hydropower energy produced by the force of running water.

hydrosphere the watery envelope surrounding Earth: a reservoir in Earth's surface system.

hydrostatic rock pressure the pressure exerted at depth in the crust by the overlying column of water within the openings of rocks.

hydrothermal reaction a chemical reaction involving hot water and minerals in a rock.

hydrothermal vent a hot water spring on the seafloor or the land through which water and chemical substances from depth exit.

hydroxyl radical (OH*) the excited chemical compound of hydrogen and oxygen in the atmosphere with an imbalance of electric charge. The hydroxyl radical is responsible for the oxidation of many chemically reduced gases emitted from the surface of Earth.

ice age a glacial epoch, especially the last Pleistocene Epoch beginning about 1.8 million years ago.

Ice House an extended period of geologic time during which Earth was cool.

ice sheet a thick and extensive blanket of ice found on land and often covering a large portion of the continent.

ice shelf an ice sheet that projects from the continent into coastal waters so that the end of the sheet is often floating.

igneous rock produced under conditions of intense heat; for example, rocks of volcanic origin or rocks crystallized from molten magma.

infrared radiation the region of the electromagnetic spectrum with wavelengths longer than visible light (about 1 micrometer) but shorter than microwaves (about 1 millimeter). Earth radiation emitted back to space is predominantly long-wave, infrared (heat) radiation.

inorganic not having the structure or organization characteristic of life and not characterized by vital processes; pertaining to compounds that are not hydrocarbons or derivatives thereof.

insolation the amount of solar radiation received at the top of Earth's atmosphere by latitude and by season.

interglacial stage a relatively warm period of time recognized during the Pleistocene Epoch during which great ice sheets retreated from the continents of the Northern Hemisphere.

Intergovernmental Panel on Climate Change (IPCC) a large group of scientists from the international community that develops consensus statements and documents on the present scientific investigations of the impact of greenhouse gases and aerosols on climate.

Intertropical Convergence Zone (ITCZ) the region in the atmosphere along which the northeast and southeast trade winds meet, generally within the weak broad trough of low pressure slightly north of the equator.

ion an electrically charged atom or group of atoms formed by the loss or gain of one or more electrons. A positive ion, the cation, is created by electron loss, and a negative ion, the anion, is created by an electron gain.

IPAT equation A relationship developed for the total impact (**I**) of humans on the environment as a function of the variables of population (**P**), affluence (**A**), and technology available (**T**).

irradiance the rate of transfer of electromagnetic radiation divided by the area through which the radiation is passing, usually given in watts per meter squared.

irreversible process a process in which the entropy change is greater than zero. After the process is complete, the system is more disordered and different from its initial state.

isobar a line along which every point has the same barometric pressure.

isotope any of two or more forms of a chemical element having the same atomic number—that is, the same number of protons in the nucleus—but having different numbers of neutrons in the nucleus, therefore having different atomic weights. There are 275 isotopes of the stable elements and over 800 radioactive isotopes. Every element has known isotopic forms.

jet stream the narrow meandering belt(s) of air with high velocity generally moving from west to east at an altitude of about 10 kilometers.

Junge layer the globally encircling layer of sulfuric acid aerosol in the lower stratosphere named after C. E. Junge; also known as the *stratospheric sulfate veil*.

Kelvin a temperature scale in which temperature is measured in degree Celsius intervals, and water freezes at 273 K and boils at 373 K. All atomic and molecular motion ceases at 0 K.

Kelvin wave an internal wave within the ocean whose dynamics are affected by the Coriolis force and whose maximum amplitude is at the equator.

The wavelength of a Kelvin wave is measured in thousands of kilometers. For a Kelvin wave to exist there must be a barrier to the right of the direction of motion in the Northern Hemisphere and to the left in the Southern Hemisphere. The barrier for coastal Kelvin waves is the coastline, and for equatorial Kelvin waves, the equator itself.

kerogen fossil organic matter dispersed throughout a rock.

kinetic expression a mathematical expression of a rate process.

La Niña the pattern of interannual climatic change that is essentially opposite to that of El Niño. In a La Niña, sea-surface temperatures in the eastern Pacific Ocean and offshore western South America are unusually cold.

laminar flow fluid flow in which the streamlines of flow remain distinct and the flow direction in the fluid at every point remains unchanged with time.

latent heat the quantity of heat gained or lost when a substance (liquid, solid, or gas) changes state at a given temperature and pressure.

leaching the process of selectively removing substances from a substrate, usually with water. For example, rainwater percolating through a soil can dissolve nitrogen from soil materials and transport it to the groundwater.

lifetime a measure of the reactivity of an atmospheric chemical compound. The more reactive the compound, the shorter its atmospheric lifetime; analogous to the term *residence time*.

limestone a sedimentary rock consisting predominately of calcium carbonate minerals.

limiting nutrient the chemical compound, generally inorganic, that limits productivity in a terrestrial or aquatic environment. Examples are nitrate, phosphate, and iron.

lithosphere the dynamic subdivision of Earth on the order of 100 km in thickness forming the outer rigid part of the planet. Also, the solid portion of Earth comprised of minerals, rocks, and soil; a reservoir in Earth's surface system.

lithostatic rock pressure the vertical pressure exerted at depth in Earth caused by the weight of the overlying rock; also known as *geostatic pressure*.

Little Ice Age the period of time from about 1350 to the mid-nineteenth century when temperatures in the Northern Hemisphere were unusually cold, especially in Western Europe.

low pressure cell (a low) a movement of air characterized by converging air flow at the surface, ascending air currents at the core of the cell, and divergent air flow above. Replacement of the ascending air at the cell's core results from surface air flowing inward toward the cell's center. Thus, the wind field created from the balance of pressure forces and the Coriolis force is such that wind flows around the low in a counter clockwise direction in the Northern Hemisphere This is called cyclonic flow. When friction is also involved, the resulting gradient wind has an inward component toward the region of lowest pressure. Lows are often associated with bad weather.

North Atlantic Oscillation (NAO) a climatic phenomenon taking place in the North Atlantic Ocean resulting from oscillations in the atmospheric pressure difference at sea level between the Icelandic low pressure region and the Azores high pressure region. This oscillatory motion controls the strength and direction of westerly winds and storm tracks crossing the North Atlantic.

magnetic anomaly a deviation from the regional average strength of today's magnetic field.

magnetic pole either of two areas near opposite ends of a magnet where the magnetic intensity is the greatest. Earth's magnetic field might appear as though a giant magnet was embedded in its core with its poles antipodal to one another.

mantle the portion of Earth between the crust and the core that is enriched in magnesium and iron and has a thickness of about 2,900 km.

mass balance equation an equation describing the balance of materials in a system on a mass basis. For example, the total mass of dissolved inorganic carbon in the ocean is equal to the sum of the masses of bicarbonate (HCO_3^-), carbonate (CO_3^{2-}), and dissolved CO_2.

Maunder sunspot minimum that period of time between about A.D. 1645 and 1715 when astronomers, observing with telescopes, found very few sunspots on the sun.

medieval climatic optimum the interval of time between A.D. 1100 and 1300 in which some regions of the Northern Hemisphere were warmer than previously and then experienced the following Little Ice Age.

meridional overturning circulation (MOC) an ocean circulation pattern in the Atlantic that carries warm, upper waters into the far-northern latitudes and returns cold, deep waters southward across the Equator. This circulation represents part of the global conveyor belt circulation of the ocean.

mesosphere the dynamic subdivision of Earth lying below the asthenosphere. It includes the lower part of the mantle and the core. Also, the region of Earth's atmosphere between the stratosphere and the thermosphere extending from about 50 to 80 km in altitude above Earth's surface.

metamorphic rock a rock formed from a preexisting rock by an increase in temperature and pressure.

metamorphism the set of processes that leads to a change in the structure or constitution of a rock due to pressure and temperature.

methane clathrate a partially frozen mixture of methane and ice.

meteorite any solid metallic or stony object that has traveled through interplanetary space and fallen to Earth's surface without being vaporized as it passed through the atmosphere.

methanogenic bacteria bacteria that are responsible for the conversion of organic material to methane.

methanotrophic bacteria bacteria that are responsible for conversion of methane to carbon dioxide.

midocean ridge any of several, seismically active, submarine mountain ranges that are found in the Atlantic, Indian, and Pacific Oceans. These ridges are the center of seafloor spreading and the source of the crustal plates.

Milankovitch hypothesis (theory) the hypothesis that orbitally controlled fluctuations in high-latitude solar radiation (insolation) during the summer control the size of ice sheets and the periodicity of glacial-interglacial stages.

mineral a naturally occurring inorganic chemical compound such as common table salt (NaCl, halite).

mixotrophy the process of utilization of both inorganic and organic compounds in metabolism.

model a simplified representative of a system or phenomenon whose description is often mathematical and portrayed pictorially.

Mohorovičić discontinuity the seismic discontinuity between the crust and the mantle of Earth occurring at depths that average about 35 km beneath the continents and about 10 km beneath the ocean floor. Also termed *Moho.*

mole one gram atomic weight of an element or one gram molecular weight of a compound. One **gram atomic weight of an element** is its atomic weight expressed in grams (e.g., the atomic weight of oxygen is 16; its gram atomic weight is 16 grams). One **gram molecular weight of a compound** is its molecular weight expressed in grams (e.g., the molecular weight of carbon dioxide is 44; its gram molecular weight is 44 grams).

molecule the smallest physical unit of an element or compound consisting of one or more similar atoms in an element and two or more different atoms in a compound.

natural resources industrial materials and capacities supplied by nature; e.g., mineral deposits and water power.

negative feedback a process or mechanism that relieves or subtracts from an initial perturbation to a system.

nekton free-swimming, aquatic animals such as whales and squid that are essentially independent of wave and current action.

net primary production (NPP) the net amount of plant material produced per unit area per unit time; photosynthesis minus respiration equals net primary production.

neuston minute organisms that float in the surface layers of water.

nitrification the conversion of ammonium to nitrite and nitrate by nitrifying bacteria.

nitrogen fixation the metabolic assimilation of atmospheric nitrogen by soil and aquatic microorganisms; also, the industrial conversion of free nitrogen into combined forms used as starting materials for fertilizers and explosives.

non-steady state the state of a system in which a variable(s) is (are) changing with time. For example, the atmospheric concentrations of carbon dioxide and other greenhouse gases are changing with time. Thus, the atmosphere is a non-steady-state system with respect to composition.

nonsulfur purple bacteria photosynthetic bacteria that synthesize organic matter from carbon dioxide and preexisting organic compounds.

North Atlantic deep water (NADW) the water mass forming at the high northern latitudes of the North Atlantic Ocean because of cold winds blowing off of North America that sinks and flows southward in the Atlantic at a depth of 2,000 to 4,000 meters.

nutricline the depth range in the ocean where the concentrations of the nutrient elements change rapidly with depth.

nutrient a substance such as phosphorus or nitrogen that supplies nutrition to a living organism.

Ocean acidification the increase in the acidity of the ocean due to emissions of CO_2 to the atmosphere from human activities and the absorption of part of this CO_2 in surface ocean water leading to a decrease in the pH of the water. This phenomenon has been called "the other CO_2 problem." Acidification will have important consequences for marine organisms from bacteria on up the trophic chain. Coral reefs are particularly at risk of degradation due to acidification of the waters bathing them.

old-growth forest primary, original forest.

open system a system that permits matter and energy to exchange with its surroundings.

organic pertaining to a class of chemical compounds that include carbon as a component; characteristic of, or derived from, living organisms.

oxidation the process by which electrons are removed from an atom or molecule. In such a process, a substance such as iron may combine with oxygen and be converted into an oxide such as iron oxide (rust).

oxidizing capacity the intrinsic ability of a system to oxidize reduced substances. In **oxidation**, electrons are removed from an atom or molecule.

ozone a form of oxygen, O_3, having a peculiar odor of weak chlorine. It can be produced by an electric spark or ultraviolet light passing through air or oxygen. It is found in both the troposphere and stratosphere as a trace gas. In the stratosphere it absorbs UV radiation, preventing much of that energy from reaching Earth's surface.

ozone layer The layer of the upper atmosphere where most of stratospheric ozone is concentrated. This layer extends from approximately 19 to 48 km above Earth's surface. The maximum ozone concentration is found at an altitude of about 30 km in tropical regions and 22 km in polar regions.

Pacific Decadal Oscillation (PDO) the climate phenomenon of inter-decadal warming and cooling of the surface waters of the Pacific Ocean north of 20 °N. During a warm or positive phase, the western Pacific Ocean becomes cool and the surface waters of parts of the eastern Pacific warm up. The opposite pattern occurs during a cool or negative phase of the PDO.

Peak Oil the point in time when the global production rate of oil is reached after which the rate of production declines.

pelagic of, relating to, or occurring in lake water or the open ocean.

Periodic Table of Elements a table listing the elements according to their properties, such as atomic number, and arranged into related groups.

permeability the capacity of a porous rock, sediment, or soil to transmit a fluid. It is a measure of the relative ease of fluid flow under unequal pressure.

petrology the science of the study of rocks.

pH the negative logarithm of the effective hydrogen ion concentration or hydrogen ion activity used in expressing both acidity and alkalinity on a scale whose values range from 0 to 14, with 7 representing neutrality; numbers less than 7 denote increasing acidity, and numbers greater than 7, increasing alkaline conditions.

photochemical pertaining to chemical reactions involving chemical compounds in the presence of radiant energy.

photolysis (photodissociation) pertaining to chemical reactions triggered by radiant energy that convert a complex compound to more simple products. The photolysis of ammonia is an example: $2NH_3 \Rightarrow N_2 + 3H_2$.

photomicrograph a photographic enlargement of a microscopic image such as a thin section of a rock or an electron microscopic image of organic tissue.

photosynthesis the process of synthesis of complex organic materials; e.g., carbohydrates, from carbon dioxide, water, and nutrients, using sunlight as a source of energy and with the aid of chlorophyll and associated pigments.

phytoplankton minute plant life that passively float atop a body of water.

planetesimal one of the small celestial bodies that fused together to form the planets of the solar system.

plankton minute plant and animal life of the ocean ranging in size from 5 micrometers to 3 centimeters. The plant plankton are the phytoplankton; the animal plankton are the zooplankton.

plate tectonics the theory of global tectonics in which the lithosphere is divided into a number of crustal plates that move on the underlying plastic asthenosphere. These plates may collide with, slide under, or move past adjacent plates in a nearly horizontal direction. The sources of the plates are the great midocean ridges of the world's oceans, where hot molten material upwells from within Earth. The plates are destroyed at subduction zones, such as the western margin of the Pacific Ocean, where the plates sink down into the underlying asthenosphere.

polar vortex the immense low-pressure region and accompanying easterly winds that encircle the globe at both polar regions.

polymorph a mineral with a specific composition that can crystallize in at least two different structural forms. A *dimorph* is a mineral that crystallizes in two distinct structural forms.

positive feedback a process or mechanism that reinforces or adds to an initial perturbation to a system.

precipitation the removal of water from the atmosphere and its deposition on Earth's surface in the form of rain, ice, and snow.

producer an organism such as a green plant that produces its own organic compounds from simple precursor substances such as carbon dioxide and inorganic nitrogen.

productivity the rate of production of organic matter by producer organisms.

prokaryote any cellular organism that has no membrane about its nucleus and no organelles in the cytoplasm except ribosomes. Its genetic material is in the form of single continuous strands forming coils or loops, characteristic of all organisms in the kingdom Monera such as bacteria and blue-green algae.

protogalaxy a small cloud of gas and dust held together by mutual gravitational attraction and isolated from similar systems that formed early in the evolution of the universe.

protostar an individual swirling mass within a protogalaxy that eventually collapsed to form an energy-generating and self-luminous celestial body.

protozoan an eukaryotic organism of the kingdom Protoctista, phylum Protozoa, with a membrane-bound nucleus and organelles within a mass of protoplasm. Planktonic foraminifera and radiolarians which secrete shells of calcium carbonate and opal, respectively, are members of this group.

proxy a quantifiable indicator of climatic or other environmental changes that is found in an historical record and precede direct instrument measurements of change.

radiation transmission and propagation of energy in the form of rays, waves, or particles.

radiant flux the rate of transfer of electromagnetic radiation, usually in units of joules per second, or watts.

radioactivity spontaneous emission of radiation, either directly from unstable atomic nuclei or as a consequence of a nuclear reaction.

radiolarian a planktonic protozoan that constructs an intricate skeleton of spicules made up of opal.

Redfield ratio the relatively constant ratio of 106:16:1 of the bioessential elements carbon, nitrogen, and phosphorus in marine plankton. The concept of the Redfield ratio has been applied to the terrestrial realm as well as to organic matter in soils and sediments.

red shift an increase in the wavelength of radiation emitted by a celestial body as a consequence of the Doppler effect. The Doppler effect is an apparent change in the frequency of radiation waves occurring when the source of the radiation and the observer are in motion relative to one another.

reduction the process by which electrons are added to an atom or molecule.

reforestation the replanting of trees in a deforested area.

remote sensing a means of collecting data from an area that is distant from the observer; satellites are often equipped with instrumentation capable of recording information concerning the surface of Earth.

reservoir or stock that part of a system that can store or accumulate and be a source of a quantity of one of the system's variables or substances. For example, the atmosphere is a reservoir of the surface system of Earth, the ecosphere, and can store water vapor released to it from the land by evapotranspiration and the ocean by evaporation, and return the water to Earth's surface as rain.

residence time the total mass of a substance in a reservoir divided by its inflow or outflow. The residence time is a measure of the reactivity of the substance in the reservoir. For example, the residence time of sodium in the ocean is very long (55 million years). Therefore, sodium does not enter into chemical or biochemical reactions that remove it very rapidly from the ocean. Conversely, the residence time of dissolved silica in the ocean is about 20,000 years. This compound is readily taken up by diatoms and radiolarians to build their skeletons of mineral silica.

respiration the physical and chemical processes by which an organism supplies its cells and tissues with the oxygen needed for metabolism and releases carbon dioxide formed in the energy-producing reactions.

reversible process a process in which the change in entropy is zero. In general, after the process is complete, the state of the system is as it was initially.

rift a fault or graben (depressed area) of regional extent.

RNA (ribonucleic acid) a polymeric constituent of all living cells and many viruses, consisting of a long, usually single-stranded chain of alternating phosphate and ribose units with the bases adenine, guanine, cytosine, and uracil bound to the ribose. The structure and the base sequence of RNA are the determining factors in protein synthesis and the transmission of genetic information.

Rossby wave an internal wave within the ocean whose dynamics are affected by the Coriolis force and whose wavelength is measured in thousands of kilometers. The speed of a Rossby wave decreases with distance from the equator. The Rossby waves associated with ENSO (El Niño Southern Oscillation) events propagate in a westerly direction.

salinity a measure of the salt content of water; for example, the salinity of seawater is 35 and that of average river water is 0.12.

salpa an organism of the kingdom Animalia, phylum Coelenterata. Heterotrophic, multicellular, jellyfish-like, soft-bodied organism of the ocean.

saturation the degree to which a solution or a gas is at equilibrium with one of its components. It is measured in several different ways. For example, a humidity of 125% would be a supersaturation of 25% with respect to water vapor in the air. Saturation of seawater with respect to the mineral calcite ($CaCO_3$) of 50% would mean that the seawater was 50% undersaturated with respect to calcite. If a lake water contained exactly enough dissolved CO_2 to be in equilibrium with the atmosphere, it would have a saturation of 100% with respect to CO_2.

saturation state of an aqueous solution (Ω) the measure of the degree of saturation of a solution with respect to a phase like a mineral. It is equal to the ion activity product or the ion concentration product of the components in the solution divided by the thermodynamic equilibrium product or the stoichiometric solubility product (a type of equilibrium product involving concentrations and not activities), respectively. For example, in a seawater at a fixed temperature, pressure, and salinity, the Ω of $CaCO_3$ = concentration of calcium ion in seawater times the concentration of carbonate ion in seawater divided by the stoichiometric solubility product $[Ca^{2+}] \times [CO_3^{2-}]$. If Ω is greater than 1, the solution is oversaturated with respect to the phase under consideration, $CaCO_3$. If less than 1, undersaturated, and if equal to 1, at equilibrium.

Scientific Method the process by which scientists, over time, attempt to construct and present an accurate picture of a phenomenon that is reliable, consistent, and non-arbitrary.

sea ice ice formed at the surface of seawater.

seafloor spreading creation and migration of new seafloor by the separation of lithospheric plates generally at midocean ridges and the filling of the gap with lava and igneous intrusive rocks.

secondary growth vegetation that replaces the primary, original plants.

sedimentary rock a rock formed from the erosion of preexisting rocks and the deposition of the eroded materials as sediment. Sedimentary rocks are also formed by inorganic or biological precipitation of minerals from natural waters.

shear wave the type of seismic wave that is transmitted because of a shearing motion that involves oscillation perpendicular to the direction of propagation of the wave; also termed *S wave*. Shear waves do not travel through liquids, nor through Earth's outer core, which is molten.

soil the upper layers of Earth's surface consisting of weathered and disintegrated rock and humus.

solar constant the rate at which solar energy is received outside Earth's atmosphere on a surface normal to the incident radiation at Earth's mean distance from the sun. The mean solar constant is 1.94 calories per minute per square centimeter.

solar radiation the electromagnetic radiation emitted by the sun. It includes energy wavelengths from the very short ultraviolet (<0.2 micrometer) to about 3 micrometers.

solar system the sun together with all the planets and other celestial bodies that revolve around it.

Southern Oscillation Index (SOI) a measure of monthly or seasonal fluctuations in the air pressure difference between Tahiti and Darwin, Australia. Sustained negative values of the SOI indicate El Niño episodes; sustained positive values, La Niña episodes. The SOI may be calculated as follows:

$$SOI = 10 \times (Pdiff - Pdiffave)/SD(Pdiff)$$

where Pdiff is (average Tahiti mean sea level pressure for the month) − (average Darwin mean sea level pressure for the month), Pdiffave is the long-term

average of the Pdiff for the month in question, and SD(Pdiff) is the long-term standard deviation of the Pdiff for the month in question.

space scale a particular three-dimensional realm or expanse in which material objects are located and events occur.

Sporer sunspot minimum a period of time between A.D. 1460 and 1550 when very few sunspots were observed with telescopes on the sun's surface.

star a large self-luminous heavenly body such as the sun.

steady state the property of a system that implies no change in space and time.

stoichiometry the quantitative relationship between reactants and products in a chemical reaction.

stony meteorite a general name for meteorites consisting chiefly of the silicate minerals olivine, pyroxene, and plagioclase. Stony meteorites constitute 90% of all meteorites that fall to Earth.

stony iron meteorite a general name for relatively rare meteorites containing at least 25 weight% of both nickel-iron and silicates such as pyroxene and olivine.

strata beds of sedimentary rocks that generally consist of one kind of material representing continuous deposition.

stratosphere the region of the upper atmosphere extending upward from the tropopause to about 50 km above Earth's surface.

stromatolite a fossil laminated structure usually found in limestones and constructed by algae and bacteria. The structure may have a rounded, columnar, or more complex form.

subduction zone the juncture of two lithospheric plates where collision of the plates results in one plate being drawn down under or overridden by another plate. This region is the sink of the crustal plates of Earth.

sunspot cycle a natural cycle of about 11 years in the number of dark spots (sunspots) observed on the surface of the sun.

system a selected set of interactive components. A simple system is an air-conditioning unit. A biogeochemical system consists of reservoirs, processes and mechanisms, and associated fluxes involving material transport. The global climate system is very complex and involves all the physical, chemical, and biological interactions that control the long-term environmental conditions of the world.

technosphere defined by Barry Commoner in 1991 as the sum of the industrial, technological, and agricultural components of human activity.

temperate zone the region between the Tropic of Cancer and the Arctic Circle or between the Tropic of Capricorn and the Antarctic Circle.

terpene any of various hydrocarbons present in essential oils, such as those from coniferous vegetation; commonly used in solvents and organic syntheses.

terrestrial of, or relating to, the land as distinct from air or water.

thermocline a layer of water at depth in the oceans or in lakes in which the temperature rapidly decreases with increasing depth.

thermodynamic threshold a threshold value (a value for a variable that separates two different behaviors of a system) for a variable of a thermodynamic nature, like temperature, free energy, an equilibrium condition, etc. A system behaves one way below the thermodynamic threshold and a different way above the threshold.

thermodynamics the scientific study of energy, its transformations and flows, and the influence of a system's energy on its variables.

thermohaline circulation that part of the global circulation pattern of the ocean driven by seawater density differences created by changes in temperature (surface heat fluxes) and salinity (freshwater fluxes); sometimes referred to as the ocean conveyor belt circulation.

thermosphere the region of the upper atmosphere in which the temperature increases continuously with altitude. This region includes essentially all of the atmosphere above the mesosphere.

time scale a particular period of time encompassing the duration of an event.

trace gas a gas present in the atmosphere in very low concentration. For example, methane, nitrous oxide, and carbon monoxide are considered trace gases.

transform fault a fault in which the movement is principally horizontal and that offsets a midocean ridge in opposing directions on either side of an axis of seafloor spreading.

transitional phenomenon a feature of a system that displays change from one state to another. Such a change may be relatively slow or more generally abrupt. In the boiling of water, the change from the state of little water motion to that of turbulence is a transitional phenomenon.

transpiration the process of excreting water in the form of a vapor through a living membrane.

transport path the process or mechanism that moves materials in and out of a reservoir. A transport path is associated with a flux for a substance.

tropical zone generally taken as the region lying between the Tropic of Cancer and the Tropic of Capricorn.

troposphere the lowest level of the atmosphere, 8 to 13 kilometers high, within which there is a steady drop in temperature with increasing altitude. It is the region where most cloud formations occur and weather conditions manifest themselves.

ultraviolet radiation the region of the electromagnetic spectrum with wavelengths longer than 0.5 nanometers but shorter than 0.5 micrometers. Solar radiation has an important component of ultraviolet radiation of varying intensity.

uniformitarianism the concept first proposed by James Hutton of Scotland that "the present is the key to the past"—in other words, that natural physical laws applied to the interpretation of the rock record are time invariant.

upwelling the upward movement of water from depths of typically 50 to 150 meters at speeds of approximately 1 to 3 meters per day. The upwelling of water generally results from the lateral movement of surface water. Upwelling zones in the ocean are found along the western margins of the continents, in equatorial regions, and at high latitudes of the Southern Hemisphere.

vascular plant a gymnosperm or angiosperm plant.

visible light the wavelength range within the electromagnetic spectrum between 0.4 and 0.7 millionths of a meter (micrometers).

volatilization the conversion of a substance into the gas or vapor state and its emission into the environment.

wash out the scavenging of particles from the atmosphere by rainfall and their subsequent deposition on Earth's surface.

water cycle the circulation of water about Earth's surface, including its movement to the oceans via rivers, to the atmosphere via evaporation from the ocean, and return of some of this evaporated water to the continents.

wavelength the distance between successive troughs and crests in a waveform like an ocean wave.

weather the state of the atmosphere in terms of temperature, precipitation, cloudiness, humidity, etc., generally on a short time scale.

weathering the chemical, physical, and biological processes that lead to the disintegration of minerals, kerogen, and rocks.

wet deposition the deposition on Earth's surface of solid particles and dissolved chemical compounds in rain.

wetland land or areas such as tidal flats and swamps that contain much soil moisture.

Younger Dryas a period of colder climate between approximately 12,800 to 11,500 years ago named after the alpine/tundra wildflower, *Dryas octopetala*. The Younger Dryas brief cold period interrupted the general trend of warming coming out of the Last Glacial Maximum (LGM) 18,000 years ago. The Younger Dryas started abruptly and ended abruptly and perhaps was caused by changes in the strength of the meridional overturning circulation, part of the great conveyor belt circulation of the ocean.

zooplankton minute animal life in a body of water that drift passively or swim very weakly.

REFERENCES

Air Quality Committee, 1994, United States–Canada air quality agreement, progress report, 1994. U. S.–Canada Air Quality Committee, International Joint-Commission, Washington, DC.

Allen, L. H., Jr., and Amthor, J. S., 1995, Plant physiological responses to elevated CO_2, temperature, air pollution, and UV-B radiation. In Biotic Feedbacks in the Global Climatic System, G. M. Woodwell and F. T. Mackenzie (eds.), Oxford University Press, New York, pp. 51–84.

Anderson, D. L., 1989, Theory of the Earth. Blackwell Scientific Publications, Boston.

Andersson, A., Mackenzie, F. T., and Lerman, A., 2005, Coastal ocean and carbonate systems in the high CO_2 world of the Anthropocene. American Journal of Science, 305, pp. 875–918.

Andreae, M. O., 1987, The oceans as a source of biogenic gases. Oceanus, v. 29, no. 4, pp. 27–35.

Atjay, G. L., Ketner, P., and Duvigneaud, P., 1979, Terrestrial primary production and phytomass. In The Global Carbon Cycle, B. Bolin, E. T. Degens, S. Kempe, and P. Ketner (eds.), SCOPE 13, John Wiley & Sons, New York, pp. 129–181.

Atkinson, T. C., Briffa, K. R., and Coope, G. R., 1987, Seasonal temperatures in Britain during the past 22,000 years. Nature, v. 325, pp. 587–592.

Bailey, R. (ed.), 1995, The True State of the Planet. The Free Press, New York.

Balling, R. C., Jr., 1992, The Heated Debate, Greenhouse Predictions versus Climate Reality. Pacific Research Institute for Public Policy, San Francisco.

Bard, E., Hamelin, B., and Fairbanks, R. G., 1990, Additional uranium series ages for Barbados corals by the mass spectrometry method. Nature, v. 346, pp. 456–458.

Bard, E., Hamelin, B., Fairbanks, R. G., and Zindler, A., 1990, Comparison between radiocarbon and uranium series ages on glacial age Barbados corals. Nature, v. 345, pp. 405–409.

Beerling, D., Berner, R. A., Mackenzie, F. T., Harfoot, M. B., and Pyle, J. A., 2009, Atmospheric methane over the past 400 million years. American Journal of Science, v. 309, pp. 97–113.

Behrenfeld, M. and Chapman, J., 1991, Our disappearing ozone shield. Currents, v. 10, no. 3, pp. 13–17.

Berner, E. K., and Berner, R. A., 1996, Global Environment: Water, Air, and Geochemical Cycles. Prentice Hall, Upper Saddle River, NJ.

Berner, R. A., 1991, A model for atmospheric CO_2 over Phanerozoic time. American Journal of Science, v. 291, pp. 339–376.

Berner, R. A., and Canfield, D. E., 1989, A new model for atmospheric oxygen over Phanerozoic time. American Journal of Science, v. 289, pp. 333–361.

Berner, R. A. and Kothavala, Z., 2001, GEOCARB III: A revised model of atmospheric CO_2 over Phanerozoic time. American Journal of Science, v. 301, pp. 182–204.

Berner, R. A., and Lasaga, A. C., 1989, Modeling the geochemical carbon cycle. Scientific American, v. 222, March, pp. 74–81.

Bindoff, N., Willebrand, J., Artale, V., Cazenave, A., Gregory, J., Gulev, S., Hanawa, K., Le Quere, C., Levitus, S., Nojiri, Y., Shum, C. K., Talley, L., and Unnikrishnan, A., 2007, Observations: oceanic climate change and sea level. In Climate Change 2007: The Physical Science Basis, Cambridge University Press, New York.

Boden, T. A., Sepanski, R. J., and Stoss, F. W., 1991, Trends '91: A Compendium of Data on Global Change. Oak Ridge National Laboratory/Carbon Dioxide Information Analysis Center/U.S. Department of Energy, Tennessee.

Boden, T. A., Marland, G., and Andres, R. J., 2009, Global, Regional, and National Fossil-Fuel CO_2 Emissions. Carbon Dioxide Information Analysis Center, Oak Ridge National Laboratory, U.S. Department of Energy, Oak Ridge, Tenn., U.S.A. doi 10.3334/CDIAC/00001

Bond, G., Heinrich, H., Broecker, W., Labeyrie, L., McManus, J., Andrews, J., Huon, S., Jantschik, R., Clasen, S., Simet, C., Tedesco, K., Klas, M., Bonani, G., and Ivy, S., 1992, Evidence for massive discharges of icebergs into the North Atlantic ocean during the last glacial period. Science, v. 360, pp. 245–249.

Boyce, J. M., and Maxwell, T., 1992, Our solar system, a geologic snapshot. Washington, DC, May.

Briggs, D. E. G., and Crowther, P. R., 1990, Paleobiology, A Synthesis. Blackwell Scientific Publications, Oxford, England.

Broecker, W. S., 1985, How to Build a Habitable Planet. Eldigio Press, Palisades, NY.

Broecker, W. S., 1993, The Glacial World According to Wally. Eldigio Press, Lamont-Doherty Earth Observatory of Columbia University, NY, 318 pp. (revised 1995).

Broecker, W. S., 1995, Chaotic climate. Scientific American, v. 273, November, pp. 62–68.

Broecker, W. S., 2004, Fossil Fuel CO_2 and the Angry Climate Beast. Compact disc.

Broecker, W.S., Barker, S., Clark, E., Hajdas, I., Bonani, G., and Stott, L., 2004, Ventilation of the glacial deep Pacific Ocean. Science, v. 306, pp. 1,169–1,172.

Bruce, J., Lee, H., and Haites, E. (eds.), 1996, Climate Change 1995—Economic and Social Dimensions of Climate Change. Contribution of Working Group III to the Second Assessment Report of the Intergovernmental Panel on Climate Change. Cambridge University Press, Cambridge.

Bryant, D. et al., 1995, Coastlines at risk: an index of potential development-related threats to coastal ecosystems. World Resources Institute Indicator Brief, Washington, DC, pp. 5–6.

Calder, N., 1983, Timescale, An Atlas of the Fourth Dimension. Viking Press, New York.

Cameron, A. G. W., 1966, Abundance of the elements. In Handbook of Physical Constants, S. P. Clark, Jr. (ed.), Geological Society of America Memoir 97, pp. 8–10.

Caraco, N. F., 1995, Influence of human populations on P transfers to aquatic systems: a regional scale study using large rivers. In Phosphorus in the Global Environment: Transfers, Cycles and Management, H. Tiessen (ed.), SCOPE 54, John Wiley & Sons, New York, pp. 235–244.

Caufield, C., 1991, In the Rainforest. University of Chicago Press, Chicago.

Charlson, R. J., and Wigley, T. M. L., 1994, Sulfate aerosol and climate change. Scientific American, v. 270, February, pp. 48–57.

Charlson, R. J., Lovelock, J. E., Ardreae, M. O., and Warren, S. G., 1987, Oceanic phytoplankton, atmospheric sulfur, cloud albedo and climate. Nature, v. 326, pp. 655–661.

Chiras, D. D., 1988, Environmental Science. Benjamin Cumming Publishing Co., Red Wood City, CA.

Christensen, J. W., 1991, Global Science. Kendall/Hunt Publishing Co., Dubuque, IA.

Church, J. A., and Gregory, J. M., 2001, Changes in sea level. In Climate Change 2001: The Scientific Basis, J. T. Houghton et al. (eds.), Cambridge University Press, Cambridge, pp. 639–693.

Clarke, F. W., 1924, Data of Geochemistry. U.S. Geological Survey Bulletin, v. 770, Washington, DC.

Clarke, R., 1987, The ozone layer. United Nations Environment Programme, UNEP GEMS Environment Library, no. 2, Nairobi, Kenya.

Clark, S. P., Jr., 1966, Composition of rocks. In Handbook of Physical Constants, S. P. Clark, Jr. (ed.), Geological Society of America Memoir 97, pp. 1–5.

CLIMAP, 1981, Seasonal reconstructions of the Earth's surface at the last glacial maximum. Geological Society of America Map and Chart Series, MC-36.

Climate Diagnostics Bulletin, July 1997, Near real-time analyses ocean/atmosphere. U.S. Department of Commerce, Climate Prediction Center, Washington, DC.

Cole, J. J., Peierls, B. L., Caraco, N. F., and Pace, M. L., 1993, Nitrogen loading of rivers as a human-driven process. In Humans as Components of Ecosystems: The Ecology of Subtle Human Effects and Populated Areas, M. J. McDonnell and S. T. A. Pickett (eds.), Springer-Verlag, Berlin, pp. 141–157.

Commoner, B., 1991, Making Peace with the Planet. Pantheon Books, New York.

Coppens, Y., 1994, East side story: the origin of humankind. Science, v. 270, pp. 88–95.

Covey, C., 1984, The earth's orbit and the ice ages. Scientific American, v. 250, February, pp. 58–66.

Cowen, R., 1995, History of Life. Blackwell Scientific Publications, Boston.

Craig, J. R., Vaughan, D. J., and Skinner, B. J., 1988, Resources of the Earth. Prentice Hall, Englewood Cliffs, NJ.

Crough, 1979, Hotspot epeirogeny. Tectonophysics, v. 61, pp. 321–333.

Crowley, T. J., 1995, Ice age terrestrial carbon changes revisited. Global Biogeochemical Cycles, v. 9, pp. 377–389.

Crowley, T. J. and Berner, R. A., 2003, CO_2 and climate change. Science, v. 292, pp. 870–872.

Curtis, H., 1983, Biology (4th ed.). Worth Publishers, New York.

Cubasch, U., and Meehl, G. A. (lead authors) et al., 2001, Projections of future climate change. In Climate Change 2001: The Scientific Basis, J. T. Houghton et al. (eds.), Cambridge University Press, Cambridge, pp. 525–582.

Darnell, W. L., Staylor, W. F., Gupta, S. K., Ritchey, N. A., and Wilber, A. C., 1992, Seasonal variation of surface radiation budget derived from international satellite cloud climatology project C1 data. Journal of Geophysical Research, v. 97, pp. 15,741–15,760.

Delmas, R. J., 1992, Environmental information from ice cores. Reviews of Geophysics, v. 30, pp. 1–21.

Dignon, J., and Hameed, S., 1989, Global emissions of nitrogen and sulfur oxides from 1860 to 1980. Journal of the Air and Waste Management Association, v. 39, pp. 180–186.

Dudley, W. C., and Lee, M., 1998, Tsunami! (2nd ed.). University of Hawaii Press, Honolulu.

Duncan, R. C., 2001, World energy production, population growth, and the road to Olduvai Gorge. Population and Environment: A Journal of Interdisciplinary Studies, v. 22, pp. 503–522.

Duplessy, J.-C., 2001, Physical and chemical properties of the glacial ocean. In Geosphere–Biosphere Interactions and Climate, L. O. Bengtsson and C. U. Hammer (eds.), Cambridge University Press, Cambridge, pp. 220–255.

Duplessy, J.-C., Labeyrie, L., Juillet-Leclerc, A., Maitre, F., Duprat, J., and Sarnthein, M., 1991, Surface salinity reconstruction of the North Atlantic Ocean during the last glacial maximum. Oceanologica Acta, v. 14, pp. 311–324.

Durning, A. T., 1993, Saving the Forest, What Will It Take? Worldwatch Paper 117, Worldwatch Institute, Washington, DC.

Eclogae Geologicae Helvetiae, 1981, v. 74, no. 2.

Ecological Society of America, 1991, The sustainable biosphere initiative, an ecological research agenda. Reprinted from Ecology, v. 72, no. 2, pp. 371–412.

Eicher, D. L., 1968, Geologic Time. Prentice Hall, Englewood Cliffs, NJ.

Electrical Power Research Institute (EPRI), 1988, Ozone, one gas, two environmental issues, an environmental briefing, EN. 3003.12.88. EPRI, Palo Alto, CA.

Elkins, J. W., Thompson, T. M., Swanson, T. H., Butler, J. H., Hall, B. D., Cummings, S. O., Fisher, D. A., and Raffo, A. G., 1993, Decrease in the growth rates of atmospheric chlorofluorocarbons 11 and 12. Nature, v. 364, pp. 780–783.

Elsom, D., 1996, Smog Alert: Managing Urban Air Quality. Earthscan Publications Limited, London.

Emiliani, C., and Shackleton, N. J., 1974, The Brunhes Epoch: isotopic paleotemperatures and geochronology. Science, v. 183, pp. 511–524.

Energy Information Administration (EIA), 1994, Annual energy review 1993. U.S. Department of Energy, Washington, DC.

Energy Information Administration (EIA), 1997, International Energy Outlook. U.S. Department of Energy, Washington, DC.

Enger, E. D., and Smith, B. F., 2002, Environmental Science: A Study of Interrelationships (8th ed.). McGraw Hill, New York, 486 pp.

Environment Canada, 1981, Downwind, the acid rain story, Cat. No. En 56-561, 1981E. Minister of Supply and Services Canada, Ottawa.

Environmental Protection Agency (EPA), 2006, http://www.epa.gov/.

Ernst, W. G., 1969, Earth Materials. Prentice Hall, Englewood Cliffs, NJ.

Fairbanks, R. G., 1989, A 17,000-year glacio-eustatic sea level record: influence of meltwater on the Younger Dryas event and ocean circulation. Nature, v. 349, pp. 637–642.

Faiz, A., Sinha, K., Walsh, M. et al., 1990, Automotive air pollution, issues and options for developing countries, Report No. 492, Policy Research and External Affairs Working Paper Series, The World Bank, Washington, DC, p. xi.

Few, A. A., 1991, System Behavior and System Modeling. Global Change Instruction Program, University Corporation for Atmospheric Research, Boulder, CO.

Foley et al., 2007, Global consequences of land use. Science, v. 309, pp. 570–574.

Folland, C. K., Rayner, N. A., Brown, S. J., Smith, T. M., Shen, S. S., Parker, D. E., Macadam, I., Jones, P. D., Jones, R. N., Nicholls, N., and Sexton, D. M. H., 2001, Global temperature change and its uncertainties since 1861. Geophysical Research Letters, v. 28, pp. 2621–2624.

Food and Agricultural Organization (FAO) of the United Nations, 1991, Major Climatic Zones. Food and Agricultural Organization, Rome, Italy.

Food and Agricultural Organization (FAO) of the United Nations, 1997, State of the World's Forests 1997. Food and Agricultural Organization of the United Nations, Rome, Italy.

Food and Agricultural Organization (FAO) of the United Nations, 2006, The State of the World's Forests 2005. FAO, Rome, Italy (on completion of this book the world's forest report for 2009 appeared).

Foukal, P. V., 1990, The variable sun. Scientific American, v. 262, February, pp. 34–41.

Foley, J. A., Monfreda, C., Ramankutty, N., and Zaks, D., 2007, Our share of the planetary pie. Proceedings of the National Academy of Sciences, v. 104, No. 31, pp. 12585–12586. Published online 2007 July 23. doi: 10.1073/pnas.0705190104.

French, H. F., 1990, Clearing the Air: A Global Agenda. Worldwatch Paper 94, Worldwatch Institute, Washington, DC.

French, H. F., 1995, Partnership for the Planet: An Environmental Agenda for the United Nations. Worldwatch Paper 126, Worldwatch Institute, Washington, DC.

Friis-Christensen, E., and Lassen, K., 1991, Length of the solar cycle, an indicator of solar activity closely associated with climate. Science, v. 254, pp. 698–700.

Frisch, J. R., 1986, Future stresses for energy resources. In World Energy Conference, Conservation Commission, Graham and Trotman, London.

Galloway, J. N., 1989, Atmospheric acidification, projection for the future. Ambio, v. 18, pp. 161–166.

Galloway, J. N., Schlesinger, W. H., Levy, H., II, Michaels, A., and Schnoor, J. L., 1995, Nitrogen fixation: anthropogenic enhancement—environmental response. Global Biogeochemical Cycles, v. 9, pp. 235–252.

Galloway, J. N., Dentener, F. J., Capone, D. G., Boyer, E. W., Howarth, R. W., Seitzinger, S. P., Asner, G. P., Cleveland, C. C., Green, P. A., Holland, E. A., Karl, D. M., Michaels, A. F., Porter, J. H., Townsend, A. R., and Vorosmarty, C. J., 2004, Nitrogen cycles: past, present, and future. Biogeochemistry, v. 70, pp. 153–226.

Gardner, G., 1996, Shrinking Fields: Cropland Loss in a World of Eight Billion People. Worldwatch Paper 131, Worldwatch Institute, Washington, DC.

Garrels, R. M., 1951, A Textbook of Geology. Harper and Row, New York.

Garrels, R. M., and Mackenzie, F. T., 1971, Evolution of Sedimentary Rocks. W. W. Norton & Co., New York.

Garrels, R. M., Mackenzie, F. T., and Hunt, C., 1975, Chemical Cycles and the Global Environment, Assessing Human Influences. William Kaufmann, Los Altos, CA.

George, T. S., 2001, Minamata: Pollution and the Struggle for Democracy in Postwar Japan. Harvard University Press, Cambridge, MA.

Glantz, M. H., 2001, Currents of Change: The Impact of El Niño and La Niña on Climate and Society. Cambridge University Press, Cambridge.

Gough, D. O., 1981, Solar Physics, v. 74, p. 21.

Graedel, T. E., and Crutzen, P. J., 1993, Atmospheric Change, An Earth System Perspective. W. H. Freeman and Co., New York.

Gregor, C. B., 1988, Prologue, cyclic processes in geology, a historical sketch. In Chemical Cycles in the Evolution of the Earth, C. B. Gregor, R. M. Garrels, F. T. Mackenzie, and J. B. Maynard (eds.), John Wiley & Sons, New York, pp. 5–16.

Gschwandtner, G., Wagner, J. K., and Husar, R. B., 1988, Comparison of historic SO2 and NOx emission data sets, EPA/600/S7-88/009. U.S. Environmental Protection Agency, Research Triangle Park, NC.

Guidry, M., and Mackenzie, F. T., 2009, unpublished.

Guinotte, J. M., Buddemeier, R. W., and Kleypas, J. A., 2003, Future coral reef habitat marginality: temporal and spatial effects of climate change in the Pacific basin, Coral Reefs v. 22, pp. 551–558.

Halpert, M. S., and Ropelewski, C. F., 1993, Fourth annual climate assessment, 1992. Climate Analysis Center/U.S. Department of Commerce, Camp Springs, MD.

Hameed, S., and Dignon, J., 1992, Global emissions of nitrogen and sulfur oxides in fossil fuel combustion 1970–1986. Journal of the Air and Waste Management Association, v. 42, pp. 159–163.

Harris, N. R. P., Ancellet, G., Bishop, L., Hofmann, D. J., Kerr, J. B., McPeters, R. D., Prendez, M., Randel, W. J., Staehelin, J., Subbaraya, B. H., Volz-Thomas, A., Zawodny, J., and Zerefos, C. S., 1997, Trends in stratospheric and free tropospheric ozone. Journal of Geophysical Research, v. 102, pp. 1,571–1,590.

Harris, S. A., 1986, Permafrost distribution, zonation, and stability along the Eastern Ranges of the Cordillera of North America. Arctic, v. 39, pp. 29–38.

Hedgpeth, J. W., 1957, Classification of marine organisms. In Treatise on Marine Ecology and Paleoecology, J. W. Hedgpeth (ed.), Geological Society of America Memoir 67, The Geological Society of America, New York, pp. 17–27.

Hedin, L. O., and Likens, G. E., 1996, Atmospheric dust and acid rain. Scientific American, v. 275, December, pp. 88–92.

Hicks, B. and Nelder, C., 2008, Profit from the Peak. John Wiley & Sons, Hoboken, New Jersey.

Hielman, B., 1989, Global warming. Chemical Engineering News, March 13, pp. 25–44.

Hindu Survey of the Environment, 1992, National Press, Madras, India.

Holland, H. D., and Petersen, U., 1995, Living Dangerously: The Earth, Its Resources, and the Environment. Princeton University Press, Princeton, NJ.

Houghton, J. T., Jenkins, G. J., and Ephraums, J. J. (eds.), 1990, Climate Change: The IPCC Scientific Assessment. Cambridge University Press, Cambridge, MA.

Houghton, J. T., Meira Filho, L. G., Callander, B. A., Harris, N., Kattenberg, A., and Maskell, K. (eds.), 1996, Climate Change: The Science of Climate Change. Cambridge University Press, Cambridge.

Houghton, J. T., Ding, Y., Griggs, D. J., Noguer, M., van der Linden, P. J., Dai, X., Maskell, K., and Johnson, C. A. (eds.), 2001, Climate Change 2001: The Scientific Basis. Cambridge University Press, Cambridge.

Houghton, R. A., 1995, Land-use change and the carbon cycle. Global Change Biology, v. 1, pp. 275–287.

Hughes, R. J., Sullivan, M. E., and Yok, D., 1991, Human-induced erosion in the highlands catchment in Papua, New Guinea: the prehistoric and contemporary records. Zeitschrift für Geomorphologie, Supplement v. 83, pp. 227–339.

Idyll, C. P., 1973, The anchovy crisis. Scientific American, v. 228, June, pp. 22–29.

Imbrie, J., and Imbrie, K. P., 1979, Ice Ages, Solving the Mystery. Macmillan, London.

Intergovernmental Panel on Climate Change (IPCC), 1992, Global climate change and the rising challenge of the sea. The Hague, the Netherlands, Ministry of Transport, Public Works and Water Management, Directorate General Rijkswaterstaat, Tidal Waters Division.

Intergovernmental Panel on Climate Change (IPCC), 2007, Climate Change 2007: The Physical Science Basis. Contribution of Working Group I to the Fourth Assessment Report of the Intergovernmental Panel on Climate Change, S. Solomon, D. Qin, M. Manning, Z. Chen, M. Marquis, K. B. Averyt, M. Tignor, and H. L. Miller (eds.). Cambridge University Press, Cambridge, UK and New York.

International Energy Agency, 1993, World Energy Outlook to the Year 2010. Organization for Economic Co-Operation and Development and International Energy Agency, Paris, France.

International Energy Agency, 1999, World Energy Outlook: Looking at Energy Subsidies: Getting the Prices Right. Organization for Economic Co-Operation and Development and International Energy Agency, Paris, France.

Ishii, M., Kimoto, M., Sakamoto, K., and Iwasaki, S.-I., 2006, Steric sea level changes estimated from historical subsurface temperature and salinity analyses. Journal of Oceanography, v. 61, pp. 155–170.

Jacobsen, J. E., 1993, Population Growth: Global Change Instruction Program. University Corporation for Atmospheric Research, Boulder, CO.

Jiang, Y., and Yung, Y. L., 1996, Concentrations of tropospheric ozone from 1979 to 1992 over tropical Pacific South America from TOMS date. Science, v. 272, pp. 714–716.

Johnson, R., 1990, The Greenhouse Effect. Lerner Publishing Co., Minneapolis, MN.

Jones, P. D., Wigley, T. M. L., and Briffa, K. R., 1994, Global and hemispheric temperature anomalies: land and marine instrumental records. In Trends '93: A Compendium of Data on Global Change (ORNL/CDIAC-65), T. A. Boden, D. P. Kaiser, R. J. Sepanski, and F. W. Stoss (eds.), Carbon Dioxide Information Analysis Center, Oak Ridge National Laboratory, Oak Ridge, TN, pp. 603–608.

Jouzel, J., Barkov, N. I., Barnola, J. M., Bender, M., Chappellaz, J., Genthon, C., Kotlyakov, V. M., Lipenkov, V., Lorius, C., Petit, J. R., Raynaud, D., Raisbeck, G., Ritz, C., Sowers, T., Stievenard, M., Yiou, G., and Yiou, P., 1993, Extending the Vostok ice-core of palaeoclimate to the penultimate glacial period. Nature, v. 364, pp. 407–411.

Karl, T. R., Nicholls, N., and Gregory, J., 1997, The coming climate. Scientific American, v. 276, May, pp. 78–83.

Kasting, J., Toon, O. B., and Pollack, J. B., 1988, How climate evolved on the terrestrial planets. Scientific American, v. 258, pp. 90–97.

Kattenberg, A., Giorgi, F., Grassl, H., Meehl, G. A., Mitchell, J. F. B., Stouffer, R. J., Tokioka, T., Weaver, A. J., and Wigley, T. M. L., 1996, Climate models—projections for future climate change. In Climate Change 1995: The Science of Climate Change, J. T. Houghton, L. G. Meira Filho, B. A. Callander, N. Harris, A. Kattenberg, and K. Maskell (eds.), Cambridge University Press, Cambridge, pp. 285–357.

Kearey, P., and Vine, F. J., 1990, Global Tectonics. Blackwell Scientific Publications, London.

Kleypas, J.A. et al., 2006. Impacts of Ocean Acidification on Coral Reefs and Other Marine Calcifiers: A Guide for Future Research, report of a workshop held April 18–20, 2005, St. Petersburg, FL, sponsored by NSF, NOAA, and the U.S. Geological Survey.

Kohout, E. J., Miller, D. J., Nieves, L. A., Rothmann, D. S., Saricks, D. L., Stodolsky, F., and Hanson, D. A., 1990, Current emission trends for nitrogen oxides, sulfur dioxide and volatile organic compounds by month and state, methodology and results. Policy and Economic Analysis Group, Environmental Assessment and Information Services Division, ANL/EAIS/TM-25, Argonne National Laboratory, Argonne, IL.

Kump, L. R., Kasting, J. F., and Crane, R. G., 1999, The Earth System. Prentice Hall, Upper Saddle River, NJ.

Kundzewicz, Z. W., Mata, L. J., Arnell, N. W., Doll, P., Jimenez, B., Miller, K., Oki, T., Sen, Z., and Shiklomanov, I., 2008, The implications of projected climate change for freshwater resources and their management. Hydrological Sciences Journal, v. 53, pp. 3–10.

Kutzbach, J. E., 1975, Diagnostic studies of past climates. In The Physical Basis of Climate and Climate Modeling, Global Atmospheric Research Program Publication Series No. 16, World Meteorological Organization, Geneva, Switzerland, pp. 119–126.

Kutzbach, J., 1998, Climate and biome simulations for the past 21,000 years. Quaternary Science Reviews, v. 17, pp. 473–506.

Laporte, L. F., 1975, Encounter with the Earth: Resources. Canfield Press, San Francisco.

Laws, E. D., 1992, El Niño and the Peruvian anchovy fishery. Global Change Instruction Program Module 104, University Corporation for Atmospheric Research, Boulder, CO.

Laws, E. A., 2000, Aquatic Pollution. John Wiley & Sons, New York.

Lee, K. N., 1993, Compass and Gyroscope, Integrating Science and Politics for the Environment. Island Press, London.

LeGrand, H. E., 1988, Drifting Continents and Shifting Theories. Cambridge University Press, Cambridge.

Lents, J. M., and Kelly, W. J., 1993, Clearing the air in Los Angeles. Scientific American, v. 269, October, pp. 32–39.

Lerman, A. and Mackenzie, F.T., 2006, CO_2 air–sea exchange due to calcium carbonate and organic matter storage, and its implications for the global carbon cycle. Aquatic Geochemistry 11:4, 345–390.

Levinton, J. S., 2001, Marine Biology: Function, Diversity, Ecology (2nd ed.). Oxford University Press, New York.

Levinton, J. S., 1992, The big bang of animal evolution. Scientific American, v. 267, pp. 84–91.

Levorsen, A. I., 1956, Geology of Petroleum. W. H. Freeman and Co., San Francisco.

Liu, L, Sundet, J. K., Liu, Y., Bernsten, T. K., and Isaksen, I. S. A., 2007, A study of tropospheric ozone over China with the 3-D Global CTM Model. Terr. Atmos. Ocean. Sci., v. 18, No. 3, pp. 515–545.

Luthi, D., Le Floch, M., Bereiter, B., Blunier, T., Barnola, J.-M., Siegenthaler, U., Raynaud, D., Jouzel, J., Fischer, H., Kawamura, K., and Stocker, T., 2008, High-resolution carbon dioxide concentration record 650,000–800,000 years before present. Nature, v. 453, pp. 379–382.

Lutz, R. A., and Kennish, M. J., 1993, Ecology of deep-sea hydrothermal vent communities. Reviews of Geophysics, v. 31, pp. 211–242.

Lutz, W., 1994, The Future Population of the World, What Can We Assume Today? International Institute for Applied Systems Analysis, Earthscan Publications Ltd., London.

Mackenzie, F. T., 1995, Climatically important biogenic gases and feedbacks on global climatic change. In Biotic Feedbacks in the Global Climatic System, Will the Warming Feed the Warming? G. M. Woodwell and F. T. Mackenzie (eds.), Oxford University Press, New York, pp. 22–46.

Mackenzie, F. T., 1999, Global Biogeochemical Cycles and the Physical Climate System. Global Change Instruction Program, University Corporation for Atmospheric Research, Boulder, CO, 69 pp.

Mackenzie, F. T., and Agegian, C. R., 1989, Biomineralization and tentative links to plate tectonics. In Origin, Evolution, and Modern Aspects of Biomineralization in Plants and Animals, R. E. Crick (ed.), Plenum Press, pp. 11–27.

Mackenzie, F. T., Bewers, J. M., Charlson, R. J., Hofmann, E. E., Knauer, G. A., Kraft, J. C., Nöthig, E. M., Quack, B., Walsh, J. J., Whitfield, M., and Wollast, R., 1991, What is the importance of ocean margin processes in global change? In Ocean Margin Processes in Global Change, John Wiley & Sons, New York, pp. 433–454.

Mackenzie, F. T. and Lerman, A., 2006, Carbon in the Geobiosphere—Earth's Outer Shell. Springer, Dordrecht, The Netherlands.

Mackenzie, F. T., Ver, L. H., and Lerman, A., 2000, Coastal-zone biogeochemical dynamics under global warming. International Geology Review, v. 42, pp. 193–206.

Mackenzie, F. T., Lerman, A., Ver, L. M. B., 2001, Recent past and future of the global carbon cycle. *In:* Geological Perspectives of Global Climate Change. Am. Assoc. Pet. Geologists Special Publication 47, L. Gerhard, B. Hanson, and W. Harrison (eds.), AAPG, Tulsa, OK, pp. 51–82.

Mackenzie, F. T., Ver, L. H., and Lerman, A., 2002, Century-scale nitrogen and phosphorus controls of the carbon cycle. Chemical Geology, v. 190, pp. 13–32.

Malle, K.-G., 1996, Cleaning up the river Rhine. Scientific American, v. 274, January, pp. 70–75.

Malm, W. C., 1989, Atmospheric haze, its sources and effects on visibility in rural areas of the continental United States. Environmental Monitoring and Assessment, v. 12, pp. 203–225.

Manabe, S., and Stouffer, R. J., 1994, Multiple-century response of a coupled ocean–atmosphere model to increase of atmospheric carbon dioxide. Journal of Climate, v. 7, pp. 5–23.

Mann, M. E., Bradley, R. S., and Hughes, M. K., 1999, Northern Hemisphere temperatures during the past millennium: inferences, uncertainties, and limitations. Geophysical Research Letters, v. 26, pp. 759–762.

Margulis, L., and Olendzenski, L. (eds.), 1992, Environmental Evolution: Effects of the Origin and Evolution of Life on Planet Earth. MIT Press, Cambridge, MA.

Margulis, L., and Sagan, D., 1995, What is Life? Simon and Schuster, New York.

Mason, B., 1966, Principles of Geochemistry (3rd ed.). John Wiley & Sons, New York.

Maurits la Rivière, J. W., 1990, Threats to the world's water. In Managing Planet Earth, W. H. Freeman and Co., New York, pp. 37–48.

Meadows, D. H., Meadows, D. L., Handers, J., and Behrens, W., III, 1974, The Limits to Growth. Universe Books, New York.

Meehl, G. A., Stocker, T. F., Collins, W.D. Friedlingstein, P. Gaye, A.T. Gregory, J.M. Kitoh, A. Knutti, R. Murphy, J.M. Noda, A. Raper, S.C.B. Watterson, I.G. Weaver A.J. and Z.-C. Zhao, 2007: Global Climate Projections. In Climate Change 2007: The Physical Science Basis. Contribution of Working Group I to the Fourth Assessment Report of the Intergovernmental Panel on Climate Change, S., Solomon, D. Qin, M. Manning, Z. Chen, M. Marquis, K. B. Averyt, M. Tignor and H. L. Miller (eds.). Cambridge University Press, Cambridge, UK and New York.

Millennium Ecosystem Assessment, 2005, Ecosystems and Human Well-being: Synthesis. Island Press, Washington, DC.

Mintz, W., 1972, Historical Geology (2nd ed.). Charles E. Merrill Publishing Co., Columbus, OH.

Mohnen, V. A., 1988, The challenge of acid rain. Scientific American, v. 259, February, pp. 30–38.

Montzka, S. A., Butler, J. H., Myers, R. C., Thompson, T. M., Swanson, T. H., Clarke, A. D., Lock, L. T., and Elkins, J. W., 1996, Decline in the tropospheric abundance of halogen from halocarbons: implications for stratospheric ozone depletion. Science, v. 272, pp. 1,318–1,322.

Mooney, H., Vitousek, P. M., and Matson, R. A., 1987, Exchange of materials between terrestrial ecosystems and the atmosphere. Science, v. 239, pp. 926–932.

Morse, J., and Mackenzie, F. T., 1998, Hadean ocean carbonate geochemistry. Aquatic Geochemistry, v. 4, pp. 301–319.

Munk, W. R., 1971, The circulation of the oceans. In Oceanography, Readings from Scientific American, San Francisco, pp. 64–69.

Myers, N., 1984, Gaia, An Atlas of Planet Management. Doubleday, New York.

National Acid Precipitation Assessment Program (NAPAP), 1990, Integrated assessment, questions 1 and 2, external review draft. Washington, DC, NAPAP.

National Atmospheric Deposition Program (NADP), 2008, https://nadp.isws.illinois.edu/.

National Aeronautics and Space Administration (NASA), 1997, The Global Biosphere. Washington, DC.

National Aeronautics and Space Administration (NASA), 1991, Ocean Color from Space (code EE). EOS Program Office, Washington, DC.

National Oceanographic and Atmospheric Administration/Climate Monitoring and Diagnostics Laboratory (NOAA/CMDL), 1997, CO_2 and other gas concentration measurements on the World Wide Web. www.cmdl.noaa.gov.

National Research Council, 1992, Policy Implications of Greenhouse Warming, Report of the Panel on Policy Implications of Greenhouse Warming. Committee on Science, Engineering, and Public Policy; National Academy of Sciences; National Academy of Engineering; and the Institute of Medicine, National Academy Press, Washington, DC.

National Research Council, 2000, Reconciling Observations of Global Temperature Change. National Academy Press, Washington, DC.

Nebel, B., 1981, Environmental Science. Prentice Hall, Englewood Cliffs, NJ.

Nebel, B.J., and Wright, R.T., 1998, Environmental Science. Prentice Hall, Upper Saddle River, NJ.

Newell, R. E., 1971, The global circulation of atmospheric pollutants. Scientific American, v. 224, pp. 32–42.

Newman, A., 1990, Tropical Rainforest. Eddison Editions, London.

Nicholls, N., Gruza, G. V., Jouzel, J., Karl, T. R., Ogallo, L. A., and Parker, D. E., 1996, Observed climate variability and change. In Climate Change 1995: The Science of Climate Change, J. T. Houghton, L. G. Meira Filho, B. A. Callander, N. Harris, A. Kattenberg, and K. Maskell (eds.), Cambridge University Press, Cambridge, pp. 132–192.

Nicholls, R., 1997, Climate Change and Its Impacts: A Global Perspective. Brittanic Crown Copyright, United Kingdom Meteorological Office, London.

Nisbet, E. G., 1990, The end of the ice age. Canadian Journal of Earth Science, v. 27, pp. 148–157.

Office of Science and Technology Policy, 1997, Climate Change: State of Our Knowledge. Executive Office of the President, Office of Science and Technology Policy (OSTP), Washington, DC.

Oldeman, L. R., van Engelen, V. W. P., and Pulles, J. H. M., 1990, The extent of human-induced soil degradation. In Annex 5 of Oldeman, L. R., Hakkeling, R. T. A., and Sombroek, W. G., World Map of the Status of Human-induced Soil Degradation, an Explanatory Note (rev. 2nd ed.). International Soil Reference and Information, Wageningen, the Netherlands.

Open University, 1989, Ocean Chemistry and Deep-Sea Sediments. Pergamon Press, Oxford, England.

Open University, 1989, Ocean Circulation. Pergamon Press, Oxford, England.

Open University, 1989, Seawater: Its Composition, Properties and Behavior. Pergamon Press, Oxford, England.

Open University, 1989, The Ocean Basins: Their Structure and Evolution. Pergamon Press, Oxford, England.

Open University, 1989, Wave, Tides and Shallow-Water Processes. Pergamon Press, Oxford, England.

Organisation for Economic Cooperation and Development (OECD), 1991, The State of the Environment. OECD, Paris.

Organisation for Economic Cooperation and Development (OECD), 1994, The State of the Environment. OECD, Paris.

Pacala, S. and Socolow, R., 2004, Stabilization wedges: solving the climate problem for the next 50 years with current technologies. Science, v. 305, pp. 969–972.

Pace, N.R., 1997, A molecular view of microbial diversity and the biosphere. Science 276:734–740.

Parry, M. et al., 1997, Climate Change and Its Impacts: A Global Perspective. Brittanic Crown Copyright, United Kingdom Meteorological Office, London.

Patullo, J., Munk, W., Revelle, R., and Strong, E., 1955, The seasonal oscillations in sea level. Journal of Marine Research, v. 14, pp. 88–156.

Petit, J. R., Jouzel, J., Raynaud, D., Barkov, N. I., Barnola, J. M., Basile, I., Bender, M., Chappellaz, J., Davis, J., Delaygue, G., Delmotte, M., Kotyakov, V. M., Legrand, M., Lipenkov, V. Y., Lorius, C., Pepin, L., Ritz, C., Saltzman, E., and Stievenard, M., 1999, Climate and atmospheric history of the past 420,000 years from the Vostok Ice Core, Antarctica. Nature, v. 399, pp. 429–436.

Pollack, H. N., Hurter, S. J., and Johnson, J. R., 1993, Heat flow from the Earth's interior: analysis of the global data set. Reviews of Geophysics, v. 31, pp. 267–280.

Postel, S., 1984, Air Pollution, Acid Rain, and the Future of Forests. Worldwatch Paper 58, Worldwatch Institute, Washington, DC.

Postel, S., 1996, Dividing the Waters: Food Security, Ecosystem Health, and the New Politics of Scarcity. Worldwatch Paper 132, Worldwatch Institute, Washington, DC.

Prentice, I. C., and Sykes, M. T., 1995, Vegetation geography and global carbon storage changes. In Biotic Feedbacks in the Global Climatic System, G. M. Woodwell and F. T. Mackenzie (eds.), Oxford University Press, New York, pp. 302–312.

Press, F., and Siever, R., 1974, Earth. W. H. Freeman and Co., San Francisco.

Prinn, R. G., Cunnold, D., Rasmussen, R., Simmonds, P., Alyea, F., Crawford, A., Fraser, P., and Rosen, R., 1990, Atmospheric emissions and trends of nitrous oxide deduced from 10 years of ALE/GAGE data. Journal of Geophysical Research, v. 95, pp. 18,369–18,385.

Rambler, M. B., Margulis, L., and Fester, R. (eds.), 1989, Global Ecology: Towards a Science of the Biosphere. Academic Press, Boston.

Raup, D. M., 1991, Extinction: Bad Genes or Bad Luck? W. W. Norton & Co., New York.

Readings from Scientific American, 1990, Managing Planet Earth. W. H. Freeman and Co., New York.

Reid, S. J., 2000, Ozone and Climate Change: A Beginner's Guide. Gordon and Breach Science Publishers, Amsterdam, the Netherlands, 210 pp.

Ringwood, A. E., 1966, Chemical evolution of the terrestrial planets. Geochimica et Cosmochmica Acta, v. 30, pp. 41–104.

Rodhe, H., Langner, J., Gallardo, L., and Kjellstrom, E., 1995, Global transport of acidifying pollutants. Water, Air and Soil Pollution, v. 85, pp. 37–50.

Ropelewski, C. F., 1992, Predicting El Niño events. Nature, v. 356, pp. 476.

Ruddiman, W. F., 2001, Earth's Climate: Past and Future. W. H. Freeman and Co., New York.

Sabine, C. L., Feely, R. A., Gruber, N., Key, R. M., Lee, K., Bullister, J. L., Wanninkhof, R., Wong, C. S., Wallace, D. W. R., Tilbrook, B., Millero, F. J., Peng, T.-H., Kozyr, A., Ono, T., and Rios, A. F., 2004, The ocean sink for anthropogenic CO_2. Science, v. 305, pp. 367–371.

Schlesinger, W. H., 1997, Biogeochemistry: An Analysis of Global Change (2nd ed.). Academic Press, New York.

Schneider, S. H., 1976, The Genesis Strategy, Climate and Global Survival. Plenum Press, New York.

Schneider, S. H., and Londer, R., 1989, The Coevolution of Climate and Life. Sierra Club Books, San Francisco.

Schopf, J. W. (ed.), 1983, Earth's Earliest Biosphere: Its Origin and Evolution. Princeton University Press, Princeton, NJ.

Schopf, J. W., and Klein, C., 1992, The Proterozoic Biosphere, A Multidisciplinary Study. Cambridge University Press, New York.

Sedjo, R. A., and Clawson, M., 1984, Global forests. In The Resourceful Earth, J. T. Simon and H. Kahn (eds.), Basil Blackwell, New York, pp. 128–170.

Siegenthaler, U., and Oeschger, H., 1987, Geospheric CO_2 emissions during the past 200 years, reconstructed by deconvolution of ice core data. Tellus, v. 39, pp. 140–154.

Skinner, B. J., and Porter, S. C., 1987, Physical Geology. John Wiley & Sons, New York.

Skinner, B. J., and Porter, S. C., 1995, The Dynamic Earth (3rd ed.). John Wiley & Sons, New York.

Smil, V., 1997, Cycles of Life: Civilization and the Biosphere. Scientific American Library, W. H. Freeman and Co., New York.

Smith, S. J., Andres, R., Conception, E., and Lurz, J. 2004, Historical sulphur dioxide emissions 1850–2000: Methods and results. Research Report prepared for the U. S. Department of Energy under the Contract DE-AC06-76RL01830, Joint Global Change Research Institute, College Park, MD, Pacific Northwest National Laboratory, PNNL-14537.

Smith, S. V., and Hollibaugh, J. T., 1993, Role of coastal ocean organic metabolism in the oceanic organic carbon cycle. Reviews of Geophysics, v. 31, pp. 75–89.

Somville, M., 1980, Étude ecophysiologique des metabolismes bacteriens dans l'estuaire de l'Escaut (Ph.D. thesis). Universite Libré de Bruxelles, Bruxelles, Belgium.

Southwick, C. H., 1996, Global Ecology in Human Perspective. Oxford University Press, New York.

Stern, D. I., 2006, Global sulfur dioxide from emissions from 1850 to 2000. Chemosphere, v. 58, pp. 163–175.

Stockton, C. W., Boggess, W. R., and Meko, D. M., 1985, Climate and tree rings. In Paleoclimate Analysis and Modeling, A. Hecht (ed.). Wiley-Interscience, New York, pp. 71–150.

Stolarski, R. S., 1988, The Antarctic ozone hole. Scientific American, v. 258, January, pp. 30–37.

Stolz, J. F., Botkin, D. B., and Dastoor, M. N., 1989, The integral biosphere. In Global Ecology, Towards a Science of the Biosphere, M. B. Rambler, L. Margulis, and R. Fester (eds.), Academic Press, Boston, pp. 31–49.

Strahler, A. N., 1963, The Earth Sciences. Harper and Row, New York.

Streete, J., 1991, The Sun–Earth System. University Corporation for Atmospheric Research, Global Change Instructional Program, Module 102, Boulder, CO.

Taylor, S. R., 1964, Abundance of chemical elements in the continental crust: a new table. Geochimica et Cosmochimica Acta, v. 28, pp. 1,273–1,285.

Tennissen, A. C., 1974, Nature of Earth Materials. Prentice Hall, Englewood Cliffs, NJ.

Travis, R. B., 1955, Classification of rocks. Quarterly of the Colorado School of Mines, Colorado School of Mines, Golden, CO, v. 50, pp. 1–98.

Trewartha, G. T., and Horn, L. H., 1980, An Introduction to Climate. McGraw-Hill, New York.

United Nations, 1987, The prospects of world urbanization, revised as of 1984–85. United Nations, New York, pp. 25–27; Table 2, p. 8.

United Nations Population Division, 1995. Concise Report on the World Population Situation in 1995. United Nations, New York, 44 pp.

United Nations Statistical Office, 1991, U. N. energy tape. United Nations, New York, May.

University Corporation for Atmospheric Research (UCAR)/Office for Interdisciplinary Earth Studies (OIES), 1991, Science Capsule, Changes in Time in the Temperature of the Earth. EarthQuest, v. 5, no. 1.

U.S. Department of Commerce, 2001, Climate Diagnostics Bulletin, November 2001. U.S. Department of Commerce, Camp Springs, MD.

U.S. Environmental Protection Agency (EPA), 1988, National air quality and emissions trends report, 1986, EPA-450/4-88-001. Office of Air Quality Planning and Standards, Research Triangle Park, NC, pp. 3–35.

Velicogna, I., and Wahr, J., 2006, Acceleration of Greenland ice mass loss in spring 2004. Nature, v. 443, pp. 329–331.

Vitousek, P. M., Aber, J. D., Howath, R. W., Likens, G. E., Matson, P. A., Schindler, D. W., Schlesinger, W. H., and Tilman, D., 1997, Human alteration of the global nitrogen cycle: sources and consequences. Ecological Applications, v. 7, pp. 737–750.

Von Arx, W. S., 1962, An Introduction to Physical Oceanography. Addison-Wesley, Reading, MA.

Warneck, P., 1988, Chemistry of the Natural Atmosphere. Academic Press, San Diego.

Warrick, R. A., Le Provost, C., Meir, M., Oerlemans, J., and Woodworth, P., 1996, Changes in sea level. In Climate Change 1995: The Science of Climate Change, J. T. Houghton, L. G. Meira Filho, B. A. Callander, N. Harris, A. Kattenberg, and K. Maskell (eds.), Cambridge University Press, Cambridge, pp. 358–405.

Watson, R. T. et al. (eds.), 2000, Land Use, Land-Use Change, and Forestry. A Special Report of the Intergovernmental Panel on Climate Change. Cambridge University Press, Cambridge.

Wayne, R. P., 1991, Chemistry of Atmospheres. Oxford University Press, New York.

Weber, P., 1993, Abandoned Seas, Reversing the Decline of the Oceans. Worldwatch Paper 116, Worldwatch Institute, Washington, DC.

Whelpdale, D., 1991, personal communication.

White, R. M., 1996, Climate science and national interests. Issues in Science and Technology, v. 13, pp. 33–38.

Whitmore, T. C., and Sayer, J. A., 1992, Tropical Deforestation and Species Extinction. Chapman and Hall, New York.

Whittaker, R. H., and Likens, G. E., 1975, The biosphere and man. In Primary Productivity of the Biosphere, H. Lieth and G. Likens (eds.). Springer-Verlag, Berlin, pp. 305–328.

Wigley, T. M. L., Richels, R., and Edmonds, J. A., 1996, Economic and environmental choices in the stabilization of atmospheric CO_2 emissions. Nature, v. 379, pp. 242–245.

Willis, J. K., Roemmich, D., and Cornuelle, B, 2004, Interannual variability in upper ocean heat content, temperature, and thermosteric expansion on global scales. Journal of Geophysical Research, v. 109, C12036, doi:10.1029/2003JC002260.

Wilson, E. O., 1992, The Diversity of Life. Belknap Press of Harvard University Press, Cambridge, MA.

Wilson, E. O., 1990, Threats to biodiversity. In Managing Planet Earth, W. H. Freeman and Co., New York, pp. 49–60.

Wilson, E. O., (ed.), 1988, Biodiversity. National Academy Press, Washington, DC.

Wilson, J. T., 1976, Continents adrift and continents aground. In Readings from Scientific American, W. H. Freeman and Co., San Francisco.

Wilson, T. R. S., 1975, Salinity and the major elements of sea water. In Chemical Oceanography, v. I (2nd ed.), J. P. Riley and G. Skirrow (eds.), Academic Press, New York, pp. 365–413.

Woese, C. R., 1987, Bacterial evolution. Microbiological Reviews, v. 51, pp. 221–271.

Wollast, R., and Mackenzie, F. T., 1989, Global biogeochemical cycles and climate. In Climate and Geosciences, A. Berger, S. Schneider, and J.-C. Duplessy (eds.), Kluver Academic Publishers, Dordrecht, the Netherlands, pp. 453–473.

Woodwell, G. M., and Mackenzie, F. T., 1995, Biotic Feedbacks in the Global Climatic System. Oxford University Press, New York.

World Bank, 1995, The World Bank Atlas 1995. International Bank for Reconstruction and Development/ World Bank, Washington, DC.

World Meteorological Organization (WMO), 1994, Scientific Assessment of Ozone Depletion: 1994. WMO, Global Ozone Research and Monitoring Project Report No. 37, Geneva, Switzerland.

World Meteorological Organization (WMO), 1998, Scientific Assessment of Ozone Depletion: 1998, Executive Summary. WMO, Global Ozone Research and Monitoring Project Report No. 44, Geneva, Switzerland.

World Meteorological Organization (WMO), 2007, Scientific Assessment of Ozone Depletion: 2006. WMO, Global Ozone Research and Monitoring Project Report No. 50, Geneva, Switzerland.

World Resources Institute, 1996, World Resources: A Guide to the Global Environment. Oxford University Press, New York, 365 pp. Also editions of 1986, 1988, 1990, 1992, and 1994.

World Resources Institute, 2000, World Resources 2000–2001. World Resources Institute, Washington, DC.

Worldwatch Institute, 1997, State of the World 1997. W. W. Norton & Company, New York.

Worldwatch Institute, 2001, Vital Signs 2001: The Trends That Are Shaping Our Future. W. W. Norton & Company, New York.

Worldwatch Institute, 2006, Peak Oil. World Watch Magazine, Worldwatch Institute, v. 119, Washington, DC.

Wyllie, P. J., 1976, The Way the Earth Works. John Wiley & Sons, New York.

Young, J. E., 1992, Mining the Earth. Worldwatch Paper 109, Worldwatch Institute, Washington, DC.

Ziemke, J. R., Chandra, S., Duncan, B. N., Froidevaux, L., Bhartia, P. K., Levelt, P. F., and Waters, J. W., 2006, Tropospheric ozone determined from Aura OMI and MLS: Evaluation of measurements and comparison with the Global Modeling Initiative's Chemical Transport Model. Journal of Geophysical Research, v. 1111, D19303, pp. 1–18.

INDEX